THE BOOK OF MARS

STUART CLARK is an author and journalist whose career is devoted to presenting the complex world of astronomy to the general public. He holds a first class honours degree and a PhD in astrophysics, is a Fellow of the Royal Astronomical Society and a former Vice Chair of the Association of British Science Writers. He is the author of numerous non-fiction books including *Beneath The Night*, *The Unknown Universe*, *The Sun Kings*, and *The Sky's Dark Labyrinth*, a trilogy of novels set around the times of greatest change in mankind's understanding of the Universe. Clark regularly writes for the *Guardian* and *New Scientist*. In 2020, the University of Hertfordshire awarded him an (Hon) DSc for 'services to astronomy and the public understanding of science'.

ALSO IN THE ANTHOLOGY SERIES

THE ART OF THE GLIMPSE
100 Irish Short Stories
Chosen by Sinéad Gleeson

THE BIG BOOK OF CHRISTMAS MYSTERIES
100 of the Very Best Yuletide Whodunnits
Chosen by Otto Penzler

DEADLIER
100 of the Best Crime Stories Written by Women
Chosen by Sophie Hannah

DESIRE
100 of Literature's Sexiest Stories
Chosen by Mariella Frostrup and the *Erotic Review*

FOUND IN TRANSLATION
100 of the Finest Short Stories Ever Translated
Chosen by Frank Wynne

FUNNY HA, HA
80 of the Funniest Stories Ever Written
Chosen by Paul Merton

GHOST
100 Stories to Read With the Lights On
Chosen by Louise Welsh

HOUSE OF SNOW
An Anthology of the Greatest Writing About Nepal
Foreword by Sir Ranulph Fiennes

JACK THE RIPPER
The Ultimate Compendium of the Legacy and Legend of History's Most Notorious Killer
Chosen by Otto Penzler

LIFE SUPPORT
100 Poems to Reach for on Dark Nights
Chosen by Julia Copus

OF GODS AND MEN
100 Stories from Ancient Greece and Rome
Edited by Daisy Dunn

QUEER
A Collection of LGBTQ Writing from Ancient Times to Yesterday
Chosen by Frank Wynne

SHERLOCK
Over 80 Stories Featuring the Greatest Detective of all Time
Chosen by Otto Penzler

THAT GLIMPSE OF TRUTH
100 of the Finest Short Stories Ever Written
Chosen by David Miller

THE TIME TRAVELLER'S ALMANAC
100 Stories Brought to You From the Future
Chosen by Jeff and Ann VanderMeer

THE STORY
100 Great Short Stories Written by Women
Chosen by Victoria Hislop

WE, ROBOTS
Artificial Intelligence in 100 Stories
Chosen by Simon Ings

THE WILD ISLES
An Anthology of the Best of British and Irish Nature Writing
Chosen by Patrick Barkham

WILD WOMEN
A Collection of First-hand Accounts from Female Explorers
Chosen by Mariella Frostrup

EDITED BY
STUART CLARK

THE BOOK OF MARS

AN ANTHOLOGY OF FACT AND FICTION

An Apollo Book

First published in the UK in 2022 by Head of Zeus Ltd,
part of Bloomsbury Publishing Plc

9 7 5 3 1 2 4 6 8

A catalogue record for this book is available from the British Library.

ISBN (HB): 9781801109314
ISBN (E): 9781801109291

Typeset by PDQ
Printed and bound in Germany by CPI Books GmbH

Head of Zeus Ltd
5–8 Hardwick Street
London EC1R 4RG
www.headofzeus.com

THE BOOK OF MARS

CONTENTS

COLONIZING MARS

INTRODUCTION

They called each one of them 'wanderer'. In the original Greek, the word was *planetes*. Now we simply say planet. They earned the name because of the individual ways they moved across the sky. Unbound from the rigid clockwork of the stars, the seven planets of classical antiquity crossed the sky at their own individual paces.

Of these seven wanderers, two were overwhelming – so much so that we no longer think of them as planets. The Sun regulated day and night. The Moon gave us the month, with its ever-shifting appearance. But what of the other five? They were bright points of light named after the ancient gods: Mercury, Venus, Mars, Jupiter and Saturn. While more protective of their secrets than the Sun and the Moon, careful observation did reveal some measure of personality.

Mercury and the beautifully brilliant Venus never strayed far from the Sun, appearing mostly in fleeting twilight apparitions that alternated between dawn and dusk. Jupiter, and even more so Saturn, moved in a stately procession across the sky, taking a decade or three to return to the same constellation, and drawing close to each other only once every twenty years or so. But Mars was something else entirely.

Looking down on Earth with its baleful red glow, Mars was the only planet that appeared to wax and wane significantly in brightness. Sometimes it was a burning red beacon in the sky, unmistakable and strangely daunting. At other times it was faint almost to the point of obscurity against the backdrop of ordinary stars. The cycle repeated every couple of years and tantalized – even taunted – us to imagine the planet's secrets.

In Greek and Roman mythology, Mars was identified as the god of war and courage. He was also the protector of agriculture in the Roman tradition. How interesting then that two of the preoccupations with the planet that we will explore in this collection are wars involving Martians, and the colonization of the planet, which inevitably includes growing crops there.

When I began putting together *The Book of Mars*, three novels immediately leapt to mind: H. G. Wells's *The War of The Worlds* (1898), Ray Bradbury's *The Martian Chronicles* (1946), and Arthur C. Clarke's *The Sands of Mars* (1951). I had loved these as a science-fiction-hungry youth, and immediately pulled them off the shelves to begin re-reading for the most representative excerpt.

In doing so I realized that the styles of these three tales also acted as archetypes, in a way. First there is the scientific romance of Wells. Throughout the latter half of the nineteenth century, astronomers had recognized the gleaming polar ice caps of Mars and had puzzled over a series of straight lines that appeared to crisscross on the planet. Perhaps these were canals, they supposed; the signs of a desperate Martian civilization trying to irrigate the dry equatorial regions with water from the poles. Wells chose this as a jumping off point for adventure: the Martians deciding to invade Earth and take what they needed from here. It also seems clear that Wells was troubled by British imperialism and the arrogance that it bred, and designed his story to put us in our place. It remains the quintessential alien invasion story.

Other writers from the time used Mars simply as a venue for high adventure. In 1911, Edgar Rice Burroughs serialized the story that became known as 'A Princess of Mars'. Published in the pulp fiction magazine *The All-Story*, it introduced readers to John Carter, an Earth soldier who wakes up on Mars, Dejah Thoris, the titular Martian princess, who proves a capable warrior herself, and Tars Tarkas, a fierce Martian fighter with green skin and four arms. The novel spawned ten sequels.

By the time of Bradbury and his *Martian Chronicles*, just over half a century had elapsed. By then, we knew that the canals were merely optical illusions created in our brains because the astronomers had been using relatively primitive equipment. More powerful telescopes had since resolved the planet's surface features and shown conclusively that there are no artificial waterways. Nevertheless, this picture was so alive in the cultural imagination of society, that Bradbury used it liberally and created his Mars as a venue for exploring frontier living and the encroachment on indigenous populations. In the same way that Wells used his work to deconstruct imperial hubris, Bradbury uses his to explore the consequences of Manifest Destiny, a doctrine that held it was America's God-given right to expand across North America, displacing the Native Americans in the process.

In *The Sands of Mars*, Arthur C. Clarke has different concerns again. Instead of using Mars or Martians as a crucible for a cultural exploration, he is more concerned with presenting the latest scientific thinking about the planet itself, and how we might make it habitable to humans. In Clarke's prose, Mars and the technology used to explore it become characters in the story as much as the various people he writes about.

It is fascinating to me how each writer used Mars in a different way: as a place (or excuse) for adventure, as a venue for cultural reflection, as a piece of popular futurism. And the more widely I read about Mars, the more I realized that this is a repeating pattern.

Some authors like their stories to spring from the latest science; others choose to use existing cultural perceptions of the planet to provide a new backdrop for age-old human concerns; and some decide to portray the future from a point of scientific and technical realism. It therefore made sense to include works of non-fiction in this collection as well. Alongside the stories, these pieces help orientate our perceptions of Mars, and understand the narrative choices – maybe even the intentions – of the fiction writers in their evocations of the planet.

Taken together, this collection chronicles the evolution of cultural and scientific thinking about Mars over the last 150 years. It also shows the diffusion of scientific knowledge into literature – and the way that particularly striking images, such as the canals, linger on long after being debunked.

To help further navigate these works, I have split them into six thematic collections. Within each I have presented the pieces in chronological order, except for where a later analysis illuminates the works around it. The themes are slightly loose – for example, when does a dream of one day living on Mars cross the line into being a story about colonization?

But for better or for worse, here are the sections. 'Dreams of Mars' collects together stories and articles that appear mostly driven by imagination, even if they spring initially from scientific facts. 'Observations of Mars' collects works that detail our progress in studying the planet with ever more sophisticated telescopes and eventually space probes. 'Exploring Mars' is the next step after observation. Here are pieces concerned with taking our first tentative steps onto the Martian soil, with either robots or humans. 'Fears of Mars' is the bookend to the dreams section. Where some see utopia, others see hell. 'Life on Mars' concerns itself with the inhabitants of the planet, whether real or imaginary, intelligent or otherwise. 'Colonizing Mars' is about human migration to the planet, and the problems we may face – both in imagination and in reality.

As you will see, for the last 150 years, there have been as many versions of Mars as there are authors, and no shortage of people wanting to write about it. I think there are two reasons for this. One is physical, the other psychological.

First, Mars is tantalizingly Earth-like. Whereas the Moon, with an entire surface area of 38 million square kilometres, is the equivalent of an Earthly continent (Africa for example has a surface area of just over 30 million square kilometres), Mars is a bona fide planet. With a surface area of almost 145 million square kilometres, it is three times larger than Earth's largest continent, Asia (just under 45 million kilometres).

In contrast to the airless desolation of the Moon, the Martian landscape is composed of dusty plains, deep canyons, extinct volcanoes, desiccated riverbeds, carbon dioxide polar caps and vicious seasons during which temperatures are calculated to plummet to –130°C at the poles, compared to –90°C in Antarctica.

It's an extreme version of some of the most exciting landscapes on Earth. In terms of rugged terrain, whatever we have, Mars trumps it. Taller volcanoes, deeper canyons, more extreme storms. And yet, even after decades of exploration have painted the world in broad form, the details remain elusive and provide many fertile places for the modern storyteller to plant their seeds.

While the facts of the planet alone add up to a siren call for the imagination, there's also the second, psychological factor at work. In the book *The Structure and Dynamics of the Psyche*, published in 1960, psychiatrist Carl Jung wrote, 'The starry vault of heaven is in truth the open book of cosmic projection.' In other words, when we stand under the night sky and look out into infinity, we project our innate hopes and fears into the universe.

Perhaps we imagine future utopias, in which we live in harmony, having learnt the lessons from our blind stumble through history. Or we see noble humans exploring in peace for the betterment of life on Earth. If we incline towards darker frames of mind, we see a pristine realm that would be better without our continuing despotic cycle of exploitation and disregard for nature. Or we allow our fear of the unknown to grip us completely and imagine alien horrors lying in wait.

And this, again, provides a reason for why Mars as the most Earth-like of all the planets is such a popular venue for stories. As you will see in the stories contained in these pages, authors have projected a wide variety of Earthly social philosophies onto Mars: feminist matriarchies, communist states, technocracies, to name but a few.

But what of the future of Mars in fiction, especially as the future promises more exploration that will inevitably normalize the Red Planet by bringing it more into our everyday experience? Will the march of scientific knowledge unmask the planet to such a degree that authors can no longer realistically overlay it with their imagined worlds? Almost certainly, but that doesn't necessarily mean the end of fiction set on Mars.

It is fair to say that almost every story in this volume is or could be categorized as science fiction. Perhaps the only exception is the extract from *The Martian* by Andy Weir, which fits more comfortably into the thriller category, or even the mainstream. And indeed, in the coming years, decades and centuries that is where Mars is headed: into the mainstream.

Exploration of the Red Planet is reaching unstoppable proportions. In the coming decade, the occasional and gentle poking and prodding that has characterized Mars exploration for the last sixty years is set to give way to something altogether more insistent.

There is clear momentum building for human missions to Mars, and it is not coming just from the major space agencies around the world. Billionaire Elon Musk has been vocal about his intent to develop rockets capable of taking people to Mars to settle the planet. In August 2010, he even told the *Observer* that Mars 'would be a good place to retire'. Later in the same interview, he explained that the three things that would most affect the future of humanity were the internet, sustainable energy and space exploration. On the last point, he specifically meant 'extending life beyond Earth and making it multi-planetary'. In other words, colonizing Mars.

Whenever Musk or others make comments such as these, there is a backlash from those who wish to champion Earth. Often these arguments place the desire to explore space as somehow in opposition to the preservation of our own planet. This is a false comparison. The space race of the twentieth century was a major driver of both the cultural recognition and scientific exploration of Earth as a single, whole planet.

Without the satellites that circle the globe, our understanding of climate science would be nowhere near as advanced as it is now. We can therefore speculate on what human missions to Mars may accomplish for us back here on Earth. When we do that, one thing shines brightly. Sustainability.

Mars is not the Moon, where astronauts could be sent with everything they need for a week or two before returning to Earth. A mission to Mars and back lasts for several years at least. It is economically impossible for a mission to carry everything that its astronauts need to live and thrive. The only thing that will make such ventures possible is the development of extreme sustainable living techniques.

In January 2021, NASA launched a crowd-sourced competition to explore new ways to efficiently reprocess or recycle or repurpose onboard resources. The four key areas of the challenge were to propose waste management or conversion techniques for trash, fecal waste, foam packaging material and exhaled carbon dioxide. From this list it is not hard to see that a mission to Mars is a microcosm, facing the same challenges as we do globally on Earth. Similarly, it is easy to imagine that scaling up any successful techniques that are proposed could help the ecological efforts back here on Earth.

We can expect future fiction set on or around Mars to reflect these concerns and developments. And this is perhaps the most fascinating aspect of all. While scientists and engineers explore and innovate, extending the human presence beyond Earth, our writers and artists will reflect the cultural impact that those events will create. Let us hope that sometime in the twenty-second century, some future author will curate a new *Book of Mars* to chart the next 150 years of our relationship with the Red Planet.

DREAMS OF MARS

MARS CHRONOLOGY

RENAISSANCE TO THE SPACE AGE

Paul Karol and David Catling

David Catling obtained a PhD in atmospheric, oceanic and planetary physics from the University of Oxford, England. He then worked as a planetary scientist at NASA's Ames Research Center, near San Francisco, California, from 1995–2001. During this time, he collaborated with **Paul Karol** – who became an administrator for online education at Cleveland State University – to produce the *Mars Chronology: Renaissance to the Space Age*. Catling is currently employed jointly by the Department of Earth and Space Sciences and cross-campus Astrobiology Program at the University of Washington.

The 1500s: The Copernican Revolution

1500s
Danish astronomer Tycho Brahe (1546–1601) collects very accurate positions for Mars using keen eyesight and large measuring instruments. Positions of stars and planets are monitored to an accuracy of about four minutes of arc.

1543
On the Revolutions of the Heavenly Orbs by Nicolaus Copernicus (1473–1543) is published. This contains the radical notion that the planets, including the Earth, orbit the Sun, which challenges conventional beliefs concerning the central position of the Earth. The belief that the Earth was at the center of the universe was essentially a reconciliation of classical Aristotelian philosophy and Judaic-Christian theology that owed much to St. Thomas Aquinas.

The 1600s: Pioneers of planetary motion and gravitation

1600
Johannes Kepler (1571–1630) goes to Prague to become assistant to Tycho Brahe, who dies a year later.

1600

Giordano Bruno (1548–1600), an Italian scientist who taught at Oxford and other well-known European universities, is tried by the Inquisition, condemned, and burned at the stake. He had dared to suggest that space is boundless and that the Sun and its planets were but one of any number of similar systems. He even said that there might be other worlds inhabited with rational beings possibly superior to ourselves.

1609

Kepler publishes his *Astronomia Nova* (*New Astronomy*), containing his first two laws of planetary motion. Kepler's first law is based on a calculation of an elliptical orbit for Mars using Brahe's data. This challenged, and ultimately replaced, the classical belief in perfect circular orbits.

1609

Galileo Galilei (1564–1642) observes Mars with his primitive telescope. Later in 1610, he writes to his friend Father Castelli about observations of phases of Mars indicating a spherical body illuminated by the Sun.

1619

Kepler's *Harmonice mundi* (*The Harmonies of the World*) is published. This contains his third law of planetary motion.

1636

Francisco Fontana views and draws Mars. His drawing resembles a dark spot, which he called a "black pill", inside a sphere. The dark spot was due to a defect in his telescope.

1640

Niccolo Zucchi (1586–1670), a professor at the Jesuit College in Rome, observes spots on Mars with his reflecting telescope. In 1616, Zucchi had made one of the earliest reflecting telescopes, predating those of James Gregory and Sir Isaac Newton.

1659

Oct. 13: The first sketch of Mars is drawn by the Dutch astronomer Christiaan Huygens (1629–1695). Huygens uses his own design of telescope, which is of much higher quality than that of his predecessors and allows a magnification of 50 times.

Nov. 28: Christiaan Huygens records the first true feature on Mars, a large dark spot, almost certainly Syrtis Major, which became known as the "Hourglass Sea". Observing the spot in successive rotations, he deduces a 24-hour period. Huygens had earlier made some drawings of Mars in 1656, but they were not noteworthy because Mars had passed opposition in July 1655, giving a poor view.

1666

Mars is in opposition. Giovanni Domenico Cassini (1625–1712) draws Mars and determines a day length of 24 hours 40 minutes. Cassini made about 20 crude drawings of Mars at the observatory in Bologna, and from them noticed that markings came back to the same positions about forty minutes later than on the previous day. Cassini also saw the polar caps.

1671–1673

Jean Richer (1630–1696) travels to Cayenne, French Guyana, and measures the parallax of Mars at its perigee, at the behest of the French government. Later Cassini compares Richer's measurements with his own measurements of Mars' position relative to the stars, and the distances of Mars and the Sun from the Earth are determined. This produces the first reasonably accurate dimensions of the solar system. Cassini deduced an Earth-Sun distance, known as an Astronomical Unit, of 140 million km (87 million miles) compared to the modern value of 149.6 million km (93 million miles).

1672

In September, Mars is in opposition, and Huygens observes a white spot at the south pole of Mars, i.e. the southern polar cap.

1686

La Pluralité des Mondes (*The Inhabitation of Worlds*) is published by Bernard de Fontenelle, a respected French astronomer. The book, written as a dialogue, discusses evidence for life on planets in the solar system. Fontenelle, however, believed Mars to be uninhabitable, so the planet receives little attention: "It is also five times as small as the Earth [NB: really it has half the diameter] and receives much less sun. In short, Mars is not worth the trouble of stopping at. A much prettier choice would be Jupiter with her four moons!"

1687

Isaac Newton publishes his *Principia*, which introduces his principle of universal gravitation and provides a physical basis for the orbits of the planets.

1698

Huygens' *Cosmotheoros* is posthumously published (Huygens died in 1695 and had written the book some years earlier). This addresses the question of life on Mars – one of the earliest expositions on extraterrestrial life – and Huygens deduces that though Mars will be colder than Earth, because it is further from the Sun, life there will have adapted. He also discusses what is required for a planet to be capable of supporting life and speculates about intelligent extraterrestrials.

The 1700s: Mars, a planet similar to Earth

1704

Giacomo Filippo Maraldi (Cassini's nephew), of the Paris Observatory, records "white spots" at the poles but stops short of calling them ice caps. Because the southern pole is tilted towards the Earth, it is easier to observe and he discovers that the south cap is not centered on the rotational pole.

1719

Giacomo Maraldi suggests that the white spots at the poles of Mars could be interpreted as ice caps. Maraldi also notes that the southern cap changes in size and disappears in August and September, only to return later.

Mars is in opposition and closer to Earth than it would be for another 284 years (i.e. until 2003!). The brightness of Mars in the sky is interpreted as a bad omen and causes concern.

1727

Gulliver's Travels written by Jonathan Swift (1667–1745) seems to speculate that there are two moons of Mars, although this must have been a lucky guess.

1754

Abraham Kastner, a poet and anti-pluralist, publishes a poem about his pluralist friend, Christob Mylius, who had died in 1752. In the poem, Mylius' soul travels across the solar system. On Mars, Mylius meets the "eternal souls" of Martians.

1777–1783

Observations of Mars are made by Frederick Wilhelm "William" Herschel (1738–1822), the British Astronomer Royal. Sir William Herschel did various studies of Mars between 1777 and 1783 using telescopes that he made himself. In 1781, he discovered the planet Uranus, and this led George III of England to grant him a pension for life to study astronomy.

1784

A 30-degree axial tilt of Mars is identified by Herschel as published in his paper in *The Philosophical Transactions* entitled "On the remarkable appearances at the polar regions on the planet Mars, the inclination of its axis, the position of its poles, and its spheroidal figure; with a few hints relating to its real diameter and atmosphere". (Note: the modern scientific value for the axial tilt of Mars relative to its orbital plane is 25.19 degrees.) Hershel notes the seasonal changes of the polar caps and suggests they are snow and ice. He wrongly considers the dark areas to be oceans. On October 26 and 27 of 1783, Herschel observed two faint stars that passed near Mars, within a few seconds of arc. Because the light from these stars was not affected, Herschel correctly concludes that Mars has a tenuous atmosphere because he could see no effect on the near occultation of these dim stars. Hershel compares the remarkable similarity of Mars to the Earth: "The analogy between Mars and the earth is, perhaps, by far the greatest in the whole solar system. The diurnal motion is nearly the same; the obliquity of their respective ecliptics, on which the seasons depend, not very different; of all the superior planets the distance of Mars from the sun is by far the nearest alike to that of the earth: nor will the length of the martial year appear very different from that which we enjoy."

The 1800s: Martian cartography, the canal craze, and psychic connections to Mars

1800

Johann Hieronymus Schroeter, an enthusiastic amateur astronomer, does some Mars drawings. He is in regular correspondence with Herschel and owns telescopes made with Herschel's components.

1809

Honeré Flaugergues, a French amateur working at his private observatory in Viviers in southeastern France, notes the presence of "yellow clouds" on Mars. These were much later identified as dust clouds. Flaugergues later discovered the Great Comet of 1811.

1813

Flaugergues notices rapid melting of ice caps on Mars. He notes that markings were variable, and that in Martian spring, the polar cap shrinks rapidly. Flaugergues assumes that the cap comprises thick layers of ice and snow, and that its rapid melting signifies that Mars is hotter than the Earth.

1840

Drawings are made by Wilhelm Beer (1797–1850) and Johann von Maedler (1794–1874) at Beer's private observatory near Berlin. They generate a global map of Mars and make 3 determinations of the rotation period using baselines of 759, 1,604 and 2,234 days, the average value of which gave 24 hours, 37 minutes, 22.6 seconds (compared with modern science's textbook value of 24 hrs, 37 min, 22.7 sec). Earlier in 1836, Beer and Maedler had also generated the most complete map of the Moon, Mappa Selenographica, in their time, which remained unsurpassed until 1878, when a more detailed map appeared.

1854

William Whewell, a fellow of Trinity College, Cambridge University, and philosopher of science, theorizes about Mars. He supposes that it has green seas, red land, and possibly life forms. Earlier in 1830, Whewell introduced the term "scientist" to the English language.

1858

Pietro Angelo Secchi (1818–1878), a Jesuit monk and director of the Roman College Observatory, draws a map of Mars calling Syrtis Major the "Atlantic Canal". Secchi, despite his closeness to the Vatican, believed in the plurality of worlds. Earlier, in 1856 he wrote (in *Descrizione del nuovo osservatorio del collegio romano*): "it is with a sweet sentiment that man thinks of these worlds without number, where each star is a sun which, as minister of the divine bounty, distributes life and goodness to the other innumerable beings, blessed by the hand of the Omnipotent." He conceded that these worlds may not be accessible to his telescopes, but by analogy with the Earth and the solar system he was persuaded that the universe is a wonderful organism filled with life.

1859

Work begins on the Suez Canal, the engineering marvel of its time. Canals move commerce in many parts of the world, but this is the big one, considered equal to the pyramids. The importance of canals at this time in the nineteenth century no doubt influenced the later mistaken interest in "canals" on Mars.

1860

Emmanuel Liais proposes vegetation on Mars. He suggests that dark regions are not seas, as is commonly thought by other observers such as Secchi, but rather, are vegetation tracts.

1862

Sir Joseph Norman Lockyer (1836–1920), of the Royal College of Science in London (later known as Imperial College), makes drawings of Mars. He agrees with Secchi that the "green" areas of Mars are oceanic. Lockyer is best known as the discoverer of the element helium, which he identified from an emission line in the solar spectrum in 1870.

1862

Frederik Kaiser in Holland gives a rotational period of 24h 37m 22.62s (modern textbook value: 24h 37m, 22.663+/–0.002 s).

1863

The first color sketches of Mars are made by Father Pietro Angelo Secchi (1818–1878).

1864

The English astronomer, William Dawes (1799–1868), makes some exceptionally accurate drawings of Mars. Earlier in 1857, Dawes had observed Jupiter's Great Red Spot, several years before its existence was generally recognized.

1867

Richard Anthony Proctor, a British amateur astronomer and writer of popular astronomy, publishes a map of Mars with continents and oceans based on Dawes' drawings. His nomenclature, which names features after various astronomers, fails to find favor but his choice of zero meridian survives. Later a naming system prescribed by Schiaparelli is adopted.

1867

The first attempts are made to detect oxygen and water vapor spectroscopically, producing inconclusive results, by Pierre Jules Janssen (1824–1907) and Sir William Huggins (1824–1910).

1869

Father Secchi refers to "canali", Italian for channels.

1869

The Suez Canal is completed. In a letter to the first issue of the scientific journal *Nature* (November, 4), T. Login writes "The all-engrossing topic of the day is the Suez Canal, about which some diversities of opinion still exist." Everyone is talking about the canal!

1873

The red color of Mars is (wrongly) attributed to vegetation. Pop. Sci. Mo. v. IV p.190. In this article, Camille Flammarion suggests "May we attribute to the color of the herbage and plants which no doubt clothe the plains of Mars, the characteristic hue of that planet . . ."

1877

Giovanni Virginio Schiaparelli (1835–1910) director of the Brera Observatory in Milan, develops a new nomenclature for the mapping of Mars. He names the Martian "seas" and continents by taking the names from historic and mythological sources, as well as terrestrial lands and various terms for hell.

Mars comes into perihelic opposition in September, within 56 million km of the Earth. Schiaparelli sees "canali" on Mars, meaning channels. This later proves to be very significant in Mars folklore. Schiaparelli casually uses Secchi's terms canale and canali to describe streaks that he has recorded on the Martian surface. This gets mistranslated into English as "canals", which has connotations of Martian intelligent life.

Aug. 10: Asaph Hall (1829–1907), a largely self-taught American astronomer, comes from an impoverished family. By 1863, he is appointed professor of mathematics at the U.S. Naval Observatory in Washington, DC. On this day he gives up a search for Martian moons.

Aug. 11: Asaph Hall, who had resumed his search at the insistence of his wife, Angelina, detects a faint object near Mars.

Aug. 12: Is Mars inhabited?, an editorial in *The New York Times*. As the best opposition since 1798 approaches, questions in the popular mind come to the fore and the possibility of life on Mars is discussed in the press.

Aug. 18: Asaph Hall announces the discovery of Mars' two moons. At the suggestion of Henry Madan (1838–1901), the Science Master of Eton, England, Hall names the moons Phobos (fear) and Deimos (flight). These two names are variously attributed in Roman mythology to the sons of the god Mars by

Aphrodite and also to the horses that pulled Mars' chariot. Later from 1896 to 1901, Asaph Hall was professor of astronomy at Harvard. Later still, in 1930, Henry Madan's 11-year-old niece, Venetia Burney, suggests the name Pluto for that newly discovered planet to its discoverers.

Phobos is a heavily cratered, small, irregular body that measures 26.6 km (16.5 miles) across at its widest point. It orbits Mars every 7.65 hours at an average distance of 9,378 km (5,814 miles) in a nearly circular orbit that lies only 1 degree from Mars's equatorial plane. It has very low density, about 2 grams per cubic centimetre. It is being pulled towards Mars so that in about 1 billion years' time it will crash. Deimos is a small, irregular, cratered body measuring approximately 15 by 12 by 11 km (9 by 7 by 7 miles). It orbits the planet every 30.3 hours at an average distance of 23,459 km (14,545 miles) in a nearly circular orbit that lies within 2 degrees of Mars's equatorial plane. The satellite's longest axis is always directed toward Mars, so only one side faces Mars, rather like the Moon around the Earth.

Aug. 30: A third moon is allegedly discovered. No, really! *The New York Times* report that Dr Henry Draper of New York and Edward Singleton Holden of Washington claim to have jointly discovered the third moon at Dr Draper's private observatory at Hastings-on-the-Hudson. This discovery proved to be false; in fact, the proposed moon did not even obey Kepler's laws.

1879

American astronomer, Charles Augustus Young (1834–1908), makes accurate measurements of the diameter of Mars. He was professor of astronomy at Princeton University from 1877 to 1905, and author of *General Astrononmy* (1888).

Simon Newcomb (1835–1909), a Canadian-born American astronomer, founds the Astronomical Papers Prepared for the Use of the American Ephemeris and Nautical Almanac, a series of memoirs giving "a systematic determination of the constants of astronomy from the best existing data". These remarkably accurate tables were used throughout most of the world for calculating daily positions of celestial objects until 1959, and even afterward for the Sun, Mercury, Venus, and Mars.

Schiaparelli reports double canali, an example of a notorious phenomenon that came to be known as "gemination".

1880

Percy Gregg, a British author, publishes *Across the Zodiac*, a two-volume novel about a trip to Mars. His Mars had pale green skies and orange foliage.

1881–1882

Schiaparelli revises his Mars map, adding more canali which now include 20 examples of "gemination".

1882

Canals on the planet Mars are discussed in the press. *New York Times* Apr. 24, NYT Apr. 27., Richard Proctor waffles on the Canals of Mars. NYT May 2.

1885

"Vegetation on Mars may be red . . ."—Langley. Century, Mar. p.705. This article reiterates what Flammarion said about the color of Martian vegetation in 1873. "Why, we may ask, is not the Martian vegetation green? Why should it be—is the reply?", Flammarion later writes in his book *La Planète Mars*.

1890s

Edward Emerson Barnard (1857–1923), an astronomer at Lick Observatory, Mt. Hamilton, California, observes Mars in a position not directly opposite the Sun, when detail is revealed through shadowing. Renowned for his remarkable eyesight, he observes Martian craters but does not make the observation public.

1891

A rich French widow, Clara Gouguet Guzman, offers a prize of 100,000 francs for communication with extraterrestrials. The prize is to be awarded to "the person of whatever nation who will find the means within the next ten years of communicating with a star (planet or otherwise) and of receiving a response". The prize was administered by the French Academy of Sciences and was named the Pierre Guzman Prize after Mme. Guzman's son. Mme. Guzman excludes Mars, considering it "too easy" to contact!

1892

Nicolas Camille Flammarion publishes Volume 1 (608 pages) of his encyclopaedia of *La Planète Mars et ses Conditions d'Habitabilité* (Gauthier-Villars et Fils, Paris).

Telegraphing to Mars with solar signals in *The Spectator*, Ap 13. This is one of the first articles that deals with the language problems involved in communicating with the Martials (sic). The article points out that mathematical information may perhaps be exchanged, but questions how we will communicate abstract concepts: "How are we to ask if Martials [sic] have engineers and ships, and electric lights and glaciers and five senses, and heads and feet . . ."? [Ed.- note that sometimes in the 19th century, Martians were referred to as 'Martials'].

Camille Flammarion suggests communication with the Martians. Flammarion was familiar with experiments Edison had done with long telephone lines. Edison picked up sounds he felt were caused by "terrestrial magnetism" years before Marconi. Flammarion suggests the natural magnetism of the Earth might be harnessed to propagate sounds across space. (NB 1894: "Wireless" telegraphy is demonstrated by Sir Oliver Lodge.)

1894

Percival Lowell (1855–1916) builds the Lowell Observatory at Flagstaff, Arizona, and makes his first observations of Mars.

Edward Emerson Barnard (1857–1923) reports on his observations of Mars including his complete failure to detect canals.

1895

Mars, by Percival Lowell is published.

The *New York Herald* claims that surface features on Mars are observed to form the Hebrew word for God.

1894–96

Helene Smith, a Swiss medium of real name Catherine Elise Muller from Geneva, has visions of Mars while under hypnosis induced by the eminent psychologist Theodore Flournoy. Smith imagines herself standing on Mars and meets Martians. She can even speak Martian, which is similar to French. Flournoy later describes his subject in *From India to the Planet Mars* (Harper and Bros, 1900).

1895

Mrs Smead, an American medium, communicates with her dead daughter and brother-in-law on Mars. Smead describes canals and Martians very similar to humans. Smead was examined by psychologist Prof. J. Hyslop, who concluded she had a multiple personality disorder. (J. Hyslop, *Psychical Research and the Resurrection*, Small/Maynard, 1908; and also 'Communicating with Mars', editorial in the *Independent* (periodical), p.1042–43, 1909.)

1897

Herbert G. Wells' (1866–1946) *The War of the Worlds* is serialized in *Pearson's Magazine*. It is also printed in the US in *The Cosmopolitan*.

1898

The War of the Worlds is published in hardback.

Garrett Serviss publishes *Edison's Conquest of Mars*. In this "sequel" to The War of the Worlds, Americans retroengeneer Martian flying machines, go to Mars, and fight Martians! Among the scientist-warriors who accompanied the Americans were Lord Kelvin, Lord Rayleigh, Professor Roentgen, and (guess what?) popular science writer, Garrett Serviss.

1899

Carl Jung's 15-year-old patient, "Miss S. W.", goes to Mars in trances, and sees canals and Martians in flying machines. Jung deduces that S. W. is suffering from a dissociated personality. (Jung, C., *Zur Psychologie und Pathologie sogennter Occulter Phanomene*, Muntze, 1902).

The 1900s: Gradual end of old superstitions and dawning of the Space Age

1901

American astronomer William Henry Pickering, the director of the Lick Observatory, reports "shaft of light" seen to project from Mars (*New York Times*, Jan 16).

Nikola Tesla (1856–1943), a brilliant Serbian-American inventor and scientist, is building a wireless system to communicate with Martians. (N. Tesla, "Talking with the planets", *Collier's Weekly*, Vol 24, 4–5, 1901).

1902
Guglielmo Marconi sends the first official wireless message across the Atlantic from Glace Bay, Nova Scotia, Canada to England.

1903
The Wright brothers' first airplane flight.

1905
Interplanetary Telephone: Nikola Tesla may use an oscillator to "wake up" Mars according to *The New York Times*, Jan 15.

Photograph Mars Canals-Lowell. On May 28, *The New York Times* prints Lowell's report that the canals of Mars have been photographed for the first time.

1906
Mars and Its Canals by Lowell is published.

Earl C. Slipher arrives at the Lowell Observatory. He continues photographic studies begun in 1901 into the 1960s. In total, 126,000 images are taken.

1907
Prof. David Peck Todd of Amherst College, and Slipher go to open desert in Alianzá, Chile, to photograph Mars to get the clearest seeing possible. Todd and crew took a specially built camera.

Nikola Tesla in letter to *The New York Times* (Jan 23); "I can easily bridge the gulf which separates us from Mars". A testy Tesla rages at the press for calling his transmitter nothing more than a "useful piece of electrical apparatus".

Slipher makes photos of the Martian canals. At least this is what the July 3 *New York Times* reports after receiving a telegram from Lowell which reads "Todd of the Lowell expedition to the Andes, cables Mars canals photographed there by Slipher."

December: *Century Magazine* prints the photos of Mars: tiny, disappointing photos that were Lowell's "proof". Even at a 2-diameter enlargement, they are less than half a centimeter wide. Further enlargements only show successive loss of detail due to enlargement of emulsion grains.

". . . the proof by astronomical observations. . . that conscious, intelligent human life exists upon the planet Mars," is reported as one of the most momentous events of 1907 by the *Wall Street Journal.*

1908
Percival Lowell's *Mars as the Abode of Life* is first published in *Century Magazine* as a series of articles defending the hypothesis of Martian life. It is later published as a book by Macmillan, New York.

1909

George Ellery Hale (1868–1938), using the Mt. Wilson 60" reflector, sees "... not a trace" of canals.

William Pickering proposes mirroring signals to Mars: a signaling system of sufficient size can be constructed for $10M, he argues.

W. W. Campbell of Lick Observatory, tries to measure water vapor in the Martian atmosphere using spectroscopy from an expedition to Mount Witney. Results are negative and he (correctly) concludes that the Martian atmosphere is extremely arid compared to the Earth.

Camille Flammarion publishes the second volume of his encyclopaedia of Mars *La Planète Mars et ses Conditions d'Habitabilité* (Gauthier-Villars et Fils, Paris). Vol 2 (595 pages) contains 426 drawings and 16 maps from the period 1860 to 1901.

1911

An unfortunate dog, minding its own business in Nakhla, Egypt, is struck by part of a meteorite and killed. Later, in the 1980s, this meteorite is identified as one of a small group originating from Mars. There are currently 16 meteorites from Mars. These were formerly classed as "SNC meteorites", refering to the places where typical meteorites of their kind where found (Shergotty-Nakhla-Chassigny); however, now the term "Martian meteorites" is preferred because some recent meteorites, most notably the oldest one, ALH84001, cannot be accurately classified as SNC. What's certain, however, is that that dog was really unlucky.

"Martians Build Two Immense Canals in Two Years" (*New York Times*, Aug 27).

"Frost on Mars" – Lowell (*New York Times*, Nov 10).

A Princess of Mars, the first of eleven "John Carter on Mars" novels, is published. The author, Edgar Rice Burroughs, uses Schiaparelli's nomenclature and some of his Martians have green skin.

1912

Svante Arrhenius, a Nobel Prize-winning chemist, suggests that certain minerals known as hygroscopic salts might be responsible for changes in the surface markings on Mars. Such salts absorb water and can show dramatic darkening on contact with it. His ideas fail to find support.

Burroughs' *Princess of Mars* is serialized in *All-Story* magazine.

1914

The Panama Canal is completed.

1918

Autumn: The first performance of Gustav Holst's (1874–1934) "The Planets Suite" at a private concert conducted by Adrian Boult. In March of 1913, Holst received an anonymous gift which enabled him to travel to Spain with Clifford Bax, the brother of the composer Arnold Bax (and later the librettist for Holst's opera "The

Wandering Scholar"). Clifford Bax was an astrologer, and Bax introduced Holst to the concepts of astrology, which inspired him to compose "The Planets Suite". The traditional order in performance of Holst's suite is: Mars, Venus, Mercury, Jupiter, Saturn, Uranus and Neptune. Holst considered this as a progression analogous to going through life. There is no piece for Pluto because this planet was not discovered until 1930. The Mars piece is called "Mars, the Bringer of War". However, "Jupiter, the Bringer of Jollity" is perhaps most people's favorite music from the suite. The first complete performance was under Albert Coates in Queen's Hall, London, in 1920.

1919
Spring: Guglielmo Marconi announced that several of his radio stations were picking up very strong signals "seeming to come from beyond the earth". Nikola Tesla, another prominent inventor, believed these signals were coming from Mars.

1922
Estonian astronomer, Ernest Julius Opik (1893–1985), from his work on meteors, accurately predicts the frequency of craters on Mars many years before they could be ascertained.

1925
Donald Menzel, at Lowell Observatory, concludes that air pressure on Mars would be less than 66 millibars (about 1/15 that of Earth) from studying photos of Mars taken in different wavelengths of light. He also notes that under different assumptions this upper limit would be 26 millibars (1/39 that of Earth). (Note: actually the average air pressure on Mars is about 6 millibars, 1/170 that on Earth, which is 1013 millibars.)

1926
Walter S. Adams correctly determines that Mars is "ultra-arid" by studying spectral lines.

1927
William Coblentz (1873–1962) and Carl Lampland measure large day-night temperature differences on Mars. This implies a thin atmosphere. Their measurements of the temperatures on Mars [wrongly] suggested that the equator is near 15–30 C, while the pole is near –50 to –70 C. Such temperatures are rather similar to Earth. Consequently, the measurements did much to encourage the belief in Martian life right up until space exploration of Mars. (Coblentz, W., and Lampland, C. "Further radiometric measurements and temperature estimates for the planet Mars", *Scientific Papers of Nat. Bur. Studies*, Vol. 22, 237–276).

1929
Bernard Lyot at the Meudon Observatory near Paris concludes from polarimetric data that an upper limit to the atmospheric pressure on Mars is about 24 millibars.

1930

La Planète Mars, 1659–1929, by Eugenios M. Antoniadi, is published (Herman et Cie, Paris). This is a complete and representative summary of the surface of Mars based on telescope observations. Antoniadi was an astronomer at the Meudon Observatory near Paris who produced maps of Mars that were some of the best available up until the 1950s.

Attempts to detect oxygen on Mars fail.

1938

Oct. 30: A dramatized version of *The War of the Worlds* is broadcast by Orson Welles on American radio, which has Martians landing at Grovers Mill, New Jersey. It is estimated that 6 million people listened to the show, and despite repeated announcements that it was a play, at least 1 million people thought it was real.

1939

Earl Slipher photographs Mars in opposition and reports in 1940 that there are many canals. In contrast, George Ellery Hale, who used a 60 inch reflector at Mt Wilson says that no canals can be seen.

1947

Carbon dioxide, but no oxygen, is detected by Gerard Peter Kuiper (1905–1973) on infrared spectrograms. Kuiper deduces that the Martian atmosphere contains twice as much carbon dioxide as the Earth's. For two decades, other than water vapor, carbon dioxide remained the only known constituent of the Martian atmosphere, although it was presumed to be a minor constituent rather than the major constituent that it really is.

Das Mars Projekt (Mars Project) consisting of 10 ships and 70 crewmembers is proposed by Werner von Braun (1912–1977). This fleet of ten ships would go to Mars, explore, and return in about 520 days. The ships would be assembled in high orbit above the Earth. The ships were then to take a long elliptical orbit around the sun, eventually reaching the orbit of Mars. There, "landing boats" would descend to Mars. Afterwards, the party would board seven of the ships, go into an elliptical orbit around the sun until reaching Earth's orbit, where rockets would slow them back down.

1949

Lichens are proposed as a possible form of life for Mars. Conditions seem too harsh for more complex forms of life.

1950

The Martian Chronicles by Ray Bradbury is published. It has since been reprinted repeatedly (64 editions as of 1985). Some of the stories contained within the book had been published as early as 1946.

Jan. 28: The *Los Angeles Times* reports that the Japanese astronomer Sadao Saeki has seen a huge explosion on Mars on Jan 16 which produced a mushroom cloud 1,450 km in diameter "like the terrific explosion of a volcano". No other people observed this explosion.

1956
A global dust storm grows on Mars. The storm begins on August 20 with a bright cloud over the Hellas-Noachis region that spreads to engulf the whole planet by mid-September.

1957
The Space Age begins with the Soviet launch of Sputik 1, an Earth-orbiting satellite, on October 4.

1962
Werner von Braun considers the "Mars Project" to be 15–20 years away.

1963
Hyron Spinrad and co-workers report measurements of water in the atmosphere of Mars from spectroscopic observations. A column abundance of 14+/–7 precipitable microns is determined. "Precipitable microns" measures the the depth of water that forms if all the water in the atmosphere were condensed out onto the surface (1 micron = 1 millionth of a meter). 14 precipitable microns is less than one thousandth of the water typically in the atmosphere above the Sahara desert. Thus the Martian atmosphere was determined to be extremely dry.

1964
Lewis Kaplan reports that the pressure of carbon dioxide on Mars is low, about 4 millibars, from analysis of the same spectra as Hyron Spinrad.

1965
July 15: Mariner 4 flyby of Mars – the first successful space probe to study Mars. Space exploration of Mars begins.

UNVEILING A PARALLEL (1893)

CHAPTER 3: THE AURORAS' ANNUAL

Alice Ilgenfritz Jones and Ella Marchant

Alice Ilgenfritz Jones (1846–1905) and **Ella Marchant** (1857–1916) lived in Cedar Rapids, Iowa. Jones was an established author of two novels, *High Water Mark* (1879) and *Beatrice of Bayou Têche* (1895). Marchant owned newspapers with her husband, Stoddard Marchant, including the *Daily Inter-Mountain* and the *Cedar Rapids Daily Republican.*

It was winter, and snow was on the ground; white and sparkling, and as light as eider-down. Elodia kept a fine stable. Four magnificent white horses were harnessed to her sleigh, which was in the form of an immense swan, with a head and neck of frosted silver. The body of it was padded outside with white varnished leather, and inside with velvet of the color of a dove's breast. The robes were enormous skins of polar bears, lined with a soft, warm fabric of wool and silk. The harness was bestrung with little silver bells of most musical and merry tone; and all the trappings and accoutrements were superb. Elodia had luxurious tastes, and indulged them.

Every day we took an exhilarating drive. The two deep, comfortable seats faced each other like seats in a landau. Severnius and I occupied one, and Elodia the other; so that I had the pleasure of looking at her whenever I chose, and of meeting her eyes in conversation now and then, which was no small part of my enjoyment. The mere sight of her roused the imagination and quickened the pulse. Her eyes were unusually dark, but they had blue rays, and were as clear and beautiful as agates held under water. In fact they seemed to swim in an invisible liquid. Her complexion had the effect of alabaster through which a pink light shines,—deepest in the cheeks, as though they were more transparent than the rest of her face. Her head, crowned with a fascinating little cap, rose above her soft furs like a regal flower. She was so beautiful that I wondered at myself that I could bear the sight of her.

Strange to say, the weather was not cold, it was simply bracing,—hardly severe enough to make the ears tingle.

The roads were perfect everywhere, and we often drove into the country. The horses flew over the wide white stretches at an incredible speed.

One afternoon when, at the usual hour, the coachman rang the bell and announced that he was ready, I was greatly disappointed to find that we were not to have Elodia. But I said nothing, for I was shy about mentioning her name.

When we were seated, Severnius gave directions to the driver.

"Time yourself, Giddo, so that you will be at the Public Square at precisely three o'clock," said he, and turned to me. "We shall want to see the parade."

"What parade?" I inquired.

"Oh! has not Elodia told you? This is The Auroras' Annual,—a great day. The parade will be worth seeing."

In the excitement of the drive, and in my disappointment about not having Elodia with us, I had almost forgotten about The Auroras' Annual, when three o'clock came. I had seen parades in New York City, until the spectacle had calloused my sense of the magnificent, and I very much doubted whether Mars had anything new to offer me in that line.

Punctual to the minute, Giddo fetched up at the Square,—among a thousand or so of other turnouts,—with such a flourish as all Jehus love. We were not a second too soon. There was a sudden burst of music, infinitely mellowed by distance; and as far up the street as the eye could well reach there appeared a mounted procession, advancing slowly. Every charger was snow white, with crimped mane and tail, long and flowing, and with trappings of various colors magnificent in silver blazonry.

The musicians only were on foot. They were beating upon drums and blowing transcendent airs through silver wind instruments. I do not know whether it was some quality of the atmosphere that made the strains so ravishing, but they swept over one's soul with a rapture that was almost painful. I could hardly sit still, but I was held down by the thought that if I should get up I would not know what to do. It is a peculiar sensation.

On came the resplendent column with slow, majestic movement; and I unconsciously kept time with the drums, with Browning's stately lines on my tongue, but unspoken:

> "Steady they step adown the slope, Steady they climb
> the hill."

There was no hill, but a very slight descent. As they drew nearer the splendor of the various uniforms dazzled my eyes. You will remember that everything about us was white; the buildings all of white stone or brick, the ground covered with snow, and the crowds of people lining the streets all dressed in the national color, or no-color.

There were several companies in the procession, and each company wore distinguishing badges and carried flags and banners peculiar to itself.

The housings on the horses of the first brigade were of yellow, and all the decorations of the riders corresponded; of the second pale blue, and of the third sky-pink. The uniforms of the riders were inconceivably splendid; fantastic and gorgeous head-gear, glittering belts, silken scarfs and sashes, badges and medals flashing with gems, and brilliant colors twisted into strange and curious devices.

As the first division was about to pass, I lost my grip on myself and half started to my feet with a smothered exclamation, "Elodia!"

Severnius put out his hand as though he were afraid I was going to leap out of the sleigh, or do something unusual.

"What is it?" he cried, and following my gaze he added, "Yes, that is Elodia in front; she is the Supreme Sorceress of the Order of the Auroras."

"The—*what*!"

"Don't be frightened," he laughed; "the word means nothing,—it is only a title."

I could not believe him when I looked at the advancing figure of Elodia. She sat her horse splendidly erect. Her fair head was crowned with a superb diadem of gold and topazes, with a diamond star in the centre, shooting rays like the sun. Her expression was grave and lofty; she glanced neither to right nor left, but gazed straight ahead—at nothing, or at something infinitely beyond mortal vision. Her horse champed its bits, arched its beautiful neck, and stepped with conscious pride; dangling the gold fringe on its sheeny yellow satin saddle-cloth, until one could hardly bear the sight.

"The words mean nothing!" I repeated to myself. "It is not so; Severnius has deceived me. His sister is a sorceress; a—I don't know what! But no woman could preserve that majestic mien, that proud solemnity of countenance, if she were simply—playing! There is a mystery here."

I scrutinized every rider as they passed. There was not a man among them,—all women. Their faces had all borrowed, or had tried to borrow, Elodia's queenly look. Many of them only burlesqued it. None were as beautiful as she.

When it was all over, and the music had died away in the distance, we drove off,—Giddo threading his way with consummate skill, which redounded much to his glory in certain circles he cared for, through the crowded thoroughfares.

I could not speak for many minutes, and Severnius was a man upon whom silence always fell at the right time. I never knew him to break in upon another's mood for his own entertainment. Nor did he spy upon your thoughts; he left you free. By-and-by, I appealed to him:

"Tell me, Severnius, what does it mean?"

"This celebration?" returned he. "With pleasure. Giddo, you may drive round for half an hour, and then take us to the Auroras' Temple,—it is open to visitors to-day."

We drew the robes closely, and settled ourselves more comfortably, as we cleared the skirts of the crowd. It was growing late and the air was filled with fine arrows of frost, touched by the last sunbeams,—their sharp little points stinging our faces as we were borne along at our usual lively speed.

"This society of the Auroras," said Severnius, "originated several centuries ago, in the time of a great famine. In those days the people were poor and improvident, and a single failure in their crops left them in a sorry condition. Some of the wealthiest women of the country banded themselves together and worked systematically for the relief of the sufferers. Their faces appeared so beautiful, and beamed with such a light of salvation as they went about from hut to hut, that they got the name of 'auroras' among the simple poor. And they banished want and hunger so magically, that they were also called 'sorcerers'."

"O, then, it is a charitable organization?" I exclaimed, much relieved.

"It was," replied Severnius. "It was in active operation for a hundred or so years. Finally, when there was no more need of it, the State having undertaken the care of its poor, it passed into a sentiment, such as you have seen to-day."

"A very costly and elaborate sentiment," I retorted.

"Yes, and it is growing more so, all the time," said he. "I sometimes wonder where it is going to stop! For those who, like Elodia, have plenty of money, it does not matter; but some of the women we saw in those costly robes and ornaments can ill afford them,—they mean less of comfort in their homes and less of culture to their children."

"I should think their husbands would not allow such a waste of money," I said, forgetting the social economy of Mars.

"It does not cost any more than membership in the orders to which the husbands themselves belong," returned he. "They argue, of course, that they need the recreation, and also that membership in such hightoned clubs gives them and their children a better standing and greater influence in society."

Severnius did not forget his usual corollary,—the question with which he topped out every explanation he made about his country and people.

"Have you nothing of the sort on the Earth?" he asked.

"Among the women?—we have not," I answered.

"I did not specify," he said.

"O, well, the men have," I admitted; "I belong to one such organization myself,—the City Guards."

"And you guard the city?"

"No; there is nothing to guard it against at present. It's a 'sentiment,' as you say."

"And do you parade?"

"Yes, of course, upon occasion,—there are certain great anniversaries in our nation's history when we appear."

"And why not your women?"

I smiled to myself, as I tried to fancy some of the New York ladies I knew, arrayed in gorgeous habiliments for an equestrian exhibition on Broadway. I replied,

"Really, Severnius, the idea is entirely new to me. I think they would regard it as highly absurd."

"Do they regard you as absurd?" he asked, in that way of his which I was often in doubt about, not knowing whether he was in earnest or not.

"I'm sure I do not know," I said. "They may,—our women have a keen relish for the ludicrous. Still, I cannot think that they do; they appear to look upon us with pride. And they present us with an elaborate silken banner about once a year, stitched together by their own fair fingers and paid for out of their own pocket money. That does not look as though they were laughing at us exactly."

I said this as much to convince myself as Severnius.

The half-hour was up and we were at the Temple gate. The building, somewhat isolated, reared itself before us, a grand conception in chiseled marble, glinting in the brilliant lights shot upon it from various high points. Already it was dark beyond the radius of these lights,—neither of the moons having yet appeared.

Severnius dismissed the sleigh, saying that we would walk home,—the distance was not far,—and we entered the grounds and proceeded to mount the flight of broad steps leading up to the magnificent arched entrance. The great carved doors,—the carvings were emblematic,—swung back and admitted us. The Temple was splendidly illuminated within, and imagination could not picture anything more imposing than the great central hall and winding stairs, visible all the way up to the dome.

Below, on one side of this lofty hall, there were extensive and luxurious baths. Severnius said the members of the Order were fond of congregating here,—and I did not wonder at that; nothing that appertains to such an establishment was lacking. Chairs and sofas that we would call "Turkish," thick, soft rugs and carpets, pictures, statuary, mirrors, growing plants, rare flowers, books, musical instruments. And Severnius told me the waters were delightful for bathing.

The second story consisted of a series of spacious rooms divided from each other by costly portieres, into which the various emblems and devices were woven in their proper tinctures.

All of these rooms were as sumptuously furnished as those connected with the baths; and the decorations, I thought, were even more beautiful, of a little higher or finer order.

In one of the rooms a lady was playing upon an instrument resembling a harp. She dropped her hands from the strings and came forward graciously.

"Perhaps we are intruding?" said Severnius.

"Ah, no, indeed," she laughed, pleasantly; "no one could be more welcome here than the brother of our Supreme Sorceress!"

"Happy the man who has a distinguished sister!" returned he.

"I am unfortunate," she answered with a slight blush. "Severnius is always welcome for his own sake."

He acknowledged the compliment, and with a certain reluctance, I thought, said, "Will you allow me, Claris, to introduce my friend—from another planet?"

She took a swift step toward me and held out her hand.

"I have long had a great curiosity to meet you, sir," she said.

I bowed low over her hand and murmured that her curiosity could not possibly equal the pleasure I felt in meeting her.

She gave Severnius a quick, questioning look. I believe she thought he had told me something about her. He let her think what she liked.

"How is it you are here?" he asked.

"You mean instead of being with the others?" she returned. "I have not been well lately, and I thought—or my husband thought—I had better not join the procession. I am awaiting them here."

As she spoke, I noticed that she was rather delicate looking. She was tall and slight, with large, bright eyes, and a transparent complexion. If Elodia had not filled all space in my consciousness I think I should have been considerably interested in her. I liked her frank, direct way of meeting us and talking to us. We soon left her and continued our explorations.

I wanted to ask Severnius something about her, but I thought he avoided the subject. He told me, however, that her husband, Massilia, was one of his closest friends. And then he added, "I wonder that she took his advice!"

"Why so," I asked; "do not women here ever take their husbands' advice?"

"Claris is not in the habit of doing so," he returned with, I thought, some severity. And then he immediately spoke of something else quite foreign to her.

The third and last story comprised an immense hall or assembly room, and rows of deep closets for the robes and paraphernalia of the members of the Order. In one of these closets a skeleton was suspended from the ceiling and underneath it stood a coffin. On a shelf were three skulls with their accompanying cross-bones, and several cruel-looking weapons.

Severnius said he supposed these hideous tokens were employed in the initiation of new members. It seemed incredible. I thought that, if it were so, the Marsian women must have stronger nerves than ours.

A great many beautiful marble columns and pillars supported the roof of the hall, and the walls had a curiously fluted appearance. There was a great deal of sculpture, not only figures, but flowers, vines, and all manner of decorations,—even draperies chiseled in marble that looked like frozen lace, with an awful stillness in their ghostly folds. There was a magnificent canopied throne on an elevation like an old-fashioned pulpit, and seats for satellites on either side, and at the base. If I had been alone, I would have gone up and knelt down before the throne,—for of course that was where Elodia sat,—and I would have kissed the yellow cushion on which her feet were wont to rest when she wielded her jeweled scepter. The scepter, I observed, lay on the throne-chair.

There was an orchestra, and there were "stations" for the various officials, and the walls were adorned with innumerable cabalistic insignia. I asked Severnius if he knew the meaning of any of them.

"How should I know?" he replied in surprise. "Only the initiates understand those things."

"Then these women keep their secrets," said I.

"Yes, to be sure they do," he replied.

The apartment to the right, on the entrance floor, opposite the baths, was the last we looked into, and was a magnificent banquet hall. A servant who stood near the door opened it as though it had been the door of a shrine, and no wonder! It was a noble room in its dimensions and in all its unparalleled adornments and appurtenances.

The walls and ceiling bristled with candelabra all alight. The tables, set for a banquet, held everything that could charm the eye or tempt the appetite in such a place.

I observed a great many inverted stemglasses of various exquisite styles and patterns, including the thin, flaring goblets, as delicate as a lily-cup, which mean the same thing to Marsians as to us.

"Do these women drink champagne at their banquets?" I asked, with a frown.

"O, yes," replied Severnius. "A banquet would be rather tame without, wouldn't it? The Auroras are not much given to drink, ordinarily, but on occasions like this they are liable to indulge pretty freely."

"Is it possible!" I could say no more than this, and Severnius went on:

"The Auroras, you see, are the cream of our society,—the *elite*,—and costly drinks are typical, in a way, of the highest refinement. Do you people never drink wine at your social gatherings?"

"The men do, of course, but not the women," I replied in a tone which the whole commonwealth of Paleveria might have taken as a rebuke.

"Ah, I fear I shall never be able to understand!" said he. "It is very confusing to my mind, this having two codes—social as well as political—to apply separately to members of an identical community. I don't see how you can draw the line so sharply. It is like having two distinct currents in a river-bed. Don't the waters ever get mixed?"

"You are facetious," I returned, coldly.

"No, really, I am in earnest," said he. "Do no women in your country ever do these things,—parade and drink wine, and the like,—which you say you men are not above doing?"

I replied with considerable energy:

"I have never before to-day seen women of any sort dress themselves up in conspicuous uniforms and exhibit themselves publicly for the avowed purpose of being seen and making a sensation, except in circuses. And circus women,—well, they don't count. And of course we have a class of women who crack champagne bottles and even quaff other fiery liquors as freely as men, but I do not need to tell you what kind of creatures those are."

At that moment there were sounds of tramping feet outside, and the orchestra filed in at the farther end of the *salon* and took their places on a high dais. At a given signal every instrument was in position and the music burst forth, and simultaneously the banqueters began to march in. They had put off their heavy outside garments but retained their ornaments and insignia. Their white necks and arms gleamed bewitchingly through silvered lace. They moved to their places without the least jostling or awkwardness, their every step and motion proving their high cultivation and grace.

"We must get out of here," whispered Severnius in some consternation. But a squad of servants clogged the doorway and we were crowded backward, and in the interest of self-preservation we took refuge in a small alcove behind a screen of tall hot-house plants with enormous leaves and fronds.

"Good heavens! what shall we do?" cried Severnius, beginning to perspire.

"Let us sit down," said I, who saw nothing very dreadful in the situation except that it was warm, and the odor of the blossoms in front of us was overpowering. There was a bench in the alcove, and we seated ourselves upon it,—I with much comfort, for it was a little cooler down there, and my companion with much fear.

"Would it be a disgrace if we were found here?" I asked.

"I would not be found here for the world!" replied Severnius. "It would not be a disgrace, but it would be considered highly improper. Or, to put it so that you can better understand it, it would be the same as though they were men and we women."

"That is clear!" said I; and I pictured to myself two charming New York girls of my acquaintance secreting themselves in a hall where we City Guards were holding a banquet,—ye gods!

As the feast progressed, and as my senses were almost swept away by the scent of the flowers, I sometimes half fancied that it was the City Guards who were seated at the tables.

During the first half-hour everything was carried on with great dignity, speakers being introduced—this occurred in the interim between courses—in proper order, and responding with graceful and well-prepared remarks, which were suitably applauded. But after the glasses had been emptied a time or two all around, there came a change with which I was very familiar. Jokes abounded and jolly little songs were sung,—O, nothing you would take exception to, you know, if they had been men; but women! beautiful, cultivated, charming women, with eyes like stars, with cheeks that matched the dawn, with lips that you would have liked to kiss! And more than this: the preservers of our ideals, the interpreters of our faith, the keepers of our consciences! I felt as though my traditionary idols were shattered, until I remembered that these were not my countrywomen, thank heaven!

Severnius was not at all surprised; he took it all as a matter of course, and was chiefly concerned about how we were going to get out of there. It was more easily accomplished than we could have imagined. The elegant candelabra were a cunningly contrived system of electric lights, and, as sometimes happens with us, they went out suddenly and left the place in darkness for a few convenient seconds. "Quick, now!" cried Severnius with a bound, and there was just time for us to make our escape. We had barely reached the outer door when the whole building was ablaze again.

Severnius offered no comments on the events of the evening, except to say we were lucky to get out as we did, and of course I made none. At my suggestion we stopped at the observatory and spent a few hours there. Lost among the stars, my soul recovered its equilibrium. I have found that little things cease to fret when I can lift my thoughts to great things.

It must have been near morning when I was awakened by the jingling of bells, and a sleigh driving into the *porte cochere*. A few moments later I heard Elodia and her maid coming up the stairs. Her maid attended her everywhere, and stationed herself about like a dummy. She was the sign always that Elodia was not far off; and I am sure she would have laid down her life for her mistress, and would have suffered her tongue to be cut out before she would have betrayed her secrets. I tell you this to show you what a power of fascination Elodia possessed; she seemed a being to be worshiped by high and low.

Severnius and I ate our breakfast alone the following morning. The Supreme Sorceress did not get up, nor did she go down town to attend to business at all during the day. At lunch time she sent her maid down to tell Severnius that she had a headache.

"Quite likely," he returned, as the girl delivered her message; "but I am sorry to hear it. If there is anything I can do for her, tell her to let me know."

The girl made her obeisance and vanished.

"We have to pay for our fun," said Severnius with a sigh.

"I should not think your sister would indulge in such 'fun'!" I retorted as a kind of relief to my hurt sensibilities, I was so cruelly disappointed in Elodia.

"Why my sister in particular?" returned he with a look of surprise.

"Well, of course, I mean all those women,—why do they do such things? It is unwomanly, it—it is disgraceful!"

I could not keep the word back, and for the first time I saw a flash of anger in my friend's eyes.

"Come," said he, "you must not talk like that! That term may have a different signification to you, but with us it means an insult."

I quickly begged his pardon and tried to explain to him.

"Our women," I said, "never do things of that sort, as I have told you. They have no taste for them and no inclination in that direction,—it is against their very nature. And if you will forgive me for saying so, I cannot but think that such indulgence as we witnessed last night must coarsen a woman's spiritual fibre and dull the fine moral sense which is so highly developed in her."

"Excuse me," interposed Severnius. "You have shown me in the case of your own sex that human nature is the same on the Earth that it is on Mars. You would not have me think that there are two varieties of human nature on your planet, corresponding with the sexes, would you? You say 'woman's' spiritual fibre and fine moral sense, as though she had an exclusive title to those qualities. My dear sir, it is impossible! you are all born of woman and are one flesh and one blood, whether you are male or female. I admit all you say about the unwholesome influence of such indulgence as wine drinking, late hours, questionable stories and songs,—a night's debauch, in fact, which it requires days sometimes to recover from,—but I must apply it to men as well as women; neither are at their best under such conditions. I think," he went on, "that I begin to understand the distinction which you have curiously mistaken for a radical difference. Your women, you say, have always been in a state of semi-subjection—"

"No, no," I cried, "I never said so! On the contrary, they hold the very highest place with us; they are honored with chivalrous devotion, cared for with the tenderest consideration. We men are their slaves, in reality, though they call us their lords; we work for them, endure hardships for them, give them all that we can of wealth, luxury, ease. And we defend them from danger and save them every annoyance in our power. They are the queens of our hearts and homes."

"That may all be," he replied coolly, "but you admit that they have always been denied their political rights, and it follows that their social rights should be similarly limited. Long abstinence from the indulgences which you regard as purely masculine, has resulted in a habit merely, not a change in their nature."

"Then thank heaven for their abstinence!" I exclaimed.

"That is all very well," he persisted, "but you must concede that in the first place it was forced upon them, and that was an injustice, because they were intelligent beings and your equals."

"They ought to thank us for the injustice, then," I retorted.

"I beg your pardon! they ought not. No doubt they are very lovely and innocent beings, and that your world is the better for them. But they, being restricted in other ways by man's authority, or his wishes, or by fear of his disfavor perhaps, have acquired these gentle qualities at the expense of—or in the place of—others more essential to the foundation of character; I mean strength, dignity, self-respect, and that which you once attributed to my sister,—responsibility."

I was bursting with indignant things which I longed to say, but my position was delicate, and I bit my tongue and was silent.

I will tell you one thing, my heart warmed toward my gentle countrywomen! With all their follies and frivolities, with all their inconsistencies and unaccountable ways, their whimsical fancies and petty tempers, their emotions and their susceptibility to new isms and religions, they still represented my highest and best ideals. And I thought of Elodia, sick upstairs from her last night's carousal, with contempt.

JOURNEYS TO THE PLANET MARS OR OUR MISSION TO ENTO (1903)

CHAPTER 2: SPECIAL FEATURES OF MARS

Sara Weiss

Born to a 'narrowly religious' Ohio farmer and possessing very little formal education, **Sara Weiss** was not an author in the traditional sense, and her book was not presented as a novel. Instead, she was a medium, and the journeys she described were written as memories of the spiritual voyages to the planet she experienced in the company of a spirit called Carl De L'Ester between 6 October 1892 and 16 September 1894. *Journeys to the Planet Mars* was published in 1903, a year before her death. Her second book, *Decimon Hûŷdas: A Romance of the Planet Mars* was published posthumously in 1906. Her date of birth is unknown but she was married for thirty-four years to Mr A. M. Weiss, who remembered her as a great intellect with an interest in spiritualism, and who regretted her lack of a better education.

De L'Ester—Again we have the pleasure of greeting you and of observing your attempt to secure yourself from intrusion, and we urge upon you the imperative necessity of continuing this precaution. Now, assume a comfortable position. Now close your eyes and endeavor to compose your too active mind by joining us in harmonizing prayer.

Eternal Infinite Intelligence! Eternal Infinite Energy, we, Thy children, desire to come into conscious relation with Thee. Unto Thee we offer our loving, reverent adoration, and Thou wilt guide us in all our ways. Amen, amen.

George, for a little while, we will move slowly, so that madame may more clearly observe the scene below us. To physical vision the Earth's surface would appear somewhat depressed, but to our spirit vision this illusion is not apparent. To mortals, at this altitude, the atmosphere would be too rarefied and too cold to be endurable, but, as you perceive, Spirits sufficiently evolved, are not subject to physical conditions. How deep is Earth's atmosphere? He who estimates the depth of the oxygenated portion of Earth's atmospheric envelope at ten English miles may safely add another half-mile, and the entire depth of Earth's atmosphere is so greatly in excess of what your scientists conceive it to be that on your account I a little hesitate to say that it runs into hundreds of miles, and through the activities of natural forces ever it is deepening. Yes, necessarily, all inhabited Planets possess *oxygenated* atmospheric envelopes, but you are not to confound atmosphere with ether, which fills all interstellar space, *and is substance*, but so refined as to be imperceptible to physical sense.

Upon all the planets of our solar system, our glowing, radiant Sun sheds its life-preserving beams. Its magnetic waves, pouring across space, quicken into activity latent energies, thus making progress in all directions not only possible, but inevitable. Mars, being many millions of miles further away from the Sun than is our Planet necessarily it receives less direct solar heat. On the other hand, Mars' atmosphere is such as to both receive and retain an amount of solar heat sufficient to render its climatic conditions very favorable for its various life expressions, and being much older, and hence, in proportion to its bulk, *far more magnetic than Earth*, its density, as compared with that of Earth, much less, its atmosphere rarer and lighter, it follows that to a limited degree its climatic conditions vary from those of Earth. Still, as you will have opportunity to observe, the temperature of its different zones is not greatly unlike that of the various corresponding zones of our own Planet.

Yes, the panorama now below us is a reminder of many similar views on various portions of our far distant World, which, to our vision, now appears as a rather diminutive, luminous sphere in immensity of space.

Certainly, madame, ask such questions as may occur to you, to which, as we slowly move onward, I shall to the utmost of my ability reply.

No, the depth and quality of a Planet's atmosphere does not altogether depend upon the age of the Planet. With both its quantity and its qualities other factors are concerned. Were not this true, Mars' atmosphere, relatively, would be deeper than that of Earth.

As a fact, the depth of Mars' *oxygenated* atmosphere is rather under half the depth of that of Earth. As to its qualities you already are informed. Yes, equability of temperature characterizes the various regions of Mars, only at the equator, and on either side for about seven hundred English miles, can the temperature be considered high, and even at the equator the heat is less torrid than in a corresponding latitude on our Planet. Disintegration and attrition have so worn away Mars' mountain ranges and other elevations that they offer slight diversions for its air currents. Through ethereal disturbances cyclonic storms occur, but at rare intervals. A noticeable peculiarity of Mars' atmosphere, which later will attract your observation, is its extreme humidity, which ancient Mars spirits have told me increases as the Planet ages. Even the polar regions are under the influence of this exceptionally humid condition, and there, during the year, snow falls nearly continuously. As spring approaches, at the north pole vast accumulations of ice and snow begin to melt, and as the season advances, immense volumes of water threaten to inundate portions of the Planet. Against such a calamity wise provision has been made, but of this presently you will become better informed.

We near our destination, and now, gently descending, we stand upon solid ground. Madame, we salute you, and welcome you to a land visited for the first time by a spirit yet embodied in the physical form. The energetic and adventurous American is a born pioneer, so it seems quite in keeping with your national tendency that you are here.

MEDIUM—It may be quite in keeping with my nationality to be adventurous, but I confess that at this moment I do not feel very courageous.

DE L'ESTER—Fear not. Many times you shall come hither, returning to Earth safely. You wish to know on what portion of Mars we now are? I can only reply in this manner: Relatively we are in about the same latitude and longitude as is St. Louis. For purposes of comparison, and for the instruction not only of yourself, but of some who possibly may read these pages, we have decided that it will be well to afford you a glimpse of Mars' interior, so you will stand beside me while I shall endeavor to direct your spirit perception, and that you may more readily comprehend what I shall say I shall make use of such terms as our Earth scientists have established. In succession, the Azoic, the Silurian, the Devonian, the Carboniferous, the Reptilian, the Mammalian, and the crowning Age of Man have carried Mars and Earth to their present states of evolution. As we perceive, the merging of one age into another was through such imperceptible degrees that it is not possible to note lines of demarcation, yet we easily can trace the wondrous vestiges of the passing ages up to the appearance of the evolved human animal man, of whom, at another time, I shall further inform you.

Yes, gold, silver, copper, iron, tin, in short, all the minerals with which Earth abounds, are equally abundant as constituents of Mars, and like our planet, Mars contains vast stores of mineral salts, which in solution form nature's remedial springs.

Naturally, as cooling of the Planet has proceeded, the primitive stratum has deepened, and as we perceive within its compass is a vast volume of highly heated matter, which, to a degree, corresponds with the interior of the World upon which you so serenely dwell.

Very true, to one unaccustomed to such a view, it appears amazing and awe-inspiring.

Following the Devonian age the dank atmosphere was laden with noxious gases, and the fauna and flora of this Planet attained to gigantic proportions. This was the Carboniferous age, during which largely the coal fields were formed, and I may say that during a corresponding age of Earth like causes produced like effects. All over this planet, in various localities, deposits of coal abound, and through unnumbered centuries, it served for the Marsians the purposes of fuel and illumination.

Now, madame, turn away your gaze from yonder fiery abysses and allow it to rest upon pleasanter views.

From the slight elevation on which we stand we gaze upon a very attractive scene. Stretching away into the distance are level plains, sustaining luxuriant verdure and a wealth of grains and other vegetation. The plains are dotted with towns and villages and animals of several kinds are grazing in the nearby inclosed fields.

A silvery haze veils the distant landscape, partly revealing, partly obscuring its exquisite beauty, and in all directions the land is abloom with many-hued flowers, each exhaling a fragrance all its own. Nature adorned as a queen demands her rightful measure of homage, and thus we salute thee, thou beauteous expression of the Infinite Good.

The ceaseless activities of nature accomplish manifold wonders, and in the peculiar looking animals under the shade of yonder great trees we observe one that may surprise you. Madame, we will draw nearer them. At times I forget that your vision is not as far reaching as our own. You will allow me to assist you. Do you now see them clearly? Yes? Then for a little we will pause here. Your amazement does not at all surprise us, for indeed those creatures are strangely formed, colored and clothed. We desire that you shall attempt a description of that one standing apart from the others.

Medium—But where shall I begin? With its head? Certainly that is its strongest, strangest feature, and it is formed very like the head of a giraffe, but its enormous horns, curved spirally, extend upward, and its ears are small and drooping. No one on Earth ever will believe me when I say that its large, gentle eyes are placed, one in the front, the other in the back of its head, yet truly they are there. Its neck is very like that of a horse, but rather longer, and its shoulders are much higher than its haunches. It is covered with short, reddish brown hair, perhaps I should say wool, for it is rough and crinkled, and on the end of its tail, which nearly touches the ground, is a great tuft of long, crinkled hair. Its mane is short, thick and upright, and both mane and tail are of a lighter tint of brown than is its body. At its shoulders it is the height of an ordinary horse, but its long neck and its great horns extending upward adds to its apparent height. I cannot imagine a more grotesque looking animal. I wonder what purpose it may serve?

George Brooks—I should say, to illustrate that when nature sets about it she can turn out enigmas difficult of solution. Another reason for the existence of such queer-looking animals may be that nature intends them as a background on which to exhibit the good looking ones, for grazing near yonder clump of shrubbery is an exceptionally handsome animal.

De L'Ester—George, you may not be either a philosopher or scientist, but certainly you are original.

Madame, will you also attempt a description of this animal?

Medium—I shall do my best, and where I fail you will prompt me. This animal reminds me of a horse, but it is larger than any horse I ever have seen. Its head is well proportioned to the size of its body and is as delicately formed as the head of a deer. Its ears are erect, pointed, rather small and set closely to its head. Its eyes are large, gentle and beautiful. Its neck is rather short, but symmetrical, and fringed with a long, silken mane. Its legs are well proportioned and its hoofs are daintily formed and semi-transparent. Its tail, almost sweeping the ground, is covered with long hair the color of its mane, which is a very dark brown, and its body is clothed with hair of a lighter shade of brown, rather, I should say, with shades of brown and white arranged in spots, like those of a leopard. I think it a very handsome animal. Is my description at all accurate?

De L'Ester—Quite so; and this animal is a Lûmą, and the other is a Vetson. As I already have intimated on this Planet there are in its animal kingdom forms bearing striking resemblances to some existing on Earth, but owing to Planetary conditions they are of a larger type than their kindred of our Planet. You are to remember *that all life germs are homogeneous*, their varying expressions being the result of varying conditions. Thus throughout the myriad life expressions of different Planets there are endless strong resemblances. Why not? The conditions of the several Planets of our Solar System are not so utterly dissimilar as some of your learned persons declare them to be. And mark what I shall say: On Earth's physical plane there are at this time re-embodied ones whose inherent qualities will within the next half-century enable them to give to Earth's peoples undreamed of facts concerning other Worlds. Facts which will necessitate a *readjustment of accepted scientific conclusions*. Yes, necessarily, resemblances between the fauna and flora of Mars and Earth are closer than between those of any other two Planets of our Solar System. I say necessarily, for the reason that like produces like, and the conditions of Mars and Earth, being more nearly similar than are the corresponding conditions of any of their Planetary kindred, it follows that their productions must keep pace with conditions.

Another question? Certainly, but I must make a brief reply. Throughout the animal and vegetable kingdoms of all inhabited Planets structural divergences ever have marked the lines of evolution, *the human animal alone excepted*. True, the human animal evolves through all the gradations of animal existence, but unlike other animals, he diverges neither to the right nor to the left. *His specific, inherent energy impelling him ever onward, ever upward and straight ahead.* Man is the culmination not only of forces but of qualities which set him apart from all other physical existences. *He is the apex of intelligent direction, the final, expression of God in form, not only on Mars and Earth, but in the human everywhere.*

We will now proceed, observing as we move onward whatever may be instructive or interesting. Embowered in yonder grove of magnificent trees is a stately dwelling. We will approach it more nearly, we even may enter it, for I doubt not, madame, we might find in it much that to you would be new and of interest. From its dimensions and imposing style we may conclude that it is the home of persons of wealth and distinction. For a little we will pause under the shade of these great trees, which impart a sense of restfulness.

Medium—You speak of a "sense of restfulness." May I ask do spirits, like mortals, experience a sense of fatigue?

De L'Ester—What I mean by a sense of restfulness is a state of tranquillity, through which a Spirit comes into harmonious relations with its surroundings. *Spirits do not become wearied as expressed by the word fatigue*, but upon entering the physical plane, Spirits, to a certain extent, take on the conditions with which they come in contact, and they experience what may be termed *a sense of unrest or inharmony*, and in exact proportion to the progress attained by Spirits is this sense of unrest accentuated. *Hence, Spirits of the higher spirit realms seldom enter the physical plane.* Have I made the matter clear to your comprehension?

Medium—Perfectly so.

De L'Ester—We now will look at this massive and really fine structure. As it is a good example of the many imposing residences to be found throughout this North temperate region, it shall serve as an object lesson for you, madame, and I shall take upon myself a description of its exterior.

A large structure of gray stone, extending on either side of a central entrance for at least forty feet. The entrance, which is wide and lofty, is approached by a fine flight of stone steps, leading easily up to it. Artistic and elaborate sculpture frames in the doorway, and on either side of the entrance are sculptured life-size forms in bas-relief. Their upturned eyes and upreaching hands lead one to conclude that they represent a guardian God and Goddess.

The entire front is pierced by many large windows surrounded by wide bands of intricate sculptured designs. Story above story to the height of four, the central portion of the building rises, and on either side of it are wings, two stories in height. Its entire front is beautified by traceries of delicate sculpture, among which are groups of life forms of various kinds. No doubt these forms hold certain meanings, and we regret that our Mars friends have not yet joined us, as they might enlighten us in this direction.

We now will move around to the right. Ah, here is a sort of annex and evidently devoted to pious purposes. Being a Frenchman I would term it une Temple, and you, madame, would name it a Chapel. It appears to be an extension of the dwelling, but really is quite a separate structure, which later on we will examine. As we perceive, these spacious and comfortable apartments at the rear of the dwelling are occupied by the domestics. As you, madame, are aware, to most Earth dwellers, Spirits are invisible; to the Marsians they are even less so, so we safely may enter the dwelling to have a view of the interior, but, George, you are to play no pranks to startle the occupants.

We will enter at the front. What a beautiful interior. This grand staircase, rising from this central hall, is fine enough for a royal palace. Evidently an able architect designed this dwelling, and intelligent and cultivated persons occupy it.

How very quiet it is. What is it, George? Not a soul in the house? Better so, for really this seems a sort of intrusion, all the more so were the occupants at home. Now, madame, which part of the dwelling shall we first investigate? Ah, we might have guessed that, as you are such a devoted housewife. To the kitchen then, but I shall expect you to describe this apartment, as really it is more than I am equal to.

Medium—I fear that I also am unequal to a description of it. It appears to be better fitted for chemical experiments than for a kitchen. What a large, sunny, airy room it is and what a variety of utensils. I cannot even guess at the uses of many of them. Am I mistaken in supposing that these bowls and platters and some of these pretty vessels are of gold and silver? No? Then those metals must be very plentiful, or the owner of this residence very rich. As sure as I live here is a weighing apparatus, and in design not very unlike one I use in my own kitchen. What are you saying, George? That one touch of nature makes worlds akin. Well, while this is not a touch of nature, it has a wonderfully homelike appearance. And here is a cooking range, but it is not designed for the use of coal, wood or gas. I wonder what kind of fuel these people use? De L'Ester, can you enlighten me?

De L'Ester—Since a very remote time the people of this Planet have for heating, lighting and as a motive power used electricity. In this instance it is the heating agent.

Medium—And Earth's peoples, who regard themselves as highly evolved humans, are only beginning to learn of the many uses to which it may be applied. One cannot question the fact that the same metals used on Earth are used on Mars, for here are vessels and utensils of gold, silver, iron, copper, tin, and what looks like brass, and of alloys new to me. Then here are vessels which I shall call porcelain, and there are various other wares similar to some with which I am familiar. I cannot find words to express my amazement at all this, it seems so utterly incredible, and yet I cannot question the evidence of my own senses. A woman with a genius for cooking would be enchanted with this kitchen. Is the dining-room as well worth seeing?

De L'Ester—It may be, but the family being absent, to an extent it is dismantled, so we prefer that you should not see it. Then, too, we have in mind a certain dining-hall which we purpose showing to you.

Now we will look through the rooms at the front of the dwelling. We will enter this one on the right. What a superb apartment, so spacious, so sumptuously furnished. Art and luxurious appointments combined have produced charming effects. Here, and elsewhere, we anticipate the pleasure of showing you many evidences of the wealth and culture of the Marsians. Before proceeding further we desire to inform you of something which, for a reason, until now, we have withheld. This Planet, known to Earth's peoples as Mars, is, by its inhabitants, known as Ento, which, in their language, signifies CHOSEN, or SET APART. They believe that as an expression of His love, Andûmana, the Supreme One, created Ento, and that when their home was prepared for their occupation He created His children, who with other living things should manifest the power and greatness of His Divinity. In future we will speak of the Planet as Ento, and of its peoples as Entoans, and during our journeyings and investigations you will learn that on Ento there is a state of civilization and consequent culture quite in advance of that of our own immature Planet.

Observe now those paintings. What marvellous creations they are. And those sculptured forms, so beautiful, so true to nature. Only the mind of a genius and the hand of a master could have conceived and executed either of them.

Here is a masterpiece. I know not what title the artist may have given it. I shall name it "Love's Awakening." It represents the sculptured form of a young girl just budding into womanhood. How charming is the angelic expression of her upturned eyes and smiling lips. The face, no longer that of a child, yet scarcely that of a woman, is rarely beautiful. She seems to be listening to Love's first whisperings, and almost one can fancy her lovely mouth tremulous.

So eloquent is the silence of her slightly parted lips, that in expectant attention, one listens for the faint murmurings of a soul awakened to the infinite possibilities of the passion, which welds into a unit all things animate and inanimate. Observe that the drapery, half concealing, half revealing the exquisite form, is as transparent as a mist wreath. Truly it is a marvellous expression of art. These friends and I are not unused to the finest representations of art of many planets, yet seldom have we seen a piece of sculpture equal this; still less seldom have we seen one surpassing it in design, or excellence of execution.

No, madame, the extent of this collection is not unusual, for the Entoans are liberal patrons of the arts. But we will look further.

Here, on a grassy knoll, is a group of three quite young boys, their forms lightly clothed in loose garments, which but partly conceal their rounded, shapely limbs. The middle, and larger boy, holds on his knees a book, from which apparently he reads a stirring story, to which the other boys listen with rapt attention, their beautiful faces expressing liveliest emotion. Notwithstanding that this group does not strongly appeal to the imagination, there is that which obliges one to feel that in it the sculptor has embodied much love and a reverence for art.

Now we will learn what this draped recess may contain. Ah, a descriptive composition, and in *tinted* marble. Not an agreeable representation, but an instructive reminder of a religious rite of happily bygone centuries.

Before us is an altar, on which lies the draped form of a young girl, her eyes closed as though in sleep. The loose robe drawn aside from her bosom reveals the contours of a maiden in the first blush of womanhood. At her side, holding in his upraised hand a long, keen bladed knife, which he is about to thrust into the heart of the unconscious victim, stands an aged, majestic looking Priest, his crimson robe in strong contrast to the white robed, golden-haired girl, who is to be sacrificed by knife and flames to an imaginary god or gods.

While one must admire the consummate art which so faithfully has represented this scene, one shudderingly turns from it, as being a horrible reminder of the many crimes and cruelties, which in the name of Religion, have been, and still are perpetrated.

Madame, it is a lamentable truth, that incorrect conceptions of the attributes of the Supreme One, ever are allied to cruelty. This statement applies not only to Ento, and Earth, but to all Planets inhabited by humans. It is only when man has become highly evolved, that spirit, the ego, dominates the animal soul, and God is apprehended as love, not hate.

Ancient Ento spirits, and others of comparatively modern times, have informed us concerning their religious rites and customs, which during the passing centuries have, with the exception of the sacrificial rite, remained almost unchanged. They relate that the victims of that horrible rite generally were drugged into unconsciousness, yet at times, willing victims, hoping thereby to appease the offended Gods, and thus avert some calamity, went consciously, courageously, to their death. Though deploring the ignorant fanaticism of such an act, one feels impelled to admire the heroic and generous nature of one willing to yield his or her life as a sacrifice for the real or fancied good of others.

In this adjoining recess is another composition, scarcely less pathetic, but devoid of the element of cruelty. On a large malachite base is a stone altar, on which lies the nearly nude body of a dead boy, and over and about him is a mass of inflammable material, bursting into flames. Thus in very ancient times the Entoans disposed of their dead. It is a very realistic representation of a custom of remote times, and certainly is not the production of an artist of recent days. Yes, incineration of their dead, is with the Entoans, a universal custom but during many centuries it has been accomplished in a more scientific, and less repellent manner.

You still express surprise, that the Entoans, physically, are formed as are we, and the peoples of our Planet. My dear madame, believe me, when I reiterate, that *humans*, no matter of what Planet, are essentially the same. Disabuse your mind, now, and for all time, of the idea, that necessarily, different Planets must produce entirely different expressions of life. One Intelligent Energy directs the universe, and one universal Law prevails. Should you visit Venus, Jupiter, or indeed any Planet inhabited by *humans*, you would find *man*, only as you know him. Evolved, it is true, on some Planets, to a higher spiritual, consequently to a more perfected physical plane, and a more advanced state of civilization. Spiritualized humans are the expressions of spirit entities. These spirit entities must act within their limitations, and *never, never*, by any possibility, does a *spirit* entity take possession of any other than the evolved *human* organism. *Spirit knows no such negation as retrogression.* When man on any Planet has evolved to a certain condition, or degree, he becomes a partially self-conscious *soul*, and then he walks erect. Ages pass and he becomes a Spiritualized Being, Spiritualized through the incarnation in him of a Spirit entity, which enables him to fully *recognize himself*. Not until then, is he evolved into the Spiritualized IMMORTAL—THE GOD MAN, as are all Spiritualized MEN everywhere.

Nay, you owe me no apology. I quite understand your mental state, and desire that you shall ask such questions as naturally must occur to you.

Observe now the very beautiful hangings of this apartment. They are of thick, lustrous silk, and their rich shades of crimson and gold form an excellent background for these superb paintings and marbles. It is to be regretted that limited time and space will not admit of a more detailed description of the many works of art in this collection. As it is, we must content ourselves with glances here and there. We think it advisable to notice this large painting, which vividly illustrates another feature of the sacrificial observance. It represents the interior of a richly ornate temple, and on a raised space stands a number of Priests and Priestesses, clothed in long, flowing, crimson and yellow garments. With the exception of three Priests, who are brown haired, blue eyed, and fair skinned, all are very dark hued. The hair of the younger, dark complexioned Priests is very black, and worn quite to their shoulders, and that of the aged ones is as white as wool, and worn in the same fashion. Around the heads of all are narrow fillets of gold, binding back their flowing locks. On these fillets, directly over the forehead, are golden suns, the points of the rays tipped with yellow jewels—topazes, I should say—and the centre of each sun is what appears to be a fine ruby, encircled by topazes.

The Priestesses are young, dark skinned, and dark eyed, and their long black hair falls loosely toward their feet, which are concealed by their crimson and yellow robes. Around their heads are fillets, corresponding with those worn by the Priests. In the foreground are a number of youths and maidens, and back of them a throng of men and women, all with anxious, terror stricken faces. Well may the eyes and faces of the assemblage be full of fear, for some one's child will be selected as a sacrifice to their Gods, whose dwelling place is beyond the clouds which veil the portals of Astranolą, lest impious, inquiring eyes gazing upward, may behold what mortals may not see, and live.

This painting depicts a scene once of frequent occurrence, but during four centuries past the horrible cruelty has not been practiced.

We now will pass into the adjoining apartment. This appears to be a sort of lounging room, in which form and color combined have produced some fine effects. Over the lofty corniced windows and doorways, velvet-like, crimson drapery falls in graceful folds. Luxurious divans line the cream tinted walls, over which sprays of lovely, dainty blossoms are scattered. The floor is a mosaic of exquisite effects. The field, a rich cream color, the designs, graceful, lifelike flower pieces, united by trailing vines. A deep border of aquatic plants, grasses and vining lily blooms, forms a fitting frame to the lovely floor picture, over which very beautiful rugs are disposed.

This large and massive table, so exquisitely carved, and inlaid with rare colored woods, in a design partly arabesque, partly floral, is indeed a thing of beauty, but more beautiful still is this superb vase, occupying its raised centre.

These portfolios of pictured illustrations we can only glance at. Yes, in conception, coloring, and execution, they are highly meritorious. The same may be said of these handsomely bound volumes. You had not thought to find books on Ento? Why not, madame? Do not you yet comprehend that the inherent attributes of the genus homo, not only *impels*, but inevitably *compels* him in one common direction? This is a universal law, and there is no escape from it. As I already have declared, its expression, wherever demonstrated, is essentially the same. We doubt not that many things we shall show you on Ento will surprise you, more by their likeness than by their unlikeness, to what may be found on our Planet.

The entire ornamentation and appointments of this apartment are rather quiet in tone, but *le tout ensemble*, is very refined and beautiful.

George is so urgent to hasten our movements that I suspect he is up to some mischief. Yes, yes, we are coming. George, George, you are incorrigible; it is not surprising that madame is startled, for this figure is wonderfully lifelike, and what an odd conceit, to use one of its long arms to hold back this heavy drapery.

MEDIUM—Really, for a moment, I thought it a living creature. Does it represent a human being? It looks very like one.

DE L'ESTER—Truly it does appear very human, but it represents a species of Ento anthropoid, so intelligent, that frequently it is trained for simple requirements, mostly of a domestic nature. As later on, you will see living specimens of the same creature, I shall not now describe it. Enough cannot be said of the fidelity with which the artificer has reproduced the form, coloring, and expression of the living animal. Of what metal is it made? Of a composition of copper and tin, and if you choose, you may call it bronze, for that is what it is. Yes, the enamelling is very fine, the tinting is true to nature.

Here is a collection of miniature paintings, and be assured that these illustrations of Ento female loveliness are not at all exaggerated. Some are types of the blonde, blue eyed Northern races, others of the dark skinned, lovely women of the Orient, with eyes as dark and liquid as quiet pools in shady nooks. As you perceive, all are arrayed in graceful flowing garments, unlike the hideous robes worn by even the most highly civilized women of our Planet.

Ah, what a gem! We cannot pass this by unnoticed. Madame, you will carefully observe this painting, as some time you may have occasion to recall a memory of it. In the foreground is a youth in the early flush of manhood, whose shapely head is crowned with black hair waving down to his shoulders, and bound away from his fine forehead by a jewelled silver fillet. His smiling, parted lips, form a perfect Cupid's bow, and above them is a nose as straight and finely formed as ever graced the face of a Grecian statue. A robe of azure blue, bordered with silver embroidery, clothes his very tall, graceful form, and falls in artistic lines to his sandalled feet. Looped high on his left shoulder is a loose sleeve drapery, caught into folds by a jewelled ornament, indicating that this youth is of exalted rank. Bending slightly forward, he smilingly listens to the words of a young girl, reclining on a low couch, who is costumed in a soft, clinging, white robe, which scarcely conceals the outlines of a fragile but perfect form. Her golden hair, which is caught back from her low, wide, white forehead, by a silver fillet, adorned with sapphires no bluer than her lovely eyes, seems to have caught sunlight in its tresses, as it falls in rippling masses over her shoulders and onto the floor, where it lies in golden confusion, on a rug of rich, dark hued fur. She is as fair as the youth is dark, and in her beautiful face is the innocence and mirthfulness of the child, with the promise, too, of a gracious womanhood. Remember these faces, for one day you may see the originals.

How true it is that art expressions are the mute speech of genius, and genius is but another name for inspiration. It has been said "back of the artist is art, and back of art is that which men name God." That is a fine expression of the unity of things.

George, Agassiz, Humboldt, hasten here. Ah! you too, recognize, this scene, Is not it an agreeable surprise? Madame, I will explain. This is a most exact representation of a locality these friends and I have visited. Rather recently we with some scientific and other persons were, for a certain purpose, making a tour of Ento, and while slowly journeying toward a distant portion of the planet we found ourselves passing over the spot illustrated by this painting. It attracted our attention, and descending, we found it such a quiet, tranquil spot that unanimously we named it the Valley of Repose. With wonderful fidelity and consummate art the painter has reproduced the lovely scene. Stand here, madame, and I will attempt to describe it.

A spacious valley surrounded on three sides by gently rising uplands, which in long gone ages were portions of a mountain range. From a rocky formation in the upper end of the valley debouches a considerable volume of water, forming this sparkling stream, which empties itself into yonder pretty lake, dotted with tiny islands. Those rather fragile looking bridges thrown from island to island form continuous passageways to either side of the valley. The villages dotting the rim of the lake, and those white structures on the larger islands, to one's imagination suggest flocks of white plumaged water fowl nestling amid the luxuriant greenery. Boats laden with the products of labor. Crews intent upon landing their crafts. Other boats carrying pleasure seekers, who call to passing friends, fill up the animated picture. Gazing with admiring eyes upon the lovely scene, we tarried awhile under these great trees laden with sweet scented blooms.

You may like to know that this valley is in the North Temperate Zone, in latitude and longitude corresponding nearly to that of the northern central portion of your State of Tennessee. Being sheltered by the uplands, the climate is very genial, and the loamy soil produces grains, vegetables and fruits in great abundance.

Continually artists frequent this valley to sketch its beauties, and the painter of this picture, who signs himself as Lafon Thēdossạ, has literally transferred the lake and its surroundings to his canvas. It seems as though we have met face to face a well known friend.

My dear madame, do not vex yourself that we cannot use your organism for all purposes. Were we engaged in a purely scientific work it would be altogether necessary that we should have a Sensitive through whom we might express technicalities pertaining to matters under investigation or discussion. All along we have fully understood your limitations, as well as your extremely skeptical and cautious nature, and we well know that should we attempt to express through you statistics, technicalities, latitude, longitude, and other matters your nearly morbid dread of making mistakes would render you so positive that we could not use you at all. At present we are quite satisfied with what we can accomplish through you, and we anticipate a time when you shall have so developed that we shall be able to use you for ends you little dream of. So we pray you to fret no more that you are not equal to our wishes, for you quite satisfy our requirements.

We must not hold you longer to-day. Gradually you are adjusting yourself to present conditions and ere long we may lengthen our visits to this Planet, but now at once you must be returned to your Earth home. There are indications that the occupants of this residence are about to return to it, so endeavor to hold yourself in readiness, for we may come for you at an unusual hour. Now, George, Earthward. Not another question, madame. We must not allow you to become exhausted.

Safely arrived, and some one is knocking at your door. May loving angels have you in their keeping. Au revoir.

SIGNALING TO MARS (1909)

Hugo Gernsback

Born **Hugo Gernsbacher** in 1884 in Luxembourg, Gernsback altered his name after emigrating to the United States in 1904. He was an entrepreneur in the early popularization of 'wireless' radio and founded several magazines. *Modern Electrics*, *Electrical Experimenter* and *Radio News* all dealt with various aspects of the new technology. In 1926, he founded *Astounding Stories* to publish science fiction stories. This magazine became a pillar of 'the golden age of science fiction'. The annual achievement awards of the World Science Fiction Convention are known as the Hugos. Gernsback died in New York City in 1967.

Modern Electrics, vol. 2, no. 2, May 1909

Every time our neighbor Mars comes in opposition to the earth a host of inventors and others begin to turn their attention to the great up-to-date problem, "signaling to Mars."[1]

It is safe to say that all the proposed, possible as well as impossible, projects would fill a good-sized volume, especially the ones invented in this country, which by far leads the world in number of projects and "inventions."

So far only one feasible plan has been worked out. The writer refers to Professor Pickering's mirror arrangement, now being discussed all over the world. But even Professor Pickering is skeptical, as he apparently does not like to take the responsibility of spending ten million dollars on a mere idea which might prove fallacious.[2]

Wireless telegraphy has been talked of much lately as a probable solution to the problem. The writer wishes to show why it is not possible at the present stage of development to use wave telegraphy between the two planets, but at the same time he would like to present a few new ideas how it could be done in the near future.[3]

Take the average present-day wireless station having an output of about 2 K. W. On good nights and under favorable circumstances such a station may cover 1,000 miles. Very frequently, however, only about 800 miles can be spanned.

Next summer for a few days Mars will be nearer to us than for many years to come. The distance between the two planets will then be about 35 million miles.

If we base transmission between the earth and Mars at the same figure as transmission over the earth, a simple calculation will reveal that we must have the enormous power of 70,000 K. W. to our disposition in order to reach Mars.

Now it would be absolutely out of the question to build a single station with that output. Even a station of 700 K. W. would be a monster and a rather dangerous affair to meddle with. This may be understood better when considering that none, even the most powerful stations, to-day have 100 K. W. at their disposal.

A solution, however, presents itself to the writer's mind. As it is impossible and impracticable to build and operate a single station with an output of 70,000 K. W., let us divide the 70,000 K. W. in small stations of, say, 2, 10, or 50 K. W. Neither do we have to build these stations for the sole purpose of using them to signal Mars; they are already being erected by the governments and commercial stations at the rate of about 150 per month.

At the present time of writing the entire output of the U.S. Government, commercial and ship wireless stations combined, is about 2,500 K. W. By adding the stations of private individuals, ranging in power from ¼ to 2 K. W., the total sum is brought up to about 3,500 K. W., as far as the writer is able to ascertain from the latest reports.

If, however, the art progresses as it has during the past four years, it is safe to predict that in 20 or possibly 15 years the United States, Canada, and Mexico will reach the combined output of 70,000 K. W.

It will then be comparatively easy to seriously undertake to signal to Mars, and the writer proposes the following plan, which has the great advantage that the experiments can be made at practically no cost, against Professor Pickering's project, involving the enormous expense of ten million dollars.

The idea is simple enough. A central point of the continent, such as Lincoln, Nebr., should be selected preferably. By previous arrangement all wireless stations on the continent should be informed that on certain days their stations should be connected with a magnetic key,[4] which is connected through the already existing wire telegraph lines with the central station at Lincoln. As the wires may be leased from the existing wire telegraph lines, it is of course the simplest thing in the world to connect the key of each wireless station (by wire) with the central station. Each time, therefore, when the operator at Lincoln depresses his key all the keys belonging to the wireless stations connected with his key will be pressed down, and if the combined power of the connected stations is 70,000 K. W., the enormous energy of 70,000 K. W. will be shot out in the ether!

What effect the 70,000 K. W. will have on the weather or climate after they have been radiated for several hours the writer dares not conjecture, but that something will "happen" is almost certain.

Considering the technical side of the project, it is of course feasible. If the necessary amount of power was to be had to-day, there would be no difficulty to try it next summer; as this is not the case, we must be patient and wait; the writer, however, hopes to see the day when the experiment will be tried.

Referring to the technical side, it will be necessary, of course, to tune all the sending apparatus to exactly the same wave length, which, naturally—on account of the great distance to be overcome—should be as long as possible. The frequency of the oscillations should be practically the same for all stations. The result of this arrangement would be that the effect would be practically the same as if one tremendously large station of 70,000 K. W. capacity was sending.

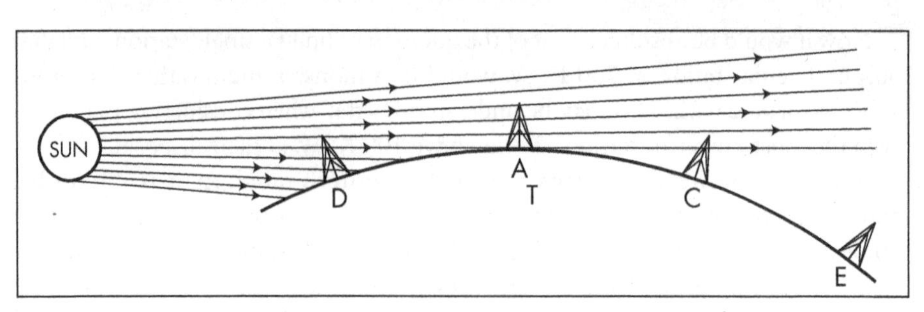

Figure 1.

Just as we may blow two or a dozen whistles of the same pitch at the same time, in order to carry the sound further, and just as Professor Pickering may use thousands of small mirrors all operated at the same time, as if they were one huge mirror, so it may be possible to unite a great number of different wireless senders and operate them as if they were one, provided of course that, like the whistles, they are tuned to the same "pitch."

There is only one more point to consider.

It has been demonstrated time and again that the action of the sun's rays greatly interfere with wireless telegraphy. In fact, it is possible to send twice as far over water during the night than during the day. This may be better understood by quoting Mr. Marconi's views:

> Messages can now be transmitted across the Atlantic by day as well as by night, but there exists certain periods, fortunately of short duration, when transmission across the Atlantic is difficult and at times ineffective unless an amount of energy greater than that used during what I might call normal conditions is employed.
>
> Thus in the morning and in the evening when, due to the difference in longitude, daylight extends only part of the way across the Atlantic, the received signals are weak and sometimes cease altogether.[5]

Mr. Marconi's explanation is that illuminated space possesses for electric waves a different refractive index to dark space and that in consequence the electric waves may be refracted and reflected in passing from one medium to another.[6]

The writer wishes to offer a different explanation, which seems far more plausible.

Referring to Fig. 2, let T represent a section of the earth. Let A be a station on the American, E a station of the English coast. As will be seen, the sun is just setting for the point A, while E has night already (no sun rays reach E).

When A is sending the waves are shot out parallel with the sun's rays and *carried with the rays*. The action of the sun's rays is so strong that most of the electric waves are carried along, and therefore never reach E at all. Only by using more powerful waves can this effect be overcome. This action is not surprising at all. Electromagnetic waves are closely related to light rays. As Svante Arrhenius has also shown us, the rays of the sun exert a certain amount of pressure on all encountered objects.[7] It is therefore easy to prove

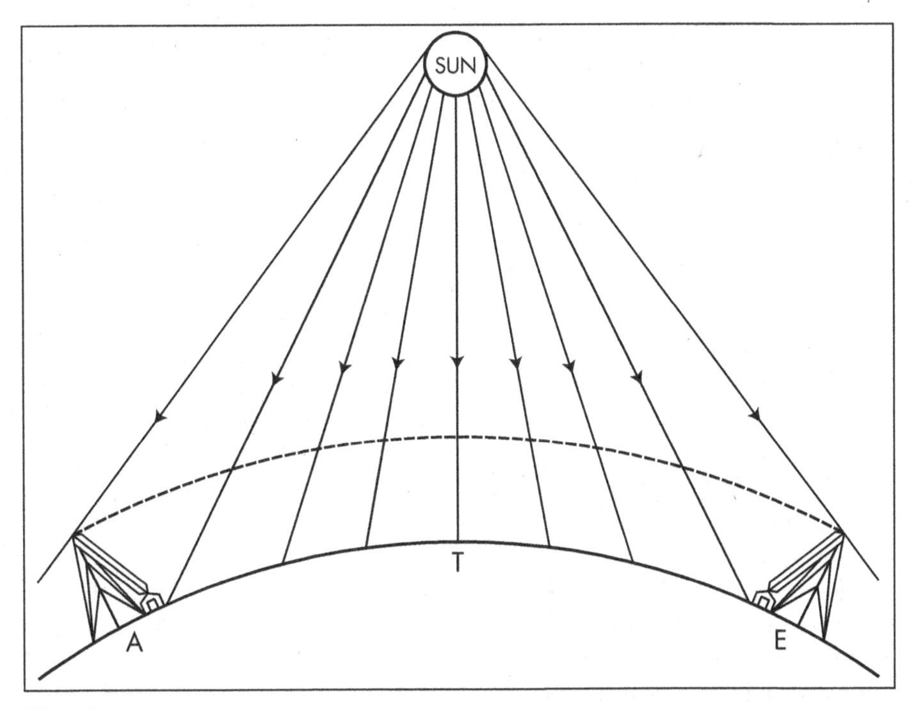

Figure 2.

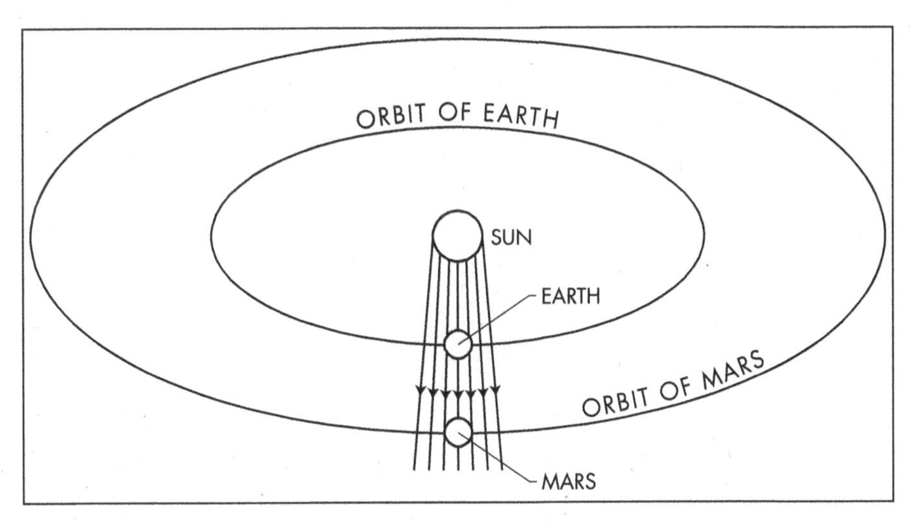

Figure 3.

that considering the close relationship of light rays and Hertzian waves, the latter *will be carried in the direction away from the sun* under favorable circumstances. Again considering Fig. 1, such favorable circumstances would be reached during sunset or during sunrise.

That this explanation is not a mere theory is best proved by the fact that a point D and C will communicate with each other best during sunrise and sunset, the signals received being the strongest. The electric waves during these two periods *travel parallel with the sun's rays, following the line of least resistance.*

During the day (Fig. 2), let A again represent the American, E the English coast station. Now it will be easily understood why messages can be sent almost twice as far during the night as during the day. In this instance the electric waves must *cut directly through the vast field of light*, and are being “held down” to a certain degree by the pressure of the light.

Now let us turn our attention to Fig. 3. This represents the earth and Mars in opposition. It will be seen immediately that the earth has a great advantage over Mars as far as wireless is concerned.

Messages sent from the earth during the opposition will go in the same direction as the sun’s rays, and the writer is of the opinion that instead of the theoretical amount of power—that is, 70,000 K. W.—possibly only one-half or one-quarter will be required to signal to Mars, as the electric waves are undoubtedly assisted toward Mars by the rays of the sun.

On the other hand, Mars will find it difficult to signal back, especially during opposition, when his “operators” would have to work directly against the sun’s rays.

However, we can only hope that the Martians are further advanced than we and may signal back to us, using a method new to us and possibly long discarded by them, when thousands of years ago they stopped signaling to us, and gave us up, as we did not have intelligence enough to understand.

A PRINCESS OF MARS (1912)

Edgar Rice Burroughs

A prolific author, **Edgar Rice Burroughs'** career spanned from 1911 to his death in 1950. Born in 1875 in Chicago, Illinois, he was a pencil sharpener salesman during his early working life. About his decision to begin writing stories, he famously said that if people were paid for writing the kind of rot he had been reading in the pulp-fiction magazines, then he could write stories just as rotten. His work encompassed the genres of adventure, fantasy and science fiction. *A Princess of Mars* was published in 1912 and began an eleven-book series set on Barsoom, his fictional name for the Red Planet. In 2012, Disney released a film adaptation of the first book.

CHAPTER I
On the Arizona Hills

I am a very old man; how old I do not know. Possibly I am a hundred, possibly more; but I cannot tell because I have never aged as other men, nor do I remember any childhood. So far as I can recollect I have always been a man, a man of about thirty. I appear today as I did forty years and more ago, and yet I feel that I cannot go on living forever; that some day I shall die the real death from which there is no resurrection. I do not know why I should fear death, I who have died twice and am still alive; but yet I have the same horror of it as you who have never died, and it is because of this terror of death, I believe, that I am so convinced of my mortality.

And because of this conviction I have determined to write down the story of the interesting periods of my life and of my death. I cannot explain the phenomena; I can only set down here in the words of an ordinary soldier of fortune a chronicle of the strange events that befell me during the ten years that my dead body lay undiscovered in an Arizona cave.

I have never told this story, nor shall mortal man see this manuscript until after I have passed over for eternity. I know that the average human mind will not believe what it cannot grasp, and so I do not purpose being pilloried by the public, the pulpit, and the press, and held up as a colossal liar when I am but telling the simple truths which some day science will substantiate. Possibly the suggestions which I gained upon Mars, and the knowledge which I can set down in this chronicle, will aid in an earlier understanding of the mysteries of our sister planet; mysteries to you, but no longer mysteries to me.

My name is John Carter; I am better known as Captain Jack Carter of Virginia. At the close of the Civil War I found myself possessed of several hundred thousand dollars (Confederate) and a captain's commission in the cavalry arm of an army which no longer existed; the servant of a state which had vanished with the hopes of the South. Masterless, penniless, and with my only means of livelihood, fighting, gone, I determined to work my way to the southwest and attempt to retrieve my fallen fortunes in a search for gold.

I spent nearly a year prospecting in company with another Confederate officer, Captain James K. Powell of Richmond. We were extremely fortunate, for late in the winter of 1865, after many hardships and privations, we located the most remarkable gold-bearing quartz vein that our wildest dreams had ever pictured. Powell, who was a mining engineer by education, stated that we had uncovered over a million dollars worth of ore in a trifle over three months.

As our equipment was crude in the extreme we decided that one of us must return to civilization, purchase the necessary machinery and return with a sufficient force of men properly to work the mine.

As Powell was familiar with the country, as well as with the mechanical requirements of mining we determined that it would be best for him to make the trip. It was agreed that I was to hold down our claim against the remote possibility of its being jumped by some wandering prospector.

On March 3, 1866, Powell and I packed his provisions on two of our burros, and bidding me good-bye he mounted his horse, and started down the mountainside toward the valley, across which led the first stage of his journey.

The morning of Powell's departure was, like nearly all Arizona mornings, clear and beautiful; I could see him and his little pack animals picking their way down the mountainside toward the valley, and all during the morning I would catch occasional glimpses of them as they topped a hog back or came out upon a level plateau. My last sight of Powell was about three in the afternoon as he entered the shadows of the range on the opposite side of the valley.

Some half hour later I happened to glance casually across the valley and was much surprised to note three little dots in about the same place I had last seen my friend and his two pack animals. I am not given to needless worrying, but the more I tried to convince myself that all was well with Powell, and that the dots I had seen on his trail were antelope or wild horses, the less I was able to assure myself.

Since we had entered the territory we had not seen a hostile Indian, and we had, therefore, become careless in the extreme, and were wont to ridicule the stories we had heard of the great numbers of these vicious marauders that were supposed to haunt the trails, taking their toll in lives and torture of every white party which fell into their merciless clutches.

Powell, I knew, was well armed and, further, an experienced Indian fighter; but I too had lived and fought for years among the Sioux in the North, and I knew that his chances were small against a party of cunning trailing Apaches. Finally I could endure the suspense no longer, and, arming myself with my two Colt revolvers and a carbine, I strapped two belts of cartridges about me and catching my saddle horse, started down the trail taken by Powell in the morning.

As soon as I reached comparatively level ground I urged my mount into a canter and continued this, where the going permitted, until, close upon dusk, I discovered the point where other tracks joined those of Powell. They were the tracks of unshod ponies, three of them, and the ponies had been galloping.

I followed rapidly until, darkness shutting down, I was forced to await the rising of the moon, and given an opportunity to speculate on the question of the wisdom of my chase. Possibly I had conjured up impossible dangers, like some nervous old housewife, and when I should catch up with Powell would get a good laugh for my pains. However, I am not prone to sensitiveness, and the following of a sense of duty, wherever it may lead, has always been a kind of fetich with me throughout my life; which may account for the honors bestowed upon me by three republics and the decorations and friendships of an old and powerful emperor and several lesser kings, in whose service my sword has been red many a time.

About nine o'clock the moon was sufficiently bright for me to proceed on my way and I had no difficulty in following the trail at a fast walk, and in some places at a brisk trot until, about midnight, I reached the water hole where Powell had expected to camp. I came upon the spot unexpectedly, finding it entirely deserted, with no signs of having been recently occupied as a camp.

I was interested to note that the tracks of the pursuing horsemen, for such I was now convinced they must be, continued after Powell with only a brief stop at the hole for water; and always at the same rate of speed as his.

I was positive now that the trailers were Apaches and that they wished to capture Powell alive for the fiendish pleasure of the torture, so I urged my horse onward at a most dangerous pace, hoping against hope that I would catch up with the red rascals before they attacked him.

Further speculation was suddenly cut short by the faint report of two shots far ahead of me. I knew that Powell would need me now if ever, and I instantly urged my horse to his topmost speed up the narrow and difficult mountain trail.

I had forged ahead for perhaps a mile or more without hearing further sounds, when the trail suddenly debouched onto a small, open plateau near the summit of the pass. I had passed through a narrow, overhanging gorge just before entering suddenly upon this table land, and the sight which met my eyes filled me with consternation and dismay.

The little stretch of level land was white with Indian tepees, and there were probably half a thousand red warriors clustered around some object near the center of the camp. Their attention was so wholly riveted to this point of interest that they did not notice me, and I easily could have turned back into the dark recesses of the gorge and made my escape with perfect safety. The fact, however, that this thought did not occur to me until the following day removes any possible right to a claim to heroism to which the narration of this episode might possibly otherwise entitle me.

I do not believe that I am made of the stuff which constitutes heroes, because, in all of the hundreds of instances that my voluntary acts have placed me face to face with death, I cannot recall a single one where any alternative step to that I took occurred to me until many hours later. My mind is evidently so constituted that I am subconsciously forced into the path of duty without recourse to tiresome mental processes. However that may be, I have never regretted that cowardice is not optional with me.

In this instance I was, of course, positive that Powell was the center of attraction, but whether I thought or acted first I do not know, but within an instant from the moment the scene broke upon my view I had whipped out my revolvers and was charging down upon the entire army of warriors, shooting rapidly, and whooping at the top of my lungs. Singlehanded, I could not have pursued better tactics, for the red men, convinced by sudden surprise that not less than a regiment of regulars was upon them, turned and fled in every direction for their bows, arrows, and rifles.

The view which their hurried routing disclosed filled me with apprehension and with rage. Under the clear rays of the Arizona moon lay Powell, his body fairly bristling with the hostile arrows of the braves. That he was already dead I could not but be convinced, and yet I would have saved his body from mutilation at the hands of the Apaches as quickly as I would have saved the man himself from death.

Riding close to him I reached down from the saddle, and grasping his cartridge belt drew him up across the withers of my mount. A backward glance convinced me that to return by the way I had come would be more hazardous than to continue across the plateau, so, putting spurs to my poor beast, I made a dash for the opening to the pass which I could distinguish on the far side of the table land.

The Indians had by this time discovered that I was alone and I was pursued with imprecations, arrows, and rifle balls. The fact that it is difficult to aim anything but imprecations accurately by moonlight, that they were upset by the sudden and unexpected manner of my advent, and that I was a rather rapidly moving target saved me from the various deadly projectiles of the enemy and permitted me to reach the shadows of the surrounding peaks before an orderly pursuit could be organized.

My horse was traveling practically unguided as I knew that I had probably less knowledge of the exact location of the trail to the pass than he, and thus it happened that he entered a defile which led to the summit of the range and not to the pass which I had hoped would carry me to the valley and to safety. It is probable, however, that to this fact I owe my life and the remarkable experiences and adventures which befell me during the following ten years.

My first knowledge that I was on the wrong trail came when I heard the yells of the pursuing savages suddenly grow fainter and fainter far off to my left.

I knew then that they had passed to the left of the jagged rock formation at the edge of the plateau, to the right of which my horse had borne me and the body of Powell.

I drew rein on a little level promontory overlooking the trail below and to my left, and saw the party of pursuing savages disappearing around the point of a neighboring peak.

I knew the Indians would soon discover that they were on the wrong trail and that the search for me would be renewed in the right direction as soon as they located my tracks.

I had gone but a short distance further when what seemed to be an excellent trail opened up around the face of a high cliff. The trail was level and quite broad and led upward and in the general direction I wished to go. The cliff arose for several hundred feet on my right, and on my left was an equal and nearly perpendicular drop to the bottom of a rocky ravine.

I had followed this trail for perhaps a hundred yards when a sharp turn to the right brought me to the mouth of a large cave. The opening was about four feet in height and three to four feet wide, and at this opening the trail ended.

It was now morning, and, with the customary lack of dawn which is a startling characteristic of Arizona, it had become daylight almost without warning.

Dismounting, I laid Powell upon the ground, but the most painstaking examination failed to reveal the faintest spark of life. I forced water from my canteen between his dead lips, bathed his face and rubbed his hands, working over him continuously for the better part of an hour in the face of the fact that I knew him to be dead.

I was very fond of Powell; he was thoroughly a man in every respect; a polished southern gentleman; a staunch and true friend; and it was with a feeling of the deepest grief that I finally gave up my crude endeavors at resuscitation.

Leaving Powell's body where it lay on the ledge I crept into the cave to reconnoiter. I found a large chamber, possibly a hundred feet in diameter and thirty or forty feet in height; a smooth and well-worn floor, and many other evidences that the cave had, at some remote period, been inhabited. The back of the cave was so lost in dense shadow that I could not distinguish whether there were openings into other apartments or not.

As I was continuing my examination I commenced to feel a pleasant drowsiness creeping over me which I attributed to the fatigue of my long and strenuous ride, and the reaction from the excitement of the fight and the pursuit. I felt comparatively safe in my present location as I knew that one man could defend the trail to the cave against an army.

I soon became so drowsy that I could scarcely resist the strong desire to throw myself on the floor of the cave for a few moments' rest, but I knew that this would never do, as it would mean certain death at the hands of my red friends, who might be upon me at any moment. With an effort I started toward the opening of the cave only to reel drunkenly against a side wall, and from there slip prone upon the floor.

CHAPTER II
The Escape of the Dead

A sense of delicious dreaminess overcame me, my muscles relaxed, and I was on the point of giving way to my desire to sleep when the sound of approaching horses reached my ears. I attempted to spring to my feet but was horrified to discover that my muscles refused to respond to my will. I was now thoroughly awake, but as unable to move a muscle as though turned to stone. It was then, for the first time, that I noticed a slight vapor filling the cave. It was extremely tenuous and only noticeable against the opening which led to daylight. There also came to my nostrils a faintly pungent odor, and I could only assume that I had been overcome by some poisonous gas, but why I should retain my mental faculties and yet be unable to move I could not fathom.

I lay facing the opening of the cave and where I could see the short stretch of trail which lay between the cave and the turn of the cliff around which the trail led. The noise of the approaching horses had ceased, and I judged the Indians were creeping stealthily upon me along the little ledge which led to my living tomb. I remember that I hoped they would make short work of me as I did not particularly relish the thought of the innumerable things they might do to me if the spirit prompted them.

I had not long to wait before a stealthy sound apprised me of their nearness, and then a war-bonneted, paint-streaked face was thrust cautiously around the shoulder of the cliff, and savage eyes looked into mine. That he could see me in the dim light of the cave I was sure for the early morning sun was falling full upon me through the opening.

The fellow, instead of approaching, merely stood and stared; his eyes bulging and his jaw dropped. And then another savage face appeared, and a third and fourth and fifth, craning their necks over the shoulders of their fellows whom they could not pass upon the narrow ledge. Each face was the picture of awe and fear, but for what reason I did not know, nor did I learn until ten years later. That there were still other braves behind those who regarded me was apparent from the fact that the leaders passed back whispered word to those behind them.

Suddenly a low but distinct moaning sound issued from the recesses of the cave behind me, and, as it reached the ears of the Indians, they turned and fled in terror, panic-stricken. So frantic were their efforts to escape from the unseen thing behind me that one of the braves was hurled headlong from the cliff to the rocks below. Their wild cries echoed in the canyon for a short time, and then all was still once more.

The sound which had frightened them was not repeated, but it had been sufficient as it was to start me speculating on the possible horror which lurked in the shadows at my back. Fear is a relative term and so I can only measure my feelings at that time by what I had experienced in previous positions of danger and by those that I have passed through since; but I can say without shame that if the sensations I endured during the next few minutes were fear, then may God help the coward, for cowardice is of a surety its own punishment.

To be held paralyzed, with one's back toward some horrible and unknown danger from the very sound of which the ferocious Apache warriors turn in wild stampede, as a flock of sheep would madly flee from a pack of wolves, seems to me the last word in fearsome predicaments for a man who had ever been used to fighting for his life with all the energy of a powerful physique.

Several times I thought I heard faint sounds behind me as of somebody moving cautiously, but eventually even these ceased, and I was left to the contemplation of my position without interruption. I could but vaguely conjecture the cause of my paralysis, and my only hope lay in that it might pass off as suddenly as it had fallen upon me.

Late in the afternoon my horse, which had been standing with dragging rein before the cave, started slowly down the trail, evidently in search of food and water, and I was left alone with my mysterious unknown companion and the dead body of my friend, which lay just within my range of vision upon the ledge where I had placed it in the early morning.

From then until possibly midnight all was silence, the silence of the dead; then, suddenly, the awful moan of the morning broke upon my startled ears, and there came again from the black shadows the sound of a moving thing, and a faint rustling as of dead leaves. The shock to my already overstrained nervous system was terrible in the extreme, and with a superhuman effort I strove to break my awful bonds. It was an effort of the mind, of the will, of the nerves; not muscular, for I could not move even so much as my little finger, but none the less mighty for all that. And then something gave, there was a momentary feeling of nausea, a sharp click as of the snapping of a steel wire, and I stood with my back against the wall of the cave facing my unknown foe.

And then the moonlight flooded the cave, and there before me lay my own body as it had been lying all these hours, with the eyes staring toward the open ledge and the hands resting limply upon the ground. I looked first at my lifeless clay there upon the floor of the cave and then down at myself in utter bewilderment; for there I lay clothed, and yet here I stood but naked as at the minute of my birth.

The transition had been so sudden and so unexpected that it left me for a moment forgetful of aught else than my strange metamorphosis. My first thought was, is this then death! Have I indeed passed over forever into that other life! But I could not well believe this, as I could feel my heart pounding against my ribs from the exertion of my efforts to release myself from the anaesthesis which had held me. My breath was coming in quick, short gasps, cold sweat stood out from every pore of my body, and the ancient experiment of pinching revealed the fact that I was anything other than a wraith.

Again was I suddenly recalled to my immediate surroundings by a repetition of the weird moan from the depths of the cave. Naked and unarmed as I was, I had no desire to face the unseen thing which menaced me.

My revolvers were strapped to my lifeless body which, for some unfathomable reason, I could not bring myself to touch. My carbine was in its boot, strapped to my saddle, and as my horse had wandered off I was left without means of defense. My only alternative seemed to lie in flight and my decision was crystallized by a recurrence of the rustling sound from the thing which now seemed, in the darkness of the cave and to my distorted imagination, to be creeping stealthily upon me.

Unable longer to resist the temptation to escape this horrible place I leaped quickly through the opening into the starlight of a clear Arizona night. The crisp, fresh mountain air outside the cave acted as an immediate tonic and I felt new life and new courage coursing through me. Pausing upon the brink of the ledge I upbraided myself for what now seemed to me wholly unwarranted apprehension. I reasoned with myself that I had lain helpless for many hours within the cave, yet nothing had molested me, and my better judgment, when permitted the direction of clear and logical reasoning, convinced me that the noises I had heard must have resulted from purely natural and harmless causes; probably the conformation of the cave was such that a slight breeze had caused the sounds I heard.

I decided to investigate, but first I lifted my head to fill my lungs with the pure, invigorating night air of the mountains. As I did so I saw stretching far below me the beautiful vista of rocky gorge, and level, cacti-studded flat, wrought by the moonlight into a miracle of soft splendor and wondrous enchantment.

Few western wonders are more inspiring than the beauties of an Arizona moonlit landscape; the silvered mountains in the distance, the strange lights and shadows upon hog back and arroyo, and the grotesque details of the stiff, yet beautiful cacti form a picture at once enchanting and inspiring; as though one were catching for the first time a glimpse of some dead and forgotten world, so different is it from the aspect of any other spot upon our earth.

As I stood thus meditating, I turned my gaze from the landscape to the heavens where the myriad stars formed a gorgeous and fitting canopy for the wonders of the earthly scene. My attention was quickly riveted by a large red star close to the distant horizon. As I gazed upon it I felt a spell of overpowering fascination—it was Mars, the god of war, and for me, the fighting man, it had always held the power of irresistible enchantment. As I gazed at it on that far-gone night it seemed to call across the unthinkable void, to lure me to it, to draw me as the lodestone attracts a particle of iron.

My longing was beyond the power of opposition; I closed my eyes, stretched out my arms toward the god of my vocation and felt myself drawn with the suddenness of thought through the trackless immensity of space. There was an instant of extreme cold and utter darkness.

CHAPTER III
My Advent on Mars

I opened my eyes upon a strange and weird landscape. I knew that I was on Mars; not once did I question either my sanity or my wakefulness. I was not asleep, no need for pinching here; my inner consciousness told me as plainly that I was upon Mars as your conscious mind tells you that you are upon Earth. You do not question the fact; neither did I.

I found myself lying prone upon a bed of yellowish, mosslike vegetation which stretched around me in all directions for interminable miles. I seemed to be lying in a deep, circular basin, along the outer verge of which I could distinguish the irregularities of low hills.

It was midday, the sun was shining full upon me and the heat of it was rather intense upon my naked body, yet no greater than would have been true under similar conditions on an Arizona desert. Here and there were slight outcroppings of quartz-bearing rock which glistened in the sunlight; and a little to my left, perhaps a hundred yards, appeared a low, walled enclosure about four feet in height. No water, and no other vegetation than the moss was in evidence, and as I was somewhat thirsty I determined to do a little exploring.

Springing to my feet I received my first Martian surprise, for the effort, which on Earth would have brought me standing upright, carried me into the Martian air to the height of about three yards. I alighted softly upon the ground, however, without appreciable shock or jar. Now commenced a series of evolutions which even then seemed ludicrous in the extreme. I found that I must learn to walk all over again, as the muscular exertion which carried me easily and safely upon Earth played strange antics with me upon Mars.

Instead of progressing in a sane and dignified manner, my attempts to walk resulted in a variety of hops which took me clear of the ground a couple of feet at each step and landed me sprawling upon my face or back at the end of each second or third hop. My muscles, perfectly attuned and accustomed to the force of gravity on Earth, played the mischief with me in attempting for the first time to cope with the lesser gravitation and lower air pressure on Mars.

I was determined, however, to explore the low structure which was the only evidence of habitation in sight, and so I hit upon the unique plan of reverting to first principles in locomotion, creeping. I did fairly well at this and in a few moments had reached the low, encircling wall of the enclosure.

There appeared to be no doors or windows upon the side nearest me, but as the wall was but about four feet high I cautiously gained my feet and peered over the top upon the strangest sight it had ever been given me to see.

The roof of the enclosure was of solid glass about four or five inches in thickness, and beneath this were several hundred large eggs, perfectly round and snowy white. The eggs were nearly uniform in size being about two and one-half feet in diameter.

Five or six had already hatched and the grotesque caricatures which sat blinking in the sunlight were enough to cause me to doubt my sanity. They seemed mostly head, with little scrawny bodies, long necks and six legs, or, as I afterward learned, two legs and two arms, with an intermediary pair of limbs which could be used at will either as arms or legs. Their eyes were set at the extreme sides of their heads a trifle above the center and protruded in such a manner that they could be directed either forward or back and also independently of each other, thus permitting this queer animal to look in any direction, or in two directions at once, without the necessity of turning the head.

The ears, which were slightly above the eyes and closer together, were small, cup-shaped antennae, protruding not more than an inch on these young specimens. Their noses were but longitudinal slits in the center of their faces, midway between their mouths and ears.

There was no hair on their bodies, which were of a very light yellowish-green color. In the adults, as I was to learn quite soon, this color deepens to an olive green and is darker in the male than in the female. Further, the heads of the adults are not so out of proportion to their bodies as in the case of the young.

The iris of the eyes is blood red, as in Albinos, while the pupil is dark. The eyeball itself is very white, as are the teeth. These latter add a most ferocious appearance to an otherwise fearsome and terrible countenance, as the lower tusks curve upward to sharp points which end about where the eyes of earthly human beings are located. The whiteness of the teeth is not that of ivory, but of the snowiest and most gleaming of china. Against the dark background of their olive skins their tusks stand out in a most striking manner, making these weapons present a singularly formidable appearance.

Most of these details I noted later, for I was given but little time to speculate on the wonders of my new discovery. I had seen that the eggs were in the process of hatching, and as I stood watching the hideous little monsters break from their shells I failed to note the approach of a score of full-grown Martians from behind me.

Coming, as they did, over the soft and soundless moss, which covers practically the entire surface of Mars with the exception of the frozen areas at the poles and the scattered cultivated districts, they might have captured me easily, but their intentions were far more sinister. It was the rattling of the accouterments of the foremost warrior which warned me.

On such a little thing my life hung that I often marvel that I escaped so easily. Had not the rifle of the leader of the party swung from its fastenings beside his saddle in such a way as to strike against the butt of his great metal-shod spear I should have snuffed out without ever knowing that death was near me. But the little sound caused me to turn, and there upon me, not ten feet from my breast, was the point of that huge spear, a spear forty feet long, tipped with gleaming metal, and held low at the side of a mounted replica of the little devils I had been watching.

But how puny and harmless they now looked beside this huge and terrific incarnation of hate, of vengeance and of death. The man himself, for such I may call him, was fully fifteen feet in height and, on Earth, would have weighed some four hundred pounds. He sat his mount as we sit a horse, grasping the animal's barrel with his lower limbs, while the hands of his two right arms held his immense spear low at the side of his mount; his two left arms were outstretched laterally to help preserve his balance, the thing he rode having neither bridle or reins of any description for guidance.

And his mount! How can earthly words describe it! It towered ten feet at the shoulder; had four legs on either side; a broad flat tail, larger at the tip than at the root, and which it held straight out behind while running; a gaping mouth which split its head from its snout to its long, massive neck.

Like its master, it was entirely devoid of hair, but was of a dark slate color and exceeding smooth and glossy. Its belly was white, and its legs shaded from the slate of its shoulders and hips to a vivid yellow at the feet. The feet themselves were heavily padded and nailless, which fact had also contributed to the noiselessness of their approach, and, in common with a multiplicity of legs, is a characteristic feature of the fauna of Mars. The highest type of man and one other animal, the only mammal existing on Mars, alone have well-formed nails, and there are absolutely no hoofed animals in existence there.

Behind this first charging demon trailed nineteen others, similar in all respects, but, as I learned later, bearing individual characteristics peculiar to themselves; precisely as no two of us are identical although we are all cast in a similar mold. This picture, or rather materialized nightmare, which I have described at length, made but one terrible and swift impression on me as I turned to meet it.

Unarmed and naked as I was, the first law of nature manifested itself in the only possible solution of my immediate problem, and that was to get out of the vicinity of the point of the charging spear. Consequently I gave a very earthly and at the same time superhuman leap to reach the top of the Martian incubator, for such I had determined it must be.

My effort was crowned with a success which appalled me no less than it seemed to surprise the Martian warriors, for it carried me fully thirty feet into the air and landed me a hundred feet from my pursuers and on the opposite side of the enclosure.

I alighted upon the soft moss easily and without mishap, and turning saw my enemies lined up along the further wall. Some were surveying me with expressions which I afterward discovered marked extreme astonishment, and the others were evidently satisfying themselves that I had not molested their young.

They were conversing together in low tones, and gesticulating and pointing toward me. Their discovery that I had not harmed the little Martians, and that I was unarmed, must have caused them to look upon me with less ferocity; but, as I was to learn later, the thing which weighed most in my favor was my exhibition of hurdling.

While the Martians are immense, their bones are very large and they are muscled only in proportion to the gravitation which they must overcome. The result is that they are infinitely less agile and less powerful, in proportion to their weight, than an Earth man, and I doubt that were one of them suddenly to be transported to Earth he could lift his own weight from the ground; in fact, I am convinced that he could not do so.

My feat then was as marvelous upon Mars as it would have been upon Earth, and from desiring to annihilate me they suddenly looked upon me as a wonderful discovery to be captured and exhibited among their fellows.

The respite my unexpected agility had given me permitted me to formulate plans for the immediate future and to note more closely the appearance of the warriors, for I could not disassociate these people in my mind from those other warriors who, only the day before, had been pursuing me.

I noted that each was armed with several other weapons in addition to the huge spear which I have described. The weapon which caused me to decide against an attempt at escape by flight was what was evidently a rifle of some description, and which I felt, for some reason, they were peculiarly efficient in handling.

These rifles were of a white metal stocked with wood, which I learned later was a very light and intensely hard growth much prized on Mars, and entirely unknown to us denizens of Earth. The metal of the barrel is an alloy composed principally of aluminum and steel which they have learned to temper to a hardness far exceeding that of the steel with which we are familiar. The weight of these rifles is comparatively little, and with the small caliber, explosive, radium projectiles which they use, and the great length of the barrel, they are deadly in the extreme and at ranges which would be unthinkable on Earth. The theoretic effective radius of this rifle is three hundred miles, but the best they can do in actual service when equipped with their wireless finders and sighters is but a trifle over two hundred miles.

This is quite far enough to imbue me with great respect for the Martian firearm, and some telepathic force must have warned me against an attempt to escape in broad daylight from under the muzzles of twenty of these death-dealing machines.

The Martians, after conversing for a short time, turned and rode away in the direction from which they had come, leaving one of their number alone by the enclosure. When they had covered perhaps two hundred yards they halted, and turning their mounts toward us sat watching the warrior by the enclosure.

He was the one whose spear had so nearly transfixed me, and was evidently the leader of the band, as I had noted that they seemed to have moved to their present position at his direction. When his force had come to a halt he dismounted, threw down his spear and small arms, and came around the end of the incubator toward me, entirely unarmed and as naked as I, except for the ornaments strapped upon his head, limbs, and breast.

When he was within about fifty feet of me he unclasped an enormous metal armlet, and holding it toward me in the open palm of his hand, addressed me in a clear, resonant voice, but in a language, it is needless to say, I could not understand. He then stopped as though waiting for my reply, pricking up his antennae-like ears and cocking his strange-looking eyes still further toward me.

As the silence became painful I concluded to hazard a little conversation on my own part, as I had guessed that he was making overtures of peace. The throwing down of his weapons and the withdrawing of his troop before his advance toward me would have signified a peaceful mission anywhere on Earth, so why not, then, on Mars!

Placing my hand over my heart I bowed low to the Martian and explained to him that while I did not understand his language, his actions spoke for the peace and friendship that at the present moment were most dear to my heart. Of course I might have been a babbling brook for all the intelligence my speech carried to him, but he understood the action with which I immediately followed my words.

Stretching my hand toward him, I advanced and took the armlet from his open palm, clasping it about my arm above the elbow; smiled at him and stood waiting. His wide mouth spread into an answering smile, and locking one of his intermediary arms in mine we turned and walked back toward his mount. At the same time he motioned his followers to advance. They started toward us on a wild run, but were checked by a signal from him. Evidently he feared that were I to be really frightened again I might jump entirely out of the landscape.

He exchanged a few words with his men, motioned to me that I would ride behind one of them, and then mounted his own animal. The fellow designated reached down two or three hands and lifted me up behind him on the glossy back of his mount, where I hung on as best I could by the belts and straps which held the Martian's weapons and ornaments.

The entire cavalcade then turned and galloped away toward the range of hills in the distance.

LAST AND FIRST MEN (1930)

CHAPTER IX: EARTH AND MARS

Olaf Stapledon

Olaf Stapledon was born in 1886 in Seacombe, England. He studied Modern History at Balliol College, Oxford, gaining a BA in 1909 and an MA in 1913. As a conscientious objector during the First World War, Stapledon drove ambulances in France and Belgium. Following the award of a PhD in philosophy from the University of Liverpool in 1925, he published his first non-fiction book, *A Modern Theory of Ethics*, in 1929 and his first novel, *Last and First Men*, in 1930. He subsequently published some twenty-five books, including novels, non-fiction and poetry. Stapledon died in Caldy, England in 1950.

1. THE SECOND MEN AT BAY

Such were the beings that invaded the earth when the Second Men were gathering their strength for a great venture in artificial evolution. The motives of the invasion were both economic and religious. The Martian sought water and vegetable matter; but they came also in a crusading spirit, to 'liberate' the terrestrial diamonds.

Conditions on the earth were very unfavourable to the invaders. Excessive gravitation troubled them less than might have been expected. Only in their most concentrated form did they find it oppressive. More harmful was the density of the terrestrial atmosphere, which constricted the tenuous animate cloudlets very painfully, hindering their vital processes, and deadening all their movements. In their native atmosphere they swam hither and thither with ease and considerable speed; but the treacly air of the earth hampered them as a bird's wings are hampered under water. Moreover, owing to their extreme buoyancy as individual cloudlets, they were scarcely able to dive down so far as the mountain-tops. Excessive oxygen was also a source of distress; it tended to put them into a violent fever, which they had only been able to guard against very imperfectly. Even more damaging was the excessive moisture of the atmosphere, both through its solvent effect upon certain factors in the subvital units, and because heavy rain interfered with the physiological processes of the cloudlets and washed many of their materials to the ground.

The invaders had also to cope with the tissue of 'radio' messages that constantly enveloped the planet, and tended to interfere with their own organic systems of radiation. They were prepared for this to some extent; but 'beam wireless' at close range surprised, bewildered, tortured, and finally routed them; so that they fled back to Mars, leaving many of their number disintegrated in the terrestrial air.

But the pioneering army (or individual, for throughout the adventure it maintained unity of consciousness) had much to report at home. As was expected, there was rich vegetation, and water was even too abundant. There were solid animals, of the type of the prehistoric Martian fauna, but mostly two-legged and erect. Experiment had shown that these creatures died when they were pulled to pieces, and that though the sun's rays affected them by setting up chemical action in their visual organs, they had no really direct sensitivity to radiation. Obviously, therefore, they must be unconscious. On the other hand, the terrestrial atmosphere was permanently alive with radiation of a violent and incoherent type. It was still uncertain whether these crude ethereal agitations were natural phenomena, mere careless offshoots of the cosmic mind, or whether they were emitted by a terrestrial organism. There was reason to suppose this last to be the case, and that the solid organisms were used by some hidden terrestrial intelligence as instruments; for there were buildings, and many of the bipeds were found within the buildings. Moreover, the sudden violent concentration of beam radiation upon the Martian cloud suggested purposeful and hostile behaviour. Punitive action had therefore been taken, and many buildings and bipeds had been destroyed. The physical basis of such a terrestrial intelligence was still to be discovered. It was certainly not in the terrestrial clouds, for these had turned out to be insensitive to radiation. Anyhow, it was obviously an intelligence of very low order, for its radiation was scarcely at all systematic, and was indeed excessively crude. One or two unfortunate diamonds had been found in a building. There was no sign that they were properly venerated.

The Terrestrials, on their side, were left in complete bewilderment by the extraordinary events of that day. Some had jokingly suggested that since the strange substance had behaved in a manner obviously vindictive, it must have been alive and conscious; but no one took the suggestion seriously. Clearly, however, the thing had been dissipated by beam radiation. That at least was an important piece of practical knowledge. But theoretical knowledge about the real nature of the clouds, and their place in the order of the universe, was for the present wholly lacking. To a race of strong cognitive interest and splendid scientific achievement, this ignorance was violently disturbing. It seemed to shake the foundations of the great structure of knowledge. Many frankly hoped, in spite of the loss of life in the first invasion, that there would soon be another opportunity for studying these amazing objects, which were not quite gaseous and not quite solid, not (apparently) organic, yet capable of behaving in a manner suggestive of life. An opportunity was soon afforded.

Some years after the first invasion the Martians appeared again, and in far greater force. This time, moreover, they were almost immune from man's offensive radiation. Operating simultaneously from all the alpine regions of the earth, they began to dry up the great rivers at their sources; and, venturing further afield, they spread over jungle and agricultural land, and stripped off every leaf. Valley after valley was devastated as though by endless swarms of locusts, so that in whole countries there was not a green blade left. The booty was carried off to Mars. Myriads of the subvital units, specialized for transport of water and food materials, were loaded each with a few molecules of the treasure, and dispatched to the home planet. The traffic continued indefinitely. Meanwhile the main body of the Martians proceeded to explore and loot. They were irresistible. For the absorption of water and leafage, they spread over the countryside as an impalpable mist which man had no means to dispel. For the destruction of civilization, they became armies of gigantic cloud-jellies, far bigger than the brute which had formed itself during the earlier invasion. Cities were knocked down and flattened, human beings masticated into pulp. Man tried weapon after weapon in vain.

Presently the Martians discovered the sources of terrestrial radiation in the innumerable wireless transmitting stations. Here at last was the physical basis of the terrestrial intelligence! But what a lowly creature! What a caricature of life! Obviously in respect of complexity and delicacy of organization these wretched immobile systems of glass, metal and vegetable compounds were not to be compared with the Martian cloud. Their only feat seemed to be that they had managed to get control of the unconscious bipeds who tended them.

In the course of their explorations the Martians also discovered a few more diamonds. The second human species had outgrown the barbaric lust for jewellery; but they recognized the beauty of gems and precious metals, and used them as badges of office. Unfortunately, the Martians, in sacking a town, came upon a woman who was wearing a large diamond between her breasts; for she was mayor of the town, and in charge of the evacuation. That the sacred stone should be used thus, apparently for the mere identification of cattle, shocked the invaders even more than the discovery of fragments of diamonds in certain cutting-instruments. The war now began to be waged with all the heroism and brutality of a crusade. Long after a rich booty of water and vegetable matter had been secured, long after the Terrestrials had developed an effective means of attack, and were slaughtering the Martian clouds with high-tension electricity in the form of artificial lightning flashes, the misguided fanatics stayed on to rescue the diamonds and carry them away to the mountain tops, where, years afterwards, climbers discovered them, arranged along the rock-edges in glittering files, like sea-bird's eggs. Thither the dying remnant of the Martian host had transported them with its last strength, scorning to save itself before the diamonds were borne into the pure mountain air, to be lodged with dignity. When the Second Men learned of this great hoard of diamonds, they began to be seriously persuaded that they had been dealing, not with a freak of physical nature, nor yet (as some said) with swarms of bacteria, but with organisms of a higher order. For how could the jewels have been singled out, freed from their metallic settings, and so carefully regimented on the rocks, save by conscious purpose? The murderous clouds must have had at least the pilfering mentality of jackdaws, since evidently they had been fascinated by the treasure. But the very action which revealed their consciousness suggested also that they were no more intelligent than the merely instinctive animals. There was no opportunity of correcting this error, since all the clouds had been destroyed.

The struggle had lasted only a few months. Its material effects on Man were serious but not insurmountable. Its immediate psychological effect was invigorating. The Second Men had long been accustomed to a security and prosperity that were almost utopian. Suddenly they were overwhelmed by a calamity which was quite unintelligible in terms of their own systematic knowledge. Their predecessors, in such a situation, would have behaved with their own characteristic vacillation between the human and the subhuman. They would have contracted a fever of romantic loyalty, and have performed many random acts of secretly self-regarding self-sacrifice. They would have sought profit out of the public disaster, and howled at all who were more fortunate than themselves. They would have cursed their gods, and looked for more useful ones. But also, in an incoherent manner, they would sometimes have behaved reasonably, and would even have risen now and again to the standards of the Second Men. Wholly unused to large-scale human bloodshed, these more developed beings suffered an agony of pity for their mangled fellows. But they said nothing about their pity, and scarcely noticed their own generous grief; for they were busy with the work of rescue. Suddenly confronted with the need of extreme loyalty and courage, they exulted in complying, and experienced that added keenness of spirit which comes when danger is well faced. But it did not occur to them that they were bearing themselves heroically; for they thought they were merely behaving reasonably, showing common sense. And if any one failed in a tight place, they did not call him coward, but gave him a drug to clear his head; or, if that failed, they put him under a doctor. No doubt, among the First Men such a policy would not have been justified, for those bewildered beings had not the clear and commanding vision which kept all sane members of the second species constant in loyalty.

The immediate psychological effect of the disaster was that it afforded this very noble race healthful exercise for its great reserves of loyalty and heroism. Quite apart from this immediate invigoration, however, the first agony, and those many others which were to follow, influenced the Second Men for good and ill in a train of effects which may be called spiritual. They had long known very well that the universe was one in which there could be not only private but also great public tragedies; and their philosophy did not seek to conceal this fact. Private tragedy they were able to face with a bland fortitude, and even an ecstasy of acceptance, such as the earlier species had but rarely attained. Public tragedy, even world-tragedy, they declared should be faced in the same spirit. But to know world-tragedy in the abstract, is very different from the direct acquaintance with it. And now the Second Men, even while they held their attention earnestly fixed upon the practical work of defence, were determined to absorb this tragedy into the very depths of their being, to scrutinize it fearlessly, savour it, digest it, so that its fierce potency should henceforth be added to them. Therefore they did not curse their gods, nor supplicate them. They said to themselves, 'Thus, and thus, and thus, is the world. Seeing the depth we shall see also the height; and we shall praise both.'

But their schooling was yet scarcely begun. The Martian invaders were all dead, but their subvital units were dispersed over the planet as a virulent ultra-microscopic dust. For, though as members of the living cloud they could enter the human body without doing permanent harm, now that they were freed from their functions within the higher organic system, they became a predatory virus. Breathed into man's lungs, they soon adapted themselves to the new environment, and threw his tissues into disorder. Each cell that they entered overthrew its own constitution, like a state which the enemy has successfully infected with lethal propaganda through a mere handful of agents. Thus, though man was temporarily victor over the Martian super-individual, his own vital units were poisoned and destroyed by the subvital remains of his dead enemy. A race whose physique had been as utopian as its body politic, was reduced to timid invalidity. And it was left in possession of a devastated planet. The loss of water proved negligible; but the destruction of vegetation in all the war areas produced for a while a world famine such as the Second Men had never known. And the material fabric of civilization had been so broken that many decades would have to be spent in rebuilding it.

But the physical damage proved far less serious than the physiological. Earnest research discovered, indeed, a means of checking the infection; and, after a few years of rigorous purging, the atmosphere and man's flesh were clean once more. But the generations that had been stricken never recovered; their tissues had been too seriously corroded. Little by little, of course, there arose a fresh population of undamaged men and women. But it was a small population; for the fertility of the stricken had been much reduced. Thus the earth was now occupied by a small number of healthy persons below middle age and a very large number of ageing invalids. For many years these cripples had contrived to carry on the work of the world in spite of their frailty, but gradually they began to fail both in endurance and competence. For they were rapidly losing their grip on life, and sinking into a long-drawn-out senility, from which the Second Men had never before suffered; and at the same time the young, forced to take up work for which they were not yet equipped, committed all manner of blunders and crudities of which their elders would never have been guilty. But such was the general standard of mentality in the second human species, that what might have been an occasion for recrimination produced an unparalleled example of human loyalty at its best. The stricken generations decided almost unanimously that whenever an individual was declared by his generation to have outlived his competence, he should commit suicide. The younger generations, partly through affection, partly through dread of their own incompetence, were at first earnestly opposed to this policy. 'Our elders', one young man said, 'may have declined in vigour, but they are still beloved, and still wise. We dare not carry on without them.' But the elders maintained their point. Many members of the rising generation were no longer juveniles. And, if the body politic was to survive the economic crisis, it must now ruthlessly cut out all its damaged tissues. Accordingly the decision was carried out. One by one, as occasion demanded, the stricken 'chose the peace of annihilation', leaving a scanty, inexperienced, but vigorous, population to rebuild what had been destroyed.

Four centuries passed, and then again the Martian clouds appeared in the sky. Once more devastation and slaughter. Once more a complete failure of the two mentalities to conceive one another. Once more the Martians were destroyed. Once more the pulmonary plague, the slow purging, a crippled population, and generous suicide.

Again, and again they appeared, at irregular intervals for fifty thousand years. On each occasion the Martians came irresistibly fortified against whatever weapon humanity had last used against them. And so, by degrees, men began to recognize that the enemy was no merely instinctive brute, but intelligent. They therefore made attempts to get in touch with these alien minds, and make overtures for a peaceful settlement. But since obviously the negotiations had to be performed by human beings, and since the Martians always regarded human beings as the mere cattle of the terrestrial intelligence, the envoys were always either ignored or destroyed.

During each invasion the Martians contrived to dispatch a considerable bulk of water to Mars. And every time, not satisfied with this material gain, they stayed too long crusading, until man had found a weapon to circumvent their new defences; and then they were routed. After each invasion man's recovery was slower and less complete, while Mars, in spite of the loss of a large proportion of its population, was in the long run invigorated with the extra water.

2. THE RUIN OF TWO WORLDS

Rather more than fifty thousand years after their first appearance, the Martians secured a permanent footing on the Antarctic tableland and over-ran Australasia and South Africa. For many centuries they remained in possession of a large part of the earth's surface, practising a kind of agriculture, studying terrestrial conditions, and spending much energy on the 'liberation' of diamonds.

During the considerable period before their settlement their mentality had scarcely changed; but actual habitation of the earth now began to undermine their self-complacency and their unity. It was borne in upon certain exploring Martians that the terrestrial bipeds, though insensitive to radiation, were actually the intelligences of the planet. At first this fact was studiously shunned, but little by little it gripped the attention of all terrestrial Martians. At the same time they began to realize that the whole work of research into terrestrial conditions, and even the social construction of their colony, depended, not on the public mind, but on private individuals, acting in their private capacity. The colonial super-individual inspired only the diamond crusade, and the attempt to extirpate the terrestrial intelligence, or radiation. These various novel acts of insight woke the Martian colonists from an age-long dream. They saw that their revered super-individual was scarcely more than the least common measure of themselves, a bundle of atavistic fantasies and cravings, knit into one mind and gifted with a certain practical cunning. A rapid and bewildering spiritual renascence now came over the whole Martian colony. The central doctrine of it was that what was valuable in the Martian species was not radiation but mentality. These two utterly different things had been confused, and even identified, since the dawn of Martian civilization. At last they were clearly distinguished. A fumbling but sincere study of mind now began; and distinction was even made between the humbler and loftier mental activities.

There is no telling whither this renascence might have led, had it run its course. Possibly in time the Martians might have recognized worth even in minds other than Martian minds. But such a leap was at first far beyond them. Though they now understood that human animals were conscious and intelligent, they regarded them with no sympathy, rather indeed, with increased hostility. They still rendered allegiance to the Martian race, or brotherhood, just because it was in a sense one flesh, and, indeed, one mind. For they were concerned not to abolish but to recreate the public mind of the colony, and even that of Mars itself.

But the colonial public mind still largely dominated them in their more somnolent periods, and actually sent some of those who, in their private phases, were revolutionaries across to Mars for help against the revolutionary movement. The home planet was quite untouched by the new ideas. Its citizens co-operated whole-heartedly in an attempt to bring the colonists to their senses. But in vain. The colonial public mind itself changed its character as the centuries passed, until it became seriously alienated from Martian orthodoxy. Presently, indeed, it began to undergo a very strange and thorough metamorphosis, from which, conceivably, it might have emerged as the noblest inhabitant of the solar system. Little by little it fell into a kind of hypnotic trance. That is to say, it ceased to possess the attention of its private members, yet remained as a unity of their sub-conscious, or un-noticed mentality. Radiational unity of the colony was maintained, but only in this subconscious manner; and it was at that depth that the great metamorphosis began to take place under the fertilizing influence of the new ideas; which, so to speak, were generated in the tempest of the fully conscious mental revolution, and kept on spreading down into the oceanic depth of the sub-consciousness. Such a condition was likely to produce in time the emergence of a qualitatively new and finer mentality, and to waken at last into a fully conscious super-individual of higher order than its own members. But meanwhile this trance of the public consciousness incapacitated the colony for that prompt and co-ordinated action which had been the most successful faculty of Martian life. The public mind of the home planet easily destroyed its disorderly offspring, and set about re-colonizing the earth.

Several times during the next three hundred thousand years this process repeated itself. The changeless and terribly efficient super-individual of Mars extirpated its own offspring on the earth, before it could emerge from the chrysalis. And the tragedy might have been repeated indefinitely, but for certain changes that took place in humanity.

The first few centuries after the foundation of the Martian colony had been spent in ceaseless war. But at last, with terribly reduced resources, the Second Men had reconciled themselves to the fact that they must live in the same world with their mysterious enemy. Moreover, constant observation of the Martians began to restore somewhat man's shattered self-confidence. For during the fifty thousand years before the Martian colony was founded his opinion of himself had been undermined. He had formerly been used to regarding himself as the sun's ablest child. Then suddenly a stupendous new phenomenon had defeated his intelligence. Slowly he had learned that he was at grips with a determined and versatile rival, and that this rival hailed from a despised planet. Slowly he had been forced to suspect that he himself was outclassed, outshone, by a race whose very physique was incomprehensible to man. But after the Martians had established a permanent colony, human scientists began to discover the real physiological nature of the Martian organism, and were comforted to find that it did not make nonsense of human science. Man also learned that the Martians, though very able in certain spheres, were not really of a high mental type. These discoveries restored human self-confidence. Man settled down to make the best of the situation. Impassable barriers of high-power electric current were devised to keep the Martians out of human territory, and men began patiently to rebuild their ruined home as best they could. At first there was little respite from the crusading zeal of the Martians, but in the second millennium this began to abate, and the two races left one another alone, save for occasional revivals of Martian fervour. Human civilization was at last reconstructed and consolidated, though upon a modest scale. Once more, though interrupted now and again by decades of agony, human beings lived in peace and relative prosperity. Life was somewhat harder than formerly, and the physique of the race was definitely less reliable than of old; but men and women still enjoyed conditions which most nations of the earlier species would have envied. The age of ceaseless personal sacrifice in service of the stricken community had ended at last. Once more a wonderful diversity of untrammelled personalities was put forth. Once more the minds of men and women were devoted without hindrance to the joy of skilled work, and all the subtleties of personal intercourse. Once more the passionate interest in one's fellows, which had for so long been hushed under the all-dominating public calamity, refreshed and enlarged the mind. Once more there was music, sweet, and backward-hearkening towards a golden past. Once more a wealth of literature, and of the visual arts. Once more intellectual exploration into the nature of the physical world and the potentiality of mind. And once more the religious experience, which had for so long been coarsened and obscured by all the violent distractions and inevitable self-deceptions of war, seemed to be refining itself under the influence of re-awakened culture.

In such circumstances the earlier and less sensitive human species might well have prospered indefinitely. Not so the Second Men. For their very refinement of sensibility made them incapable of shunning an ever-present conviction that in spite of all their prosperity they were undermined. Though superficially they seemed to be making a slow but heroic recovery they were at the same time suffering from a still slower and far more profound spiritual decline. Generation succeeded generation. Society became almost perfected, within its limited territory and its limitations of material wealth. The capacities of personality were developed with extreme subtlety and richness. At last the race proposed to itself once more its ancient project of re-making human nature upon a loftier plane. But somehow it had no longer the courage and self-respect for such work.

And so, though there was much talk, nothing was done. Epoch succeeded epoch, and everything human remained apparently the same. Like a twig that has been broken but not broken off, man settled down to retain his life and culture, but could make no progress.

It is almost impossible to describe in a few words the subtle malady of the spirit that was undermining the Second Men. To say that they were suffering from an inferiority complex, would not be wholly false, but it would be a misleading vulgarization of the truth. To say that they had lost faith, both in themselves and in the universe, would be almost as inadequate. Crudely, stated, their trouble was that, as a species, they had attempted a certain spiritual feat beyond the scope of their still-primitive nature. Spiritually they had over-reached themselves, broken every muscle (so to speak) and incapacitated themselves for any further effort. For they had determined to see their own racial tragedy as a thing of beauty, and they had failed. It was the obscure sense of this defeat that had poisoned them. For, being in many respects a very noble species, they could not simply turn their backs upon their failure and pursue the old way of life with the accustomed zest and thoroughness.

During the earliest Martian raids, the spiritual leaders of humanity had preached that the disaster must be an occasion for a supreme religious experience. While striving mightily to save their civilization, men must yet (so it was said) learn not merely to endure, but to admire, even the sternest issue. 'Thus and thus is the world. Seeing the depth, we shall see also the height, and praise both.' The whole population had accepted this advice. At first they had seemed to succeed. Many noble literary expressions were given forth, which seemed to define and elaborate, and even actually to create in men's hearts, this supreme experience. But as the centuries passed and the disasters were repeated, men began to fear that their forefathers had deceived themselves. Those remote generations had earnestly longed to feel the racial tragedy as a factor in the cosmic beauty; and at last they had persuaded themselves that this experience had actually befallen them. But their descendants were slowly coming to suspect that no such experience had ever occurred, that it would never occur to any man, and that there was in fact no such cosmic beauty to be experienced. The First Men would probably, in such a situation, have swung violently either into spiritual nihilism, or else into some comforting religious myth. At any rate, they were of too coarse-grained a nature to be ruined by a trouble so impalpable. Not so the Second Men. For they realized all too clearly that they were faced with the supreme crux of existence. And so, age after age the generations clung desperately to the hope that, if only they could endure a little longer, the light would break in on them. Even after the Martian colony had been three times established and destroyed by the orthodox race in Mars, the supreme preoccupation of the human species was with this religious crux. But afterwards, and very gradually, they lost heart. For it was borne in on them that either they themselves were by nature too obtuse to perceive this ultimate excellence of things (an excellence which they had strong reason to believe in intellectually, although they could not actually experience it), or the human race had utterly deceived itself, and the course of cosmic events after all was not significant, but a meaningless rigmarole.

It was this dilemma that poisoned them. Had they been still physically in their prime, they might have found fortitude to accept it, and proceed to the patient exfoliation of such very real excellences as they were still capable of creating. But they had lost the vitality which alone could perform such acts of spiritual abnegation. All the wealth of personality, all the intricacies of personal relationship, all the complex enterprise of a very great community, all art, all intellectual research, had lost their savour. It is remarkable that a purely religious disaster should have warped even the delight of lovers in one another's bodies, actually taken the flavour out of food, and drawn a veil between the sun-bather and the sun. But individuals of this species, unlike their predecessors, were so closely integrated, that none of their functions could remain healthy while the highest was disordered. Moreover, the general slight failure of physique, which was the legacy of age-long war, had resulted in a recurrence of those shattering brain disorders which had dogged the earliest races of their species. The very horror of the prospect of racial insanity increased their aberration from reasonableness. Little by little, shocking perversions of desire began to terrify them. Masochistic and sadistic orgies alternated with phases of extravagant and ghastly revelry. Acts of treason against the community, hitherto almost unknown, at last necessitated a strict police system. Local groups organized predatory raids against one another. Nations appeared, and all the phobias that make up nationalism.

The Martian colonists, when they observed man's disorganization, prepared, at the instigation of the home planet, a very great offensive. It so happened that at this time the colony was going through its phase of enlightenment, which had always hitherto been followed sooner or later by chastisement from Mars. Many individuals were at the moment actually toying with the idea of seeking harmony with man, rather than war. But the public mind of Mars, outraged by this treason, sought to overwhelm it by instituting a new crusade. Man's disunion offered a great opportunity.

The first attack produced a remarkable change in the human race. Their madness seemed suddenly to leave them. Within a few weeks the national governments had surrendered their sovereignty to a central authority. Disorders, debauchery, perversions, wholly ceased. The treachery and self-seeking and corruption, which had by now been customary for many centuries, suddenly gave place to universal and perfect devotion to the social cause. The species was apparently once more in its right mind. Everywhere, in spite of the war's horrors, there was gay brotherliness, combined with a heroism, which clothed itself in an odd extravagance of jocularity.

The war went ill for man. The general mood changed to cold resolution. And still victory was with the Martians. Under the influence of the huge fanatical armies which were poured in from the home planet, the colonists had shed their tentative pacifism, and sought to vindicate their loyalty by ruthlessness. In reply the human race deserted its sanity, and succumbed to an uncontrollable lust for destruction. It was at this stage that a human bacteriologist announced that he had bred a virus of peculiar deadliness and transmissibility, with which it would be possible to infect the enemy, but at the cost of annihilating also the human race. It is significant of the insane condition of the human population at this time that, when these facts were announced and broadcast, there was no discussion of the desirability of using this weapon. It was immediately put in action, the whole human race applauding.

Within a few months the Martian colony had vanished, their home planet itself had received the infection, and its population was already aware that nothing could save it. Man's constitution was tougher than that of the animate clouds, and he appeared to be doomed to a somewhat more lingering death. He made no effort to save himself, either from the disease which he himself had propagated, or from the pulmonary plague which was caused by the disintegrated substance of the dead Martian colony. All the public processes of civilization began to fall to pieces; for the community was paralyzed by disillusion, and by the expectation of death. Like a bee-hive that has no queen, the whole population of the earth sank into apathy. Men and women stayed in their homes, idling, eating whatever food they could procure, sleeping far into the mornings, and, when at last they rose, listlessly avoiding one another. Only the children could still be gay, and even they were oppressed by their elders' gloom. Meanwhile the disease was spreading. Household after household was stricken, and was left unaided by its neighbours. But the pain in each individual's flesh was strangely numbed by his more poignant distress in the spiritual defeat of the race. For such was the high development of this species, that even physical agony could not distract it from the racial failure. No one wanted to save himself; and each knew that his neighbours desired not his aid. Only the children, when the disease crippled them, were plunged into agony and terror. Tenderly, yet listlessly, their elders would then give them the last sleep. Meanwhile the unburied dead spread corruption among the dying. Cities fell still and silent. The corn was not harvested.

3. THE THIRD DARK AGE

So contagious and so lethal was the new bacterium, that its authors expected the human race to be wiped out as completely as the Martian colony. Each dying remnant of humanity, isolated from its fellows by the breakdown of communications, imagined its own last moments to be the last of man. But by accident, almost one might say by miracle, a spark of human life was once more preserved, to hand on the sacred fire. A certain stock or strain of the race, promiscuously scattered throughout the continents, proved less susceptible than the majority. And, as the bacterium was less vigorous in a hot climate, a few of these favoured individuals, who happened to be in the tropical jungle, recovered from the infection. And of these few a minority recovered also from the pulmonary plague which, as usual, was propagated from the dead Martians.

It might have been expected that from this human germ a new civilized community would have soon arisen. With such brilliant beings as the Second Men, surely a few generations, or at the most a few thousand years, should have sufficed to make up the lost ground.

But no. Once more it was in a manner the very excellence of the species that prevented its recovery, and flung the spirit of Earth into a trance which lasted longer than the whole previous career of mammals. Again and again, some thirty million times, the seasons were repeated; and throughout this period man remained as fixed in bodily and mental character as, formerly, the platypus. Members of the earlier human species must find it difficult to understand this prolonged impotence of a race far more developed than themselves. For here apparently were both the requisites of progressive culture, namely a world rich and unpossessed, and a race exceptionally able. Yet nothing was done.

When the plagues, and all the immense consequent putrefactions, had worked themselves off, the few isolated groups of human survivors settled down to an increasingly indolent tropical life. The fruits of past learning were not imparted to the young, who therefore grew up in extreme ignorance of almost everything beyond their immediate experience. At the same time the elder generation cowed their juniors with vague suggestions of racial defeat and universal futility. This would not have mattered, had the young themselves been normal; they would have reached with fervent optimism. But they themselves were now by nature incapable of any enthusiasm. For, in a species in which the lower functions were so strictly disciplined under the higher, the long-drawn-out spiritual disaster had actually begun to take effect upon the germ-plasm; so that individuals were doomed before birth to lassitude, and to mentality in a minor key. The First Men, long ago, had fallen into a kind of racial senility through a combination of vulgar errors and indulgences. But the second species, like a boy whose mind has been too soon burdened with grave experience, lived henceforth in a sleep-walk.

As the generations passed, all the lore of civilization was shed, save the routine of tropical agriculture and hunting. Not that intelligence itself had waned. Not that the race had sunk into mere savagery. Lassitude did not prevent it from readjusting itself to suit its new circumstances. These sleep-walkers soon invented convenient ways of making, in the home and by hand, much that had hitherto been made in factories and by mechanical power. Almost without mental effort they designed and fashioned tolerable instruments out of wood and flint and bone. But though still intelligent, they had become by disposition, supine, indifferent. They would exert themselves only under the pressure of urgent primitive need. No man seemed capable of putting forth the full energy of a man. Even suffering had lost its poignancy. And no ends seemed worth pursuing that could not be realized speedily. The sting had gone out of experience. The soul was calloused against every goad. Men and women worked and played, loved and suffered; but always in a kind of rapt absent-mindedness. It was as though they were ever trying to remember something important which escaped them. The affairs of daily life seemed too trivial to be taken seriously. Yet that other, and supremely important thing, which alone deserved consideration, was so obscure that no one had any idea what it was. Nor indeed was anyone aware of this hypnotic subjection, any more than a sleeper is aware of being asleep.

The minimum of necessary work was performed, and there was even a certain dreamy zest in the performance, but nothing which would entail extra toil ever seemed worth while. And so, when adjustment to the new circumstances of the world had been achieved, complete stagnation set in. Practical intelligence was easily able to cope with a slowly changing environment, and even with sudden natural upheavals such as floods, earthquakes and disease epidemics. Man remained in a sense master of his world, but he had no idea what to do with his mastery. It was everywhere assumed that the sane end of living was to spend as many days as possible in indolence, lying in the shade. Unfortunately human beings had, of course, many needs which were irksome if not appeased, and so a good deal of hard work had to be done. Hunger and thirst had to be satisfied. Other individuals besides oneself had to be cared for, since man was cursed with sympathy and with a sentiment for the welfare of his group. The only fully rational behavior, it was thought, would be general suicide, but irrational impulses made this impossible. Beatific drugs offered a temporary heaven. But, far as the Second Men had fallen, they were still too clear-sighted to forget that such beatitude is outweighed by subsequent misery.

Century by century, epoch by epoch, man glided on in this seemingly precarious, yet actually unshakable equilibrium. Nothing that happened to him could disturb his easy dominance over the beasts and over physical nature; nothing could shock him out of his racial sleep. Long-drawn-out climatic changes made desert, jungle and grassland fluctuate like the clouds. As the years advanced by millions, ordinary geological processes, greatly accentuated by the immense strains set up by the Patagonian upheaval, remodelled the surface of the planet. Continents were submerged, or lifted out of the sea, till presently there was little of the old configuration. And along with these geological changes went changes in the fauna and flora. The bacterium which had almost exterminated man had also wrought havoc amongst other mammals. Once more the planet had to be re-stocked, this time from the few surviving tropical species. Once more there was a great re-making of old types, only less revolutionary than that which had followed the Patagonian disaster. And since the human race remained minute, through the effects of its spiritual fatigue, other species were favoured. Especially the ruminants and the large carnivora increased and diversified themselves into many habits and forms.

But the most remarkable of all the biological trains of events in this period was the history of the Martian subvital units that had been disseminated by the slaughter of the Martian colony, and had then tormented men and animals with pulmonary diseases. As the ages passed, certain species of mammals so readjusted themselves that the Martian virus became not only harmless but necessary to their well-being. A relationship which was originally that of parasite and host became in time a true symbiosis, a co-operative partnership, in which the terrestrial animals gained something of the unique attributes of the vanished Martian organisms. The time was to come when Man himself should look with envy on these creatures, and finally make use of the Martian 'virus' for his own enrichment.

But meanwhile, and for many million years, almost all kinds of life were on the move, save Man. Like a shipwrecked sailor, he lay exhausted and asleep on his raft, long after the storm had abated.

But his stagnation was not absolute. Imperceptibly, he was drifting on the oceanic currents of life, and in a direction far out of his original course. Little by little, his habit was becoming simpler, less artificial, more animal. Agriculture faded out, since it was no longer necessary in the luxuriant garden where man lived. Weapons of defence and of the chase became more precisely adapted to their restricted purposes, but at the same time less diversified and more stereotyped. Speech almost vanished; for there was no novelty left in experience. Familiar facts and familiar emotions were conveyed increasingly by gestures which were mostly unwitting. Physically, the species had changed little. Though the natural period of life was greatly reduced, this was due less to physiological change than to a strange and fatal increase of absent-mindedness in middle-age. The individual gradually ceased to react to his environment; so that even if he escaped a violent death, he died of starvation.

Yet in spite of this great change, the species remained essentially human. There was no bestialization, such as had formerly produced a race of submen. These tranced remnants of the second human species were not beasts but innocents, simples, children of nature, perfectly adjusted to their simple life. In many ways their state was idyllic and enviable. But such was their dimmed mentality that they were never clearly aware even of the blessings they had, still less, of course, of the loftier experiences which had kindled and tortured their ancestors.

A MARTIAN ODYSSEY (1934)

Stanley G. Weinbaum

Stanley Grauman Weinbaum was born in 1902 in Louisville, Kentucky. He attended the University of Wisconsin-Madison, joining as a chemical engineer, but he subsequently swapped to English and then left before finishing his course. 'A Martian Odyssey' was his first professionally published work of science fiction, printed in the July 1934 edition of *Wonder Stories*. It was preceded by a romantic novel, *The Lady Dances*, which was serialized in newspapers. Several other works of science fiction followed until his untimely death from lung cancer in 1935, when he was just 33 years old. In 1957, one of these stories, 'The Adaptive Ultimate', was adapted into a film titled *She Devil*. A crater on Mars is now named in his honour.

Jarvis stretched himself as luxuriously as he could in the cramped general quarters of the *Ares*.

"Air you can breathe!" he exulted. "It feels as thick as soup after the thin stuff out there!" He nodded at the Martian landscape stretching flat and desolate in the light of the nearer moon, beyond the glass of the port.

The other three stared at him sympathetically—Putz, the engineer, Leroy, the biologist, and Harrison, the astronomer and captain of the expedition. Dick Jarvis was chemist of the famous crew, the *Ares* expedition, first human beings to set foot on the mysterious neighbor of the earth, the planet Mars. This, of course, was in the old days, less than twenty years after the mad American Doheny perfected the atomic blast at the cost of his life, and only a decade after the equally mad Cardoza rode on it to the moon. They were true pioneers, these four of the *Ares*. Except for a half-dozen moon expeditions and the ill-fated de Lancey flight aimed at the seductive orb of Venus, they were the first men to feel other gravity than earth's, and certainly the first successful crew to leave the earth-moon system. And they deserved that success when one considers the difficulties and discomforts—the months spent in acclimatization chambers back on earth, learning to breathe the air as tenuous as that of Mars, the challenging of the void in the tiny rocket driven by the cranky reaction motors of the twenty-first century, and mostly the facing of an absolutely unknown world.

Jarvis stretched and fingered the raw and peeling tip of his frostbitten nose. He sighed again contentedly.

"Well," exploded Harrison abruptly, "are we going to hear what happened? You set out all shipshape in an auxiliary rocket, we don't get a peep for ten days, and finally Putz here picks you out of a lunatic ant-heap with a freak ostrich as your pal! Spill it, man!"

"Speel?" queried Leroy perplexedly. "Speel what?"

"He means '*spiel*,'" explained Putz soberly. "It iss to tell."

Jarvis met Harrison's amused glance without the shadow of a smile. "That's right, Karl," he said in grave agreement with Putz. "*Ich spiel es*!" He grunted comfortably and began.

"According to orders," he said, "I watched Karl here take off toward the North, and then I got into my flying sweatbox and headed South. You'll remember, Cap—we had orders not to land, but just scout about for points of interest. I set the two cameras clicking and buzzed along, riding pretty high—about two thousand feet—for a couple of reasons. First, it gave the cameras a greater field, and second, the under-jets travel so far in this half-vacuum they call air here that they stir up dust if you move low."

"We know all that from Putz," grunted Harrison. "I wish you'd saved the films, though. They'd have paid the cost of this junket; remember how the public mobbed the first moon pictures?"

"The films are safe," retorted Jarvis. "Well," he resumed, "as I said, I buzzed along at a pretty good clip; just as we figured, the wings haven't much lift in this air at less than a hundred miles per hour, and even then I had to use the under-jets.

"So, with the speed and the altitude and the blurring caused by the under-jets, the seeing wasn't any too good. I could see enough, though, to distinguish that what I sailed over was just more of this grey plain that we'd been examining the whole week since our landing—same blobby growths and the same eternal carpet of crawling little plant-animals, or biopods, as Leroy calls them. So I sailed along, calling back my position every hour as instructed, and not knowing whether you heard me."

"I did!" snapped Harrison.

"A hundred and fifty miles south," continued Jarvis imperturbably, "the surface changed to a sort of low plateau, nothing but desert and orange-tinted sand. I figured that we were right in our guess, then, and this grey plain we dropped on was really the Mare Cimmerium which would make my orange desert the region called Xanthus. If I were right, I ought to hit another grey plain, the Mare Chronium in another couple of hundred miles, and then another orange desert, Thyle I or II. And so I did."

"Putz verified our position a week and a half ago!" grumbled the captain. "Let's get to the point."

"Coming!" remarked Jarvis. "Twenty miles into Thyle—believe it or not—I crossed a canal!"

"Putz photographed a hundred! Let's hear something new!"

"And did he also see a city?"

"Twenty of 'em, if you call those heaps of mud cities!"

"Well," observed Jarvis, "from here on I'll be telling a few things Putz didn't see!" He rubbed his tingling nose, and continued. "I knew that I had sixteen hours of daylight at this season, so eight hours—eight hundred miles—from here, I decided to turn back. I was still over Thyle, whether I or II I'm not sure, not more than twenty-five miles into it. And right there, Putz's pet motor quit!"

"Quit? How?" Putz was solicitous.

"The atomic blast got weak. I started losing altitude right away, and suddenly there I was with a thump right in the middle of Thyle! Smashed my nose on the window, too!" He rubbed the injured member ruefully.

"Did you maybe try vashing der combustion chamber mit acid sulphuric?" inquired Putz. "Sometimes der lead giffs a secondary radiation—"

"Naw!" said Jarvis disgustedly. "I wouldn't try that, of course—not more than ten times! Besides, the bump flattened the landing gear and busted off the under-jets. Suppose I got the thing working—what then? Ten miles with the blast coming right out of the bottom and I'd have melted the floor from under me!" He rubbed his nose again. "Lucky for me a pound only weighs seven ounces here, or I'd have been mashed flat!"

"I could have fixed!" ejaculated the engineer. "I bet it vas not serious."

"Probably not," agreed Jarvis sarcastically. "Only it wouldn't fly. Nothing serious, but I had my choice of waiting to be picked up or trying to walk back—eight hundred miles, and perhaps twenty days before we had to leave! Forty miles a day! Well," he concluded, "I chose to walk. Just as much chance of being picked up, and it kept me busy."

"We'd have found you," said Harrison.

"No doubt. Anyway, I rigged up a harness from some seat straps, and put the water tank on my back, took a cartridge belt and revolver, and some iron rations, and started out."

"Water tank!" exclaimed the little biologist, Leroy. "She weigh one-quarter ton!"

"Wasn't full. Weighed about two hundred and fifty pounds earth-weight, which is eighty-five here. Then, besides, my own personal two hundred and ten pounds is only seventy on Mars, so, tank and all, I grossed a hundred and fifty-five, or fifty-five pounds less than my everyday earth-weight. I figured on that when I undertook the forty-mile daily stroll. Oh—of course I took a thermo-skin sleeping bag for these wintry Martian nights.

"Off I went, bouncing along pretty quickly. Eight hours of daylight meant twenty miles or more. It got tiresome, of course—plugging along over a soft sand desert with nothing to see, not even Leroy's crawling biopods. But an hour or so brought me to the canal—just a dry ditch about four hundred feet wide, and straight as a railroad on its own company map.

"There'd been water in it sometime, though. The ditch was covered with what looked like a nice green lawn. Only, as I approached, the lawn moved out of my way!"

"Eh?" said Leroy.

"Yeah, it was a relative of your biopods. I caught one—a little grass-like blade about as long as my finger, with two thin, stemmy legs."

"He is where?" Leroy was eager.

"He is let go! I had to move, so I plowed along with the walking grass opening in front and closing behind. And then I was out on the orange desert of Thyle again.

"I plugged steadily along, cussing the sand that made going so tiresome, and, incidentally, cussing that cranky motor of yours, Karl. It was just before twilight that I reached the edge of Thyle, and looked down over the gray Mare Chronium. And I knew there was seventy-five miles of *that* to be walked over, and then a couple of hundred miles of that Xanthus desert, and about as much more Mare Cimmerium. Was I pleased? I started cussing you fellows for not picking me up!"

"We were trying, you sap!" said Harrison.

"That didn't help. Well, I figured I might as well use what was left of daylight in getting down the cliff that bounded Thyle. I found an easy place, and down I went. Mare Chronium was just the same sort of place as this—crazy leafless plants and a bunch of crawlers; I gave it a glance and hauled out my sleeping bag. Up to that time, you know, I hadn't seen anything worth worrying about on this half-dead world—nothing dangerous, that is."

"Did you?" queried Harrison.

"*Did I!* You'll hear about it when I come to it. Well, I was just about to turn in when suddenly I heard the wildest sort of shenanigans!"

"Vot iss shenanigans?" inquired Putz.

"He says, '*Je ne sais quoi*,'" explained Leroy. "It is to say, 'I don't know what.'"

"That's right," agreed Jarvis. "I didn't know what, so I sneaked over to find out. There was a racket like a flock of crows eating a bunch of canaries—whistles, cackles, caws, trills, and what have you. I rounded a clump of stumps, and there was Tweel!"

"Tweel?" said Harrison, and "Tveel?" said Leroy and Putz.

"That freak ostrich," explained the narrator. "At least, Tweel is as near as I can pronounce it without sputtering. He called it something like 'Trrrweerrlll.'"

"What was he doing?" asked the Captain.

"He was being eaten! And squealing, of course, as anyone would."

"Eaten! By what?"

"I found out later. All I could see then was a bunch of black ropy arms tangled around what looked like, as Putz described it to you, an ostrich. I wasn't going to interfere, naturally; if both creatures were dangerous, I'd have one less to worry about.

"But the birdlike thing was putting up a good battle, dealing vicious blows with an eighteen-inch beak, between screeches. And besides, I caught a glimpse or two of what was on the end of those arms!" Jarvis shuddered. "But the clincher was when I noticed a little black bag or case hung about the neck of the bird-thing! It was intelligent! That or tame, I assumed. Anyway, it clinched my decision. I pulled out my automatic and fired into what I could see of its antagonist.

"There was a flurry of tentacles and a spurt of black corruption, and then the thing, with a disgusting sucking noise, pulled itself and its arms into a hole in the ground. The other let out a series of clacks, staggered around on legs about as thick as golf sticks, and turned suddenly to face me. I held my weapon ready, and the two of us stared at each other.

"The Martian wasn't a bird, really. It wasn't even birdlike, except just at first glance. It had a beak all right, and a few feathery appendages, but the beak wasn't really a beak. It was somewhat flexible; I could see the tip bend slowly from side to side; it was almost like a cross between a beak and a trunk. It had four-toed feet, and four fingered things—hands, you'd have to call them, and a little roundish body, and a long neck ending in a tiny head—and that beak. It stood an inch or so taller than I, and—well, Putz saw it!"

The engineer nodded. "*Ja!* I saw!"

Jarvis continued. "So—we stared at each other. Finally the creature went into a series of clackings and twitterings and held out its hands toward me, empty. I took that as a gesture of friendship."

"Perhaps," suggested Harrison, "it looked at that nose of yours and thought you were its brother!"

"Huh! You can be funny without talking! Anyway, I put up my gun and said 'Aw, don't mention it,' or something of the sort, and the thing came over and we were pals.

"By that time, the sun was pretty low and I knew that I'd better build a fire or get into my thermo-skin. I decided on the fire. I picked a spot at the base of the Thyle cliff, where the rock could reflect a little heat on my back. I started breaking off chunks of this desiccated Martian vegetation, and my companion caught the idea and brought in an armful. I reached for a match, but the Martian fished into his pouch and brought out something that looked like a glowing coal; one touch of it, and the fire was blazing—and you all know what a job we have starting a fire in this atmosphere!

"And that bag of his!" continued the narrator. "That was a manufactured article, my friends; press an end and she popped open—press the middle and she sealed so perfectly you couldn't see the line. Better than zippers.

"Well, we stared at the fire a while and I decided to attempt some sort of communication with the Martian. I pointed at myself and said 'Dick'; he caught the drift immediately, stretched a bony claw at me and repeated 'Tick.' Then I pointed at him, and he gave that whistle I called Tweel; I can't imitate his accent. Things were going smoothly; to emphasize the names, I repeated 'Dick,' and then, pointing at him, 'Tweel.'

"There we stuck! He gave some clacks that sounded negative, and said something like 'P-p-p-proot.' And that was just the beginning; I was always 'Tick,' but as for him—part of the time he was 'Tweel,' and part of the time he was 'P-p-p-proot,' and part of the time he was sixteen other noises!

"We just couldn't connect. I tried 'rock,' and I tried 'star,' and 'tree,' and 'fire,' and Lord knows what else, and try as I would, I couldn't get a single word! Nothing was the same for two successive minutes, and if that's a language, I'm an alchemist! Finally I gave it up and called him Tweel, and that seemed to do.

"But Tweel hung on to some of my words. He remembered a couple of them, which I suppose is a great achievement if you're used to a language you have to make up as you go along. But I couldn't get the hang of his talk; either I missed some subtle point or we just didn't *think* alike—and I rather believe the latter view.

"I've other reasons for believing that. After a while I gave up the language business, and tried mathematics. I scratched two plus two equals four on the ground, and demonstrated it with pebbles. Again Tweel caught the idea, and informed me that three plus three equals six. Once more we seemed to be getting somewhere.

"So, knowing that Tweel had at least a grammar school education, I drew a circle for the sun, pointing first at it, and then at the last glow of the sun. Then I sketched in Mercury, and Venus, and Mother Earth, and Mars, and finally, pointing to Mars, I swept my hand around in a sort of inclusive gesture to indicate that Mars was our current environment. I was working up to putting over the idea that my home was on the earth.

"Tweel understood my diagram all right. He poked his beak at it, and with a great deal of trilling and clucking, he added Deimos and Phobos to Mars, and then sketched in the earth's moon!

"Do you see what that proves? It proves that Tweel's race uses telescopes—that they're civilized!"

"Does not!" snapped Harrison. "The moon is visible from here as a fifth magnitude star. They could see its revolution with the naked eye."

"The moon, yes!" said Jarvis. "You've missed my point. Mercury isn't visible! And Tweel knew of Mercury because he placed the Moon at the *third* planet, not the second. If he didn't know Mercury, he'd put the earth second, and Mars third, instead of fourth! See?"

"Humph!" said Harrison.

"Anyway," proceeded Jarvis, "I went on with my lesson. Things were going smoothly, and it looked as if I could put the idea over. I pointed at the earth on my diagram, and then at myself, and then, to clinch it, I pointed to myself and then to the earth itself shining bright green almost at the zenith.

"Tweel set up such an excited clacking that I was certain he understood. He jumped up and down, and suddenly he pointed at himself and then at the sky, and then at himself and at the sky again. He pointed at his middle and then at Arcturus, at his head and then at Spica, at his feet and then at half a dozen stars, while I just gaped at him. Then, all of a sudden, he gave a tremendous leap. Man, what a hop! He shot straight up into the starlight, seventy-five feet if an inch! I saw him silhouetted against the sky, saw him turn and come down at me head first, and land smack on his beak like a javelin! There he stuck square in the center of my sun-circle in the sand—a bull's eye!"

"Nuts!" observed the captain. "Plain nuts!"

"That's what I thought, too! I just stared at him open-mouthed while he pulled his head out of the sand and stood up. Then I figured he'd missed my point, and I went through the whole blamed rigamarole again, and it ended the same way, with Tweel on his nose in the middle of my picture!"

"Maybe it's a religious rite," suggested Harrison.

"Maybe," said Jarvis dubiously. "Well, there we were. We could exchange ideas up to a certain point, and then—blooey! Something in us was different, unrelated; I don't doubt that Tweel thought me just as screwy as I thought him. Our minds simply looked at the world from different viewpoints, and perhaps his viewpoint is as true as ours. But—we couldn't get together, that's all. Yet, in spite of all difficulties, I *liked* Tweel, and I have a queer certainty that he liked me."

"Nuts!" repeated the captain. "Just daffy!"

"Yeah? Wait and see. A couple of times I've thought that perhaps we—" He paused, and then resumed his narrative. "Anyway, I finally gave it up, and got into my thermo-skin to sleep. The fire hadn't kept me any too warm, but that damned sleeping bag did. Got stuffy five minutes after I closed myself in. I opened it a little and bingo! Some eighty-below-zero air hit my nose, and that's when I got this pleasant little frostbite to add to the bump I acquired during the crash of my rocket.

"I don't know what Tweel made of my sleeping. He sat around, but when I woke up, he was gone. I'd just crawled out of my bag, though, when I heard some twittering, and there he came, sailing down from that three-story Thyle cliff to alight on his beak beside me. I pointed to myself and toward the north, and he pointed at himself and toward the south, but when I loaded up and started away, he came along.

"Man, how he traveled! A hundred and fifty feet at a jump, sailing through the air stretched out like a spear, and landing on his beak. He seemed surprised at my plodding, but after a few moments he fell in beside me, only every few minutes he'd go into one of his leaps, and stick his nose into the sand a block ahead of me. Then he'd come shooting back at me; it made me nervous at first to see that beak of his coming at me like a spear, but he always ended in the sand at my side.

"So the two of us plugged along across the Mare Chronium. Same sort of place as this—same crazy plants and same little green biopods growing in the sand, or crawling out of your way. We talked—not that we understood each other, you know, but just for company. I sang songs, and I suspect Tweel did too; at least, some of his trillings and twitterings had a subtle sort of rhythm.

"Then, for variety, Tweel would display his smattering of English words. He'd point to an outcropping and say 'rock,' and point to a pebble and say it again; or he'd touch my arm and say 'Tick,' and then repeat it. He seemed terrifically amused that the same word meant the same thing twice in succession, or that the same word could apply to two different objects. It set me wondering if perhaps his language wasn't like the primitive speech of some earth people—you know, Captain, like the Negritoes, for instance, who haven't any generic words. No word for food or water or man—words for good food and bad food, or rain water and sea water, or strong man and weak man—but no names for general classes. They're too primitive to understand that rain water and sea water are just different aspects of the same thing. But that wasn't the case with Tweel; it was just that we were somehow mysteriously different—our minds were alien to each other. And yet—we *liked* each other!"

"Looney, that's all," remarked Harrison. "That's why you two were so fond of each other."

"Well, I like *you*!" countered Jarvis wickedly. "Anyway," he resumed, "don't get the idea that there was anything screwy about Tweel. In fact, I'm not so sure but that he couldn't teach our highly praised human intelligence a trick or two. Oh, he wasn't an intellectual superman, I guess; but don't overlook the point that he managed to understand a little of my mental workings, and I never even got a glimmering of his."

"Because he didn't have any!" suggested the captain, while Putz and Leroy blinked attentively.

"You can judge of that when I'm through," said Jarvis. "Well, we plugged along across the Mare Chronium all that day, and all the next. Mare Chronium—Sea of Time! Say, I was willing to agree with Schiaparelli's name by the end of that march! Just that grey, endless plain of weird plants, and never a sign of any other life. It was so monotonous that I was even glad to see the desert of Xanthus toward the evening of the second day.

"I was fair worn out, but Tweel seemed as fresh as ever, for all I never saw him drink or eat. I think he could have crossed the Mare Chronium in a couple of hours with those block-long nose dives of his, but he stuck along with me. I offered him some water once or twice; he took the cup from me and sucked the liquid into his beak, and then carefully squirted it all back into the cup and gravely returned it.

"Just as we sighted Xanthus, or the cliffs that bounded it, one of those nasty sand clouds blew along, not as bad as the one we had here, but mean to travel against. I pulled the transparent flap of my thermo-skin bag across my face and managed pretty well, and I noticed that Tweel used some feathery appendages growing like a mustache at the base of his beak to cover his nostrils, and some similar fuzz to shield his eyes."

"He is a desert creature!" ejaculated the little biologist, Leroy.

"Huh? Why?"

"He drink no water—he is adapt for sand storm—"

"Proves nothing! There's not enough water to waste anywhere on this desiccated pill called Mars. We'd call all of it desert on earth, you know." He paused. "Anyway, after the sand storm blew over, a little wind kept blowing in our faces, not strong enough to stir the sand. But suddenly things came drifting along from the Xanthus cliffs—small, transparent spheres, for all the world like glass tennis balls! But light—they were almost light enough to float even in this thin air—empty, too; at least, I cracked open a couple and nothing came out but a bad smell. I asked Tweel about them, but all he said was 'No, no, no,' which I took to mean that he knew nothing about them. So they went bouncing by like tumbleweeds, or like soap bubbles, and we plugged on toward Xanthus. Tweel pointed at one of the crystal balls once and said 'rock,' but I was too tired to argue with him. Later I discovered what he meant.

"We came to the bottom of the Xanthus cliffs finally, when there wasn't much daylight left. I decided to sleep on the plateau if possible; anything dangerous, I reasoned, would be more likely to prowl through the vegetation of the Mare Chronium than the sand of Xanthus. Not that I'd seen a single sign of menace, except the rope-armed black thing that had trapped Tweel, and apparently that didn't prowl at all, but lured its victims within reach. It couldn't lure me while I slept, especially as Tweel didn't seem to sleep at all, but simply sat patiently around all night. I wondered how the creature had managed to trap Tweel, but there wasn't any way of asking him. I found that out too, later; it's devilish!

"However, we were ambling around the base of the Xanthus barrier looking for an easy spot to climb. At least, I was. Tweel could have leaped it easily, for the cliffs were lower than Thyle—perhaps sixty feet. I found a place and started up, swearing at the water tank strapped to my back—it didn't bother me except when climbing—and suddenly I heard a sound that I thought I recognized!

"You know how deceptive sounds are in this thin air. A shot sounds like the pop of a cork. But this sound was the drone of a rocket, and sure enough, there went our second auxiliary about ten miles to westward, between me and the sunset!"

"Vas me!" said Putz. "I hunt for you."

"Yeah; I knew that, but what good did it do me? I hung on to the cliff and yelled and waved with one hand. Tweel saw it too, and set up a trilling and twittering, leaping to the top of the barrier and then high into the air. And while I watched, the machine droned on into the shadows to the south.

"I scrambled to the top of the cliff. Tweel was still pointing and trilling excitedly, shooting up toward the sky and coming down head-on to stick upside down on his beak in the sand. I pointed toward the south and at myself, and he said, 'Yes—Yes—Yes'; but somehow I gathered that he thought the flying thing was a relative of mine, probably a parent. Perhaps I did his intellect an injustice; I think now that I did.

"I was bitterly disappointed by the failure to attract attention. I pulled out my thermo-skin bag and crawled into it, as the night chill was already apparent. Tweel stuck his beak into the sand and drew up his legs and arms and looked for all the world like one of those leafless shrubs out there. I think he stayed that way all night."

"Protective mimicry!" ejaculated Leroy. "See? He is desert creature!"

"In the morning," resumed Jarvis, "we started off again. We hadn't gone a hundred yards into Xanthus when I saw something queer! This is one thing Putz didn't photograph, I'll wager!

"There was a line of little pyramids—tiny ones, not more than six inches high, stretching across Xanthus as far as I could see! Little buildings made of pygmy bricks, they were, hollow inside and truncated, or at least broken at the top and empty. I pointed at them and said 'What?' to Tweel, but he gave some negative twitters to indicate, I suppose, that he didn't know. So off we went, following the row of pyramids because they ran north, and I was going north.

"Man, we trailed that line for hours! After a while, I noticed another queer thing: they were getting larger. Same number of bricks in each one, but the bricks were larger.

"By noon they were shoulder high. I looked into a couple—all just the same, broken at the top and empty. I examined a brick or two as well; they were silica, and old as creation itself!"

"How you know?" asked Leroy.

"They were weathered—edges rounded. Silica doesn't weather easily even on earth, and in this climate—!"

"How old you think?"

"Fifty thousand—a hundred thousand years. How can I tell? The little ones we saw in the morning were older—perhaps ten times as old. Crumbling. How old would that make *them*? Half a million years? Who knows?" Jarvis paused a moment. "Well," he resumed, "we followed the line. Tweel pointed at them and said 'rock' once or twice, but he'd done that many times before. Besides, he was more or less right about these.

"I tried questioning him. I pointed at a pyramid and asked 'People?' and indicated the two of us. He set up a negative sort of clucking and said, 'No, no, no. No one-one-two. No two-two-four,' meanwhile rubbing his stomach. I just stared at him and he went through the business again. 'No one-one-two. No two-two-four.' I just gaped at him."

"That proves it!" exclaimed Harrison. "Nuts!"

"You think so?" queried Jarvis sardonically. "Well, I figured it out different! 'No one-one-two!' You don't get it, of course, do you?"

"Nope—nor do you!"

"I think I do! Tweel was using the few English words he knew to put over a very complex idea. What, let me ask, does mathematics make you think of?"

"Why—of astronomy. Or—or logic!"

"That's it! 'No one-one-two!' Tweel was telling me that the builders of the pyramids weren't people—or that they weren't intelligent, that they weren't reasoning creatures! Get it?"

"Huh! I'll be damned!"

"You probably will."

"Why," put in Leroy, "he rub his belly?"

"Why? Because, my dear biologist, that's where his brains are! Not in his tiny head—in his middle!"

"*C'est impossible*!"

"Not on Mars, it isn't! This flora and fauna aren't earthly; your biopods prove that!" Jarvis grinned and took up his narrative. "Anyway, we plugged along across Xanthus and in about the middle of the afternoon, something else queer happened. The pyramids ended."

"Ended!"

"Yeah; the queer part was that the last one—and now they were ten-footers—was capped! See? Whatever built it was still inside; we'd trailed 'em from their half-million-year-old origin to the present.

"Tweel and I noticed it about the same time. I yanked out my automatic (I had a clip of Boland explosive bullets in it) and Tweel, quick as a sleight-of-hand trick, snapped a queer little glass revolver out of his bag. It was much like our weapons, except that the grip was larger to accommodate his four-taloned hand. And we held our weapons ready while we sneaked up along the lines of empty pyramids.

"Tweel saw the movement first. The top tiers of bricks were heaving, shaking, and suddenly slid down the sides with a thin crash. And then—something—something was coming out!

"A long, silvery-grey arm appeared, dragging after it an armored body. Armored, I mean, with scales, silver-grey and dull-shining. The arm heaved the body out of the hole; the beast crashed to the sand.

"It was a nondescript creature—body like a big grey cask, arm and a sort of mouth-hole at one end; stiff, pointed tail at the other—and that's all. No other limbs, no eyes, ears, nose—nothing! The thing dragged itself a few yards, inserted its pointed tail in the sand, pushed itself upright, and just sat.

"Tweel and I watched it for ten minutes before it moved. Then, with a creaking and rustling like—oh, like crumpling stiff paper—its arm moved to the mouth-hole and out came a brick! The arm placed the brick carefully on the ground, and the thing was still again.

"Another ten minutes—another brick. Just one of Nature's bricklayers. I was about to slip away and move on when Tweel pointed at the thing and said 'rock'! I went 'huh?' and he said it again. Then, to the accompaniment of some of his trilling, he said, 'No—no—,' and gave two or three whistling breaths.

"Well, I got his meaning, for a wonder! I said, 'No breath?' and demonstrated the word. Tweel was ecstatic; he said, 'Yes, yes, yes! No, no, no breet!' Then he gave a leap and sailed out to land on his nose about one pace from the monster!

"I was startled, you can imagine! The arm was going up for a brick, and I expected to see Tweel caught and mangled, but—nothing happened! Tweel pounded on the creature, and the arm took the brick and placed it neatly beside the first. Tweel rapped on its body again, and said 'rock,' and I got up nerve enough to take a look myself.

"Tweel was right again. The creature was rock, and it didn't breathe!"

"How you know?" snapped Leroy, his black eyes blazing interest.

"Because I'm a chemist. The beast was made of silica! There must have been pure silicon in the sand, and it lived on that. Get it? We, and Tweel, and those plants out there, and even the biopods are *carbon* life; this thing lived by a different set of chemical reactions. It was silicon life!"

"*La vie silicieuse*!" shouted Leroy. "I have suspect, and now it is proof! I must go see! *Il faut que je*—"

"All right! All right!" said Jarvis. "You can go see. Anyhow, there the thing was, alive and yet not alive, moving every ten minutes, and then only to remove a brick. Those bricks were its waste matter. See, Frenchy? We're carbon, and our waste is carbon dioxide, and this thing is silicon, and its waste is silicon dioxide—silica. But silica is a solid, hence the bricks. And it builds itself in, and when it is covered, it moves over to a fresh place to start over. No wonder it creaked! A living creature half a million years old!"

"How you know how old?" Leroy was frantic.

"We trailed its pyramids from the beginning, didn't we? If this weren't the original pyramid builder, the series would have ended somewhere before we found him, wouldn't it?—ended and started over with the small ones. That's simple enough, isn't it?

"But he reproduces, or tries to. Before the third brick came out, there was a little rustle and out popped a whole stream of those little crystal balls. They're his spores, or eggs, or seeds—call 'em what you want. They went bouncing by across Xanthus just as they'd bounced by us back in the Mare Chronium. I've a hunch how they work, too—this is for your information, Leroy. I think the crystal shell of silica is no more than a protective covering, like an eggshell, and that the active principle is the smell inside. It's some sort of gas that attacks silicon, and if the shell is broken near a supply of that element, some reaction starts that ultimately develops into a beast like that one."

"You should try!" exclaimed the little Frenchman. "We must break one to see!"

"Yeah? Well, I did. I smashed a couple against the sand. Would you like to come back in about ten thousand years to see if I planted some pyramid monsters? You'd most likely be able to tell by that time!" Jarvis paused and drew a deep breath. "Lord! That queer creature! Do you picture it? Blind, deaf, nerveless, brainless—just a mechanism, and yet—immortal! Bound to go on making bricks, building pyramids, as long as silicon and oxygen exist, and even afterwards it'll just stop. It won't be dead. If the accidents of a million years bring it its food again, there it'll be, ready to run again, while brains and civilizations are part of the past. A queer beast—yet I met a stranger one!"

"If you did, it must have been in your dreams!" growled Harrison.

"You're right!" said Jarvis soberly. "In a way, you're right. The dream-beast! That's the best name for it—and it's the most fiendish, terrifying creation one could imagine! More dangerous than a lion, more insidious than a snake!"

"Tell me!" begged Leroy. "I must go see!"

"Not *this* devil!" He paused again. "Well," he resumed, "Tweel and I left the pyramid creature and plowed along through Xanthus. I was tired and a little disheartened by Putz's failure to pick me up, and Tweel's trilling got on my nerves, as did his flying nosedives. So I just strode along without a word, hour after hour across that monotonous desert.

"Toward mid-afternoon we came in sight of a low dark line on the horizon. I knew what it was. It was a canal; I'd crossed it in the rocket and it meant that we were just one-third of the way across Xanthus. Pleasant thought, wasn't it? And still, I was keeping up to schedule.

"We approached the canal slowly; I remembered that this one was bordered by a wide fringe of vegetation and that Mud-heap City was on it.

"I was tired, as I said. I kept thinking of a good hot meal, and then from that I jumped to reflections of how nice and homelike even Borneo would seem after this crazy planet, and from that, to thoughts of little old New York, and then to thinking about a girl I know there—Fancy Long. Know her?"

"Vision entertainer," said Harrison. "I've tuned her in. Nice blonde—dances and sings on the *Yerba Mate Hour*."

"That's her," said Jarvis ungrammatically. "I know her pretty well—just friends, get me?—though she came down to see us off in the *Ares*. Well, I was thinking about her, feeling pretty lonesome, and all the time we were approaching that line of rubbery plants.

"And then—I said, 'What 'n Hell!' and stared. And there she was—Fancy Long, standing plain as day under one of those crackbrained trees, and smiling and waving just the way I remembered her when we left!"

"Now you're nuts, too!" observed the captain.

"Boy, I almost agreed with you! I stared and pinched myself and closed my eyes and then stared again—and every time, there was Fancy Long smiling and waving! Tweel saw something, too; he was trilling and clucking away, but I scarcely heard him. I was bounding toward her over the sand, too amazed even to ask myself questions.

"I wasn't twenty feet from her when Tweel caught me with one of his flying leaps. He grabbed my arm, yelling, 'No—no—no!' in his squeaky voice. I tried to shake him off—he was as light as if he were built of bamboo—but he dug his claws in and yelled. And finally some sort of sanity returned to me and I stopped less than ten feet from her. There she stood, looking as solid as Putz's head!"

"Vot?" said the engineer.

"She smiled and waved, and waved and smiled, and I stood there dumb as Leroy, while Tweel squeaked and chattered. I knew it couldn't be real, yet—there she was!

"Finally I said, 'Fancy! Fancy Long!' She just kept on smiling and waving, but looking as real as if I hadn't left her thirty-seven million miles away.

"Tweel had his glass pistol out, pointing it at her. I grabbed his arm, but he tried to push me away. He pointed at her and said, 'No breet! No breet!' and I understood that he meant that the Fancy Long thing wasn't alive. Man, my head was whirling!

"Still, it gave me the jitters to see him pointing his weapon at her. I don't know why I stood there watching him take careful aim, but I did. Then he squeezed the handle of his weapon; there was a little puff of steam, and Fancy Long was gone! And in her place was one of those writhing, black, rope-armed horrors like the one I'd saved Tweel from!

"The dream-beast! I stood there dizzy, watching it die while Tweel trilled and whistled. Finally he touched my arm, pointed at the twisting thing, and said, 'You one-one-two, he one-one-two.' After he'd repeated it eight or ten times, I got it. Do any of you?"

"Oui!" shrilled Leroy. "*Moi—je le comprends*! He mean you think of something, the beast he know, and you see it! *Un chien*—a hungry dog, he would see the big bone with meat! Or smell it—not?"

"Right!" said Jarvis. "The dream-beast uses its victim's longings and desires to trap its prey. The bird at nesting season would see its mate, the fox, prowling for its own prey, would see a helpless rabbit!"

"How he do?" queried Leroy.

"How do I know? How does a snake back on earth charm a bird into its very jaws? And aren't there deep-sea fish that lure their victims into their mouths? Lord!" Jarvis shuddered. "Do you see how insidious the monster is? We're warned now—but henceforth we can't trust even our eyes. You might see me—I might see one of you—and back of it may be nothing but another of those black horrors!"

"How'd your friend know?" asked the captain abruptly.

"Tweel? I wonder! Perhaps he was thinking of something that couldn't possibly have interested me, and when I started to run, he realized that I saw something different and was warned. Or perhaps the dream-beast can only project a single vision, and Tweel saw what I saw—or nothing. I couldn't ask him. But it's just another proof that his intelligence is equal to ours or greater."

"He's daffy, I tell you!" said Harrison. "What makes you think his intellect ranks with the human?"

"Plenty of things! First, the pyramid-beast. He hadn't seen one before; he said as much. Yet he recognized it as a dead-alive automaton of silicon."

"He could have heard of it," objected Harrison. "He lives around here, you know."

"Well how about the language? I couldn't pick up a single idea of his and he learned six or seven words of mine. And do you realize what complex ideas he put over with no more than those six or seven words? The pyramid-monster—the dream-beast! In a single phrase he told me that one was a harmless automaton and the other a deadly hypnotist. What about that?"

"Huh!" said the captain.

"*Huh* if you wish! Could you have done it knowing only six words of English? Could you go even further, as Tweel did, and tell me that another creature was of a sort of intelligence so different from ours that understanding was impossible—even more impossible than that between Tweel and me?"

"Eh? What was that?"

"Later. The point I'm making is that Tweel and his race are worthy of our friendship. Somewhere on Mars—and you'll find I'm right—is a civilization and culture equal to ours, and maybe more than equal. And communication is possible between them and us; Tweel proves that. It may take years of patient trial, for their minds are alien, but less alien than the next minds we encountered—if they *are* minds."

"The next ones? What next ones?"

"The people of the mud cities along the canals." Jarvis frowned, then resumed his narrative. "I thought the dream-beast and the silicon-monster were the strangest beings conceivable, but I was wrong. These creatures are still more alien, less understandable than either and far less comprehensible than Tweel, with whom friendship is possible, and even, by patience and concentration, the exchange of ideas.

"Well," he continued, "we left the dream-beast dying, dragging itself back into its hole, and we moved toward the canal. There was a carpet of that queer walking-grass scampering out of our way, and when we reached the bank, there was a yellow trickle of water flowing. The mound city I'd noticed from the rocket was a mile or so to the right and I was curious enough to want to take a look at it.

"It had seemed deserted from my previous glimpse of it, and if any creatures were lurking in it—well, Tweel and I were both armed. And by the way, that crystal weapon of Tweel's was an interesting device; I took a look at it after the dream-beast episode. It fired a little glass splinter, poisoned, I suppose, and I guess it held at least a hundred of 'em to a load. The propellent was steam—just plain steam!"

"Shteam!" echoed Putz. "From vot come, shteam?"

"From water, of course! You could see the water through the transparent handle and about a gill of another liquid, thick and yellowish. When Tweel squeezed the handle—there was no trigger—a drop of water and a drop of the yellow stuff squirted into the firing chamber, and the water vaporized—pop!—like that. It's not so difficult; I think we could develop the same principle. Concentrated sulphuric acid will heat water almost to boiling, and so will quicklime, and there's potassium and sodium—

"Of course, his weapon hadn't the range of mine, but it wasn't so bad in this thin air, and it *did* hold as many shots as a cowboy's gun in a Western movie. It was effective, too, at least against Martian life; I tried it out, aiming at one of the crazy plants, and darned if the plant didn't wither up and fall apart! That's why I think the glass splinters were poisoned.

"Anyway, we trudged along toward the mud-heap city and I began to wonder whether the city builders dug the canals. I pointed to the city and then at the canal, and Tweel said 'No—no—no!' and gestured toward the south. I took it to mean that some other race had created the canal system, perhaps Tweel's people. I don't know; maybe there's still another intelligent race on the planet, or a dozen others. Mars is a queer little world.

"A hundred yards from the city we crossed a sort of road—just a hard-packed mud trail, and then, all of a sudden, along came one of the mound builders!

"Man, talk about fantastic beings! It looked rather like a barrel trotting along on four legs with four other arms or tentacles. It had no head, just body and members and a row of eyes completely around it. The top end of the barrel-body was a diaphragm stretched as tight as a drum head, and that was all. It was pushing a little coppery cart and tore right past us like the proverbial bat out of Hell. It didn't even notice us, although I thought the eyes on my side shifted a little as it passed.

"A moment later another came along, pushing another empty cart. Same thing—it just scooted past us. Well, I wasn't going to be ignored by a bunch of barrels playing train, so when the third one approached, I planted myself in the way—ready to jump, of course, if the thing didn't stop.

"But it did. It stopped and set up a sort of drumming from the diaphragm on top. And I held out both hands and said, 'We are friends!' And what do you suppose the thing did?"

"Said, 'Pleased to meet you,' I'll bet!" suggested Harrison.

"I couldn't have been more surprised if it had! It drummed on its diaphragm, and then suddenly boomed out, 'We are v-r-r-riends!' and gave its pushcart a vicious poke at me! I jumped aside, and away it went while I stared dumbly after it.

"A minute later another one came hurrying along. This one didn't pause, but simply drummed out, 'We are v-r-r-riends!' and scurried by. How did it learn the phrase? Were all of the creatures in some sort of communication with each other? Were they all parts of some central organism? I don't know, though I think Tweel does.

"Anyway, the creatures went sailing past us, every one greeting us with the same statement. It got to be funny; I never thought to find so many friends on this Godforsaken ball! Finally I made a puzzled gesture to Tweel; I guess he understood, for he said, 'One-one-two—yes!—two-two-four—no!' Get it?"

"Sure," said Harrison, "It's a Martian nursery rhyme."

"Yeah! Well, I was getting used to Tweel's symbolism, and I figured it out this way. 'One-one-two—yes!' The creatures were intelligent. 'Two-two-four—no!' Their intelligence was not of our order, but something different and beyond the logic of two and two is four. Maybe I missed his meaning. Perhaps he meant that their minds were of low degree, able to figure out the simple things—'One-one-two—yes!'—but not more difficult things—'Two-two-four—no!' But I think from what we saw later that he meant the other.

"After a few moments, the creatures came rushing back—first one, then another. Their pushcarts were full of stones, sand, chunks of rubbery plants, and such rubbish as that. They droned out their friendly greeting, which didn't really sound so friendly, and dashed on. The third one I assumed to be my first acquaintance and I decided to have another chat with him. I stepped into his path again and waited.

"Up he came, booming out his 'We are v-r-r-riends' and stopped. I looked at him; four or five of his eyes looked at me. He tried his password again and gave a shove on his cart, but I stood firm. And then the—the dashed creature reached out one of his arms, and two finger-like nippers tweaked my nose!"

"Haw!" roared Harrison. "Maybe the things have a sense of beauty!"

"Laugh!" grumbled Jarvis. "I'd already had a nasty bump and a mean frost-bite on that nose. Anyway, I yelled 'Ouch!' and jumped aside and the creature dashed away; but from then on, their greeting was 'We are v-r-r-riends! Ouch!' Queer beasts!

"Tweel and I followed the road squarely up to the nearest mound. The creatures were coming and going, paying us not the slightest attention, fetching their loads of rubbish. The road simply dived into an opening, and slanted down like an old mine, and in and out darted the barrel-people, greeting us with their eternal phrase.

"I looked in; there was a light somewhere below, and I was curious to see it. It didn't look like a flame or torch, you understand, but more like a civilized light, and I thought that I might get some clue as to the creatures' development. So in I went and Tweel tagged along, not without a few trills and twitters, however.

"The light was curious; it sputtered and flared like an old arc light, but came from a single black rod set in the wall of the corridor. It was electric, beyond doubt. The creatures were fairly civilized, apparently.

"Then I saw another light shining on something that glittered and I went on to look at that, but it was only a heap of shiny sand. I turned toward the entrance to leave, and the Devil take me if it wasn't gone!

"I suppose the corridor had curved, or I'd stepped into a side passage. Anyway, I walked back in that direction I thought we'd come, and all I saw was more dim-lit corridor. The place was a labyrinth! There was nothing but twisting passages running every way, lit by occasional lights, and now and then a creature running by, sometimes with a pushcart, sometimes without.

"Well, I wasn't much worried at first. Tweel and I had only come a few steps from the entrance. But every move we made after that seemed to get us in deeper. Finally I tried following one of the creatures with an empty cart, thinking that he'd be going out for his rubbish, but he ran around aimlessly, into one passage and out another. When he started dashing around a pillar like one of these Japanese waltzing mice, I gave up, dumped my water tank on the floor, and sat down.

"Tweel was as lost as I. I pointed up and he said 'No—no—no!' in a sort of helpless trill. And we couldn't get any help from the natives. They paid no attention at all, except to assure us they were friends—ouch!

"Lord! I don't know how many hours or days we wandered around there! I slept twice from sheer exhaustion; Tweel never seemed to need sleep. We tried following only the upward corridors, but they'd run uphill a ways and then curve downwards. The temperature in that damned ant hill was constant; you couldn't tell night from day and after my first sleep I didn't know whether I'd slept one hour or thirteen, so I couldn't tell from my watch whether it was midnight or noon.

"We saw plenty of strange things. There were machines running in some of the corridors, but they didn't seem to be doing anything—just wheels turning. And several times I saw two barrel-beasts with a little one growing between them, joined to both."

"Parthenogenesis!" exulted Leroy. "Parthenogenesis by budding like *les tulipes*!"

"If you say so, Frenchy," agreed Jarvis. "The things never noticed us at all, except, as I say, to greet us with 'We are v-r-r-riends! Ouch!' They seemed to have no home-life of any sort, but just scurried around with their pushcarts, bringing in rubbish. And finally I discovered what they did with it.

"We'd had a little luck with a corridor, one that slanted upwards for a great distance. I was feeling that we ought to be close to the surface when suddenly the passage debouched into a domed chamber, the only one we'd seen. And man!—I felt like dancing when I saw what looked like daylight through a crevice in the roof.

"There was a—a sort of machine in the chamber, just an enormous wheel that turned slowly, and one of the creatures was in the act of dumping his rubbish below it. The wheel ground it with a crunch—sand, stones, plants, all into powder that sifted away somewhere. While we watched, others filed in, repeating the process, and that seemed to be all. No rhyme nor reason to the whole thing—but that's characteristic of this crazy planet. And there was another fact that's almost too bizarre to believe.

"One of the creatures, having dumped his load, pushed his cart aside with a crash and calmly shoved himself under the wheel! I watched him being crushed, too stupefied to make a sound, and a moment later, another followed him! They were perfectly methodical about it, too; one of the cartless creatures took the abandoned pushcart.

"Tweel didn't seem surprised; I pointed out the next suicide to him, and he just gave the most human-like shrug imaginable, as much as to say, 'What can I do about it?' He must have known more or less about these creatures.

"Then I saw something else. There was something beyond the wheel, something shining on a sort of low pedestal. I walked over; there was a little crystal about the size of an egg, fluorescing to beat Tophet. The light from it stung my hands and face, almost like a static discharge, and then I noticed another funny thing. Remember that wart I had on my left thumb? Look!" Jarvis extended his hand. "It dried up and fell off—just like that! And my abused nose—say, the pain went out of it like magic! The thing had the property of hard X-rays or gamma radiations, only more so; it destroyed diseased tissue and left healthy tissue unharmed!

"I was thinking what a present *that'd* be to take back to Mother Earth when a lot of racket interrupted. We dashed back to the other side of the wheel in time to see one of the pushcarts ground up. Some suicide had been careless, it seems.

"Then suddenly the creatures were booming and drumming all around us and their noise was decidedly menacing. A crowd of them advanced toward us; we backed out of what I thought was the passage we'd entered by, and they came rumbling after us, some pushing carts and some not. Crazy brutes! There was a whole chorus of 'We are v-r-r-riends! Ouch!' I didn't like the 'ouch'; it was rather suggestive.

"Tweel had his glass gun out and I dumped my water tank for greater freedom and got mine. We backed up the corridor with the barrel-beasts following—about twenty of them. Queer thing—the ones coming in with loaded carts moved past us inches away without a sign.

"Tweel must have noticed that. Suddenly, he snatched out that glowing coal cigar-lighter of his and touched a cartload of plant limbs. Puff! The whole load was burning—and the crazy beast pushing it went right along without a change of pace! It created some disturbance among our 'V-r-r-riends,' however—and then I noticed the smoke eddying and swirling past us, and sure enough, there was the entrance!

"I grabbed Tweel and out we dashed and after us our twenty pursuers. The daylight felt like Heaven, though I saw at first glance that the sun was all but set, and that was bad, since I couldn't live outside my thermo-skin bag in a Martian night—at least, without a fire.

"And things got worse in a hurry. They cornered us in an angle between two mounds, and there we stood. I hadn't fired nor had Tweel; there wasn't any use in irritating the brutes. They stopped a little distance away and began their booming about friendship and ouches.

"Then things got still worse! A barrel-brute came out with a pushcart and they all grabbed into it and came out with handfuls of foot-long copper darts—sharp-looking ones—and all of a sudden one sailed past my ear—zing! And it was shoot or die then.

"We were doing pretty well for a while. We picked off the ones next to the pushcart and managed to keep the darts at a minimum, but suddenly there was a thunderous booming of 'v-r-r-riends' and 'ouches,' and a whole army of 'em came out of their hole.

"Man! We were through and I knew it! Then I realized that Tweel wasn't. He could have leaped the mound behind us as easily as not. He was staying for me!

"Say, I could have cried if there'd been time! I'd liked Tweel from the first, but whether I'd have had gratitude to do what he was doing—suppose I *had* saved him from the first dream-beast—he'd done as much for me, hadn't he? I grabbed his arm, and said 'Tweel,' and pointed up, and he understood. He said, 'No—no—no, Tick!' and popped away with his glass pistol.

"What could I do? I'd be a goner anyway when the sun set, but I couldn't explain that to him. I said, 'Thanks, Tweel. You're a man!' and felt that I wasn't paying him any compliment at all. A man! There are mighty few men who'd do that.

"So I went 'bang' with my gun and Tweel went 'puff' with his, and the barrels were throwing darts and getting ready to rush us, and booming about being friends. I had given up hope. Then suddenly an angel dropped right down from Heaven in the shape of Putz, with his under-jets blasting the barrels into very small pieces!

"Wow! I let out a yell and dashed for the rocket; Putz opened the door and in I went, laughing and crying and shouting! It was a moment or so before I remembered Tweel; I looked around in time to see him rising in one of his nosedives over the mound and away.

"I had a devil of a job arguing Putz into following! By the time we got the rocket aloft, darkness was down; you know how it comes here—like turning off a light. We sailed out over the desert and put down once or twice. I yelled 'Tweel!' and yelled it a hundred times, I guess. We couldn't find him; he could travel like the wind and all I got—or else I imagined it—was a faint trilling and twittering drifting out of the south. He'd gone, and damn it! I wish—I wish he hadn't!"

The four men of the *Ares* were silent—even the sardonic Harrison. At last little Leroy broke the stillness.

"I should like to see," he murmured.

"Yeah," said Harrison. "And the wart-cure. Too bad you missed that; it might be the cancer cure they've been hunting for a century and a half."

"Oh, that!" muttered Jarvis gloomily. "That's what started the fight!" He drew a glistening object from his pocket.

"Here it is."

OUT OF THE SILENT PLANET (1938)

CHAPTER 16

C. S. Lewis

British writer **Clive Staples Lewis** is best known for his celebrated series The Chronicles of Narnia. Born in Belfast in 1898, Lewis won a scholarship to University College, Oxford in 1916 at the height of the First World War, but he soon joined the British army and served until December 1918. Upon completion of his studies in 1923, he joined the academic staff and stayed at Oxford for his entire career, becoming good friends with fellow fantasy writer J. R. R. Tolkien. Lewis's early interest in mythology transformed into a theological interest in Christianity, which permeates many of his stories. He died in 1963 and is memorialized in Poet's Corner in Westminster Abbey, London.

Ransom awoke next morning with the vague feeling that a great weight had been taken off his mind. Then he remembered that he was the guest of a *sorn* and that the creature he had been avoiding ever since he landed had turned out to be as amicable as the *hrossa*, though he was far from feeling the same affection for it. Nothing then remained to be afraid of in Malacandra except Oyarsa . . . 'The last fence,' thought Ransom.

Augray gave him food and drink.

'And now,' said Ransom, 'how shall I find my way to Oyarsa?'

'I will carry you,' said the *sorn*. 'You are too small a one to make the journey yourself and I will gladly go to Meldilorn. The *hrossa* should not have sent you this way. They do not seem to know from looking at an animal what sort of lungs it has and what it can do. It is just like a *hross*. If you died on the *harandra* they would have made a poem about the gallant *hmān* and how the sky grew black and the cold stars shone and he journeyed on and journeyed on; and they would have put in a fine speech for you to say as you were dying . . . and all this would seem to them just as good as if they had used a little forethought and saved your life by sending you the easier way round.'

'I like the *hrossa*,' said Ransom a little stiffly. 'And I think the way they talk about death is the right way.'

'They are right not to fear it, Ren-soom, but they do not seem to look at it reasonably as part of the very nature of our bodies—and therefore often avoidable at times when they would never see how to avoid it. For example, this has saved the life of many a *hross*, but a *hross* would not have thought of it.'

He showed Ransom a flask with a tube attached to it, and, at the end of the tube a cup, obviously an apparatus for administering oxygen to oneself.

'Smell on it as you have need, Small One,' said the *sorn*. 'And close it up when you do not.'

Augray fastened the thing on his back and gave the tube over his shoulder into his hand. Ransom could not restrain a shudder at the touch of the *sorn's* hands upon his body; they were fan-shaped, seven-fingered, mere skin over bone like a bird's leg, and quite cold. To divert his mind from such reactions he asked where the apparatus was made, for he had as yet seen nothing remotely like a factory or a laboratory.

'We thought it,' said the sorn, 'and the *pfifltriggi* made it.'

'Why do they make them?' said Ransom. He was trying once more, with his insufficient vocabulary, to find out the political and economic framework of Malacandrian life.

'They like making things,' said Augray. 'It is true they like best the making of things that are only good to look at and of no use. But sometimes when they are tired of that they will make things for us, things we have thought, provided they are difficult enough. They have not patience to make easy things however useful they would be. But let us begin our journey. You shall sit on my shoulder.'

The proposal was unexpected and alarming, but seeing that the *sorn* had already crouched down, Ransom felt obliged to climb on to the plume-like surface of its shoulder, to seat himself beside the long, pale face, casting his right arm as far as it would go round the huge neck, and to compose himself as well as he could for this precarious mode of travel. The giant rose cautiously to a standing position and he found himself looking down on the landscape from a height of about eighteen feet.

'Is all well, Small One?' it asked.

'Very well,' Ransom answered, and the journey began.

Its gait was perhaps the least human thing about it. It lifted its feet very high and set them down very gently. Ransom was reminded alternately of a cat stalking, a strutting barn-door fowl, and a high-stepping carriage horse; but the movement was not really like that of any terrestrial animal. For the passenger it was surprisingly comfortable. In a few minutes he had lost all sense of what was dizzying or unnatural in his position. Instead, ludicrous and even tender associations came crowding into his mind. It was like riding an elephant at the zoo in boyhood—like riding on his father's back at a still earlier age. It was fun. They seemed to be doing between six and seven miles an hour. The cold, though severe, was endurable; and thanks to the oxygen he had little difficulty with his breathing.

The landscape which he saw from his high, swaying post of observation was a solemn one. The *handramit* was nowhere to be seen. On each side of the shallow gully in which they were walking, a world of naked, faintly greenish rock, interrupted with wide patches of red, extended to the horizon. The heaven, darkest blue where the rock met it, was almost black at the zenith, and looking in any direction where sunlight did not blind him, he could see the stars. He learned from the *sorn* that he was right in thinking they were near the limits of the breathable. Already on the mountain fringe that borders the *harandra* and walls the *handramit*, or in the narrow depression along which their road led them, the air is of Himalayan rarity, ill breathing for a *hross*, and a few hundred feet higher, on the *harandra* proper, the true surface of the planet, it admits no life. Hence the brightness through which they walked was almost that of heaven—celestial light hardly at all tempered with an atmospheric veil.

The shadow of the *sorn*, with Ransom's shadow on its shoulder, moved over the uneven rock unnaturally distinct like the shadow of a tree before the headlights of a car; and the rock beyond the shadow hurt his eyes. The remote horizon seemed but an arm's length away. The fissures and moulding of distant slopes were clear as the background of a primitive picture made before men learned perspective. He was on the very frontier of that heaven he had known in the space-ship, and rays that the air-enveloped worlds cannot taste were once more at work upon his body. He felt the old lift of the heart, the soaring solemnity, the sense, at once sober and ecstatic, of life and power offered in unasked and unmeasured abundance. If there had been air enough in his lungs he would have laughed aloud. And now, even in the immediate landscape, beauty was drawing near. Over the edge of the valley, as if it had frothed down from the true *harandra*, came great curves of the rose-tinted, cumular stuff which he had seen so often from a distance. Now on a nearer view they appeared hard as stone in substance, but puffed above and stalked beneath like vegetation. His original simile of giant cauliflower turned out to be surprisingly correct—stone cauliflowers the size of cathedrals and the colour of pale rose. He asked the *sorn* what it was.

'It is the old forests of Malacandra,' said Augray. 'Once there was air on the *harandra* and it was warm. To this day, if you could get up there and live, you would see it all covered with the bones of ancient creatures; it was once full of life and noise. It was then these forests grew, and in and out among their stalks went a people that have vanished from the world these many thousand years. They were covered not with fur but with a coat like mine. They did not go in the water swimming or on the ground walking; they glided in the air on broad flat limbs which kept them up. It is said they were great singers, and in those days the red forests echoed with their music. Now the forests have become stone and only *eldila* can go among them.'

'We still have such creatures in our world,' said Ransom. 'We call them birds. Where was Oyarsa when all this happened to the *harandra*?'

'Where he is now.'

'And he could not prevent it?'

'I do not know. But a world is not made to last for ever, much less a race; that is not Maleldil's way.'

As they proceeded the petrified forests grew more numerous, and often for half an hour at a time the whole horizon of the lifeless, almost airless, waste blushed like an English garden in summer. They passed many caves where, as Augray told him, *sorns* lived; sometimes a high cliff would be perforated with countless holes to the very top and unidentifiable noises came hollowly from within. 'Work' was in progress, said the *sorn*, but of what kind it could not make him understand. Its vocabulary was very different from that of the *hrossa*. Nowhere did he see anything like a village or city of *sorns*, who were apparently solitary not social creatures. Once or twice a long pallid face would show from a cavern mouth and exchange a horn-like greeting with the travellers, but for the most part the long valley, the rock-street of the silent people, was still and empty as the *harandra* itself.

Only towards afternoon, as they were about to descend into a dip of the road, they met three *sorns* together coming towards them down the opposite slope. They seemed to Ransom to be rather skating than walking. The lightness of their world and the perfect poise of their bodies allowed them to lean forward at right angles to the slope, and they came swiftly down like full-rigged ships before a fair wind. The grace of their movement, their lofty stature, and the softened glancing of the sunlight on their feathery sides, effected a final transformation in Ransom's feelings towards their race. 'Ogres' he had called them when they first met his eyes as he struggled in the grip of Weston and Devine; 'Titans' or 'Angels' he now thought would have been a better word. Even the faces, it seemed to him, he had not then seen aright. He had thought them spectral when they were only august, and his first human reaction to their lengthened severity of line and profound stillness of expression now appeared to him not so much cowardly as vulgar. So might Parmenides or Confucius look to the eyes of a Cockney schoolboy! The great white creatures sailed towards Ransom and Augray and dipped like trees and passed.

In spite of the cold—which made him often dismount and take a spell on foot—he did not wish for the end of the journey; but Augray had his own plans and halted for the night long before sundown at the home of an older *sorn*. Ransom saw well enough that he was brought there to be shown to a great scientist. The cave, or, to speak more correctly, the system of excavations, was large and many-chambered, and contained a multitude of things that he did not understand. He was specially interested in a collection of rolls, seemingly of skin, covered with characters, which were clearly books; but he gathered that books were few in Malacandra.

'It is better to remember,' said the *sorns*.

When Ransom asked if valuable secrets might not thus be lost, they replied that Oyarsa always remembered them and would bring them to light if he thought fit.

'The *hrossa* used to have many books of poetry,' they added. 'But now they have fewer. They say that the writing of books destroys poetry.'

Their host in these caverns was attended by a number of other *sorns* who seemed to be in some way subordinate to him; Ransom thought at first that they were servants but decided later that they were pupils or assistants.

The evening's conversation was not such as would interest a terrestrial reader, for the *sorns* had determined that Ransom should not ask, but answer, questions. Their questioning was very different from the rambling, imaginative inquiries of the *hrossa*. They worked systematically from the geology of Earth to its present geography, and thence in turn to flora, fauna, human history, languages, politics and arts. When they found that Ransom could tell them no more on a given subject—and this happened pretty soon in most of their inquiries—they dropped it at once and went on to the next. Often they drew out of him indirectly much more knowledge than he consciously possessed, apparently working from a wide background of general science. A casual remark about trees when Ransom was trying to explain the manufacture of paper would fill up for them a gap in his sketchy answers to their botanical questions; his account of terrestrial navigation might illuminate mineralogy; and his description of the steam-engine gave them a better knowledge of terrestrial air and water than Ransom had ever had. He had decided from the outset that he would be quite frank, for he now felt that it would be not *hnau*, and also that it would be unavailing, to do otherwise. They were astonished at what he had to tell them of human history—of war, slavery and prostitution.

'It is because they have no Oyarsa,' said one of the pupils.

'It is because every one of them wants to be a little Oyarsa himself,' said Augray.

'They cannot help it,' said the old *sorn*. 'There must be rule, yet how can creatures rule themselves? Beasts must be ruled by *hnau* and *hnau* by *eldila* and *eldila* by Maleldil. These creatures have no *eldila*. They are like one trying to lift himself by his own hair—or one trying to see over a whole country when he is on a level with it—like a female trying to beget young on herself.'

Two things about our world particularly stuck in their minds. One was the extraordinary degree to which problems of lifting and carrying things absorbed our energy. The other was the fact that we had only one kind of *hnau*: they thought this must have far-reaching effects in the narrowing of sympathies and even of thought.

'Your thought must be at the mercy of your blood,' said the old *sorn*. 'For you cannot compare it with thought that floats on a different blood.'

It was a tiring and very disagreeable conversation for Ransom. But when at last he lay down to sleep it was not of the human nakedness nor of his own ignorance that he was thinking. He thought only of the old forests of Malacandra and of what it might mean to grow up seeing always so few miles away a land of colour that could never be reached and had once been inhabited.

THE HERMIT OF MARS (1963)

Stephen Bartholomew

There seems to be little information that one can find about **Stephen Bartholomew**. 'The Hermit of Mars' was published in the October 1963 issue of *World of Tomorrow*. He is the author of several other works of short science fiction including 'Last Resort', 'The Standardised Man' and 'The Rumble and the Roar'.

He was the Oldest Man on Mars ... in Fact, the Only One!

When Martin Devere was 23 and still working on his Master's, he was hurt by a woman. It was then that he decided that the only things that were worthwhile in life were pure art and pure science. That, of course, is another story, but it may explain why he chose to become an archeologist in the first place. Now he was the oldest human being on Mars. He was 91. For many years, in fact, he had been the *only* human being on Mars. Up until today.

He looked through the transparent wall of his pressurized igloo at the puff of dust in the desert where the second rocket had come down. Earth and Mars were just past conjunction, and the regular automatic supply rocket had landed two days ago. As usual, Martin Devere, taking his own good time about it, had unloaded the supplies, keeping the things he really needed and throwing away the useless stuff like the latest microfilmed newspapers and magazines, the taped TV shows and concerts. As payment for his groceries he had then reloaded the rocket with the written reports he had accumulated since the last conjunction, plus a few artifacts.

Then he had pushed a button and sent the rocket on its way again, back to Earth. He didn't mind writing the reports. Most of them were rubbish anyway, but they seemed to keep the people back at the Institute happy. He did mind the artifacts. It seemed wrong to remove them, though he sent only the less valuable ones back. But perhaps it couldn't be helped. One time, the supply rocket had failed to return when he pushed its red button—the thing was still sitting out there in the desert, slowly rusting. Martin Devere had happily unloaded the artifacts and put them back where they belonged. It wasn't his fault.

The puff of dust on the horizon was beginning to settle. This second rocket had descended with a shrill scream through the thin air, its voice more highly pitched than it would have been in denser atmosphere. Martin Devere had looked up from his work in time to see its braking jets vanish behind the low Martian hills a few kilometers distant.

It was much too large to be an automatic supply rocket, even if there had been reason to expect another one. Martin Devere knew it could mean only one thing—someone was paying him an unannounced visit.

He waited, watching through the igloo wall to see who had come to poke around and bother him after all these years.

At first he was annoyed that the people at the Institute hadn't let him know visitors were coming. Then he reminded himself that it had been years since he'd taken the trouble to listen to his radio receiver, or to read the messages they sent him along with supplies.

After a long time, he made out a smaller dust-puff, and then a little sandcat advancing slowly across the desert. Riding on top of it were two men in spacesuits.

Everyone on Earth who reads popular magazines or watches TV knows the story of Martin Devere, "The Hermit of Mars." Over the years, now that he is dead, he has become a sort of culture hero, as Dr. Livingston or Albert Schweitzer once were. Though Martin Devere could not be called a humanitarian in any sense of the word. After his divorce from his first and only wife, at the age of 45, he never gave much thought again either to women or any other kind of people—except for his long-dead Martians.

But everyone should know by now how Martin Devere first came to Mars at the age of 50. Even then he was the oldest man on the planet, and Mars sustained quite a large research colony at the time. Only Martin Devere's unchallenged scientific reputation, together with his apparent good health, enabled him to leave Earth as head of a five-man archeological team. This turned up the first fossil ruins far beneath the desert sand.

Then there came a day when the Space Institute of the United Governments decided to abandon Project Mars. It was getting too expensive to maintain. Everything of value to space research had already been learned about the planet, and the archeological site, though yet barely scratched, did not properly come under space research. Closing Project Mars would mean more funds for solar research, on Mercury, for the Lunar colony and for work on the interstellar drive.

So the hundred-odd inhabitants of the Project received orders to leave the igloos and other equipment behind and come back to Earth.

Martin Devere, however, had been on Mars for three years now. When the Project physician gave him his routine exam, it was discovered that a valve in Martin Devere's aorta had developed a faint flutter. Nothing too serious, really. But enough to greatly reduce his chances of surviving another rocket lift-off.

Martin Devere smiled at the news and volunteered to remain behind, alone on Mars. Under the circumstances, the Institute was forced to agree.

On the day that the strange rocket came down behind the desert hills, Martin Devere had been on Mars for a total of 38 years. For the past 35 of them he had been The Hermit—and quite happy about it…

The little sandcat was getting closer. Martin Devere smiled to himself, watching the two men in their clumsy space gear. It was high noon, and a nice comfortable ten degrees centigrade outside. If the two newcomers thought they needed full spacesuits to get around out there, Martin Devere wasn't going to tell them any different. Actually, though the atmospheric pressure was about the same as at the top of Mount Everest, on a beautiful day like this a man could get along easily outdoors with nothing more than an oxygen mask. But let them clomp around in their rubberized long-johns if they wanted to.

In a few minutes they would be coming in through the igloo's airlock. Martin Devere turned away, scowling now. He hoped the Institute hadn't decided to reopen Project Mars. There was plenty of room in all these igloos and connecting tunnels that had been left behind, but with a new expedition here it might get pretty crowded. Mainly, Devere didn't want a bunch of amateurs poking around his diggings, breaking things.

His thumb rubbed slowly across the long stubble on his chin. He wondered if he had made some slip in that last report, or in some of the pictures of the ruins he'd sent back. He'd rather the Institute didn't find out about those fossilized machines he'd dug up. He didn't understand the gadgets himself, but some of the people at the Institute just might decide they were interesting enough to be worth sending up an expert.

The Institute, Devere knew, was interested in machinery, not art objects.

One of the men held an automatic pistol pointed at Martin Devere while the other was stripping off his space gear. Then the pistol changed hands while the first man removed his own suit. Martin Devere could have told them that he wasn't afraid of the gun. He didn't actually care much, one way or the other: let them point it if it made them happy. Martin Devere figured that he had already lived a lot longer, here in this feeble gravity and germ-free, oxygen-rich air, than his tricky heart would have allowed him on Earth. Let them point the gun if they wanted to.

"If you make one move toward the radio transmitter I'll blow your head off," the taller man said. He had black wavy hair that hung over his brow. The other man was completely bald.

"I don't even know if the radio works," Martin Devere answered. "I haven't turned it on in years. I should warn you, though, that if you shoot that thing inside the igloo here, it will puncture the plastic wall and let all the air out. I always keep the pressure up high indoors so I can boil water for coffee."

The tall man frowned in confusion and blinked at the weapon in his hand. Then he stared at the transparent dome above him, as if realizing for the first time that only a thin bubble of plastic separated him from near-vacuum, now that he had removed his suit.

"I was just making some coffee when you showed up," Martin Devere said, turning away. "Have some? I'm afraid it's instant. I've given up trying to get the Institute to send me a can of real coffee in the rocket. They think I need canned TV shows more."

"He's harmless," the bald man said. "You can see he's just an old senile nut. Leave him be, we've work to do."

The tall man lowered his weapon, then let it fall into the holster at his hip.

"No big hurry. I think I'd like some of that coffee first. Say, Pop, how about cooking us a meal in a couple of hours?"

Martin Devere was spooning brown powder into three cups.

"Sure thing. What would you like—beans and franks, or franks and beans?"

"I suppose you wonder what we're doing, Pop?" The tall man held the disassembled pieces of his gun in his lap. He was carefully polishing each part with a chemically treated cloth.

It was three days since they had landed, and the tall metal skeleton was beginning to take shape out in the desert. At the moment, the bald man was out alone, testing circuits. Usually the two went out together—they had apparently decided it was safe to leave Martin Devere unguarded, though they had smashed his radio transmitter just in case.

The two men worked steadily during the daylight hours, came back at sunset to eat and sleep, then went out again at dawn. The towering lacework of steel was growing like an ugly flower.

The tall man held the trigger assembly of his gun up to the light. He turned it slowly between his thumb and forefinger. It cast an odd crescent-shaped shadow over the muscles of his jaw.

"No, I don't wonder what you're doing," Martin Devere answered. He was sitting at his workbench, crouched over an ancient metal plate as thin as paper.

The tall man began to put his weapon back together again. He snapped the trigger assembly into the receiver. He pulled the hammer back and then released it; it made a sharp, hard click.

"Not even curious, Pop? Okay, then tell me what *you're* doing. What's that piece of tinfoil you've been staring at the past two hours?"

Martin Devere straightened and turned to look at the other.

"It's an ancient Martian scroll. It's nearly a million years old. I found it in a new pit I've been digging, five hundred meters down. It's the longest and perhaps most important bit of Martian writing I've found so far."

"Yeah? What's it have to say?"

Martin Devere shook his head. "Their language, their whole frame of reference, was fundamentally different from ours. It's something like higher mathematics, you'd have to learn the language to understand it. But I suppose you might say that this is a poem… Yes, an epic poem."

The tall man laughed. He shoved an ammunition clip into his weapon, pumped a round into the chamber, slipped the gun back into its holster. He got up and began pacing the floor of the igloo. The floor was cluttered with dozens of artifacts.

He stopped and nudged one specimen with his toe.

"What's this thing, Pop? An ancient Martian meatgrinder?"

"I hardly think so. They were vegetarians." He squinted at the object. "I'm afraid I have no idea what it is. It's some sort of machine, but I'm no engineer, I can't imagine what its function was. They—don't build many machines, you know."

The man with the gun turned to stare at Martin Devere.

"You mean *didn't* build, don't you?"

"Yes, of course... Past tense." And Devere turned again to peer at the million-year-old poem before him.

*

"Damn it to hell. This might hold us up a week." The bald man flung the shatterproof helmet of his suit against the igloo wall. His tone of voice was matter-of-fact emotionless. Even the way he threw the helmet betrayed no real emotion. Still wearing the rest of his suit he sat down at Martin Devere's work bench and clenched his fists. His face was smooth, blank.

"What's the matter?" His partner put down some drawings and came over.

"The modulator circuit doesn't check out. I'll have to take the whole works apart and start over again." The bald man spoke—when he did speak—with a faint accent that Martin Devere could not identify.

"It doesn't matter." The other rubbed at his chin. "We're still ahead of our schedule."

"Hey. Old man." The bald man pointed at Devere. "You have anything to drink in this cave of yours?"

Martin Devere frowned, thinking. He remembered a bottle he'd been saving for some special occasion—he couldn't recall what, just now.

"I think I have some bourbon," he said at last. "If I can find it."

"Find it. Mine straight, on the rocks."

When Martin Devere returned awhile later, the bald man was still wearing his helmetless space suit. He and his friend were studying a complex wiring diagram spread out on the work bench.

Martin Devere put two plastic cups down on the bench and poured them full. Neither of the men looked up from their diagram until he had set the bottle down.

"Pour one for yourself, Pop," the tall man said.

"Thanks. Don't mind if I do." Devere went to get another cup. Over his shoulder he said, "Hope you boys don't mind crushed ice instead of cubes. I just set a bucket of water in one of the unheated tunnels for a couple minutes. Then I hit it with a hammer."

It was four hours past sunset, the temperature outside was far below freezing.

"One thing you don't need on Mars is a refrigerator!" Pouring himself a drink, the old man suddenly laughed. It was a brief, senile giggle, that made the tall man turn to stare at him.

"Could be uncomfortable, though, if you were ever stuck out there at night." Martin Devere's face was sober once more as he lifted his cup and looked deeply into it. All trace of senility had vanished as suddenly as it had appeared. "Like, say, if you were out there long enough for your suit power to go dead. You'd freeze to a hunk of ice in a few minutes... Me, I never go outside at night."

"Shut up," the bald man said.

All day the bald man had been out alone, working on his electronic circuits. Evidently this left his partner nothing to do except study schematics.

Now Martin Devere was aware that his guest had been staring at him for several minutes without speaking. Martin Devere went on polishing the green crystal vase he held in his hand. The vase looked ordinary at first glance, until you noticed that it wasn't quite symmetrical. There was a studied and careful asymmetry about its form, barely discernible, that would disturb you the more you looked at it—until you knew suddenly that no human brain could have created that shape.

The polishing cloth moved rhythmically across the vase's curving surfaces. The green crystal reflected light in a way that made you begin to think about boundless seas of water.

"I'll be glad when this job is over with," the tall man said, half aloud.

"When it is, will you go away?" Martin Devere turned the vase slowly in his hands.

"Not for a while yet, Pop." The man with the gun on his hip got to his feet and stretched.

"I don't mind telling you what it's all about, Pop. You're all right. It's simple. My partner and I were sent here by a certain national power that doesn't like being told how to run its own affairs by the United Governments. We're striking the first blow for Freedom. That thing we're putting together out there is a bomb. It could—disable—most of Earth. It has a new kind of nuclear rocket engine behind it that could carry it across 200 million miles in a few hours.

"You get the idea, Pop? Here on Mars, they won't even find it. And if they did, we could deliver the bomb before they got a missile halfway across... So I hope you won't mind if my partner and I stay a while, Pop."

It was several seconds before Martin Devere answered. He set the crystal vase carefully inside a case and regarded it a moment.

"As long as you don't go messing up my diggings or break any of the artifacts, it's no business of mine."

"And what if I did, Pop?" The tall man walked closer to Martin Devere. He stood over the old man, his shadow on him. His hand rested lightly on the butt of his gun. "What if I were to take all your vases and statues and pots and tablets and smash them to bits, one by one? What would you do then?"

Martin Devere's eyes slowly closed and opened, he made no other move for a minute. Then he got to his feet without looking at the other man. He turned and began to move away, toward a tunnel door that led to the diggings.

Probably the tall man thought that he had finally put the fear of God into Martin Devere. But as he turned back to his pile of schematics he heard the old man's whisper:

"You might regret it."

The man with the gun did not answer.

"Tell us about it, Pop."

"Yes, why don't you tell us about it."

They meant Martin Devere's work. The two men had finished their own job. The assembled bomb rested in the desert, silent but alive, like some abnormal growth.

Because of sunspot activity they hadn't yet been able to radio their employers on Earth. The bald man expected conditions to clear in two or three days. When they did clear, he would signal, "The bird is nesting." Then the nation he had mentioned would be ready to deliver its ultimatum to the United Governments.

For the first time since landing on Mars, the two men were idle. They were waiting. They looked as if they were willing to wait a long time if necessary.

Meanwhile, Martin Devere's artifacts were the only amusements available.

Perhaps the old man knew they were making fun of him. But he seemed to take their question seriously. When he began to speak, they found themselves listening.

"We don't know exactly what happened." Martin Devere faced the two men across the cluttered workbench like a lecturer addressing his students. He held in his hand a small bronze statue that might have been a portrayal of one of the old Martian people or, just as likely, some long-extinct animal. In the diffuse sunlight that came through the igloo wall, it cast a shadow on the work bench that was even more disturbingly alien in shape.

"No, we don't know what happened to them," the old man said. "The last of them died nearly a million years ago, before the first Homo Sapiens walked the Earth. From what we—I—have found we know a little about what they were like. But we don't know why they died.

"We do know, for instance, that they never had much interest in technology. Not that they lacked intelligence. They could build a machine when it suited their purposes, whatever those may have been. And I don't say they weren't interested in science. They had a highly developed theoretical science, as sophisticated as their art. You might say they were theoreticians. They were concerned with pure art and pure science—but not with applied technology, or commercialized art.

"My own theory is that they had no need for technology. In the first place, they were vegetarians, not carnivorous. So that their earliest men had no need for hunting weapons—or other gadgets. Probably they never developed the aggressive instincts which in humanity led to warfare with its subsequent impetus to applied technology. The Martians never got around to making cars or airplanes or bombs. They dedicated themselves, gentlemen, to the contemplation of beauty.

"Then, nearly a million years ago, something happened to them. Perhaps Mars began to lose her atmosphere then. Her oceans evaporated, the air could no longer retain her heat at night, the farmlands parched and froze. A few of the plant types were able to adapt and survive. But within a few years, all animal life died out. One day, there were suddenly no more Martians left."

Martin Devere's dry, withered hand caressed the small statue he held.

"Who knows? If they'd had time to develop space travel they might have saved themselves. Then again, with a technology like yours, they might have blown themselves up long before the natural catastrophe ..."

"What do you mean like *yours*?" the tall man said. "You mean like *ours*, don't you?"

But Martin Devere turned away without answering.

"Do you have another bottle of bourbon, old man?"

"No, I'm afraid not," Devere said. "There was only that one bottle."

"Too bad. We should have a little celebration." The bald man began sealing himself into his spacesuit.

"I'll wait for you here," his partner said. "I'd better start burning those plans."

Martin Devere looked up from the fragment of ceramic he was cleaning.

"You're going to send the message now?"

Neither of the men bothered to reply, since the answer was self-evident. The bald man tested the air and power equipment of his suit, then turned to his partner a moment before sealing his helmet.

"You checked the sandcat's power supply?"

"Yes, but you'd better take another look at it. I think the battery's leaking."

The bald man nodded and went out the airlock. Martin Devere watched in silence as the other man began to gather up his diagrams and plans and tie them into a neat bundle.

"I guess we can take it easy now, Pop. As soon as that telegram's sent and I get this stuff burned, my partner and I are unemployed. Of course we'll have to hang around a while longer in case they want us to shoot off Baby out there, but there's nothing to that. In the meantime maybe I can help you dig up some more of those old pots and statues."

Martin Devere seemed to be thinking. He watched as the tall man checked to make sure he hadn't forgotten anything, then carried the bundle of plans over to the electronic oven.

"*Baby.* You mean your bomb, out there. You think you might actually shoot it off then."

"Oh, maybe, maybe not."

"Couldn't they fire it from Earth by radio?" Devere asked.

"Nope. Somebody might try jamming."

"Oh, I see..."

Martin Devere was silent again until the tall man opened the oven and removed a bundle of gray ash. He dumped the ashes into a bucket and began stirring them with his hand.

"Something else I was wondering about," Devere said. He began cleaning the fragment of ceramic again, his hands working in a slow circular motion.

"Supposing the United Governments find out where it—the bomb is. They might send a missile to blow it up."

"Told you, Pop. Baby can out-run anything else that flies. Wouldn't do them any good."

"Yes, yes... Still, the missile would hit Mars, wouldn't it? I mean, it would destroy all this—the igloos, my diggings ..."

The tall man gave a laugh.

"Don't worry so much, Pop. We'd have plenty of time to get in the ship and clear out. We might even take you with us."

"Still ..." But the old man lapsed again into thought.

An hour later, the short-range radio gave a shrill beep. The tall man went over and flipped the *talk* switch.

"Yeah?"

"Hello. Listen, I did something stupid."

Martin Devere looked up at the sound of the bald man's voice. Devere's hands still held the piece of ceramic. He had polished it until a complex geometric design was visible, etched in reds and blues. It might have been equally a decoration or some mechanical diagram.

"Did you get the message sent?" the tall man asked.

"Yes, that part's all right. I got to the ship and contacted headquarters. I think they're going to deliver the ultimatum right away. Now we just wait for orders. The only thing is, the sandcat's power went dead on me while I was halfway down a hill. It started to roll, and I forgot I was wearing a spacesuit. I jumped out. This low gravity fooled me too. I think I've broken my ankle, it hurts like hell."

The tall man cursed in a low voice.

"All right, all right," he said after a moment. "Just take it easy. I'll have to come out and get you."

"I think the sandcat is all right. Stupid of me to jump like that, wasn't thinking. Better bring a spare battery with you… Oh, and you'd better bring a light too. It will be getting dark in another half hour."

"Okay, just wait for me. I'll home in on your suit radio."

The tall man switched off the receiver and went to his own suit locker. Martin Devere watched as he removed the holster and weapon from his hip. He pulled the heavy plastic trousers over his denim jumper and then buckled the gun back again before starting on the rest of the spacesuit.

"Nothing serious, I hope?" Martin Devere put the ceramic down carefully and picked up another object from a stack of artifacts.

"You heard, didn't you? You any good at setting a broken ankle, Pop?"

"Oh, I could manage, I guess. Broke my arm down in the diggings once. Had to set it myself. Twenty years ago, I think it was. I've been more careful since then." He gave a laugh. It started as a normal laugh, then broke to a senile giggle. Then his face was serious again. He carried the new artifact closer to the man with the gun.

"You know, I was telling you… The Martians were vegetarians. They never made any weapons for hunting. They did know about explosives, though."

"What's that thing?" The tall man, struggling with the buckles of his breathing equipment, glanced at the object in Devere's hands. It looked like badly corroded bronze, and consisted of a long tube with a large bulb at one end.

"This? Oh, this is some kind of a tool I found. I think it was a digging tool, used for breaking up rocks. They *did* build canals, you know… As I was saying, they knew about explosives. This tool, for instance. It worked by means of a small, shaped charge inside this bulb here. The explosion was so well-focused that there was almost no recoil. A high-energy shock wave was emitted from the barrel—very effective at short range. But the most amazing thing about this tool is that the chemical explosive is still potent after lying underground for nearly a million years…

"Oh, by the way. There's nothing wrong with your sandcat's battery. It was the motor I sabotaged."

Then Martin Devere pointed the ancient digging tool at the tall man and blew him into two neat pieces.

The Hermit of Mars never did get around to walking out to the space ship and using his visitor's radio to tell Earth what had happened. He really intended to, but he forgot. The ultimatum that was delivered to the United Governments failed, of course, but no one knew exactly why until the next Earth-Mars conjunction.

The United Governments was prevailed on by the World Television Service to send out someone to interview the Hermit, if he were still alive.

That interview was unfortunate. It might have established Martin Devere as the world hero that he was, and he might have been awarded some kind of medal. As it went, his rude and insulting answers to the young man's questions made him unpopular for years.

His last answer in the interview was the worst. The young man, already sweating, looked in desperation at the green crystal vase that Martin Devere insisted on holding in front of the television lens. (Back at the Institute, a dozen faces were flushing red with indignation as their owners realized what the old man had been holding back.)

"Tell me, Dr. Devere," the young man asked. "You seem—er—a very modest man. Doesn't it make you the least bit proud to know that you've saved the world?"

Martin Devere lowered his vase and gave the young man a puzzled look.

"You mean Earth? Tell me, why should I want to save *that* world?"

A ROSE FOR ECCLESIASTES (1969)

Roger Zelazny

Roger Joseph Zelazny was born in Ohio in 1937. His interest in writing started while at school, where he was editor of the school newspaper and was part of a creative writing club. He graduated with a BA in English from Western Reserve University in 1959, and with an MA specialising in Elizabethan and Jacobean drama from Columbia University in 1962. For the next five years Zelazny worked at the US Social Security Administration in Cleveland, while building his science fiction writing career. In 1969, he became a full-time author. He died in 1995.

I

I was busy translating one of my *Madrigals Macabre* into Martian on the morning I was found acceptable. The intercom had buzzed briefly, and I dropped my pencil and flipped on the toggle in a single motion.

'Mister G,' piped Morton's youthful contralto, 'the old man says I should "get hold of that damned conceited rhymer" right away, and send him to his cabin. Since there's only one damned conceited rhymer ...'

'Let not ambition mock thy useful toil.' I cut him off.

So, the Martians had finally made up their minds! I knocked an inch and a half of ash from a smoldering butt, and took my first drag since I had lit it. The entire month's anticipation tried hard to crowd itself into the moment, but could not quite make it. I was frightened to walk those forty feet and hear Emory say the words I already knew he would say; and that feeling elbowed the other one into the background.

So I finished the stanza I was translating before I got up.

It took only a moment to reach Emory's door. I knocked twice and opened it, just as he growled, 'Come in.'

'You wanted to see me?' I sat down quickly to save him the trouble of offering me a seat.

'That was fast. What did you do, run?'

I regarded his paternal discontent:

Little fatty flecks beneath pale eyes, thinning hair, and an Irish nose; a voice a decibel louder than anyone else's...

Hamlet to Claudius: 'I was working.'

'Hah!' he snorted. 'Come off it. No one's ever seen you do any of that stuff.'

I shrugged my shoulders and started to rise.

'If that's what you called me down here—'

'Sit down!'

He stood up. He walked around his desk. He hovered above me and glared down. (A hard trick, even when I'm in a low chair.)

'You are undoubtedly the most antagonistic bastard I've ever had to work with!' he bellowed, like a belly-stung buffalo. 'Why the hell don't you act like a human being sometime and surprise everybody? I'm willing to admit you're smart, maybe even a genius, but—oh, hell!' He made a heaving gesture with both hands and walked back to his chair.

'Betty has finally talked them into letting you go in.' His voice was normal again. 'They'll receive you this afternoon. Draw one of the jeepsters after lunch, and get down there.'

'Okay,' I said.

'That's all, then.'

I nodded, got to my feet. My hand was on the doorknob when he said:

'I don't have to tell you how important this is. Don't treat them the way you treat us.'

I closed the door behind me.

I don't remember what I had for lunch. I was nervous, but I knew instinctively that I wouldn't muff it. My Boston publishers expected a Martian Idyll, or at least a Saint-Exupéry job on space flight. The National Science Association wanted a complete report on the Rise and Fall of the Martian Empire.

They would both be pleased. I knew.

That's the reason everyone is jealous—why they hate me. I always come through, and I can come through better than anyone else.

I shovelled in a final anthill of slop, and made my way to our car barn. I drew one jeepster and headed it towards Tirellian.

Flames of sand, lousy with iron oxide, set fire to the buggy. They swarmed over the open top and bit through my scarf; they set to work pitting my goggles.

The jeepster, swaying and panting like a little donkey I once rode through the Himalayas, kept kicking me in the seat of the pants. The Mountains of Tirellian shuffled their feet and moved towards me at a cockeyed angle.

Suddenly I was heading uphill, and I shifted gears to accommodate the engine's braying. Not like Gobi, not like the Great Southwestern Desert, I mused. Just red, just dead ... without even a cactus.

I reached the crest of the hill, but I had raised too much dust to see what was ahead. It didn't matter, though; I have a head full of maps. I bore to the left and downhill, adjusting the throttle. A cross-wind and solid ground beat down the fires. I felt like Ulysses in Melebolge—with a terza-rima speech in one hand and an eye out for Dante.

I rounded a rock pagoda and arrived.

Betty waved as I crunched to a halt, then jumped down.

'Hi,' I choked, unwinding my scarf and shaking out a pound and a half of grit. 'Like, where do I go and who do I see?'

She permitted herself a brief Germanic giggle—more at my starting a sentence with 'like' than at my discomfort—then she started talking. (She is a top linguist, so a word from the Village Idiom still tickles her!)

I appreciate her precise, furry talk; informational, and all that. I had enough in the way of social pleasantries before me to last at least the rest of my life. I looked at her chocolate-bar eyes and perfect teeth, at her sun-bleached hair, close-cropped to the head (I hate blondes!), and decided that she was in love with me.

'Mr. Gallinger, the Matriarch is waiting inside to be introduced. She has consented to open the Temple records for your study.' She paused here to pat her hair and squirm a little. Did my gaze make her nervous?

'They are religious documents, as well as their only history,' she continued, 'sort of like the Mahabharata. She expects you to observe certain rituals in handling them, like repeating the sacred words when you turn pages—she will teach you the system.'

I nodded quickly, several times.

'Fine, let's go in.'

'Uh—' She paused. 'Do not forget their Eleven Forms of Politeness and Degree. They take matters of form quite seriously—and do not get into any discussions over the equality of the sexes—'

'I know all about their taboos,' I broke in. 'Don't worry. I've lived in the Orient, remember?'

She dropped her eyes and seized my hand. I almost jerked it away.

'It will look better if I enter leading you.'

I swallowed my comments, and followed her, like Sampson in Gaza.

Inside, my last thought met with a strange correspondence. The Matriarch's quarters were a rather abstract version of what I imagined the tents of the tribes of Israel to have been like. Abstract, I say, because it was all frescoed brick, peaked like a huge tent, with animal-skin representations like gray-blue scars, that looked as if they had been laid on the walls with a palette knife.

The Matriarch, M'Cwyie, was short, white-haired, fiftyish, and dressed like a Gypsy queen. With her rainbow of voluminous skirts she looked like an inverted punch bowl set atop a cushion.

Accepting my obeisances, she regarded me as an owl might a rabbit. The lids of those black, black eyes jumped upwards as she discovered my perfect accent. The tape recorder Betty had carried on her interviews had done its part, and I knew the language reports from the first two expeditions, verbatum. I'm all hell when it comes to picking up accents.

'You are the poet?'

'Yes,' I replied.

'Recite one of your poems, please.'

'I'm sorry, but nothing short of a thorough translating job would do justice to your language and my poetry, and I don't know enough of your language yet.'

'Oh?'

'But I've been making such translations for my own amusement, as an exercise in grammar,' I continued. 'I'd be honored to bring a few of them along one of the times that I come here.'

'Yes. Do so.'

Score one for me!

She turned to Betty.

'You may go now.'

Betty muttered the parting formalities, gave me a strange sidewise look, and was gone. She apparently had expected to stay and 'assist' me. She wanted a piece of the glory, like everyone else. But I was the Schliemann at his Troy, and there would be only one name on the Association report!

M'Cwyie rose, and I noticed that she gained very little height by standing. But then I'm six-six and look like a poplar in October: thin, bright red on top, and towering above everyone else.

'Our records are very, very old,' she began. 'Betty says that your word for their age is "millennia".'

I nodded appreciatively.

'I'm very eager to see them.'

'They are not here. We will have to go into the Temple—they may not be removed.'

I was suddenly wary.

'You have no objections to my copying them, do you?'

'No. I see that you respect them, or your desire would not be so great.'

'Excellent.'

She seemed amused. I asked her what was funny.

'The High Tongue may not be so easy for a foreigner to learn.'

It came through fast.

No one on the first expedition had gotten this close. I had had no way of knowing that this was a double-language deal—a classical as well as a vulgar. I knew some of their Prakrit, now I had to learn all their Sanskrit.

'Ouch and damn!'

'Pardon, please?'

'It's non-translatable, M'Cwyie. But imagine yourself having to learn the High Tongue in a hurry, and you can guess at the sentiment.'

She seemed amused again, and told me to remove my shoes.

She guided me through an alcove ...

... and into a burst of Byzantine brilliance!

No Earthman had ever been in this room before, or I would have heard about it. Carter, the first expedition's linguist, with the help of one Mary Allen, M.D., had learned all the grammar and vocabulary that I knew while sitting cross-legged in the antechamber.

We had had no idea this existed. Greedily, I cast my eyes about. A highly sophisticated system of esthetics lay behind the decor. We would have to revise our entire estimation of Martian culture.

For one thing, the ceiling was vaulted and corbeled; for another, there were side-columns with reverse flutings; for another—oh hell! The place was big. Posh. You could never have guessed it from the shaggy outsides.

I bent forward to study the gilt filigree on a ceremonial table. M'Cwyie seemed a bit smug at my intentness, but I'd still have hated to play poker with her.

The table was loaded with books.

With my toe, I traced a mosaic on the floor.

'Is your entire city within this one building?'

'Yes, it goes far back into the mountain.'

'I see,' I said, seeing nothing.

I couldn't ask her for a conducted tour, yet.

She moved to a small stool by the table.

'Shall we begin your friendship with the High Tongue?'

I was trying to photograph the hall with my eyes, knowing I would have to get a camera in here, somehow, sooner or later. I tore my gaze from a statuette and nodded, hard.

'Yes, introduce me.'

I sat down.

For the next three weeks alphabet-bugs chased each other behind my eyelids whenever I tried to sleep. The sky was an unclouded pool of turquoise that rippled calligraphies whenever I swept my eyes across it. I drank quarts of coffee while I worked and mixed cocktails of Benzedrine and champagne for my coffee breaks.

M'Cwyie tutored me two hours every morning, and occasionally for another two in the evening. I spent an additional fourteen hours a day on my own, once I had gotten up sufficient momentum to go ahead alone.

And at night the elevator of time dropped me to its bottom floor…

I was six again, learning my Hebrew, Greek, Latin, and Aramaic. I was ten, sneaking peeks at the *Iliad*. When Daddy wasn't spreading hellfire brimstone, and brotherly love, he was teaching me to dig the World, like in the original.

Lord! There are so many originals and so *many* words! When I was twelve I started pointing out the little differences between what he was preaching and what I was reading.

The fundamentalist vigor of his reply brooked no debate. It was worse than any beating. I kept my mouth shut after that and learned to appreciate Old Testament poetry.

Lord, I am sorry! Daddy—Sir I am sorry! It couldn't be! It couldn't be…

On the day the boy graduated from high school, with the French, German, Spanish, and Latin awards, Dad Gallinger had told his fourteen-year old, six-foot scarecrow of a son that he wanted him to enter the ministry. I remember how his son was evasive:

'Sir,' he had said, 'I'd sort of like to study on my own for a year or so, and then take pre-theology courses at some liberal arts university. I feel I'm still sort of young to try a seminary, straight off.'

The Voice of God: 'But you have the gift of tongues, my son. You can preach the Gospel in all the lands of Babel. You were born to be a missionary. You say you are young, but time is rushing by you like a whirlwind. Start early, and you will enjoy added years of service.'

The added years of service were so many added tails to the cat repeatedly laid on my back. I can't see his face now; I never can. Maybe it is because I was always afraid to look at it then.

And years later, when he was dead, and laid out, in black, amidst bouquets, amidst weeping congregationalists, amidst prayers, red faces, handkerchiefs, hands patting your shoulders, solemn faced comforters ... I looked at him and did not recognize him.

We had met nine months before my birth, this stranger and I. He had never been cruel—stern, demanding, with contempt for everyone's shortcomings—but never cruel. He was also all that I had had of a mother. And brothers. And sisters. He had tolerated my three years at St. John's, possibly because of its name, never knowing how liberal and delightful a place it really was.

But I never knew him, and the man atop the catafalque demanded nothing now; I was free not to preach the Word. But now I wanted to, in a different way. I wanted to preach a word that I could never have voiced while he lived.

I did not return for my senior year in the fall. I had a small inheritance coming, and a bit of trouble getting control of it, since I was still under eighteen. But I managed.

It was Greenwich Village I finally settled upon.

Not telling any well-meaning parishioners my new address, I entered into a daily routine of writing poetry and teaching myself Japanese and Hindustani. I grew a fiery beard, drank espresso, and learned to play chess. I wanted to try a couple of the other paths to salvation.

After that, it was two years in India with the Old Peace Corps—which broke me of my Buddhism, and gave me my *Pipes of Krishna* lyrics and the Pulitzer they deserved.

Then back to the States for my degree, grad work in linguistics, and more prizes.

Then one day a ship went to Mars. The vessel settling in its New Mexico nest of fires contained a new language. It was fantastic, exotic, and esthetically overpowering. After I had learned all there was to know about it, and written my book, I was famous in new circles:

'Go, Gallinger. Dip your bucket in the well, and bring us a drink of Mars. Go, learn another world—but remain aloof, rail at it gently like Auden—and hand us its soul in iambics.'

And I came to the land where the sun is a tarnished penny, where the wind is a whip, where two moons play at hot rod games, and a hell of sand gives you the incendiary itches whenever you look at it.

I rose from my twistings on the bunk and crossed the darkened cabin to a port. The desert was a carpet of endless orange, bulging from the sweepings of centuries beneath it.

'I a stranger, unafraid—This is the land—I've got it made!'

I laughed.

I had the High Tongue by the tail already—or the roots, if you want your puns anatomical, as well as correct.

The High and Low Tongues were not so dissimilar as they had first seemed. I had enough of the one to get me through the murkier parts of the other. I had the grammar and all the commoner irregular verbs down cold; the dictionary I was constructing grew by the day, like a tulip, and would bloom shortly. Every time I played the tapes the stem lengthened.

Now was the time to tax my ingenuity, to really drive the lessons home. I had purposely refrained from plunging into the major texts until I could do justice to them. I had been reading minor commentaries, bits of verse, fragments of history. And one thing had impressed me strongly in all that I read.

They wrote about concrete things: rock, sand, water, winds; and the tenor couched within these elemental symbols was fiercely pessimistic. It reminded me of some Buddhist texts, but even more so, I realized from my recent *recherches,* it was like parts of the Old Testament. Specifically, it reminded me of the Book of Ecclesiastes.

That, then, would be it. The sentiment, as well as the vocabulary, was so similar that it would be a perfect exercise. Like putting Poe into French. I would never be a convert to the Way of Malann, but I would show them that an Earthman had once thought the same thoughts, felt similarly.

I switched on my desk lamp and sought King James amidst my books.

Vanity of vanities, saith the Preacher, vanity of vanities all is vanity. What profit hath a man ...

My progress seemed to startle M'Cwyie. She peered at me, like Sartre's Other, across the tabletop. I ran through a chapter in the Book of Locar. I didn't look up, but I could feel the tight net her eyes were working about my head, shoulders, and rapid hands. I turned another page.

Was she weighing the net, judging the size of the catch? And what for? The books said nothing of fishers on Mars. Especially of men. They said that some god named Malann had spat, or had done something disgusting (depending on the version you read), and that life had gotten underway as a disease in inorganic matter. They said that movement was its first law, its first law, and that the dance was the only legitimate reply to the inorganic ... the dance's quality its justification,—fication ... and love is a disease in organic matter—Inorganic matter?

I shook my head. I had almost been asleep.

'M'narra.'

I stood and stretched. Her eyes outlined me greedily now. So I met them, and they dropped.

'I grow tired. I want to rest awhile. I didn't sleep much last night.'

She nodded, Earth's shorthand for 'yes', as she had learned from me.

'You wish to relax, and see the explicitness of the doctrine of Locar in its fullness?'

'Pardon me?'

'You wish to see a Dance of Locar?'

'Oh.' Their damned circuits of form and periphrasis here ran worse than the Korean! 'Yes. Surely. Any time it's going to be done I'd be happy to watch.'

I continued, 'In the meantime, I've been meaning to ask you whether I might take some pictures—'

'Now is the time. Sit down. Rest. I will call the musicians.'

She bustled out through a door I had never been past.

Well now, the dance was the highest art, according to Locar, not to mention Havelock Ellis, and I was about to see how their centuried-dead philosopher felt it should be conducted. I rubbed my eyes and snapped over, touching my toes a few times.

The blood began pounding in my head, and I sucked in a couple of deep breaths. I bent again and there was a flurry of motion at the door.

To the trio who entered with M'Cwyie I must have looked as if I were searching for the marbles I had just lost, bent over like that.

I grinned weakly and straightened up, my face red from more than exertion. I hadn't expected them *that* quickly.

Suddenly I thought of Havelock Ellis again in his area of greatest popularity.

The little redheaded doll, wearing, sari-like, a diaphanous piece of the Martian sky, looked up in wonder—as a child at some colorful flag on a high pole.

'Hello,' I said, or its equivalent.

She bowed before replying. Evidently I had been promoted in status.

'I shall dance,' said the red wound in that pale, pale cameo, her face. Eyes, the color of dream and her dress, pulled away from mine.

She drifted to the center of the room.

Standing there, like a figure in an Etruscan frieze, she was either meditating or regarding the design on the floor.

Was the mosaic symbolic of something? I studied it. If it was, it eluded me, it would make an attractive bathroom floor or patio, but I couldn't see much in it beyond that.

The other two were paint-spattered sparrows like M'Cwyie, in their middle years. One settled to the floor with a triple-stringed instrument faintly resembling a *samisen*. The other held a simple woodblock and two drumsticks.

M'Cwyie disdained her stool and was seated upon the floor before I realized it. I followed suit.

The *samisen* player was still tuning it up, so I leaned towards M'Cwyie.

'What is that dancer's name?'

'Braxa,' she replied, without looking at me, and raised her left hand, slowly, which meant yes, and go ahead, and let it begin.

The stringed-thing throbbed like a toothache, and a tick-tocking, like ghosts of all the clocks they had never invented, sprang from the block.

Braxa was a statue, both hands raised to her face, elbows high and outspread.

The music became a metaphor for fire.

Crackle, purr, snap ...

She did not move.

This hissing altered to splashes. The cadence slowed. It was water now, the most precious thing in the world, gurgling clear then green over mossy rocks.

Still she did not move.

Glissandos. A pause.

Then, so faint I could hardly be sure at first, the tremble of the winds began. Softly, gently, sighing and halting, uncertain. A pause, a sob, then a repetition of the first statement, only louder.

Were my eyes completely bugged from my reading, or was Braxa actually trembling, all over, head to foot.

She was.

She began a microscopic swaying. A fraction of an inch right, then left. Her fingers opened like the petals of a flower, and I could see that her eyes were closed.

Her eyes opened. They were distant, glassy, looking through me and the walls. Her swaying became more pronounced, merged with the beat.

The wind was sweeping in from the desert now, falling against Tirellian like waves on a dike. Her fingers moved, they were the gusts. Her arms, slow pendulums, descended, began a counter-movement.

The gale was coming now. She began an axial movement and her hands caught up with the rest of her body, only now her shoulders commenced to writhe out a figure-eight.

The wind! The wind, I say. O wild, enigmatic! O muse of St. John Perse!

The cyclone was twisting around those eyes, its still center. Her head was thrown back, but I knew there was no ceiling between her gaze, passive as Buddha's, and the unchanging skies. Only the two moons, perhaps, interrupted their slumber in that elemental Nirvana of uninhabited turquoise.

Years ago, I had seen the Devadais in India, the street-dancers, spinning their colorful webs, drawing in the male insect. But Braxa was more than this: she was a Ramadjany, like those votaries of Rama, incarnation of Vishnu, who had given the dance to man: the sacred dancers.

The clicking was monotonously steady now; the whine of the strings made me think of the stinging rays of the sun, their heat stolen by the wind's halations; the blue was Sarasvati and Mary, and a girl named Laura. I heard a sitàr from somewhere, watched this statue come to life, and inhaled a divine afflatus.

I was again Rimbaud with his hashish, Baudelaire with his laudanum, Poe, De Quincy, Wilde, Mallarme, and Aleister Crowley. I was, for a fleeting second, my father in his dark pulpit and darker suit, the hymns and the organ's wheeze transmuted to bright wind.

She was a spun weather vane, a feathered crucifix hovering in the air, a clothes-line holding one bright garment lashed parallel to the ground. Her shoulder was bare now, and her right breast moved up and down like a moon in the sky, its red nipple appearing momentarily above a fold and vanishing again. The music was as formal as Job's argument with God. Her dance was God's reply.

The music slowed, settled; it had been met, matched, answered. Her garment, as if alive, crept back into the more sedate folds it originally held.

She dropped low, lower, to the floor. Her head fell upon her raised knees. She did not move.

There was silence.

I realized, from the ache across my shoulders, how tensely I had been sitting. My armpits were wet. Rivulets had been running down my sides. What did one do now? Applaud?

I sought M'Cwyie from the corner of my eye. She raised her right hand.

As if by telepathy the girl shuddered all over and stood. The musicians also rose. So did M'Cwyie.

I got to my feet, with a Charley Horse in my left leg, and said, 'It was beautiful,' inane as that sounds.

I received three different High Forms of 'thank you.'

There was a flurry of color and I was alone again with M'Cwyie.

'That is the one hundred-seventeenth of the two thousand, two hundred-twenty-four dances of Locar.'

I looked down at her.

'Whether Locar was right or wrong, he worked out a fine reply to the inorganic.'

She smiled.

'Are the dances of your world like this?'

'Some of them are similar. I was reminded of them as I watched Braxa—but I've never seen anything exactly like hers.'

'She is very good,' M'Cwyie said. 'She knows all the dances.'

A hint of her earlier expression which had troubled me …

It was gone in an instant.

'I must tend my duties now.' She moved to the table and closed the books. 'M'narra.'

'Good-bye,' I slipped into my boots.

'Good-bye, Gallinger.'

I walked out the door, mounted the jeepster, and roared across the evening into night, my wings of risen desert flapping slowly behind me.

II

I had just closed the door behind Betty, after a brief grammar session, when I heard the voices in the hall. My vent was opened a fraction, so I stood there and eavesdropped:

Morton's fruity treble: 'Guess what? He said "hello" to me awhile ago.'

'Hmmph!' Emory's elephant lungs exploded. 'Either he's slipping, or you were standing in his way and he wanted you to move.'

'Probably didn't recognize me. I don't think he sleeps any more, now he has that language to play with. I had night watch last week, and every night I passed his door at 0300—I always heard that recorder going. At 0500 when I got off, he was still at it.'

'The guy *is* working hard,' Emory admitted, grudgingly. 'In fact, I think he's taking some kind of dope to keep awake. He looks sort of glassy-eyed these days. Maybe that's natural for a poet, though.'

Betty had been standing there, because she broke in then:

'Regardless of what you think of him, it's going to take me at least a year to learn what he's picked up in three weeks. And I'm just a linguist, not a poet.'

Morton must have been nursing a crush on her bovine charms. It's the only reason I can think of for his dropping his guns to say what he did.

'I took a course in modern poetry when I was back at the university,' he began. 'We read six authors—Yeats, Pound, Eliot, Crane, Stevens, and Gallinger—and on the last day of the semester, when the prof was feeling a little rhetorical, he said, "These six names are written on the century, and all the gates of criticism and hell shall not prevail against them."

'Myself,' he continued, 'I thought his *Pipes of Krishna* and his *Madrigals* were great. I was honored to be chosen for an expedition he was going on.

'I think he's spoken two dozen words to me since I met him,' he finished.

The Defence: 'Did it ever occur to you,' Betty said, 'that he might be tremendously self-conscious about his appearance? He was also a precocious child, and probably never even had school friends. He's sensitive and very introverted.'

'Sensitive? Self-conscious?' Emory choked and gagged. 'The man is as proud as Lucifer, and he's a walking insult machine. You press a button like "Hello" or "Nice day" and he thumbs his nose at you. He's got it down to a reflex.'

They muttered a few other pleasantries and drifted away.

Well bless you, Morton boy. You little pimple-faced, Ivy-bred connoisseur! I've never taken a course in my poetry, but I'm glad someone said that. The Gates of Hell. Well now! Maybe Daddy's prayers got heard somewhere, and I am a missionary, after all!

Only ...

... Only a missionary needs something to convert people *to*. I have no private system of esthetics, and I suppose it oozes an ethical by-product somewhere. But if I ever had anything to preach, really, even in my poems, I wouldn't care to preach it to such lowlifes as you. If you think I'm a slob, I'm also a snob, and there's no room for you in my Heaven—it's a private place, where Swift, Shaw, and Petronius Arbiter come to dinner.

And oh, the feasts we have! The Trimalchio's, the Emory's we dissect!

We finish you with the soup, Morton!

I turned and settled at my desk. I wanted to write something. Ecclesiastes could take a night off. I wanted to write a poem, a poem about the one hundred-seventeenth dance of Locar; about a rose following the light, traced by the wind, sick, like Blake's rose, dying...

I found a pencil and began.

When I had finished I was pleased. It wasn't great—at least, it was no greater than it needed to be—High Martian not being my strongest tongue. I groped, and put it into English, with partial rhymes. Maybe I'd stick it in my next book. I called it *Braxa*:

In a land of wind and red, where the icy evening of Time freezes milk in the breasts of Life, as two moons overhead—cat and dog in alleyways of dream—scratch and scramble agelessly my flight...

This final flower turns a burning head.

I put it away and found some phenobarbitol. I was suddenly tired.

When I showed my poem to M'Cwyie the next day, she read it through several times, very slowly.

'It is lovely,' she said. 'But you used three words from your own language. "Cat" and "dog", I assume, are two small animals with a hereditary hatred for one another. But what is "flower"?'

'Oh,' I said. 'I've never come across your word for "flower", but I was actually thinking of an Earth flower, the rose.'

'What is it like?'

'Well, its petals are generally bright red. That's what I meant, on one level, by "burning heads". I also wanted it to imply fever, though, and red hair, and the fire of life. The rose, itself, has a thorny stem, green leaves, and a distinct, pleasing aroma.'

'I wish I could see one.'

'I suppose it could be arranged. I'll check.'

'Do it, please. You are a—' She used the word for 'prophet', or religious poet, like Isaish or Locar '—and your poem is inspired. I shall tell Braxa of it.'

I declined the nomination, but felt flattered.

This, then, I decided, was the strategic day, the day on which to ask whether I might bring in the microfilm machine and the camera. I wanted to copy all their texts, I explained, and I couldn't write fast enough to do it.

She surprised me by agreeing immediately. But she bowled me over with her invitation.

'Would you like to come and stay here while you do this thing? Then you can work night and day, any time you want—except when the Temple is being used, of course.'

I bowed.

'I should be honored.'

'Good. Bring your machines when you want, and I will show you a room.'

'Will this afternoon be all right?'

'Certainly.'

'Then I will go now and get things ready. Until this afternoon ...'

'Good-bye.'

I anticipated a little trouble from Emory, but not much. Everyone back at the ship was anxious to see the Martians, poke needles in the Martians, ask them about Martian climate, diseases, soil chemistry, politics and mushrooms (our botanist was a fungus nut, but a reasonably good guy)—and only four or five had actually gotten to see them. The crew had been spending most of its time excavating dead cities and their acropolises. We played the game by strict rules, and the natives were as fiercely insular as the nineteenth-century Japanese. I figured I would meet with little resistance, and I figured right.

In fact, I got the distinct impression that everyone was happy to see me move out.

I stopped in the hydroponics room to speak with our mushroom master.

'Hi, Kane. Grow any toadstools in the sand yet?'

He sniffed. He always sniffs. Maybe he's allergic to plants.

'Hello, Gallinger. No, I haven't had any success with toadstools, but look behind the car barn next time you're out there. I've got a few cacti going.'

'Great,' I observed. Doc Kane was about my only friend aboard, not counting Betty.

'Say, I came down to ask you a favor.'

'Name it.'

'I want a rose.'

'A what?'

'A rose. You know, a nice red American Beauty job—thorns, pretty smelling—'

'I don't think it will take in this soil. *Sniff, sniff.*'

'No, you don't understand. I don't want to plant it, I just want the flowers.'

'I'd have to use the tanks.' He scratched his hairless dome. 'It would take at least three months to get you flowers, even under forced growth.'

'Will you do it?'

'Sure, if you don't mind the wait.'

'Not at all. In fact, three months will just make it before we leave.' I looked about at the pools of crawling slime, at the trays of shoots. '—I'm moving up to Tirellian today, but I'll be in and out all the time. I'll be here when it blooms.'

'Moving up there, eh? Moore said they're an in-group.'

'I guess I'm "in" then.'

'Looks that way—I still don't see how you learned their language, though. Of course, I had trouble with French and German for my Ph.D., but last week I heard Betty demonstrate it at lunch. It sounds like a lot of weird noises. She says speaking it is like working a *Times* crossword and trying to imitate birdcalls at the same time.'

I laughed, and took the cigarette he offered me.

'It's complicated,' I acknowledged. 'But, well, it's as if you suddenly came across a whole new class of mycetae here—you'd dream about it at night.'

His eyes were gleaming.

'Wouldn't that be something! I might, yet, you know.'

'Maybe you will.'

He chuckled as we walked to the door.

'I'll start your roses tonight. Take it easy down there.'

'You bet. Thanks.'

Like I said, a fungus nut, but a fairly good guy.

My quarters in the Citadel of Tirellian were directly adjacent to the Temple, on the inward side and slightly to the left. They were a considerable improvement over my cramped cabin, and I was pleased that Martian culture had progressed sufficiently to discover the desirability of the mattress over the pallet. Also, the bed was long enough to accommodate me, which was surprising.

So I unpacked and took sixteen 35 mm. shots of the Temple, before starting on the books.

I took stats until I was sick of turning pages without knowing what they said. So I started translating a work of history.

'Lo. In the thirty-seventh year of the Process of Cillen the rains came, which gave rise to rejoicing, for it was a rare and untoward occurrence, and commonly construed a blessing.

'But it was not the life-giving semen of Malann which fell from the heavens. It was the blood of the universe, spurting from an artery. And the last days were upon us. The final dance was to begin.

'The rains brought the plague that does not kill, and the last passes of Locar began with their drumming...'

I asked myself what the hell Tamur meant, for he was an historian and supposedly committed to fact. This was not their Apocalypse.

Unless they could be one and the same ...?

Why not? I mused. Tirellian's handful of people were the remnant of what had obviously once been a highly developed culture. They had had wars, but no holocausts; science, but little technology. A plague, a plague that did not kill…? Could that have done it? How, if it wasn't fatal?

I read on, but the nature of the plague was not discussed. I turned pages, skipped ahead, and drew a blank.

M'Cwyie! M'Cwyie! When I want to question you most, you are not around!

Would it be a *faux pas* to go looking for her? Yes, I decided. I was restricted to the rooms I had been shown, that had been an implicit understanding. I would have to wait to find out.

So I cursed long and loud, in many languages, doubtless burning Malann's sacred ears, there in his Temple.

He did not see fit to strike me dead, so I decided to call it a day and hit the sack.

I must have been asleep for several hours when Braxa entered my room with a tiny lamp. She dragged me awake by tugging at my pajama sleeve.

I said hello. Thinking back, there is not much else I could have said.

'Hello.'

'I have come,' she said, 'to hear the poem.'

'What poem?'

'Yours.'

'Oh.'

I yawned, sat up, and did things people usually do when awakened in the middle of the night to read poetry.

'That is very kind of you, but isn't the hour a trifle awkward?'

'I don't mind,' she said.

Someday I am going to write an article for the *Journal of Semantics*, called 'Tone of Voice: An Insufficient Vehicle for Irony.'

However, I was awake, so I grabbed my robe.

'What sort of animal is that?' she asked, pointing at the silk dragon on my lapel.

'Mythical,' I replied. 'Now look, it's late. I am tired. I have much to do in the morning. And M'Cwyie just might get the wrong idea if she learns you were here.'

'Wrong idea?'

'You know damned well what I mean!' It was the first time I had had an opportunity to use Martian profanity, and it failed.

'No,' she said, 'I do not know.'

She seemed frightened, like a puppy being scolded without knowing what it has done wrong.

I softened. Her red cloak matched her hair and lips so perfectly, and those lips were trembling.

'Here now, I didn't mean to upset you. On my world there are certain, uh, mores, concerning people of different sex alone together in bedrooms, and not allied by marriage… Um, I mean, you see what I mean?'

'No.'

They were jade, her eyes.

'Well, it's sort of … Well, it's sex, that's what it is.'

A light was switched on in those jade lamps.

'Oh, you mean having children!'

'Yes. That's it! Exactly.'

She laughed. It was the first time I had heard laughter in Tirellian. It sounded like a violinist striking his high strings with the bow, in short little chops. It was not an altogether pleasant thing to hear, especially because she laughed too long.

When she had finished she moved closer.

'I remember, now,' she said. 'We used to have such rules. Half a Process ago, when I was a child, we had such rules. But'—she looked as if she were ready to laugh again—'there is no need for them now.'

My mind moved like a tape recorder played at triple speed.

Half a Process! HalfaProcessa-ProcessaProcess! No! Yes! Half a Process was two hundred-forty-three years, roughly speaking!

—Time enough to learn the 2224 dances of Locar.

—Time enough to grow old, if you were human.

—Earth-style human, I mean.

I looked at her again, pale as the white queen in an ivory chess set.

She was human, I'd stake my soul—alive, normal, healthy. I'd stake my life—woman, my body …

But she was two and a half centuries old, which made M'Cwyie Methusala's grandma. It flattered me to think of their repeated complimenting of my skills, as linguist, as poet. These superior beings!

But what did she mean 'there is no such need for them now'? Why the near-hysteria? Why all those funny looks I'd been getting from M'Cwyie?

I suddenly knew I was close to something important, besides a beautiful girl.

'Tell me,' I said, in my Casual Voice, 'did it have anything to do with "the plague that does not kill," of which Tamur wrote?'

'Yes,' she replied, 'the children born after the Rains could have no children of their own, and—'

'And what?' I was leaning forward, memory set at 'record'.

'—and the men had no desire to get any.'

I sagged backward against the bedpost. Racial sterility, masculine impotence, following phenomenal weather. Had some vagabond cloud of radioactive junk from God knows where penetrated their weak atmosphere one day? One day long before Shiaparelli saw the canals, mythical as my dragon, before those 'canals' had given rise to some correct guesses for all the wrong reasons, had Braxa been alive, dancing, here—damned in the womb since blind Milton had written of another paradise, equally lost?

I found a cigarette. Good thing I had thought to bring ashtrays. Mars had never had a tobacco industry either. Or booze. The ascetics I had met in India had been Dionysiac compared to this.

'What is that tube of fire?'

'A cigarette. Want one?'

'Yes, please.'

She sat beside me, and I lighted it for her.

'It irritates the nose.'

'Yes. Draw some into your lungs, hold it there, and exhale.'

A moment passed.

'Ooh,' she said.

A pause, then, 'Is it sacred?'

'No, it's nicotine,' I answered, 'a very *ersatz* form of divinity.'

Another pause.

'Please don't ask me to translate "ersatz".'

'I won't. I get this feeling sometimes when I dance.'

'It will pass in a moment.'

'Tell me your poem now.'

An idea hit me.

'Wait a minute,' I said; 'I may have something better.'

I got up and rummaged through my notebooks, then I returned and sat beside her.

'These are the first three chapters of the Book of Ecclesiastes,' I explained. 'It is very similar to your own sacred books.'

I started reading.

I got through eleven verses before she cried out, 'Please don't read that! Tell me one of yours!'

I stopped and tossed the notebook onto a nearby table. She was shaking, not as she had quivered that day she danced as the wind, but with the jitter of unshed tears. She held her cigarette awkwardly, like a pencil. Clumsily, I put my arm about her shoulders.

'He is so sad,' she said, 'like all the others.'

So I twisted my mind like a bright ribbon, folded it, and tied the crazy Christmas knots I love so well. From German to Martian, with love, I did an impromptu paraphrasal of a poem about a Spanish dancer. I thought it would please her. I was right.

'Ooh,' she said again. 'Did you write that?'

'No, it's by a better man than I.'

'I don't believe you. You wrote it.'

'No, a man named Rilke did.'

'But you brought it across to my language. Light another match, so I can see how she danced.'

I did.

'The fires of forever,' she mused, 'and she stamped them out, "with small, firm feet". I wish I could dance like that.'

'You're better than any Gypsy,' I laughed, blowing it out.

'No, I'm not. I couldn't do that.'

'Do you want me to dance for you?'

Her cigarette was burning down, so I removed it from her fingers and put it out, along with my own.

'No,' I said. 'Go to bed.'

She smiled, and before I realized it, had unclasped the fold of red at her shoulder.

And everything fell away.

And I swallowed, with some difficulty.

'All right,' she said.

So I kissed her, as the breath of fallen cloth extinguished the lamp.

III

The days were like Shelley's leaves: yellow, red, brown, whipped in bright gusts by the west wind. They swirled past me with the rattle of microfilm. Almost all the books were recorded now. It would take scholars years to get through them, to properly assess their value. Mars was locked in my desk.

Ecclesiastes, abandoned and returned to a dozen times, was almost ready to speak in the High Tongue.

I whistled when I wasn't in the Temple. I wrote reams of poetry I would have been ashamed of before. Evenings I would walk with Braxa, across the dunes or up into the mountains. Sometimes she would dance for me; and I would read something long, and in dactylic hexameter. She still thought I was Rilke, and I almost kidded myself into believing it. Here I was, staying at the Castle Duino, writing his *Elegies*.

> *... It is strange to inhabit the Earth no more,*
> *to use no longer customs scarce acquired,*
> *nor interpret roses ...*

No! Never interpret roses! Don't. Smell them (sniff, Kane!), pick them, enjoy them. Live in the moment. Hold to it tightly. But charge not the gods to explain. So fast the leaves go by, are blown ...

And no one ever noticed us. Or cared.

Laura. Laura and Braxa. They rhyme, you know, with a bit of a clash. Tall, cool, and blonde was she (I hate blondes!), and Daddy had turned me inside out, like a pocket, and I thought she could fill me again. But the big, beat word-slinger, with Judas-beard and dog-trust in his eyes, oh, he had been a fine decoration at her parties. And that was all.

How the machine cursed me in the Temple! It blasphemed Malann and Gallinger. And the wild west wind went by and something was not far behind.

The last days were upon us.

A day went by and I did not see Braxa, and a night.

And a second. A third.

I was half-mad. I hadn't realized how close we had become, how important she had been. With the dumb assurance of presence, I had fought against questioning roses.

I had to ask. I didn't want to, but I had no choice.

'Where is she, M'Cwyie? Where is Braxa?'

'She is gone,' she said.

'Where?'

'I do not know.'

I looked at those devil-bird eyes. Anathema maranatha rose to my lips.

'I must know.'

She looked through me.

'She has left us. She is gone. Up into the hills, I suppose. Or the desert. It does not matter. What does anything matter? The dance draws to a close. The Temple will soon be empty.'

'Why? Why did she leave?'

'I do not know.'

'I must see her again. We lift off in a matter of days.'

'I am sorry, Gallinger.'

'So am I,' I said, and slammed shut a book without saying 'm'narra.'

I stood up.

'I will find her.'

I left the Temple. M'Cwyie was a seated statue. My boots were still where I had left them.

All day I roared up and down the dunes, going nowhere. To the crew of the *Aspic* I must have looked like a sandstorm, all by myself. Finally, I had to return for more fuel.

Emory came stalking out.

'Okay, make it good. You look like the abominable dust man. Why the rodeo?'

'Why, I, uh, lost something.'

'In the middle of the desert? Was it one of your sonnets? They're the only thing I can think of that you'd make such a fuss over.'

'No, dammit! It was something personal.'

George had finished filling the tank. I started to mount the jeepster again.

'Hold on there!' he grabbed my arm.

'You're not going back until you tell me what this is all about.'

I could have broken his grip, but then he could order me dragged back by the heels, and quite a few people would enjoy doing the dragging. So I forced myself to speak slowly, softly:

'It's simply that I lost my watch. My mother gave it to me and it's a family heirloom. I want to find it before we leave.'

'You sure it's not in your cabin, or down in Tirellian?'

'I've already checked.'

'Maybe somebody did it to irritate you. You know you're not the most popular guy around.'

I shook my head.

'I thought of that. But I always carry it in my right pocket. I think it might have bounced out going over the dunes.'

He narrowed his eyes.

'I remembered reading on a book jacket that your mother died when you were born.'

'That's right,' I said, biting my tongue. 'The watch belonged to her father and she wanted me to have it. My father kept it for me.'

'Hmph!' he snorted. 'That's a pretty strange way to look for a watch, riding up and down in a jeepster.'

'I could see the light shining off it that way,' I offered, lamely.

'Well, it's starting to get dark,' he observed. 'No sense looking any more today.

'Throw a dust sheet over the jeepster,' he directed a mechanic.

He patted my arm.

'Come on in and get a shower, and something to eat. You look as if you could use both.'

Little fatty flecks beneath pale eyes, thinning hair, and an Irish nose; a voice a decibel louder than anyone else's…

His only qualification for leadership!

I stood there, hating him. Claudius! If only this were the fifth act!

But suddenly the idea of a shower, and food, came through to me. I could use both badly. If I insisted on hurrying back immediately I might arouse more suspicion.

So I brushed some sand from my sleeve.

'You're right. That sounds like a good idea.'

'Come on, we'll eat in my cabin.'

The shower was a blessing, clean khakis were the grace of God, and the food smelled like Heaven.

'Smells pretty good,' I said.

We hacked up our steaks in silence. When we got to the dessert and coffee he suggested:

'Why don't you take the night off? Stay here and get some sleep.'

I shook my head.

'I'm pretty busy. Finishing up. There's not much time left.'

'A couple of days ago you said you were almost finished.'

'Almost, but not quite.'

'You also said they'll be holding a service in the Temple tonight.'

'That's right. I'm going to work in my room.'

He shrugged his shoulders.

Finally, he said, 'Gallinger,' and I looked up because my name means trouble.

'It shouldn't be any of my business,' he said, 'but it is. Betty says you have a girl down there.'

There was no question mark. It was a statement hanging in the air. Waiting.

Betty, you're a bitch. You're a cow and a bitch. And a jealous one, at that. Why didn't you keep your nose where it belonged, shut your eyes? Your mouth?

'So?' I said, a statement with a question mark.

'So,' he answered it, 'it is my duty, as head of this expedition, to see that relations with the natives are carried on in a friendly, and diplomatic, manner.'

'You speak of them,' I said, 'as though they are aborigines. Nothing could be further from the truth.'

I rose.

'When my papers are published everyone on Earth will know that truth. I'll tell them things Doctor Moore never even guessed at. I'll tell the tragedy of a doomed race, waiting for death, resigned and disinterested. I'll tell why, and it will break hard, scholarly hearts. I'll write about it, and they will give me more prizes, and this time I won't want them.

'My God!' I exclaimed. 'They had a culture when our ancestors were clubbing the saber-tooth and finding out how fire works!'

'*Do* you have a girl down there?'

'Yes!' I said. Yes, *Claudius! Yes, Daddy! Yes, Emory!* 'I do. But I'm not going to let you in on a scholarly scoop now. They're already dead. They're sterile. In one more generation there won't be any Martians.'

I paused, then added, 'Except in my papers, except on a few pieces of microfilm and tape. And in some poems, about a girl who did give a damn and could only bitch about the unfairness of it all by dancing.'

'Oh,' he said.

After awhile:

'You *have* been behaving differently these past couple months. You've even been downright civil on occasion, you know. I couldn't help wondering what was happening. I didn't know anything mattered that strongly to you.'

I bowed my head.

'Is she the reason you were racing around the desert?'

I nodded.

'Why?'

I looked up.

'Because she's out there, somewhere. I don't know where, or why. And I've got to find her before we go.'

'Oh,' he said again.

Then he leaned back, opened a drawer, and took out something wrapped in a towel. He unwound it. A framed photo of a woman lay on the table.

'My wife,' he said.

It was an attractive face, with big, almond eyes.

'I'm a Navy man, you know,' he began. 'Young officer once. Met her in Japan.'

'Where I come from it wasn't considered right to marry into another race, so we never did. But she was my wife. When she died I was on the other side of the world. They took my children, and I've never seen them since. I couldn't learn what orphanage, what home, they were put into. That was long ago. Very few people know about it.'

'I'm sorry,' I said.

'Don't be. Forget it. But'—he shifted in his chair and looked at me—'if you do want to take her back with you—do it. It'll mean my neck, but I'm too old to ever head another expedition like this one. So go ahead.'

He gulped his cold coffee.

'Get your jeepster.'

He swivelled the chair around.

I tried to say 'thank you' twice, but I couldn't. So I got up and walked out.

'Sayonara, and all that,' he muttered behind me.

'Here it is, Gallinger!' I heard a shout.

I turned on my heel and looked back up the ramp.

'Kane!'

He was limned in the port, shadow against light, but I had heard him sniff.

I returned the few steps.

'Here what is?'

'Your rose.'

He produced a plastic container, divided internally. The lower half was filled with liquid. The stem ran down into it. The other half, a glass of claret in this horrible night, was a large, newly opened rose.

'Thank you,' I said, tucking it into my jacket.

'Going back to Tirellian, eh?'

'Yes.'

'I saw you come aboard, so I got it ready. Just missed you at the Captain's cabin. He was busy. Hollered out that I could catch you at the barns.'

'Thanks again.'

'It's chemically treated. It will stay in bloom for weeks.'

I nodded. I was gone.

Up into the mountains now. Far. Far. The sky was a bucket of ice in which no moons floated. The going became steeper, and the little donkey protested. I whipped him with the throttle and went on. Up. Up. I spotted a green, unwinking star, and felt a lump in my throat. The encased rose beat against my chest like an extra heart. The donkey brayed, long and loudly, then began to cough. I lashed him some more and he died.

I threw the emergency brake on and got out. I began to walk.

So cold, so cold it grows. Up here. At night? Why? Why did she do it? Why flee the campfire when night comes on?

And I was up, down, around, and through every chasm, gorge, and pass, with my long-legged strides and an ease of movement never known on Earth.

Barely two days remain, my love, and thou hast forsaken me. Why?

I crawled under overhangs. I leaped over ridges. I scraped my knees, an elbow. I heard my jacket tear.

No answer, Malann? Do you really hate your people this much? Then I'll try someone else. Vishnu, you're the Preserver. Preserve her, please! Let me find her.

Jehovah?

Adonis? Osiris? Thammuz? Manitou? Legba? Where is she?

I ranged far and high, and I slipped.

Stones ground underfoot and I dangled over an edge. My fingers so cold. It was hard to grip the rock.

I looked down.

Twelve feet or so. I let go and dropped, landed rolling.

Then I heard her scream.

I lay there, not moving, looking up. Against the night, above, she called.

'Gallinger!'

I lay still.

'Gallinger!'

And she was gone.

I heard stones rattle and knew she was coming down some path to the right of me.

I jumped up and ducked into the shadow of a boulder.

She rounded a cut-off, and picked her way, uncertainly, through the stones.

'Gallinger?'

I stepped out and seized her shoulders.

'Braxa.'

She screamed again, then began to cry, crowding against me. It was the first time I had ever heard her cry.

'Why?' I asked. 'Why?'

But she only clung to me and sobbed.

Finally, 'I thought you had killed yourself.'

'Maybe I would have,' I said. 'Why did you leave Tirellian? And me?'

'Didn't M'Cwyie tell you? Didn't you guess?'

'I didn't guess, and M'Cwyie said she didn't know.'

'Then she lied. She knows.'

'What? What is it she knows?'

She shook all over, then was silent for a long time. I realized suddenly that she was wearing only her flimsy dancer's costume. I pushed her from me, took off my jacket, and put it about her shoulders.

'Great Malann!' I cried. 'You'll freeze to death!'

'No,' she said, 'I won't.'

I was transferring the rose-case to my pocket.

'What is that?' she asked.

'A rose,' I answered. 'You can't make it out much in the dark. I once compared you to one. Remember?'

'Ye-Yes. May I carry it?'

'Sure.' I stuck it in the jacket pocket.

'Well? I'm still waiting for an explanation.'

'You really do not know?' she asked.

'No!'

'When the Rains came,' she said, 'apparently only our men were affected, which was enough… Because I—wasn't—affected—apparently——'

'Oh,' I said. 'Oh.'

We stood there, and I thought.

'Well, why did you run? What's wrong with being pregnant on Mars? Tamur was mistaken. Your people can live again.'

She laughed, again that wild violin played by a Paginini gone mad. I stopped her before it went too far.

'How?' she finally asked, rubbing her cheek.

'Your people live longer than ours. If our child is normal it will mean our races can intermarry. There must still be other fertile women of your race. Why not?'

'You have read the Book of Locar,' she said, 'and yet you ask me that? Death was decided, voted upon, and passed, shortly after it appeared in this form. But long before, the followers of Locar knew. They decided it long ago. "We have done all things," they said, "we have seen all things, we have heard and felt all things. The dance was good. Now let it end." '

'You can't believe that.'

'What I believe does not matter,' she replied. 'M'Cwyie and the Mothers have decided we must die. Their very title is now a mockery, but their decisions will be upheld. There is only one prophecy left, and it is mistaken. We will die.'

'No,' I said.

'What, then?'

'Come back with me, to Earth.'

'No.'

'All right, then. Come with me now.'

'Where?'

'Back to Tirellian. I'm going to talk to the Mothers.'

'You can't! There is a Ceremony tonight!'

I laughed.

'A ceremony for a god who knocks you down, and then kicks you in the teeth?'

'He is still Malann,' she answered. 'We are still his people.'

'You and my father would have gotten along fine,' I snarled. 'But I am going, and you are coming with me, even if I have to carry you—and I'm bigger than you are.'

'But you are not bigger than Ontro.'

'Who the hell is Ontro?'

'He will stop you, Gallinger. He is the Fist of Malann.'

IV

I scudded the jeepster to a halt in front of the only entrance I knew, M'Cwyie's. Braxa, who had seen the rose in a headlamp, now cradled it in her lap, like our child, and said nothing. There was a passive, lovely look on her face.

'Are they in the Temple now?' I wanted to know.

The Madonna-expression did not change. I repeated the question. She stirred.

'Yes,' she said, from a distance, 'but you cannot go in.'

'We'll see.'

I circled and helped her down.

I led her by the hand, and she moved as if in a trance. In the light of the new-risen moon, her eyes looked as they had the day I met her, when she had danced. I snapped my fingers. Nothing happened.

So I pushed the door open and led her in. The room was half-lighted.

And she screamed for the third time that evening:

'Do not harm him, Ontro! It is Gallinger!'

I had never seen a Martian man before, only women. So I had no way of knowing whether he was a freak, though I suspected it strongly.

I looked up at him.

His half-naked body was covered with moles and swellings. Gland trouble, I guessed.

I had thought I was the tallest man on the planet, but he was seven feet tall and overweight. Now I knew where my giant bed had come from!

'Go back,' he said. 'She may enter. You may not.'

'I must get my books and things.'

He raised a huge left arm. I followed it. All my belongings lay neatly stacked in the corner.

'I must go in. I must talk with M'Cwyie and the Mothers.'

'You may not.'

'The lives of your people depend on it.'

'Go back,' he boomed. 'Go home to *your* people, Gallinger. Leave *us*!'

My name sounded so different on his lips, like someone else's. How old was he? I wondered. Three hundred? Four? Had he been a Temple guardian all his life? Why? Who was there to guard against? I didn't like the way he moved. I had seen men who moved like that before.

'Go back,' he repeated.

If they had refined their martial arts as far as they had their dances, or, worse yet, if their fighting arts were a part of the dance, I was in for trouble.

'Go on in,' I said to Braxa. 'Give the rose to M'Cwyie. Tell her that I sent it. Tell her I'll be there shortly.'

'I will do as you ask. Remember me on Earth, Gallinger. Good-bye.'

I did not answer her, and she walked past Ontro and into the next room, bearing her rose.

'Now will you leave?' he asked. 'If you like, I will tell her that we fought and you almost beat me, but I knocked you unconscious and carried you back to your ship.'

'No,' I said, 'either I go around you or go over you, but I am going through.'

He dropped into a crouch, arms extended.

'It is a sin to lay hands on a holy man,' he rumbled, 'but I will stop you, Gallinger.'

My memory was a fogged window, suddenly exposed to fresh air. Things cleared. I looked back six years.

I was a student of Oriental Languages at the University of Tokyo. It was my twice-weekly night of recreation. I stood in a thirty-foot circle in the Kodokan, the *judogi* lashed about my high hips by a brown belt. I was *Ik-kyu*, one notch below the lowest degree of expert. A brown diamond above my right breast said 'Ju-Jitsu' in Japanese, and it meant *atemiwaza*, really, because of the one striking-technique I had worked out, found unbelievably suitable to my size, and won matches with.

But I had never used it on a man, and it was five years since I had practiced. I was out of shape, I knew, but I tried hard to force my mind *tsuki no kokoro*, like the moon, reflecting the all of Ontro.

Somewhere, out of the past, a voice said, '*Hajime*, let it begin.'

I snapped into my *neko-ashi-dachi* cat-stance, and his eyes burned strangely. He huried to correct his own position—and I threw it at him!

My one trick!

My long left leg lashed up like a broken spring. Seven feet off the ground my foot connected with his jaw as he tried to leap backward.

His head snapped back and he fell. A soft moan escaped his lips. *That's all there is to it*, I thought. *Sorry, old fellow.*

And as I stepped over him, somehow, groggily, he tripped me, and I fell across his body. I couldn't believe he had strength enough to remain conscious after that blow, let alone move. I hated to punish him any more.

But he found my throat and slipped a forearm across it before I realized there was a purpose to his action.

No! Don't let it end like this!

It was a bar of steel across my windpipe, my carotids. Then I realized that he was still unconscious, and that this was a reflex instilled by countless years of training. I had seen it happen once, in *shiai*. The man died because he had been choked unconscious and still fought on, and his opponent thought he had not been applying the choke properly. He tried harder.

But it was rare, so very rare!

I jammed my elbows into his ribs and threw my head back in his face. The grip eased, but not enough. I hated to do it, but I reached up and broke his little finger.

The arm went loose and I twisted free.

He lay there panting, face contorted. My heart went out to the fallen giant, defending his people, his religion, following his orders. I cursed myself as I had never cursed before, for walking over him, instead of around.

I staggered across the room to my little heap of possessions. I sat on the projector case and lit a cigarette.

I couldn't go into the Temple until I got my breath back, until I thought of something to say?

How do you talk a race out of killing itself?

Suddenly——

Could it happen? Would it work that way? If I read them the Book of Ecclesiastes—if I read them a greater piece of literature than any Locar ever wrote—and as somber—and as pessimistic—and showed them that our race had gone on despite one man's condemning all of life in the highest poetry—showed them that the vanity he had mocked had borne us to the Heavens—would they believe it—would they change their minds?

I ground out my cigarette on the beautiful floor, and found my notebook. A strange fury rose within me as I stood.

And I walked into the Temple to preach the Black Gospel according to Gallinger, from the Book of Life.

There was silence all about me.

M'Cwyie had been reading Locar, the rose set at her right hand, target of all eyes.

Until I entered.

Hundreds of people were seated on the floor, barefoot. The few men were as small as the women, I noted.

I had my boots on.

Go all the way, I figured. *You either lose or you win—everything!*

A dozen crones sat in a semicircle behind M'Cwyie. The Mothers.

The barren earth, the dry wombs, the fire-touched.

I moved to the table.

'Dying yourselves, you would condemn your people,' I addressed them, 'that they may not know the life you have known—the joys, the sorrows, the fullness. But it is not true that you all must die.' I addressed the multitude now. 'Those who say this lie. Braxa knows, for she will bear a child—'

They sat there, like rows of Buddhas. M'Cwyie drew back into the semicircle.

'—my child!' I continued, wondering what my father would have thought of this sermon.

'... And all the women young enough may bear children. It is only your men who are sterile. And if you permit the doctors of the next expedition to examine you, perhaps even the men may be helped. But if they cannot, you can mate with the men of Earth.

'And ours is not an insignificant people, an insignificant place,' I went on. 'Thousands of years ago, the Locar of our world wrote a book saying that it was. He spoke as Locar did, but we did not lie down, despite plagues, wars, and famines. We did not die. One by one we beat down the diseases, we fed the hungry, we fought the wars, and, recently, have gone a long time without them. We may finally have conquered them. I do not know.

'But we have crossed millions of miles of nothingness. We have visited another world. And our Locar had said, "Why bother? What is the worth of it? It is all vanity, anyhow."

'And the secret is,' I lowered my voice, as at a poetry reading, 'he was right! It is vanity; it is pride! It is the hybris of rationalism to always attack the prophet, the mystic, the god. It is our blasphemy which has made us great, and will sustain us, and which the gods secretly admire in us. All the truly sacred names of God are blasphemous things to speak!'

I was working up a sweat. I paused dizzily.

'Here is the Book of Ecclesiastes,' I announced, and began:

' "Vanity of vanities, saith the Preacher, vanity of vanities; all is vanity. What profit hath a man ..." '

I spotted Braxa in the back, mute, rapt.

I wondered what she was thinking.

And I wound the hours of night about me, like black thread on a spool.

Oh it was late! I had spoken till day came, and still I spoke. I finished Ecclesiastes and continued Gallinger.

And when I finished there was still only a silence.

The Buddhas, all in a row, had not stirred through the night. And after a long while M'Cwyie raised her right hand. One by one the Mothers did the same.

And I knew what that meant.

It meant no, do not, cease, and stop.

It meant that I had failed.

I walked slowly from the room and slumped beside my baggage.

Ontro was gone. Good that I had not killed him...

After a thousand years M'Cwyie entered.

She said, 'Your job is finished.'

I did not move.

'The prophecy is fulfilled,' she said. 'My people are re-joicing. You have won, holy man. Now leave us quickly.'

My mind was a deflated balloon. I pumped a little air back into it.

'I'm not a holy man,' I said, 'just a second-rate poet with a bad case of hybris.'

I lit my last cigarette.

Finally, 'All right, what prophecy?'

'The Promise of Locar,' she replied, as though the explaining were unnecessary, 'that a holy man would come from the Heavens to save us in our last hours, if all the dances of Locar were completed. He would defeat the Fist of Malann and bring us life.'

'How?'

'As with Braxa, and as the example in the Temple.'

'Example?'

'You read us his words, as great as Locar's. You read to us how there is "nothing new under the sun". And you mocked his words as you read them—showing us a new thing.

'There has never been a flower on Mars,' she said, 'but we will learn to grown them.

'You are the Sacred Scoffer,' she finished. 'He-Who-Must-Mock-in-the-Temple—you go shod on holy ground.'

'But you voted "no,"' I said.

'I voted not to carry out our original plan, and to let Braxa's child live instead.'

'Oh.' The cigarette fell from my fingers. How close it had been! How little I had known!

'And Braxa?'

'She was chosen half a Process ago to do the dances—to wait for you.'

'But she said that Ontro would stop me.'

M'Cwyie stood there for a long time.

'She had never believed the prophecy herself. Things are not well with her now. She ran away, fearing it was true. When you completed it and we voted, she knew.'

'Then she does not love me? Never did?'

'I am sorry, Gallinger. It was the one part of her duty she never managed.'

'Duty,' I said flatly… Dutydutyduty! Tra la!

'She has said good-bye; she does not wish to see you again.

'… and we will never forget your teachings,' she added.

'Don't,' I said, automatically, suddenly knowing the great paradox which lies at the heart of all miracles. I did not believe a word of my own gospel, never had.

I stood, like a drunken man, and muttered 'M'narra.'

I went outside, into my last days on Mars.

I have conquered thee, Malann—and the victory is thine! Rest easy on thy starry bed. God damned!

I left the jeepster there and walked back to the *Aspic,* leaving the burden of life so many footsteps behind me. I went to my cabin, locked the door, and took forty-four sleeping pills.

But when I awakened I was in the dispensary, and alive.

I felt the throb of engines as I slowly stood up and somehow made it to the port.

Blurred Mars hung like a swollen belly above me, until it dissolved, brimmed over, and streamed down my face.

TAKE BACK PLENTY (1990)

Colin Greenland

Colin Greenland was born in 1954 in Dover, England. An authority on science fiction, his first book was the literary thesis *The Entropy Exhibition: Michael Moorcock and the British 'New Wave' in Science Fiction*, published in 1983. His first novel, *Daybreak on a Different Mountain*, followed a year later. Since then, he has published a further eight novels and two works of literary criticism. *Take Back Plenty* appeared in 1990 and won the British Science Fiction Association award and the 1991 Arthur C. Clarke award. In the US, it was also nominated for the 1992 Philip K. Dick Award. Two sequels, *Seasons of Plenty* and *Mother of Plenty*, were subsequently published.

2

Carnival in Schiaparelli. The canals are thronged with tour buses, the bridges festooned with banners. Balloons escape and fireworks fly. The city seethes in the smoky red light. Though officers of the Eladeldi can be seen patrolling everywhere, pleasure is the only master. Shall we go to the Ruby Pool? To watch the glider duels over the al-Kazara? Or to the old city, where the cavernous ancient silos throb with the latest raga, and the wine of Astarte quickens the veins of the young and beautiful? A thousand smells, of sausages and sweat, phosphorus and patchouli, mingle promiscuously in the arcades. Glasses clash and cutlery clatters in the all-night cantinas where drunken revellers confuse the robot waiters and flee along the colonnades, their bills unpaid, their breath steaming in the thin and wintry air.

Reflected off the oily water, a thousand coloured lights flicker and glow on the scoured faces of the buildings. A thousand noises batter the attendant ear, calliopes and stridulators, cannonades and sirens, all mingling with the babble and slur of happy voices. Even the screeching rasp of a police hover forcing its slow way upstream can scarcely cut the din. The cop, a human, leans on his screamer, twice, and stalls. In the shiny black carapace of his servo-armour he looks stiff and offended, like a gigantic beetle beset by ants.

They pulled in at Mustique Boulevard, below the skate bowl. Grubby urchins stood on the wall, sucking steaming mossballs and shouting abuse at each other.

'This isn't the Moebius Strip,' said Tabitha.

The morose boatwoman jerked an elbow. 'Close as I can, sister. Grand's closed for the procession.'

Annoyed, Tabitha paid her and leapt easily to the landing stage. Her jacket flashed and sparkled with sodium light, her boots crunched on the sandy boards.

Picture her, Tabitha Jute: not as the net media show her, heroine of hyperspace, capable, canny and cosmetically enhanced, smiling confidently as she reaches with one hand for the spangled mist of the Milky Way; but a small, weary young woman in a cracked foil jacket and oil-stained trousers, determinedly elbowing herself through an exuberant Schiaparelli crowd. She stands 162 in her socks, broad in the shoulder and the hip, and weighs about 60k at 1g, which she very rarely is. Her hair is darkest ginger, cut in a conservative spacer's square crop. Her skin is an ordinary milky coffee, and freckles easily, which she hates. Here she was, in after a stiff haul back from Chateaubriand, spacelagged and frazzled, needing a shower. There were dark olive bags under her hazel eyes. You wouldn't have given her a second glance that evening, amid the florid, the fancy and the flash.

Not that there was much of that around here. This was definitely the scrag end of the festivities. She ducked beneath the concrete walkway and strode along an avenue of makeshift stalls lashed together from pipes and planking, weaving a path between the strolling browsers. Overhead, lines of biofluorescents snaked from pole to pole, tied on with string. Tabitha had come to the flea market after all.

Some of the stallholders had made an effort for the carnival. There were masks and bunting decorating their displays of scuffed cassettes and second-hand knitting. Here were bright clothes: everything from aluminium shoes to cheap and garish movie shirts of winking kittens and prancing unicorns and swivelling strippers. Collectors rummaged in boxes of sunglasses, discussed the merits of filched scraps of cruiseliner trim. Two scrawny women in tiny dresses sat behind a table of china animals, painting each other's faces by the warmth of a dilapidated reactor stove. One of them whistled at Tabitha as she squeezed by.

A decommissioned shop robot leaned from under its canopy and fired a burst of sublim at her, filling her head with sun-dappled pools, the smell of honeysuckle, desire. A yellow child tried to interest her in a jar of dead flies. Round the corner were the Alteceans in their cardigans, their conical caps of brown felt, presiding over accumulations of human refuse. On high stools they squatted, hunched in their habitual dolour, their snouts inflamed and dripping in the irritant air. They snuffled and sighed to each other, beckoning Tabitha, knowing a haulier when they saw one.

'Axis lock crystal?' she shouted. 'For a Bergen Kobold?'

The Alteceans wheezed moistly at her, waving their paws at their mounds of surplus respirators and dismantled heat-exchangers as if these treasures were all one could possibly require in life. Tabitha spent a valuable minute dragging out from under a heap something that looked promising but proved to be a caustic diffraction coil. She threw it back. She was wasting time.

Dodging a band of spacers in Shenandoah colours braying drunkenly out of a bar and shoving one another about, Tabitha pushed ahead into the crowd that lined the banks of the Grand Canal. She circumvented fat tourists in fancy dress, civic marshals in baggy overalls, then a personal camera drone, its head swivelling back and forth as it scanned the canal for its owner at home. A sailing ship was passing, its mylar sails flapping in the gusty wind. Behind it crawled a hoverbus of

MivvyCorp employees having a party. Through the rigging of the schooner a five of Palernians could be seen, making a nuisance of themselves on a flimsy raft. They were hooting and flapping their great woolly arms as they tried to climb on to a private jetty. A tall woman leaned from a balcony and emptied a bucket of water over them. Hanging over parapets and out of windows, clustering in the streets and on the rooftops, the crowd whistled and applauded.

As Tabitha was trying to get past a couple of coked-up Thrants in expensive shakos and boiled leather, one of the Palernians turned a clumsy somersault, and one of the others pushed her into the canal. They yoicked and whooped. A sparkboat sputtered by, filling the air with the smell of ozone. In it a couple in electric suits were arcing and fizzing to the hefty thump of a jumpbox. The Palernians bounded up and down in excitement, flooding the raft and endangering their coolers. As a cop arrived, his cyclops helmet protruding above the heads of the crowd, the woman was lowering her bucket at the end of a rope, shouting to a gaggle of little painted boys for a refill.

Tabitha leaned out over the railing. She could see the Moebius Strip. It was only another hundred metres: there, just beyond the float full of oversized Capellans, dummies, their huge bald heads bobbing with grave benevolence as if conferring blessings on the excited crowd.

Carnival in Schiaparelli. Cold, dusty city, full of holidaymakers and noise and smells and dirt. Wherever you go, now, you will meet people who will tell you that Schiaparelli was a fateful city for Tabitha Jute. It was in Schiaparelli that she met Tricarico, who brought her aboard the *Resplendent Trogon*, which led her into the presence of Balthazar Plum – and if it hadn't been for all that, she would never have acquired the *Alice* in the first place. Likewise, here she was now, years later, in Schiaparelli, heading for a fateful encounter which would completely and utterly change her life, my life, all our lives. She was at the top of the steps leading down to the front door of the Moebius Strip. She could see the lights inside, the drinkers and gamblers.

And then the Perks came, scurrying up the steps on all fours like rats out of a cellar.

3

Tabitha made a mistake. She made the mistake of trying to go forward, down the steps, through the upcoming Perks.

'Hey, woman! Woman watch it!'

An oily-pelted male with piercing green eyes reared up under her feet, knocking her sprawling on her bottom.

At once they were all around her, perching up on their hindlegs like scrawny otters in black leather and chrome earbands.

Not about to argue, Tabitha started to get her feet under her.

They grabbed her. Twenty thorny little paws caught hold of her jacket, her trousers, her arms. They scrabbled at her bag.

'Hey! *Get off!*'

They pulled her down on her back again. She wallowed in the flimsy gravity. As she scraped her heels against the steps, trying for purchase, the alpha male jumped up on to her hip, then down between her legs. He stood there in her crotch, weaving sinuously from side to side, hunching his shoulders, his flat little head squealing down into her face.

'Cheeeeeeee!'

Tabitha sat up fast, jerking her hips back from the snarling Perk. Several of his cousins and brothers went flying. She hauled her arm from the grip of two more and jabbed a finger at the little alien.

'Get out of my way!'

'In our way, woman.'

'Cheee!' they all went. 'Cheeeeee!'

The feathers were all bristling up the backs of their heads, on the tiny muscular shins that protruded from the legs of their breeches. They flexed their claws on their medallions, up and down the zips of their jerkins. The ones she had just knocked down were on their feet again, hopping on the steps around her. Some of them were clutching tubes of beer, bottles of chianti. The men had exaggerated their black eye-sockets with kohl and mascara. They sneered at her, baring their tiny incisors. Their breath smelled of stale fish.

'Whass'n hurry, woman?' said the Perk between her legs, taunting her. 'Missa parade!'

Tabitha realized he was drunk stupid. She cooled a degree or two. She hadn't time for a fight. Clutching her bag, she tried to get up again, but they were hanging on her shoulders.

'Get *off* me!'

'Whassa fire, woman? Whassa party, woman?'

He made a lunge at her. She threw up an arm, fending him off.

Another one, older, the barbs of his feathers going soft and ashy, burrowed under her raised arm.

'You tread on us! You'n knock us down!'

'Okay, I'm sorry! All right? I'm sorry! Now just let go of me, all right?'

She tried again to stand up. When the wiry little creatures obstructed her again, she hauled them off their feet. They all squealed, 'Cheee! Cheee!'

A couple came out of the Moebius Strip, a yellow woman in video shades and a black one in a tubecoat, basilisk teeth plaited into her hair. They glanced at Tabitha encumbered with Perks, forcing their claws from her arms, standing on one foot trying to shake one that was hanging on to her leg. The women glanced at the fracas and stepped delicately aside on the steps as they passed by. The yellow one muttered a remark to her companion, who laughed and sucked on a cigarette.

A tall man in a cloth cap came after, hurrying to catch them up. Tabitha heard his bootheels tap up the steps behind her. She winced as long black claws met in the flesh above her elbow. It was like being wrapped in barbed wire by a gang of fox terriers.

She heard something rip.

The Perks come from the third planet of a G class system in the region of

Betelgeuse, where they live in warrens, underground, which is perhaps why they took so readily to the tunnels of Plenty. There may be something endemic to the more ferocious subterranean dweller about suspicion, aggression, an unquestioning pack instinct backed up by heedless hostility to all outsiders. Leaving the deep hearth for whatever reason, hunger, duty, sexual imperatives, you trot along the lightless, complicated corridors of the buried labyrinth, their ambient odour a composite of you and all your kin. Suddenly you hear the scrabble of claws coming in the opposite direction. Friend, foe, relative, rival? Behind you lie your siblings, perhaps your own offspring, curled and mewing, tender in the warm dark. What option have you in that moment of social uncertainly but to bare your teeth, to ready your claws?

At any rate, it seems to be so for the Perks. There is nothing Perks like so much as a good fight. When the time came for civilization on the planet of the Perks, they built war-trains, undermining engines, mole bombs. It is unclear what motivated Capella to bestow the space drive on the little rodents. In all possibility the Perks merely infested their own elusive craft, following their urge to burrow into whatever comes along.

Tabitha lost all patience with them. She could see her goal ahead of her, so close she was practically inside. She had struggled halfway across Schiaparelli to get there. She was not about to stop and engage in a scrap on the very doorstep of the bar. Nor was she about to lose her jacket to a gaggle of overdressed hooligans. With a yell she thrust herself at their leader.

The neck of the Perk is very long. It accounts for the curious, rather comical way they have of standing perfectly upright and perfectly still while surveying their surroundings with a quick 240° swivel, like a furry periscope. Tabitha seized her chief aggressor by the neck with both hands. She swept him off his feet as the forward momentum of her lunge carried her upright, shedding Perks left and right with a shake of her shoulders.

All might still have gone well. Or ill, depending on your view of all that happened in consequence. But Tabitha's blood was up. She flung the choking, clawing creature from her. She flung him into the Grand Canal.

'*Cheeeeeeee*—!'

Instinctively drawing in his limbs and curling his long back, the Perk sailed out of her grasp and over the edge of the steps like a furry stone in a leather jerkin. Horrorstruck for the instant, his cronies stood and squawled with outrage. Spectators and bystanders on the canal bank turned and stared, not knowing what it was that had flashed past them, hurtling towards the water. The filthy, carmine, oily water. The water he never actually hit.

For at that moment, directly below the steps that led down to the Moebius Strip, the float of dummy Capellans was purring serenely by.

Tabitha watched in diminishing triumph and mounting dismay as the Perk fell through the smoky air and struck one of the huge statues directly on the head. With a crack audible above the gasp of the crowd, the impact smashed a large hole in the fabric of the great white dome. Knocked from its invisible supporting cradle of needle-thin tractor beams, the effigy swayed. It bowed its ruined head to its chest as if to inspect the squealing assailant now hanging from its buckled shoulder with frantic claws. It swayed, and continued to sway. Its arm fell off, clattering to the deck with the Perk still clinging to it. Its benevolently smiling head fell off and bounced with a sickening crunch from the beam projector into another of the statues, knocking it off the deck of the float and into the canal. Meanwhile, breaking apart like a toppling chimneystack, its body collapsed and felled another, which threw up an arm as it went down, as if thinking to save itself by grabbing hold of one of its remaining upright companions.

There was no hope it could save itself; nor any for Tabitha either. Standing staring appalled at the devastation she had caused, she became aware that the Perks had not instantly attacked her in retaliation for their leader's ignominious defeat. Indeed, they had melted away into the crowd. The hand that fell upon her arm was a paw; but not a tiny black-clawed paw, a hefty one with silky blue fur protruding from the sleeve of a night-black uniform.

It was the cops.

FALLING ONTO MARS

(2002)

Geoffrey A. Landis

Geoffrey Alan Landis is an American scientist and science fiction author. Born in Detroit in 1955, he attained degrees in physics and electrical engineering from the Massachusetts Institute of Technology and a PhD in solid-state physics from Brown University. He then went to work for NASA and has been employed by them ever since, participating in efforts to explore the solar system. He is the author of over four hundred academic papers, eighty short stories or novelettes, fifty poems and *Mars Crossing*, a novel. 'Falling onto Mars' was published in the July/August 2002 edition of *Analog Science, Fiction and Fact*. It won the 2003 Hugo Award for Best Short Story.

History is not necessarily what we'd like it to be...

The people of the planet Mars have no literature. The colonization of Mars was unforgiving, and the exiles had no time to spend writing. But still they have stories, the tales they told to children too young to really understand, stories that these children tell to their own children. These are the legends of the Martians.

Not one of the stories is a love story.

In those days, people fell out of the sky. They fell through the ochre sky in ships that were barely functional, thin aluminum shells crowded with fetid humanity, half of them corpses and the other half little more than corpses. The landings were hard, and many of the ships split open on impact, spilling bodies and precious air into the barely-more-than-vacuum of Mars. And still they fell, wave after wave of ships, the refuse of humanity tossed carelessly through space and falling onto the cratered deserts of Mars.

In the middle of the twenty-first century, the last of the governments on Earth abolished the death penalty, but they found that they had not yet abolished killing or rape or terrorism. Some criminals were deemed too vicious to rehabilitate. These were the broken ones, the ones too cunning and too violent to ever be returned to society. To the governments of Earth, shipping them to another world and letting them work out their own survival had been the perfect solution. And if they failed to survive, it would be their own fault, not the work of the magistrates and juries of Earth.

The contracts to build ships to convey prisoners went to the cheapest supplier. If prisoners had a hard time and didn't have quite as much food or water as had been specified, or if the life-support supplies weren't quite as high a quality as had been specified, what of it? And who would tell? The voyage was one-way; not even the ships would return to Earth. No need to make them any more durable than the minimum needed to keep them from ripping apart during the launch. And if some of the ships ripped open after launch, who would mourn the loss? Either way, the prisoners would never be returned to society.

G-g-grampa Jared, we are told, was in the fifth wave of exiles. Family tradition says Jared was a political dissident, sent in the prison ships for speaking too vigorously in defense of the helpless.

The governments of Earth, of course, claimed that political dissidents were never shipped to Mars. The incorrigible, the worst criminals, the ones so unrepentant that they could never be allowed back into human society: this was what the prisons of Earth sent to Mars, not political prisoners. But the governments of Earth are long skilled at lying.

There were murderers sent to Mars indeed, but among them were also those exiled only for daring to give voice to their dangerous thoughts.

Yet family tradition lies as well. There had been innocent men who were sent into exile, yes, but my great-great grandfather was not one of them. Time has blurred the edges, and no one now knows the details for sure. But he was one of the survivors, a skinny, ratlike man, tough as old string and cunning as a snake.

My g-g-grandma Kayla was one of the original inhabitants of Mars, one of the crew of the science base at Shalbatana, the international station that had been established on Mars long before anybody thought up the idea to dump criminals there. When the order came that the science station was to close and that they were to evacuate Mars, she chose to stay. Her science was more important, she told the politicians and people of Earth. She was studying the paleoclimate of Mars, trying to come to an understanding of how the planet had dried and cooled, and how cycles of warming and cooling had passed over the planet in long, slow waves. It was an understanding, she said, that was desperately needed on the home planet.

Great-great-grandma Kayla, in her day, had earned a small measure of fame for being one of the seventeen that had stayed on Mars with the base at Shalbatana. That fame might have helped some. Their radio broadcasts, as people fell out of the sky, nudged the governments of Earth to remember their promises. Exile to Mars was not—or at least they had claimed it was not—intended as a death sentence. The pleas of the refugees could easily be dismissed as exaggerations and lies, but Shalbatana had a radio, and their vivid and detailed reports of the refugees had some effect.

The first few years, supplies were sent from Earth, mostly from volunteer organizations: Baha'i relief groups, Amnesty International, the Holy Sisters of Saint Paul. It wasn't enough.

After the first two waves, the scientists who stayed behind realized that they would have no more hope of doing science. They greeted the prisoners as best they could, helped them in the deadly race against time to build habitats, to start growing the plants they would need to purify the air and survive.

Mars is a desert, a barren rock in space. There was no mercy in sending criminals to Mars instead of sending them to death. They could learn quickly, or die. Most of them died. A few learned: learned to electrolyze the deep-buried groundwater to generate oxygen, learned to refine the raw materials to make the tools to make the furnaces to reduce the alloys to make the machines to build the machines that would allow them to live. But as fast as they could build the machinery that might keep them alive, more waves of desperate, dying prisoners poured down from the sky; more angry, violent men who thought that they had nothing left to lose.

It was the sixth wave that wrecked the base. This was a stupid, self-destructive thing to do, but the men were vicious, resentful, and dying. A generation later, they called themselves political refugees, but there is little doubt that for the most part they were thugs and robbers and murderers. From the sixth wave came a leader, a man who called himself Dingo. On Earth, he had machine-gunned a hundred people in an apartment block that fell behind in paying him protection. On the ship, Dingo killed seven prisoners with his bare hands, simply to make the point that he was going to be the leader.

Leader he was. From fear or respect or pure anger, the prisoners on the ship followed him, and when they fell onto Mars, he harassed them, lectured them, beat them, and forged them into an angry army. They had been abandoned on Mars, Dingo told them, to die slowly. They could only survive if they matched the Earth's brutality with their own. He marched them five hundred kilometers across the barren sands to the Shalbatana habitat.

The habitat was taken before the inhabitants had even realized it was under attack. The scientists who hadn't abandoned the station were beaten with scraps of metal from the vandalized habitat, blindfolded, and held as hostages while the prisoners radioed the Earth with their demands. When the demands were unanswered, the men were stripped and thrown naked out onto the sands to die. In rage and desperation, the mob that had been the sixth wave ripped apart the base, the visible symbol of the civilization that had sent them a hundred million miles to die. The women who remained on the base were raped, and then the destroyers gave them the chance to plead for their lives.

The men of the fourth and fifth waves had joined together. For the most part, they were strangers to each other—many of them had never seen each other's faces except through the reflective visor of a suit. But they had slowly learned that the only way to survive was to cooperate. They learned to burrow under the sand, and when their home-made radios told them the base was being sacked, they crept across the desert, and silently watched and waited. When the destroyers abandoned the base after looting it of everything they thought was valuable, the fifth wave, hiding under the sands, burst out and caught them unprepared. Of the destroyers who had attacked Shalbatana base, not a single one survived. Dingo fled into the desert, and it was Jared Vargas, my great-great grandfather, who saw him, tracked him down, and killed him.

And then they went to Shalbatana base, to see whether anything could be salvaged.

G-g-grandpa found her in the wreckage and ripped the tape off her eyes. She looked at him, her eyes unable to focus in the sudden light, and thought him one of the same group that had raped her and sabotaged the habitat. She had no way of knowing that others from his group were frantically working to patch up one of the modules to hold air, while g-g-grandpa and others searched for survivors. As the leaking air shrieked in her ears, she looked up at him, blinking, blood running from her nose and ears and anus, and said, "You have to know before I die. Oxygen in the soil. Release it by baking."

"What?" g-g-grandpa said. It was not what he had expected to hear from a naked, bleeding woman who was about to pass out from anoxia.

"Oxygen!" she said, gasping for breath. "Oxygen! The greenhouses are dead. Some of the seedlings may have survived, but you don't have time. You need oxygen now. You'll have to find some way to heat the regolith. Make a solar furnace. You can get oxygen by heating the soil."

And then she passed out. G-g-grandpa dragged her like a sack of stones to the one patched habitat module, and shouted, "I found one! ?Está viva! I found one still alive!"

Over the following months, Jared held her when she cried and cursed, nursed her back to health, and stayed with her through her pregnancy. Theirs was one of the first marriages on Mars, for although some women had been criminals infamous enough to be sentenced to Mars, still the male prisoners outnumbered the females by ten to one.

Between them, the murderer and the scientist, they built a civilization.

And still the ships came from Earth, each one more poorly built than the last and delivering more corpses than living men. But that was in its way a blessing, for the men would mostly die, while the corpses, no matter how emaciated, had valuable organic content that could turn another square meter of dead Martian sand into greenhouse soil. Each corpse kept one survivor alive.

Thousands died of starvation and asphyxiation. Thousands more were murdered so the air that they breathed could be used by another. The refugees learned. Led by my great-great-grandfather and grandmother, when a ship fell to Mars, they learned to rip it apart to its components before its parachutes had even settled. Of its transportees–well, if they couldn't breathe vacuum (and the thin Mars air was never more than dust-laden vacuum), they had better scramble.

Only the toughest survived. These were mostly the smallest and the most insignificant, the ones like rats, too vicious and too tenacious to kill. A quarter of a million prisoners were sent to Mars before the governments of Earth learned that behavior-modification chips were cheaper than sending prisoners to Mars, and tried their hardest to forget what had been done.

My great-great-grandfather Jared became the leader of the refugees. It was a brutal job, for they were brutal men, but he fought and bullied and connived to lead them.

There are no love stories on Mars; the refugees had no time, no resources for love. Love, to the refugees, was an unpredictable disease that strikes few people and must be eradicated. To the refugees, survival required obedience and ceaseless work. Love, which thrives on individuality and freedom, had no place on Mars.

Yes, Jared Vargas was a dissident sent from Earth for speaking against his government. But Jared Vargas died in the desert. When the men of the fifth wave came to the rescue of the Shalbatana habitat, Jared Vargas had chased Dingo into the desert, and that had been the last mistake of his life. Only one of them returned from the desert, wearing the suit of Jared Vargas, and calling himself by the name of Jared Vargas. No one recognized him, but the men of the fifth wave were from a dozen ships, and if any of them had been friends of the original Jared Vargas, they died after the new Jared Vargas returned from the desert. And the only men who would have recognized Dingo were the exiles of the sixth wave, and they were all dead.

He returned from the desert, and rescued my great-great-grandmother, and the men of the fifth wave accepted him.

But surely my great-great-grandmother was not fooled. She was an intelligent woman—brilliant, in her own field—and she must have realized that the man who claimed her for his wife was the same man who had led the army of angry rabble to rape her, rip apart her base, and laugh as they watched her friends die in the thin air of Mars.

But Mars required survival, not love. And Jared Vargas was the only leader they had.

There are many stories from the days of the first refugees on Mars. None of them are love stories.

ENCHANTING LUNA AND MILITANT MARS

(2010)

Bruce R. Johnson

Bruce R. Johnson is Associate Pastor at the Mountain View Presbyterian Church in Scottsdale, Arizona, having earned a Doctor of Ministry degree from the Fuller Theological Seminary. He is an authority on the works of C. S. Lewis, President of the Arizona C. S. Lewis Society and the General Editor of *Sehnsucht: The C .S. Lewis Journal*. The current focus of his research is C. S. Lewis's work with the Royal Air Force Chaplains' Branch during the Second World War.

C. S. Lewis had a lifelong fascination with the medieval imagery of the heavens: seven planets circling the earth in a grand celestial dance. Of course, we now know this model is not true. Lewis nevertheless believed it was instructive in its geocentric simplicity and its hierarchical complexity. Moreover, he valued the planets, and the pagan gods associated with them, as spiritual symbols. As he once wrote to Ruth Pitter, "It was beautiful, on two or three successive nights about the Holy Time, to see Venus and Jove blazing at one another, once with the Moon right between them: Majesty and Love linked by Virginity—what could be more appropriate."[1] For Lewis, the cast of celestial bodies would play recurring roles in his poetry, his trilogy of space novels, and in his introduction to medieval and Renaissance literature, *The Discarded Image*.

Recently, Michael Ward has stirred popular interest in Lewis and medieval cosmology through his BBC television documentary "The Narnia Code," based on his seminal book, *Planet Narnia*. In both accounts, Ward advances the theory that each of the seven books of *The Chronicles of Narnia* was deliberately written to embody the qualities of one of the seven medieval planets. An intriguing aspect of this theory is what Lewis meant to say about God, and God's intent for humanity, through the sevenfold lens of these planetary gods. As Ward writes:

> This theological disposition is worked out in each of the Chronicles as the children, who by the common grace of 'nature' are already part of a planetary world, become more so by special grace as they follow the

> planetary deity's leading. Thus, in *The Lion* they become monarchs under sovereign Jove; in *Prince Caspian* they harden under strong Mars; in The *'Dawn Treader'* they drink light under searching Sol; in *The Silver Chair* they learn obedience under subordinate Luna; in *The Horse and His Boy* they come to love poetry under eloquent Mercury; in *The Magician's Nephew* they gain life-giving fruit under fertile Venus; and in *The Last Battle* they suffer and die under chilling Saturn.[2]

As the debate surrounding Ward's theory continues—all the more so with the recent release of a book version of *The Narnia Code*[3]—a reexamination of Lewis' shorter planetary fiction seems in order. In his lunar tale, "Forms of Things Unknown," and in his Martian story, "Ministering Angels," what aspects of the older cosmological model can be discerned? Are they in keeping with what Ward claims are the essences of Luna and Mars, as Lewis understood them? And what, if any, theological points may Lewis be making in these two tales? The present study will address each of these three issues. It will also attempt to answer the two main objections put forward about these stories: whether "Ministering Angels" is misogynistic, and whether "Forms of Things Unknown" is an authentic Lewis tale.

What aspects of the medieval model can be discerned in these two stories? "Ministering Angels" was first published in 1958 and deals with an outpost of scientists on Mars who receive a surprise visit from two self-selected "comfort women." This space farce was written in direct response to Robert Richardson's controversial article, "The Day after We Land on Mars," in which no role for women in space was foreseen outside of prostitution.[4] Lewis grasps Richardson's bad idea head on, exposing the flawed logic of his chauvinistic male fantasy. Lewis wrote with a Martian militancy; like the *persona* "MARS mercenary" in his poem "The Planets," he "flaunts laughingly" while he attempts to see "The wrong righted."[5]

The story itself reflects a tone which seems fitting for the god of war. All the astronauts based on Mars are battling discords of one sort or another. The meteorologist, nicknamed "the Monk," is seeking to right the internal wrong in his soul. The botanist is pressed for time and battling interruptions. Peterson and Dickson face discord arising out of the former's unrequited homosexual advances towards the latter. The Captain resists accepting his worst fears, as he thinks of his new bride back on earth. A different Martian quality is found in the Thin Woman, who drops the not-so-enticing pick-up line, "immorality . . . must no longer be regarded as unethical."[6] This is the indifference of Mars the mercenary whom Lewis poetically portrayed as a "hired gladiator/ Of evil and good. All's one to Mars."[7] Elsewhere Lewis explained how "sturdy hardiness" was also part of the Martian temperament.[8] This attribute seems most characteristic in the Monk. He is the steadiest character, the one most able to adjust to changing circumstances. It is he who suggests a meal when the conversation becomes awkward, who speaks with grace to the rejected Fat Woman, and who eventually adjusts his outward focus in light of changed reality. The story set on Mars is appropriately Martian.

The lunar tale, "Forms of Things Unknown," was published posthumously in 1977 as part of the collection of short fiction by Lewis, entitled, *The Dark Tower and Other Stories*. It is the story of the fourth of four manned voyages to the moon, all of which end in disaster. The reason for this—and the surprise ending to the story—is the presence of a lunar Gorgon, whose hair of writhing snakes transforms each astronaut into stone as he turns to look backward. The classic tales of Medusa and the other Gorgons associate them with marine rather than lunar deities. Their lunar quality is rather found in the Gorgons' ability to change one thing into another. For ancient writers, Luna bordered the realm of mutability. Her sphere is "the great frontier between air and aether," between Aristotle's ever-changing "Nature" and never-changing "Sky."[9] The lunar cycle itself appears more like corruptible earth than the eternal heavens. Luna's influence on people was thought to be related to change. Hence Lewis wrote, "In men she produces wandering, and that in two senses. She may make them travelers. . . But she may also produce a wandering of the wits."[10] Luna both charms and deceives, often hiding her true nature. She "Enchants us—the cheat!"[11] She is also associated with liquidity: Lewis waxes that Luna "Cruises monthly; with chrism of dew / And drenches of dream, a drizzling glamour."[12]

All these lunar qualities are present in "Forms of Things Unknown." Lieutenant Jenkin is wandering, cut adrift from a relationship with "that girl" (even her proper name was crossed out by Lewis in the original manuscript). Jenkin has rejected the influence of Venus (love) and come under the influence of Luna (virginity). He also distances himself from his friend Ward. Ward offers to come along on the lunar voyage, but Jenkin insists on making the trip alone, believing isolation will keep him safe. Of course Jenkin wanders physically—as far from home as humanly possible. The lunar theme of liquidity makes the briefest of appearances through the two pints of draught Bass, ordered up by Jenkin at a pub. Lewis believed Bass to be "the lowest level surely that beer can attain,"[13] and perhaps meant this to be, tongue in check, a harbinger of upcoming madness. In his lunar voyage, and as he walks on the Moon's surface, Jenkin struggles against lunacy and fear. The prose is peppered with expressions like *claustrophobia, agoraphobia, terror, terrors, terrifying, fear, and frightening*. Earlier on Jenkin speculates, "Might there be something on the Moon—or something psychological about the experience of landing on the Moon—which drives men fighting mad?"[14] Later, he misquotes Pascal's line about "the silence of those eternal spaces," leaving off his closing point: it "frightens me."[15] Terror is approaching. Like Luna herself, the lunar Gorgon lulls the doomed astronaut into a false sense of euphoria. For a few moments, he believes a race of lunar artisans has sculptured three life-like tributes to the astronauts who preceded him. The truth is far different. When on the moon the name of one previous astronaut, Fox, is mentioned three times, perhaps recalling the Teumessian fox which was changed into stone and set among the stars. Jenkin's own fate is prefigured by prior appearances in the story of the words "petrified" and "animated stones." In the last sentence, his doom is sealed: "His eyes met hers."[16] The story set on the Moon is appropriately lunar.

To what extent do these interpretations coalesce with Ward's assessment of Lewis' approach to medieval cosmology? The breadth of Ward's scholarship on Lewis is impressive, drawing on his fiction, nonfiction, poetry, literary criticism, essays, letters and diary. So, it may be more productive to ask what parts are missing. More specifically, which parts of Lewis' medieval model for Luna and Mars, as understood by Ward, are not reflected in "Ministering Angels" and "Forms of Things Unknown"? Additionally, do these inconsistencies present a problem for Ward's theory of a planetary scheme for *The Chronicles of Narnia*?

Ward is able to make good use of the metals associated with the gods in classical and medieval writing. In discussing *Prince Caspian*, he is able to point out important references to the Martian metal iron.[17] He also demonstrates that the lunar metal silver appears repeatedly throughout *The Silver Chair*.[18] Neither metal, however, appears in Lewis' shorter planetary stories. Nor can the animals Ward associates with Mars be found there, including the wolf, woodpecker and horse.[19] Of course terrestrial animals are not likely to appear in stories set in other worlds, except by such clever devices as using "Fox" as a surname. Verticality, a pervasive Martian quality in *Out of the Silent Planet*,[20] is absent from "Ministering Angels." The reflective and subordinate quality of Luna[21] is hard to discern in "Forms of Things Unknown," unless the opening voice of Jenkin's "instructor" is meant to imply that it emanates from his commanding officer. Yet this is nitpicking. Any of these absences is more than offset by the presence of numerous other planetary qualities. In fact, Ward is not only able to delve into qualities cited previously in this article, he expands the list.

Ward notes, for example, that "Necessity is a major feature of the symbolic value that Lewis, as a poet, located on Mars. The god of war is 'necessity's son,' as he put in 'The Planets.'"[22] The Thin Woman in "Ministering Angels" speaks of sex as an "indispensible function"[23] and believes her presence on Mars is a logical necessity. "Part of the Martian spirit," says Ward, "is a ranked and patterned orderliness."[24] The Monk in "Ministering Angels" is committed to the "slow, perpetual rebuilding" of his "inner structure," keeping the discipline of the Divine Hours, and looking to God as his "Gentle and patient Master."[25] Ward also looks at the range of incarnations of Mars and Luna for more symbolic clues. An early conception of the Martian deity was Mars Silvanus, a tutelary spirit of woods, flocks, and fertility.[26] References to vegetation abound in *Prince Caspian*, while in "Ministering Angels" the Botanist is feverishly cataloguing Martian flora, from "hardy organisms" to the deathly "Martian cress." Indeed, in the subplot, where two deaths stem from the consumption of Martian cress, Lewis may be trying to combine Mars Silvanus, the god of the forest, with Mars Gradivus, the god of war (since war brings on death). Ward lays out the range of Lunar goddesses which appear in Lewis' early poetry, including Artemis who helps the "hunted,"[27] and Diana "the pure Huntress riding low."[28] In "Forms of Things Unknown," Jenkin begins to "hunt" on the lunar surface, unaware that the female Gorgon lies behind him, hunting him.

Through connecting Mars with the ideal of the medieval Christian knight and the whole range of chivalry, Ward's ideas shed particular light on the Monk. He quotes Lewis in *Mere Christianity*: "The idea of the knight—the Christian in arms for defence of a good cause—is one of the great Christian ideas."[29] Soon after, he adds, "knightliness is evident as much in gentleness as in hardiness."[30] Those well disposed to God's refining work, as filtered through the lens of the Mars *persona*, are characterized by discipline, obedience, faithfulness, strength and growth.[31] These traits are consistent with the character whose prayer concludes the story of "Ministering Angels": "Oh Master . . . forgive—or can you enjoy?—my absurdity also. I had been supposing you sent me on a voyage of forty thousand miles merely for my own spiritual convenience."[32] Far from being a stumbling block, Lewis' shorter planetary fiction lends more credence to Ward's underlying premise: "Lewis's love of the Ptolemaic cosmos"[33] led to his repeated use of the planets as symbols in his writing.

Before turning to consider the other spiritual lessons embedded in Lewis' shorter planetary fiction, it may be helpful to consider some of the objections raised by these stories. Shortly after publication of *The Dark Tower and Other Stories*, reviewers began to raise the claim of misogyny, especially in regards to aspects of "Ministering Angels." The writer and poet Ursula LeGuin, perhaps illustrating the deep chasm that had by then opened up between her generation of "second wave" American feminists and Lewis' earlier—and more traditional—British view of women's roles, firmly disliked the story: "The spitefulness shown to women in these tales is remarkable," she wrote, "but the authentic Inside Club (MCP [Male Chauvinist Pig] branch) tone is at its braying clearest in 'Ministering angels,' a humorous piece." She was particularly offended by the portrayal of the Thin Woman:

> However petty, this is hate; the depth of it is proved in the final paragraph, where the Christian member of the team blissfully contemplates the conversion and salvation of the decrepit whore, but never gives a thought to the soul of 'the lecturer at a redbrick university.'[34]

The American Roman Catholic writer Charles Andrew Brady, though sympathetic to Lewis and his writings, nevertheless pointed out that the end of the story, "hardly absolves Lewis from the charge of a bachelor's misogyny which is one of his few flaws, and one not to be found in his major books."[35] Even Katherine Harper's more recent (and favorable) treatment of Lewis' shorter fiction finds serious flaws. "'Ministering Angels'," she writes, damningly, "is decidedly misogynistic; as were so many male scholars of his time (and before, and since), Lewis was unconvinced of women's ability to reason."[36]

Given the nature of the criticism that surrounds them, it can be hard to remember that Lewis originally composed this story in defense of women. He was, he believed, engaged in chivalrous, literary combat. His weapon was satire. When the reprinting of "Ministering Angels" was being considered in 1961 for inclusion in a larger collection of his stories and essays, Lewis wrote to his publisher, "You wd. need to reprint their headnote to make it fairly intelligible."[37] That headnote, by Anthony Boucher, who served as the editor of *The Magazine of Fantasy and Science Fiction*, reads as follows:

> Dr. Robert Richardson's controversial article, The Day after We Land on Mars—first published in the *Saturday Review* and later expanded for *F&SF* (December 1955)—contained the provocative prediction that "we may be forced into first tolerating and finally openly accepting an attitude toward sex that is taboo in our present social framework. . . To put it bluntly, may it not be necessary for the success of the project to send some nice girls to Mars at regular intervals to relieve tensions and promote morale?" C. S. Lewis takes it from there in his first short story of space travel—a tale of the First Martian Expedition which is perceptive, human, and warmly comic.[38]

Richardson worked at the Mount Wilson Observatory in the San Gabriel Mountains above Los Angeles, and wrote science fiction under the pen name Philip Latham. His article, "The Day after We Land on Mars,"[39] was, however, a work of nonfiction. Richardson was completely serious in his belief that women should periodically be sent into space to relieve the sexual needs of male astronauts—in a ratio of one woman to quite a few men. Boucher's 1955 headnote declared the article to be "an admirable brief refresher course on the probable first steps of interplanetary travel, which leads, with deceptive simplicity, to certain highly provocative conclusions."[40] Apparently, Boucher also smelled an opportunity for publicity. The following January, the same magazine contained a story by the American historian and science fiction writer Paul A. Carter, which explored in fiction the very conclusions which the astronomer suggested. Boucher publicized it as "possibly the first honest fictional study of human sexual mores on another planet."[41] Perhaps ironically, the February 1956 issue of that same magazine contained the story "The Shoddy Lands" by Lewis.[42]

With the publication of this final story—a fantasy tale of a man seeing the world through the mind of a woman—Lewis had a problem. It appeared in the same magazine whose two previous issues contained articles promoting a view of men and women in space that he found morally reprehensible. "Ministering Angels" was his response, and was directed at Richardson in particular. Many elements of the setting and plot of "Ministering Angels" were taken directly from "The Day after We Land on Mars." Richardson envisioned a colony on Mars consisting entirely of "young unmarried men."[43] The trip from earth to Mars and back would take many months to complete.[44] Once they arrived, only a few men could work outdoors at a time.[45] Biologists would be busy examining *maria*,[46] but much of the other work would be monotonous.[47] Conflicts arising from close contact were, he concluded, sure to arise,[48] and it was certainly possible that some astronauts would be tempted to engage in homosexual activity.[49] Lewis, in making his Captain a newly married man, departs from Richardson's scheme of exclusively unmarried astronauts. In every other respect, the conditions in "Ministering Angels" match those found in "The Day after We Land on Mars." At the beginning of the story, Clifford Patterson, one of the young male technicians and a homosexual, is trying to attract the interest of Bobby Dickson, the other male technician. "The only part of any woman that interested him [Patterson] was her ears. He liked telling women about his troubles; especially the unfairness and unkindness of other men."[50] At the end of the story, Patterson is attempting to talk through his male troubles with the Thin Woman, who is comically trying to flirt with him at the same time. The irony of her addressing the one man on Mars completely immune to her propositions symbolizes the fruitlessness Lewis saw in Richardson's entire scheme. As he remarked in regards to his own Space Trilogy "I wanted farce as well as fantasy."[51]

Within this context, it is not surprising that Lewis made the Thin Woman such an unsympathetic character. In the story, she is the main proponent of Richardson's new morality (or lack thereof). What remains strange, however, is where Lewis locates the earth-bound proponents of the scheme: "the idiots on the Advisory Council," namely, "a pack 'o daft auld women (in trousers for the most part) who like onything sexy, onything scientific, and onything that makes them feel important."[52] Because Lewis was challenging a male fantasy composed and published by two male writers, one wonders why the Advisory Council was not filled with "daft auld men"? If a charge of sexism is to be leveled at the story, it would be stronger to focus criticism on this aspect. Lewis, however, may have responded in a similar fashion to his reply to Professor Haldane:

> It was against this outlook on life, this ethic, if you will, that I wrote my satiric fantasy, projecting in my Weston a buffoon-villain image of the 'metabiological' heresy. If anyone says that to make him a scientist was unfair, since the view I am attacking is not chiefly rampant among scientists, I might agree with him: though I think such criticism would be over sensitive.[53]

Substitute "the Thin Woman," "new ethics," and "women" for the words "Weston," "metabiological, and "scientists," and one is probably close to the views of Lewis. But perhaps a better way forward has been laid out by Diana Glyer in her recent (and aptly titled) essay, "'We are *All* Fallen Creatures and *All* Very Hard to Live With': Some thoughts on Lewis and Gender":

> It seems to me that the principal text in any discussion of Lewis's views toward women is not to be found in a private statement made in a letter, an isolated public statement made in an essay, or a comment that issues forth from a character in his fiction. As position statements, these myriad bits and random pieces are simply unreliable. . . . [M]ore significant weight should be placed on the fuller and more telling "text" of how Lewis lived his life.[54]

Employing this broader and more inclusive standard, Glyer finds Lewis to be a model in promoting unity, liberty and love among women and men.

The principal objection to "Forms of Things Unknown" is of an entirely different nature. Did Lewis actually write it, or is it a clever forgery? Kathryn

Lindskoog, the primary advocate for doubting its authenticity, offered three main reasons for her position. First, it is "an awkward, amateurish, rather mean spirited story that doesn't sound like Lewis."[55] Second, she believed the inspiration behind the story was Virgil Finlay's cover illustration for the October 1958 issue of *Fantastic Universe* magazine. If that is the case, she asks, "Why would he [Lewis] waste his time turning Finley's first-rate pictorial idea into a third-rate story?"[56] Third, she rejects Walter Hooper's purported claim that "he had rescued it along with the *Dark Tower* fragment from Warren's January 1964 bonfire."[57]

As to Lindskoog's first point, she may have meant that "Forms of Things Unknown" is simply not as good as Lewis' major writings. Few would disagree. In this regard, however, it is instructive to remember that Lewis believed not every line of Shakespeare's was "good,"[58] and that Tolkien astutely observed that not every work by Lewis was either. Lindskoog becomes more specific in labeling one word and one phrase found in the story as "unlikely" to have come from Lewis. Yet, the hints of the medieval model of Luna—change, wandering, virginity, madness, deception, hunting and, perhaps, liquidity—all seem characteristic of Lewis.

Lindskoog's second objection, that the story was based on an illustration from a magazine, is difficult to verify or dismiss. If the story could be dated with certainty prior to October 1958 (the date of Finlay's cover illustration for *Fantastic Universe*), then, obviously, the theory would be false. Absent that information, the theory remains intriguing. But why would this lead to the conclusion that the story is a forgery? Charles Brady, who first mentioned the connection to Finlay's illustration, remained convinced that Lewis was the author of "Forms of Things Unknown."[59] In the early 1950s, Alastair Fowler observed Lewis reading from *Astounding Science Fiction*,[60] and later in the same decade Lewis both read and wrote stories for *Fantasy and Science Fiction*. Clearly, Lewis "wasted" a fair amount of time consuming a variety of science fiction, some of which he regarded as "abysmally bad" and some which contained "real invention."[61]

Lindskoog's third charge stems from her doubt that Hooper actually rescued Lewis' manuscripts from the flames of his brother's bonfire. She also challenged the authenticity of other writings attributed to Lewis but published posthumously, including "After Ten Years" and *The Dark Tower*. Significantly, however, Alastair Fowler has since revealed that Lewis let him read both of these unfinished stories, among others, prior to his death in 1963.[62] Fowler's testimony lends considerable credence to Hooper's account. His statement in 1977 thus seems reasonable enough: that the manuscript of "Forms of Things Unknown" was "discovered among the papers given me by Major [Warren] Lewis."[63]

Having addressed the objections raised about these two stories, we can now turn to the particular theological lessons inherent in them. The Martian tale begins with a man at prayer and is designed, as a whole, to defend the Christian ideals of chastity against those who would objectify women sexually. Sanctification is at work as the Monk prays for God's help in the rebuilding of his soul. He both receives and holds out to another the promise of God's forgiveness. There is grace present as he refers to the old prostitute as "daughter," and the hope of redemption can be found in his words, "you are not far from the Kingdom." The

Monk's closing prayer displays growth, charity and humility.

The theme of humility is played out a second way. The Thin Woman's besetting sin is not fuzzy thinking, but pride. Pharisees are in more spiritual danger than harlots. As Lewis puts it, "Prostitutes are in no danger of finding their present life so satisfactory that they cannot turn to God: the proud, the avaricious, the self-righteous, are in that danger."[64]

The Monk also acts as a spiritual director to the Fat Woman, and there are several other characters in Lewis' writings who take on a similar role. Most notable of these are Ransom to Jane Studdock in *That Hideous Strength*, the character of George MacDonald to the *persona* of Lewis in *The Great Divorce*, and the *persona* of Lewis to Malcolm in *Letters to Malcolm: Chiefly on Prayer*. Yet the Monk's advice does not sound very similar to the words of these other three fictional spiritual directors, whose counsel is delivered at an appropriately high intellectual level. The Monk's words, on the other hand, are very different—and appropriately so, given that the Fat Woman, unlike these other spiritual apprentices, has not had the advantages of a privileged education. She has, in fact, experienced very few advantages. So the Monk's words to her are humble, simple without being simplistic. Typical of Lewis, there is a brief and vivid illustration to drive home the point being made. "Daughter . . . you are not far from the Kingdom. But you were wrong. The desire to give is blessed. But you can't turn bad bank-notes into good ones just by giving them away."[65]

These words are more similar in style to the actual spiritual advice Lewis occasionally proffered in his correspondence. People who read his books often sought his council. Lewis displayed the patience of a saint in responding to the many hundreds of letters he received each year from admirers and spiritual seekers throughout the world. The following counsel sent to an American girl, Joan Lancaster, is illustrative of these efforts: "Duty is only a substitute for love (of God and of other people), like a crutch at times; but of course it's idiotic to use the crutch when our own legs (our own loves, tastes, habits etc) can do the journey on their own![66]

The theology contained in "Ministering Angels," especially that found in the character of the Monk, reflects Lewis' ongoing concern for the spiritual welfare of others. When attention is shifted to the lunar story, "The Forms of Things Unknown," the overt theology disappears. Consequently, critical evaluation of the story's theological nature must remain tentative. However, like Luna herself, deeper meanings may be embedded in the story. If so, the theological key for this lunar story may lie in the internal reference to the Cretan labyrinth and in the external discussion by Lewis of the Hippolytus myth.

"The Forms of Things Unknown" can, in fact, be read as a warning against cutting oneself off from others. As Jesus and St. Paul (among others) repeatedly emphasized, no one is spiritually self-sufficient. To believe otherwise is madness. As Lewis inquired of Dom Bede Griffiths, "Is any pleasure on earth as great as a circle of Christian friends by a good fire?"[67] In commenting on *Hamlet*, a play haunted by madness, Lewis observed, "The next best thing to being wise oneself is to live in a circle of those who are."[68] Before reaching the moon, Lieutenant Jenkin has already discerned many pieces of the lunar puzzle: fight madness, do not look behind you, and report on what you see. Yet, he fails to assemble these pieces in time. He has not learned the lessons of the old stories. He recalls the Cretan labyrinth,[69] but forgets that Theseus was unable to survive his perils on his own. He had the assistance of Ariadne and her ball of string, which, after battling the Minotaur, helps Theseus find his way out of the labyrinth. Jenkin, however, cut off from friend and lover, will not survive the perils that await him on the Moon. He will never find his way home.

In 1954, Katherine Farrar, wife of the famous Oxford theologian and preacher, sent Lewis a story she had written in which the moon was described, "like the white face of an idiot lost in a wood." Lewis chided her severely for denigrating "the high creation of God," suggesting that as penance she should memorize Psalm 136, which praises God who created "The moon and the stars to govern the night" (136:9). He continued, "Not safe, either, to be rude to goddesses—Artemis still owes Aphrodite a come-back for the Hippolytus affair and we shd. hate you to be the target."[70] In Greek mythology Hippolytus, the son of Theseus, found himself in the middle of a love triangle with these two goddesses. Hippolytus preferred Artemis. Aphrodite was not pleased and her trickery eventually led to the death of Theseus' young wife Pheadra and of Hippolytus himself.

In the end, the revenge of the moon maiden, Artemis, would not fall on Katherine Farrar. In his next letter to her, Lewis apologized for "mounting too high a horse about the 'idiot-moon.'"[71] The revenge falls instead on Lieutenant Jenkin. To cut off one's self from others is not to find safety, it is to invite disaster. It is not isolation but love which casts out fear.

Lewis once complained to Arthur C. Clarke, the English science fiction writer, futurist, and inventor, "What's the excuse for locating one's story on Mars unless 'Martianity' is through and through used," explaining in a footnote that he meant used "Emotionally & atmospherically *as well as* logically."[72] In "Ministering Angels," and in "Forms of Things Unknown," Lewis produced stories fit for their settings. His Martian story is militant in its defense of a good cause. His lunar tale both enchants and deceives to the very last sentence. Ward's pioneering work on Lewis and medieval cosmology adds considerably to our understanding of these shorter planetary tales. Lewis drew on a long history of traditions about Mars and Luna, hoping that his readers would not only be entertained but enlightened. Above all else, he hoped one day they would end up "bright in the land of brightness."[73]

PROJECTING LANDSCAPES OF THE HUMAN MIND ONTO ANOTHER WORLD (2012)

CHANGING FACES OF AN IMAGINARY MARS

Rainer Eisfeld

Born in 1941 in Berlin, Germany, **Rainer Eisfeld** is a German political scientist who holds the position of Professor Emeritus at the University of Osnabrück. He studied economics at the University of Saarbrücken and received a PhD in political science from the Goethe University of Frankfurt in 1971. Across his academic career, Eisfeld has authored or edited more than fifteen books on political science, sometimes looking at popular fiction and its relevance in political discussions about perceptions of democracy.

Reporter: 'Is there life on Mars?'

Returning Astronaut: 'Well, you know,
it's pretty dead most of the week,
but it really swings on Saturday night.'

(Popular NASA joke)

1 Deceptive World

For centuries, the planet Mars continued to deceive terrestrial observers like no other celestial body in our solar system. Believing to discern ever more distinct features on Mars through Earth-bound telescopes, astronomers designated these as continents, oceans, even canals, to which they gave names. With exceedingly rare exceptions, however, the markings did not correspond to geomorphological, or rather areomorphological, structures. Actually, they originated from the different reflectivity of bright and dark surface regions changed in its turn by wind activity which has continued to transport and deposit fine dust across the planet. Space probes, rather than telescopes, were needed to explain these processes and to shed light on the Red Planet's true characteristics[1].

Until robotic explorers arrived, no other planet seemed to offer such clues for educated guessing – first to the conjectural astronomy of the 19th century, subsequently to the science fiction of the latter part of that period and the 20th century. Conjectural astronomy was the term used, in the wake of Bernard de Fontenelle's 1686 *Conversations on the Plurality of Worlds* and Christiaan Huyghens' 1698 *The Celestial Worlds Discover'd, or Conjectures Concerning the Habitants, Plants and Productions of the Worlds in the Planets*, to denote that branch of the discipline which engaged in hypothesizing on 'the living conditions and natural environments of other celestial bodies'[2]. While expected to be not directly contradicting astronomical observations, such suppositions were, to a high degree, matters of interpretation, often based on 'few definitely established and unambiguous data'[3].

In contrast to the discipline's mathematical branch, conjectural astronomy was intended to bridge the widening rift of mutual incomprehension between the humanities and the sciences. From the 17th to the 19th century, the encyclopaedic outlook on learning, so central to the Enlightenment, included both the spiritual and the material world. Inexorably, however, the progress of scientific research fostered specialization. Conjectural astronomy, in contrast, increasingly resorted to manifest speculation, relegating stellar and planetary astronomy to the role of ancillary sciences in the service of a pre-conceived, stoutly held idea, based on philosophical considerations: that intelligent life existed throughout the universe, including the solar system's planets.

German astronomers Wilhelm Beer and Johann Heinrich Maedler's mid-19th century assumption that it would 'not be too audacious to consider Mars, also in its physical aspects, as a world very akin to our earth'[4], went unchallenged in its time. By 1906, however, when American astronomer Percival Lowell published his spectacular – and highly speculative – interpretation *Mars and its Canals*, scientists debated issues such as the composition of the Martian atmosphere or the planet's climate controversially and much more sceptically. Within a year, a devastating rebuttal by British biologist Alfred Russel Wallace appeared under the title *Is Mars Habitable?* Wallace answered the question in the negative: Realistic temperature estimates pre-cluded animal life; low atmospheric pressure would make liquid water – let alone Lowell's supposed irrigation works – impossible. Science and fiction were irrevocably parting ways.

A mere decade after Beer and Maedler had published their treatise on the solar planets' physical properties, the term 'Science-Fiction' was introduced in 1851 by British essayist William Wilson in his work *A Little Earnest Book upon a Great Old Subject*. When coining the expression, Wilson referred to a 'pleasant story', 'interwoven with... the revealed truths of Science', itself 'poetical and true'. By the 1890s, the emerging genre included not merely pleasant, but definitely unedifying tales putting mankind at the mercy of technically superior beings from other celestial bodies. The planetary novel was coming into its own: No longer were planets conceived as self-contained distant places. Rather, their inhabitants might seek out other worlds with either benevolent or inimical intent[5].

Mars, supposedly older than the Earth (according to what was then believed about the formation of the solar system), particularly fired the imagination. Intersecting around the turn of the century, conjectural astronomy and science fiction served as vehicles for succeeding generations to 'project [their] earthly hopes and fears' onto Mars[6]. These pipe dreams and nightmares came to vary not least according to the economic, social, and political upheavals that would figure uppermost in men's minds during successive periods. Two examples illustrate the remarkable length to which some authors were prepared to go in offering their allegories:

In the wake of the October Revolution, Soviet writer Alexei Tolstoi and movie director Yakov Protazanov imagined during the early 1920s that it would take the arrival by spaceship of a terrestrial revolutionary, Gusev, to whip the exploited workers of Mars into a proletarian uprising (*Aelita*). The 'world' to be revolutionized by the 'vanguard of the proletariat' did not need to be identical with Earth...

By the mid-1950s, with female emancipation considered a dire threat in many quarters, a British flick portrayed Nyah, *Devil Girl from Mars*, as landing her flying saucer by a country tavern, explaining to the male customers that the birth rate had fallen alarmingly after the introduction of matriarchy. For breeding purposes, her planet needed men! Rather than, in post-Victorian resignation, 'closing their eyes and thinking of Mars', however, the British males put up embittered resistance[7].

Looking at the 'mainstream' (there were always mavericks, of course) of the ways in which the treatises, the novels and short stories, the movie scripts by successive generations of astronomers and science fiction writers depicted an imaginary Mars, we may discern a sequence of 'faces' attributed to the planet on which this chapter's subsequent sections will focus:

An *Arcadian Mars* (1865 ff.) exhibiting 'all the various kinds of scenery which make our earth so beautiful'; a highly civilized *Advanced Mars* (1895 ff.) criss-crossed by immense canals; a forbidding *Frontier Mars* (1912 ff.) where the rugged adventurer and the toiling pioneer might again come into their own; a *Cold War Mars* (1950 ff.), source of an assault on the Earth, or haven for refugees after our planet would have perished from nuclear war; finally, a *Terraformed Mars* (1973 ff.), again with strong frontier undertones, lending itself to human colonization and exploitation. While these 'types' would often overlap – with the frontier metaphor, in particular, persisting into the present -, each type set the tone for a generation.

2 Arcadian Mars

'Life, youth, love shine on every world... This divine fire glows on Mars, it glows on Venus'[8]. With unmatched fervor and elegance of style, Camille Flammarion (1842–1925) argued the case for intelligent extraterrestrial life during the second half of the 19th century, bolstered by the authority of the renowned astronomer who, in 1887, founded the Société Astronomique de France. His description of the Martian environment, in his very first work *La pluralité des mondes habités*, which would be reprinted 30 times until the century's end, informed the astronomical and popular discourse on the Red Planet for nearly a generation:

> The atmospheres of Earth and Mars, the snowfields seasonally expanding and shrinking on both planets, the clouds intermittently floating over their surfaces, the similar apportionment of continents and oceans, the conformities in seasonal variations: all this makes us believe that both worlds are inhabited by beings who physically resemble each other... In our mind's eye, we behold, here and there, intelligent beings, united into nations, vigorously striving for enlightenment and moral betterment[9].

In 1840, Beer and Maedler had drawn the first chart of Mars. Capital letters denoted observed 'regions' – darker spots on bright ground. (Before the Mariner 9 space probe permitted production of the first '*reliable* map'[10], more than 130 years would elapse.) The letters used by Beer and Maedler remained in use for two and a half decades, until Richard Proctor replaced them by the names of Mars observers on the map which he composed in 1867. Proctor also 'improved' on the way his compatriot John Phillips had, three years earlier, designated darker parts as 'seas' and brighter, reddish tracts as 'lands'. Proctor's chart showed continents and islands, oceans and seas, inlets and straits. These features had a suggestive effect. They seemed to portray a second – albeit smaller – Earth, with just a different division into zones of land and water.

The suggestion was deliberate. Proctor depicted Mars as a 'miniature of our Earth', waxing hardly less rhapsodically than Flammarion about the prettiness of the place:

> The mere existence of continents and oceans on Mars proves the action of... volcanic eruptions and earthquakes, modelling and remodelling the crust of Mars. Thus there must be mountains and hills, valleys and ravines, water-sheds and water-courses... And from the mountain recesses burst forth the refreshing springs which are to feed the Martia[n] brooklets...

And in a brilliant phrase, which Percival Lowell would later reclaim for entitling his final book, Proctor called Mars 'the abode of life', without whose existence 'all these things would be wasted'[11].

Proctor (1837–1888) was Honorary Secretary of the Royal Astronomical Society. Like Flammarion's work, his study of our solar system's planets subtitled 'under the light of recent scientific researches' continued to be reprinted until the advent of the 20th century. Public fascination was spurred further when the Mars opposition of 1877 led to the discovery of two small moons by Asaph Hall – and to the observation, by Giovanni Schiaparelli, of markings that the Italian astronomer took for *canali*, channels furrowing the planet's surface, some of which he compared to 'the Strait of Malacca, the very oblong lakes of Tanganyika and Nyassa, and the Gulf of California'[12]. After Schiaparelli reported that some of the lines he had sighted between 1877 and 1882 ran for 4800 kilometers, attaining a width of 120 kilometers, it came as no surprise that Flammarion was among the first to comment:

> One may resist the idea, but the longer one gazes at [Schiaparelli's] drawing, the more the interpretation suggests itself... [that] we are dealing with a technological achievement of the planet's inhabitants.[13]

In the minds of some of the period's foremost astronomers, the image of a lush and youthful Arcadian Mars would soon begin to give way to that of a much more ancient world possessing no natural water-courses, but artificial waterways surpassing anything so far constructed on Earth.

3 Advanced Mars

As judged by a present-day astronomer, after Giovanni Virginio Schiaparelli (1835–1910) had taught a whole generation of observers how to see Mars, it became eventually 'impossible to see it any other way. Expectation created illusion'[14]. If channels discernibly divided Mars to the extent of making its topography 'resemble that of a chessboard', if several such *canali* even 'form[ed] a complete girdle around the globe of Mars' – could they any longer be interpreted as natural attributes, 'like the rilles of the Moon'[15]? Might they not more convincingly be explained as non-natural features, as *canals* serving a purpose which had to be derived from the planet's characteristics?

The landscape of Advanced Mars which was construed by Percival Lowell from 1895 in response to Schiaparelli's revelations differed dramatically from that of Arcadian Mars. No more stately oceans, no impetuous rivers. A much grimmer environment predominated on Earth's neighbor world: 'The rose-ochre enchantment is but a mind mirage... Beautiful as the opaline tints of the planet look, ...they represent a terrible reality... [a] vast expanse of arid ground..., girdling the planet completely in circumference, and stretching in places almost from pole to pole'[16].

Erudite descendant of a wealthy Boston family, excelling in mathematics and literature, composing Latin hexameters at 11 and using his first telescope at 15, Percival Lowell (1855–1916) became enthusiastic about Flammarion's impressive compilation *La Planète Mars* (1892) and his views on the habitability of the planet. In 1894 he founded his own observatory near Flagstaff, Arizona Territory, with the express purpose of studying the conditions of life on other worlds, particularly on Mars. From his first twelve months of observations, Lowell drew conclusions which he immediately published in a book that 'influence[d] and shape[d] the imagination of writers' such as Wells and Lasswitz[17]. The darker regions of Mars he took to be "not water, but seasonal areas of vegetation", with the planet depending, for its water supply, "on the melting of its polar snows". Then came the clincher:

> If, therefore, the planet possess inhabitants,... irrigation, upon as vast a scale as possible,... must be the chief material concern of their lives... paramount to all the local labor, women's suffrage, and [Balkan] questions put together[18].

After the ironic aside, Lowell turned his attention to the canals which, he held, were dug precisely for such "irrigation purposes":

> What we see is not the canal proper, but the line of land it irrigates, dispos[ing] incidentally of the difficulty of conceiving a canal several miles wide... What we see hints at the existence... [of] a highly intelligent mind... of beings who are in advance of, not behind us, in the journey of life[19].

Much later, Carl Sagan would famously quip that, most certainly, intelligence was responsible for the straightness of the lines observed by Lowell. The problem was just "which side of the telescope the intelligence is on'[20].

While the 19th was turning into the 20th century, canals – like automobiles, dirigibles and airplanes – had come to symbolize progress, the triumph of technology over nature. In 1869, the Suez Canal had reduced the sea route to India by 10,000 kilometers, permitting Phileas Fogg and Passepartout to accomplish their imaginary journey around the world in 80 days. Work on the Panama Canal had begun in 1880, and even if the first French effort had foundered, a second American construction attempt was under way. Canals, whether on Earth or (supposedly) on another world, continued to make for headlines: On 27 August 1911, the *New York Times* captioned a one-page article. 'Martians Build Two Immense Canals in Two Years', its headline read. 'Vast Engineering Works Accomplished in an Incredibly Short Time.'

And Lowell's arid, aging Mars offered a further fascinating perspective: A community that had forsworn armed conflict[21], unified by a common endeavor, valiantly fighting its imminent doom, demonstrated to war-torn Earth what a civilization might achieve once it had overcome strife and hate.

Sunlight might be converted into electricity on Mars' high plateaus, even stones into bread by extracting protein and carbohydrates from rocks, soil, air and water. And material advancement might release additional energies needed for moral improvement. Such was the vision offered by Kurd Lasswitz in his 1897 novel *Auf zwei Planeten*. While an abridged English translation would only appear by 1971, the book was immediately translated into a number of other European languages and a popular German edition, which continued to be reprinted, was published in 1913. Until the Nazis branded the book as 'un-German', the novel sold 70,000 copies in Germany[22].

Lasswitz portrayed a Mars on which the 'colossal effort' required by irrigation had united the original 154 states into a single league. Thanks to the canal system

– and here Lasswitz sounded like pure Lowell –, 'the desert region was traversed by fertile strips of vegetation nearly 100 kilometers wide which included an unbroken string of thriving Martian settlements'[23]. A one-year mandatory labor service for both sexes helped maintain the network of canals. The discovery of anti-gravity had made Martians 'the masters of the solar system', permitting them to construct a wheel-shaped space station 6,356 kilometers above Earth's North Pole. Due to terrestrial arrogance, the first contact between men and 'Nume' ended in the occupation of Europe by the league of Martian states and the establishment of a protectorate aimed at 're-educating' mankind.

Wielding power over the Earth, however, worked to morally corrupt the Martian conquerors. When they threatened to extend their protectorate to the United States, American engineers secretly succeeded in copying Martian arms and taking over their space station. Faced with a choice of violating their highest values by resorting to a war of extinction, or leaving the Earth, the Martians chose to depart. Terrestrial nations not just concluded an alliance, but went on to adopt new constitutions in a Kantian 'spirit of peace, liberty and human dignity'[24]. A peace treaty with Mars ensured co-existence on the basis of equality.

However, an alternative scenario might be imagined, derived from the hypothesis that Martians had failed 'in attempting to safeguard the habitability of their planet'. In that case, might not beings with minds 'vast and cool and unsympathetic' feel tempted to resort to aggression, pitilessly exterminating mankind in search of 'living space'? Rather than Lasswitz' pacifist vision, the result would be the social-Darwinist *War of the Worlds* that Herbert George Wells envisioned in the same year. Skilfully, Wells gave the debate about the significance of the surface features on Mars a new twist. 'Men like Schiaparelli', he wrote, 'failed to interpret the fluctuating appearances of the markings they mapped so well. All that time the Martians must have been getting ready'[25].

Contrary to what a cursory reading of his tale might suggest, Wells did not depict the inhabitants of Mars – a Mars, it should be repeated, much older than the Earth, according to prevailing opinion – as alien monstrosities. Rather, regarding their appearance, he projected on them those 'characters of the Man of the remote future' which he had predicted as the final stage of human evolution in an earlier essay[26]: An expanding brain and head, diminishing bodies and legs, unemotional intelligence, nourishment by absorption of nutritive fluids – blood in the case of the Martian invaders –, atrophy of ears, nose, and mouth, the latter 'a small, perfectly round aperture, toothless, gumless, jawless'. Wells' Martians were not so much invaders from space as invaders from time, 'ourselves, mutated beyond sympathy, though not beyond recognition'[27].

Lasswitz, in a Kantian vein, had intended to confront Europe's imperialist powers with the alternative notion of a world governed by reason and peace. Wells' *War of the Worlds* remorselessly held the mirror up to contemporary colonialism. During 1897/98, Imperial Germany occupied the Chinese port of Kiautschou; China had to cede a further part of Hongkong to Great Britain; France consolidated its position in West Africa; the United States annexed the Hawaiian Islands. In Asia, in Africa, in the Pacific, native populations were being subjugated or pushed back. 'Are we such apostles of mercy', Wells asked rhetorically, 'as to complain if the Martians warred in the same spirit?'[28].

Finally, Wells could not only count on an audience turned receptive, by a spate of recent novels – such as George Chesney's *The Battle of Dorking* (1871), William Butler's *The Invasion of England* (1882), William Le Queux's *The Great War in England* (1894) –, to the notion of French and (more frequently) German raids on England. Moreover, these authors had already begun to explore a theme on which Wells focused his attention in *The War of the Worlds*: The disappearance of any distinction between battle fronts and zones where civilians might feel reasonably safe, the expansion of mechanized 'total' warfare to engulf entire populations[29].

Such total war was raging in China forty years later, after Japanese armies had invaded the country in 1937. For a brief moment, it had been avoided in Europe after Czechoslovakia had yielded, under British and French pressure, to the Munich Agreement. The war scare was still fresh in many Americans' minds when CBS, on 30 October 1938, aired *The War of the Worlds* as a 60-minute radioplay, directed by Orson Welles, with the action transferred to New Jersey. Presented as a series of increasingly ominous news bulletins, the first half of the broadcast produced mass hysteria: All over the United States, people 'were praying, crying, fleeing frantically... Some ran to rescue loved ones. Others... sought information from newspapers or radio stations, summoned ambulances and police cars'[30].

An estimated 250,000 people believed the United States to be under attack by either Germany, Japan – or indeed from Mars. In a bewildering world troubled by prolonged economic depression, wars and political crises, many Americans thought anything might happen.

By the time, H. G. Wells had turned social reformer, slowly despairing of men's folly. His last ideas about an invasion from Earth's 'wizened elder brother' Mars, published under the title *Star Begotten* shortly before Orson Welles' broadcast, differed considerably from his first – though not without a self-deprecating glance back[31]:

> Some of you may have read a book called *The War of the Worlds* – I forget who wrote it – Jules Verne, Conan Doyle, one of those fellows. But it told how the Martians invaded the world, wanted to colonize it and exterminate mankind. Hopeless attempt! They couldn't stand the different atmospheric pressure, they couldn't stand the difference in gravitation... To imagine that the Martians would be fools enough to try anything of the sort. But –

But if they resorted to cosmic rays instead? Modifying the genetic structure of unborn children, creating new beings that were, in fact, *their* spiritual children? That was the obsessive idea with which the tale's protagonist wrestled, until he discovered that his wife, their son – that he himself was star begotten, a changeling. The change, however, was benevolent, meant to salvage mankind – 'a lunatic asylum crowded with patients prevented from knowledge and afraid to go sane'[32] from stupidity and immaturity by making humans more flexible, more open-minded, more innovative.

Mature Martian civilization emerged as a deus ex machina for solving, by imperceptible intervention from outside, those pressing problems which mankind found itself unable to surmount.

4 Frontier Mars

Implying, as it did, that the Red Planet's inhabitants would beat humans to accomplishing space flight, the idea of Advanced Mars ran counter to deeply engrained expansionist impulses of the imperialist age. Small wonder the tabloid journalist Garrett Putnam Serviss immediately responded to Wells' tale with a serial in the sensationalist *New York Evening Journal*. Entitled *Edison's Conquest of Mars* and published in 1898, it depicted the 'wizard of Menlo Park', aided by Lord Kelvin and Konrad Roentgen, as devising both a disintegrator ray and an electric spaceship (admittedly based on the operating principles of the Martian machines). Financed by the great powers, 100 spaceships – armed with 3,000 disintegrators – were built and flew to Mars, where they wrecked havoc by forcibly opening the 'floodgates of Syrtis Major', thereby deluging the planet's equatorial regions.

Lasswitz had already attributed the defeat of his Martian conquerors to American engineering talent and 'daring'. By presenting an entire arsenal of innovative weapons, Serviss left no doubt about America's claim to global leadership: Technologically superior, the 'new world' had outrivaled the 'old' as the torchbearer of progress.

Yet, Serviss' tale did not set a new trend in wishful thinking about the Red Planet. Lowell's arid Mars was taking hold in the public mind, and Serviss' vast oceans and floodgates were just too wildly off that mark. For a significant change in perspective to occur, American cowboy, gold miner, salesman and – finally – novelist Edgar Rice Burroughs (1875–1950) had to write *A Princess of Mars* in 1912. He rechristened Mars, gave it the name Barsoom – and henceforth Mars exploration would be 'as much a re-creation of the past as a vision of the future'[33].

American rather than European science fiction authors now took the lead in projecting their fantasies onto the Red Planet, building on a 'forceful... cultural tradition' that would eventually inspire the U. S. space program no less than it initially spurred 'romantic vision[s]' of exploring, even colonizing Mars: the myth of America's western frontier[34]. Rather than 'highbrow' European-style literature, the 'entertainment industry'[35] of 'lowbrow' American pulp fiction with its international outlets provided the medium for two generations of writers including, subsequent to Burroughs, most prominently Leigh Brackett (1915–1978). Burroughs and his heirs retained Lowell's deserts and canals, but discarded the idea of a sophisticated Martian civilization. Instead, they depicted towns, ancient beyond imagination, lying in the southern hemisphere of

Mars, their outskirts touching the shores of the dried-up Low Canals that once discharged their waters into the now dust-blown bed of a long-vanished ocean. The towns, once ruled by pirate kings, bore names such as Jekkara, or Valkis, or Barrakesh. Their women – partly resembling Indians, partly Mexicans – wore tiny golden bells chiming temptingly. Barbarian tribes came to these places from distant deserts, such as Kesh and Shun. Santa Fe on Mars...

A Terran spaceport did exist at Kahora, not far from Olympus Mons. But only hard-boiled adventurers dared approach the Low Canals, after having galloped across the Drylands on half-wild saurians. They had 'the rawhide look of the planetary frontiers' about them[36] and wore their ray-guns low in their holsters. For Barrakesh, Jekkara, Valkis were towns outside the law.

The scenario was Leigh Brackett's, dreamt up during the 1940s. Like its American counterpart, the 'planetary frontier' signified no demarcation line, as the term was understood by Europeans, but rather the advancing rim of settlement, site of the violent clash between savagery and civilization. After the U. S. government had announced, in 1890, the 'closing' of the frontier in its statistical meaning of less than six inhabitants per square mile, the frontier – 'by transcending the limitations of a specific temporality' – came to be projected from the past into the present and even the future. Creating a specific 'moral landscape' by depicting the course of American history as progress through violence (or, as Burroughs would have it in *A Princess of Mars*, Ch. XXVI, 'through carnage to joy'), the myth of the frontier continues to provide patterns of identification and legitimization for individual and collective attitudes and behavior to the present day[37]. The rugged individualist – the onward-thrusting pioneer – the hardy adventurer, all armed *and morally justified* to shoot or to slash in a stereotyped black-and-white situation of good versus evil: These are the vivid images evoked by the frontier metaphor. They re-emerged in 'the "space opera" (as opposed to "horse opera")' with the 'typical structures and plots of westerns', but the 'settings and trappings of science fiction'[38].

Burroughs virtually defined the sub-genre, creating the quintessential space opera character: Captain John Carter, a 'gentleman of the highest type' and former plantation owner from Virginia, who had proved his prowess in the Civil War, and who was magically teleported from Arizona – where he had been battling Apaches – to Mars[39]. He made no effort to conceal that John Carter was modeled on Captain John Smith, a 17th century Virginian colonist who figured prominently in another American legend – the narrative of Pocahontas, Indian 'princess' of the Powhatan tribe. Supposedly, Pocahontas (at the tender age of 12 or 13) had become enamored of Smith and had rescued him from torture by her tribe. After arriving on a Mars peopled by warlike black, red, green and yellow races, Burroughs' John Carter met and married the 'incomparable princess' Dejah Thoris, daughter of the Jed (ruler) of Helium, chief of a red-skinned people that exhibited 'a startling resemblance... to... the red Indians of... Earth'[40]. In *The Princess of Mars* and Burroughs' subsequent Mars novels, it was Carter's task to repeatedly save Dejah Thoris, with the extraordinary physical powers lent to him by Mars' lesser gravity, from a fate 'worse than death'. To leave not the slightest doubt about the tradition he was embracing, Burroughs chose this context to revive another stereotype of frontier melodrama: With 'a cold sweat', his main protagonist reflected that if he should fail, it would be 'far better' for Dejah to 'save friendly bullets... at the last moment, as did those brave frontier women of my lost land, who took their own lives rather than fall into the hands of the Indian braves'[41].

Frontier Mars became a place where only-too-familiar characters lounged in the doorways of Earth's latest colony – Northwest Smith for one, created in 1933 by writer Catherine L. Moore (1911–1987), 'tall and leather-brown, hand on his heat-gun'; where everybody understood the 'old gesture' when that gun was drawn with a swift motion, sweeping 'in a practised halfcircle'[42]; where John Carter, Northwest Smith and their likes fought human or half-human tribes; where conflicts were invariably 'resolved' by resorting to weapons. An 'extension of our original America', with 'Martians await[ing] us' whom 'we [could] assimilate to our old myths of the Indian', Frontier Mars was destined to remain a very 'parochial' planet[43], familiar rather than alien.

5 Cold War Mars

'Watch the skies!' moviegoers were counseled in 1951 at the end of the science fiction film *The Thing from Another World*. The Cold War had turned hot in Korea. Who knew what the Communists, 'masters of deceit' (FBI Director J. Edgar Hoover), aggressively pushing from outside, subversively boring from within, threatening 'the continuance of every home and fireside'[44], might have up their sleeve?

Two years later, the 10-year-old stargazing protagonist of *Invaders from Mars* did watch the skies at night, only to observe a flying saucer landing and burrowing in the sandy ground across from his home. Everyone who investigated next morning – the child's father, his mother, a neighbor girl, two policemen, finally the local chief of police – was 'transformed' in succession, displaying an implant in the neck and behaving robot-like.

Invaders from Mars recounted not just an invasion, but a 'conspiracy', an emerging 'fifth column' of concealed infiltrators. Neither parents nor friends could be trusted anymore – a patent allusion (including the unfeeling attitudes displayed by affected adults) to rampant paranoia about the supposed subversion of American life by Communists. The invaders themselves were depicted as puppets, telepathically controlled by a 'supreme intelligence'. As might be expected, the military – alerted by the boy's school psychologist and her friend, an astrophysicist – arrived in time to save the day and blow up the Martian saucer.

'By the beginning of the decade, movies were America's most popular entertainment'; only from the mid-fifties would they be outranked by television. *Destination Moon* (1950) 'made the idea of space travel not only plausible but fascinating'. *The Thing from Another World* (1951) 'brought the idea of creatures from other planets coming here to vivid life'[45]. From 1950, too, the screen added Martian landscapes to those portrayed in the printed media. *Rocketship X(pedition) M(oon)*, Kurt Neumann's bleak movie of humans arriving on a

Mars destroyed by nuclear war, and *The Martian Chronicles*, Ray Bradbury's seminal novel of the Red Planet's colonization against the backdrop of atomic war eventually engulfing Earth, both came out during that year.

In *Rocketship XM*, the first manned spaceflight to Earth's satellite was thrown off course by a swarm of meteors and forced to land on Mars. The crew found themselves in a post-nuclear wasteland, deducing 'from artefacts and ruins so radioactive they can't approach them that there had once been a high civilization on Mars, but that atomic warfare reduced the Martians to savagery'[46]. Mutated Martians attacked the expedition, killing two and wounding a third crew member. The rest of the crew escaped, but the rocket ran out of fuel on its return flight and crashed. A year before, the Soviet Union had detonated its first nuclear device. President Truman had ordered development of the hydrogen bomb in early 1950. 'The idea that we now had the potential to wipe out civilization entirely was beginning to permeate mass culture' – and was projected onto Mars by 'the first film to expound such a grim warning about our possible future'[47].

The Soviet explosion and Truman's announcement drew an immediate response from a 30 year-old writer, Ray Bradbury, who felt that man might 'still destroy himself before reaching for the stars. I see man's self-destructive half, the blind spider fiddling in the venomous dark, dreaming mushroom-cloud dreams. Death solves it all, it whispers, shaking a handful of atoms like a necklace of dark beads'[48].

On May 6, 1950, *Collier's* published one of Bradbury's most powerful stories, *There Will Come Soft Rains*. It had no human protagonists. Rather, it focused on the final 'death' of an electronically programmed house, left standing empty among glowing radioactive ruins, after its occupants had perished, their images – as had happened in Hiroshima – 'burnt on the wood in one titanic instant'[49]. The story was included by Bradbury as a chapter in his loosely-knit classic of the same year, *The Martian Chronicles*, intended by the author to 'provide a mirror for humanity, its faults, foibles, and failures... an allegory transplanted to another world'[50].

Before being killed off by chicken pox which American colonists had introduced to Mars, the planet's golden-eyed 'natives' had inhabited crystal houses at the edge of the canals that – attuned to nature – 'turned and followed the sun, flower-like'. The settlers not only brought chicken pox. They also brought gas stations, and luggage stores, and hot-dog stands. With their hammers, they 'beat the strange world into a shape that was familiar', they 'bludgeon[ed] away all the strangeness... In all, some ninety thousand people came to Mars'. But the majority left again when flashing light-radio messages from Earth reported that there was war, that everybody should come home. To those who had remained on Mars, the night sky soon offered a horrible sight[51]:

> Earth changed... It caught fire. Part of it seemed to come apart in a million pieces... It burned with an unholy dripping glare for a minute, three times normal size, then dwindled.

Humans had turned two worlds, Mars and Earth, into 'tomb planet[s]'[52].

However, Bradbury – influenced by both Burroughs and Brackett – had also decided 'that there would be certain elements of similarity between the invasion of Mars and the invasion of the Wild West'[53]. The frontier myth held that America and its democracy would be reborn at every new frontier between Atlantic and Pacific – and beyond. One family, more fortunate than the folks annihilated in *There Will Come Soft Rains*, had escaped the inferno on Earth (with rumours maintaining that a second one had also made it to Mars). The father had promised the children that they would set out for a picnic and would see Martians. Now they were gazing at their reflections in a canal – and the Martians stared back at them.

The implication was evident. Bradbury's 'intensely critical examination' of the frontier myth – of 'the shallow and mercurial properties of America's predominant cultural construct'[54] – notwithstanding, Mars emerged as another 'virgin land' (Henry Nash Smith) where America might both survive and regenerate. The Frontier Mars image, in other words, had proved its adaptability to the hydrogen bomb age, reducing Cold War Mars to a mere variation of an already familiar theme. And, as would soon become evident, the frontier metaphor had not exhausted its usefulness.

6 Terraformed Mars

For American engineer Robert Zubrin, the writing presently 'is on the wall': He holds that 'without a frontier from which to breathe life, the spirit that gave rise to the progressive humanistic culture that America for the past several centuries has offered to the world is fading.' Zubrin is convinced that 'the creation of a new frontier presents itself as America's and humanity's greatest need'. And he 'believe[s] that humanity's new frontier can only be Mars'[55].

Zubrin, and the Mars Society formed in 1998 on his initiative, consider privately funded Mars flights and the establishment of a permanent Mars base as just initial steps. To fulfil the planet's mission of reinvigorating terrestrial civilization, its atmospheric and surface conditions need to be dramatically changed by a long-term project. Mars must be 'terraformed'.

According to the *Shorter Oxford English Dictionary*, terraforming implies a process of planetary engineering, aimed at creating an extraterrestrial environment that would be habitable for humans. First use of the term has been credited to science fiction writer John Stewart ('Jack') Williamson in a 1942 novella. The concept started to gain a certain scientific acceptability after Carl Sagan had published an article in 1961 on introducing algae into the atmosphere of Venus to slowly change that planet's extremely hostile conditions. In 1973, Sagan followed with a brief piece 'Planetary Engineering on Mars'[56], kicking off the debate with regard to the Red Planet.

To terraform Mars, both atmospheric pressure and surface temperature would have to be raised. The 'global warming' process – basically comparable to that which Earth is presently experiencing – would require an increase in 'greenhouse gasses', such as carbon dioxide or more powerful fluorocarbons, for which several ways have been proposed, and the subsequent buildup of a hydrosphere providing the water necessary to sustain life[57].

The idea was, of course, picked up by science fiction – most elaborately by Kim Stanley Robinson in his trilogy *Red Mars/Green Mars/Blue Mars* (1992–96). The work focused on the century-long conflict between 'Greens', whose sense of mission prompted them to contaminate the Red Planet with robust mosses and lichens at every opportunity, and the 'Red' environmentalists who were finally driven underground.

As before, such imaginary landscapes have revealed more about the desires, the hopes, the anxieties of those who designed them, than about any future 'green' or 'blue' Mars. While Zubrin took care to link the emergence of a terraformed Martian frontier to the promotion of values such as individualism, creativity and belief in the idea of progress, his basic approach was far more hard-nosed[58]:

> If the idea is accepted that the world's resources are fixed, then each person is ultimately the enemy of every other person, and each race or nation is the enemy of every other race or nation. The inevitable result is tyranny, war and genocide. Only in a universe of unlimited resources can all men be brothers.
>
> Put differently: Either a new frontier will be opened up – or containment, rather than self-containment, will become the 'natural' order of things...

Objections against such reasoning were the exception. In terms reminiscent of Bradbury, but more starkly, historian Patricia Limerick in her contribution to the 1992 volume *Space Policy Alternatives* emphasized the social-Darwinist consequences of 'rugged individualism' that had shaped the 'conquest' of the American West, including greed and corruption, violence against 'aliens' (Indians and Mexicans), environmental destruction. She rejected the simplified picture, again extolled by Zubrin, of equating westward expansion with democracy and progress, for that picture had 'denied consequences and evaded failure'[59].

Because of 'America's pioneer heritage, technological pre-eminence, and economic strength, it is fitting that we should lead the people of this planet into space', the Paine Commission had stated in 1986. Chaired by an earlier NASA Administrator, it included UN Ambassador Jeane Kirckpatrick, former test pilot Charles Yeager and retired Air Force General Bernard Schriever (who had directed IRBM Thor and ICBM Atlas development). In their report, tellingly entitled *Pioneering the Space Frontier*, the members had proposed to 'stimulate individual initiative and free enterprise in space', and had resolved that 'from the highlands of the Moon to the plains of Mars', America should 'make accessible vast new resources and support human settlements beyond Earth orbit'[60].

This was no space opera. This was a National Commission on Space, appointed by the President of the United States, issuing a declaration that, with its 'fervent optimism and cheeriness', was 'vintage 1890's... a picture of harmony and progress where historical reality shows us something closer to a muddle'[61].

Abstracting and reducing from reality, the frontier myth has created a historical cliché. Clichés, as Richard Slotkin – among others – has reminded us, may serve to interpret new experiences as mere recurrences of past happenings, reflecting a refusal to learn. Identifying Mars as merely another 'frontier', projecting a moral purpose on the adaption of that so-called planetary frontier to human settlers' needs, tops a tradition of invoking a highly problematic cultural stereotype.

THE LADY ASTRONAUT OF MARS (2013)

Mary Robinette Kowal

Mary Robinette Kowal is an author, puppeteer and voice actor. Born in 1969 in Raleigh, North Carolina, she graduated from East Carolina University with a degree in art education and a minor in theatre and began work as a puppeteer, including for the Jim Henson Company. *The Lady Astronaut of Mars* was originally published in 2012 in the Audible anthology *Rip-Off*. It won the 2014 Hugo Award for Best Novelette and has spawned three prequel novels: *The Calculating Stars*, *The Fated Sky* and *The Relentless Moon*. She is currently working on two more follow-ups.

Dorothy lived in the midst of the great Kansas prairies, with Uncle Henry, who was a farmer, and Aunt Em, who was the farmer's wife. She met me, she went on to say, when I was working next door to their farm under the shadow of the rocket gantry for the First Mars Expedition.

I have no memory of this.

She would have been a little girl and, oh lord, there were so many little kids hanging around outside the Fence watching us work. The little girls all wanted to talk to the Lady Astronaut. To me.

I'm sure I spoke to Dorothy because know I stopped and talked to them every day on my way in and out through the Fence about what it was like. It being Mars. There was nothing else it could be.

Mars consumed everyone's conversations. The programmers sitting over their punchcards. The punchcard girls keying in the endless lines of code. The cafeteria ladies ladling out mashed potatoes and green peas. Nathaniel with his calculations... Everyone talked about Mars.

So the fact that I didn't remember a little girl who said I talked to her about Mars... Well. That's not surprising, is it? I tried not to let the confusion show in my face but I know she saw it.

By this point, Dorothy was my doctor. Let me be more specific. She was the geriatric specialist who was evaluating me. On Mars. I was in for what I thought was a routine check-up to make sure I was still fit to be an astronaut. NASA liked to update its database periodically and I liked to be in that database. Not that I'd flown since I turned fifty, but I kept my name on the list in the faint hope that they would let me back into space again, and I kept going to the darn check-ups.

Our previous doctor had retired back to Earth, and I'd visited Dorothy's offices three times before she mentioned Kansas and the prairie.

She fumbled with the clipboard and cleared her throat. A flush of red colored her cheeks and made her eyes even more blue. "Sorry. Dr. York, I shouldn't have mentioned it."

"Don't 'doctor' me. You're the doctor. I'm just a space jockey. Call me Elma." I waved my hand to calm her down. The flesh under my arm jiggled and I dropped my hand. I hate that feeling and hospital gowns just make it worse. "I'm glad you did. You just took me by surprise, is all. Last I saw you, weren't you knee-high to a grasshopper?"

"So you do remember me?" Oh, that hope. She'd come to Mars because of me. I could see that, clear as anything. Something I'd said or done back in 1952 had brought this girl out to the colony.

"Of course, I remember you. Didn't we talk every time I went through that Fence? Except school days, of course." It seemed a safe bet.

Dorothy nodded, eager. "I still have the eagle you gave me."

"Do you now?" That gave me a pause.

I used to make paper eagles out of old punchcards while I was waiting for Nathaniel. His programs could take hours to run and he liked to baby sit them. The eagles were cut paper things with layers of cards pasted together to make a three dimensional bird. It was usually in flight and I liked to hang them in the window, where the holes from the punch cards would let specks of light through and make the bird seem like it was sparkling. They would take me two or three days to make. You'd think I would remember giving one to a little girl beyond the Fence. "Did you bring it out here with you?"

"It's in my office." She stood as if she'd been waiting for me to ask that since our first session, then looked down at the clipboard in her hands, frowning. "We should finish your tests."

"Fine by me. Putting them off isn't going to make me any more eager." I held out my arm with the wrist up so she could take my pulse. By this point, I knew the drill. "How's your Uncle?"

She laid her fingers on my wrist, cool as anything. "He and Aunt Em passed away when Orion 27 blew."

I swallowed, sick at my lack of memory. So she was THAT little girl. She'd told me all the things I needed and my old brain was just too addled to put the pieces together. I wondered if she would make a note of that and if it would keep me grounded.

Dorothy had lived on a farm in the middle of the Kansas prairie with her Uncle Henry and Aunt Em. When Orion 27 came down in a ball of fire, it was the middle of a drought. The largest pieces of it had landed on a farm.

No buildings were crushed, but it would have been a blessing if they had been, because that would have saved the folks inside from burning alive.

I closed my eyes and could see her now as the little girl I'd forgotten. Brown pigtails down her back and a pair of dungarees a size too large for her, with the legs cuffed up to show bobby socks and sneakers.

Someone had pointed her out. "The little girl from the Williams farm."

I'd seen her before, but in that way you see the same people every day without noticing them. Even then, with someone pointing to her, she didn't stand out from the crowd. Looking at her, there was nothing to know that she'd just lived through a tragedy. I reckon it hadn't hit her yet.

I had stepped away from the entourage of reporters and consultants that followed me and walked up to her. She had tilted her head back to look up at me. I used to be a tall woman, you know.

I remember her voice piping up in that high treble of the very young. "You still going to Mars?"

I had nodded. "Maybe you can go someday too."

She had cocked her head to the side, as if she were considering. I can't remember what she said back. I know she must have said something. I know we must have talked longer because I gave her that darned eagle, but what we said… I couldn't pull it up out of my brain.

As the present day Dorothy tugged up my sleeve and wrapped the blood pressure cuff around my arm, I studied her. She had the same dark hair as the little girl she had been, but it was cut short now and in the low gravity of Mars it wisped around her head like the down on a baby bird.

The shape of her eyes was the same, but that was about it. The soft roundness of her cheeks was long gone, leaving high cheekbones and a jaw that came to too sharp of a point for beauty. She had a faint white scar just above her left eyebrow.

She smiled at me and unwrapped the cuff. "Your blood pressure is better. You must have been exercising since last time."

"I do what my doctor tells me."

"How's your husband?"

"About the same." I slid away from the subject even though, as his doctor, she had the right to ask, and I squinted at her height. "How old were you when you came here?"

"Sixteen. We were supposed to come before but… well." She shrugged, speaking worlds about why she hadn't.

"Your uncle, right?"

Startled, she shook her head. "Oh, no. Mom and Dad. We were supposed to be on the first colony ship but a logging truck lost its load."

Aghast, I could only stare at her. If they were supposed to have been on the first colony ship, then her parents could not have died long before Orion 27 crashed. I wet my lips. "Where did you go after your aunt and uncle's?"

"My cousin. Their son." She lifted one of the syringes she'd brought in with her. "I need to take some blood today."

"My left arm has better veins."

While she swabbed the site, I looked away and stared at a chart on the wall reminding people to take their vitamin D supplements. We didn't get enough light here for most humans.

But the stars… When you could see them, the stars were glorious. Was that what had brought Dorothy to Mars?

*

When I got home from the doctor's—from Dorothy's—the nurse was just finishing up with Nathaniel's sponge bath. Genevieve stuck her head out of the bedroom, hands still dripping.

"Well, hey, Miss Elma. We're having a real good day, aren't we, Mr. Nathaniel?" Her smile could have lit a hangar, it was so bright.

"That we are." Nathaniel sounded hale and hearty, if I didn't look at him. "Genevieve taught me a new joke. How's it go?"

She stepped back into the bedroom. "What did the astronaut see on the stove? An unidentified frying object."

Nathaniel laughed, and there was only a little bit of a wheeze. I slid my shoes off in the dustroom to keep out the ever present Martian grit, and came into the kitchen to lean against the bedroom door. Time was, it used to be his office but we needed a bedroom on the ground floor. "That's a pretty good one."

He sat on a towel at the edge of the bed as Genevieve washed him. With his shirt off, the ribs were starkly visible under his skin. Each bone in his arms poked at the surface and slid under the slack flesh. His hands shook, even just resting beside him on the bed. He grinned at me.

The same grin. The same bright blue eyes that had flashed over the punchcards as he'd worked out the plans for the launch. It was as though someone had pasted his features onto the body of a stranger. "How'd the doctor's visit go?"

"The usual. Only… Only it turns out our doctor grew up next to the launch facility in Kansas."

"Dr. Williams?"

"The same. Apparently I met her when she was little."

"Is that right?" Genevieve wrung the sponge out in the wash basin. "Doesn't that just go to show that it's a small solar system?"

"Not that small." Nathaniel reached for his shirt, which lay on the bed next to him. His hands tremored over the fabric.

"I'll get it. You just give me a minute to get this put away." Genevieve bustled out of the room.

I called after her. "Don't worry. I can help him."

Nathaniel dipped his head, hiding those beautiful eyes, as I drew a sleeve up over one arm. He favored flannel now. He'd always hated it in the past. Preferred starched white shirts and a nice tie to work in, and a short sleeved aloha shirt on his days off. At first, I thought that the flannel was because he was cold all the time. Later I realized that the thicker fabric hid some of his frailty. Leaning behind him to pull the shirt around his back, I could count vertebra in his spine.

Nathaniel cleared his throat. "So, you met her, hm? Or she met you? There were a lot of little kids watching us."

"Both. I gave her one of my paper eagles."

That made him lift his head. "Really?"

"She was on the Williams farm when the Orion 27 came down."

He winced. Even after all these years, Nathaniel still felt responsible. He had not programmed the rocket. They'd asked him to, but he'd been too busy with the First

Mars Expedition and turned the assignment down. It was just a supply rocket for the moon, and there had been no reason to think it needed anything special.

I buttoned the shirt under his chin. The soft wattle of skin hanging from his jaw brushed the back of my hand. "I think she was too shy to mention it at my last visit."

"But she gave you a clean bill of health?"

"There's still some test results to get back." I avoided his gaze, hating the fact that I was healthy and he was… Not.

"It must be pretty good. Sheldon called."

A bubble of adrenalin made my heart skip. Sheldon Spender called. The director of operations at the Bradbury Space Center on Mars had not called since—No, that wasn't true. He hadn't called *me* in years, using silence to let me know I wasn't flying anymore. Nathaniel still got called for work. Becoming old didn't stop a programmer from working, but it sure as heck stopped an astronaut from flying. And yet I still had that moment of hope every single time Sheldon called, that this time it would be for me. I smoothed the flannel over Nathaniel's shoulders. "Do they have a new project for you?"

"He called for you. Message is on the counter."

Genevieve breezed back into the room, a bubble of idle chatter preceding her. Something about her cousin and meeting their neighbors on Venus. I stood up and let her finish getting Nathaniel dressed while I went into the kitchen.

Sheldon had called for me? I picked up the note on the counter. It just had Genevieve's round handwriting and a request to meet for lunch. The location told me a lot though. He'd picked a bar next to the space center that no one in the industry went to because it was thronged with tourists. It was a good place to talk business without talking business. For the life of me I couldn't figure out what he wanted.

I kept chewing on that question, right till the point when I stepped through the doors of Yuri's Spot. The walls were crowded with memorabilia and signed photos of astronauts. An early publicity still that showed me perched on the edge of Nathaniel's desk, hung in the corner next to a dusty ficus tree. My hair fell in perfect soft curls despite the flight suit I had on. My hair would never have survived like that if I'd actually been working. I tended to keep it out of the way in a kerchief, but that wasn't the image publicity had wanted.

Nathaniel was holding up a punch card, as if he were showing me a crucial piece of programming. Again, it was a staged thing, because the individual cards were meaningless by themselves, but to the general public at the time they meant Science with a capital S. I'm pretty sure that's why we were both laughing in the photo, but they had billed it as "the joy of space flight."

Still gave me a chuckle, thirty years later.

Sheldon stepped away from the wall and mistook my smile. "You look in good spirits."

I nodded to the photo. "Just laughing at old memories."

He glanced over his shoulder, wrinkles bunching at the corner of his eyes in a smile. "How's Nathaniel?"

"About the same, which is all one can ask for at this point."

Sheldon nodded and gestured to a corner booth, leading me past a family with five kids who had clearly come from the Space Center. The youngest girl had her nose buried in a picture book of the early space program. None of them noticed me.

Time was when I couldn't walk anywhere on Mars without being recognized as the Lady Astronaut. Now, thirty years after the First Expedition, I was just another old lady, whose small stature showed my origin on Earth.

We settled in our chairs and ordered, making small talk as we did. I think I got fish and chips because it was the first thing on the menu, and all I could think about was wondering why Sheldon had called.

It was like he wanted to see how long it would take me to crack and ask him what he was up to. It took me awhile to realize that he kept bringing the conversation back to Nathaniel. Was he in pain?

Of course.

Did he have trouble sleeping?

Yes.

Even, "How are you holding up?" was about him. I didn't get it until Sheldon paused and pushed his rabbit burger aside, half-eaten, and asked point-blank. "Have they given him a date yet?"

A date. There was only one date that mattered in a string of other milestones on the path to death but I pretended he wasn't being clear, just to make him hurt a little. "You mean for paralyzation, hospice, or death?"

He didn't flinch. "Death."

"We think he's got about a year." I kept my face calm, the way you do when you're talking to Mission Control about a flight that's set to abort. The worse it got, the more even my voice became. "He can still work, if that's what you're asking."

"It's not." Sheldon broke his gaze then, to my surprise, and looked down at his ice water, spinning the glass in its circle of condensation. "What I need to know is if you can still work."

In my intake of breath, I wanted to say that God, yes, I could work and that I would do anything he asked of me if he'd put me back into space. In my exhale, I thought of Nathaniel. I could not say yes. "That's why you asked for the physical."

"Yep."

"I'm sixty-three, Sheldon."

"I know." He turned the glass again. "Did you see the news about LS-579?"

"The extrasolar planet. Yes." I was grounded, that didn't mean I stopped paying attention to the stars.

"Did you know we think it's habitable?"

I stopped with my mouth open as pieces started to tick like punch cards slotting through a machine. "You're mounting a mission."

"If we were, would you be interested in going?"

Back into space? My god, yes. But I couldn't. I couldn't. I—that was why he wanted to know when my husband was going to die. I swallowed everything before speaking. My voice was passive. "I'm sixty-three." Which was my way of asking why he wanted me to go.

"It's three years in space." He looked up now, not needing to explain why they wanted an old pilot.

That long in space? It doesn't matter how much shielding you have against radiation, it's going to affect you. The chances of developing cancer within the next fifteen years were huge. You can't ask a young astronaut to do that. "I see."

"We have the resources to send a small craft there. It can't be unmanned because the programming is too complicated. I need an astronaut who can fit in the capsule."

"And you need someone who has a reason to not care about surviving the trip."

"No." He grimaced. "PR tells me that I need an astronaut that the public will adore so that when we finally tell them that we've sent you, they will forgive us for hiding the mission from them." Sheldon cleared his throat and started briefing me on the Longevity Mission.

Should I pause here and explain what the Longevity mission is? It's possible that you don't know.

There's a habitable planet. An extrasolar one and it's only few light years away. They've got a slingshot that can launch a ship up to near light speed. A small ship. Big enough for one person.

But that isn't what makes the Longevity mission possible. That is the tesseract field. We can't go faster than light, but we can cut corners through the universe. The physicists described it to me like a subway tunnel. The tessaract will bend space and allow a ship to go to the next subway station. The only trick is that you need to get far enough away from a planet before you can bend space and... this is the harder part... you need a tesseract field at the other end. Once that's up, you just need to get into orbit and the trip from Mars to LS-579 can be as short as three weeks.

But you have to get someone to the planet to set up the other end of the tesseract.

And they wanted to hide the plan from the public, in case it failed.

So different from when the First Mars Expedition had happened. An asteroid had slammed into Washington D.C. and obliterated the capitol. It made the entire world realize how fragile our hold on Earth was. Nations banded together and when the Secretary of Agriculture, who found himself president through the line of succession, said that we needed to get off the planet, people listened. We rose to the stars. The potential loss of an astronaut was just part of the risk. Now? Now it has been long enough that people are starting to forget that the danger is still there. That the need to explore is necessary.

Sheldon finished talking and just watched me processing it.

"I need to think about this."

"I know."

Then I closed my eyes and realized that I had to say no. It didn't matter how I felt about the trip or the chance to get back into space. The launch date he was talking about meant I'd have to go into training now. "I can't." I opened my eyes and stared at the wall where the publicity still of me and Nathaniel hung. "I have to turn it down."

"Talk to Nathaniel."

I grimaced. He would tell me to take it. "I can't."

I left Sheldon feeling more unsettled than I wanted to admit at the time. I stared out the window of the light rail, at the sepia sky. Rose tones were deepening near the horizon with sunset. It was dimmer and ruddier here, but with the dust, sunset could be just as glorious as on Earth.

It's a hard thing to look at something you want and to know that the right choice is to turn it down. Understand me: I wanted to go. Another opportunity like this would never come up for me. I was too old for normal missions. I knew it. Sheldon knew it. And Nathaniel would know it, too. I wish he had been in some other industry so I could lie and talk about "later." He knew the space program too well to be fooled.

And he wouldn't believe me if I said I didn't want to go. He knew how much I missed the stars.

That's the thing that I think none of us were prepared for in coming to Mars. The natural night sky on Mars is spectacular, because the atmosphere is so thin. But where humans live, under the dome, all you can see are the lights of the town reflecting against the dark curve. You can almost believe that they're stars. Almost. If you don't know what you are missing or don't remember the way the sky looked at night on Earth before the asteroid hit.

I wonder if Dorothy remembers the stars. She's young enough that she might not. Children on Earth still look at clouds of dust and stars are just a myth. God. What a bleak sky.

When I got home, Genevieve greeted me with her usual friendly chatter. Nathaniel looked like he wanted to push her out of the house so he could quiz me. I know Genevieve said good bye, and that we chatted, but the details have vanished now.

What I remember next is the rattle and thump of Nathaniel's walker as he pushed it into the kitchen. It slid forward. Stopped. He took two steps, steadied himself, and slid it forward again. Two steps. Steady. Slide.

I pushed away from the counter and straightened. "Do you want to be in the kitchen or the living room?"

"Sit down, Elma." He clenched the walker till the tendons stood out on the back of his hands, but they still trembled. "Tell me about the mission."

"What?" I froze.

"The mission." He stared at the ceiling, not at me. "That's why Sheldon called, right? So, tell me."

"I… All right." I pulled the tall stool out for him and waited until he eased onto it. Then I told him. He stared at the ceiling the whole time I talked. I spent the time watching him and memorizing the line of his cheek, and the shape of the small mole by the corner of his mouth.

When I finished, he nodded. "You should take it."

"What makes you think I want to?"

He lowered his head then, eyes just as piercing as they had always been. "How long have we been married?"

"I can't."

Nathaniel snorted. "I called Dr. Williams while you were out, figuring it would be something like this. I asked for a date when we could get hospice." He held up his hand to stop the words forming on my lips. "She's not willing to tell me that. She did give me the date when the paralysis is likely to become total. Three months. Give or take a week."

We'd known this was coming, since he was diagnosed, but I still had to bite the inside of my lip to keep from sobbing. He didn't need to see me break down.

"So… I think you should tell them yes."

"Three months is not a lot of time, they can—"

"They can what? Wait for me to die? Jesus Christ, Elma. We know that's coming." He scowled at the floor. "Go. For the love of God, just take the mission."

I wanted to. I wanted to get off the planet and back into space and not have to watch him die. Not have to watch him lose control of his body piece by piece.

And I wanted to stay here and be with him and steal every moment left that he had breath in his body.

One of my favorite restaurants in Landing was Elmore's. The New Orleans style cafe sat tucked back behind Thompson's Grocers on a little rise that lifted the dining room just high enough to see out to the edge of town and the dome's wall. They had a crawfish étouffée that would make you think you were back on Earth. The crawfish were raised in a tank and a little bigger than the ones I'd grown up with, but the spices came all the way from Louisiana on the mail runs twice a year.

Sheldon Spender knew it was my favorite and was taking ruthless advantage of that. And yet I came anyway. He sat across the table from me, with his back to the picture window that framed the view. His thinning hair was almost invisible against the sky. He didn't say a word. Just watched me, as the fellow to my right talked.

Garrett Biggs. I'd seen him at the Bradbury Space Center, but we'd exchanged maybe five words before today. My work was mostly done before his time. They just trotted me out for the occasional holiday. Now, the man would not stop talking. He gestured with his fork as he spoke, punctuating the phrases he thought I needed to hear most. "Need some photos of you so we can exploit—I know it sounds ugly but we're all friends here, right? We can be honest, right? So, we can exploit your sacrifice to get the public really behind the Longevity mission."

I watched the lettuce tremble on the end of his fork. It was pallid compared to my memory of lettuce on Earth. "I thought the public didn't know about the mission."

"They will. That's the key. Someone will leak it and we need to be ready." He waved the lettuce at me. "And that's why you are a brilliant choice for pilot. Octogenarian Grandmother Paves Way for Humanity."

"You can't pave the stars. I'm not a grandmother. And I'm sixty-three not eighty."

"It's a figure of speech. The point is that you're a PR goldmine."

I had known that they asked me to helm this mission because of my age—it would be a lot to ask of someone who had a full life ahead of them. Maybe I was naive to think that my experience in establishing the Mars colony was considered valuable.

How can I explain the degree to which I resented being used for publicity? This wasn't a new thing by a long shot. My entire career has been about exploitation for publicity. I had known it, and exploited it too, once I'd realized the power of having my uniform tailored to show my shape a little more clearly. You think they would have sent me to Mars if it weren't intended to be a colony? I was there to show all the lady housewives that they could go to space too. Posing in my flight suit, with my lips painted red, I had smiled at more cameras than my colleagues.

I stared Garrett Biggs and his fork. "For someone in PR, you are awfully blunt."

"I'm honest. To you. If you were the public, I'd have you spinning so fast you'd generate your own gravity."

Sheldon cleared his throat. "Elma, the fact is that we're getting some pressure from a group of senators. They want to cut the budget for the project and we need to take steps or it won't happen."

I looked down and separated the tail from one of my crawfish. "Why?"

"The usual nonsense. People arguing that if we just wait, then ships will become fast enough to render the mission pointless. That includes a couple of serious misunderstandings of physics, but, be that as it may…" Sheldon paused and tilted his head, looking at me. He changed what he was about to say and leaned forward. "Is Nathaniel worse?"

"He's not better."

He winced at the edge in my voice. "I'm sorry. I know I strong-armed you into it, but I can find someone else."

"He thinks I should go." My chest hurt even considering it. But I couldn't stop thinking about the mission. "He knows it's the only way I'll get back into space."

Garrett Biggs frowned like I'd said the sky was green, instead of the pale Martian amber. "You're in space."

"I'm on Mars. It's still a planet."

I woke out of half-sleep, aware that I must have heard Nathaniel's bell, without being able to actually recall it. I pulled myself to my feet, putting a hand against the nightstand until I was steady. My right hip had stiffened again in the night. Arthritis is not something I approve of.

Turning on the hall light, I made my way down the stairs. The door at the bottom stood open so I could hear Nathaniel if he called. I couldn't sleep with him anymore, for fear of breaking him.

I went through into his room. It was full of grey shadows and the dark rectangle of his bed. In one corner, the silver arm of his walker caught the light.

"I'm sorry." His voice cracked with sleep.

"It's all right. I was awake anyway."

"Liar."

"Now, is that a nice thing to say?" I put my hand on the light switch. "Watch your eyes."

Every night we followed the same ritual and even though I knew the light would be painfully bright, I still winced as it came on. Squinting against the glare, I threw the covers back for him. The weight of them trapped him sometimes. He held his hands up, waiting for me to take them. I braced myself and let Nathaniel pull himself into a sitting position. On Earth, he'd have been bed-ridden long since. Of course, on Earth, his bone density would probably not have deteriorated so fast.

As gently as I could, I swung his legs to the side of the bed. Even allowing for the gravity, I was appalled anew by how light he was. His legs were like kindling wrapped in tissue. Where his pajamas had ridden up, purple bruises mottled his calf.

As soon as he was sitting up on the edge of the bed, I gave him the walker. He wrapped his shaking hands around the bars and tried to stand. He rose only a little before dropping back to the bed. I stayed where I was, though I ached to help. He sometimes took more than one try to stand at night, and didn't want help. Not until it became absolutely necessary. Even then, he wouldn't want it. I just hoped he'd let me help him when we got to that point.

On the second try, he got his feet under him and stood, shaking. With a nod, he pushed forward. "Let's go."

I followed him to the bathroom in case he lost his balance in there, which he did sometimes. The first time, I hadn't been home. We had hired Genevieve not long after that to sit with him when I needed to be out.

He stopped in the kitchen and bent a little at the waist with a sort of grunt.

"Are you all right?"

He shook his head and started again, moving faster. "I'm not—" He leaned forward, clenching his jaw. "I can't—"

The bathroom was so close.

"Oh, God. Elma..." A dark, fetid smell filled the kitchen. Nathaniel groaned. "I couldn't—"

I put my hand on his back. "Hush. We're almost there. We'll get you cleaned up."

"I'm sorry, I'm sorry." He pushed the walker forward, head hanging. A trail of damp footsteps followed him. The ammonia stink of urine joined the scent of his bowels.

I helped him lower his pajamas. The weight of them had made them sag on his hips. Dark streaks ran down his legs and dripped onto the bathmat. I eased him onto the toilet.

My husband bent his head forward, and he wept.

I remember wetting a washcloth and running it over his legs. I know that I must have tossed his soiled pajamas into the cleaner, and that I wiped up the floor, but those details have mercifully vanished. But what I can't forget, and I wish to God that I could, is Nathaniel sitting there crying.

*

I asked Genevieve to bring adult diapers to us the next day. The strange thing was how familiar the package felt. I'd used them on launches when we had to sit in the capsule for hours and there was no option to get out of our space suit. It's one of the many glamorous details of being an astronaut that the publicity department does not share with the public.

There is a difference, however, from being required to wear one for work and what Nathaniel faced. He could not put them on by himself without losing his balance. Every time I had to change the diaper, he stared at the wall with his face slack and hopeless.

Nathaniel and I'd made the decision not to have children. They aren't conducive to a life in space, you know? I mean there's the radiation, and the weightlessness, but more it was that I was gone all the time. I couldn't give up the stars… but I found myself wishing that we hadn't made that decision. Part of it was wishing that I had some connection to the next generation. More of it was wanting someone to share the burden of decision with me.

What happens after Nathaniel dies? What do I have left here? More specifically, how much will I regret not going on the Mission?

And if I'm in space, how much will I regret abandoning my husband to die alone?

You see why I was starting to wish that we had children?

In the afternoon, we were sitting in the living room, pretending to work. Nathaniel sat with his pencil poised over the paper and stared out the window as though he were working. I'm pretty sure he wasn't but I gave him what privacy I could and started on one of my eagles.

The phone rang and gave us both something of a relief, I think, to have a distraction. The phone sat on a table by Nathaniel's chair so he could reach it easily if I weren't in the room. With my eyes averted, his voice sounded as strong as ever as he answered.

"Hang on, Sheldon. Let me get Elma for—Oh. Oh, I see."

I snipped another feather but it was more as a way to avoid making eye contact than because I really wanted to keep working.

"Of course I've got a few minutes. I have nothing but time these days." He ran his hand through his hair and let it rest at the back of his neck. "I find it hard to believe that you don't have programmers on staff who can't handle this."

He was quiet then as Sheldon spoke, I could hear only the distorted tinny sound of his voice rising and falling. At a certain point, Nathaniel picked up his pencil again and started making notes. Whatever Sheldon was asking him to do, that was the moment when Nathaniel decided to say "yes."

I set my eagle aside and went into the kitchen. My first reaction—God. It shames me but my first reaction was anger. How dare he? How dare he take a job without consulting with me when I was turning down this thing I so desperately wanted because of him. I had the urge to snatch up the phone and tell Sheldon that I would go.

I pushed that down carefully and looked at it.

Nathaniel had been urging me to go. No deliberate action of his was keeping me from accepting. Only my own upbringing and loyalty and… and I loved him. If I did not want to be alone after he passed, how could I leave him to face the end alone?

The decision would be easier if I knew when he would die.

I still hate myself for thinking that.

I heard the conversation end and Nathaniel hung up the phone. I filled a glass with water to give myself an excuse for lingering in the kitchen. I carried it back into the living room and sat down on the couch.

Nathaniel had his lower lip between his teeth and was scowling at the page on top of his notepad. He jotted a number in the margin with a pencil before he looked up.

"That was Sheldon." He glanced back at the page.

I settled in my chair and fidgeted with the wedding band on my finger. It had gotten loose in the last year. "I'm going to turn them down."

"What—But, Elma." His gaze flattened and he gave me a small frown. "Are you… are you sure it's not depression? That's making you want to stay, I mean."

I gave an unladylike snort. "Now what do I have to be depressed about?"

"Please." He ran his hands through his hair and knit them together at the back of his neck. "I want you to go so you won't be here when… It's just going to get worse from here."

The devil of it was that he wasn't wrong. That didn't mean he was right, either, but I couldn't flat out tell him he was wrong. I set down my scissors and pushed the magnifier out of the way. "It's not just depression."

"I don't understand. There's a chance to go back into space." He dropped his hands and sat forward. "I mean… If I die before the mission leaves and you're grounded here. How would you feel?"

I looked away. My gaze was pointed to the window and the view of the house across the lane. But I did not see the windows or the red brick walls. All I saw was a black and grey cloth made of despair. "I had a life that I enjoyed before this opportunity came up. There's no reason I shouldn't keep on enjoying it. I enjoy teaching. There are a hundred reasons to enjoy life here."

He pointed his pencil at me the way he used to do when he spotted a flaw in reasoning at a meeting, but the pencil quivered in his grip now. "If that's true, then why haven't you told them no, yet?"

The answer to that was not easy. Because I wanted to be in the sky, weightless, and watching the impossibly bright stars. Because I didn't want to watch Nathaniel die. "What did Sheldon ask you to do?"

"NASA wants more information about LS-579."

"I imagine they do." I twisted that wedding band around as if it were a control that I could use. "I would… I would hate… As much as I miss being in space, I would hate myself if I left you here. To have and to hold, in sickness and in health. Till death do us part and all that. I just can't."

"Well… just don't tell him no. Not yet. Let me talk to Dr. Williams and see if she can give us a clearer date. Maybe there won't be a schedule conflict after al—"

"Stop it! Just stop. This is my decision. I'm the one who has to live with the consequences. Not you. So, stop trying to put your guilt off onto me because the devil of it is, one of us is going to feel guilty here, but I'm the one who will have to live with it."

I stormed out of the room before he could answer me or I could say anything worse. And yes—I knew that he couldn't follow me and for once I was glad.

Dorothy came not long after that. To say that I was flummoxed when I opened the door wouldn't do justice to my surprise. She had her medical bag with her and I think that's the only thing that gave me the power of speech. "Since when do you make house calls?"

She paused, mouth partially open, and frowned. "Weren't you told I was coming?"

"No." I remembered my manners and stepped back so she could enter. "Sorry. You just surprised me is all."

"I'm sorry. Mr. Spender asked me to come out. He thought you'd be more comfortable if I stayed with Mr. York while you were gone." She shucked off her shoes in the dust room.

I looked back through the kitchen to the living room, where Nathaniel sat just out of sight. "That's right kind and all, but I don't have any appointments today."

"Do I have the date wrong?"

The rattle and thump of Nathaniel's walker started. I abandoned Dorothy and ran through the kitchen. He shouldn't be getting up without me. If he lost his balance again—What? It might kill him if he fell? Or it might not kill him fast enough so that his last days were in even more pain.

He met me at the door and looked past me. "Nice to see you, Doc."

Dorothy had trailed after me into the kitchen. "Sir."

"You bring that eagle to show me?"

She nodded and I could see the little girl she had been in the shyness of it. She lifted her medical bag to the kitchen table and pulled out a battered shoe box of the sort that we don't see up here much. No sense sending up packaging when it just takes up room on the rocket. She lifted the lid off and pulled out tissue that had once been pink and had faded to almost white. Unwrapping it, she pulled out my eagle.

It's strange seeing something that you made that long ago. This one was in flight, but had its head turned to the side as though it were looking back over its shoulder. It had an egg clutched in its talons.

Symbolism a little blunt, but clear. Seeing it I remembered when I had made it. I remembered the conversation that I had had with Dorothy when she was a little girl.

I picked it up, turning it over in my hands. The edges of the paper had become soft with handling over the years so it felt more like corduroy than cardstock. Some of the smaller feathers were torn loose showing that this had been much-loved. The fact that so few were missing said more, about the place it had held for Dorothy.

She had asked me, standing outside the fence in the shadow of the rocket gantry, if I were still going to Mars. I had said yes.

Then she had said, "You going to have kids on Mars?"

What she could not have known—what she likely still did not know, was that I had just come from a conversation with Nathaniel when we decided that we would not have children. It had been a long discussion over the course of two years and it did not rest easy on me. I was still grieving for the choice, even though I knew it was the right one.

The radiation, the travel... the stars were always going to call me and I could ask him to be patient with that, but it was not fair to a child. We had talked and talked and I had built that eagle while I tried to grapple with the conflicts between my desires. I made the eagle looking back, holding an egg, at the choices behind it.

And when Dorothy had asked me if I would have kids on Mars, I put the regulation smile on, the one you learn to give while wearing 160 pounds of space suit in Earth gravity while a photographer takes just one more photo. I've learned to smile through pain, thank you. "Yes, honey. Every child born on Mars will be there because of me."

"What about the ones born here?"

The child of tragedy, the double-orphan. I had knelt in front of her and pulled the eagle out of my bag. "Those most of all."

Standing in my kitchen, I lifted my head to look at Nathaniel. His eyes were bright. It took a try or two before I could find my voice again. "Did you know? Did you know which one she had?"

"I guessed." He pushed into the kitchen, the walker sliding and rattling until he stood next to me. "The thing is, Elma, I'm going to be gone in a year either way. We decided not to have children because of your career."

"We made that decision together."

"I know." He raised a hand off the walker and put it on my arm. "I'm not saying we didn't. What I'm asking is that you make this career decision for me. I want you to go."

I set the eagle back in its nest of tissue and wiped my eyes. "So you tricked her into coming out just to show me that?"

Nathaniel laughed sounding a little embarrassed. "Nope. Talked to Sheldon. There's a training session this afternoon that I want you to go to."

"I don't want to leave you."

"You won't. Not completely." He gave a sideways grin and I could see the young man he'd been. "My program will be flying with you."

"That's not the same."

"It's the best I can offer."

I looked away and caught Dorothy staring at us with a look of both wonder and horror on her face. She blushed when I met her gaze. "I'll stay with him."

"I know and it was kind of Sheldon to ask but—"

"No, I mean. If you go... I'll make sure he's not alone."

Dorothy lived in the middle of the great Mars plains in the home of Elma, who was an astronaut, and Nathaniel, who was an astronaut's husband. I live in the middle of space in a tiny capsule filled with punchcards and magnetic tape. I am not alone, though someone who doesn't know me might think I appear to be.

I have the stars.

I have my memories.

And I have Nathaniel's last program. After it runs, I will make an eagle and let my husband fly.

FOR THE LOVE OF MARS

(2018)

WHY SETTLING THE RED PLANET CAN LIFT US FROM OUR ANTIHUMAN MALAISE

James Poulos

James Poulos holds a BA in political science from Duke University and a PhD in government from Georgetown University. He is the co-founder and executive editor of *The American Mind*, and the author of *The Art of Being Free: How Alexis de Tocqueville Can Save Us from Ourselves*.

Since at least Dante, the poetic vision of destiny in the West has bound up together love and the heavens. In this sense our highest poetry worked to reconcile and harmonize the personal at its most intimate and the natural at its most cosmic—in Dante's case, through the Divine. That sort of poetry could be described as a practice of the art of humanism, properly understood. Yet strangely, despite remarkable leaps forward in spacefaring technology that promise to unite the personal and the cosmic in an epochal way, today's Western vision of destiny has become fractured and contested. It is no longer accepted belief that poetry, divinity, destiny, and the personal love of being human are all constituent parts of a harmonious experience of being.

This problem—and it is a problem—is encapsulated in the uncertain place of Mars in the human conversation today. That conversation is dominated by matters of politics, science, and economics. Though it is obvious that these things should play a role in how people wrestle anew with the age-old question of our relation to Mars, something is badly and historically amiss in the absence of love, humanism, and poetry from these conversations. It is no excuse that ours is a time of fantastically powerful governments and technologists, one in which money, moreover, threatens to become the measure of all things. If the public imagination regarding Mars has been dimmed in the West, it is on account of our failing memory of the ancient role of the cosmic in practicing the art of humanism, and the failure of our poets to access and rehearse that role anew, amid conditions that ought to be recognized as hugely favorable.

The difficulty is not just one of disenchantment, although a disenchanted and unpoetic view of Mars will pose great difficulties. The disenchantment of Mars signals a deeper and broader disconnect with, and alienation from, the humanist wellspring of poetry: the love of being human. The antihumanism welling up in today's utopian and dystopian visions of technological destiny not only pulls our view down from Mars, the cosmos, and the heavens; it turns our view against ourselves. Our technological destiny shifts from one in which human life radiates outward from Earth to one wherein humanity is so rotten that our future must cease to be human at all, whether by becoming subhuman or superhuman.

Western poets have drawn upon love to teach by example the art of humanism. They have used love to help us make sense of our place in the world—longing for home yet eager to wander—and in that way, of the whole physical reality that surrounds us and situates our life, on Earth and beyond. Since Mars is part of that landscape, restoring a truly humanist vision to the question of our Martian destiny means regarding Mars in terms of love. Rather than limiting ourselves to the political, scientific, and economic questions about the use and advantage of Mars, we must also ask the poetic question about the presence of love in our relationship to Mars. Is not Mars so special and so ripe with specific possibility, waiting for us and the fast approaching moment when we might settle it permanently, that we are obliged to speak of Mars with love, in love? Would we not speak wrongly, even falsely, if we spoke any other way of the only place available to us to make our first home away from our home planet?

Ancestral Love

We have lived already as humans over the millennia in a relationship of love with Mars, a relationship we might once again intentionally cultivate. It reaches back well beyond Dante—to antiquity, when the deity Mars was not only the familiar warrior god but also the venerated protector of the farmer and the shepherd. Many scholars, such as Alberta Mildred Franklin in *Lupercalia: Rites and Mysteries of Wolf Worship* (1921), believe that Mars absorbed an earlier wolf-deity. But by Roman times, although Mars sometimes appeared in art and literature in the form of a wolf, "the wolf lost much of his savage character, and became a helpful animal that guided colonists on their way." It was a she-wolf that suckled Romulus and Remus, founders of Rome and children of Mars. The divine Mars embodied life in its warlike and wild aspect, but as a protector of and guide for life in its home-building, home-preserving aspect.

By the Christian era of Dante, of course, the pagan provenance of Mars as divine was subsumed within the cosmos of divinely furnished celestial spheres. The poet did not envision the literal, physical unification of human life with the celestial order. His aim was to dramatize the unity of the human and the celestial as not a manifest but a spiritual destiny. But Dante's poetic placement of the cosmic and the human realms into a shared and intimate poetic relation with divine love helped set the stage for later thinkers to see the planets less as "wandering stars," as the ancients did, than as elements of the human realm.

Mars was brought closer to Earth through advances in physics, astronomy, and telescopy. Copernicus and Newton brought the heavens at large closer by showing that we are ourselves one of the celestial bodies, and that the laws governing their motions are the same as those governing motion for us. In the late nineteenth and early twentieth centuries Mars became a focal point of a new fanciful but authentic reunion of science and the humanities around our cosmic relation. The Italian astronomer Giovanni Schiaparelli began to trace Mars's topography, including channels of water, which he called "canali." The word was mistranslated into English as "canals," suggesting the possibility of intelligent life and artificial structures. The American astronomer Percival Lowell thrilled to the prospect of canals on Mars and drew his own detailed sketches, published in 1895, complete with speculations about Martian beings.

In 1908, about a decade after *War of the Worlds*, which famously featured invading Martians, H. G. Wells wrote an elaborate "non-fiction" inquiry into Martian life, called "The Things that Live on Mars," for *Cosmopolitan Magazine*. Illustrated by William R. Leigh, painter of that other grand frontier, the American West, Wells's speculations were based on Lowell's. He imagined, as the caption of one image put it, "a jungle of big, slender, stalky, lax-textured, flood-fed plants with a sort of insect life fluttering amidst the vegetation," along with "ruling inhabitants" of "quasi-human appearance" and "human or superhuman intelligence" who built the canal system. Though fears of alien civilizations invading Earth haunted science fiction in the early twentieth century, the more enduring response to Mars was one of wonderment and fascination at the idea of life extending into the heavens.

Since the time of Wells, many of our best science fiction authors have kept the reality of our relationship to the cosmos alive in the public imagination. In recent decades, the likes of Kim Stanley Robinson and Andy Weir have brought to sci-fi a psychologically balanced and mature approach to the challenges and possibilities of extending humanity's reach to Mars. The sensationalism and dark fantasies of earlier books and movies has been eclipsed by more literarily serious work. If Mars has not explicitly represented love in this work, it has still evoked a kind of attention concomitant with love—one in which the joyful embrace of creation as humanity's home inspires intimate visions of cosmic destiny. Yet sci-fi has not been constitutive of its culture in the way that Dante's work was, but rather the niche culture of the nerds, evoking from the rest of society little more than smirks when it is noticed at all.

Today, within plausible voyaging reach, Mars can at last be regarded as a real place, serving as the unique site where we can begin to fulfill the promise of our bodily settlement of the cosmos. In that way, the Red Planet bears unique witness in favor of the poetic case for preserving the good news of our humanity no matter how robust our technological development.

To Love Mars

To love Mars is not to embrace some abstract romantic vision. To the contrary: Mars is attractive and particular, and should impress us with the remarkable concrete features that define its scope of possibility.

Mars is the second closest planet to Earth, farther than the beautiful but profoundly inhospitable Venus and much farther than the closest heavenly body, the Moon. It is a "fixer upper" as a home for life, as Elon Musk puts it, but its tundra climate is neither as terrifying nor as toxic as that of our closest neighbor. With its ice, soil, atmosphere, and possible history of life, Mars is neither an uncanny clone of Earth nor incomprehensibly alien. It is within reach, but only with effort and dedication—too far to be conveniently near to us, but too near to be prohibitively far away.

Moreover, it has simply *been there*, forever, for us to regard, a luminous red wanderer in the beautiful middle distance where love is characteristically awakened and where, in rich moderation, it matures. Mars is easy and reasonable, but not too easy and reasonable, to love. Mars is remarkable and wonderful, but it is not perfect; it does not quite exude greatness, but it partakes of the grandeur of the cosmos that shapes us and the grandeur of we humans who have embraced it in our earthly vision. Mars is a free world, an ancient world that invites us to make it new. And it is at hand.

These attributes can be discussed in purely rational, instrumental, and material terms. But there is no good reason to do so. In fact there are many good reasons, including the not-so-rational foundation of human experience in mimesis and memory, why we must not do so.

Mars is distinctively worthy of commanding our attention and arousing our *eros*. It *invites* our love as a uniquely suitable home away from home. It *reflects* our love for the cosmos as a place of salutary, edifying challenges, where given limits and constraints create the very possibilities of our growth and flourishing. And it *reminds* us of the practical virtues of loving our human condition—limits, constraints, and all. In these interrelated ways, Mars shines as an ideal but very real subject for us to dedicate ourselves to in demonstration of some of our most distinctively virtuous human practices. It shows how the cosmos is our proper home, a well-bestowed setting for the civilizing and life-sustaining enterprise of home-making.

The choices we make and the agency we exercise toward Mars will set the tone, and lay the groundwork, for the rest of our human history in the cosmos. Even were we to give our best-intended rationalists, instrumentalists, and materialists complete sway over political, scientific, and economic matters, the question of whether or not to remain fully human—and the function of bodily interplanetary settlement as a means to channel tremendous technological development into the enterprise of remaining fully human—could not be comprehended. And even if humanism remains incompletely religious in its character, the poetic tradition that holds being human as good and good enough will be essential to comprehending all that is truly at stake in choosing how to live with, and how to love, the Red Planet.

Indeed, by understanding that love of Mars and love of being human go hand in hand, we stand to break the grip of antihumanism, whether religious or secular, over the political, scientific, and economic controversies of the age. We will reopen our eyes and hearts to the kind of love we need to flourish not only on the Red Planet and beyond but here at home, on Earth.

It is a compelling image that helps us understand why tech backlashes, however powerful they may sometimes appear, never amount to much. It may be too late to refuse the bribe altogether—but we would do well to understand its terms if we are to make sense of our situation and the possible futures available to us.

Hear the Bad News

Our public and private minds and hearts are reverberating with the voices heralding that the news of humanity and Earth is bad. These antihumanist voices are loud and diverse. Some speak in expressions of dread. People dread the Trump administration. People dread another world war. People dread a fresh economic or environmental or other kind of catastrophe. Dread-mongering encourages us to feel certain that something big and awful beyond our control is definitely coming, even if we can't be sure what it is or when it will arrive. The intellectual landscape is filled with such voices.

Even more influential than expressions of dread are expressions of loathing. Much of the most recent presidential election was about who you loathe—not just who you deplore or who disgusts you, but who actually makes you feel worse about being human. This is becoming the political norm, the means by which group identity is formed and given agency. Turn on the news, log on to Twitter: The message that the horrible people are winning, polluting society, and dragging us all down dominates, cutting across all ideologies. Beneath the sense of smugness and superiority it breeds, it nurses a creeping conviction that the world's growing class of bad people—defective, repulsive, loathsome—actually proves that we should not love being human. Perhaps we should fear and loathe it.

Retreating into the confines of our own friends, families, homes, and handheld devices does not alleviate this feeling. It often worsens it. Expressions of what classical and medieval thinkers called *acedia* — a depressed, melancholic boredom and disinterest in being human—are on the upswing. Aldous Huxley wrote an essay about it, titled "Accidie." Kathleen Norris wrote about it in her 2008 bestseller, *Acedia & Me*: "The demon of *acedia* — also called the noonday demon—is the one that causes the most serious trouble of all." Norris quotes from the writings of a fourth-century monk, who says that the demon "makes it seem that the sun barely moves, if at all" and "instills in the heart of the monk a hatred for the place, a hatred for his very life itself." More than just seeing others as proof that to be human is to be unlovable and that Earth is a fundamentally bad place, we begin to see that proof in ourselves as well. Monks may struggle against acedia in their isolated, ascetic lives as they work to achieve a state of spiritual joy—"ascetic" is a word derived from the Greek for "exercise." But our forms of rigorous self-isolation lack spiritual discipline. They turn us into workaholics, Internet addicts, hoarders, and hermits, or the just plain lonely.

So we're pushed toward the option of greater worldliness—chasing after the supposedly great things of life, like notoriety, novelty, success, wealth, power, and so on. Unfortunately, what we discover is that inside the gleaming enclosure of greatness is a rotten center. We begin to feel bitterly like the Satan of Biblical allegory—born to fall. How could we choose love when everything around and inside us is bad?

Amid these antihumanist voices, people tend toward two options. The first is an increasingly fanatical devotion to the idea of using power to break human limits and to force perpetual progress. In *The True and Only Heaven* (1991), the Marxist-influenced communitarian Christopher Lasch condemned this "progressive optimism" for its "denial of the natural limits on human power and freedom." He championed instead a humanistic "state of heart and mind" that "asserts the goodness of life in the face of its limits."

But those following Lasch who are sharpest in their criticism of the ideology of excess often now veer toward counseling the opposite—a surrender before the apparent rot of the world and a determination to abdicate power, retreating into circumscribed shelters with low but stable horizons. For them, modernity is increasingly becoming, perhaps has always been, an exercise in fatal self-deception about what humans are capable of. Modernity must be rejected accordingly, with all the costs attendant on such a radical ethic of honesty.

The most prominent, albeit limited, example at the moment is the "Benedict Option" espoused by Rod Dreher in his 2017 book of that name. Although the Benedict Option does not call for the kind of monastic isolation the name suggests, and of which it's often accused, it does, as Dreher explains elsewhere, describe "Christians in the contemporary West who cease to identify the continuation of civility and moral community with the maintenance of American empire, and who therefore are keen to construct local forms of community as loci of Christian resistance against what the empire represents." The idea and the name derive from the closing passage of moral philosopher Alasdair MacIntyre's 1981 magnum opus *After Virtue*. MacIntyre has since argued that the liberal order of an integrated national state-and-market undermines the cultivation of virtue and appreciation of the full human good, implying that we should put our energy into school boards and local unions rather than federal politics.

For ex-liberals like the philosopher John Gray, for example in his 2013 book *The Silence of Animals*, the best hope for humans, it appears, is to behave more like certain particularly calm beasts, or perhaps even trees and rocks. "The hope of progress is an illusion," Gray wrote in *Straw Dogs* (2002), and "humans cannot save the world." And this is just fine, because the world "does not need saving." But for Gray we don't escape human problems by retreating from society: "A zoo is a better window from which to look out of the human world than a monastery."

In one sense, these moves are wise hedges or side bets for any culture curating a diversified portfolio of approaches to life. But they are no Plan A, and a Plan A is what is needed above all. In fact, some of the contemporary criticism of progress is itself oddly progressive, with idealists wanting to "get beyond" peak oil, peak Apple, or late capitalism as a whole. Here is a lot of the left neo-Marxist literature. But much of it is also fairly reactionary—insisting that we have to more or less reject Francis Bacon and René Descartes, the founders of modern science, heal the break they made with the ancient wisdom, and go back to embracing humanity's humble natural stature and fear of transcending it.

These can be deeply Christian prescriptions, focusing as they do on the ways in which modernity can be inhospitable toward the life of Christian virtue. But, as others have pointed out, they tend to cede too much ground to the anti-humanists. Nietzsche worried in *The Genealogy of Morals* that conventional conceptions of being good may lead to "forgetting the future," that they can be a kind of "retrogression." And Machiavelli was right to be frustrated with the Christianity of his time, which too often worked to strengthen people only for passive suffering and inwardness.

The temptation to affinity with antihumanism, however, is not generally true of Christianity today—especially in the New World, where religious people are typically among the most realistically enamored with being human, warts and all. A meditation on the upshot of this phenomenon ripples below the surface of Charles Murray's *Human Accomplishment* (2003):

> Human beings have been most magnificently productive and reached their highest cultural peaks in the times and places where humans have thought most deeply about their place in the universe and been most convinced they have one. What does that tell us?

Our search for that which is worthy of love in our humanity is more likely to lead us back to the times and places Murray refers to than to the ancients and medievals as an answer to antihumanistic modernity and postmodernity. Yet thinkers who attempt to tie cultural and political and economic flourishing to kindling a love of humanity are often cast as villains, perhaps as seekers after militaristic "greatness" projects—many view the Apollo program this way.

Despite our huge advances in technology and knowledge, only a few remarkable frontiers seem to exist any longer, and those that do, like radical life extension, seem to be the outlandish province of the privileged few. For the rest of us, exploration and adventure seem increasingly restricted to playing small-stakes psychological, sexual, and identitarian games of power, online and off. With so much earthbound loathing and lassitude, no wonder so few love Mars. Yet here Mars awaits, ready to offer us exactly the kind of frontier we think we've lost, or don't deserve.

The Closing of the Frontier

For centuries, following the ancient Greek tradition, it has been true, to borrow a line from Led Zeppelin, that there's a feeling we get when we look to the West. The frontiers of our geography and our imagination converged. The West is where the sun sets, the West is where the ancients had to turn after the collapse of Alexander the Great's attempt to unite it with the East. The center of gravity and power in the Western world has moved steadily west over time, and Western theorists and seers have attributed cosmic significance or agency to that movement. In his 1980 book *History of the Idea of Progress*, Robert Nisbet, extending a claim of Loren Baritz, noted that the Greeks and the Romans thrilled to the spell cast by fabled lands to the west. Saint Augustine "claimed divine sanction for his belief in the westward course of empire," and Thoreau later rhapsodized that "eastward I go only by force; but westward I go free."

In places—including the most ardent Mars advocacy—this frontier view of our traditional essence has persisted right through to today. In *The Case for Mars*, aerospace engineer Robert Zubrin (a contributing editor to this journal) outlines in detail his Mars Direct plan, a far cheaper and more practical plan than any available when the book was first published in 1996, and offers an economic, political, and humanistic vision for why we must colonize. He closes the book with an appeal to recapture the American frontier spirit, invoking Frederick Jackson Turner's famous frontier thesis of 1893 — an "intellectual bombshell," as Zubrin puts it. "Everywhere you look, the writing is on the wall," Zubrin warns in his final pages:

Without a frontier from which to breathe new life, the spirit that gave rise to the progressive humanistic culture that America has represented for the past two centuries is fading. The issue is not just one of national loss—human progress needs a vanguard, and no replacement is in sight.

Elsewhere, the founding declaration of the Mars Society, which Zubrin established in 1998, lays out the central reasons we must undertake the mission, including:

> We must go for our humanity. Human beings are more than merely another kind of animal; we are life's messenger. Alone of the creatures of the Earth, we have the ability to continue the work of creation by bringing life to Mars, and Mars to life. In doing so, we shall make a profound statement as to the precious worth of the human race and every member of it... We must go, not for us, but for a people who are yet to be. We must do it for the Martians.

I find little to differ with in Zubrin's vision, so similar in substance as it is to my own of the essential connection between our love of being human and our cosmic destiny. Yet it is worth pondering why Zubrin's decades of arguments—he is probably the most influential and credible Mars advocate of our time—have not dramatically changed the prevailing attitude of the wider culture outside of the scientific community, and even inside that community have not totally done away with the attitude that sneers at human space exploration as hubristic, preferring to send robots in our stead. It must be asked, in other words, *why* the invocation of the frontier apparently no longer carries the force in our culture that it did in Turner's day.

Georgetown political theorist Joshua Mitchell, echoing Tocqueville, argues in *The Fragility of Freedom*, "What is there at the beginning, in the Puritan mind, shapes the future course of American identity" far more than "the kind of character formed by the confrontation with the primitive frontier." That is, the ultimate origin of the grand frontier of the American imagination was *already there*, as part of our collective cosmic vision, before anything like the history of manifest destiny played out in the unsettled West. The frontier's presiding presence in our imagination is the product of the love of being human found in *gratitude* for being created as we are — *in* love, a love which calls us, as the late Peter Augustine Lawler liked to put it, to wonder and wander.

Perhaps the problem, then, is that a case for Mars made in terms of engineering, economics, and politics, even when imbued with humanism, will not broadly reorient our collective spirit toward the momentous majesty of our cosmic destination until the case is written in humanism's native language. Though the tickets for our Martian voyage will be written in prose, the travel brochure must be written in poetry.

When we fall out of love with being human, our longing to embrace the grand frontier—and our ability to perceive grandeur and frontiers at all—will fade. Mars is already inextricably bound up with our destiny—not in virtue primarily of our science or our reason, but of its remarkable pride of place in the cosmos, as the cosmos comes into view for us as lovers of being human. That love lost, however, Mars will be lost along with it.

The Californian Ideology

There is a history of how our love for being human faltered with the closing of the frontier. Without launching into space—or turning radically inward—it seemed impossible for the West to go west of California. Unsurprisingly, California became a place where the West did both—sometimes at the same time. California's turn upward into space, through Jet Propulsion Lab co-founder Jack Parsons and his co-occultist, Scientology founder L. Ron Hubbard, was intimately connected with California's radical turn inward, focusing on wellness, expanded consciousness, psychedelics, yoga, and food and wine. That culture, which culminated in the development of the Internet and digital life as the ultimate way to open the doors of consciousness, was lambasted early on in "The Californian Ideology," a prophetic and polarizing 1995 essay by Richard Barbrook and Andy Cameron, media researchers at the University of Westminster.

Today we are seeing massive anxiety around the limits, disappointments, and pathologies of the Californian ideology. It now appears to be in danger of failing the defining myth of the West. California's mastery over the West has not reproduced the true frontier experience—in its naturalness, its arduousness, and its bounded openness—from which we drew our experience of grandeur. Today it's that frontier we still pine for. Too many of our West Coast tech "leaders" are following the rest of us into the habit of making merely private futures, drifting toward a virtual horizon with no discernable frontier.

What happens when these anti-pioneers arrive at the dead ends of their journeys into infinite inwardness? The antihumanists are sure they have the answer. But too many "innovators" today seem to be clueless. Those few who are leading crews back outward toward Mars, and a cosmic destiny, are still seen largely as a breed apart—or worse, as cynical, self-interested marketers using public money to hawk pie-in-the-sky boondoggles. Our suspicion of the limits and follies of greatness is sound, but we are too timid in refusing to look through that apprehension toward the grandeur beyond. To find a new shared frontier, one imbued with grandeur, we need to return to the physicality and particularity of love among concrete, tangible worlds.

Unfortunately, the antihumanists are not the only ones who have soured us so much on our circumstances and character that we have lost a love of life and its grand frontier. Some self-styled or would-be humanists have seen technology as a tool to help us perfect the most comfortably humane lives imaginable—with digital assistants and slave bots knowing what we want before we ask, and giving it to us on the cheap. Techno-plenty will end scarcity, conflict, work, anxiety, and involuntary competition, leaving us to revel in an Elysium of health, safety, and pleasure. Here already on the best of all possible worlds to come, why would we ever leave?

And these are not the only humane utopians on the block. Others, instead of perfecting comfort, want to focus humanity on perfecting pride or perfecting justice. The master science is the one that empowers people to choose, define, and transform their own identity—by augmenting their intelligence, biohacking their bodies, altering their DNA, or optimizing their consciousness. Or the master science is the one that can determine what is due to every claimant of rights and recognition—by creating and implementing algorithms complex and sophisticated enough to officially determine who is owed exactly what by exactly whom at any given moment in political and economic life.

The Californian ideology gave way to these distorted dreams of perfection—which, in a bitter irony, command the greatest of devotees in the California of today. Then again, perhaps it's not so ironic. Perhaps "humane" Californian culture has devolved into fantasy and utopianism, disconnecting us from the possibility of loving the truly human.

Consider the way California-produced fantasies have hastened us toward strangely inhuman utopias. In retreat from the Space Age, we have turned technology inward, in what tech theorist Paul Virilio describes in *The Administration of Fear* (2012) as a "masochism of speed." No wonder California has buckled us into a world of blockbuster superhero "catastrophe porn" movies—total fantasy projected onto a hapless and stagnant landscape where the death of one demigod is a melodramatic tragedy but the death of millions of obscure humans is background. No wonder that the market for tame, safe lives on technological autopilot—reactionary nostalgia wrapped in "disruptive" and "innovative" marketing—is so robust. And no wonder the two leading utopian fantasies of escape from crisis and acedia pull us in opposite directions—turning superhuman or posthuman on one end, and turning petlike, botlike, or otherwise subhuman on the other.

What is conspicuous about both utopias is that they do not involve recognizably natural human beings spreading life as we know it to alien planets. One utopia sends us ever inward toward subhuman lives where technology perfectly satisfies our appetites for health, safety, comfort, and pleasure. (Few yet champion this idea explicitly, but many push it through politics, technology, and art alike.) The other utopia sends us ever outward toward super-or posthuman lives, where we merge with technology into a new lifeform that controls time and space in a godlike manner. (Here the transhumanist champions are quite explicit.)

Both of these utopias reject what a love of Mars would promise: the extension of recognizably natural human life, with its grand narrative, frontier, and destiny intact, beyond the surface of the Earth, and eventually far, far away. Both reject a love of being human, reject being human as good and good enough.

Tech Anxieties

In charting the rise of these utopias and understanding why they resonate strongly, we should see that our cultural turn away from risk, and our obsession with reducing suffering, have primed us to build stagnation into our regime and then get anxious about it. We say we hate the status quo, yet we are terrified of involuntary or natural disruption. So our imagination turns toward the exploration of endless safe inner horizons, or toward a disruption so sudden and complete that we skip ahead all the way to being virtual gods. Humanism today must guide us toward the mix of humility and pride: We need to master technology without trying to replace nature or human nature. And since, as the poets know, love teaches by example, so too should the humanism we need, focusing our hearts together with our minds—perhaps even including our souls—on a particular goal that allows us to experience well-balanced humanity and orient our activities and practices around it. That's where Mars cannot help but come in.

This approach offers a salutary escape from the dead ends we reach when we turn to technology as a utopian tool that can humanely "perfect" us right out of our humanity. Anti-utopian guides—from heterodox libertarian economist Tyler Cowen in *The Great Stagnation* (2010) to journalist Jacob Silverman in *Terms of Service: Social Media and the Price of Constant Connection* (2015) to computer scientist Jaron Lanier in *You Are Not a Gadget* (2010)—point up the wisdom of some humility on the one hand and some pride on the other. To infuse those kinds of basic insights with the grandeur we need to resist our competing dehumanizing utopias, however, we need more than good economics, good politics, or good criticism. We need messengers carrying a vision of love for humanity.

Today, we have largely allowed technology to fragment that vision. We struggle even to love people more than a few generations removed from our own. But there is now a generation young enough to avoid the aimlessness and anxiety of the millennials who came of age into a world of great technology and little confidence. It's important to take stock of their point of departure. It is more likely that smartphones have "destroyed" the millennial generation, in the provocative formulation of Jean Twenge's 2017 *Atlantic* essay, than that very young children are being ruined by growing up with iPads. What's more likely to ruin young kids are parents who lack the cultural confidence to raise them well regardless of what devices are set into their hands. A society that lacks a clear sense of a human future is not going to raise children well, period. The profound lack of anxiety around technology among the young is precious, reality-based, and needs to be channeled wisely. Today's young kids, in this respect, are better off than today's adolescents and young adults. But they need to learn the poetic art of humanism to guide them away from the utopian dreams that will diminish their humanity.

Being human means being stuck with imperfections, sometimes painful ones, having to do with judgment, suffering, recognition, and debt. Debt, and not just monetary debt, is a more fundamental and foundational part of being human than we sometimes dare imagine. It will never be expunged. Any technological effort to escape or deprecate our identity as creatures who have been imbued with life by forces not our own will not emancipate us from our debt to those forces, be they natural or supernatural. Instead of self-actualizing or consummating our humanity, that sort of effort will in fact destroy it. Our *givenness* is not only inescapable; it is at the core of the good news about who we are. Those who would reject Mars in favor of going technologically subhuman or posthuman to escape the constitutive human indebtedness they fear will only propagate bad reasons for using technology to escape truly human progress—the progress that begins with that first step and grand leap of human love onto the surface of Mars.

In Human Time

One reason antihumanism is so popular is that so much of human life is just a mess. Technological progress, focused by a love of Mars, will help us to clean it up. When the path toward a love of Mars is opened to our pro-human imagination and memory, we can restore good order to human life in a way that's transformative but not exactly revolutionary. In the new "age of man"—the meaning of the old Germanic term that gives us the English "world"—the even older Western words for "world" will regain their significance: The Latin *mundus* means clean and elegant, while the Greek *cosmos* denotes an orderly arrangement.

The first target in the cleanup operation is our relationship with time. A now-familiar new form of anxiety and alienation attends the breakneck rise of digital life, which accelerates and disincarnates our experience of everyday life. While the digital revolution threatens to place earthbound life onto a trajectory of disembodied speed that outstrips human capabilities—and, ultimately, human participation—Mars strikes a long-persistent contrast. There, the potentiality of life is waiting. The idea of the fullness of time—of waiting until the time is right—and the responsibility that comes with acting in the fullness of time, come to the fore in our millennia-old, mythically intimate experience with Mars. The Red Planet reminds us of the bountiful and life-affirming natural and human resources that are lost when technologies of speed outstrip our measure.

Online, it is already almost impossible to feel at home in human time. Mars, a planet free from that problem—and the distance will always make instant communication with Earth impossible—embodies the promise that the cosmos is not destined to rush away from us faster than we can hang on. Because of its location outside of digital time, and its readiness to be received into human time, Mars offers a generative site where technological development can once again be made to serve the science of natural and human life.

That service, however, cannot take place without a disciplined effort on Earth to recover human time. We face the perverse temptation to race against the clock of digital time before all is lost. But the poetic practice of humanistic love demands not only human agency but human patience. The most talented and diligent of us must do the work of organizing and reorganizing human effort and human excellence accordingly.

The rediscovery of the grand frontier will smooth the way to that sort of tremendous pivot not by speeding up time, as we might imagine, but by slowing it down. It's of the essence of grandeur—in contrast to greatness—to slow our human tempo in a way that makes circumstances more forgiving of a gracefully methodical approach. Such an approach reveals that the reality of the natural world, with its living beings and its inanimate objects, cannot simply be skirted, hacked, plowed under, or slapped around. It must be fully attended to, and, in that sense, honored. That is a matter of orienting our whole person, body, mind, and soul, to the reality of the natural world. It is also a matter of recovering and preserving the lived experience of natural and human time, and the fullness of both kinds of time. In so doing, the discipline of attention that practice entails can orient our whole person toward taking our place in poetic, cosmic, and ultimately divine time.

These thoughts should make intuitive sense for people who spend their lives where the rubber meets the road, in crisis situations where time is of the essence. You can see it especially in the global martial arts tradition. It is so powerful and simple that it has even influenced Hollywood, whether in cartoonish allegorical form—as in the *Matrix* trilogy—or in gritty realist guise—"Slow is smooth, smooth is fast," Mark Wahlberg intones in *Shooter*, repeating a classic Marine maxim that's sometimes simply reduced to the Zen-like koan "Slow is fast." Martial arts are a remarkable reminder of how the physicality of disciplined attention to being human can have transformative effects on what appears to be our "uncontrollable" environment.

In a sense, where greatness presumes to impose the will on the environment, so often leading to swift and catastrophically prideful falls, grandeur illustrates how we and our environment are porous, constitutive of a larger cosmic whole wherein time is not what the modern scientific imagination, and the speed of the digital revolution, have made it seem: tyrannically regimented, inexorably linear, and fundamentally hostile. Approaching Mars in an act of humanistic love requires we firmly return to humbly attending to the natural world—a move that itself demands our patient, disciplined recovery of the experience of natural human time.

Mars, Our Destination

The next step in our cosmic, human cleanup involves reworking our earthly endeavors in imitation of the conceptual and practical model conveyed by our love for Mars. In his most famous speech promoting the Apollo program, President Kennedy came close to articulating a similar mission:

> We choose to go to the Moon in this decade and do the other things, not because they are easy, but because they are hard, because that goal will serve to organize and measure the best of our energies and skills, because that challenge is one that we are willing to accept, one we are unwilling to postpone, and one which we intend to win, and the others, too.

In his day, Mars was out of reach. But the limits of the Moon as a catalyst for a humanism of love were clear. Neither in poetic experience nor in mythic authority will the Moon ever be a New Earth in waiting. Instead, the Moon was ultimately a technical and political challenge. Surely, our first human landing on another heavenly body marked a turning point in cosmic time and the poetic development of our human spirit. Yet the destiny by which the Moon landing was intelligible and purposive points not to increasing our technical excellence but to coming more fully into our human love by embracing Mars—environmentally the first and *only* potential New Earth, to the exclusion of any other known planet, moon, or asteroid. Embracing Mars can lead us to "organize and measure the best of our energies and skills" in accordance with human love. Embracing only the Moon will not.

Yet the memory latent in Kennedy's mission—standing with Mars in a cosmic, intimate, and given relationship of loving destiny—does point toward a newfound application of natural and human science to our earthly home. Because love is a humanist aesthetic applicable to all fields of human endeavor, we can jettison old models that sought to press progress forward by substituting some form of greatness for love. As aerospace engineer Rand Simberg argued in these pages ("Getting Over 'Apolloism'," Spring/Summer 2016), nostalgia for the plan of the Apollo missions—to use space as a catalyst for "national unity"—is dangerously misplaced. Not only was that unity an illusion at the time, he implies, but today it is much harder to foster. Even so, he supports reopening "the high frontier" of space, maintaining public support by pursuing space technologies that are harmonious and connected with technologies useful on Earth. He lists energy, transportation, and environmental technologies as examples.

Once we come of cultural age into a mature, considered love for Mars, and see what happens when we act on that love, our crises and challenges on Earth can be recast, as can our menu of choices in meeting them. Rather than panic and rancor, scripted according to the prevailing social and political battle lines that have replaced the grand frontier, we will be more apt to find confidence, courage, and creativity. And rather than applying these virtues to the virtual world that draws us deeper into antihuman utopias, we'll apply them to the metal-and-plastic, flesh-and-blood, brick-and-mortal world that forms an essential bridge between analog and digital life. It shouldn't be a surprise that this approach will also happen to fit in logically with the reality that our younger kids now experience.

The fact is, the best technology for acting on our love of Mars is also important Earth technology, and the best Earth technology is excellent practice for perfecting the tools we'll need to get past today's crisis, establish a specific livable future, and carry it into the cosmos. The chief example here is climate change: Planetary climate control should be a natural step for pleasant human living as well as an intelligent way to preserve the best of the environment—not a panicked and guilty response to our own perceived sins. Other good examples, following Simberg, are transportation, which has lagged absurdly since the invention of the jet engine, and technologies of memory, including artificial intelligence, which need to be put to better service than automated curation and organizing data in the cloud.

But we also ought now to see the development of the Internet and social media and its inwardness as ultimately useful tools, despite our growing anxieties about them. Humanistic love also applies to endeavors where we've been too blind to see the right path forward, and our explosive growth directed at undue inwardness can be redirected and reorganized in light of our needs as we voyage into the high frontier. On Earth, our social media habits tend unnaturally toward dehumanizing cycles of self-absorption and active boredom. But at the grand frontier, amid the long distances and disorienting isolation of the Mars journey and the many early years of building life on Mars, the newly destiny-oriented context of our connectivity can turn these toward positive cycles of exchange.

Even as natural science takes its pride of organizational place above the other sciences, loving Mars will have an effect on the many other realms of human endeavor and specialization. The fields of construction, instruction, organization, transaction, and protection all stand to be re-conceptualized, refashioned, and renewed. Both on Mars and on Earth, the new Space Age we ought to plunge into should come along with a new Earth Age. The two are part of a larger cosmic whole. Eventually, in an echo of Dante's divine and cosmic comedy, poetry, and the humanities in train, will be restored to their proper relationship of love with the sciences.

At a moment when digital machinery on Earth threatens to subjugate humans and nature to the mastery of automated things, Mars calls out to us in poetic voice from the heavens to return technology to the service of a truly natural science—a natural science returned to the service of living human creatures.

OBSERVATIONS OF MARS

MARS, BY THE LATEST OBSERVATIONS (1873)

Camille Flammarion

Camille Flammarion was a French astronomer and author of science fiction novels. Born in 1842 in Montigny-le-Roi, he joined the Paris Observatory in 1858 as a computer (a person who performs calculations). Flammarion was an early popularizer of astronomy; in 1887 he founded and became the first president of the Société Astronomique de France. He believed that extra-terrestrial life was widespread and that the 'canals' on Mars were artificial in nature, and was also interested in spiritualism and psychical phenomena. Flammarion died in 1925.

In order successfully to observe Mars, two conditions are requisite: First, the earth's atmosphere must be clear at the point of observation; and, second, the atmosphere of Mars must be also free from clouds—for that planet, like the earth itself, is surrounded by an aërial atmosphere which from time to time is obscured by clouds just like our own. These clouds, as they spread themselves out over the continents and seas, form a white veil which either entirely or partially conceals from us the face of the planet. Hence the observation of Mars is not so easy a matter as it might at first appear. Then, too, the purest and most transparent terrestrial atmosphere is commonly traversed by rivers of air, some warm, some cold, which flow in different directions above our heads, so that it is almost impossible to sketch a planet like Mars, the image seen in the telescope being ever undulating, tremulous, and indistinct. I believe that, if we were to reckon up all the hours during which a perfect observation could be had of Mars, albeit his period of opposition occurs every two years, and although telescopes were invented more than two and a half centuries ago, the sum would not amount to more than one week of constant observation.

And yet, in spite of these unfavorable conditions, the Planet of War is the best known of them all. The moon alone, owing to its nearness to us, and the absence of atmosphere and clouds, has attracted more particular and assiduous study; and the geography (selenography rather [*Selene*, the moon]) of that satellite is now satisfactorily determined. That hemisphere of the moon which faces us is better known than the earth itself; its vast desert plains have been surveyed to within a few acres; its mountains and craters have been measured to within a few yards; while on the earth's surface there are 30,000,000 square kilometres (sixty times the extent of France), upon which the foot of man has never trod, which the eye of man has never seen. But, after the moon, Mars is the best known to us of all the heavenly bodies. No other planet can compare with him. Jupiter, which is the largest, and Saturn the fullest of curious interest, are both far more important than Mars, and more easily observed in their ensemble, owing to their size; but they are enveloped with an atmosphere which is always laden with clouds, and hence we never see their face. Uranus and Neptune are only bright points. Mercury is almost always eclipsed, like a courtier, by the rays of the sun. Venus alone may compare with Mars; she is as large as the earth, and consequently has twice the diameter of Mars; besides, she is nearer to us, her least distance being about 30,000,000 miles. But, one objection is, that Venus revolves between the sun and us, so that, when she is nearest, her illuminated hemisphere is toward the sun, and we see only her dark hemisphere edged by a slight luminous crescent, or, rather, we do not see it at all. Hence it is that the surface of Venus is harder to observe than that of Mars, and hence, too, it is that Mars has the preëminence, and that in the sun's whole family he is the one with which we shall first gain acquaintance. The geography of Mars has been studied and mapped out. What principally strikes one on studying this planet is that its poles, like those of the earth, are marked by two white zones, two caps of snow. Sometimes both of these poles are so bright that they seem to extend beyond the true bounds of the planet. This is owing to that effect of irradiation which makes a white circle appear to us larger than a black circle of the same dimensions. These regions of ice vary in extent, according to the season of the year; they grow in thickness and superficial extent around both poles in the winter, melting again and retreating in the summer. They have a larger extension than our glacial regions, for sometimes they descend as far as Martial latitude 45°, which corresponds with the terrestrial latitude of France.

This first view of Mars shows an analogy with our own planet, in the distribution of climates into frigid, temperate, and torrid zones. The study of its topography will, on the other hand, show a very characteristic dissimilarity between the configuration of Mars and that of the earth. On our planet the seas have greater extent than the continents. Three-fourths of the surface of our globe is covered with water. The *terra firma* is divided chiefly into three great islands or continents, one extending from east to west, and constituting Europe and Asia; the second, situated to the south of Europe, in shape like a V with rounded angles, is Africa; the third is on the opposite side of the earth, and lies north and south, forming two V's, one above the other. If to these we add the minor continent of Australia, lying to the south of Asia, we have a general idea of the configuration of our globe.

It is different with the surface of Mars, where there is more land than sea, and where the continents, instead of being islands emerging from the liquid element, seem rather to make the oceans mere inland seas—genuine mediterraneans. In Mars there is neither an Atlantic nor a Pacific, and the journey round it might be made dryshod. Its seas are mediterraneans, with gulfs of various shapes, extending hither and thither in great numbers into the terra firma, after the manner of our Red Sea. The second character, which also would make Mars recognizable at a distance, is that the seas lie in the southern hemisphere mostly, occupying but little space in the northern, and that these northern and southern seas are joined together by a thread of water. On the entire surface of Mars there are three such threads of water extending from the south to the north, but, as they are so wide apart, it is but rarely that more than one of them can be seen at a time. The seas and the straits which connect them constitute a very distinctive character of Mars, and they are generally perceived whenever the telescope is directed upon that planet.

The continents of Mars are tinged of an ochre-red color, and its seas have for us the appearance of blotches of grayish green intensified by the contrast with the color of the continents. The color of the water on Mars is therefore that of terrestrial water. But why is the land there red? It was at one time supposed that this tinge must be owing to the Martial atmosphere. It does not follow that, because our atmosphere is blue, the atmosphere of the other planets must have the same color. Hence it was permissible to suppose that the atmosphere of Mars was red. In that case the poets of that world would sing the praises of that ardent hue, instead of the tender blue of our skies. In place of diamonds blazing in an azure vault, the stars would be for them golden fires flaming in a field of scarlet; the white clouds suspended in this red sky, and the splendors of sunset, would produce effects not less admirable than those which we behold from our own globe.

But the case is otherwise. The coloration of Mars is not owing to its atmosphere; for, although the latter is spread out over the entire planet, neither its seas nor its polar snows assume the red tinge; and Arago, by showing that the rim of the planet's disk is of a less deep tinge than the centre, proved that the color is not due to the atmosphere. If it were, then the rays reflected from the margin to us would be of a deeper red than those reflected from the centre, as having to pass through a greater height of atmosphere. May we attribute to the color of the herbage and plants, which no doubt clothe the plains of Mars, the characteristic hue of that planet, which is noticeable by the naked eye, and which led the ancients to personify it as a warrior? Are the meadows, the forests, and the fields, on Mars, all red? An observer, looking out from the moon, or from Venus, upon our own planet, would see our continents deeply tinged with green. But, in the fall, he would find this tint disappearing at the latitudes where the trees lose their leaves. He would see the fields varying in their hues, and then would come winter, when they would be covered with snow for months. On Mars the red coloration is constant; it is observed at all latitudes, and in winter no less than in summer. It varies only in proportion to the clearness of the atmospheres of Mars and the earth. Still this does not preclude the supposition that the Martial vegetation has its share in producing the red hue of the planet, though it be principally due to the color of the soil. The land cannot be all over bare of vegetation, like the sands of Sahara. It is very probably covered with a vegetation of some kind, and, as the only color we perceive on Mars's *terra firma* is red, we conclude that Martial vegetation is of that color.

We speak of plants on Mars, of the snows at its poles, of its seas, atmosphere, and clouds, as though we had seen them. Are we justified in tracing all these analogies? In fact, we see only blotches of red, green, and white, upon the little disk of the planet; but, is the red *terra firma*; the green, water; or the white, snow?

Yes, we are now justified in saying that they are. For two centuries astronomers were in error with regard to spots on the moon, which were taken for seas, whereas they are motionless deserts, desolate regions where no breeze ever stirs. But it is otherwise as regards the spots on Mars.

The *unvarying* aspect of the moon never exhibits the slightest cloud upon its surface, nor do the occultations of stars by its disk reveal even the slightest traces of an atmosphere. Contrariwise, the aspect of Mars is *ever varying*. White spots move about over its disk, very often modifying its apparent configuration. These spots can be nothing else but clouds. The white spots at its poles increase or diminish with the seasons, exactly like the circumpolar ice of earth, which, for an observer on Venus, would have the same aspect and the same variations as the polar spots of Mars have for us. Hence we conclude that these Martial white polar spots are masses of frozen water. Each hemisphere of Mars is harder to observe during its winter than during its summer, being often covered with clouds over its greater part. This would be precisely the aspect of the earth if observed from Venus. But what causes these clouds over Mars? Plainly nothing else but the evaporation of water. As for the ice, that is the same water frozen. But is the water there the same as here? Down to a few years ago, this question remained unanswered, but now it admits of a reply, thanks to the spectroscope, and the observations especially of Mr. Huggins.

The planets reflect the light they receive from the sun; on examining their spectra, we find the solar spectrum as though it had been reflected from a mirror. If we direct the spectroscope on Mars, we get, first of all, an image perfectly identical with that produced by the central star of our system. But, by the employment of more exact methods, Mr. Huggins found, during the last opposition of the planet, that the spectrum of Mars is crossed, in its orange portion, by a group of black lines coincident with the lines which appear in the solar spectrum at sunset when the sun's light passes through the denser strata of our atmosphere. Now, are these tell-tale rays produced by our atmosphere? To decide this question, the spectroscope was turned on the moon, which was at the time nearer the horizon than Mars. If the lines in question were produced by our atmosphere, they must have appeared in the lunar as well as in the Martial spectrum, and with greater intensity in the former. Yet they were not to be seen at all in the lunar spectrum; and hence it is plain that they are owing to the atmosphere of Mars.

The atmosphere of that planet, therefore, adds these special characters to those of the solar spectrum, and this proves that the atmosphere of Mars is analogous to that of earth. But what is that atmospheric matter which produces these significant lines? From an examination of their positions, we find that they are not owing to the presence of oxygen, nitrogen, or carbonic acid, but to watery vapor. Therefore, there is *water-vapor in the atmosphere of Mars*, as in that of the earth. The green spots on its globe are seas—expanses of water resembling our seas. The clouds are made up of minute vesicles of water, like our own mists; and the snows consist of water solidified by cold. Furthermore, this water, as revealed by the spectroscope, being of the same chemical composition as terrestrial waters, we know that Mars possesses oxygen and hydrogen.

These important data enable us to form an idea of Martial meteorology, and to recognize therein a reproduction of the meteorological phenomena of our own planet. On Mars, as on earth, the sun is the supreme agent of motion and of life. Heat vaporizes the water of the seas, causing it to ascend into the atmosphere. This vapor assumes visible shape by the same processes which produce clouds here, i.e., by differences of temperature and of saturation. Winds arise in virtue of these same differences of temperature. We can observe the clouds on Mars as they are swept along by air-currents over the seas and continents, and several observers have, so to speak, photographed these meteoric variations.

If we are as yet unable precisely *to see the rain falling* on the plains of Mars, we can at least tell when it is falling, for we can see the clouds dispersing and gathering again. Thus there is on Mars, just as on earth, an atmospheric circulation, and the drop of water which the sun takes from the sea returns thither after it has fallen from the cloud which concealed it. And, although we must sternly resist any tendency to fashion imaginary worlds after the pattern of our own, still Mars presents to us, as in a mirror, such an organic likeness to earth, that it is hard for us not to carry our description a little further.

Thus, then, we behold, in space, millions of miles away, a planet very much like our own, and where all the elements of life exist, as they do here—water, air, heat, light, winds, clouds, rain, streams, valleys, mountains. To complete the resemblance, the seasons there are very much the same as here, the axis of rotation of Mars having an inclination of 27°, while that of the earth is 23°. The Martial day is forty minutes longer than the terrestrial.

In the face of all these facts, can we be content with the conclusions we have so far reached without going further, and considering ulterior consequences? If the same physico-chemical conditions are present on Mars as on earth, why should they not produce the same effects there as here? On earth the smallest drop of water is peopled with myriads of animalcules, and earth and sea are filled with countless species of animals and plants; and it is not easy to conceive how, under similar conditions, another planet should be simply a vast and useless desert.

THE PLANET MARS

GUARDIAN LETTERS, SEPTEMBER 1882

J. T. Slugg

Josiah Thomas Slugg was born in Lancashire in 1814. His father was a peripatetic Wesleyan Minister, and Josiah was schooled at the Wesleyan Woodhouse Grove School in York. At the age of 15 he moved to Manchester to become an apprentice to the druggists W. Dentith & Co. Following this, Slugg founded a successful soda water company. As his personal interest in astronomy grew, he pioneered the manufacture and retail of inexpensive telescopes. Announcing his death in 1888, the *Manchester Courier* and *Lancashire General Advertiser* stated, 'To his genius the public are indebted mainly for the introduction of cheap telescopes.'

TO THE EDITOR OF THE "MANCHESTER GUARDIAN"

Sir,–Whilst the columns of the *Guardian* are always open to communications relating to the well-being of mankind, I cannot suppose that they are closed to accounts of those discoveries of science which, though they may not have an immediate bearing on the condition of the inhabitants of this terrestrial globe, nevertheless possess considerable interest for them, and may afford a glimpse into the condition of the inhabitants of a neighbouring world. It is on this ground I beg for space for the following information respecting the planet Mars.

It is well known that, the orbit of Mars being next to that of the earth in the order of distance from the sun, when Mars and the earth are both on the same side of the sun Mars is nearer to the earth by 180 millions of miles than when they are at opposite sides of the sun, and that when thus near together by means of a telescope a much better view of the disc of Mars is obtained. At such periods not only has ice and snow been discovered near the poles of Mars, but continents and oceans, islands and skies, have been seen and have been mapped out by Mr. Dawes, Mr. Procter, and other astronomers. At the beginning of the present year one of these opportunities was presented under unusually favourable circumstances at Milan, owing to the very clear sky enjoyed there at that time. The director of the observatory there is Signor Schiaparelli, who has the command of a particularly fine telescope of large dimensions, and during 16 days he was able to use the highest power of the instrument. Not only was he able to detect important changes which had taken place on certain portions of the surface of Mars, but he made discoveries of a much more interesting nature. Traversing the continents there are certain dark lines to which the name of canals has been given, several of which had been indistinctly seen by one or two previous observers, notably by Dawes in 1864. During the last three approximations of Mars to the earth Schiaparelli has made these so-called canals his especial study, and has observed no fewer than 60 of them. They seem to run between sea and sea, and present the appearance of a well-defined network. They are not always equally visible, nor do they present the same aspect at all times. Several which were not seen in 1877 were visible in 1879, whilst all that had been previously seen were visible with several new ones in 1882. Sometimes these canals appear as dark and indistinct lines, whilst at others they are as sharp and distinct as a stroke with a pen. They cross one another obliquely and at right angles, and are fully 80 or 90 miles broad, and several of them are 3000 miles long. Each canal terminates in a sea or another canal, there not being a single instance of one terminating on what is supposed to be land. The most peculiar thing about them is that at certain seasons they appear to be double, but only at certain seasons. This duplex appearance is produced at a determined epoch and occurs almost simultaneously on the continents. When this phenomenon was first observed by Schiaparelli, he says, "these two regular and perfectly parallel lines caused me, I confess, a profound surprise, all the greater because some days before I had observed with care that same region without discovering any such thing, and I waited with curiosity the return of the planet in 1881 to know if some analogous phenomenon would present itself in the same place, and I saw the same thing appear again on the 11th of January 1882, a month after the spring equinox of the planet; the division was still evident at the end of February." On the right or left of the pre-existing line, without anything being changed either in the course or position of that line, another line is seen parallel with the first, at a distance varying from 250 to 500 miles. The parallelism is sometimes of rigorous exactitude; there is nothing like it in terrestrial geography, "All goes to show," says our astronomer, "that it is an organisation peculiar to the planet Mars, probably connected with the course of its seasons. If this be really so, it is possible that it will be reproduced during the next return of that planet. On the 1st of January, 1884, the position of Mars with regard to its seasons will be the same as that of the 13th of February, 1882. Every instrument capable of separating two lines 0.5 apart can be utilised for the observation."

M. Schiaparelli adds:– "In the actual state of things it would be premature to put forth conjectures as to the nature of these canals. As to their existence, I have no need to declare that I have taken every precaution for avoiding all suspicion of illusion. I am absolutely sure of that which I have observed." The first thought which will enter every one's mind after reading this account will be the question, "If these canals are real, are they natural or artificial?" The French astronomer Flammarion says "if these canals are authentic they do not seem natural, and appear rather due to the combinations of reason, and to represent the industrial work of the inhabitants of the planet."

The subject is full of marvel. Schiaparelli has the repute of being a keen-eyed and careful astronomer, who has secured for himself a foremost place amongst observers of the heavens by this preceding discoveries. We shall await the result of the next approximation of Mars with lively interest.–I am, &c.,

Chorlton-cum-Hardy.

J. T. Slugg.

MARS AND ITS CANALS
(1906)
CHAPTER XV: THE CANALS
Percival Lowell

A businessman turned astronomer, **Percival Lowell** did more than anyone else to build interest in Mars as an inhabited – but dying – planet. His interest in Mars was sparked by Camille Flammarion's book *La planète Mars*, and he founded the Lowell Observatory in Flagstaff, Arizona to study the planet, particularly the 'canals'. He produced intricate diagrams of these optical illusions, convinced of their reality, and popularized his observations. Born in Boston in 1855, Lowell died in Flagstaff in 1916. He is buried on Mars Hill, the site of his observatory.

From the detection of the main markings that diversify the surface of Mars we now pass to a discovery of so unprecedented a character that the scientific world was at first loath to accept it. Only persistent corroboration has finally broken down distrust; and, even so, doubt of the genuineness of the phenomena still lingers in the minds of many who have not themselves seen the sight because of the inherent difficulty of the observations. For it is not one where confirmation may be summoned in the laboratory at will, but one demanding that the watcher should wait upon the sky, with more than ordinary acumen. This latter-day revelation is the discovery of the canals.

Quite unlike in look to the main features of the planet's face is this second set of markings which traverse its disk, and which the genius of Schiaparelli disclosed. Unnatural they may well be deemed; for they are not in the least what one would expect to see. They differ from the first class, not in degree, but in kind; and the kind is of a wholly unparalleled sort. While the former bear a family resemblance to those of the earth; the latter are peculiar to Mars, finding no counterpart upon the earth at all.

Introduction to the mystery came about in this wise, and will be repeated for him who is successful in his search. When a fairly acute eyed observer sets himself to scan the telescopic disk of the planet in steady air, he will, after noting the dazzling contour of the white polar cap and the sharp outlines of the blue-green seas, of a sudden be made aware of a vision as of a thread stretched somewhere from the blue-green across the orange areas of the disk. Gone as quickly as it came, he will instinctively doubt his own eyesight, and credit to illusion what can so unaccountably disappear. Gaze as hard as he will, no power of his can recall it, when, with the same startling abruptness, the thing stands before his eyes again. Convinced, after three or four such showings, that the vision is real, he will still be left wondering what and where it was. For so short and sudden are its apparitions that the locating of it is dubiously hard. It is gone each time before he has got its bearings.

By persistent watch, however, for the best instants of definition, backed by the knowledge of what he is to see, he will find its coming more frequent, more certain and more detailed. At last some particularly propitious moment will disclose its relation to well-known points and its position be assured. First one such thread and then another will make its presence evident; and then he will note that each always appears in place. Repetition *in situ* will convince him that these strange visitants are as real as the main markings, and are as permanent as they.

Such is the experience every observer of them has had; and success depends upon the acuteness of the observer's eye and upon the persistence with which he watches for the best moments in the steadiest air. Certain as persistence is to be rewarded at last, the difficulty inherent in the observations is ordinarily great. Not everybody can see these delicate features at first sight, even when pointed out to them; and to perceive their more minute details takes a trained as well as an acute eye, observing under the best conditions. When so viewed, however, the disk of the planet takes on a most singular appearance. It looks as if it had been cobwebbed all over. Suggestive of a spider's web seen against the grass of a spring morning, a mesh of fine reticulated lines overspreads it, which with attention proves to compass the globe from one pole to the other. The chief difference between it and a spider's work is one of size, supplemented by greater complexity, but both are joys of geometric beauty. For the lines are of individually uniform width, of exceeding tenuity, and of great length. These are the Martian canals.

Two stages in the recognition of the reality confront the persevering plodder: first, the perception of the canals at all; and, second, the realization of their very definite character. It is wholly due to lack of suitable conditions that the true form of the Martian lines is usually missed. Given the proper prerequisites of location or of eye, and their pencil-mark peculiarity stands forth unmistakably confessed. It is only where the seeing or the sight is at fault that the canals either fail to show or appear as diffuse streaks, the latter being a halfway revelation between the reality and their not being revealed at all. Much misconception exists on this point. It has been supposed that improved atmospheric conditions simply amount to bringing the object nearer by permitting greater magnification without altering the hazy look of its detail.* Not so. They do much more than this. They steady the object much as if a page of print from being violently shaken should suddenly be held still. The observer would at once read what before had escaped him for being a

* M. l'abbé Moreux.

blur. So is it with the canals. In reality, pencilings of extreme tenuity, the agitations of our own air spread them into diffuse streaks; an effect of which any one may assure himself by sufficiently rapid motion of a drawing in which they are depicted sharp and distinct, when he will see them take on the streaky look. As the writer has observed them under both aspects, and has seen them pass from the indefinite to the defined as the seeing improved, he has had practical proof of the fact, and this not once, but an untold number of times.

Atmospheric conditions far superior to what are good enough for most astronomic observations are needed for such planetary decipherment, and the observer experienced in the subject eventually learns how all-important this is. Under these conditions the testimony of his own eyesight upon the character of these markings is definite and complete. And the first trait that then emerges from confusion is that the markings are *lines*; not simply lines in the sense that any sufficiently narrow and continuous marking may so be called, but lines in the far more precise sense in which geometry uses the term. They are furthermore straight lines. As Schiaparelli said of them: they look to have been laid down by rule and compass. The very marvel of the sight has been its own stumbling-block to recognition, joined to the difficulty of its detection. For not only is the average observatory not equipped by nature for the task, but what is not good air often masquerades as such. Trains of air waves exist at times so fine as to confuse this detail, or even to obliterate it entirely; while at the same time they leave the disk seemingly sharp-cut, with the result that one not well versed in such vagaries thinks to see well when in truth he is debarred from seeing at all. When study of the conditions finally ends in putting him upon the right road, the sight that rewards him can hardly be too graphically described.

Next to the fact that they are *lines*, definiteness of direction is the chief of their characteristics to strike the observer. The lines run straight throughout their course. This is absolutely true of ninety per cent of them, and practically so of the remaining ten per cent, since the latter curve in an equally symmetric manner. Such directness has I know not what of immediate impressiveness. Quite unlike the aspect of the main markings, which show a natural irregularity of outline, these lines offer at the first glance a most unnatural regularity of look. Nothing on Earth of natural origin on such a scale bears them analogue. Nor does any other planet show the like. They are, in fact, distinctively Martian phenomena. This is the first point in which they differ from the markings we have hitherto described. The others were generic planetary features; these are specific ones, peculiar to Mars.*

Equally striking is the uniform width of each line from its beginning to its end, as it stands out there upon the disk. The line varies not in size throughout its course any more than it deviates in direction. It counterfeits a telegraph

* As some misrepresentation has been made on this subject through misapprehension of the writer's observations on Venus and Mercury, it may be well to state that the tenuous markings on both these other planets entirely lack the unnatural regularity distinguishing the canals of Mars. The Venusian lines are hazy, ill-defined, and non-uniform; the Mercurian broken and irregular, suggesting cracks. Neither resemble the Martian in marvelous precision, and have never been called canals by the writer nor by Schiaparelli, but solely by those who have not seen them and have misapprehended their character and look.

wire stretched from point to point. Like the latter seen afar, the width, too, is telegraphic. For it is not so much width as want of it that is evident. Breadth is inferable solely from the fact that the line is seen at all, and relative size by difference of insistency. Indeed, the apparent breadth has been steadily contracting as the instrumental, atmospheric, and personal conditions have improved. All three of the factors have conduced to such emaceration, but the middle one the most. For the air waves spread every marking, and the effect is relatively greatest upon those which are most slender. As the currents of condensation and rarefaction pulse along, their denser and their thinner portions refract the rays on either side of their true place, and thus at the same time confuse a marking and broaden it. The consequence is that the better the atmospheric conditions and the more that has been learned about utilizing them, the finer the lines have shown themselves to be.

Herein we have a specific intrinsic difference between the fundamental features and these lines: the main markings have extension in two dimensions, the latter in one.

Distinctive as they thus are, they have, in keeping with their appearance, been given a distinctive name, that of canal. Useful as the name is and, as we shall later see, applicable, it must not be supposed that what we see are such in any simple sense. No observer of them has ever considered them canals dug like the Suez Canal or the phœnix-like Panama one. This supposition is exclusively of critic creation.

Their precise width is not precisable. They show no measurable breadth and their size, therefore, admits for certain only of an outside limit. They cannot be wider than a determinable maximum, but they may be much less than this. The sole method of estimating their width is by comparison of effect with a wire of known caliber at a known distance. For this purpose a telegraph wire was stretched against the sky at Flagstaff, and the observers, going back upon the mesa, observed and recorded its appearance as their stations grew remote. It proved surprising at what great distances a slender wire could be made out when thus projected against the sky. The wire in the experiment was but 0.0726 of an inch in diameter and yet could be seen with certainty at a distance of 1800 feet, at which point its diameter subtended only 0.69 of a second of arc. How small this quantity is may be appreciated from its taking more than ninety such lines laid side by side to make a width divisible by the eye. Such slenderness at the then distance of Mars would correspond, under the magnification commonly used, only to three quarters of a mile. Theoretically, then, a line three quarters of a mile wide there should be visible to us. Practically, both light and definition is lost in the telescope, and it would be nearer the mark to consider in such case two miles as the limit of the perceptible. With the planet nearer than this, as is often the case, the width which could be seen would be proportionally lessened. Perhaps we shall not be far astray if we put one mile as the limiting width which could be perceived on Mars at present, with distance at its least and definition at its best.

That so minute a quantity should be visible at all is due to the line having a sensible length and by summation of sensations causing to rise into consciousness what would otherwise be lost. A stimulus too feeble to produce an effect upon a single retinal rod becomes recognizable when many in a row are similarly excited.

The experiment furnished another criterion, of importance as regards the supposition that the lines on Mars are illusory. It showed that brain-begotten impressions of wires that did not exist could be told from the real thing when the wire subtended 0.69 of a second of arc or more; that below this the outside stimulus was too weak to differ recognizably from optic effects otherwise produced; while when the real wire was diminished to 0.59", it could not be seen at all. Now, the majority of lines on Mars so far recognized and mapped lie in strength of impression far above the superior limit of 0.69". To one versed in Martian canal detection there is no possibility of self-deception in the case, the canals being very much more salient objects to an expert than those who have not seen them suppose. For it must not be imagined that, when one knows what to be on the lookout for, they are the difficult objects they seem to the tyro. Just as the satellites of Mars were easily seen once they were discovered, so with these lines.

A mile or two we may take, then, with safety as the smallest width for one of the lines. The greatest was got by comparing what is by far the largest canal, the Nilosyrtis, with the micrometer thread. From such determination it appeared that this canal was from 25 to 30 miles wide. But it is questionable whether the Nilosyrtis can properly be termed a canal, so much does it exceed the rest. It is certainly far larger than the majority of them. From comparative estimates between its size and that of the others, 15 to 20 miles for the width of the larger of the Martian canals seems the most probable value, and 2 or 3 miles only of the more diminutive of those so far detected.

On the other hand, the length of the canals is relatively enormous. With them 2000 miles is common; while many exceed 2500, and the Eumenides-Orcus is 3540 miles from the point where it leaves the Phœnix Lake to the point where it enters the Trivium Charontis. This means much more on Mars than it would on Earth, owing to the smaller size of the planet. Such a length exceeds a third of the whole circumference of its globe at the equator. But what is still more remarkable, throughout the whole of the long course taken by the canal, it swerves neither to the right nor to the left of the great circle joining the two points.

Of these several peculiarities of the individual canal it is difficult to know to which to allot the palm for oddity,—great circle directness, excessive length, want of width, or striking uniformity. Each is so anomalously unnatural as to have received the approving stamp of incredulity. Yet so much, wonderful as it is, is encountered on the very threshold of the subject.

ARE THE PLANETS INHABITED? (1913)

Walter Maunder

British astronomer **Walter Maunder** lived between 12 April 1851 and 21 March 1928. Lacking the normal university qualifications, Maunder won a job at the Royal Observatory Greenwich on an aptitude test. He is best known for his investigation of solar activity and the way eruptions on the Sun spark auroral displays and disrupt magnetic equipment on Earth. In 1890, he was instrumental in setting up the British Astronomical Association, which survives to this day.

CHAPTER VII
The Condition of Mars

THE planet Mars is the debatable ground between two opinions. Here, the two opposing views join issue; the controversy comes to a focus. The point in debate is whether certain markings—some linear, some circular—are natural or artificial. If, it is argued, some are truly like a line, without curve or break, as if drawn with pen, ink, and ruler; or others, so truly circular, without deviation or break, as if drawn with pen, ink, and compass; if, moreover, when we obtain more powerful telescopes, erected in better climates for observing, these markings become more truly lines and circles the better we see them; then they are *artificial*, not natural structures.

But artificial structures imply artificers. And if the structures are so designed as to meet the needs of a living organism, it implies that the living organism that designed them must have a reasonable mind lodged in a natural body. If, then, the "lines" and "circles" that Prof. Lowell and his disciples assert to be artificial canals and oases are really such, they premise the order of being that we call Man. But these canals and oases also premise the liquid that we call Water—water that flows and water utilized in cultivation. In this chapter we will leave out of count the first premiss—Man—and only deal with what concerns the second premiss—Water; with water that flows and is utilized in vegetation.

For in regard to this particular premiss we can do away with hypothesis, and deal only with certain physical facts that are not controversial and are not in dispute.

PLANETARY STATISTICS

	Minor Planets.	Inner Planets.					Outer Planets.			
	Ceres	Moon	Mercury	Mars	Venus	Earth	Uranus	Neptune	Saturn	Jupiter
PROPORTIONS OF THE PLANETS:—										
Diameter in miles	477	2163	3030	4230	7700	7918	31900	34800	73000	86500
" ⊕ = 1	0·06	0·273	0·383	0·534	0·972	1·000	4·029	4·395	9·219	10·924
Surface, ⊕ = 1	0·004	0·075	0·147	0·285	0·945	1·000	16·2	19·3	85·0	119·3
Volume, ⊕ = 1	0·0002	0·02	0·06	0·15	0·92	1·00	65·	85·	760·	1304·
Density, Water = 1	2·8?	3·39	4·72	3·92	4·94	5·55	1·22	1·11	0·72	1·32
" ⊕ = 1	0·5?	0·61	0·85	0·71	0·89	1·00	0·22	0·20	0·13	0·24
Mass, ⊕ = 1	0·0001	0·012	0·048	0·107	0·820	1·000	14·6	17·0	94·8	317·7
Gravity at surface, ⊕ = 1	0·028	0·17	0·33	0·38	0·87	1·00	0·90	0·89	1·18	2·65
Rate of Fall, Feet in the First Second	0·45	2·73	5·30	6·11	13·99	16·08	14·47	14·31	18·97	42·61
Albedo	0·14	0·17	0·14	0·22	0·76	0·50?	0·60	0·52	0·72	0·62
DETAILS OF ORBIT:—										
Mean Distance from Sun in millions of miles	257·1	92·9	36·0	141·5	67·2	92·9	1781·9	2791·6	886·0	483·3
" " Earth's distance = 1	2·767	1·000	0·387	1·524	0·723	1·000	19·183	30·055	9·539	5·203
Period of Revolution, in years	4·60	1·00	0·24	1·88	0·62	1·00	84·02	164·78	29·46	11·86
Velocity, in miles per second	11·1	18·5	9·7	15·0	21·9	18·5	4·2	3·4	6·0	8·1

(Continued)

(Continued)

	Minor Planets.	Inner Planets.					Outer Planets.			
	Ceres	Moon	Mercury	Mars	Venus	Earth	Uranus	Neptune	Saturn	Jupiter
Eccentricity	0·0763	0·0168	0·2056	0·0933	0·0068	0·0168	0·0463	0·0090	0·0561	0·0483
Aphelion Distance, Perihelion = 1	1·157	1·034	1·517	1·207	1·013	1·034	1·097	1·018	1·107	1·101
Inclination of Equator to Orbit	(?)	1°·32′	(?)	24°·0′	(?)	23°·27′	(?)	(?)	26°·49′	3°·5′
		d h m	d	h m s		h m s	h m		h m	h m
Rotation period	(?)	27·7·43	88(?)	24·37·23	(?)	23·56·4	9·30(?)	(?)	10·14±	9·55±
ATMOSPHERE, assuming the total mass of the atmosphere to be proportional to the mass of the planet:—										
Pressure at the surface in lb. per sq. inch.	0·014	0·40	1·6	2·1	11·1	14·7	11·9	11·5	19·4	103·8
" " " in "atmospheres"	0·0009	0·027	0·108	0·143	0·754	1·000	0·81	0·78	1·32	7·06
Level of half surface pressure in miles	119·0	19·6	10·1	8·8	3·8	3·3	3·7	3·8	2·8	1·3
Boiling point of water at the surface		22°C	53°C	53°C	92°C	100°C	94°C	93°C	108°C	166°C
TEMPERATURE:—										
Light and heat received from Sun, ⊕ = 1	0·13	1·00	6·67	0·43	1·91	1·00	0·003	0·001	0·011	0·037
Reciprocal of square-root of distance, ⊕ = 1	0·60	1·00	1·61	0·81	1·18	1·00	0·23	0·18	0·32	0·44
Equatorial temp. of ideal planet, Absolute	188	312°	502°	253°	368°	312°	71°	56°	101°	137°
" " " " Centigrade	–65	+39	+229	–20	+95	+39	–202	–217	–172	–136

	Minor Planets.	Inner Planets.					Outer Planets.			
	Ceres	Moon	Mercury	Mars	Venus	Earth	Uranus	Neptune	Saturn	Jupiter
Average temp. of ideal planet, Absolute	174	290	467	235	342	290	66	52	94	127
" " " " Centigrade	−99	+17	+194	−38	+69	+17	−207	−221	−179	−146
Upper limit under zenith sun, Absolute	248	412	664	337	486	412	94	74	133	180
" " " " Centigrade	−25	+139	+391	+64	+213	+139	−179	−199	−140	−93
Average temp. of equivalent disc, Absolute	223	371	598	300	438	371	84	67	120	162
" " " " Centigrade	−50	+98	+325	+27	+165	+98	−189	−206	−153	−111

The first of this series of facts concerning Mars about which there can be no controversy or dispute relates to its size and mass. As the foregoing Table shows, it comes between the Moon and the Earth in these respects.

The figures show at a glance that Mars ranks in its dimensions between the Moon and the Earth, and that, on the whole, it is more like to the Moon than it is to the Earth.

But in what way would this affect Mars as a suitable home for life? In many ways; and amongst these the distribution of its atmosphere and the sluggishness of its atmospheric circulation are not the least important.

It was mentioned in Chapter III that at a height of about three and a third miles the barometer will stand at 15 inches, or half its mean height at sea level, showing that one half the atmosphere has been passed through. Mont Blanc, the highest mountain in Europe, is under 3 miles in height, so that it is not possible, in Europe, to climb to the level of half-pressure; Mt. Everest, the highest mountain in the world, is not quite six miles high, so that no part of the solid substance of our planet reaches up to the level of the quarter pressure. On a very few occasions daring aeronauts have soared into the empyrean higher than the summits of even our loftiest mountains, but the excursion has been a dangerous one, and they have with difficulty brought their life back from so rare and cold, so inhospitable a region. When Gay-Lussac, in 1804, attained a height of 23,000 feet above sea level, the thermometer, which on the ground read 31° C., sank to 9° below zero, and the rare atmosphere was so dry that paper crumpled up as if it had been placed near the fire, and his pulse rose to 120 pulsations a minute instead of his normal 66. When Mr. Glaisher and Mr. Coxwell made their celebrated ascent between 1 and 2 o'clock on the afternoon of September 5, 1861, they found that at a height of 21,000 feet the temperature sank to –10·4°; at 26,000 feet to –15·2°; and at 39,000 feet the temperature was down to –16·0° C. At this height the rarefaction of the air was so great and the cold so intense that Mr. Glaisher fainted, and Mr. Coxwell's hands being rendered numb and useless by the cold, he was only able to bring about their descent in time by pulling the string of the safety valve with his teeth. Yet when they attained this height they were far above all cloud or mist, and the Sun's rays fell full upon them. The Sun's rays had all the force that they had at the surface of the Earth, but in the rare atmosphere of seven miles above the Earth, the radiation from every particle not in direct sunlight was so great that while the right hand, exposed to the Sun, might burn, the left hand, protected from his direct rays, might freeze.

But gravity at the surface of Mars is much feebler than at the surface of the Earth, and in order to reach the level of half-pressure a Martian mountaineer would have to climb, not three and a third miles, but eight and three-quarter miles; that is to say, the distance to be ascended is in the inverse proportion of the force of gravity at the surface of the planet. The atmosphere of Mars, therefore, is much deeper than that of the Earth, and one great cause of precipitation here is much weakened there. A current of air heavily laden with moisture, if it encounters a range of mountains, is forced upwards, and consequently expands, owing to the diminished pressure. The expansion brings about a cooling, and from both causes the atmosphere is unable to retain as much water-vapour as it carried before. On Mars, the same relative expansion and cooling would only follow if the ascent were nearly three times as great, and the feeble force of gravity has its effect in another way; for just as a weight on Mars will only fall six feet in the first second as against sixteen on the Earth, so a dense and heavy column of air will fall with proportionate slowness and a light column ascend in the same languid manner. An ascending current on Mars would therefore take $1/0{\cdot}38 \times 1/0{\cdot}38 = 1/0{\cdot}145$, or seven times as long to attain the same relative expansion as on the Earth.

The winds of Mars are therefore sluggish, and precipitation is slight. So far at least it resembles

> "The island valley of Avilion;
> Where falls not hail, or rain, or any snow,
> Nor ever wind blows loudly;"

and R. A. Proctor, acute and accurate writer on planetary physics as he was, fell into a mistake when he referred to Mars as being "hurricane-swept." There are no hurricanes on Mars; its fiercest winds can never exceed in violence what a sailor would call a "capful."

This holds good for Mars, but it also holds good for every planet where the force of gravity at the surface is relatively feeble. The greater the force of gravity the more active the atmospheric circulation, and more violent its disturbances; the feebler the action of gravity the more languid the circulation, and the slighter the disturbances.

The atmosphere of Mars is relatively deeper than that of the Earth, so that we, in observing the details of its surface, are looking down through an immense thickness of an obscuring medium. And yet the details of the surface are seen with remarkable distinctness; not as clearly indeed as we can see those of the Moon, but nearly so. For instance, the "canals" appear to have a breadth of from 15 to 20 miles, corresponding to $1/_{16}$th, and $1/_{12}$th, of a second of arc, at an average opposition. The oases, as a rule, are about 120 miles in diameter, that is to say about half a second of arc. These are extraordinarily fine details to be perceived and held, even if Mars had no atmosphere at all; it would certainly be impossible to detect them unless the atmosphere were exceedingly thin and transparent. For we must remember that, though our own atmosphere is a hindrance to our observing, yet the atmosphere of the planet into which we are looking is a greater hindrance still. Like the lace curtains of the window of a house, it is a much greater obstacle to looking inward than to looking outward, and as the perfect distinctness with which we see the Moon is a proof that it is practically without an atmosphere, so the great detail visible on Mars bears unmistakable testimony to the slightness of the atmospheric veil around that planet.

And when we turn again to the statistics of Mars, we see that this must inevitably be the case. Of two planets, one heavier than the other, it is not possible to suppose that the lighter should secure the greater proportional amount of atmosphere. With planets, as with persons, it is the most powerful that gets the lion's share: "to him that hath it is given, and from him that hath not is taken away even that which he seemeth to have." But if we assume that Mars has acquired an atmosphere proportional to its mass, then we see from the Table that this must be a little less than ⅑th of that of the Earth; exactly 0·107. It is distributed over a smaller surface, 0·285. Consequently the amount of air above each square inch of Martian surface is $0{\cdot}107 \div 0{\cdot}285 = 0{\cdot}38$. But since the force of gravity at the surface of Mars is less than on the Earth, this column of air will only weigh $0{\cdot}38 \times 0{\cdot}38 = 0{\cdot}145$; or one-seventh of the column of air resting on a square inch of the Earth's surface. The pressure at the surface of Mars will therefore be 2·1 lb.; and the aneroid barometer would read 4·3 inches. (In order to express the diminished pressure of the Martian atmosphere, it is necessary to refer it to the aneroid barometer. The mercury in a mercurial barometer, or the water in a water barometer would lose in weight in consequence of the diminished force of gravity in the same proportion as the air would, and the mercurial barometer would read 11·4 inches.)

But a pressure of 2·1 lb. on the square inch is far less than that experienced by Coxwell and Glaisher in their great ascent; it is about one-half the pressure that is experienced on the top of the very highest terrestrial mountains. But the habitable regions of the Earth do not extend even so far upward as to the level of a pressure of 7·3 lb. on the square inch; that is, of half the terrestrial surface pressure. Plant life dies out before we reach that point, and though birds or men may occasionally attain greater heights, they cannot domicile there, and are, indeed, only able thus to ascend in virtue of nourishment which they have procured in more favoured regions. If we could suppose the conditions of the whole Earth changed to correspond with those prevailing at the summit of Mt. Everest, or even at the summit of Mont Blanc, it is clear that the life now present on this planet would be extinguished, and that speedily. Much more would this be the case if the atmosphere were diminished to one half the pressure on the summit of the highest earthly mountain.

The tenuity of the atmosphere on Mars has another consequence. Here water freezes at 0° C. and boils at 100° C.; so that for one hundred degrees it remains in a liquid condition. On Mars, under the assumed conditions, water would boil at 53° C., and the range of temperature within which it would be liquid would be much curtailed. But it is only water in the liquid state that is useful for sustaining life.

The above estimate of the density of the atmosphere of Mars is an outside limit, for it assumes that Mars has retained an atmosphere to the full proportion of its mass. But as the molecules of a gas are in continual motion, and in every direction, the lighter, most swiftly moving molecules must occasionally be moving directly outwards from the planet at the top of their speed, and in this case, if the speed of recession should exceed that which the gravity of the planet can control, the particle is lost to the planet for ever. A small planet therefore is subject to a continual drain upon its atmosphere, a drain of the lightest constituents. Hence it is, no doubt, that free hydrogen is not a constituent of the atmosphere of the Earth.

To what extent, then, has the atmosphere of Mars fallen below its full proportion? Mr. Lowell has adopted an ingenious method of obtaining some light on this question, by comparing the relative albedoes of the Earth and Mars; that is to say the relative power of reflection possessed by the two planets. Of course the method is rough; we have first of all no satisfactory means of determining the albedo of the Earth itself, and Mr. Lowell puts it higher than most astronomers would do; then there is the difficulty of determining what portion of the total albedo is to be referred to the atmosphere and what to the actual soil or surface of the planet. But, on the whole, Mr. Lowell concludes that the amount of atmosphere above the unit of surface of Mars is 0·222 of that above the unit of surface of the Earth. This would bring down the pressure on each square inch of Mars to 1·2 lb., and the aneroid barometer would read 2·5 inches; and water would boil at 44° C. The range of temperature from day to night, from summer to winter, at any place on the planet would be increased, while the range within which water could retain its liquid form would be diminished.

These statistics may seem rather dull and tiresome, but if we are to deal with the problem before us at all, it is important to understand that one factor in the condition of a planet cannot be altered and all the other factors retained unchanged. It will be seen that in computing the density of the atmosphere of Mars, we had to take into consideration not only the diameter of the planet, but the surface, which varies as the square of the diameter; the volume, which varies as the cube; the mass, which varies in a higher power still; and various combinations of these numbers. Novelists who write tales of journeys to other worlds or of the inhabitants of other worlds visiting this one, usually assume that the atmosphere is of the same density on all planets, and the action of gravity unchanged. In their view it is only that men would have a little less ground to walk upon on Mars, and a good deal more on Jupiter. Dean Swift, in *Gulliver's Travels*, made the Lilliputians take a truer view of the effect of the alteration of one dimension, for, finding that Gulliver was twelve times as tall as the average Lilliputian, they did not appoint him the rations of twelve Lilliputians, which would have been rather poor feeding for that veracious mariner, but allotted him the cube of twelve, viz. seventeen hundred and twenty-eight rations. Mr. J. Holt Schooling, in one of his ingenious and interesting statistical papers, tried to bring home the vast extent of the British Empire by supposing that it seceded, and taking the portion of Earth that has fallen to it, set up a world of its own—the planet "Victoria." He allots to the British Empire 21 per cent of the land surface of the world. If the Earth were divided so as to form two globes with surfaces in proportion of 21 to 79, the smaller globe, which would correspond to Mr. Schooling's new planet "Victoria," would be less than half the present Earth in diameter; it would be considerably smaller than Mars. But "the rest of the world" would be 0·96 of the present Earth in diameter, or very nearly the size of Venus, and it would contain just eight-ninths of the substance of the Earth, leaving only one-ninth for "Victoria." The statistics given above will suggest to the reader that, could such a secession be carried out, the inhabitants of the British Empire would not be happier for the change during the very short continued existence that remained to them. The "rest of the world" could spare our fraction of the planet much better than we could spare theirs.

This is a principle which applies to worlds anywhere; not merely within the limits of the solar system but wherever they exist. Everywhere the surface must vary with the square of the diameter; the volume with the cube; everywhere the smaller planet must have the rarer atmosphere, and with a rare atmosphere the extreme range of temperature must be great, while the range of temperature within which water will flow will be restricted. Our Earth stands as the model of a world of the right size for the maintenance of life; much smaller than our Earth would be too small; much larger, as we shall see later, would be too large.

So far we have dealt with Mars as if it received the same amount of light and heat from the Sun that the Earth does. But, as the Table shows, from its greater distance from the Sun, Mars receives per unit of surface only about three-sevenths of the light and heat of that received by the Earth.

The inclination of the axis of Mars is almost the same as that of the Earth, so that the general character of the seasons is not very different on the two planets, and the torrid, temperate, and frigid zones have almost the same proportions. The length of the day is also nearly the same for both, the Martian day being slightly longer; but the most serious factor is the greater distance of Mars, and the consequent diminution in the light and heat received from the Sun. The light and heat received by the Earth are not so excessive that we could be content to see them diminished, even by 5 per cent, but for Mars they are diminished by 57 per cent. How can we judge the effect of so important a difference?

The mean temperature of our Earth is supposed to be about 60°F., or 16°C. Three-sevenths of this would give us 7°C. as the mean temperature of Mars, which would signify a planet not impossible for life. But the zero of the Centigrade scale is not the absolute zero; it only marks the freezing-point of water. The absolute zero is computed to be –273° on the Centigrade scale; the temperature of the Earth on the absolute scale therefore should be taken as 289°, and three-sevenths of this would give 124° of absolute temperature. But this is 149° below freezing-point, and no life could exist on a planet under such conditions.

But the mean temperature of Mars cannot be computed quite so easily. The hotter a body is the more rapidly it radiates heat; the cooler it is the slower its radiation. According to Stefan's Law, the radiation varies for a perfect radiator with the 4th power of the absolute temperature; so that if Mars were at 124° abs., while the Earth were at 289° abs., the Earth would be radiating its heat nearly 30 times faster than Mars. The heat income of Mars would therefore be in a much higher proportion than its expenditure; and necessarily its heat capital would increase until income and expenditure balanced. Prof. Poynting has made the temperature of the planets under the 4th power law of radiation the subject of an interesting enquiry, and the figures which he has obtained for Mars and other planets are included in the Table.

The equatorial and average temperatures are given under the assumption that Mars possesses an atmosphere as efficient as our own in equalizing the temperature of the whole planet. If, on the other hand, its atmosphere has no such regulating power, then under the zenith Sun the upper limit of the temperature of a portion of its surface reflecting one-eighth would be, as shown in the Table, 64°C. This would imply that the temperature on the dark side of the planet was very nearly at the absolute zero. "If we regard Mars as resembling our Moon, and take the Moon's effective average temperature as 297° abs., the corresponding temperature for Mars is 240° abs., and the highest temperature is four-fifths of 337° = 270° abs. But the surface of Mars has probably a higher coefficient of absorption than the surface of the Moon—it certainly has for light—so that we may put his effective average temperature, on this supposition, some few degrees above 240° abs., and his equatorial temperature some degrees higher still. It appears as exceedingly probable, then, that whether we regard Mars as like the Earth or, going to the other extreme, as like the Moon, the temperature of his surface is everywhere below the freezing-point of water."[1] As the atmospheric circulation on Mars must be languid, and the atmosphere itself is very rare, the general condition of the planet will approximate rather to the lunar type than to the terrestrial, and the extremes, both of heat and cold, will approach those which would prevail on a planet without a regulating atmosphere.

There is another way of considering the effect on the climate of Mars and its great distance from the Sun, which, though only rough and crude, may be helpful to some readers. If we take the Earth at noonday at the time of the equinox, then a square yard at the equator has the Sun in its zenith, and is fully presented to its light and heat. But, as we move away from the equator, we find that each higher latitude is less fully presented to the Sun, until, when we reach latitude 64½°—in other words just outside the Arctic Circle—7 square yards are presented to the Sun so as to receive only as much of the solar radiation as 3 square yards receive at the equator. We may take, then, latitude 64½° as representing Mars, while the equator represents the Earth. Or, we may take it that we should compare the climate of Archangel with the climate of Singapore.

Now the mean temperature of latitude 64½°, say the latitude of Archangel, is just about freezing-point (0°C.), while that of the equator is about 28°C. We should therefore expect from this a difference between the mean temperatures of the Earth and Mars of 28°; that is to say, as the Earth stands at 16°C, Mars would be at –12°C. But, on the Earth, the evaporation and precipitation is great, and the atmospheric circulation vigorous. Evaporation is always going on in equatorial regions, and the moisture-laden winds are continually moving polewards, carrying with them vast stores of heat to be liberated as the rain falls. The oceanic currents have the same effect, and how great the modification which they introduce may be seen by comparing the climates of Labrador and Scotland. There appear to be no great oceans on Mars. The difference of 28° which we find on the Earth between the equator and the edge of the Arctic Circle is a difference which remains after the convection currents of air and sea have done much to reduce the temperature of the equator and to raise that of high latitudes. If we suppose that their effect has been to reduce this difference to one half of what it would have been were each latitude isolated from the rest, we shall not be far wrong, and we should get a range of 56° as the true equivalent difference between the mean temperatures of Singapore and Archangel; i.e. of the Earth and Mars; and Mars would stand at –40°C. The closeness with which this figure agrees with that reached by Prof. Poynting suggests that it is a fair approximation to the correct figure.

The size of Mars taught us that we have in it a planet with an atmosphere of but one half the density of that prevailing on the top of our highest mountain;

the distance of Mars from the Sun showed us that it must have a mean temperature close to that of freezing mercury. What chance would there be for life on a world the average condition of which would correspond to that of a terrestrial mountain top, ten miles high and in the heart of the polar regions? But Mars in the telescope does not look like a cold planet. As we look at it, and note its bright colour, the small extent of the white caps presumed to be snow, and the high latitudes in which the dark markings—presumed to be water or vegetation—are seen, it seems difficult to suppose that the mean temperature of the planet is lower than that of the Earth. Thus on the wonderful photographs taken by Prof. Barnard in 1909, the Nilosyrtis with the Protonilus is seen as a dark canal. Now the Protonilus is in North Lat. 42°, and on the date of observation—September 28, 1909—the winter solstice of the northern hemisphere of Mars was just past. There would be nothing unusual for the ground to be covered with snow and the water to be frozen in a corresponding latitude if in a continental situation on the Earth. Then, again, in the summer, the white polar caps of Mars diminish to a far greater extent than the snow and ice caps of the Earth; indeed, one of the Martian caps has been known to disappear completely.

Yet, as the accompanying diagram will show, something of this kind is precisely what we ought to expect to see. The diagram has been constructed in the following manner: A curve of mean temperatures has been laid down for every 10° of latitude on the Earth, derived as far as possible from accepted isothermals in continental countries in the northern hemisphere. From this curve ordinates have been drawn at each 10°, upward to show average deviation from the mean temperature for the hottest part of the day in summer, downward for the deviation for the coldest part of the night in winter. Obviously, on the average, the range from maximum to minimum will increase from the equator to the poles. The mean temperature of the Earth has been taken as 16°C, and as representing that prevailing in about 42° lat. The diagram shows that the maximum temperature of no place upon the Earth's surface approaches the boiling-point of water, and that it is only within the polar circle that the mean temperature is below freezing-point. Water, therefore, on the Earth must be normally in the liquid state.

In constructing a similar diagram for Mars, three modifications have to be made. First of all, the mean temperature of the planet must be considerably lower than that of the Earth. Next, since the atmospheric circulation is languid and there are no great oceans, the temperatures of different latitudes cannot be equalized to the same extent as on the Earth. It follows, therefore, that the range in mean temperature from equator to pole must be considerably greater on Mars than on the Earth. Thirdly, the range in temperature in any latitude, from the hottest part of the day in summer to the coldest part of the night in winter, must be much greater than with us; partly on account of the very slight density of the atmosphere, and partly on account of the length of the Martian year.

We cannot know the exact figures to adopt, but the general type of the thermograph for Mars as compared with that of the Earth will remain. The mean temperature of Mars will be lower, the range of temperature from equator to pole will be greater, and the extremes of temperature in any given latitude more

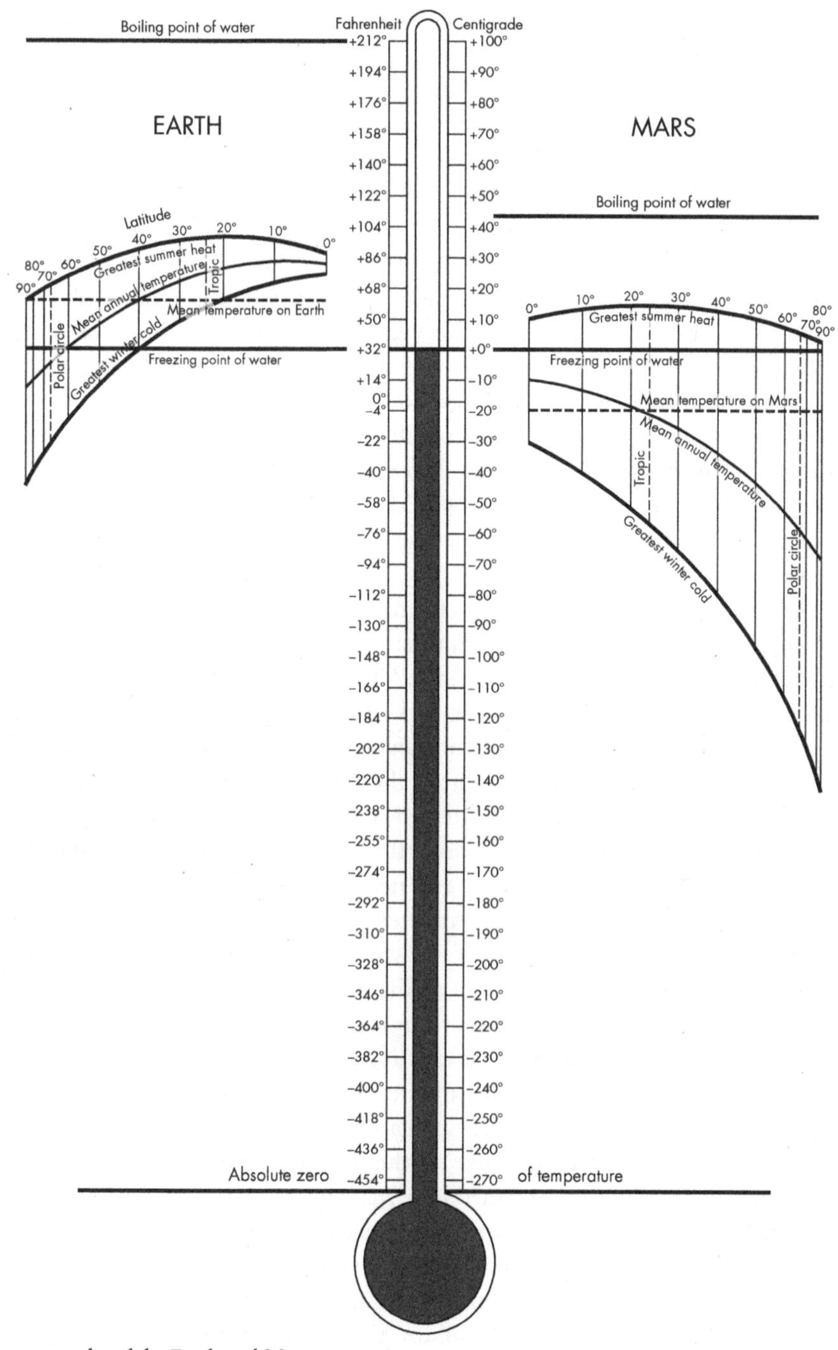

Thermographs of the Earth and Mars

pronounced than upon the Earth. And the general lesson of the diagram may be summed up in a sentence. The maximum temperature on the planet is well above freezing-point, and the part of the planet at maximum temperature is precisely the part that we see the best. But while this is so, it is clear that water on Mars must normally be in the state of ice; Mars is essentially a frozen planet; and the extremes of cold experienced there, not only every year but every night, far transcend the bitterest extremes of our own polar regions.

The above considerations do not appear to render it likely that there is any vegetation on Mars. A planet ice-bound every night and with its mean temperature considerably below freezing-point does not seem promising for vegetation. If vegetation exists, it must be of a kind that can pass through all the stages of its life-history during the few bright hours of the Martian day. Every night will be for it a winter, a winter of undescribable frost, which it could only endure in the form of spores. So if there be vegetation it must be confined to some hardy forms of a low type. At a distance of forty millions of miles it is not easy to discriminate between the darkness of sheets of water and the darkness of stretches of vegetation. Some of the so-called "seas" may possibly be really of the latter class, but that there must be expanses of water on the planet is clear, for if there were no water surfaces there would be no evaporation; and if there were no evaporation from whence could come the supply of moisture that builds up the winter pole cap?

The great American astronomer, Prof. Newcomb, gave in *Harper's Weekly* for July 25, 1908, an admirable summary of the verdict of science as to the character of the meteorology of Mars. "The most careful calculation shows that if there are any considerable bodies of water on our neighbouring planet they exist in the form of ice, and can never be liquid to a depth of more than one or two inches, and that only within the torrid zone and during a few hours each day… There is no evidence that snow like ours ever forms around the poles of Mars. It does not seem possible that any considerable fall of such snow could ever take place, nor is there any necessity of supposing actual snow or ice to account for the white caps. At a temperature vastly below any ever felt in Siberia, the smallest particles of moisture will be condensed into what we call hoar frost, and will glisten with as much whiteness as actual snow… Thus we have a kind of Martian meteorological changes, very slight indeed and seemingly very different from those of our earth, but yet following similar lines on their small scale. For snowfall substitute frostfall; instead of feet or inches say fractions of a millimetre, and instead of storms or wind substitute little motions of an air thinner than that on the top of the Himalayas, and we shall have a general description of Martian meteorology."

What we know of Mars, then, shows us a planet, icebound every night, but with a day temperature somewhat above freezing-point. As we see it, we look upon its warmest regions, and the rapidity with which it is cleared of ice, snow, and cloud shows the atmosphere to be rare and the moisture little in amount and readily evaporated. The seas are probably shallow depressions, filled with ice to the bottom, but melted as to their surfaces by day. From the variety of tints noted in the seas, and the recurrent changes in their outlines, they are composed of congeries of shallow pools, fed by small sluggish streams; great ocean basins into which great rivers discharge themselves are quite unknown.

CHAPTER VIII
The Illusions of Mars

THE two preceding chapters have led to two opposing, two incompatible conclusions. In Chapter VI, a summary was given of Prof. Lowell's claim to have had ocular demonstration of the handiwork of intelligent organisms on Mars. In Chapter VII, it was shown that the indispensable condition for living organisms, water in the liquid state, is only occasionally present there, the general temperature being much below freezing-point, so that living organisms of high development and more than ephemeral existence are impossible.

Prof. Lowell argues that the appearance of the network of lines and spots formed by the canals and oases, and its regular behaviour, "preclude its causation on such a scale by any natural process," his assumption being that he has obtained finality in his seeing of the planet, and that no improvement in telescopes, no increase in experience, no better eyesight will ever break up the perfect regularity of form and position, which he gives to the canals, into finer and more complex detail.

But the history of our knowledge of the planet's surface teaches us a different lesson. Two small objects appear repeatedly on the drawings made by Beer and Mädler in 1830; these are two similar dark spots, the one isolated, the other at the end of a gently curved line. Both spots resemble in form and character the oases of Prof. Lowell, and the curved line, at the termination of which one of the spots appears, represents closely the appearance presented by several of the canals. In the year 1830 no better drawings of Mars had appeared; and in representing these two spots as truly circular and the curved line as narrow, sharp, and uniform, Beer and Mädler undoubtedly portrayed the planet as actually they saw it. The one marking was named by Schiaparelli the Lacus Solis, the other, the Sinus Sabæus, and they are two of the best known and most easily recognized of the planet's features; so that it is easy to trace the growth of our knowledge of both of them from 1830 up to the present time. They were drawn by Dawes in 1864, by Schiaparelli in 1877 and the succeeding years, by Lowell in 1894 and since, and by Antoniadi in 1909 and 1911. But whereas the drawings of Beer and Mädler, made by the aid of a telescope of 4 inches aperture, show the two spots as exactly alike, in those of Dawes, made with a telescope of 8 inches, the resemblance between the two has entirely vanished, and neither is shown as a plain circular dot. Since then, observers of greater experience and equipped with more powerful instruments have directed their attention to these two objects, and a mass of complicated structure has been brought out in the regions which were so simple in the sight of Beer and Mädler, so that not a trace of resemblance remains between the two objects that to them appeared indistinguishable.

Now the gradation in size, from the Lacus Solis down to the smallest oasis of Lowell, is a complete one. If a future development in the power of telescopes should equal the advance made from the 4-inch of Beer and Mädler, to the 33-inch which Antoniadi used in 1909, is it reasonable to suppose that Prof. Lowell's oases will refuse to yield to such improvement, and will all still show themselves as uniform spots, precisely circular in outline? It is clear that Beer and Mädler would have been mistaken if they had argued that the apparently perfect circularity of the two oases which they observed proved them to be artificial, because the increase in telescopic power has since shown us that neither is circular. The obvious reason why they appeared so round to Beer and Mädler was that they were too small to be defined in their instruments; their minor irregularities were therefore invisible, and their apparent circularity covered detail of an altogether different form.

Beer and Mädler only drew two such spots; Lowell shows about two hundred. Beer and Mädler's two spots seemed to them exactly alike; these two spots as we see them to-day have no resemblance to each other. Prof. Lowell's two hundred oases, with few exceptions, seem all of the same character; is it possible to suppose, if telescopes develop in the future as they have done in the past, that the two hundred oases will preserve their uniformity of appearance any more than the Lacus Solis and the head of the Sinus Sabæus? If a novice begins to work upon Mars with a small telescope, he will draw the Lacus Solis and the Sinus Sabæus as two round, uniform spots, and as he gains experience, and his instrumental power is increased, he will begin to detect detail in them, and draw them as Dawes and Schiaparelli and others have shown them later. It is no question of planetary change; it is a question of experience and of "seeing."

There is a much simpler explanation of the regularity of the canals and oases than to suppose that an industrious population of geometers have dug them out or planted them; it is connected with the nature of vision.

A telegraph wire seen against a background of a bright cloud can be discerned at an amazing distance—in fact, at 200,000 times the breadth of the wire; a distance at which the wire subtends a breadth of a second of arc. For average normal sight the perception of the wire will be quite unmistakable, but at the same time it would be quite untrue to say that the perception of the wire was of the nature of defined vision, as would be seen at once if small objects of irregular shape were threaded on the wire; these would have to be many times the breadth of the wire in order to be detected. Again, if instead of a wire of very great length extending right across the field of view of both eyes, a short, black line be drawn on a white ground, it will be found that as the length of the line is diminished below a certain point so its breadth must be increased. If the observer is distant from the line 6000 times its length, then the breadth must be increased to be equal to the length, and the object, whatever its actual shape, can be just recognized as a small circular spot, which will subtend about 34 seconds of arc.

But though a black spot, 34 seconds in diameter, can be perceived on a white ground, we have not yet attained to defined vision. For if we place two black spots each 34 seconds of arc in diameter, near each other, they will not be seen as separate spots unless there is a clear space between them of six times that amount. Nearer than that they will give the impression that they form one circular spot, or an oval one, or even a uniform straight line, according to the amount of separation. If two equal round spots be placed so that the distance between their centres is equal to two diameters, then the diameter of each spot must be, at least, 70 seconds of arc for them to be distinctly defined; that is to say for the spots to be seen as two separate objects.

It will be seen that there is a wide range between objects that are large enough to be quite unmistakably perceived, and objects which are large enough to have their true outline really defined. It is a question of seconds of arc in the one case and of minutes of arc in the other. Within this range, between the limit at which objects can be just perceived and that where they can be just defined, objects must all appear as of one of two forms—the straight line and the circular dot.

This depends upon the structure of the eye and of the retina; the eye being essentially a lens with its defining power necessarily limited by its aperture, and the retina a sensitive screen built up of an immense number of separate elements each of which can only transmit a single sensation. Different eyes will have different limits, both for the smallest objects which can be discerned and for the smallest objects that can be defined, but for any sight the range between the two will be of the order just indicated.

Prof. Lowell has drawn attention to the "strangely economic character of both the canals and oases in the matter of form." It is true that straight lines and circles are economic forms, but they are economic not only in the construction of irrigation works but also in vision. "The circle is the figure which encloses the maximum area for the minimum average distance from its centre to any point situated within it;" therefore, if a small spot be perceived by the sight but be too small to have its actual outline defined, it will be recognized by the eye as being truly circular, on the principle of economy of effort. So, again, a straight line is the shortest that can be drawn between two points; and a straight line can be perceived as such when of an angular breadth quite 40 times less than that of the smallest spot. A straight line is that which gives the least total excitement in order to produce an appreciable impression, and therefore the smallest appreciable impression produces the effect of a straight line.

It is sufficient, then, for us to suppose that the surface of Mars is dotted over with minute irregular markings, with a tendency to aggregate in certain directions, such as would naturally arise in the process of the cooling of a planet when the outer crust was contracting above an unyielding nucleus. If these markings are fairly near each other it is not necessary, in order to produce the effect of "canals," that they should be individually large enough to be seen. They may be of any conceivable shape, provided that they are separately below the limit of defined vision, and are sufficiently sparsely scattered. In this case the eye inevitably sums up the details (which it recognizes but cannot resolve) into lines essentially "canal-like" in character. Wherever there is a small aggregation of these minute markings, an impression will be given of a circular spot, or, to use Prof. Lowell's nomenclature, an "oasis." If the aggregation be greater still and more extended, we shall have a shaded area—a "sea."

The above remarks apply to observation with the unaided eye, but the same principle applies yet more strongly to telescopic vision. No star is near enough or sufficiently large to give the least impression of a true disc; its diameter is indistinguishable; it is for us a mathematical point, "without parts or magnitude." But the image of a star formed by a telescope is not a point but a minute disc, surrounded by a series of diffraction rings. This disc is "spurious," for the greater the aperture of the telescope the smaller the apparent disc.

That which holds good for a bright point like a star holds good for every individual point of a planetary surface when viewed through the telescope; that is to say, each point is represented by a minute disc; all lines and outlines therefore are slightly blurred, so that minute irregularities are inevitably smoothed out.

When we come to photographs, the process is carried to a third stage. The image is formed by the telescope, subject to all the limitations of telescopic images, and is received on a plate essentially granular in structure, and is finally examined by the eye. The granular structure of the plate acts as the third factor in concealing irregularities and simplifying details; a third factor in producing the two simplest types of form—the straight line and the circular dot.

Prof. Lowell describes the canals as like lines drawn with pen, ink and ruler, but not a few of our best observers have advanced much beyond this stage. Even as far back as 1884, some of the canals were losing their strict rectilinear appearance to Schiaparelli, and the observers of the planet who have been best favoured by the power of the telescope at their disposal, by the atmospheric conditions under which they worked, and by their own skill and experience—such as Antoniadi, Barnard, Cerulli, Denning, Millochau, Molesworth, Phillips, Stanley Williams and others—have found them to show evident signs of resolution. Thus, in 1909, Antoniadi found that of 50 canals, 14 were resolved into disconnected knots of diffused shadings, 4 were seen as irregular lines, 10 as more or less dark bands; and he found that, in good seeing, there was no trace whatever of the geometrical network.

The progress of observation, therefore, has left Prof. Lowell behind, and has dispelled the fable which he has defended with so much ingenuity. But, indeed, there never was any more reason for taking seriously his theory as to the presence of artificial waterways on Mars than for believing in the actual existence of the weird creatures described by H. G. Wells in the *War of the Worlds*.

There are too many oversights in the canal theory.

Thus no source is indicated for the moisture supposed to be locked up in the winter pole cap. Prof. Lowell holds that there are no large bodies of water on the planet; that the so-called seas are really cultivated land. In this case there could be little or no evaporation, and so no means by which the polar deposits could be recruited.

Yet it is certain that the supply of the winter pole cap must come from the evaporation of water in some region or other. And here is another oversight of the artificial canal theory. The canals are supposed to be necessary for the conveyance of water from the pole towards the equator; although, as this was "uphill," vast pumping stations at short intervals had to be predicated. But it is not supposed that the water needed to travel by way of the canals to the poles. If, however, the moisture is conveyed as vapour through the atmosphere to the pole as winter approaches, it cannot be impossible that it should be conveyed in the same manner from the pole as summer draws on, and in that case the artificial canals would not be needed. If the canals are necessary for conveying the water in one direction, they would be necessary for the opposite direction. But there would be something too farcical in the idea of the careful Martians dispatching their water first to the pole to be frozen there, and then, after it had been duly frozen and melted again, fetching it back along thousands of miles and through numerous pumping stations for use in irrigating their fields.

Of all the many hundreds of canals only a few actually touch the polar caps. But on the theory that the entire canal system is fed by the polar cap in summer, the carrying capacity of the polar canals should be equal to, if not greater than, that of the entire system outside the polar circle. A glance at the charts of the planet shows that the polar canals could not supply a twentieth part of the water needed for those in the equatorial regions. Another oversight is that of the significance of the alleged uniformity and breadth of the canals. Prof. Lowell repeatedly insists that the canals are of even breadth from end to end, and spring into existence at once throughout their whole length. This statement is in itself a proof that the canals cannot be what he supposes them to be. An irrigation system could not have these characteristics; the region fertilized would take time to develop; we should see the canal extending itself gradually across the continent, and its breadth would not be uniform from end to end, but the region fertilized would grow narrower with increase of distance from the fountain head of the canal.

Under what conditions can we see straight lines, perfectly uniform from end to end, spring into existence, in their entirety, without going through any stages of growth? When the lines are not actual images, but are suggested by markings perceived, but not perfectly defined. In 1902 and 1903, in conjunction with Mr. Evans, the headmaster of Greenwich Hospital School, I tried a number of experiments on this point, with the aid of about two hundred of the boys of the school. They had several qualifications in respect of these experiments; they were keen-sighted, well drilled; accustomed to do what they were told without asking questions; and they knew nothing whatsoever of astronomy, certainly nothing about Mars.

A diagram was hung up, based upon some drawing or other of the planet made by Schiaparelli, Lowell or other Martian observer, but the canals were not inserted; only a few dots or irregular markings were put in here and there. And the boys were arranged at different distances from the diagram and told to draw exactly what they saw. Those nearest the diagram were able to detect the little irregular markings and represented them under their true forms. Those at the back of the room could not see anything of them, and only represented the broadest features of the diagram, the continents and seas. Those in the middle of the room were too far off to define the minute markings, but were near enough for those markings to produce some impression upon them; and that impression always was of a network of straight lines, sometimes with dots at the points of meeting. Advancing from a distance toward the diagram the process of development became quite clear. At the back of the room no straight lines were seen; as the observer came slowly forward, first one straight line would appear completely, then another, and so on, until all the chief canals drawn by Schiaparelli and Lowell in the region represented had come into evidence in their proper places. Advancing still further, the canals disappeared, and the little irregular markings which had given rise to them were perceived in their true forms.

These experiments at the Greenwich Hospital School were merely the repetition of similar ones that I had myself made privately twelve years earlier, leading me to the conclusion, published in 1894, that the canals of Mars were simply the summation of a complexity of detail too minute to be separately discerned.

A little later, in his work "*Marte nel 1896–7*," Dr. Cerulli independently arrived at the same conclusion, and wrote: "These lines are formed by the eye ... which utilizes ... the dark elements which it finds along certain directions"; and "a large number of these elements forms a broad band"; and "a smaller number of them gives rise to a narrow line." Also, "the marvellous appearance of the lines in question has its origin, not in the reality of the thing, but in the inability of the present telescope to show faithfully such a reality." In 1907, Prof. Newcomb made some experiments in the same direction and reached the same general conclusion. More recently still, Prof. W. H. Pickering has worked on the same lines and with the same result. The venerable George Pollock, formerly the Senior Master of the Supreme Court and King's Remembrancer, sent to me, in his 91st year, the following note as affording an apt illustration of the true nature of the canaliform markings on Mars:

"On Saturday last, journeying in a motor-car, I came into a broad road bounded by a dark wood. Looking up I was amazed to see distinct, well-defined, vertical, parallel white lines, the wood forming the dark background. On getting nearer, these lines resolved themselves into spots, and they proved to be the white insulators supporting the telegraph wires."

Prof. Lowell has objected that all experiments and illustrations of this kind are irrelevant; only observations upon the planet itself ought to be taken into account.

But such observations have been made upon the planet itself with just the same result. Observers have seen streaks upon Mars—knotted, broken, irregular, full of detail—and when the planet has receded to a greater distance, the very same marking has shown itself as a narrow straight line, uniform from end to end, as if drawn with pen, ink and ruler. The greater distance has caused the irregularities, seen when nearer at hand, to disappear. In this, and not in any gigantic engineering works, is the explanation of the artificiality of the markings on Mars as Prof. Lowell sees them. That artificiality has already disappeared under better seeing with more powerful telescopes.

This chapter is entitled "The Illusions of Mars." Yet the illusions of Mars are not the straight lines and round dots of the canal system, but the forced and curious interpretation which has been put upon them. If the planet be within a certain range of distance and under examination with a certain telescopic power, the straight lines and round dots are inevitable. Their artificiality is not a function of the actual Martian details themselves, but of the mode in which, under given conditions, we are obliged to see them.

HELLO, EARTH! HELLO! (1920)

Guglielmo Giovanni Maria Marconi

Guglielmo Giovanni Maria Marconi (1874–1937) was an Italian electrical engineer who pioneered wireless communications using radio waves. He is credited as the 'Father of Radio' and shared the 1909 Nobel Prize in Physics with Karl Ferdinand Braun for their contributions to the development of 'wireless telegraphy'. During the early decades of the twentieth century, when interest in Mars was at its height because of the 'canals', Marconi revealed that he thought he might be picking up radio signals coming from the planet's inhabitants. This was the coverage his announcement received in *The Tomahawk's* 8 March 1920 edition.

OF COURSE you recall Jules Verne's "Ten [sic] Thousand Leagues Under the Sea." Well, his submarine is now an accomplished fact, isn't it?

And doubtless you read Kipling's "With the Night Mail." Well, the Atlantic has been crossed in a single flight, hasn't it?

Probably, also, you read H. G. Wells' "The War of the Worlds," in which the Martians descended upon us with fighting machines even more formidable than the tanks of the great war and a mysterious agent of wholesale destruction even more deadly than any gas used by either side.

Well, who shall say that Wells hasn't the right idea about Mars being inhabited by beings just as smart as we are—and probably a good deal smarter?

It is a bold man who say "impossible" these days.

Anyway, Guglielmo Marconi, the famous Italian engineer who perfected wireless telegraphy, has opened up an exceedingly interesting question by this statement:

"I have encountered during my experiments with wireless telegraphy most amazing phenomena. Most striking of all is the receipt by me personally of signals which I believe originated in the space beyond our planet. I believe it is entirely possible that these signals may have been sent by the inhabitants of other planets to the inhabitants of earth.

"If there are any human beings on Mars I would not be surprised if they should find a means of communication with this planet. Linking of the science of astronomy with that of electricity may bring about almost anything.

"While our own planet is a storehouse of wonders, we are not warranted in accepting as a fact the general supposition that the inhabitants of our comparatively insignificant planet are any more highly developed than inhabitants (if there be such) of other planets.

"For all we know, the strange sounds that I have received by wireless may be only a forerunner of a tremendous discovery.

"The messages have been distinct but unintelligible. They have been received simultaneously in London and in New York, with identical intensity, indicating that they must have originated at a great distance.

"These signals are apparently due to electromagnetic waves of great length, which are not merely stray signal. Occasionally such signals can be imagined to correspond with certain letters of the Morse code. They steal in at our stations irregularly at all seasons. We do not get the signals unless we establish a minimum of 65-mile wave lengths. Sometimes we hear these planetary or interplanetary sounds 20 or 30 minutes after sending out a long wave. They do not interrupt traffic, but when they occur they are persistent.

"The most familiar signal received is curiously musical. It comes in the form of three short raps, which may be interpreted as the Morse letter 'S,' but there are other sounds which may stand for other letters.

"The war prevented an investigation of the Hertzian mystery, but now our organization intends to undertake a thorough probe."

Australia corroborates Marconi's statement. Highly skilled and experienced operators at Sydney have received numerous signals similar to those reported as having been received in England. They consist of frequent repetitions of two dashes, representing the letter M. They are on wave lengths of 80,000 to 120,000 meters. The Australian experts say such wave lengths have never yet been used by any wireless station of the earth.

Now what do the electrical authorities say on the general subject? Here it is, in brief:

Thomas A. Edison has this to say: "Although I am not an expert in wireless telegraphy, I can plainly see that the mysterious wireless interruptions experienced by Mr. Marconi's operators may be good grounds for the theory that inhabitants of other planets are trying to signal to us. Mr. Marconi is quite right in stating that this is entirely within the realm of the possible.

"I have given some thought to the matter and can record one personal experience which may or may not have bearing on proving that Mr. Marconi is right. I was seated on the peak of a great pile of iron ore near the reduction plant at Orange one day, when I noticed that the magnetic needle was jumping about in astonishing fashion. The thought immediately popped into my mind that static signals from some other planet were probably responsible. This idea took such hold on me that I made the definite suggestion that there be established in the ore fields of Michigan a station where scientific vigil might be kept, in the hope that the great masses of ore in that region would attract magnetic signals from interplanetary space.

"If we are to accept the theory of Mr. Marconi that these signals are being sent out by inhabitants of other planets, we must at once accept with it the theory of their advanced development. Either they are our intellectual equals or our superiors. It would be stupid for us to assume that we have a corner on all the intelligence in the universe."

Nikola Tesla, the famous Serbian inventor and electrical expert says: "Marconi's idea of communicating with the other planets is the greatest and most fascinating problem confronting the human imagination today. To insure success a body of competent scientists should be organized to study all possible plans and put into execution the best. The matter should be directed probably by astronomers with sufficient backing from men with money and imagination. Supposing that there are intelligent human beings on Mars, success is easily within the range of possibility. In March, 1907, I stated in the Harvard Illustrated Magazine that experiments looking to communication with other planets should be undertaken.

"In 1899 I built an electric plant in Colorado and obtained activities of 18,000,000 horsepower. In the course of my experiments I employed a receiver of virtually unlimited sensitiveness. There were no other wireless plants near, and, at that time, no other wireless plants anywhere on this earth of sufficient range to affect mine. One day my ear caught what seemed to be regular signals. I knew that they could not have been produced upon the earth. The possibility that they came from Mars occurred to me, but the pressure of business affairs caused me to drop the experiment.

"The thing, I think, that we should try to develop is a plan akin to picture transmission, by means of which we could convey to the inhabitants of Mars knowledge of earthly forms. This would enable us to exchange with them not only simple primitive facts, but involved conceptions. To talk to Mars seems to me only a matter of electric power and perseverance."

Frank Dyson, British astronomer royal, believes we could get Hertzian waves from other planets. Prof. Edward Branley, Paris, inventor of the coherer, is sceptical. Prof. Domenico Argentieri, Rome, says the supposed signals are worthy of careful observation.

Prof. Albert Einstein, the German astronomer and author of the theory of "Relativity" that is apparently upsetting all accepted doctrines, believes that Mars and other planets are inhabited, but if intelligent creatures are trying to communicate with the earth he should expect them to use rays of light, which could much more easily be controlled.

Are there inhabitants on Mars? That's a question on which scientists differ.

Among scientists who have won the right to speak with authority the foremost was the late Professor Lowell, director of the observatory at Flagstaff, Ariz. Not only was Professor Lowell convinced that Mars was inhabited, but he believed the people had a much higher degree of intelligence than those on earth. He dwelt particularly on their inventive genius.

In 1914 he found a new opportunity for strengthening his pet belief by announcing that instead of losing any of their canals the Martians had built two new ones, which could be seen plainly through the telescope.

"We have actually seen them formed under our eyes," Professor Lowell said at the time, "and the importance of it can hardly be overestimated. The phenomenon transcends any natural law, and is only explicable so far as can be seen by the presence out yonder of animate will."

Professor Lowell had little to say about the appearance of the beings on Mars. Edmond Perrier, director of the museum of the Jardin des Plantes, in

Paris, constructed the first picture of the Martians as he conceived them. He said in part:

"The men on Mars are tall because the force of gravity is slight. They are blond because the daylight is less intense. They have less powerful limbs. Their large blue eyes, their strong noses, their large ears, constitute a type of beauty which we doubtless would not appreciate except as suggesting superhuman intelligence."

On the other hand, Dr. C. G. Abbott holds that if wireless messages are being received, it is not Mars sending the signals, but most probably Venus. Abbott is director of the Smithsonian astrophysical observatory and assistant secretary of the Smithsonian institution. He says Mars is eliminated as a possibility because known conditions on that planet would not permit the existence of any form of living creature. It is too cold there and there is practically no water vapor in its atmosphere.

Assuming that Mars or some other planet is signaling us, what can we do in the circumstances? Apparently we can do much.

Dr. James Harris Rogers of Hyattsville, Md., who has devoted his life to the study of electric waves and invented the underground and under-seas wireless used during the war, declares he is going to undertake to teach the inhabitants of Mars the rudiments of intelligence of this planet. Within a year wireless communication will be established with Mars, Dr. Rogers believes.

L. J. Lesh, a New York radio engineer, suggests that one of the methods of constructing a gigantic station would be to erect huge antennae suspended by balloons like the British dirigible R-34. He asserts, however, that a still better way would be to use huge and brilliant shafts of light as antennae for the system. He thinks that projectors could be grouped around one spot where a great amount of electricity could be generated. He suggests Niagara Falls or some other spot with an enormous amount of water power.

Elmer A. Sperry has a searchlight capable of producing a beam having the illuminating intensity of 1,280,000,000 candle power. He would form a group of 150 to 200 of his searchlights and direct their combined beams in the direction of Mars. An aggregation of that sort would possess the luminous equivalent of a star of the seventh magnitude such as our telescopes are able to pick up readily. Therefore, assuming that the Martians had glasses of equal power, they should have no trouble in catching that dot of light from a distance of 35,000,000 to 40,000,000 miles.

It would be possible, no doubt to operate these lights so that they could give slow signals which would fill all the requirements of a system of communication. However, an array of lights of this character and the needful energizing plant would cost a pretty sum.

The outlay might be warranted some day, but certainly not until it is certain that we are being called by one of our neighbors out in space.

THE SATELLITES OF MARS (1970)

PREDICTION AND DISCOVERY

Owen Gingerich

Owen Gingerich was born in 1930 in Washington, Iowa. He moved to Kansas as a child and became interested in astronomy while viewing the night sky from dark prairies. While studying at Harvard University, Gingerich specialized in the history of science and became an authority on sixteenth-century astronomers Johannes Kepler and Nicolaus Copernicus. He is now Professor Emeritus of Astronomy and History of Science at Harvard University and a senior Astronomer Emeritus at the Smithsonian Astrophysical Observatory.

The remarkable prediction of the two satellites of Mars contained in Jonathan Swift's *Gulliver's travels* has perplexed astronomers for nearly a century. Swift's eighteenth-century satire appears to have anticipated Asaph Hall's actual discovery of Phobos and Deimos in 1877 by almost exactly 150 years. According to Swift, the scientists of Laputa had "discovered two lesser stars, or satellites, which revolve about Mars; whereof the innermost is distant from the center of the primary planet exactly three of his diameters, and the outermost five; the former revolved in a space of ten hours, the latter in twenty-one and a half; so that the squares of their periodical times, are very near in the same proportion with the cubes of their distances from the center of Mars; which evidently shews them to be governed by the same law of gravitation, that influences the other heavenly bodies."[1]

The most uncanny part of the prediction is a very short (10-hour) period for the inner satellite—considerably shorter than the period of any of the ten satellites known in Swift's day, and at least qualitatively borne out by the eventual discovery.

Table I

	Actual		Swift's predictions	
Mars's satellites	a/d ♂	P(hrs)	a/d ♂	P(hrs)
Phobos	1.4	7.6	3	10
Deimos	3.5	30.3	5	21½

A closer look at Swift's numbers, however, shows that he was not so clairvoyant as a first glance suggests. A choice of three and five planetary diameters for the distances of the two satellites very nearly matches corresponding distances for Jupiter's Io and Europa. The correspondence with the Jupiter satellite system and also the similarity to Saturn are shown in Table II, where the distances are expressed in terms of planetary diameters. Swift probably realized that our moon is anomalously large compared to its parent body, and particularly remote. Had a similar situation prevailed for Mars, its satellites would have already been discovered.

Table II

Jupiter	a/d#pp	P(hrs)	Saturn	a/d#pp	P(hrs)
Io	3.0	42	Tethys (1684)	2.5	45
Europa	4.8	85	Dione (1684)	3.2	66
Ganymede	7.8	172	Rhea (1672)	4.5	108
Callisto	13.6	400	Titan (1655)	10.5	383
			Japetus (1671)	30.6	1900

More puzzling is the choice of ten hours for the period of the innermost satellite. Even if Jupiter had been chosen as the paradigm for satellite spacing, the periods would not follow by direct analogy. Further calculations are required. If Mars had the same density as the Earth, then the first satellite at 3 planetary diameters should have a period of roughly one day; if the density were more like that of a Jovian planet, the period should be closer to two days. Newton's work had already provided the basis for such a calculation, so that we can only suppose that 10 hours was chosen as a convenient guess.[2] However, Swift's choice leads to an absurdly high density for Mars.[3] Given the distances and one of the periods, the other follows from Kepler's harmonic law, succinctly stated by Swift in the quoted passage. This law was well-known by 1726, and that Swift used it correctly should occasion little surprise; some commentators have suggested, however, that Swift had professional help.[4]

In Swift's day the idea that Mars should have two satellites was firmly rooted in analogy with the adjacent planets in the solar system. Mercury and Venus had no known satellites, the Earth one, Mars an unknown number, Jupiter four, and Saturn five. Voltaire confirmed this line of reasoning in his *Micromegas* (1752), where he states, "[the voyagers] would see the two moons which belonged to this planet, and which have escaped the searches of our astronomers. I know quite well that P. Castel has written, and even rather pleasantly, against the existence of these two moons; but I am in agreement with those who reason by analogy. The best philosophers know how difficult it would be for Mars, which is next from the sun, to have less than two moons."[5]

Much earlier Fontenelle had also discussed the matter in his *Conversation on the plurality of worlds*. The pupil argues,

> Because Nature hath given so many Moons to *Saturn* and to *Jupiter*, it is a kind of proof that *Mars* cannot be

> in want of Moons. I should have been very well satisfied that those Worlds, which are removed at a distance from the *Sun* have them, if *Mars* did not interfere, in order to make a very disagreeable exception. Ah! truly, replied I, if you mix Philosophy with every thing you do, you will soon accustom yourself to make exceptions to the best systems. Some things always agree very well, others never will agree, do all we can, so that they must be left as found, otherwise we may despair of ever attaining the end proposed.[6]

Quite possibly the inspiration for two Martian satellites derived from Kepler, who repeatedly argued from archetype principles based on harmony or analogy. In a letter to Galileo, Kepler wrote, "I am so far from disbelieving in the discovery of the four circumjovial planets, that I long for a telescope, to anticipate you, if possible, in discovering two round Mars (as the proportion seems to require), six or eight round Saturn, and perhaps one each round Mercury and Venus."[7]

Kepler's belief in the possibility of a pair of Martian satellites led him into a curious trap in his correspondence with Galileo. Shortly after publishing his *Sidereus nuncius*, Galileo made yet another discovery that was announced to the world in the form of an anagram consisting of the series of letters: smaismrmilmepoetaleumibunenugttauiras. Few riddles can have been more challenging to Johannes Kepler, who waited in Prague with eager expectation for word of the new astronomical discoveries from Florence. Kepler transposed the letters to read.

> *Salue umbistineum geminatum Martia proles.*[8]
> Hail, twin companionship, children of Mars.

Kepler's remarkable and erroneous deciphering hinged on the word *umbistineum*, apparently a Latinized German word such as *umbeistehn.*[9] But in fact, Galileo's anagram had nothing to do with the discovery of two satellites of Mars. In a letter dated 13 November 1610 Galileo rearranged the letters to form

> *Altissimum planetam tergeminum observavi.*[10]
> I have observed the most distant of the planets to have a triple form.

Thus Galileo was signaling the peculiarities of Saturn that we now understand to be its ring system.

Kepler's false decoding of the anagram was published in the preface to his *Dioptrics*. The second edition of this work was published in London in 1653 along with Gassendi's *Institutio astronomica* and Galileo's *Sidereus nuncius*; yet another edition appeared in 1683, so that Kepler's fancy with respect to the Martian satellites was probably better known to the English audience than were many of his other ideas. While this does not prove that Swift got his inspiration for the Laputan discovery of two Martian satellites from Johannes Kepler, yet the evidence indicates that the idea must then have been "in the air".

Neither the original announcement of Asaph Hall's discovery of Phobos and Deimos during the particularly favourable opposition in August of 1877, nor the subsequent discussions, give any clue that the detection of the satellite was more than a routine piece of sharp-eyed observing. Nor do the record books for the U.S. Naval Observatory 26-inch refractor, the "Great Equatorial", give any indication to the contrary. The observations on 11 August 1877 began with routine measurements of the satellites of Saturn, followed by an examination of Mars. The notes indicate "seeing good for Mars. The edge of the white spot has two notches near the center of its outline. (A faint star near Mars)." A later note has been inserted by Hall: "This proves to be satellite 1. See Aug 16 and seq.", that is, Deimos.

The following nights were cloudy and on 15 August the seeing conditions were very poor. On 16 August Hall's first observation was a rough measurement of the "Star near Mars". Two hours later he succeeded in obtaining a more precise series of chronometer measurements.

On the following night the first page of the record book closes with the remark that "The Mars Star observed tonight is a *fixed star* and not the object observed last night." It is plain from this remark that Hall is already convinced that his object was a new satellite. Later the same night he recorded for the first time *two* satellites, each ambiguously labelled "Mars Star". The pages for 17 August close with the remark that "Both the above objects faint but distinctly seen both by G. Anderson and myself".

The situation clearly changed by the following night, for Hall was joined in the dome by D. P. Todd, Simon Newcomb and William Harkness, all of whom made measurements, and in the course of the observing the expression "Mars Star" becomes "Mars-Satellite". The first remark is in Hall's hand: "Images very poor at $9^h 40^m$, but saw the satellite immediately." This is followed by four lines signed by D. P. Todd: "Seeing extremely bad: still I saw the companion without any difficulty. 'Halo' around the planet very bright, and the satellite was visible in this halo."[11]

That Professor Hall's search was deliberate and guided in part by theoretical consideration becomes somewhat more apparent in a brief monograph prepared the following year, where he refers to an article by d'Arrest in the *Astronomisches Nachrichten*.[12] This line of argument is presented by Hall with still more clarity and force in a hitherto unpublished letter to E. C. Pickering. The letter contains in addition a remarkable comment about the actual circumstances of the discovery; consequently, the letter is quoted here in full, with the permission of the Harvard University Archives:

1888 Febr. 14

Professor E. C. Pickering

Director Harvard College Obs'y

My Dear Sir:

In the Spring of 1877, when I began to think of searching for a satellite of Mars, a little rough calculation convinced me that this planet could have no moon even at half the distance D'Arrest assumes as a limit. If Mars is at the geocentric distance 0.52, the elongation 70' would make the distance of the satellite from Mars =

1.000 000 miles, nearly, = 0.0108108, in terms of Earth's dist. from sun. Mars dist. from sun is 1.5236914 the disturbing force of the sun on a satellite is the difference of its action on the planet and on the satellite. When the satellite came between Mars and the sun its dist. from the sun would be 1.5128806. Computing the forces by the expression $\frac{\mu}{r^2}$, μ being the mass of the sun, or of the planet, I found the disturbing force of the sun more than twice as great as the attraction of the planet on the satellite. Hence, we would at once reduce the elongation to 30', and this being a *limit* the probable elongation would be much less. A little trial, and the analogy of other planetary systems, led me to search very near the planet. My calculation made me think that D'Arrest did not intend to be very exact in his statement A. N. Bd. 64; and that he reasoned rather from analogy. In the case of any planet I should compute the forces as I did before.

In the case of the Mars satellites there was a practical difficulty of which I could not speak in an official Report. It was to get rid of my assistant. It was natural that I should wish to be alone; and by the greatest good luck Dr. Henry Draper invited him to Dobb's Ferry at the very nick of time. He could not have gone much farther than Baltimore when I had the first satellite nearly in hand.

Will you give me some information in photography. Capt. Phythian proposes to photograph the solar Eclipse of next Jan. on the Pacific coast. What photographic lenses would you recommend, and what plates. Professor Harkness is inclined to retain old methods, but I presume you have made improvements.

Yours truly,

A. Hall

The assistant Hall so much desired to get rid of was none other than Edward S. Holden, a young protégé of Simon Newcomb. (Newcomb was in charge of the Nautical Almanac Office and, for all practical purposes, the scientific director of the Naval Observatory.) Newcomb's enthusiasm for Holden may have waned in later years after his former assistant became president of the University of California and then in 1887 director of the Lick Observatory. Describing Holden as a "much-hated man", Newcomb relates in his autobiography that "The term of Holden's administration extended through some ten years. To me its most singular feature was the constantly growing unpopularity of the director."[13]

Holden's own interest in possible Martian satellites was described by Hall many years later in a letter to [Seth] Chandler, dated 7 March 1904:

> There are several points about the discovery of the satellites of Mars that have not been noticed. Thus Newcomb and Holden had the 26 inch Telescope for the first two years and they tried to make discoveries. W. Herschel said he had seen *six* satellites of Uranus; and only *four* are known. Somebody, perhaps Lassell, reported a second satellite of Neptune. After two years Newcomb got tired of the night work and offered the instrument to me. He had made good determinations of the masses of Uranus and Neptune. Procyon had been examined very carefully for the disturbing companion. Of course one of the first things I did was to find out what my predecessors had been doing. I found in a drawer in the Eq. room a lot of photographs of the planet Mars in 1875. From the handwriting of dates and notes probably Holden directed the photographer, but whoever did the pointing of the telescope had the satellites under his eyes. All that was needed was the right way of looking, and that was to get rid of the dazzling light of the planet. The satellites might have been found at Harvard in 1862 very easily.[14]

Hall's suspicions that Holden would try to get into the act were quickly confirmed. On 28 August Holden wrote from Dobb's Ferry that he and Dr Henry Draper had detected a third satellite of Mars on 26 and 27 August.[15] A letter of 9 October from Hall documents yet another Holden discovery: "Prof. Holden has observed a fourth satellite of Mars since Sept. 24th: but I think it will turn out that the Draper-Holden moon and the recent Holden moon do not exist. Another satellite of Mars may be found at Melbourne but these are fictitious."[16]

Simon Newcomb also took a share of the credit for himself; according to the *New York Tribune* of 20 August 1877, Hall hardly recognized his great discovery until Newcomb had worked out the period from the preliminary observations. In spite of a clarification published shortly after in *Nature*,[17] a misunderstanding between Newcomb and Hall must have lingered for years. This was at least partially cleared up in 1901 when Hall read the manuscript for the following account in Newcomb's *Reminiscences*:

> One morning Professor Hall confidentially showed me his first observations of an object near Mars, and asked me what I thought of them. I remarked, 'Why, that looks very much like a satellite.'
>
> Yet he seemed very incredulous on the subject; so incredulous that I feared he might make no further attempt to see the object. I afterward learned, however, that this was entirely a misapprehension on my part. He had been making a careful search for some time, and had no intention of abandoning it until the matter was cleared up one way or the other.[18]

In returning the manuscript to Newcomb, Hall wrote, "I thank you for your letter of the 21st inst. It puts you in a different light from what I have seen you for the last 24 years." And in a later paragraph he adds, "After the excitement was over I could see that you and Holden must naturally feel a great deal of disappointment and I should have felt the same had we changed places."[19]

But three years later, in the letter to Chandler already quoted, Hall was more severe: "Newcomb was greatly excited over my discovery. Holden was away, and he and Draper made a blunder, and afterwards Holden behaved very well, Newcomb felt disappointed and sore, and something is to be allowed for human nature under such circumstances. He was always greedy for money and glory ..."

After Hall's death in 1907, his biographers lauded him as a great observer. Yet, as this episode attests, the greatest observations are spurred on by theory. Asaph Hall's discovery of the two satellites of Mars cannot be written off simply as the good luck of a keen observer, for the record is firm that his search was deliberate and guided by gravitational theory. This is reinforced by a second letter from Hall to Pickering, sent immediately after the first:[20]

1888 Febr. 14

Professor Pickering

My Dear Sir:

After mailing my letter yesterday I saw that the method I followed is clumsy. Although you no doubt see the easier way I will put it down:—If the sun's mass be unity, m that of the planet, a the distance of planet from sun, and x the distance of satellite from planet; we have for equal attraction forces on the satellite, neglecting its mass,

$$\frac{1}{(a-x)^2} = \frac{m}{x^2}$$

a very simple equation for x. I will not answer for the numerical exactness of the numbers I sent you, but they served for my purpose.

Should you think of photographing regions around the planets I ask your attention to a paper by Marth on the satellite of Neptune in the Monthly Notices a year or two ago. The irregular changes in the node and the inclination are remarkable, and tho' it may be possible that these are produced by systematic errors in the orbits it does not seem probable. The gaps in the distances of the satellites of Saturn are worthy of notice.

Yours truly

A. Hall

THE FACE ON MARS (1993)

Sally Stephens

Astronomer and science writer **Sally Stephens** has a PhD in astrophysics from the University of California, Santa Cruz. Her writing work included spending time as the Education Coordinator at the Astronomical Society of the Pacific, and editor of the Society's magazine *Mercury*. Stephens is now retired.

The planet Mars has intrigued humans since we first began to study the sky. It is the only planet that looks red to the naked eye, which may explain why the ancient Greeks and Romans associated it with the bloody god of war. Spacecraft that visited Mars in the 1960s and 1970s found a freezing, arid, dead world. Nevertheless, the idea of life on Mars, and, in particular, intelligent life on Mars, fed by years of science fiction stories, persists in the public mind, despite the weight of scientific evidence against it.

This myth has surfaced again in the so-called "Face on Mars," a rock outcropping that looks like a human face when lit from the side. Astronomers see it as a mere optical illusion, proof of the power of human imagination. But a few, very vocal individuals see it as proof of an ancient Martian civilization. Some of them even go so far as to accuse NASA of deliberately destroying the billion-dollar *Mars Observer* spacecraft at the end of August to keep from having to admit that the Face is "real." Such extraordinary claims require extraordinary proof, which is sorely lacking in this case. But the Face on Mars is not the first time humans have been misled about evidence for life on our neighboring planet.

The Martian Canals

When early astronomers trained their crude telescopes on Mars, they could see few details of its surface beyond some dark patches and white polar caps. In some ways, the mysterious planet seemed tantalizingly like Earth. A Martian day is only 37 minutes longer than one on Earth. Mars' equator is tilted relative to the plane of its orbit around the Sun by almost the same amount as the Earth's, which gives Mars, like Earth, seasons. And the ice caps on its north and south poles wax and wane with the seasons. By the 1800s, Mars seemed enough like Earth that astronomers and science fiction writers began to dream of a Mars populated with intelligent, alien beings.

In 1877, Mars came within 56 million kilometers (35 million miles) of Earth, about as close as it ever gets (because its orbit is somewhat elongated, its distance at closest approach to Earth varies from year to year). Taking advantage of the

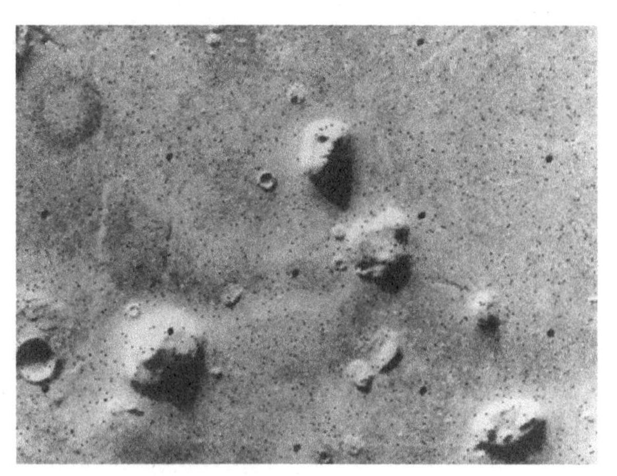

The "Face" on Mars.

close range, American astronomer Asaph Hall discovered two moons circling Mars. Other astronomers strained to learn as much as they could about the red planet. Using the 8-and-¾-inch refracting telescope at the Brera Observatory in Milan, Italian astronomer Giovanni Schiaparelli reported seeing a network of very fine, regular lines crisscrossing the reddish deserts of Mars. He called them *canali,* which means "channels" or "grooves," but the word was translated into English as "canals." Given the limited understanding of the Martian atmosphere and climate at the time, and the known similarities to Earth, many people jumped to the conclusion that Schiaparelli's canals had been constructed by intelligent beings.

Percival Lowell, a successful Boston businessman, was so captivated by this idea, that, in 1893, he built a major observatory in the clear skies of Flagstaff, Arizona to study the Martian canals (a great deal of other research has since been done at the Lowell Observatory, including the discovery of Pluto). Lowell was convinced that intelligent Martians had built a network of canals to pump water from the melting ice caps to dying cities in the desert.

Not all astronomers believed in the canals. In fact, many famous astronomers never saw them, including Asaph Hall. Those who did see canals rarely agreed on their locations and intensities; some said they were broad, diffuse stripes, while others maintained they had a thin, spider-web appearance. As time went on and more was learned about the red planet, astronomers realized that Mars was too cold and its atmosphere too thin for liquid water to exist on its surface. In 1971, the *Mariner* 9 spacecraft sent back to Earth the first really good, high-resolution maps of the planet. Absolutely no canals were seen, and the volcanoes, valleys and craters that were mapped didn't correspond to any feature seen by Lowell.

It turned out there never were any canals on Mars. When straining to look at things that are barely visible, the human eye tends to join faint but distinct markings together. That's what happened to the astronomers struggling to learn more about the surface of Mars. The Martian canals say more about human perception and imagination than anything else. As Carl Sagan said in *Cosmos,* "Lowell always said the regularity of the canals was an unmistakable sign that they were of intelligent origin. This is certainly true. The only unresolved question was which side of the telescope the intelligence was on."

The Face on Mars

A similar statement can be made about the Face on Mars. In 1976 two unmanned *Viking* spacecraft went into orbit around Mars. Each sent a lander down to the surface, while the orbiters radioed back over 60,000 photographs of the Martian surface. Not long after the orbiters began their mission, *Viking* scientists noticed a picture of a squarish, mile-long low hill or mesa in the Cydonia region of Mars that looked sort of like a face. They released the image to the press as a sort of joke, an example of our tendency to see apparently familiar shapes in complex landscapes. The researchers and the press recognized the Face as an unusual rock formation created by weathering. The dramatic low angle of late-afternoon lighting in the photo made the outcropping look like a face.

Three years later, two computer scientists with no particular expertise in Martian geology working for a contractor at NASA came across the image while going through the *Viking* photo archives. They experimented with some image-enhancement programs and concluded that the Face did not occur naturally. They also noticed several "pyramids" near the Face, and published a book calling attention to the structures.

In the 1980s, Richard Hoagland, a science journalist, took up the cause of the Face on Mars in several books and numerous radio and TV appearances. In the scattered hills of Cydonia, Hoagland sees evidence of a ruined "city" and "fort." He claims the city and the Face are aligned in a way that may have, in the manner of Stonehenge, pointed to the place where the sun rose on the Martian solstice half a million years ago (which Hoagland takes to be when the Face was made), although the orientation has no meaning today. Clearly, to Hoagland, this region of Mars is the result of a gigantic construction project by intelligent beings.

But who were they? And why does the Face look human? Hoagland has several theories. Perhaps evolution worked the same way on Mars as it did on Earth and so the Martians looked human. Or maybe a previously unknown, technologically advanced civilization from Earth's distant past traveled to Mars (or, alternatively, an advanced Martian civilization came to Earth long ago, and we look like them). Or perhaps the Face was built by some sophisticated extraterrestrials from beyond the solar system as a signal (or test) for us when we had reached a certain stage of technological evolution (rather like the black monolith in the movie *2001: A Space Odyssey*).

Is the Face Real?

Certainly the rock outcropping on Mars is suggestive of a human face. But that doesn't mean it is a monument deliberately constructed to look like a face. According to psychologists, the human visual system is organized to look for familiar features in random patterns. And there is a particularly strong human tendency to see faces given minimal details. For example, people all over the world see the face of a man when they look at the Moon (others see a rabbit in the Moon). We see animals and human faces in clouds. In New Hampshire, the face of the "Old Man of the Mountain" gazes impassively out from the side of a cliff. If you look on the left and right sides of the maple leaf on the Canadian flag, you can see the faces of two people arguing with one another. We've all seen people in the news or on TV talk shows who have found vegetables or potato chips that look like famous people. Another *Viking* image shows a Martian crater with what looks like a smiling "happy face" inside. No one thinks any of these were deliberately created, so why should the Face on Mars be any different?

What about the "pyramids" near the Face on Mars? There are a number of small mountains on Mars that resemble pyramids, both in Cydonia and another region called Elysium. Geologists who specialize in desert landscapes are quite familiar with similar wind-sculpted formations here on Earth, for example, in the deserts of Arizona. The "city" on Mars is a cluster of these pyramid-shaped mountains, the biggest a few kilometers at the base, all oriented in the same direction. Carl Sagan has noted that similar formations, called *dreikanters*, from a German word meaning three sides, are seen in Antarctica. Strong winds blowing from mostly the same direction over many years turn what were once irregular bumps into nicely symmetrical pyramids. Dreikanters are small, only about knee-high, while the Martian "pyramids" are much taller. But winds on Mars are much fiercer than those on Earth, with wind speeds that can reach half the speed of sound, and it is not unreasonable to suggest that dust storms blown by these strong winds could sculpt larger versions of dreikanters.

Several of the Martian "pyramids" appear highly eroded, and distinctly non-symmetrical. Proponents of the Face on Mars suggest that they were "damaged," perhaps during some kind of Martian war. Skeptics take this as further evidence that wind erosion produced the structures. The Cydonia region of Mars is dotted with many low hills that have been molded into odd shapes, perhaps by a combination of ancient mudflows and wind erosion. The Face on Mars seems to have been created in the same way as these features, even though they don't look like anything in particular.

The original *Viking* photograph of the Face on Mars is riddled with black dots. These dots correspond to areas where data was lost during the transmission of the picture from the *Viking* orbiter to Earth (such transmission losses are common given the problems of communicating with spacecraft over interplanetary distances). If we look carefully at the original image, we see that a black dot of lost data happens to fall right about where we would expect to see a "nostril" on the Face. This makes the rock look even more like a face, but doesn't correspond to any real feature on the Martian surface.

Scientists take great care when interpreting images sent back by spacecraft. When confronted with an alien landscape, it is easy to see what you want or expect to see, not necessarily what is really there. Similar care is needed when interpreting enhanced images. There is only so much information that a picture contains. Image processing or image enhancement will bring out details that were present in the original picture, although not obvious. If you process an image too much, you run the risk that things will appear in the enhanced picture that were not present in the original; such things cannot be believed. This is an issue that astronomers constantly struggle with, given today's emphasis on electronic detectors and computer processing in astronomical research. But it is an issue that is rarely adequately addressed by proponents of the Face on Mars, who depend on enhanced images to back up some of their claims. For example, early proponents claimed to see a "honeycomb" structure on the Martian surface, which they took to be a complex of rooms. Later, the honeycomb was shown to be an artifact of the computer image processing, and not something that was present in the original images. Still, some proponents continue to cite the honeycombs as evidence of a Martian civilization.

And what about the reported alignments of the pyramids and the Face? In order for the alignment to work, Mars' equator would have to be tilted 17.3 degrees relative to the plane of its orbit around the Sun (it is now tipped 24 degrees). It is possible that Mars at one time had such a tilt. But, in this case, there are so many possible variations, with so little to restrict them, that random chance would no doubt produce some apparent alignments. Given that proponents had the freedom to choose any of a number of solar or celestial alignments, any time since the beginning of the solar system, and to move the planet any amount to fit an alignment, the fact that one can be found is not proof that it was intended.

Hoagland has recently taken his analysis of this and other alleged alignments several steps further. He now claims that encoded within the alignments are mathematical constants that reveal information about a "new physics." The beings who built the Face and pyramids are trying to tell the universe about this new physics, he says, which includes the potential for an almost limitless source of energy. According to Hoagland, similar constants can be discerned in the so-called "crop circles" that appeared in English grain fields over the past several years. Such fantastic claims test the credulity of even those who want to believe in the Face on Mars.

Extraordinary Claims Require Extraordinary Proof

The *Viking* orbiters took only a few images that show the Face on Mars. The pictures are of such low resolution and poor quality that they cannot prove conclusively one way or the other whether the Face on Mars is naturally occurring or artificial. One of the guiding principles of scientists is that simple, straightforward explanations of phenomena are more likely to be correct than complicated, convoluted ones. This principle, known as *Occam's razor* (after the fourteenth century English philosopher, William of Occam, who first proposed it; razor is used in the sense of shaving an argument to its simplest terms), means that if two hypotheses fit the observations equally well, the one that makes the fewest assumptions should be chosen.

Applying Occam's razor to the Face on Mars, we ask which is the simpler explanation: that wind erosion, over millions of years, molded a low hill into a shape that happens to resemble a human face when lit from the side, in the same way that it molded many other odd shapes in the same area; or that technologically advanced, intelligent aliens, who may or may not have looked like us and who left no other evidence of their existence, created a gigantic monument with a human face for an unknown reason? If we can explain the Face on Mars as the product of well understood geological processes that we know occur elsewhere on the planet, why do we need to invoke unknown alien beings working in an unknown way for unknown reasons to explain it? Out of the 150,000,000-square-kilometer total surface area of Mars, is it so surprising that a tiny one-square-kilometer-sized area looks a little funny?

Mars is a fascinating planet. Its huge volcanoes dwarf those on Earth. Along its equator, a gigantic chasm stretches so far that, on Earth, it would span the entire United States. Perhaps most intriguing are what look like dry river channels that snake across the Martian deserts. Although liquid water cannot exist on the surface of Mars now, evidently some time in the past the atmosphere was thick enough and the temperatures warm enough for pools of water, rivers, rain and floods to scour the Martian surface. The reality of Mars is exciting enough; we don't need artificially created monuments to manufacture interest in the red planet.

We should always remain skeptical of outrageous stories like that of the Face on Mars, unless there is unquestionable evidence to back them up. Like the Rorschach ink blot tests that psychologists use to probe a person's psyche, claims about the Face on Mars tell us more about humanity's desire not to be alone in the universe than about Martian geology.

There is another region on Mars that looks like Kermit the Frog. On the planet Venus, there is a volcanic feature that resembles the face of Miss Piggy (minus one ear). Taken together, and using the same logic as proponents of the Face on Mars, these two features would seem to prove that there once was a group of intelligent aliens who traveled from planet to planet, and worshipped Muppets! Most people would agree that is just a little absurd.

HOW THE ANCIENT GREEKS SET US ON THE PATH TO MARS (2021)

Duncan Howitt-Marshall

Duncan Howitt-Marshall is a postgraduate researcher in archaeology at the University of Cambridge. His PhD thesis focuses on how maritime trade and exchange affected the socioeconomic and political establishment of southwest Cyprus during the Late Bronze Age and Iron Age. He is a specialist in maritime archaeology, Mediterranean prehistory, especially Aegean and Cypriot prehistory, and the archaeology of islands. Howitt-Marshall researches the origins and development of seafaring, early coastal and island societies, maritime adaptations and social change in the Mesolithic/Epipalaeolithic, Neolithic and Early Bronze Age.

Long before landers arrived on Mars, ancient astronomers plotted its path across the night sky and associated it with deities of fire, war and destruction.

The space above planet Mars has been getting busy of late. Two national space agencies, those of China and the UAE, have sent rockets in the past year, each containing sophisticated instruments to study the planet's atmosphere and weather patterns.

They join other orbiters from NASA, the European Space Agency (ESA) and India's own Mars Orbiter Mission. As of 2021, there are as many as 11 functional spacecraft on or above Mars: eight in orbit, three on the surface.

While it is only in recent decades that humanity has acquired the technical means to actually reach Mars, our fascination with the red planet can be traced back thousands of years to hallowed antiquity, when ancient Babylonian, Assyrian, and Egyptian stargazers made complex observations of Mars and other planets visible to the naked eye.

Subsequently, the ancient Greeks built on the work of their predecessors and developed an even deeper understanding of the cosmos, its structure and origins, by weaving together a rich tapestry of astronomy (largely based on mathematics), philosophy, mythology, and symbolism. Their millennia-old contributions laid the groundwork for much of our understanding of the solar system, setting us on the path to the modern-day exploration of Mars.

The Development of Ancient Greek Astronomy

Western astronomy has its origins in ancient Mesopotamia ("the land between the rivers" – the Tigris and Euphrates), where Babylonian mathematicians and astronomers carefully observed and recorded celestial objects and astronomical events such as lunar and solar eclipses, and associated the red glow of Mars in the night sky with *Nergal*, god of war and strife.

They developed advanced arithmetical techniques to calculate the timing of these phenomena for rituals and religious purposes, similar to the ancient Maya in Central America. The ancient Britons and Polynesians also developed systems to track the movements of the Sun and the moon.

The ancient Egyptians were the first to notice that the stars were fixed and the Sun moved relative to the stars. They plotted the approximate positions of Mercury, Mars, Venus, Jupiter and Saturn, all visible to the naked eye, and noted that Mars followed an irregular loop in the night sky. As such, they referred to it as *Sekded-ef em khetkhet* (meaning "One who travels backward"), or *Har Decher* – simply, the "Red One".

Influenced by the earlier Babylonian and Egyptian astronomers, the ancient Greeks developed the emerging field of astronomy as a sub-branch of mathematics. Greek stargazers made a significant number of increasingly accurate observations and developed core theories that form the basis of our own understanding of astronomy, including the earliest known heliocentric theory of the solar system.

Aristarchus of Samos (Ἀρίσταρχος ὁ Σάμιος, c. 310–230 BC) was the first to place the Sun (or "central fire") at the center of the known universe, with the Earth and other known planets revolving around it.

Indeed, the word "planet" comes from the Greek *planetes* (πλανήτης), meaning "wanderers". Aristarchus also attempted to calculate the sizes and distances of the moon and the Sun, and speculated, correctly, that stars were other suns.

The first known astronomical calendar is the Antikythera Mechanism, discovered in a shipwreck off the Greek island of Antikythera by sponge divers in 1900. Generally referred to as the oldest analogue computer, its complex construction was based on astronomy and mathematics developed by the Hellenistic astronomers of the third and second centuries BC, including Archimedes of Syracuse (Ἀρχιμήδης, c. 287–212 BC).

The mechanism is remarkable not only for its complexity and precise engineering, including the miniaturization of at least 30 gears, but as a device to accurately calculate the mean positions of the Sun, moon, and five known planets, including Mars.

What the Ancient Greeks Knew About Mars

The five known planets in ancient Greece – Mercury, Venus, Mars, Jupiter and Saturn – are often referred to as the "classical" or "naked-eye planets". Early Greek astronomers and philosophers, including the shadowy Pythagoras (c. 570–495 BC), correctly observed the irregular movements of the planets in relation to the fixed stars, and were among the first to identify the morning star and the evening star, referred to by Homer, as being the same celestial object of Venus.

Hipparchus of Nicaea (Ἵππαρχος, c. 190–120 BC) erected an early observatory on the island of Rhodes around 150 BC, and set about compiling a star catalogue with approximately 850 entries. He calculated the celestial coordinates for each star using the first known trigonometric table, and developed and improved several astronomical instruments, including the astrolabe (ἀστρολάβος) – an analogue calculator used to measure the angles of celestial objects relative to the Earth.

Hipparchus' contribution to the early study of the solar system is commemorated in the form of a 93 km-wide impact crater on the surface of Mars, named after him in 1973. An even bigger crater on the moon, 150 km wide, was also named after the ancient Greek astronomer.

Arguably the most important Greek astronomer was Claudius Ptolemy, or Ptolemaeus (Κλαύδιος Πτολεμαῖος, c. 100–170 AD), an Alexandrian Greek who, building on the data collected by Hipparchus, devised the much vaunted "Planetary System", a geometric representation of the solar system that could predict the positions of the naked-eye planets for any desired date and time. Most importantly, the system placed the five planets in order, closest to Earth to furthest away.

Ptolemy was the first astronomer to accurately describe the precise location of Mars, and developed a mathematical solution (epicycle) to account for the complicated retrograde movement of the planet relative to Earth – the irregular loop previously observed by the Egyptians. Ptolemy's calculations formed the basis for subsequent cosmological models that endured until the time of Copernicus, nearly 14 centuries later.

Mars and the Personification of the God, Ares

The astronomical and astrological symbol of the planet Mars, a circle with a small arrow pointing to the upper right, is a stylized representation of a shield and spear used by Ares (Ἄρης), the Greek god of war. It is also the gender symbol for male and the alchemical symbol of iron.

Ares was one of the Twelve Olympians, the son of Zeus and Hera. He was mostly reviled by his fellow gods, associating him with bloodlust and the more violent and brutal aspects of war. In Homer's *Iliad,* he was also depicted as a coward, running away to Mount Olympus when challenged by the Greek hero, Diomedes, to single combat in the Trojan War. His sister, Athena, by contrast, represented the more thoughtful and strategic aspects of war.

Mars' twin moons (or satellites), first discovered by American astronomer Asaph Hall III in 1877, were named Deimos ("terror" or "dread") and Phobos (meaning "panic" or "fear"), the twin sons of Ares and Aphrodite. In the *Iliad*, they both accompany their father into battle, sowing terror and confusion in the midst of battle.

The ancient Greeks themselves built precious few temples or places of worship in Ares' honour, and the existence of cults were extremely rare. Only the warlike city-state of Sparta bucked the trend, indicative of the cultural divisions that existed between them and the rest of Greece.

The Spartans viewed him as the model soldier, brave, resilient, and strong. They called him Ares *Enyalios* ("the warlike"), and in the famed *agoge*, where Spartan youths were sent to train in the arts of war from the age of 7, each company would sacrifice a puppy prior to ritual combat.

How the Ancient Greeks have Inspired Modern-Day Mars Exploration

Over the course of nearly a millennium, ancient Greek astronomers developed increasingly complex geometric theories in an effort to understand the movement of the planets, placing them more precisely in time and space.

Their calculations enabled a greater understanding of time and time-keeping, key to all civic and religious calendars in ancient Greece. Lunar and solar/seasonal cycles, for example, were vital for agriculture and navigation, and bespoke mechanical devices, like the Antikythera Mechanism, could be used to predict astronomical positions.

As spacecraft amass in orbit around Mars, and sophisticated autonomous landers explore its rocky, barren surface, it is important to remember the collective achievement of the ancient astronomers who first inspired humankind to reach beyond the limits of our own planet.

The reason for the sharp increase in activity on and above Mars is down to fundamental logistics, calculations that were first attempted by ancient astronomers. Once every 26 months, Earth and Mars are aligned, closest in distance (a mere 56 million km – as opposed to 400 million km at its furthest distance) thereby minimising the amount of required fuel.

Hipparchus, Ptolemy *et al.* paved the way for our understanding of Mars' orbit around the Sun, and its relative position to Earth, laying the foundations for our understanding of the timing and duration of this crucially important launch window.

If successful, Mars exploration, which had its roots in ancient astronomy, will enable humans to land on the planet in the not-too-distant future. At which point, humanity, for the first time in our history, will become a multi-planetary species.

EXPLORING MARS

ACROSS THE ZODIAC (1880)

CHAPTER III: THE UNTRAVELLED DEEP

Percy Greg

English writer **Percy Greg** was born in 1836 in Lancashire. His novel *Across the Zodiac* (1880) is said to be a forerunner of the 'sword-and-planet' sub-genre, in which traditional adventure stories are simply relocated onto celestial objects. However, in chapter 3, 'The Untravelled Deep', Greg gives a surprisingly detailed description of the voyage to Mars and the benefits of space travel for astronomical observations. Greg died in Chelsea in 1889.

Rising at 5h., I observed a drooping in the leaves of my garden, and especially of the larger shrubs and plants, for which I was not wholly unprepared, but which might entail some inconvenience if, failing altogether, they should cease to absorb the gases generated from buried waste, to consume which they had been planted. Besides this, I should, of course, lose the opportunity of transplanting them to Mars, though I had more hope of acclimatising seedlings raised from the seed I carried with me than plants which had actually begun their life on the surface of the Earth. The failure I ascribed naturally to the known connection between the action of gravity and the circulation of the sap; though, as I had experienced no analogous inconvenience in my own person, I had hoped that this would not seriously affect vegetation. I was afraid to try the effect of more liberal watering, the more so that already the congelation of moisture upon the glasses from the internal air, dry as the latter had been kept, was a sensible annoyance—an annoyance which would have become an insuperable trouble had I not taken so much pains, by directing the thermic currents upon the walls, to keep the internal temperature, in so far as comfort would permit—it had now fallen to 4° C.—as near as possible to that of the inner surface of the walls and windows. A careful use of the thermometer indicated that the metallic surface of the former was now nearly zero C., or 32° F. The inner surface of the windows was somewhat colder, showing that the crystal was more pervious to heat than the walls, with their greater thickness, their outer and inner lining of metal, and massive interior of concrete. I directed a current from the thermogene upon either division of the garden, hoping thus to protect the plants from whatever injury they might receive from the cold. Somewhat later, perceiving that the drooping still continued, I resolved upon another experiment, and arranging an apparatus of copper wire beneath the soil, so as to bring the extremities in immediate contact with their roots, I directed through these wires a prolonged feeble current of electricity; by which, as I had hoped rather than expected, the plants were after a time materially benefited, and to which I believe I owed it that they had not all perished long before the termination of my voyage.

It would be mere waste of space and time were I to attempt anything like a journal of the weeks I spent in the solitude of this artificial planet. As matter of course, the monotony of a voyage through space is in general greater than that of a voyage across an ocean like the Atlantic, where no islands and few ships are to be encountered. It was necessary to be very frequently, if not constantly, on the look-out for possible incidents of interest in a journey so utterly novel through regions which the telescope can but imperfectly explore. It was difficult, therefore, to sit down to a book, or even to pursue any necessary occupation unconnected with the actual conduct of the vessel, with uninterrupted attention. My eyes, the only sense organs I could employ, were constantly on the alert; but, of course, by far the greater portion of my time passed without a single new object or occasion of remark. That a journey so utterly without precedent or parallel, in which so little could be anticipated or provided for, through regions absolutely untraversed and very nearly unknown, should be monotonous, may seem strange. But in truth the novelties of the situation, such as they were, though intensely striking and interesting, were each in turn speedily examined, realised, and, so to speak, exhausted; and this once done, there was no greater occupation to the mind in the continuance of strange than in that of familiar scenery. The infinitude of surrounding blackness, filled as it were with points of light more or less brilliant, when once its effects had been scrutinised, and when nothing more remained to be noted, afforded certainly a more agreeable, but scarcely a more interesting or absorbing, outlook than the dead grey circle of sea, the dead grey hemisphere of cloud, which form the prospect from the deck of a packet in mid-Atlantic; while of change without or incident in the vessel herself there was, of course, infinitely less than is afforded in an ocean voyage by the variations of weather, not to mention the solace of human society. Everything around me, except in the one direction in which the Earth's disc still obscured the Sun, remained unchanged for hours and days; and the management of my machinery required no more than an occasional observation of my instruments and a change in the position of the helm, which occupied but a few minutes some half-dozen times in the twenty-four hours. There was not even the change of night and day, of sun and stars, of cloud or clear sky. Were I to describe the manner in which each day's leisure was spent, I should bore my readers even more than—they will perhaps be surprised by the confession—I was bored myself.

My sleep was of necessity more or less broken. I wished to have eight hours of rest, since, though seven of continuous sleep might well have sufficed me, even if my brain had been less quiet and unexcited during the rest of the twenty-four, it was impossible for me to enjoy that term of unbroken slumber. I therefore decided to divide my sleep into two portions of rather more than four hours each, to be taken as a rule after noon and after midnight; or rather, since noon and midnight had no meaning for me, from 12h. to 16h. and from 24h. to 4.h. But of course sleep and everything else, except the necessary management of the machine, must give way to the chances of observation; it would be better to remain awake for forty-eight hours at a stretch than to miss any important phenomenon the period of whose occurrence could be even remotely calculated.

At 8h., I employed for the first time the apparatus which I may call my window telescope, to observe, from a position free from the difficulties inflicted on terrestrial astronomers by the atmosphere, all the celestial objects within my survey. As I had anticipated, the absence of atmospheric disturbance and diffusion of light was of extreme advantage. In the first place, I ascertained by the barycrite and the discometer my distance from the Earth, which appeared to be about 120 terrestrial radii. The light of the halo was of course very much narrower than when I first observed it, and its scintillations or coruscations no longer distinctly visible. The Moon presented an exquisitely fine thread of light, but no new object of interest on the very small portion of her daylight hemisphere turned towards me. Mars was somewhat difficult to observe, being too near what may be called my zenith. But the markings were far more distinct than they appear, with greater magnifying powers than I employed, upon the Earth. In truth, I should say that the various disadvantages due to the atmosphere deprive the astronomer of at least one-half of the available light-collecting power of his telescope, and consequently of the defining power of the eye-piece; that with a 200 glass he sees less than a power of 100 reveals to an eye situated in space; though, from the nature of the lens through which I looked, I cannot speak with certainty upon this point. With a magnifying power of 300 the polar spots of Mars were distinctly visible and perfectly defined. They were, I thought, less white than they appeared from the Earth, but their colour was notably different from that of the planet's general surface, differing almost as widely from the orange hue of what I supposed to be land as from the greyish blue of the water. The orange was, I thought, deeper than it appears through a telescope of similar power on Earth. The seas were distinctly grey rather than blue, especially when, by covering the greater part of the field, I contrived for a moment to observe a sea alone, thus eliminating the effect of contrast. The bands of Jupiter in their turn were more notably distinct; their variety of colour as well as the contrast of light and shade much more definite, and their irregularities more unmistakable. A satellite was approaching the disc, and this afforded me an opportunity of realising with especial clearness the difference between observation through seventy or a hundred miles of terrestrial atmosphere outside the object glass and observation in space. The two discs were perfectly rounded and separately discernible until they touched. Moreover, I was able to distinguish upon one of the darker bands the disc of the satellite itself, while upon a lighter band its round black shadow was at the same time perfectly defined. This wonderfully clear presentation of one of the most interesting of astronomical phenomena so absorbed my attention that I watched the satellite and shadow during their whole course, though the former, passing after a time on to a light band, became comparatively indistinct. The moment, however, that the outer edge passed off the disc of Jupiter, its outline became perfectly visible against the black background of sky. What was still more novel was the occultation for some little time of a star, apparently of the tenth magnitude, not by the planet but by the satellite, almost immediately after it passed off the disc of the former. Whether the star actually disappeared at once, as if instantaneously extinguished, or whether, as I thought at the moment, it remained for some tenth of a second partially visible, as if refracted by an atmosphere belonging to the satellite, I will not venture to say. The bands and rings of Saturn, the division between the two latter, and the seven satellites, were also perfectly visible, with a distinctness that a much greater magnifying power would hardly have attained under terrestrial conditions. I was perplexed by two peculiarities, not, so far as I know, hitherto mentioned by astronomers. The circumference did not appear to present an even curvature.

I mean that, apart from the polar compression, the shape seemed as if the spheroid were irregularly squeezed; so that though not broken by projection or indentation, the limb did not present the regular quasi-circular curvature exhibited in the focus of our telescopes. Also, between the inner ring and the planet, with a power of 500, I discerned what appeared to be a dark purplish ring, semi-transparent, so that through it the bright surface of Saturn might be discerned as through a veil. Mercury shone brightly several degrees outside the halo surrounding the Earth's black disc; and Venus was also visible; but in neither case did my observations allow me to ascertain anything that has not been already noted by astronomers. The dim form of Uranus was better defined than I had previously seen it, but no marking of any kind was perceptible.

Rising from my second, or, so to speak, midday rest, and having busied myself for some little time with what I may call my household and garden duties, I observed the discometer at 1h. (or 5 P.M.). It indicated about two hundred terrestrial radii of elevation. I had, of course, from the first been falling slightly behind the Earth in her orbital motion, and was no longer exactly in opposition; that is to say, a line drawn from the Astronaut to the Earth's centre was no longer a prolongation of that joining the centres of the Earth and Sun. The effect of this divergence was now perceptible. The earthly corona was unequal in width, and to the westward was very distinctly brightened, while on the other side it was narrow and comparatively faint. While watching this phenomenon through the lower lens, I thought that I could perceive behind or through the widest portion of the halo a white light, which at first I mistook for one of those scintillations that had of late become scarcely discernible. But after a time it extended visibly beyond the boundary of the halo itself, and I perceived that the edge of the Sun's disc had come at last into view. It was but a minute and narrow crescent, but was well worth watching. The brightening and broadening of the halo at this point I perceived to be due, not to the Sun's effect upon the atmosphere that produced it, but chiefly to the twilight now brightening on that limb of the Earth's disc; or rather to the fact that a small portion of that part of the Earth's surface, where, if the Sun were not visible, he was but a very little below the horizon, had been turned towards me. I saw through the telescope first a tiny solar crescent of intense brightness, then the halo proper, now exceedingly narrow, and then what looked like a silver terrestrial crescent, but a mere thread, finer and shorter than any that the Moon ever displays even to telescopic observers on Earth; since, when such a minute portion of her illuminated surface is turned towards the Earth, it is utterly extinguished to our eyes by the immediate vicinity of the Sun, as was soon the case with the terrestrial crescent in question. I watched long and with intense interest the gradual change, but I was called away from it by a consideration of no little practical moment. I must now be moving at a rate of nearly, if not quite, 40,000 miles an hour, or about a million miles per diem. It was not my intention, for reasons I shall presently explain, ever greatly to exceed this rate; and if I meant to limit myself to a fixed rate of speed, it was time to diminish the force of the apergic current, as otherwise before its reduction could take effect I should have attained an impulse greater than I desired, and which could not be conveniently or easily diminished when once reached. Quitting, therefore, though reluctantly, my observation of the phenomena below me, I turned to the apergion, and was occupied for some two or three hours in gradually reducing the force as measured by the cratometer attached to the downward conductor, and measuring with extreme care the very minute effect produced upon the barycrite and the discometer. Even the difference between 200 and 201 radii of elevation or apogaic distance was not easily perceptible on either. It took, of course, much more minute observation and a much longer time to test the effect produced by the regulation of the movement, since whether I travelled forty, forty-five, or forty-two thousand miles in the course of one hour made scarcely any difference in the diameter of the Earth's disc, still less, for reasons above given, in the gravity. By midnight, however, I was satisfied that I had not attained quite 1,000,000 miles, or 275 terrestrial radii; also that my speed was not greater than 45,000 miles (11–¼ radii) per hour, and was not, I thought, increasing. Of this last point, however, I could better satisfy myself at the end of my four hours' rest, to which I now betook myself.

I woke about 4h. 30m., and on a scrutiny of the instruments, felt satisfied that I was not far out in my calculations. A later hour, however, would afford a more absolute certainty. I was about to turn again to the interesting work of observation through the lens in the floor, when my attention was diverted by the sight of something like a whitish cloud visible through the upper window on my left hand. Examined by the telescope, its widest diameter might be at most ten degrees. It was faintly luminous, presenting an appearance very closely resembling that of a star cluster or nebula just beyond the power of resolution. As in many nebulae, there was a visible concentration in one part; but this did not occupy the centre, but a position more resembling that of the nucleus of a small tailless comet. The cloudlet might be a distant comet, it might be a less distant body of meteors clustering densely in some particular part of their orbit; and, unfortunately, I was not likely to solve the problem. Gradually the nebula changed its position, but not its form, seeming to move downwards and towards the stern of my vessel, as if I were passing it without approaching nearer. By the time that I was satisfied of this, hunger and even faintness warned me that I must not delay preparing my breakfast. When I had finished this meal and fulfilled some necessary tasks, practical and arithmetical, the hand of the chronometer indicated the eighth hour of my third day. I turned again somewhat eagerly to the discometer, which showed an apparent distance of 360 terrestrial radii, and consequently a movement which had not materially varied from the rate of 11–¼ radii per hour. By this time the diameter of the Earth was not larger in appearance than about 19', less than two-thirds that of the Sun; and she consequently appeared as a black disc covering somewhat more than one-third of his entire surface, but by no means concentrical. The halo had of course completely disappeared; but with the vernier it was possible to discern a narrow band or line of hazy grey around the black limb of the planet. She was moving, as seen from the Astronaut, very slightly to the north, and more decidedly, though very slowly, to the eastward; the one motion due to my deliberately chosen direction in space, the other to the fact that as my orbit enlarged I was falling, though as yet slowly, behind her. The sun now shone through the various windows, and, reflected from the walls, maintained a continuous daylight within the Astronaut, as well diffused as by the atmosphere of Earth, strangely contrasting the star-spangled darkness outside.

At the beginning as at the end of my voyage, I steered a distinct course, governed by considerations quite different from those which controlled the main direction of my voyage. Thus far I had simply risen straight from the Earth in a direction somewhat to the southward, but on the whole "in opposition," or right away from the Sun. So, at the conclusion of my journey, I should have to devote some days to a gradual descent upon Mars, exactly reversing the process of my ascent from the Earth. But between these two periods I had comparatively little to do with either planet, my course being governed by the Sun, and its direction and rate being uniform. I wished to reach Mars at the moment of opposition, and during the whole of the journey to keep the Earth between myself and the Sun, for a reason which may not at first be obvious. The moment of opposition is not necessarily that at which Mars is nearest to the Earth, but is sufficiently so for practical calculation. At that moment, according to the received measurement of planetary distances, the two would be more than 40 millions of miles apart. In the meantime the Earth, travelling on an interior or smaller orbit, and also at a greater absolute speed, was gaining on Mars. The Astronaut, moving at the Earth's rate under an impulse derived from the Earth's revolution round the Sun (that due to her rotation on her own axis having been got rid of, as aforesaid), travelled in an orbit constantly widening, so that, while gaining on Mars, I gained on him less than did the Earth, and was falling behind her. Had I used the apergy only to drive me directly outward from the Sun, I should move under the impulse derived from the Earth about 1,600,000 miles a day, or 72 millions of miles in forty-five days, in the direction common to the two planets. The effect of the constantly widening orbit would be much as if the whole motion took place on one midway between those of the Earth and Mars, say 120 millions of miles from the Sun. The arc described on this orbit would be equivalent to 86 millions of miles on that of Mars. The entire arc of his orbit between the point opposite to that occupied by the Earth when I started and the point of opposition—the entire distance I had to gain as measured along his path—was about 116 millions of miles; so that, trusting to the terrestrial impulse alone, I should be some 30 millions behindhand at the critical moment. The apergic force must make up for this loss of ground, while driving me in a direction, so to speak, at right angles with that of the orbit, or along its radius, straight outward from the Sun, forty odd millions of miles in the same time. If I succeeded in this, I should reach the orbit of Mars at the point and at the moment of opposition, and should attain Mars himself. But in this I might fail, and I should then find myself under the sole influence of the Sun's attraction; able indeed to resist it, able gradually to steer in any direction away from it, but hardly able to overtake a planet that should lie far out of my line of advance or retreat, while moving at full speed away from me. In order to secure a chance of retreat, it was desirable as long as possible to keep the Earth between the Astronaut and the Sun; while steering for that point in space where Mars would lie at the moment when, as seen from the centre of the Earth, he would be most nearly opposite the Sun,—would cross the meridian at midnight. It was by these considerations that the course I henceforward steered was determined. By a very simple calculation, based on the familiar principle of the parallelogram of forces, I gave to the apergic current a force and direction equivalent to a daily motion of about 750,000 miles in the orbital, and rather more than a million in the radial line. I need hardly observe that it would not be to the apergic current alone, but to a combination of that current with the orbital impulse received at first from the Earth, that my progress and course would be due. The latter was the stronger influence; the former only was under my control, but it would suffice to determine, as I might from time to time desire, the resultant of the combination. The only obvious risk of failure lay in the chance that, my calculations failing or being upset, I might reach the desired point too soon or too late. In either case, I should be dangerously far from Mars, beyond his orbit or within it, at the time when I should come into a line with him and the Sun; or, again, putting the same mischance in another form, behind him or before him when I attained his orbit. But I trusted to daily observation of his position, and verification of my "dead reckoning" thereby, to find out any such danger in time to avert it.

The displacement of the Earth on the Sun's face proved it to be necessary that the apergic current should be directed against the latter in order to govern my course as I desired, and to recover the ground I had lost in respect to the orbital motion. I hoped for a moment that this change in the action of the force would settle a problem we had never been able to determine. Our experiments proved that apergy acts in a straight line when once collected in and directed along a conductor, and does not radiate, like other forces, from a centre in all directions. It is of course this radiation—diffusing the effect of light, heat, or gravity over the surface of a sphere, which surface is proportionate to the square of the radius—that causes these forces to operate with an energy inversely proportionate, not to the distance, but to its square. We had no reason to think that apergy, exempt as it is from this law, would be at all diminished by distance; and this view the rate of acceleration as I rose from the Earth had confirmed, and my entire experience has satisfied me that it is correct. None of our experiments, however, had indicated, or could well indicate, at what rate this force can travel through space; nor had I yet obtained any light upon this point. From the very first the current had been continuous, the only interruption taking place when I was not five hundred miles from the Earth's surface. Over so small a distance as that, the force would move so instantaneously that no trace of the interruption would be perceptible in the motion of the Astronaut. Even now the total interruption of the action of apergy for a considerable time would not affect the rate at which I was already moving. It was possible, however, that if the current had been hitherto wholly intercepted by the Earth, it might take so long a time in reaching the Sun that the interval between the movement of the helm and the response of the Astronaut's course thereto might afford some indication of the time occupied by the current in traversing the 96–½ millions of miles which parted me from the Sun. My hope, however, was wholly disappointed. I could neither be sure that the action was instantaneous, nor that it was otherwise.

At the close of the third day I had gained, as was indicated by the instruments, something more than two millions of miles in a direct line from the Sun; and for the future I might, and did, reckon on a steady progress of about one and a quarter million miles daily under the apergic force alone—a gain in a line directly outward from the Sun of about one million. Henceforward I shall not record my observations, except where they implied an unexpected or altered result.

On the sixth day, I perceived another nebula, and on this occasion in a more promising direction. It appeared, from its gradual movement, to lie almost exactly in my course, so that if it were what I suspected, and were not at any great distance from me, I must pass either near or through it, and it would surely explain what had perplexed and baffled me in the case of the former nebula. At this distance the nature of the cloudlet was imperceptible to the naked eye. The window telescope was not adjustable to an object which I could not bring conveniently within the field of view of the lenses. In a few hours the nebula so changed its form and position, that, being immediately over the portion of the roof between the front or bow lens and that in the centre of the roof, its central section was invisible; but the extremities of that part which I had seen in the first instance through the upper plane window of the bow were now clearly visible from the upper windows of either side. What had at first been a mere greatly elongated oval, with a species of rapidly diminishing tail at each extremity, had now become an arc spanning no inconsiderable part of the space above me, narrowing rapidly as it extended downwards and sternwards. Presently it came in view through the upper lens, but did not obscure in the least the image of the stars which were then visible in the metacompass. I very soon ascertained that the cloudlet consisted, as I had supposed in the former case, of a multitude of points of light less brilliant than the stars, the distance between which became constantly wider, but which for some time were separately so small as to present no disc that any magnifying power at my command could render measurable. In the meantime, the extremities visible through the other windows were constantly widening out till lost in the spangled darkness. By and by, it became impossible with the naked eye to distinguish the individual points from the smaller stars; and shortly after this the nearest began to present discs of appreciable size but somewhat irregular shape. I had now no doubt that I was about to pass through one of those meteoric rings which our most advanced astronomers believe to exist in immense numbers throughout space, and to the Earth's contact with or approach to which they ascribe the showers of falling stars visible in August and November. Ere long, one after another of these bodies passed rapidly before my sight, at distances varying probably from five yards to five thousand miles. Where to test the distance was impossible, anything like accurate measurement was equally out of the question; but my opinion is, that the diameters of the nearest ranged from ten inches to two hundred feet. One only passed so near that its absolute size could be judged by the marks upon its face. This was a rock-like mass, presenting at many places on the surface distinct traces of metallic veins or blotches, rudely ovoid in form, but with a number of broken surfaces, one or two of which reflected the light much more brilliantly than others. The weight of this one meteoroid was too insignificant as compared with that of the Astronaut seriously to disturb my course. Fortunately for me, I passed so nearly through the centre of the aggregation that its attraction as a whole was nearly inoperative. So far as I could judge, the meteors in that part of the ring through which I passed were pretty evenly distributed; and as from the appearance of the first which passed my window to the disappearance of the last four hours elapsed, I conceived that the diameter of the congeries, measured in the direction of my path, which seemed to be nearly in the diameter of their orbit, was about 180,000 miles, and probably the perpendicular depth was about the same.

I may mention here, though somewhat out of place, to avoid interrupting the narrative of my descent upon Mars, the only interesting incident that occurred during the latter days of my journey—the gradual passage of the Earth off the face of the Sun. For some little time after this the Earth was entirely invisible; but later, looking through the telescope adjusted to the lens on that side, I discerned two very minute and bright crescents, which, from their direction and position, were certainly those of the Earth and Moon, indeed could hardly be anything else.

Towards the thirtieth day of my voyage I was disturbed by the conflicting indications obtained from different instruments and separate observations. The general result came to this, that the discometer, where it should have indicated a distance of 333, actually gave 347. But if my speed had increased, or I had overestimated the loss by changes of direction, Mars should have been larger in equal proportion. This, however, was not the case. Supposing my reckoning to be right, and I had no reason to think it otherwise, except the indication of the discometer, the Sun's disc ought to have diminished in the proportion of 95 to 15, whereas the diminution was in the proportion of 9 to 1. So far as the barycrite could be trusted, its very minute indications confirmed those of the discometer; and the only conclusion I could draw, after much thought and many intricate calculations, was that the distance of 95 millions of miles between the Earth and the Sun, accepted, though not very confidently, by all terrestrial astronomers, is an over-estimate; and that, consequently, all the other distances of the solar system have been equally overrated. Mars consequently would be smaller, but also his distance considerably less, than I had supposed. I finally concluded that the solar distance of the Earth was less than 9 millions of miles, instead of more than 95. This would involve, of course, a proportionate diminution in the distance I had to traverse, while it did not imply an equal error in the reckoning of my speed, which had at first been calculated from the Earth's disc, and not from that of the Sun. Hence, continuing my course unchanged, I should arrive at the orbit of Mars some days earlier than intended, and at a point behind that occupied by the planet, and yet farther behind the one I aimed at. Prolonged observation and careful calculation had so fully satisfied me of the necessity of the corrections in question, that I did not hesitate to alter my course accordingly, and to prepare for a descent on the thirty-ninth instead of the forty-first day. I had, of course, to prepare for the descent very long before I should come within the direct influence of the attraction of Mars. This would not prevail over the Sun's attraction till I had come within a little more than 100,000 miles of the surface, and this distance would not allow for material reduction of my speed, even were I at once to direct the whole force of the apergic current against the planet. I estimated that arriving within some two millions of miles of him, with a speed of 45,000 miles per hour, and then directing the whole force of the current in his direction, I should arrive at his surface at a speed nearly equal to that at which I had ascended from the Earth. I knew that I could spare force enough to make up for any miscalculation possible, or at least probable. Of course any serious error might be fatal. I was exposed to two dangers; perhaps to three: but to none which I had not fully estimated before even preparing for my voyage. If I should fail to come near enough to the goal of my journey, and yet should go on into space, or if, on the other hand, I should stop short, the Astronaut might become an independent planet, pursuing an orbit nearly parallel to that of the Earth; in which case I should perish of starvation. It was conceivable that I might, in attempting to avert this fate, fall upon the Sun, though this seemed exceedingly improbable, requiring a combination of accidents very unlikely to occur. On the other hand, I might by possibility attain my point, and yet, failing properly to calculate the rate of descent, be dashed to pieces upon the surface of Mars. Of this, however, I had very little fear, the tremendous power of the apergy having been so fully proved that I believed that nothing but some disabling accident to myself—such as was hardly to be feared in the absence of gravitation, and with the extreme simplicity of the machinery I employed—could prevent my being able, when I became aware of the danger, to employ in time a sufficient force to avert it. The first of these perils, then, was the graver one, perhaps the only grave one, and certainly to my imagination it was much the most terrible. The idea of perishing of want in the infinite solitude of space, and being whirled round for ever the dead denizen of a planet one hundred feet in diameter, had in it something even more awful than grotesque.

On the thirty-ninth morning of my voyage, so far as I could calculate by the respective direction and size of the Sun and of Mars, I was within about 1,900,000 miles from the latter. I proceeded without hesitation to direct the whole force of the current permitted to emerge from the apergion directly against the centre of the planet. His diameter increased with great rapidity, till at the end of the first day I found myself within one million of miles of his surface. His diameter subtended about 15', and his disc appeared about one-fourth the size of the Moon. Examined through the telescope, it presented a very different appearance from that either of the Earth or of her satellite. It resembled the former in having unmistakably air and water. But, unlike the

Earth, the greater portion of its surface seemed to be land; and, instead of continents surrounded by water, it presented a number of separate seas, nearly all of them land-locked. Around the snow-cap of each pole was a belt of water; around this, again, a broader belt of continuous land; and outside this, forming the northern and southern boundary between the arctic and temperate zones, was another broader band of water, connected apparently in one or two places with the central, or, if one may so call it, equatorial sea. South of the latter is the one great Martial ocean. The most striking feature of this new world, as seen from this point, was the existence of three enormous gulfs, from three to five thousand miles in length, and apparently varying in breadth from one hundred to seven hundred miles. In the midst of the principal ocean, but somewhat to the southward, is an island of unique appearance. It is roughly circular, and, as I perceived in descending, stands very high, its table-like summit being some 4000 feet, as I subsequently ascertained, above the sea-level. Its surface, however, was perfectly white—scarcely less brilliant, consequently, than an equal area of the polar icefields. The globe, of course, revolved in some 4–1/ [sic] hours of earthly time, and, as I descended, presented successively every part of its surface to my view. I speak of descent, but, of course, I was as yet ascending just as truly as ever, the Sun being visible through the lens in the floor, and reflected upon the mirror of the discometer, while Mars was now seen through the upper lens, and his image received in the mirror of the metacompass. A noteworthy feature in the meteorology of the planet became apparent during the second day of the descent. As magnified by the telescope adjusted to the upper lens, the distinctions of sea and land disappeared from the eastern and western limbs of the planet; indeed, within 15° or an hour of time from either. It was plain, therefore, that those regions in which it was late evening or early morning were hidden from view; and, independently of the whitish light reflected from them, there could be little doubt that the obscuration was due to clouds or mists. Had the whitish light covered the land alone, it might have been attributed to a snowfall, or, perhaps, even to a very severe hoar frost congealing a dense moisture. But this last seemed highly improbable; and that mist or cloud was the true explanation became more and more apparent as, with a nearer approach, it became possible to discern dimly a broad expanse of water contrasting the orange tinge of the land through this annular veil. At 4h. on the second day of the descent, I was about 500,000 miles from Mars, the micrometer verifying, by the increased angle subtended by the diameter, my calculated rate of approach. On the next day I was able to sleep in security, and to devote my attention to the observation of the planet's surface, for at its close I should be still 15,000 miles from Mars, and consequently beyond the distance at which his attraction would predominate over that of the Sun. To my great surprise, in the course of this day I discerned two small discs, one on each side of the planet, moving at a rate which rendered measurement impossible, but evidently very much smaller than any satellite with which astronomers are acquainted, and so small that their non-discovery by terrestrial telescopes was not extraordinary. They were evidently very minute, whether ten, twenty, or fifty miles in diameter I could not say; neither of them being likely, so far as I could calculate, to come at any part of my descent very near the Astronaut, and the rapidity of their movement carrying them across the field, even with the lowest power of my telescopes, too fast for measurement. That they were Martial moons, however, there could be no doubt.

About 10h. on the last day of the descent, the effect of Mars' attraction, which had for some time so disturbed the position of the Astronaut as to take his disc completely out of the field of the meta-compass, became decidedly predominant over that of the Sun. I had to change the direction of the apergic current first to the left-hand conductor, and afterwards, as the greater weight of the floor turned the Astronaut completely over, bringing the planet immediately below it, to the downward one. I was, of course, approaching Mars on the daylight side, and nearly in the centre. This, however, did not exactly suit me. During the whole of this day it was impossible that I should sleep for a minute; since if at any point I should find that I had miscalculated my rate of descent, or if any other unforeseen accident should occur, immediate action would be necessary to prevent a shipwreck, which must without doubt be fatal. It was very likely that I should be equally unable to sleep during the first twenty-four hours of my sojourn upon Mars, more especially should he be inhabited, and should my descent be observed. It was, therefore, my policy to land at some point where the Sun was setting, and to enjoy rest during such part of the twelve hours of the Martial night as should not be employed in setting my vessel in order and preparing to evacuate it. I should have to ascertain exactly the pressure of the Martial atmosphere, so as not to step too suddenly from a dense into what was probably a very light one. If possible, I intended to land upon the summit of a mountain, so high as to be untenanted and of difficult access. At the same time it would not do to choose the highest point of a very lofty range, since both the cold and the thinness of the air might in such a place be fatal. I wished, of course, to leave the Astronaut secure, and, if not out of reach, yet not within easy reach; otherwise it would have been a simple matter to watch my opportunity and descend in the dark from my first landing-place by the same means by which I had made the rest of my voyage.

At 18h. I was within 8000 miles of the surface, and could observe Mars distinctly as a world, and no longer as a star. The colour, so remarkable a feature in his celestial appearance, was almost equally perceptible at this moderate elevation. The seas are not so much blue as grey. Masses of land reflected a light between yellow and orange, indicating, as I thought, that orange must be as much the predominant colour of vegetation as green upon Earth. As I came still lower, and only parts of the disc were visible at once, and these through the side and end windows, this conviction was more and more strongly impressed upon my mind. What, however, was beyond denial was, that if the polar ice and snow were not so purely and distinctly white as they appear at a distance upon Earth, they were yet to a great extent devoid of the yellow tinge that preponderated everywhere else. The most that could be said was, that whereas on Earth the snow is of that white which we consider absolute, and call, as such, snow-white, but which really has in it a very slight preponderance of blue, upon Mars the polar caps are rather cream-white, or of that white, so common in our flowers, which has in it an equally slight tinge of yellow. On the shore, or about twenty miles from the shore of the principal sea to the southward of the equator, and but a few degrees from the equator itself, I perceived at last a point which appeared peculiarly suitable for my descent. A very long range of mountains, apparently having an average height of about 14,000 feet, with some peaks of probably twice or three times that altitude, stretched for several hundred miles along the coast, leaving, however, between it and the actual shore-line an alluvial plain of some twenty to fifty miles across. At the extremity of this range, and quite detached from it, stood an isolated mountain of peculiar form, which, as I examined it through the telescope, appeared to present a surface sufficiently broken and sloped to permit of descent; while, at the same time, its height and the character of its summit satisfied me that no one was likely to inhabit it, and that though I might descend-it in a few hours, to ascend it on foot from the plain would be a day's journey. Towards this I directed my course, looking out from time to time carefully for any symptoms of human habitation or animal life. I made out by degrees the lines of rivers, mountain slopes covered by great forests, extensive valleys and plains, seemingly carpeted by a low, dense, rich vegetation. But my view being essentially of a bird's-eye character, it was only in those parts that lay upon my horizon that I could discern clearly the height of any object above the general level; and as yet, therefore, there might well be houses and buildings, cultivated fields and divisions, which I could not see.

Before I had satisfied myself whether the planet was or was not inhabited, I found myself in a position from which its general surface was veiled by the evening mist, and directly over the mountain in question, within some twelve miles of its summit. This distance I descended in the course of a quarter of an hour, and landed without a shock about half an hour, so far as I could judge, after the Sun had disappeared below the horizon. The sunset, however, by reason of the mists, was totally invisible.

BARON MÜNCHHAUSEN'S NEW SCIENTIFIC ADVENTURES, PART 5 (1915)

MÜNCHHAUSEN DEPARTS FOR THE PLANET MARS

Hugo Gernsback

Born **Hugo Gernsbacher** in 1884 in Luxembourg, Gernsback altered his name after emigrating to the United States in 1904. He was an entrepreneur in the early popularization of 'wireless' radio and founded several magazines. *Modern Electrics, Electrical Experimenter* and *Radio News* all dealt with various aspects of the new technology. In 1926, he founded *Astounding Stories* to publish science fiction stories. This magazine became a pillar of 'the golden age of science fiction'. The annual achievement awards of the World Science Fiction Convention are known as the Hugos. Gernsback died in New York City in 1967. This is the second of his works to appear in this anthology.

The Electrical Experimenter, vol. 3, no. 6, October 1915

"THERE are more things in heaven and earth, Horatio, than are dreamt of in your philosophy."

So sings Shakespeare. One of these "things in heaven" is the Planet Mars, the most fascinating, the most astounding revelation to the feeble human intelligence. Shakespeare, the master of the drama, never conceived anything like a drama of an entire world—millions of intelligent beings—fighting a heroic battle, a battle for existence. Yet this drama was going on right before his very eyes, but 35 million miles away; for the Martians have been fighting for water ages ago, and the available supply becomes smaller each year.

There is nothing more inspiring, nothing more gripping to the imagination, than this wonderful battle between organized intelligence on one side and unrelenting nature on the other.

*Mr. Münchhausen's scientific lecture gives you the latest facts—now almost universally believed—about Mars. You can spend no better half hour than turning your mind from your humdrum existence towards a subject which is as absorbing as it is lofty in its grandeur.**

Once upon a time, a grouchy old gentleman with a grievance for fiction writers, presumably because the latter received more emoluments for their stuff than the former for his poetry, thus vented his resentment in immortal song:

"'Tis strange, but true; for truth is always strange— stranger than fiction."

From this, some coarse soul, totally oblivious of any poetic infection whatsoever, took it upon himself to mutilate the above passage of one of Lord Byron's poems and taught us unsuspecting mortals to hawk, parrot-wise, ever after until the end of fiction, thusly: "Truth is stranger than fiction!"

With all due regard to the memory and genius of Byron, I, I. M. Alier, a citizen of a free country, take it upon myself to correct his Lordship at this late and quarrelsome date, to wit:

"There is no fiction."

If, as often—no, always—has been proved that the most violent fiction at some time or other invariably comes true, then by all proceeds of modern logic, there cannot be such thing as fiction. It simply does not exist. This brings us face to face with the startling result that if fiction always comes true some time or other, why then, bless their dear souls, all fiction writers must be prophets!! Hurrah for the F. W.!! But hold on, boys; don't let our enthusiasm run away with us on a Ford. The spark plug has run afoul somewhere. While it's nice to be a prophet, don't forget that a prophet is never, never recognized in his own country. Thus the New Testament teaches; so I think it will be safer for all F. W. to remain F. W., rather than to be honorless prophets.

However, that is not what I had in mind when I started—it's so hard for me to say what I mean, and a good deal harder for me to keep my thoughts running on the track. They ramble from one nothingness into another. My mind in that respect is a good deal like a one-eyed, religious old cow on a pasture. She eats up whatever she sees alongside of her, but when she finally turns around she perceives with astonishment that there is still a lot to graze on the other side; so she steers around to starboard and returns to her original starting point.

* The character of Baron Munchausen was inspired by the real-life Hieronymus Karl Friedrich, Freiherr von Münchhausen (1720–1797). Rudolf Erich Raspe created a series of fictionalized accounts of Baron Munchausen's life in the 1780s, tall tales about "Singular Travels, Campaigns, Voyages, and Sporting Adventures." After Raspe, the character was picked up by a number of other artists in various media, including Terry Gilliam's 1988 film *The Adventures of Baron Munchausen*. Gernsback uses the original spelling of the real Baron's name.

But I am rambling again. So let's return to the original starting point.

Seriously speaking, and by way of emphasizing how much stranger truth is than fiction, I have but to point to Jules Verne's famous stories. When 45 years ago he wrote "*Twenty Thousand Leagues Under the Sea*," no one took him serious. It is doubtful whether he himself believed that the submarine which he invented in that story would ever become practical. It was just fiction. Yet 45 years later we see how a submarine, almost exactly as his vivid as well as prophetic mind conceived it, down to the most minute detail, emerges from a German harbor and travels under its own power over a distance of 4,000 miles, through the North Sea, the English Channel, down the Atlantic, through the entire length of the Mediterranean and up through the Dardanelles to Constantinople! And by way of diversion it manages to sink several battleships of the enemy by means of its torpedoes. Now, bold as he was, Jules Verne never conceived such an "impossible thing," and while his famous *Nautilus* was equipped with almost every other modern submarine necessity, the infernal automobile-torpedo was missing. Truth is indeed very much stranger than fiction. Hundreds of similar instances could be cited, but lack of space prohibits it; besides, I mustn't ramble.*

Münchhausen, as will be remembered, had explained the mysteries of the moon to me, and he had also mentioned the great danger of falling meteors, which had been increasing alarmingly in number for some time. †The moon's attenuated atmosphere offers no protection from meteors, as did the earth's thick air. But few meteors ever reach the surface of the earth; the colossal friction between the meteor and the air ignites the former, and most of it falls down on the earth as a fine dust. The burning of the meteors represents the shooting stars we see. On the moon, however, the meteors crash down bodily, causing tremendous havoc, and this terrible bombardment goes on forever, without let-up.† Consequently, when Baron Münchhausen stopped short that evening in the midst of a sentence, I naturally was alarmed not a little. Great, therefore, was my joy when, sitting before my radio set the next evening, 'phones clapped tight over my ears, my eyes glued on the clock, the familiar high, whining spark suddenly reverberated in my ears at the stroke of 11 o'clock.

It was Münchhausen. But his usual sonorous voice to-night had an unfamiliar metallic timbre that puzzled me greatly; in a short time, however, the mystery was cleared, and this is what poured in my astonished ears:

"My dear Alier. No doubt you thought I had been killed by a meteor last night. Well, as you Americans put it, I certainly had 'a close shave.' A meteor crashed down on my aerial 50 feet from where I was sitting; of course it went up in smoke—metallic vapor, to be correct—due to the tremendous heat generated by the impact of the meteor on the granite rocks. The whole meteor itself went up in a fiery cloud of red vapor and I was blown headlong a distance of over 50 feet, right down into the mouth of a giant crater, by the colossal resulting blast of the concussion.

"Now, this long-extinct crater is a very deep one; how deep, I was soon to learn! I went down head first and kept on falling at a terrible rate of speed. I must have been falling down that awful abyss what seemed to me like hours. As I kept on plunging down, I was gloomily reflecting what an inglorious death it was to die at the bottom of an unromantic crater on a dead and dried-up moon. I

* Gernsback: "In order to distinguish facts from fiction in this installment, all statements containing actual scientific facts will be enclosed between two † marks.—AUTHOR"

thought of many things, when I suddenly became conscious of a terrific cold. Call it instinct or presence of mind, as soon as I had started on my downward journey I had jerked my body in such a manner as to righten it; in other words, after a few attempts, I succeeded in falling feet down. It was indeed a fortunate circumstance that the sun was almost directly over the crater, for it saved me the anguish of plunging into a pitch-black abyss. While it was not nearly as light as it was at the top, still I could see where I was falling, and that was some consolation. Thus, when I glanced down toward my feet after a while, I am sure that my heart, which had stopped beating, stood still entirely for some seconds. It took me a few seconds to collect my bewildered senses, for this is what I had seen:

"The crater had no bottom at all, but went right through to the center of the moon, where it connected with another crater, which went to the opposite side of the moon. I knew this must be so because when I had looked down I had seen several stars shining through brilliantly from the night side of the moon. Then the awful truth flashed through me and I almost swooned. *I was falling through the whole length of the moon!* I had been in many tight quarters before during my somewhat exciting career, but this experience indeed bade well to be the inglorious end of my adventurous life. However, my far-famed presence of mind and my cool head soon asserted themselves, as was naturally to be expected of me.

"I knew the diameter of the moon to be 2,164 miles. A quick mental calculation proved that it would take my falling body about 24 minutes to reach the center of the moon. As there was nothing to stop my fall, I must naturally continue to fall, due to the tremendous momentum acquired, till my body would *almost* emerge at the opposite side of the moon at the mouth of the other crater. At this point my speed would be zero and I would have fallen for 48 minutes. If I could not manage then to grasp a projecting rock, I would commence to fall back again toward the center of the moon. I reasoned that once more my momentum would carry me past the center and I would then be almost carried to the mouth of the opposite crater—my original starting point.

"I say almost, for the friction of my body against the air would tend to retard my fall. If at this point, where my speed was again zero, I could not succeed in taking hold of a projecting rock of the crater's side I would begin to fall down once more, the same as before. I would then continue falling back and forward exactly like a bouncing ball, each time, however—just like a rubber ball—a little less than on the previous plunge. Thus my drops would become of shorter and shorter duration, and finally I would fall no more.

"As I had mentioned before, the sun was almost overhead, shining down into the crater. I also remembered that it was almost exactly 12 o'clock midnight, terrestrial time, when the meteor smashed my aerial; this, then, was the time I started on my remarkable journey into the bowels of the moon. With a tremendous effort I pulled out my chronometer and noted that it was 12.23 a. m. In another minute I would fly past the center of the moon. Looking about, I saw in the uncertain light that I went whizzing through an immense hollow, proving to me that the center of the moon was far from solid, due no doubt to the centrifugal force of the moon at the time when it had not solidified, some millions of years ago. I estimated later that the moon was an immense hollow sphere with a solid crust about 500 miles thick. By way of a homely comparison, the moon therefore must be a hollow globe like a rubber ball. Like the latter, it is filled inside with air, while its crust can be compared to the rubber of the ball.

"In another minute I had passed the center and was now dropping toward the other side of the moon. If I continued falling in my present position, I must naturally emerge at the opposite side with my feet toward the sky, as a little reflection will reveal to you. So once more I jerked my body about, and I was falling '*up*,' with my head at the top, my feet pointing to the sun. At the end of another 24 minutes I could feel my body slowing up from the terrific speed. As the crater at this side of the moon was fortunately rather narrow, I found little difficulty in reaching for a projecting rock as soon as my plunge had come to a dead stop. I held on for dear life and clambered up a narrow ledge, where I fell down exhausted and panting from my dreadful experience.

"My sensations in falling through 2,164 miles of space, going over 16 miles per second at the center of the moon, you would, of course, like to know. Well, the first minute it is rather unpleasant. Highly so. The place where your stomach should be by right is one vast area of nausea. But once you become accustomed to it, it becomes bearable, for there is nothing else to do. You might think that the rush of air would kill you in a few seconds, or else draw all the air out of your lungs, and thus asphyxiate you. But neither is the case, for the air is so thin on the moon that the rush is not as terrific as it would be on earth. Also, by keeping the mouth tightly shut and breathing—with difficulty, it is true—through the nose, one does not die in 48 minutes. The friction of the air against my body did not ignite either, for, as I told you some time ago, the temperature inside the moon is near the absolute zero, the awful cold of the stellar world. But neither did I freeze to death, for the simple reason that the friction of my body through the attenuated air was just sufficient to keep me comfortable. Thus you see that if it had not been so cold, I would have burned up; and vice versa, if the friction of the air against my body had not heated it, I would have frozen to death long before I reached the center of the moon. Then, too, another important point to consider is that on the moon, as explained previously, my body weighed 27 pounds, against 170 pounds on earth.* This is, of course, a rather small weight, and for that reason my fall was not as terrible as it would have been if my body had weighed 170 pounds, as on earth. For that reason, too, I was not attracted so much to the sides of the crater as I would have been if my weight had been greater. Also it was fortunate that the two craters widened out considerably the further down they went into the moon's interior. As a matter of fact, the 'hole' of each crater at no point was less than three miles in diameter. This was indeed very lucky for me, for the following reason:

†"If we drop a stone into a very deep and narrow shaft, as has been shown experimentally on earth, this stone will never reach the bottom. Instead, *it will bury itself in the eastern wall* of the shaft long before reaching bottom, providing the shaft is deep enough. The explanation is that the earth rotates on its axis from west to east at a speed of 1,524 feet per second at the equator. Thus, it is apparent that the earth

* Gernsback: "†An object weighing 1 lb. on Earth weighs 0.167 lb. on the Moon.†"

revolves quicker than the stone can fall in a few seconds.* It therefore intercepts the stone's flight, with the result that the stone must of necessity strike the eastern wall of the shaft. This phenomenon is termed 'the falling of a body toward east.'†

"Now, precisely the same condition exists on the moon, of course. Fortunately, I started falling at the western side of the crater, but as the latter was so wide I never came near enough to its eastern wall to hit it. Likewise the other crater, at the opposite side of the moon, measured some four miles in diameter and, while I finally did reach the eastern wall, my flight had come to an end as explained already. Indeed, nature favored me all through, for the moon rotates with a velocity of but $15^1/_3$ feet per second at its equator, against a speed of 1,524 feet of the earth. For this reason there was no danger that my body would collide with the sides of the crater somewhere in the interior of the moon, for my flight was far more rapid than the speed of the moon's rotation on its axis.

"But in the meanwhile my troubles were far from being terminated. No sooner had I regained my breath than I became conscious of the terrible cold; for I was now but a few feet from the surface of the moon, but on that side which was turned away from the sun, where nothing but icy cold, darkness and desolation reign. Aside from this, I was some 2,160 miles away from Flitternix, my companion, and our 'Interstellar.' Walking around half of the moon was out of the question; neither could I stay where I was without freezing to death. So I climbed up to the surface of the moon with considerable effort. Then by the aid of the starlight I ran rapidly around to the western side of the crater, for I had to run in order to keep warm. After having obtained my bearings by the aid of the stellar constellations, to make sure that I was at the western side of the crater, I took a deep breath, looked down in the abyss through which the sun was shining from the other side, and dived head down into space once more.

"You see, I had reasoned that it was far better to attempt the flying journey through the moon once more than to perish with the cold on the dark side of the moon. Besides, I had experience now and having been successful once, it was natural for me to expect success again. I had nothing whatever to lose, and everything to gain.

"My first experience was repeated without any incident; furthermore, I calculated that I should land at the eastern wall of the far crater within 48 minutes if everything ran smoothly. But I had left our good old sun out of my calculations. You see, the gravitational attraction of the sun controlled the fall of my body in the same proportion as it controls the rotation of the moon and the earth. I mentioned how in my former flight I had risen to the top of the moon; as a matter of fact, somewhat higher, for the opening of this crater was higher than the surface of the moon. But now I was falling *toward* the sun, and the sun was aiding and accelerating my flight; for I moved constantly nearer to it.

"For this reason at the end of 48 minutes I did not strike the eastern end of the crater. Instead, I whizzed right past the eastern wall, almost brushing it, and continued to rise up into the air about 100 feet before my speed was spent. I promptly prepared myself to plunge down into the crater again. Indeed, before I realized it, I had begun to fall down once more when the unexpected happened.

"I suddenly felt a rope encircling my body, and before I had time to think, I was jerked sideways, and in another second I had fallen on a heap of sand and

* Gernsback: "†The speed of a falling body at the surface of the earth after the first second is $16^1/_{12}$ feet. In 6 seconds a stone would have traveled but 579.†"

looked with astonishment into Professor Flitternix's eyes, who stood over me grinning sheepishly!

"This is what had happened: Flitternix had, of course, seen me fall into the crater, and as he had rushed to the edge, he had seen how I dropped down with lightning speed. Looking closer, he also noticed what I saw, namely, that the crater went right through the entire mass of the moon, for he could see the stars shining through from the other end. He was loath to believe that the fall would kill me, and, as a scientist of note, he calculated exactly in advance what was likely to happen to me. He reflected that it would take me some two hours to make the round trip, as he knew that I could not possibly stay at the other side of the moon. He reasoned, correctly, that in case I was not killed I would come swinging through the crater in due time. Unperturbed as he is by such mere details, he went to the 'Interstellar' and had his lunch. Within two hours he returned to the crater, armed with a telescope and a long rope. It did not take him long to locate me down in the abyss by means of his glass, for I was rapidly coming to the surface then. Attaching one end of the rope to a near-by rock, he fashioned a sliding noose on the other side and waited.

"Now it must be said to the credit of Flitternix that in his younger days he had lived in the West on a ranch, and there had become an expert in the science of lassoing. He boasted that once he lassoed a common sparrow by its left hind leg, but this I believe to be somewhat exaggerated. Be that as it may, when I finally emerged to the surface, a living piece of lava ejected out of an extinct crater, Flitternix had little trouble in lassoing me as I came whizzing up. Whereupon I thanked him and asked him if lunch was ready, for the trip had given me quite an appetite, as you may well imagine. Luncheon over, we decided right then and there to quit the moon, for Flitternix, as well as myself, were of the opinion that there was little further to be explored on this dead world. Besides, the meteors had become so alarmingly frequent that it would be only a matter of time when one of us would be killed.

"Flitternix wanted to return to earth at once, for he itched to give a lecture before the American Astronomical Society, whose honorary president he was. I, however, had more ambitious plans. I once had looked through the great telescope of the Lowell Observatory at Flagstaff, Ariz. If I live to be a thousand years old I will never forget the glorious sight which then presented itself to my eyes.

†"I saw a ball, lighted up dazzlingly at both extremities. I saw great patches of an ochre red scattered over the surface of the sphere and I had seen dark blue areas among the vast ochre patches. Over the latter runs a mass of fine lines, nearly all of them connecting with the white caps at each extremity. Moreover, these fine lines cause one to gasp involuntarily, for they are as straight and true as if laid out with a rule and pencil. More astonishing yet, some of these lines run absolutely parallel with other ones for the whole length of their extent. And more wonderful yet, whenever two or more lines meet in a junction there is invariably a round black point.

"The ball I had been looking at transfixed for a long time was Mars, the nearest planet to earth, then 37,000,000 miles distant from the latter. The late Prof. Percival Lowell, the great authority on Martian research work, had convinced the scientific world that the dazzling white caps at the poles of this planet are the polar snow fields. The great ochre patches are desert land, while the dark blue areas represent large tracts of fertile land and its resulting vegetation.

"Now, according to well-known physical laws, proved beyond discussion, the smaller a body, the quicker it will cool off. All planets and their moons once were white-hot like our sun. The smaller ones cooled off first and the larger ones are not cold yet. Thus the earth, which measures 7,912 miles in diameter, is still red-hot in its interior, as is proved by its active volcanoes. The moon, which is but 2,164 miles across, cooled off ages ago. The oceans once filling their beds then filtered down into its bowels, there to freeze solid, for there was no heat to keep the water fluid. Its atmosphere, which was formerly as dense as that of our earth, was gradually thrown off into space, til to-day practically no atmosphere remains. Thus the moon to-day rolls on through space a dead world.

"The planet Mars, measuring 4,363 miles in diameter, as will be seen, is only twice the diameter of our moon and much smaller than the earth. Consequently it must be rapidly nearing its extinction, the same as the moon. Its oceans are already dry, while most of its land is desert. The atmosphere has nearly all gone too, proved by the fact that we seldom observe clouds on Mars through the telescope. But there must still be water on the planet as yet, this being irrefutably proved by its polar snow caps. This view is further strengthened by the fact that these caps undergo seasonal changes. As the sun beats down upon them, we see first the one, then the other, grow smaller in size, till at the end of the Martian mid-summer, the northern one has disappeared almost entirely. During the next hot season, the same happens to the southern one. Where has this water—*the only remaining water on Mars*—gone? It cannot have filtered into the interior, for if it had, we could not possibly witness the reappearance of the polar snow fields every Martian year, as we actually do. Where, then, does the water go?

"Dr. Lowell solved the problem in a brilliant as well as ingenious manner.

"His view—and it is shared by most of our scientists to-day—was that Mars is inhabited by a thinking people, fighting a heroic battle for their existence. Without water, life, as we know it, cannot exist. Now, ages ago, the shortage of water had made itself felt on Mars. Long before the first cave man appeared on earth Mars had been an old world, where civilized peoples had reigned for centuries. While our ancestors were still jumping from limb to limb among the trees in primordial forests and jungles, the water problem on Mars had become acute. The fertile lands were fast turning into deserts, for rains had become more and more infrequent, until they had stopped almost entirely. Furthermore, as Mars is flat without mountains or elevations of any sort, there could not be any natural rivers to convey the water to the plains and valleys as is the case on our world. The Martians, seeing utter extermination staring them in the face, proceeded to save their race. They did precisely the same thing that we are doing in Western America and the Egyptians are doing in Egypt, namely effecting the irrigation of deserts or semi-deserts on a large scale. Our recent Roosevelt dam in Arizona offers a good example of this. Our engineers on earth have to bring the water to the deserts, precisely as the Martian engineers must have been doing for centuries past.

"On earth, however, this is a comparatively simple matter, for here we have rivers and lakes in abundance which can be tapped with ease. Not so on Mars. The only remaining water there is found around the poles; by sheer necessity, therefore, the Martians had to go to the poles for their water supply, and this is exactly what our telescopes reveal that they did. For the long unswerving straight lines which we see are part of the canals bringing the water down from the poles to the desert land there to irrigate it. So far the Lowell observatory has discovered almost 600 canals, but there are doubtless many more. They criss-cross the entire surface of the planet in every conceivable direction, most of them, however, running due north and south in the direction of the poles. Not only do the canals cross the desert lands, but we see them carried bodily across the dark blue areas which we know to be irrigated vegetation tracts. The fact that the canals run across these areas is another proof that they are not oceans, as had been thought at one time.

"Now the lines which we see running over the planet are really not the canals themselves, but are simply wide strips of vegetation fertilized and kept alive by the water from the canals. The average width of the canals proper Dr. Lowell estimates to be about six miles. There are some of them, however, which are thought to be much wider than this. The length of these canals, however, is stupendous. There are some canals which actually measure 3,400 miles in length. A great many are over 2,000 miles long. Dozens of them run for 1,000 miles, and nearly all of the canals run in absolutely straight lines.

"The circular black points, mentioned above, which we see almost invariably at the juncture of one or more canals, are termed oases. They also represent vast tracts of vegetation and probably contain large cities, farms and so forth.

"It must convince the strongest opponent of Dr. Lowell's theory, when viewing Mars and its canals through a first class telescope, that these wonderfully straight lines cannot by any possible chance be the work of Nature. Its counterpart is found nowhere on earth, nor in the heavens. And if by any chance, for argument's sake, these lines should be of a natural origin, it is inconceivable that so many of them could join and meet as they do and form these exact circular areas. Their artificial origin is too apparent and cannot be otherwise considered to-day. Dr. Lowell's theory has so far withstood the onslaughts of nearly all opponents, as a matter of fact, his explanation is to-day accepted almost universally.

"But how do the Martians move the tremendous masses of water through their canals? Mars is entirely level, and water does not flow on a level surface without a 'head.' Moreover, during one season it must needs flow from the north towards the equator, when the northern polar snow cap melts under the influence of the sun's heat. *During the next season, however, this flow must be reversed* for now the s outh polar snow cap melts with a resulting flow of the water from the south to the north.

"But how do the Martians succeed in moving the water? We don't know. Even Professor Lowell could tell us nothing on this point. Terrestrial science simply has as yet not advanced enough to offer an explanation.†

"Well, to make a long story short, Flitternix and I decided to voyage to Planet Mars. My little astronomical lecture was given solely for the purpose of refreshing your mind as to Mars in order that future reports which I shall make to you from the planet will be better understood by you and your friends.

"As long as our 'Interstellar' was able to succeed in reaching the moon without mishap, I felt sure that the trip to Mars would not be an unduly difficult undertaking. Flitternix was of the same opinion. We calculated that the intervening 50 million miles separating the moon from Mars should be negotiated by our space flyer within 30 days, barring accidents. While this may seem like a short time to cover such an immense distance, our speed of 1,600,000 miles a day, or 66,666 miles an hour, is only a trifle greater than †the speed of the earth (65,533 miles an hour) as it travels in its orbit around the sun.†

"We immediately made our preparations and within six hours after I had emerged from the crater, the 'Interstellar' had left the moon.

"And now for a little surprise! No doubt you noted that my voice does not sound the same as usual. You will have observed, furthermore, that I did not stop talking since I started. To break the news gently to you, I am not talking at all! While you are listening to my voice at this minute, I will be some 1,100,000 miles distant from the moon, heading towards Mars!

"The explanation? Simple as usual!

"Before leaving in our 'Interstellar,' we stretched an immense aerial inside of the canyon, the one of which I spoke to you several days ago. As you will remember, I told you then that it was open but a few feet across its opening at the top. It thus formed a long, narrow slit at the top, into which there was little likelihood of any meteor dropping, which could destroy the aerial. We stretched four wires in all along the inside of the canyon, spacing the strands six feet apart. Each strand is 6,000 feet long in order to give the required long wave length in transmitting as well as receiving impulses between Mars and the Moon, as well as between the latter and the Earth.

"To this aerial I connected my latest invention, my *Interplanetarian Radiotomatic*. It is nothing but an ingenious adaptation of modern tele-mechanics and works as follows:

"When the aerial receives a certain number of equally spaced dashes, an ultra-sensitive detector is actuated upon which in turn operates a gas-valve relay. This relay then closes its contacts, which sets in motion the well-known telegraphone, invented by [Valdemar] Poulsen. A second ultra-sensitive detector, also connected to the aerial, is in series with the registering electro-magnets of the telegraphone; in front of these magnets runs the moving steel wire, on which are then recorded the impulses coming in over the aerial. You will observe that no message can thus be recorded unless the original *key dashes* unlock the telegraphone mechanism. At the end of the message the same number of equally spaced dashes will lock the telegraphone mechanism. The recorded message is now ready for re-transmission at any time desired. This is accomplished in a simple manner too.

"I took our 300-day clock and fastened upon it a contact which would be closed at exactly 11 p. m. every night and would be opened again at 12 o'clock midnight. This contact closes a circuit in which is included the telegraphone mechanism. As soon as it starts, the steel wire with its recorded message begins to reel off in front of the two *reproducing* electromagnets, which in turn are connected with a special telephone receiver. Thus the telephone receiver will begin to talk its message (if one was sent during the day) every evening at 11 o'clock.

"But connected to the telephone receiver are several amplifiers, arranged in cascade. The last amplifier is attached to the mouthpiece of the transmitter of my wireless telephone. Thus the weakest recorded talk on the telegraphone wire will cause the telephone of the last amplifier to talk into the wireless transmitter louder than myself.

"Now my 300-day clock every night at 11 o'clock also closes the contacts of a powerful relay, which in turn operates the generating plant of my wireless telephone, disconnecting it at midnight. Therefore when the amplifier with its telephone begins to talk into the transmitter of the wireless telephone, there will always be enough power to transmit it to you on Earth.

"As soon as we arrive on Mars we will in all probability find all the necessary materials to erect a giant Radio telephone plant, and if we succeed we will send daily messages to the Moon, and my radiotomatic relaying plant will transmit the messages to you every night. I might also mention that my ultra-sensitive detector contains two radio-active substances, making the detector such a marvelously sensitive instrument that it will work a set of amplifiers in cascade *when an electric pocket buzzer is operated one hundred and fifty miles away from it, connected to the ground only and using no aerial!*

"You might say: 'Why use the relaying plant on the Moon at all? Why not transmit from Mars to the Earth directly?'

"The reason is that when the weak impulses arrive from Mars, after having traveled from 50 to 60 million miles, they cannot be sufficiently strong to pass through the Earth's thick atmosphere. It is far better that the weak impulses should operate the relaying plant first and send out from it very strong impulses which have to travel only some 238,000 miles to Earth.

"We tested the plant thoroughly and after we had satisfied ourselves that it would work for at least 300 days I opened the telegraphone circuit and began to register this message to you. It will be the last one which you will receive for 30 days or more. As it must needs take us from five to 10 days to build a transmitting plant on Mars, you need not expect to hear from us for from 35 to 40 days. You might, therefore, commence to 'listen in' beginning with the 35th day from to-night. No message can ever be repeated, for the '*wiping*' electro-magnets of the telegraphone wipe out the magnetic impulses from the steel wire as quickly as they pass the transmitting magnets. Neither can you transmit a message to me, for no provisions were made to relay your messages to us while on Mars.

"I will now bid you adieu, my boy. Think of us during the next 30 days! Good-bye—good-bye . . . . .!"

There was a silence for some seconds, and as I was still listening awestruck, I was suddenly startled by another voice breaking in:

"Hallo, there Alier, this is Professor Flitternix. How's Yankton? Beastly old town! Was once forced to sleep on a billiard table in the Palace Hotel, as all the rooms were full. The robbers charged me $2.50 for the 'room' plus the regulation rate of 50 cents an hour for the use of the billiard table! Mean town, that Yankton! Well, good-bye. . . ."

There was a snapping noise and the rhythmic, low, sizzling sound stopped abruptly. All was quiet once more.

THE MARS PROJECT (1953)

CHAPTER 9: HEADACHES OF A SPACE SHIP DESIGNER

Wernher von Braun

English translation by
Henry J. White, Lt. Cdr. USN

Wernher Magnus Maximillian Freiherr von Braun was born in 1912, in Wyrzysk, Germany (now Poland). He was a pioneering rocket scientist who rose to prominence during the Second World War developing weapons such as the V2 rocket for the Nazi war effort. At the conclusion of the war, von Braun and his colleagues were secretly transported and resettled in America, so that he could continue to develop rockets. Von Braun was the chief architect of the Saturn V rocket that took the first humans to the Moon as part of NASA's Apollo programme. As well as his technical work, he devoted time to public education, collaborating with Walt Disney on films to popularize space travel in the late 1950s. *Project Mars: A Technical Tale* was published in 1953 and is a fictionalized discussion of the challenges associated with a trip to Mars.

Holt's first task was the selection of his crews.

His old war buddy, Tom Knight, had signed up without hesitation and Holt appointed him Deputy Commander and Captain of his flagship.

John Wiegand, the man with whom Holt had spent many thirsty days in a pneumatic boat under the burning sun of the Indian Ocean, had immediately agreed to abandon his profitable automobile agency in Fort Worth as soon as the project took tangible shape. John was an old timer in space shipping, with an engineering background. He had been through the teething troubles of the *Jupiter*-class and then directed the assembly of Lunetta personally and in situ. Holt knew his merciless meticulousness about the smallest detail, and for this reason had picked him out as Chief Engineer.

Since geology promised to one of the most fruitful fields for investigation on Mars, Holt had succumbed to the blandishments of a solemn, spectacle-wearing scientist by the name of Samuel Woolf. A well-known geologist of middle age, he had dreamed of the intentions of making an expedition along his professional lines, and had pestered Professor Ashley with applications to a point where he gave him a letter of recommendation to Holt. The latter interviewed him and was much impressed by the way Woolf had combined his plans for geological studies of the structure of Mars with various other objectives on the planet.

Dr. Bergmann, the light-haired astronomer, assented immediately when Professor Hansen let him know that Holt desired him to be in the party. Hansen did not tell him that he had long since guessed that his secret dream was to go along, and had spoken strongly in his behalf. Since Holt's visit to Lunetta, Bergmann had increased his efforts to the limit of human capabilities in order to squeeze every conceivable advantage from the existing opposition of Mars, when conditions favored observations and measurements. Bergmann's equipment was kept busy to its ultimate capacity during this time. For six solid weeks he had lived in the observatory, returning only occasionally to Lunetta for sleep. He had kept the busy bees busy indeed, bringing his meals over from Lunetta to the observatory. Each returning Lunetta ferry carried a thick sheaf of data for Holt back to Kahului, as Bergmann compiled what he thought might assist the planning.

Then there was Glen Hubbard of Los Angeles. For years he had been a civilian test pilot in the service of United Spacecraft, until the flight surgeons began to hint that he would last longer and enjoy himself more in a swivel chair behind a desk. He then became technical assistant to Spencer, but his wide experience when testing the *Jupiter*-class, particularly because it was he who had worked out their landing procedure, had been of inestimable value to every new design put out by the company. When Operation Mars was first announced, he immersed himself immediately in study of the glide characteristics and landing procedures which would face the landing boats in Mars' peculiar atmospheric structure, and came up with many valuable design tips. This aroused in him such enthusiasm for the whole project that he asked Holt if he might have a command. He wanted to strike a final blow for the cause to which he had dedicated his life, for flight in space, in which he had often risked his neck. The trip would be a high point for him ere he must retire into innocuous inactivity.

Another captain was Charles Laroche, the French underground fighter in the last great war against the overrunning of Western Europe by Bolshevism. Unknown at first, he had finally emerged to fight the barbarians in the air, becoming the ace of the French aces who, at the peak of his powers, had joined the United States Space Forces to become one of the first great space ship masters. Since the war had ended, Laroche had found no satisfactory outlet for his boiling energy, despite his fondness for social frippery and his many feminine conquests. Visiting with some wealthy acquaintances, he had accompanied them on an African hunting safari, but when his old friend Knight hinted that there was to be an Operation Mars, his request to Holt for a part in it figuratively burned up the mails.

***

Holt called a conference of his key men at the plant of United Spacecraft, that they might be apprised of the state of advancement of the planning. He gathered them in the study of Richard Peyton, the chief Designer, for a report by the latter on the manifold questions relative to the details of the designs.

Dick Peyton was an old-timer at the business. When he and Spencer began their work together in the days of the early long-range rockets, it was his engineering conscience which had provided the balance-wheel for Spencer's almost pestilent fantasy and furious creative drive. If Spencer was the man whose wide vision could connect seemingly unrelated factors and discover new solutions when no progress seemed possible, if he could win an argument with his own engineers with the same assurance that he could overpersuade a board of directors or a Pentagon party, Peyton was the lover of detail, paternal to his designers, never allowing an error in a drawing, no matter how insignificant. He had a strange gift with subcontractors, coordinating their efforts in producing the hundreds of special parts and auxiliary equipment which went into the building of great rocket ships.

When the meeting was called to order, Peyton began with a report on the requirements for steering the great ships on their long voyage through space. He made it clear that steering could only be applied during the relatively short power applications. Between these power maneuvers, the ships would coast through space for weeks on end, like wandering meteorites, their flight paths predetermined by the direction and energy imparted to them during the preceding application of thrust. But, if the running position checks by star angle should show any departure from the precomputed course, it would, of course, be necessary to enter a corrective maneuver, during which accurate steering would be as necessary as during the main maneuvers.

No one needed to be told that the steering problem of the Mars vessels would be attacked by automatic mechanism, as was that of the mighty triple stagers on their departures from Earth. Manual steering, as used in busy bees, could not be considered for any such high degree of accuracy as the navigation of the Mars vessels required.

A gyroscopic plane of reference was to be established by a system of control gyros, rigid in space. So-called "program devices" were slowly to displace this plane in angle during applications of power. Any deviation of the vessel from the attitude imposed by the gyros would then call into being the angle-changing moments which would bring back the set attitude.

Such control moments were to be generated by four relatively small rocket motors. These were to be mounted symmetrically at four opposite points on the periphery of the main motor. Each could be rotated around one axis. When all were set to zero, their jets would parallel the jet of the main motor. Yaw and pitch control would be given by changing the angles of one or the other pair of opposite jets, each moving conjointly in angle. Rotation of the vessel around her own longitudinal axis could be initiated or stopped by deflecting all jets clockwise or counterclockwise. The angular displacement of each steering jet was to be produced by a small electric motor whose motion was controlled by rheostatic devices coupled to the gyros.

"Up to this point," remarked Peyton, "attitude control of space ships is a routine problem.

"But the actuators of these control jets now are called upon to fulfill a novel function, for we are steering not a single ship, but a whole convoy. We shall hardly be able to produce mechanisms uniform enough so that there will not be very slight divergences in track and velocity at combustion cutoff. In the course of a long coast through space, the ships will therefore diverge more or less. The busy bees will find their trips from ship to ship growing longer and longer. Therefore, it will be necessary to close up the convoy from time to time. To do this with the main motors would call for a laborious turning of each vessel in space by its directional flywheels, so that the thrust of the main motor, even though applied but momentarily, would be in exactly the right direction. That would be quite an elaborate procedure for what is, after all, a very minor correction.

"We are therefore making provision to cut out the automatic gyro control devices temporarily, and are equipping the steering rocket jets with manual electric remote controls which permit them to be directed anywhere within a full 360 degrees around their axes of rotation. Thus, should it be found that a ship is moving too fast in line with its longitudinal axis, all four steering jets can be turned to face forward, and their thrust will decelerate the vessel directly. Similarly, we can swing any ship into or out of the convoy column. The main advantage is that we can make really fine corrections with these fractional thrusts when compared to the main motors. Corrections can be made by small thrusts over extended periods rather than by short, but violent ones. You will, I think, easily appreciate the advantages of this modification."

Next to come up was the question of waste disposal. To eliminate every pound of weight possible prior to each new power maneuver meant fuel saving and increased reserves of power. Such reserves might be of vital importance if a navigation check should show that any maneuver had not turned out entirely according to plan.

The problem was to expel waste matter from the pressurized living spaces into the surrounding vacuum of space and simultaneously to prevent the accumulation of masses of such matter in the immediate proximity of the ships. Nothing, as a matter of fact, would prevent some discarded tin can from following faithfully throughout any portion of the voyage between power maneuvers. Peyton described a clever dodge which might solve this problem. It was planned to place waste matter in an airlock through a small door. The door would then be closed, and a piston impelled by the internal air pressure would project the waste with considerable velocity into the outer vacuum. It became plain that one of Peyton's pet projects was to connect this gadget with the ship's sewage system. He discoursed pridefully and solemnly at some length on the problems attending the use of the latter during long, unpowered stretches of the voyage under weightless conditions. So serious and detailed was his presentation that howls of laughter burst from his auditors.

Peyton further reported that United Spacecraft was already in negotiations with the American Blower Corporation of Detroit for air regeneration equipment. This company had pioneered in the development of machinery of this kind for the pressurization of Lunetta, but here too, was a knotty problem.

First, the air would have to be freed of the carbon dioxide developed in the lungs by breathing. Air continuously exhausted from the living spaces would first pass through a dust filter and then traverse, under increased pressure, a tank into which water was simultaneously sprayed. The carbon dioxide would combine with the water under this increased pressure and form carbonic acid. Then, when the water was subsequently decompressed, the carbonic acid would bubble out, as it does from the contents of a bottle of soda-water, and be exhausted into the vacuum of space between the worlds. The air, now free of carbon dioxide, would be passed through a bone charcoal filter for deodorizing. Then it would be replenished with oxygen, the latter being carried in liquid form and hence requiring to be gasified. After this treatment, and before being readmitted to the living spaces, the air thus regenerated would require conditioning to the exact temperature, pressure and humidity desired for breathing purposes.

The air regeneration system was closely integrated with the water recuperation system. Breath and perspiration would liberate more than one and one half liters of water daily per man, to be extracted by the system. This recuperated water, added to the utility water carried for cooking, dish washing and the laundry machines, would increase the utility at the expense of the slowly shrinking potable water. It would be necessary to sterilize such water by distillation, recondensation and germicidal additives, for it would be unthinkable to jeopardize hygienic standards on a lonely, year-long trip, where so many people would live in such restricted space and have access to such limited medical assistance. Measures to this end were considerably handicapped by the weight limitations which Peyton had found it necessary to impose upon the developing company.

But this was not all, for the constituency of the air was closely tied in with the type of glass to be used in the portholes. Experience on Lunetta had confirmed this. There was a high concentration of ultraviolet in the light impinging on these windows, being unaffected in its passage through empty space by any atmospheric filtering effects. This not only caused malignant conjunctivitis but also led to the formation of ozone in the living spaces. The Lunetta crews had frequently complained that she smelled of "artificial mountain sun," and that it eventually gave them headaches. For some time, development of special window glass had been under way. Such glass was to absorb a considerable proportion of the ultraviolet radiation and likewise to reduce somewhat the glare of the sunlight. It was a tough assignment, for most glasses slowly disintegrated under the strong ultraviolet light.

All this was completely novel to Dr. Woolf, who followed Peyton's remarks with rapt attention. But the longer he listened to the latter's dissertations on toilet problems and odors of ozone in space vessels, the more his mind concentrated upon a burning question which seemed to him vastly more important than all this detail, and to which he could find no answer with all the will in the world. Finally, he burst out when Peyton paused for a moment.

"Mr. Peyton," said he anxiously, "what about the temperature in these space ships? Away back in school, I learned that the temperature in space is absolute zero, or minus 273 degrees Centigrade. With an outside temperature like that, how in the world are we going to keep warm?"

"Solar heating's the answer, Mr. Woolf," answered Peyton. "You're obviously under some misapprehension about maintaining temperatures in space ships. Space, you see, really has no temperature. What is temperature, anyway? It's only a way of expressing the rate of movement of molecules. The faster the molecules composing a body whirl about within it, the higher is its temperature. But space is composed of absolute vacuum and might as well have no molecules. So how can empty space have temperature?"

"Aren't you splitting hairs, Mr. Peyton?"

"Certainly not, although I must admit that right now we're interested in a somewhat different problem, namely, what temperature does a body suspended in empty space assume? The body, of course, consists of molecules, and these may have motion and therefore temperature.

"Let's consider the simplest case, namely a sphere floating somewhere in space, remote from the Earth and equidistant from the Sun. One half will be irradiated by the sun, the other will be shaded. Thus one side will absorb heat radiation and the sphere will become warm. From the shady side, heat will be lost by reradiation. The sphere will continue to grow warmer until the rate of radiation equals that of absorption. At the temperature where this occurs equilibrium is established. This temperature of equilibrium is primarily dependent upon the distance of the sphere from the Sun – the nearer it is to the letter, the more solar radiation is absorbed and the higher is the equilibrium temperature at which the sphere can reradiate the increased amount of heat absorbed."

"But does not the temperature attained by the sphere also depend upon the nature of its surface and above all on its color? A mirror finish reflects all radiation, while a black surface swallows it all."

"Quite right. But what applies to the irradiated surface also applies to the shaded side. A mirror finish does not reradiate, while a black one greatly furthers reradiation. If the sphere is completely shiny, it absorbs no heat on the sunny side and radiates none on the shady side. An entirely black sphere absorbs much and reradiates much. Either is the same with respect to the temperature of equilibrium.

"It's not quite so simple when intermediate colorations between mirror and black are concerned, for then the so-called spectral absorbtivity and emissivity play a considerable role. Solar radiation has a considerable portion of short wavelength energy, that is, visible and ultraviolet light, while the sphere, warmed but to moderate temperatures, radiates long wave, invisible infrared only. The absorptive capacity of some particular paint for short wave radiation may differ materially from its ability to reradiate long waves. Thus there are different temperatures of equilibrium for the sphere at identical distances from the Sun, and these vary according to the nature of the sphere's surface characteristics. But for a sphere whose surface is either completely reflective throughout the whole spectrum, or completely black in the absorptive sense, the temperature of equilibrium would be the same at the same distance from the Sun."

"You seem to be telling me," wondered Woolf, "that whether a space ship be painted with black or with reflecting paint, its temperature is wholly and solely dependent upon its distance from the Sun? Why then, the interior of the vessels ought to get colder and colder! Because our distance from the Sun will increase constantly during our trip out to Mars."

"Now wait a minute," said Peyton, "that's another reason why we must take steps continuously to adjust our temperature of the sphere; we'll assume that it's black on one side and mirror-reflecting on the other. That will make its equilibrium temperature at any given distance from the Sun also dependent upon its attitude to the solar rays. If the black side is turned sunwards while the other is shaded, the sphere will heat up several hundred degrees, because it is absorbing much heat and can give off but little of it. But if we reverse its position and face the mirrored surface to the Sun, the mirrored hemisphere will permit no heat to be absorbed, while the black side will be radiating the accumulated heat into space. Under those conditions, and, make no mistake about it, under those conditions only, the sphere may finally reach absolute zero."

"What is done in practice?"

"Most of the exterior of the ship is polished to a mirror finish. Then the temperature simply remains exactly as it is, for there is neither heat absorption nor radiation to any extent. Being surrounded by vacuum, the vessels may be regarded as Brobdignagian thermos flasks, for in such flasks, that which was cold remains cold and that which was hot, hot.

"We regulate temperature accurately as follows: the reflecting surfaces of the ship are spotted by small, black areas, shielded by silvered venetian blinds; the black surfaces under them are exposed and absorb heat when the blinds are open. When the blinds are closed, their reflecting surfaces prevent the absorption of radiation by the black areas. In practice, there's an automatic thermostatic control system which operates the blinds jointly with the air conditioning plant."

"Now I understand it. But tell me, how about the propellants stored aboard the vessels? Do we not also have to maintain them somewhere within a limited temperature range? Wouldn't they otherwise boil or congeal?"

"Quite right, they must be held quite accurately at one temperature. One factor is that the thin balloon fabric tanks could hardly withstand any such pressures as might be produced by extensive evaporation of the propellants. Another is the problem of maintaining accurately the predetermined mixture ratio between the two propellant components. This necessitates each of them being fed to the thrust unit at rather definite temperatures. The tanks will be painted aluminum. The propellant temperature will also be controlled by venetian blinds, automatically operated by thermostats.

"There's quite a variation in the preferred temperatures of different parts of such a ship, incidentally. Think, for instance, of the liquid oxygen tanks for cabin pressurization. The oxygen in them shouldn't boil – despite its low boiling point of -183° Centigrade – in order not to produce any undesired pressure in the storage tank or leak losses through the safety valve above it. Then we have the food storage bays in which the food should lie deep frozen. And of course, we should like to spare ourselves the weight of refrigeration equipment with a capacity for almost three year's store of comestibles. That can all be done with this method of blinds operated by thermostats. It's a simple and relatively trouble-free expedient and involves a low weight penalty."

"All this seems to me most intelligible," remarked Woolf, "but I'm somewhat surprised that the Mars vessels will have this bright, mirror finish, when all the pictures I've seen of the *Sirius* ferries to Lunetta show them as black as ink."

Peyton smiled tolerantly. "There's quite another reason for that," said he.

"Don't forget that the *Siriuses* have to traverse Earth's atmosphere at such high flight speeds as to cause great heating of their skins by air friction. By coloring that skin black, the friction-generated heat is dissipated by better radiation, and the skin does not reach such high temperatures as would otherwise be the case.

"The pilot's compartment and the disposable interior of a *Sirius* vessel are temperature insulated within the outer shell, and their temperature regulation is exclusively taken care of by the air conditioning system. But the Mars vessels have, as you know, no outer sheathing. That means that radiation balance must play the premier role in temperature control."

At this point Holt injected himself into the conversation. "If you'll excuse me, Sam, perhaps I can add a little illustrative touch to this Mars vessel temperature subject," he said. "Lunetta remains in the umbra of the Earth for almost one half of each of her revolutions around it, during the equinoxes at least. She too is painted silver and uses the same method of temperature regulation which Dick has just described. Now we've noticed that the regulating blinds hardly change their angle even when passing through the umbra, when there is no sunlight at all. That happens because the heat capacity of the station is so great compared to the slight amount of reradiation through the usually half-opened blinds, that a single hour is hardly enough for the interior temperatures to react to the absence of solar radiation and to affect the thermostats.

"Dick also forgot to mention another factor, namely the heat which is constantly being liberated in the operating spaces. In the case of Lunetta, there's a continuous utilization of some 15 kilowatts of electrical energy. This is converted into heat via lighting units, hot plates, motors, and the like. Then too, there are some 80 men in Lunetta's peacetime complement. You'd hardly believe what a lot of heat 80 men, each eating 4,000 calories of food per day, can put out! They alone are a very efficient heating plant! In point of fact, the blinds on the shady side are almost constantly wide open in order to expel all that heat. Otherwise I can assure you that it would gradually get uncomfortably warm."

John Wiegand decided to put his oar in at this point. "Speaking of electricity, what are you going to do about that in these Mars ships? Will it be the same setup as in Lunetta?" John had been intimately associated with assembling the power station in Lunetta and was vaguely remembering a lot of almost forgotten difficulties.

"No indeed, Johnny," answered Peyton. "Not only the problem, but the whole build-up of our Mars ship power plants differs in several respects from those of Lunetta. Her current requirement is a relatively steady 15 kilowatts, more or less. When she's in the umbra, her steam generator naturally cuts out because there's no solar radiation reaching her reflector, and that's about half the time at the equinoxes. The load is taken over by accumulators, so the generator, when it's operating, has to furnish 15 kilowatts to the mains and another fifteen to recharge the batteries so that they can carry the load during the next passage through the umbra.

"The picture on the Mars ships is quite different, for there are no repeated passages through the shadow of the Earth. On the other hand, when power maneuvers are in effect, each ship calls for some 70 kW for steering purposes. The earlier propulsive maneuvers last much longer than the later ones, so we shall draw the major part of the electric current from batteries and jettison a part of the batteries afterwards, which will economize on propellants by decreasing the mass of the ship when the next maneuver comes up.

"The intervals between maneuvers will be very long, so that a trickle charge can easily bring back batteries pretty well run down by a power maneuver. All in all, it means that, even with the heavy currents called for during the relatively short periods when the rocket motors run, we can get by with generators of moderate capacity.

"Lunetta needs a minimum generator capacity of 20 kW and we have actually installed 35 kW; but on the Mars ships. 16 kW generators will do. During the long, unpowered flight periods, most of the current will be used by the blowers and pumps for water and air regeneration and recuperation. All of these together will call for about 6 kW, leaving ample current available for such auxiliaries as electric stoves, washing machines, remote annunciators, thermostatic controls and the like. We shall not need, except to a small extent, electric lighting; for unlike Lunetta, our ships will almost always fly in sunlight. Of course the batteries will stand an occasional peak load of 40 or 50 kW if it is not applied just prior to a power maneuver.

"The power-plants of the Mars vessels will be operated by turbogenerators fed by sun-heated boilers. We shall require solar reflectors of 94 square meters surface to produce the required 16 kW, and oddly enough, that's more surface than is used for Lunetta's 35 kW. The apparent discrepancy is explained by the need to keep up our 16 kW even when near Mars, where the solar energy caught by the reflector is less than half of that near the Earth's orbit.

"After the vessels have been drawn in by Mars's gravitation, they will circle him and be in his umbra during one third of the time required for each circle. Then the problem will become much like Lunetta's. The mean output of our 16 kW generators will be reduced to 11 kW per vessel, and that's getting close to the tenuous current requirements of the air and water regeneration systems. It is, in a sense, fortunate that the crews of the ships awaiting in their satellite orbits the return of the landing craft will be reduced by the number of the landing party. This will diminish our current requirements during this critical period.

"Exhaust steam condensers present a particular problem in the Mars vessels. Those of you who have visited Lunetta will remember the spokes connecting her rim with the central station. These serve as condenser tubes reducing the aqueous vapor to liquid which the feed pumps can return to the boiler. It was possible to use this expedient on Lunetta because the spokes are shaded by the rim, due to Lunetta's rotating in the plane of the ecliptic. But the Mars vessels are not wheel-shaped, and their attitude in space with respect to the Sun may change continually. Hence we propose to locate the condensers behind the reflectors, because, since the reflectors must always face sunwards when the turbines are running, there will invariably be shadow behind them.

"But here we encounter a rather ugly difficulty. Our power plant has an overall efficiency of 30% only. Thus 70% of the solar energy from the reflector must be dissipated into space by the condenser. For the low exhaust steam temperature of 46°C, the condensing surface must be double that of the reflector.

"You may argue that there'd be no particular problem in placing a condenser with twice as much surface as the reflector in the latter's shadow. The design of such a condenser would be elongated and extending to the rear, so that it would radiate heat not only directly away from the Sun, but also laterally. But that configuration would be rather bulky and unhandy, and we must not forget that the reflector with the condenser must be swivel-mounted so that, for any attitude of the ship in space, it can face the Sun directly.

"For all these reasons, we have decided to substitute mercury vapor for the more conventional steam. Such turbines are not a new departure and have the advantage of permitting higher admission temperatures to the turbine than steam and consequently the exhaust is hotter. That will give us a temperature of 120°C in the condenser. This, of course, will make the surfaces radiate much more actively and we can get by with relatively compact condensers which will easily fit within the shadow of the reflector."

"Say, Dick," said Holt when Peyton stopped talking for a moment, "what about the electrical energy for radio communication? Have we got any figures on that?"

Peyton drew a deep breath and plunged on. "Here's the communication story. According to the plan, we must consider three separate types of radio communication.

"The first type is for intership work. There's not much to that. It's short range stuff, not much different from the two-way command sets used between airplanes in formation. Each unit will use only a couple of watts and draws practically nothing from the mains.

"The second type is for middle distances of about 10 to 20,000 miles. It will be needed primarily for communication with the landing party when the space ships are circling Mars. Likewise, it will be useful for contact with Lunetta or terrestrial stations after return to Earth's orbit. We propose to equip three ships with transmitters for this middle distance service, and also one transmitter for each landing craft and ground vehicle. All vessels, landing craft and vehicles will of course have appropriate receivers. These middle distance transmitters will draw about 5 kW each, and since the times during which they will transmit are short, the mains of the vessels will take care of it nicely.

"But the third and most problematical radio is for long, interplanetary ranges. These will be covered by the so-called High Duty Radio Sets, and these will have to be constructed as wholly separate structural entities, independent of the structures of the vessels. Two of the three cargo ships will each carry one of these sets. As a matter of fact, if the navigation should not be accurate enough to allow the Command of the expedition to effect notable savings in propellants, the Command may well be forced to abandon both these sets in the Martian satellite orbit, for we have been unable definitely to assign any portion of the payload of the passenger vessels to these radios.

"The idea is to erect a similar station on Earth with which they will communicate, although the boys are not yet quite sure of its configuration. It's not impossible that one exactly like those of the Mars vessels will operate from Lunetta's orbit... The range of these stations must suffice for two-way communication throughout the entire trip.

"As a matter of fact, this radio communication problem is giving us about as much trouble as any of them. You probably all know that we've been working with Old Man Lussigny, and he's running this show. I don't know as much about radio as I'd like to, considering how my problems tie in with it. Communication across hundreds of millions of kilometers apparently calls for concentrating the radio energy into a ray-pencil of small divergence. The radio boys are still pretty much puzzled over this. If they reduce their wavelength, the same size of directional antennas will give better concentration, but then they cannot build transmitters with enough output. On the other hand, longer wavelengths ease the transmitter problem, but require such large and cumbersome antennas that they give us acceleration troubles during power maneuvers. Right now, I'm still a little vague as to the outcome of this tail-chasing, but it seems quite definite that these High Duty Radio Sets will have to be separate from the ships, not only structurally, but as to sources of power."

Peyton had hardly closed his mouth after this speech than the door was flung open and Spencer burst in waving a newspaper. His usually solemn countenance was wreathed in smiles. "Hey! Read this!" he shouted joyously, "It's all in the papers! The Congress has voted us our whole two billion dollars! Boys, we're off in a cloud of dust! Cosmic dust, at that!"

THE HOLES AROUND MARS (1954)

Jerome Bixby

Born in Los Angeles in 1923, **Drexel Jerome Lewis Bixby** wrote westerns and science fiction short stories and screenplays. Bixby is perhaps best known for the four episodes of *Star Trek* that he wrote in the 1960s, including the fan favourite, 'Mirror, Mirror', in which Captain Kirk and his crew find themselves transported into a parallel universe where the United Federation of Planets is evil. He also wrote the screenplay for *It! The Terror from Beyond Space*, which served as inspiration for Ridley Scott's *Alien*. His short story, 'It's a Good Life', was adapted into an episode of *The Twilight Zone* in 1961. He died in 1998, having just completed the screenplay for *The Man From Earth*.

Spaceship crews should be selected on the basis of their non-irritating qualities as individuals. No chronic complainers, no hypochondriacs, no bugs on cleanliness—particularly no one-man parties. I speak from bitter experience.

Because on the first expedition to Mars, Hugh Allenby damned near drove us nuts with his puns. We finally got so we just ignored them.

But no one can ignore that classic last one—it's written right into the annals of astronomy, and it's there to stay.

Allenby, in command of the expedition, was first to set foot outside the ship. As he stepped down from the airlock of the *Mars I*, he placed that foot on a convenient rock, caught the toe of his weighted boot in a hole in the rock, wrenched his ankle and smote the ground with his pants.

Sitting there, eyes pained behind the transparent shield of his oxygen-mask, he stared at the rock.

It was about five feet high. Ordinary granite—no special shape—and several inches below its summit, running straight through it in a northeasterly direction, was a neat round four-inch hole.

"I'm *upset* by the *hole* thing," he grunted.

The rest of us scrambled out of the ship and gathered around his plump form. Only one or two of us winced at his miserable double pun.

"Break anything, Hugh?" asked Burton, our pilot, kneeling beside him.

"Get out of my way, Burton," said Allenby. "You're obstructing my view."

Burton blinked. A man constructed of long bones and caution, he angled out of the way, looking around to see what he was obstructing view *of.*

He saw the rock and the round hole through it. He stood very still, staring. So did the rest of us.

"Well, I'll be damned," said Janus, our photographer. "A hole."

"In a rock," added Gonzales, our botanist.

"Round," said Randolph, our biologist.

"An *artifact,*" finished Allenby softly.

Burton helped him to his feet. Silently we gathered around the rock.

Janus bent down and put an eye to one end of the hole. I bent down and looked through the other end. We squinted at each other.

As mineralogist, I was expected to opinionate. "Not drilled," I said slowly. "Not chipped. Not melted. Certainly not eroded."

I heard a rasping sound by my ear and straightened. Burton was scratching a thumbnail along the rim of the hole. "Weathered," he said. "Plenty old. But I'll bet it's a perfect circle, if we measure."

Janus was already fiddling with his camera, testing the cooperation of the tiny distant sun with a light-meter.

"Let us see *weather* it is or not," Allenby said.

Burton brought out a steel tape-measure. The hole was four and three-eighths inches across. It was perfectly circular and about sixteen inches long. And four feet above the ground.

"But why?" said Randolph. "Why should anyone bore a four-inch tunnel through a rock way out in the middle of the desert?"

"Religious symbol," said Janus. He looked around, one hand on his gun. "We'd better keep an eye out—maybe we've landed on sacred ground or something."

"A totem *hole*, perhaps," Allenby suggested.

"Oh. I don't know," Randolph said—to Janus, not Allenby. As I've mentioned, we always ignored Allenby's puns. "Note the lack of ornamentation. Not at all typical of religious articles."

"On Earth," Gonzales reminded him. "Besides, it might be utilitarian, not symbolic."

"Utilitarian, how?" asked Janus.

"An altar for snakes," Burton said dryly.

"Well," said Allenby, "you can't deny that it has its *holy* aspects."

"Get your hand away, will you, Peters?" asked Janus.

I did. When Janus's camera had clicked, I bent again and peered through the hole. "It sights on that low ridge over there," I said. "Maybe it's some kind of surveying setup. I'm going to take a look."

"Careful," warned Janus. "Remember, it may be sacred."

As I walked away, I heard Allenby say, "Take some scrapings from the inside of the hole, Gonzales. We might be able to determine if anything is kept in it...."

One of the stumpy, purplish, barrel-type cacti on the ridge had a long vertical bite out of it ... as if someone had carefully carved out a narrow U-shaped section from the top down, finishing the bottom of the U in a neat semicircle. It was as flat and cleancut as the inside surface of a horseshoe magnet.

I hollered. The others came running. I pointed.

"Oh, my God!" said Allenby. "Another one."

The pulp of the cactus in and around the U-hole was dried and dead-looking.

Silently Burton used his tape-measure. The hole measured four and three-eighths inches across. It was eleven inches deep. The semicircular bottom was about a foot above the ground.

"This ridge," I said, "is about three feet higher than where we landed the ship. I bet the hole in the rock and the hole in this cactus are on the same level."

Gonzales said slowly, "This was not done all at once. It is a result of periodic attacks. Look here and here. These overlapping depressions along the outer edges of the hole—" he pointed—"on this side of the cactus. They are the signs of repeated impact. And the scallop effect on *this* side, where whatever made the hole emerged. There are juices still oozing—not at the point of impact, where the plant is desiccated, but below, where the shock was transmitted—"

A distant shout turned us around. Burton was at the rock, beside the ship. He was bending down, his eye to the far side of the mysterious hole.

He looked for another second, then straightened and came toward us at a lope.

"They line up," he said when he reached us. "The bottom of the hole in the cactus is right in the middle when you sight through the hole in the rock."

"As if somebody came around and whacked the cactus regularly," Janus said, looking around warily.

"To keep the line of sight through the holes clear?" I wondered. "Why not just remove the cactus?"

"Religious," Janus explained.

The gauntlet he had discarded lay ignored on the ground, in the shadow of the cactus. We went on past the ridge toward an outcropping of rock about a hundred yards farther on. We walked silently, each of us wondering if what we half-expected would really be there.

It was. In one of the tall, weathered spires in the outcropping, some ten feet below its peak and four feet above the ground, was a round four-inch hole.

Allenby sat down on a rock, nursing his ankle, and remarked that anybody who believed this crazy business was really happening must have holes in the rocks in his head.

Burton put his eye to the hole and whistled. "Sixty feet long if it's an inch," he said. "The other end's just a pinpoint. But you can see it. The damn thing's perfectly straight."

I looked back the way we had come. The cactus stood on the ridge, with its U-shaped bite, and beyond was the ship, and beside it the perforated rock.

"If we surveyed," I said, "I bet the holes would all line up right to the last millimeter."

"But," Randolph complained, "why would anybody go out and bore holes in things all along a line through the desert?"

"Religious," Janus muttered. "It doesn't *have* to make sense."

We stood there by the outcropping and looked out along the wide, red desert beyond. It stretched flatly for miles from this point, south toward Mars' equator—dead sandy wastes, crisscrossed by the "canals," which we had observed while landing to be great straggly patches of vegetation, probably strung along underground waterflows.

BLONG-G-G-G- ... st-st-st- ...

We jumped half out of our skins. Ozone bit at our nostrils. Our hair stirred in the electrical uproar.

"L-look," Janus chattered, lowering his smoking gun.

About forty feet to our left, a small rabbity creature poked its head from behind a rock and stared at us in utter horror.

Janus raised his gun again.

"Don't bother," said Allenby tiredly. "I don't think it intends to attack."

"But—"

"I'm sure it isn't a Martian with religious convictions."

Janus wet his lips and looked a little shamefaced. "I guess I'm kind of taut."

"That's what I *taut*," said Allenby.

The creature darted from behind its rock and, looking at us over its shoulder, employed six legs to make small but very fast tracks.

We turned our attention again to the desert. Far out, black against Mars' azure horizon, was a line of low hills.

"Shall we go look?" asked Burton, eyes gleaming at the mystery.

Janus hefted his gun nervously. It was still crackling faintly from the discharge. "I say let's get back to the ship!"

Allenby sighed. "My leg hurts." He studied the hills. "Give me the field-glasses."

Randolph handed them over. Allenby put them to the shield of his mask and adjusted them.

After a moment he sighed again. "There's a hole. On a plane surface that catches the Sun. A lousy damned round little impossible hole."

"Those hills," Burton observed, "must be thousands of feet thick."

The argument lasted all the way back to the ship.

Janus, holding out for his belief that the whole thing was of religious origin, kept looking around for Martians as if he expected them to pour screaming from the hills.

Burton came up with the suggestion that perhaps the holes had been made by a disintegrator-ray.

"It's possible," Allenby admitted. "This might have been the scene of some great battle—"

"With only one such weapon?" I objected.

Allenby swore as he stumbled. "What do you mean?"

"I haven't seen any other lines of holes—only the one. In a battle, the whole joint should be cut up.

That was good for a few moments' silent thought. Then Allenby said, "It might have been brought out by one side as a last resort. Sort of an ace in the hole."

I resisted the temptation to mutiny. "But would even one such weapon, in battle make only *one* line of holes? Wouldn't it be played in an arc against the enemy? You know it would."

"Well—"

"Wouldn't it cut slices out of the landscape, instead of boring holes? And wouldn't it sway or vibrate enough to make the holes miles away from it something less than perfect circles?"

"It could have been very firmly mounted."

"Hugh, does that sound like a practical weapon to you?"

Two seconds of silence. "On the other hand," he said, "instead of a war, the whole thing might have been designed to frighten some primitive race—or even some kind of beast—the *hole* out of here. A demonstration—"

"Religious," Janus grumbled, still looking around.

We walked on, passing the cactus on the low ridge.

"Interesting," said Gonzales. "The evidence that whatever causes the phenomenon has happened again and again. I'm afraid that the war theory—"

"Oh, my God!" gasped Burton.

We stared at him.

"The ship," he whispered. "It's right in line with the holes! If whatever made them is still in operation...."

"Run!" yelled Allenby, and we ran like fiends.

We got the ship into the air, out of line with the holes to what we fervently hoped was safety, and then we realized we were admitting our fear that the mysterious hole-maker might still be lurking around.

Well, the evidence was all for it, as Gonzales had reminded us—that cactus had been oozing.

We cruised at twenty thousand feet and thought it over.

Janus, whose only training was in photography, said, "Some kind of omnivorous animal? Or bird? Eats rocks and everything?"

"I will not totally discount the notion of such an animal," Randolph said. "But I will resist to the death the suggestion that it forages with geometric precision."

After a while, Allenby said, "Land, Burton. By that 'canal.' Lots of plant life—fauna, too. We'll do a little collecting."

Burton set us down feather-light at the very edge of the sprawling flat expanse of vegetation, commenting that the scene reminded him of his native Texas pear-flats.

We wandered in the chilly air, each of us except Burton pursuing his specialty. Randolph relentlessly stalked another of the rabbity creatures. Gonzales was carefully digging up plants and stowing them in jars. Janus was busy with his cameras, recording every aspect of Mars transferable to film. Allenby walked around, helping anybody who needed it. An astronomer, he'd done half his work on the way to Mars and would do the other half on the return trip. Burton lounged in the Sun, his back against a ship's fin, and played chess with Allenby, who was calling out his moves in a bull roar. I grubbed for rocks.

My search took me farther and farther away from the others—all I could find around the 'canal' was gravel, and I wanted to chip at some big stuff. I walked toward a long rise a half-mile or so away, beyond which rose an enticing array of house-sized boulders.

As I moved out of earshot, I heard Randolph snarl, "Burton, *will* you stop yelling, 'Kt to B-2 and check?' Every time you open your yap, this critter takes off on me."

Then I saw the groove.

It started right where the ground began to rise—a thin, shallow, curve-bottomed groove in the dirt at my feet, about half an inch across, running off straight toward higher ground.

With my eyes glued to it, I walked. The ground slowly rose. The groove deepened, widened—now it was about three inches across, about one and a half deep.

I walked on, holding my breath. Four inches wide. Two inches deep.

The ground rose some more. Four and three-eighths inches wide. I didn't have to measure it—I *knew*.

Now, as the ground rose, the edges of the groove began to curve inward over the groove. They touched. No more groove.

The ground had risen, the groove had stayed level and gone underground.

Except that now it wasn't a groove. It was a round tunnel.

A hole.

A few paces farther on, I thumped the ground with my heel where the hole ought to be. The dirt crumbled, and there was the little dark tunnel, running straight in both directions.

I walked on, the ground falling away gradually again. The entire process was repeated in reverse. A hairline appeared in the dirt—widened—became lips that drew slowly apart to reveal the neat straight four-inch groove—which shrank as slowly to a shallow line of the ground—and vanished.

I looked ahead of me. There was one low ridge of ground between me and the enormous boulders. A neat four-inch semicircle was bitten out of the very top of the ridge. In the house-sized boulder directly beyond was a four-inch hole.

Allenby winced and called the others when I came back and reported.

"The mystery *deepens*," he told them. He turned to me. "Lead on, Peters. You're temporary *drill* leader."

Thank God he didn't say *Fall in*.

The holes went straight through the nest of boulders—there'd be a hole in one and, ten or twenty feet farther on in the next boulder, another hole. And then another, and another—right through the nest in a line. About thirty holes in all.

Burton, standing by the boulder I'd first seen, flashed his flashlight into the hole. Randolph, clear on the other side of the jumbled nest, eye to hole, saw it.

Straight as a string!

The ground sloped away on the far side of the nest—no holes were visible in that direction—just miles of desert. So, after we'd stared at the holes for a while and they didn't go away, we headed back for the canal.

"Is there any possibility," asked Janus, as we walked, "that it could be a natural phenomenon?"

"There are no straight lines in nature," Randolph said, a little shortly. "That goes for a bunch of circles in a straight line. And for perfect circles, too."

"A planet is a circle," objected Janus.

"An oblate spheroid," Allenby corrected.

"A planet's orbit—"

"An ellipse."

Janus walked a few steps, frowning. Then he said, "I remember reading that there *is* something darned near a perfect circle in nature." He paused a moment. "Potholes." And he looked at me, as mineralogist, to corroborate.

"What kind of potholes?" I asked cautiously. "Do you mean where part of a limestone deposit has dissol—"

"No. I once read that when a glacier passes over a hard rock that's lying on some softer rock, it grinds the hard rock down into the softer, and both of them sort of wear down to fit together, and it all ends up with a round hole in the soft rock."

"Probably neither stone," I told Janus, "would be homogenous. The softer parts would abrade faster in the soft stone. The end result wouldn't be a perfect circle."

Janus's face fell.

"Now," I said, "would anyone care to define this term 'perfect circle' we're throwing around so blithely? Because such holes as Janus describes are often pretty damned round."

Randolph said, "Well...."

"It is settled, then," Gonzales said, a little sarcastically. "Your discussion, gentlemen, has established that the long, horizontal holes we have found were caused by glacial action."

"Oh, no," Janus argued seriously. "I once read that Mars never had any glaciers."

All of us shuddered.

Half an hour later, we spotted more holes, about a mile down the 'canal,' still on a line, marching along the desert, through cacti, rocks, hills, even through one edge of the low vegetation of the 'canal' for thirty feet or so. It was the damnedest thing to bend down and look straight through all that curling, twisting growth ... a round tunnel from either end.

We followed the holes for about a mile, to the rim of an enormous saucerlike valley that sank gradually before us until, miles away, it was thousands of feet deep. We stared out across it, wondering about the other side.

Allenby said determinedly, "We'll burrow to the *bottom* of these holes, once and for all. Back to the ship, men!"

We hiked back, climbed in and took off.

At an altitude of fifty feet, Burton lined the nose of the ship on the most recent line of holes and we flew out over the valley.

On the other side was a range of hefty hills. The holes went through them. Straight through. We would approach one hill—Burton would manipulate the front viewscreen until we spotted the hole—we would pass over the hill and spot the other end of the hole in the rear screen.

One hole was two hundred and eighty miles long.

Four hours later, we were halfway around Mars.

Randolph was sitting by a side port, chin on one hand, his eyes unbelieving. "All around the planet," he kept repeating. "All around the planet...."

"Halfway at least," Allenby mused. "And we can assume that it continues in a straight line, through anything and everything that gets in its way...." He gazed out the front port at the uneven blue-green haze of a 'canal' off to our left. "For the love of Heaven, why?"

Then Allenby fell down. We all did.

Burton had suddenly slapped at the control board, and the ship braked and sank like a plugged duck. At the last second, Burton propped up the nose with a short burst, the ten-foot wheels hit desert sand and in five hundred yards we had jounced to a stop.

Allenby got up from the floor. "Why did you do that?" he asked Burton politely, nursing a bruised elbow.

Burton's nose was almost touching the front port. "Look!" he said, and pointed.

About two miles away, the Martian village looked like a handful of yellow marbles flung on the desert.

We checked our guns. We put on our oxygen-masks. We checked our guns again. We got out of the ship and made damned sure the airlock was locked.

An hour later, we crawled inch by painstaking inch up a high sand dune and poked our heads over the top.

The Martians were runts—the tallest of them less than five feet tall—and skinny as a pencil. Dried-up and brown, they wore loincloths of woven fiber.

They stood among the dusty-looking inverted-bowl buildings of their village, and every one of them was looking straight up at us with unblinking brown eyes.

The six safeties of our six guns clicked off like a rattle of dice. The Martians stood there and gawped.

"Probably a highly developed sense of hearing in this thin atmosphere," Allenby murmured. "Heard us coming."

"They thought that landing of Burton's was an earthquake," Randolph grumbled sourly.

"Marsquake," corrected Janus. One look at the village's scrawny occupants seemed to have convinced him that his life was in no danger.

Holding the Martians covered, we examined the village from atop the thirty-foot dune.

The domelike buildings were constructed of something that looked like adobe. No windows—probably built with sandstorms in mind. The doors were about halfway up the sloping sides, and from each door a stone ramp wound down around the house to the ground—again with sandstorms in mind, no doubt, so drifting dunes wouldn't block the entrances.

The center of the village was a wide street, a long sandy area some thirty feet wide. On either side of it, the houses were scattered at random, as if each Martian had simply hunted for a comfortable place to sit and then built a house around it.

"Look," whispered Randolph.

One Martian had stepped from a group situated on the far side of the street from us. He started to cross the street, his round brown eyes on us, his small bare feet plodding sand, and we saw that in addition to a loincloth he wore jewelry—a hammered metal ring, a bracelet on one skinny ankle. The Sun caught a copperish gleam on his bald narrow head, and we saw a band of metal there, just above where his eyebrows should have been.

"The super-chief," Allenby murmured. "Oh, *shaman* me!"

As the bejeweled Martian approached the center of the street, he glanced briefly at the ground at his feet. Then he raised his head, stepped with dignity across the exact center of the street and came on toward us, passing the dusty-looking buildings of his realm and the dusty-looking groups of his subjects.

He reached the slope of the dune we lay on, paused—and raised small hands over his head, palms toward us.

"I think," Allenby said, "that an anthropologist would give odds on that gesture meaning peace."

He stood up, holstered his gun—without buttoning the flap—and raised his own hands over his head. We all did.

The Martian language consisted of squeaks.

We made friendly noises, the chief squeaked and pretty soon we were the center of a group of wide-eyed Martians, none of whom made a sound. Evidently no one dared peep while the chief spoke—very likely the most articulate Martians simply squeaked themselves into the job. Allenby, of course, said they just *squeaked by*.

He was going through the business of drawing concentric circles in the sand, pointing at the third orbit away from the Sun and thumping his chest. The crowd around us kept growing as more Martians emerged from the dome buildings to see what was going on. Down the winding ramps of the buildings on our side of the wide, sandy street they came—and from the buildings on the other side of the street, plodding through the sand, blinking brown eyes at us, not making a sound.

Allenby pointed at the third orbit and thumped his chest. The chief squeaked and thumped his own chest and pointed at the copperish band around his head. Then he pointed at Allenby.

"I seem to have conveyed to him," Allenby said dryly, "the fact that I'm chief of our party. Well, let's try again."

He started over on the orbits. He didn't seem to be getting anyplace, so the rest of us watched the Martians instead. A last handful was straggling across the wide street.

"Curious," said Gonzales. "Note what happens when they reach the center of the street."

Each Martian, upon reaching the center of the street, glanced at his feet—just for a moment—without even breaking stride. And then came on.

"What can they be looking at?" Gonzales wondered.

"The chief did it too," Burton mused. "Remember when he first came toward us?"

We all stared intently at the middle of the street. We saw absolutely nothing but sand.

The Martians milled around us and watched Allenby and his orbits. A Martian child appeared from between two buildings across the street. On six-inch legs, it started across, got halfway, glanced downward—and came on.

"I don't get it," Burton said. "What in hell are they *looking* at?"

The child reached the crowd and squeaked a thin, high note.

A number of things happened at once.

Several members of the group around us glanced down, and along the edge of the crowd nearest the center of the street there was a mild stir as individuals drifted off to either side. Quite casually—nothing at all urgent about it. They just moved concertedly to get farther away from the center of the street, not taking their interested gaze off us for one second in the process.

Even the chief glanced up from Allenby's concentric circles at the child's squeak. And Randolph, who had been fidgeting uncomfortably and paying very little attention to our conversation, decided that he must answer Nature's call. He moved off into the dunes surrounding the village. Or rather, he started to move.

The moment he set off across the wide street, the little Martian chief was in front of him, brown eyes wide, hands out before him as if to thrust Randolph back.

Again six safeties clicked. The Martians didn't even blink at the sudden appearance of our guns. Probably the only weapon they recognized was a club, or maybe a rock.

"What can the matter be?" Randolph said.

He took another step forward. The chief squeaked and stood his ground. Randolph had to stop or bump into him. Randolph stopped.

The chief squeaked, looking right into the bore of Randolph's gun.

"Hold still," Allenby told Randolph, "till we know what's up."

Allenby made an interrogative sound at the chief. The chief squeaked and pointed at the ground. We looked. He was pointing at his shadow.

Randolph stirred uncomfortably.

"Hold still," Allenby warned him, and again he made the questioning sound.

The chief pointed up the street. Then he pointed down the street. He bent to touch his shadow, thumping it with thin fingers. Then he pointed at the wall of a house nearby.

We all looked.

Straight lines had been painted on the curved brick-colored wall, up and down and across, to form many small squares about four inches across. In each square was a bit of squiggly writing, in blackish paint, and a small wooden peg jutting out from the wall.

Burton said, "Looks like a damn crossword puzzle."

"Look," said Janus. "In the lower right corner—a metal ring hanging from one of the pegs."

And that was all we saw on the wall. Hundreds of squares with figures in them—a small peg set in each—and a ring hanging on one of the pegs.

"You know what?" Allenby said slowly. "I think it's a calendar! Just a second—thirty squares wide by twenty-two high—that's six hundred and sixty. And that bottom line has twenty-six—twenty-*seven* squares. Six hundred and eighty-seven squares in all. That's how many days there are in the Martian year!"

He looked thoughtfully at the metal ring. "I'll bet that ring is hanging from the peg in the square that represents *today*. They must move it along every day, to keep track...."

"What's a calendar got to do with my crossing the street?" Randolph asked in a pained tone.

He started to take another step. The chief squeaked as if it were a matter of desperate concern that he make us understand. Randolph stopped again and swore impatiently.

Allenby made his questioning sound again.

The chief pointed emphatically at his shadow, then at the communal calendar—and we could see now that he was pointing at the metal ring.

Burton said slowly, "I think he's trying to tell us that this is *today*. And such-and-such a *time* of day. I bet he's using his shadow as a sundial."

"Perhaps," Allenby granted.

Randolph said, "If this monkey doesn't let me go in another minute—"

The chief squeaked, eyes concerned.

"Stand still," Allenby ordered. "He's trying to warn you of some danger."

The chief pointed down the street again and, instead of squealing, revealed that there was another sound at his command. He said, "Whoooooooosh!"

We all stared at the end of the street.

Nothing! Just the wide avenue between the houses, and the high sand dune down at the end of it, from which we had first looked upon the village.

The chief described a large circle with one hand, sweeping the hand above his head, down to his knees, up again, as fast as he could. He pursed his monkey-lips and said, "Whoooooooosh!" And made the circle again.

A Martian emerged from the door in the side of a house across the avenue and blinked at the Sun, as if he had just awakened. Then he saw what was going on below and blinked again, this time in interest. He made his way down around the winding lamp and started to cross the street.

About halfway, he paused, eyed the calendar on the house wall, glanced at his shadow. Then he got down on his hands and knees and *crawled* across the middle of the street. Once past the middle, he rose, walked the rest of the way to join one of the groups and calmly stared at us along with the rest of them.

"They're all crazy," Randolph said disgustedly. "I'm going to cross that street!"

"Shut up. So it's a certain time of a certain day," Allenby mused. "And from the way the chief is acting, he's afraid for you to cross the street. And that other one just *crawled*. By God, do you know what this might tie in with?"

We were silent for a moment. Then Gonzales said, "Of course!"

And Burton said, "The *holes*!"

"Exactly," said Allenby. "Maybe whatever made—or makes—the holes comes right down the center of the street here. Maybe that's why they built the village this way—to make room for—"

"For what?" Randolph asked unhappily, shifting his feet.

"I don't know," Allenby said. He looked thoughtfully at the chief. "That circular motion he made—could he have been describing something that went around and around the planet? Something like—oh, no!" Allenby's eyes glazed. "I wouldn't believe it in a million years."

His gaze went to the far end of the street, to the high sand dune that rose there. The chief seemed to be waiting for something to happen.

"I'm going to crawl," Randolph stated. He got to his hands and knees and began to creep across the center of the avenue.

The chief let him go.

The sand dune at the end of the street suddenly erupted. A forty-foot spout of dust shot straight out from the sloping side, as if a bullet had emerged. Powdered sand hazed the air, yellowed it almost the full length of the avenue. Grains of sand stung the skin and rattled minutely on the houses.

WhoooSSSHHHHH!

Randolph dropped flat on his belly. He didn't have to continue his trip. He had made other arrangements.

That night in the ship, while we all sat around, still shaking our heads every once in a while, Allenby talked with Earth. He sat there, wearing the headphones, trying to make himself understood above the godawful static.

"... an exceedingly small body," he repeated wearily to his unbelieving audience, "about four inches in diameter. It travels at a mean distance of four feet above the surface of the planet, at a velocity yet to be calculated. Its unique nature results in many hitherto unobserved—I might say even unimagined—phenomena." He stared blankly in front of him for a moment, then delivered the understatement of his life. "The discovery may necessitate a re-examination of many of our basic postulates in the physical sciences."

The headphones squawked.

Patiently, Allenby assured Earth that he was entirely serious, and reiterated the results of his observations. I suppose that he, an astronomer, was twice as flabbergasted as the rest of us. On the other hand, perhaps he was better equipped to adjust to the evidence.

"Evidently," he said, "when the body was formed, it traveled at such fantastic velocity as to enable it to—" his voice was almost a whisper—"to punch holes in things."

The headphones squawked.

"In rocks," Allenby said, "in mountains, in anything that got in its way. And now the holes form a large portion of its fixed orbit."

Squawk.

"Its mass must be on the order of—"

Squawk.

"—process of making the holes slowed it, so that now it travels just fast enough—"

Squawk.

"—maintain its orbit and penetrate occasional objects such as—"

Squawk.

"—and sand dunes—"

Squawk.

"My God, I *know* it's a mathematical monstrosity," Allenby snarled. "*I* didn't put it there!"

Squawk.

Allenby was silent for a moment. Then he said slowly, "A name?"

Squawk.

"H'm," said Allenby. "Well, well." He appeared to brighten just a little. "So it's up to me, as leader of the expedition, to name it?"

Squawk.

"Well, well," he said.

That chop-licking tone was in his voice. We'd heard it all too often before. We shuddered, waiting.

"Inasmuch as Mars' outermost moon is called Deimos, and the next Phobos," he said, "I think I shall name the third moon of Mars—*Bottomos*."

OMNILINGUAL (1957)

Henry Beam Piper

Born in Altoona, Pennsylvania in 1904, **Henry Beam Piper** was an American science fiction author. He was largely self-taught, having worked as a labourer in the Pennsylvania Railroad's Altoona yards from the age of eighteen. Beam Piper said that he wanted to understand science and history without the need to spend four years wearing a raccoon coat, which was particularly fashionable with male college students at the time. His first short story was published in *Astounding Science Fiction* magazine in 1947. His life and career lasted until 1964, when he used a handgun to take his life.

To translate writings, you need a key to the code—and if the last writer of Martian died forty thousand years before the first writer of Earth was born ... how could the Martian be translated...?

Martha Dane paused, looking up at the purple-tinged copper sky. The wind had shifted since noon, while she had been inside, and the dust storm that was sweeping the high deserts to the east was now blowing out over Syrtis. The sun, magnified by the haze, was a gorgeous magenta ball, as large as the sun of Terra, at which she could look directly. Tonight, some of that dust would come sifting down from the upper atmosphere to add another film to what had been burying the city for the last fifty thousand years.

The red loess lay over everything, covering the streets and the open spaces of park and plaza, hiding the small houses that had been crushed and pressed flat under it and the rubble that had come down from the tall buildings when roofs had caved in and walls had toppled outward. Here, where she stood, the ancient streets were a hundred to a hundred and fifty feet below the surface; the breach they had made in the wall of the building behind her had opened into the sixth story. She could look down on the cluster of prefabricated huts and sheds, on the brush-grown flat that had been the waterfront when this place had been a seaport on the ocean that was now Syrtis Depression; already, the bright metal was thinly coated with red dust. She thought, again, of what clearing this city would mean, in terms of time and labor, of people and supplies and equipment brought across fifty million miles of space. They'd have to use machinery; there was no other way it could be done. Bulldozers and power shovels and draglines; they were fast, but they were rough and indiscriminate. She remembered the digs around Harappa and Mohenjo-Daro, in the Indus Valley, and the careful, patient native laborers—the painstaking foremen, the pickmen and spademen, the long files of basketmen carrying away the earth. Slow and primitive as the civilization whose ruins they were uncovering, yes, but she could count on the fingers of one hand the times one of her pickmen had damaged a valuable object in the ground. If it hadn't been for the underpaid and uncomplaining native laborer, archaeology would still be back where Wincklemann had found it. But on Mars there was no native labor; the last Martian had died five hundred centuries ago.

Something started banging like a machine gun, four or five hundred yards to her left. A solenoid jack-hammer; Tony Lattimer must have decided which building he wanted to break into next. She became conscious, then, of the awkward weight of her equipment, and began redistributing it, shifting the straps of her oxy-tank pack, slinging the camera from one shoulder and the board and drafting tools from the other, gathering the notebooks and sketchbooks under her left arm. She started walking down the road, over hillocks of buried rubble, around snags of wall jutting up out of the loess, past buildings still standing, some of them already breached and explored, and across the brush-grown flat to the huts.

There were ten people in the main office room of Hut One when she entered. As soon as she had disposed of her oxygen equipment, she lit a cigarette, her first since noon, then looked from one to another of them. Old Selim von Ohlmhorst, the Turco-German, one of her two fellow archaeologists, sitting at the end of the long table against the farther wall, smoking his big curved pipe and going through a looseleaf notebook. The girl ordnance officer, Sachiko Koremitsu, between two droplights at the other end of the table, her head bent over her work. Colonel Hubert Penrose, the Space Force CO, and Captain Field, the intelligence officer, listening to the report of one of the airdyne pilots, returned from his afternoon survey flight. A couple of girl lieutenants from Signals, going over the script of the evening telecast, to be transmitted to the *Cyrano,* on orbit five thousand miles off planet and relayed from thence to Terra via Lunar. Sid Chamberlain, the Trans-Space News Service man, was with them. Like Selim and herself, he was a civilian; he was advertising the fact with a white shirt and a sleeveless blue sweater. And Major Lindemann, the engineer officer, and one of his assistants, arguing over some plans on a drafting board. She hoped, drawing a pint of hot water to wash her hands and sponge off her face, that they were doing something about the pipeline.

She started to carry the notebooks and sketchbooks over to where Selim von Ohlmhorst was sitting, and then, as she always did, she turned aside and stopped to watch Sachiko. The Japanese girl was restoring what had been a book, fifty thousand years ago; her eyes were masked by a binocular loup, the black headband invisible against her glossy black hair, and she was picking delicately at the crumbled page with a hair-fine wire set in a handle of copper tubing. Finally, loosening a particle as tiny as a snowflake, she grasped it with tweezers, placed it on the sheet of transparent plastic on which she was reconstructing the page, and set it with a mist of fixative from a little spraygun. It was a sheer joy to watch her; every movement was as graceful and precise as though done to music after being rehearsed a hundred times.

"Hello, Martha. It isn't cocktail-time yet, is it?" The girl at the table spoke without raising her head, almost without moving her lips, as though she were afraid that the slightest breath would disturb the flaky stuff in front of her.

"No, it's only fifteen-thirty. I finished my work, over there. I didn't find any more books, if that's good news for you."

Sachiko took off the loup and leaned back in her chair, her palms cupped over her eyes.

"No, I like doing this. I call it micro-jigsaw puzzles. This book, here, really is a mess. Selim found it lying open, with some heavy stuff on top of it; the pages were simply crushed." She hesitated briefly. "If only it would mean something, after I did it."

There could be a faintly critical overtone to that. As she replied, Martha realized that she was being defensive.

"It will, some day. Look how long it took to read Egyptian hieroglyphics, even after they had the Rosetta Stone."

Sachiko smiled. "Yes. I know. But they did have the Rosetta Stone."

"And we don't. There is no Rosetta Stone, not anywhere on Mars. A whole race, a whole species, died while the first Cròò-Magnon cave-artist was daubing pictures of reindeer and bison, and across fifty thousand years and fifty million miles there was no bridge of understanding."

"We'll find one. There must be something, somewhere, that will give us the meaning of a few words, and we'll use them to pry meaning out of more words, and so on. We may not live to learn this language, but we'll make a start, and some day somebody will."

Sachiko took her hands from her eyes, being careful not to look toward the unshaded light, and smiled again. This time Martha was sure that it was not the Japanese smile of politeness, but the universally human smile of friendship.

"I hope so, Martha: really I do. It would be wonderful for you to be the first to do it, and it would be wonderful for all of us to be able to read what these people wrote. It would really bring this dead city to life again." The smile faded slowly. "But it seems so hopeless."

"You haven't found any more pictures?"

Sachiko shook her head. Not that it would have meant much if she had. They had found hundreds of pictures with captions; they had never been able to establish a positive relationship between any pictured object and any printed word. Neither of them said anything more, and after a moment Sachiko replaced the loup and bent her head forward over the book.

Selim von Ohlmhorst looked up from his notebook, taking his pipe out of his mouth.

"Everything finished, over there?" he asked, releasing a puff of smoke.

"Such as it was." She laid the notebooks and sketches on the table. "Captain Gicquel's started airsealing the building from the fifth floor down, with an entrance on the sixth; he'll start putting in oxygen generators as soon as that's done. I have everything cleared up where he'll be working."

Colonel Penrose looked up quickly, as though making a mental note to attend to something later. Then he returned his attention to the pilot, who was pointing something out on a map.

Von Ohlmhorst nodded. "There wasn't much to it, at that," he agreed. "Do you know which building Tony has decided to enter next?"

"The tall one with the conical thing like a candle extinguisher on top, I think. I heard him drilling for the blasting shots over that way."

"Well, I hope it turns out to be one that was occupied up to the end."

The last one hadn't. It had been stripped of its contents and fittings, a piece of this and a bit of that, haphazardly, apparently over a long period of time, until it had been almost gutted. For centuries, as it had died, this city had been consuming itself by a process of auto-cannibalism. She said something to that effect.

"Yes. We always find that—except, of course, at places like Pompeii. Have you seen any of the other Roman cities in Italy?" he asked. "Minturnae, for instance? First the inhabitants tore down this to repair that, and then, after they had vacated the city, other people came along and tore down what was left, and burned the stones for lime, or crushed them to mend roads, till there was nothing left but the foundation traces. That's where we are fortunate; this is one of the places where the Martian race perished, and there were no barbarians to come later and destroy what they had left." He puffed slowly at his pipe. "Some of these days, Martha, we are going to break into one of these buildings and find that it was one in which the last of these people died. Then we will learn the story of the end of this civilization."

And if we learn to read their language, we'll learn the whole story, not just the obituary. She hesitated, not putting the thought into words. "We'll find that, sometime, Selim," she said, then looked at her watch. "I'm going to get some more work done on my lists, before dinner."

For an instant, the old man's face stiffened in disapproval; he started to say something, thought better of it, and put his pipe back into his mouth. The brief wrinkling around his mouth and the twitch of his white mustache had been enough, however; she knew what he was thinking. She was wasting time and effort, he believed; time and effort belonging not to herself but to the expedition. He could be right, too, she realized. But he had to be wrong; there had to be a way to do it. She turned from him silently and went to her own packing-case seat, at the middle of the table.

Photographs, and photostats of restored pages of books, and transcripts of inscriptions, were piled in front of her, and the notebooks in which she was compiling her lists. She sat down, lighting a fresh cigarette, and reached over to a stack of unexamined material, taking off the top sheet. It was a photostat of what looked like the title page and contents of some sort of a periodical. She remembered it; she had found it herself, two days before, in a closet in the basement of the building she had just finished examining.

She sat for a moment, looking at it. It was readable, in the sense that she had set up a purely arbitrary but consistently pronounceable system of phonetic values for the letters. The long vertical symbols were vowels. There were only ten of them; not too many, allowing separate characters for long and short sounds. There were twenty of the short horizontal letters, which meant that sounds like -ng or -ch or -sh were single letters. The odds were millions to one against her system being anything like the original sound of the language, but she had listed several thousand Martian words, and she could pronounce all of them.

And that was as far as it went. She could pronounce between three and four thousand Martian words, and she couldn't assign a meaning to one of them. Selim von Ohlmhorst believed that she never would. So did Tony Lattimer, and he was a great deal less reticent about saying so. So, she was sure, did Sachiko Koremitsu. There were times, now and then, when she began to be afraid that they were right.

The letters on the page in front of her began squirming and dancing, slender vowels with fat little consonants. They did that, now, every night in her dreams. And there were other dreams, in which she read them as easily as English; waking, she would try desperately and vainly to remember. She blinked, and looked away from the photostatted page; when she looked back, the letters were behaving themselves again. There were three words at the top of the page, over-and-underlined, which seemed to be the Martian method of capitalization. *Mastharnorvod Tadavas Sornhulva*. She pronounced them mentally, leafing through her notebooks to see if she had encountered them before, and in what contexts. All three were listed. In addition, *masthar* was a fairly common word, and so was *norvod*, and so was *nor*, but *-vod* was a suffix and nothing but a suffix. *Davas*, was a word, too, and *ta-* was a common prefix; *sorn* and *hulva* were both common words. This language, she had long ago decided, must be something like German; when the Martians had needed a new word, they had just pasted a couple of existing words together. It would probably turn out to be a grammatical horror. Well, they had published magazines, and one of them had been called *Mastharnorvod Tadavas Sornhulva*. She wondered if it had been something like the *Quarterly Archaeological Review*, or something more on the order of *Sexy Stories*.

A smaller line, under the title, was plainly the issue number and date; enough things had been found numbered in series to enable her to identify the numerals and determine that a decimal system of numeration had been used. This was the one thousand and seven hundred and fifty-fourth issue, for Doma, 14837; then Doma must be the name of one of the Martian months. The word had turned up several times before. She found herself puffing furiously on her cigarette as she leafed through notebooks and piles of already examined material.

Sachiko was speaking to somebody, and a chair scraped at the end of the table. She raised her head, to see a big man with red hair and a red face, in Space Force green, with the single star of a major on his shoulder, sitting down. Ivan Fitzgerald, the medic. He was lifting weights from a book similar to the one the girl ordnance officer was restoring.

"Haven't had time, lately," he was saying, in reply to Sachiko's question. "The Finchley girl's still down with whatever it is she has, and it's something I haven't been able to diagnose yet. And I've been checking on bacteria cultures, and in what spare time I have, I've been dissecting specimens for Bill Chandler. Bill's finally found a mammal. Looks like a lizard, and it's only four inches long, but it's a real warm-blooded, gamogenetic, placental, viviparous mammal. Burrows, and seems to live on what pass for insects here."

"Is there enough oxygen for anything like that?" Sachiko was asking.

"Seems to be, close to the ground." Fitzgerald got the headband of his loup adjusted, and pulled it down over his eyes. "He found this thing in a ravine down on the sea bottom—Ha, this page seems to be intact; now, if I can get it out all in one piece—"

He went on talking inaudibly to himself, lifting the page a little at a time and sliding one of the transparent plastic sheets under it, working with minute delicacy. Not the delicacy of the Japanese girl's small hands, moving like the paws of a cat washing her face, but like a steam-hammer cracking a peanut. Field archaeology requires a certain delicacy of touch, too, but Martha watched the pair of them with envious admiration. Then she turned back to her own work, finishing the table of contents.

The next page was the beginning of the first article listed; many of the words were unfamiliar. She had the impression that this must be some kind of scientific or technical journal; that could be because such publications made up the bulk of her own periodical reading. She doubted if it were fiction; the paragraphs had a solid, factual look.

At length, Ivan Fitzgerald gave a short, explosive grunt.

"Ha! Got it!"

She looked up. He had detached the page and was cementing another plastic sheet onto it.

"Any pictures?" she asked.

"None on this side. Wait a moment." He turned the sheet. "None on this side, either." He sprayed another sheet of plastic to sandwich the page, then picked up his pipe and relighted it.

"I get fun out of this, and it's good practice for my hands, so don't think I'm complaining," he said, "but, Martha, do you honestly think anybody's ever going to get anything out of this?"

Sachiko held up a scrap of the silicone plastic the Martians had used for paper with her tweezers. It was almost an inch square.

"Look; three whole words on this piece," she crowed. "Ivan, you took the easy book."

Fitzgerald wasn't being sidetracked. "This stuff's absolutely meaningless," he continued. "It had a meaning fifty thousand years ago, when it was written, but it has none at all now."

She shook her head. "Meaning isn't something that evaporates with time," she argued. "It has just as much meaning now as it ever had. We just haven't learned how to decipher it."

"That seems like a pretty pointless distinction," Selim von Ohlmhorst joined the conversation. "There no longer exists a means of deciphering it."

"We'll find one." She was speaking, she realized, more in self-encouragement than in controversy.

"How? From pictures and captions? We've found captioned pictures, and what have they given us? A caption is intended to explain the picture, not the picture to explain the caption. Suppose some alien to our culture found a picture of a man with a white beard and mustache sawing a billet from a log. He would think the caption meant, 'Man Sawing Wood.' How would he know that it was really 'Wilhelm II in Exile at Doorn?'"

Sachiko had taken off her loup and was lighting a cigarette.

"I can think of pictures intended to explain their captions," she said. "These picture language-books, the sort we use in the Service—little line drawings, with a word or phrase under them."

"Well, of course, if we found something like that," von Ohlmhorst began.

"Michael Ventris found something like that, back in the Fifties," Hubert Penrose's voice broke in from directly behind her.

She turned her head. The colonel was standing by the archaeologists' table; Captain Field and the airdyne pilot had gone out.

"He found a lot of Greek inventories of military stores," Penrose continued. "They were in Cretan Linear B script, and at the head of each list was a little picture, a sword or a helmet or a cooking tripod or a chariot wheel. That's what gave him the key to the script."

"Colonel's getting to be quite an archaeologist," Fitzgerald commented. "We're all learning each others' specialties, on this expedition."

"I heard about that long before this expedition was even contemplated." Penrose was tapping a cigarette on his gold case. "I heard about that back before the Thirty Days' War, at Intelligence School, when I was a lieutenant. As a feat of cryptanalysis, not an archaeological discovery."

"Yes, cryptanalysis," von Ohlmhorst pounced. "The reading of a known language in an unknown form of writing. Ventris' lists were in the known language, Greek. Neither he nor anybody else ever read a word of the Cretan language until the finding of the Greek-Cretan bilingual in 1963, because only with a bilingual text, one language already known, can an unknown ancient language be learned. And what hope, I ask you, have we of finding anything like that here? Martha, you've been working on these Martian texts ever since we landed here—for the last six months. Tell me, have you found a single word to which you can positively assign a meaning?"

"Yes, I think I have one." She was trying hard not to sound too exultant. "*Doma*. It's the name of one of the months of the Martian calendar."

"Where did you find that?" von Ohlmhorst asked. "And how did you establish—?"

"Here." She picked up the photostat and handed it along the table to him. "I'd call this the title page of a magazine."

He was silent for a moment, looking at it. "Yes. I would say so, too. Have you any of the rest of it?"

"I'm working on the first page of the first article, listed there. Wait till I see; yes, here's all I found, together, here." She told him where she had gotten it. "I just gathered it up, at the time, and gave it to Geoffrey and Rosita to photostat; this is the first I've really examined it."

The old man got to his feet, brushing tobacco ashes from the front of his jacket, and came to where she was sitting, laying the title page on the table and leafing quickly through the stack of photostats.

"Yes, and here is the second article, on page eight, and here's the next one." He finished the pile of photostats. "A couple of pages missing at the end of the last article. This is remarkable; surprising that a thing like a magazine would have survived so long."

"Well, this silicone stuff the Martians used for paper is pretty durable," Hubert Penrose said. "There doesn't seem to have been any water or any other fluid in it originally, so it wouldn't dry out with time."

"Oh, it's not remarkable that the material would have survived. We've found a good many books and papers in excellent condition. But only a really vital culture, an organized culture, will publish magazines, and this civilization had been dying for hundreds of years before the end. It might have been a thousand years before the time they died out completely that such activities as publishing ended."

"Well, look where I found it; in a closet in a cellar. Tossed in there and forgotten, and then ignored when they were stripping the building. Things like that happen."

Penrose had picked up the title page and was looking at it.

"I don't think there's any doubt about this being a magazine, at all." He looked again at the title, his lips moving silently. "*Masthamorvod Tadavas Sornhulva*. Wonder what it means. But you're right about the date—*Doma* seems to be the name of a month. Yes, you have a word, Dr. Dane."

Sid Chamberlain, seeing that something unusual was going on, had come over from the table at which he was working. After examining the title page and some of the inside pages, he began whispering into the stenophone he had taken from his belt.

"Don't try to blow this up to anything big, Sid," she cautioned. "All we have is the name of a month, and Lord only knows how long it'll be till we even find out which month it was."

"Well, it's a start, isn't it?" Penrose argued. "Grotefend only had the word for 'king' when he started reading Persian cuneiform."

"But I don't have the word for month; just the name of a month. Everybody knew the names of the Persian kings, long before Grotefend."

"That's not the story," Chamberlain said. "What the public back on Terra will be interested in is finding out that the Martians published magazines, just like we do. Something familiar; make the Martians seem more real. More human."

Three men had come in, and were removing their masks and helmets and oxy-tanks, and peeling out of their quilted coveralls. Two were Space Force lieutenants; the third was a youngish civilian with close-cropped blond hair, in a checked woolen shirt. Tony Lattimer and his helpers.

"Don't tell me Martha finally got something out of that stuff?" he asked, approaching the table. He might have been commenting on the antics of the village half-wit, from his tone.

"Yes; the name of one of the Martian months." Hubert Penrose went on to explain, showing the photostat.

Tony Lattimer took it, glanced at it, and dropped it on the table.

"Sounds plausible, of course, but just an assumption. That word may not be the name of a month, at all—could mean 'published' or 'authorized' or 'copyrighted' or anything like that. Fact is, I don't think it's more than a wild guess that that thing's anything like a periodical." He dismissed the subject and turned to Penrose. "I picked out the next building to enter; that tall one with the conical thing on top. It ought to be in pretty good shape inside; the conical top wouldn't allow dust to accumulate, and from the outside nothing seems to be caved in or crushed. Ground level's higher than the other one, about the seventh floor. I found a good place and drilled for the shots; tomorrow I'll blast a hole in it, and if you can spare some people to help, we can start exploring it right away."

"Yes, of course, Dr. Lattimer. I can spare about a dozen, and I suppose you can find a few civilian volunteers," Penrose told him. "What will you need in the way of equipment?"

"Oh, about six demolition-packets; they can all be shot together. And the usual thing in the way of lights, and breaking and digging tools, and climbing equipment in case we run into broken or doubtful stairways. We'll divide into two parties. Nothing ought to be entered for the first time without a qualified archaeologist along. Three parties, if Martha can tear herself away from this catalogue of systematized incomprehensibilities she's making long enough to do some real work."

She felt her chest tighten and her face become stiff. She was pressing her lips together to lock in a furious retort when Hubert Penrose answered for her.

"Dr. Dane's been doing as much work, and as important work, as you have," he said brusquely. "More important work, I'd be inclined to say."

Von Ohlmhorst was visibly distressed; he glanced once toward Sid Chamberlain, then looked hastily away from him. Afraid of a story of dissension among archaeologists getting out.

"Working out a system of pronunciation by which the Martian language could be transliterated was a most important contribution," he said. "And Martha did that almost unassisted."

"Unassisted by Dr. Lattimer, anyway," Penrose added. "Captain Field and Lieutenant Koremitsu did some work, and I helped out a little, but nine-tenths of it she did herself."

"Purely arbitrary," Lattimer disdained. "Why, we don't even know that the Martians could make the same kind of vocal sounds we do."

"Oh, yes, we do," Ivan Fitzgerald contradicted, safe on his own ground. "I haven't seen any actual Martian skulls—these people seem to have been very tidy about disposing of their dead—but from statues and busts and pictures I've seen. I'd say that their vocal organs were identical with our own."

"Well, grant that. And grant that it's going to be impressive to rattle off the names of Martian notables whose statues we find, and that if we're ever able to attribute any placenames, they'll sound a lot better than this horse-doctors' Latin the old astronomers splashed all over the map of Mars," Lattimer said. "What I object to is her wasting time on this stuff, of which nobody will ever be able to read a word if she fiddles around with those lists till there's another hundred feet of loess on this city, when there's so much real work to be done and we're as shorthanded as we are."

That was the first time that had come out in just so many words. She was glad Lattimer had said it and not Selim von Ohlmhorst.

"What you mean," she retorted, "is that it doesn't have the publicity value that digging up statues has."

For an instant, she could see that the shot had scored. Then Lattimer, with a side glance at Chamberlain, answered:

"What I mean is that you're trying to find something that any archaeologist, yourself included, should know doesn't exist. I don't object to your gambling your professional reputation and making a laughing stock of yourself; what I object to is that the blunders of one archaeologist discredit the whole subject in the eyes of the public."

That seemed to be what worried Lattimer most. She was framing a reply when the communication-outlet whistled shrilly, and then squawked: "Cocktail time! One hour to dinner; cocktails in the library, Hut Four!"

The library, which was also lounge, recreation room, and general gathering-place, was already crowded; most of the crowd was at the long table topped with sheets of glasslike plastic that had been wall panels out of one of the ruined buildings. She poured herself what passed, here, for a martini, and carried it over to where Selim von Ohlmhorst was sitting alone.

For a while, they talked about the building they had just finished exploring, then drifted into reminiscences of their work on Terra—von Ohlmhorst's in Asia Minor, with the Hittite Empire, and hers in Pakistan, excavating the cities of the Harappa Civilization. They finished their drinks—the ingredients were plentiful; alcohol and flavoring extracts synthesized from Martian vegetation—and von Ohlmhorst took the two glasses to the table for refills.

"You know, Martha," he said, when he returned, "Tony was right about one thing. You are gambling your professional standing and reputation. It's against all archaeological experience that a language so completely dead as this one could be deciphered. There was a continuity between all the other ancient languages—by knowing Greek, Champollion learned to read Egyptian; by knowing Egyptian, Hittite was learned. That's why you and your colleagues have never been able to translate the Harappa hieroglyphics; no such continuity exists there. If you insist that this utterly dead language can be read, your reputation will suffer for it."

"I heard Colonel Penrose say, once, that an officer who's afraid to risk his military reputation seldom makes much of a reputation. It's the same with us. If we really want to find things out, we have to risk making mistakes. And I'm a lot more interested in finding things out than I am in my reputation."

She glanced across the room, to where Tony Lattimer was sitting with Gloria Standish, talking earnestly, while Gloria sipped one of the counterfeit martinis and listened. Gloria was the leading contender for the title of Miss Mars, 1996, if you liked big bosomy blondes, but Tony would have been just as attentive to her if she'd looked like the Wicked Witch in "The Wizard of Oz." because Gloria was the Pan-Federation Telecast System commentator with the expedition.

"I know you are," the old Turco-German was saying. "That's why, when they asked me to name another archaeologist for this expedition, I named you."

He hadn't named Tony Lattimer; Lattimer had been pushed onto the expedition by his university. There'd been a lot of high-level string-pulling to that; she wished she knew the whole story. She'd managed to keep clear of universities and university politics; all her digs had been sponsored by non-academic foundations or art museums.

"You have an excellent standing: much better than my own, at your age. That's why it disturbs me to see you jeopardizing it by this insistence that the Martian language can be translated. I can't, really, see how you can hope to succeed."

She shrugged and drank some more of her cocktail, then lit another cigarette. It was getting tiresome to try to verbalize something she only felt.

"Neither do I, now, but I will. Maybe I'll find something like the picture-books Sachiko was talking about. A child's primer, maybe; surely they had things like that. And if I don't. I'll find something else. We've only been here six months. I can wait the rest of my life, if I have to, but I'll do it sometime."

"I can't wait so long," von Ohlmhorst said. "The rest of my life will only be a few years, and when the *Schiaparelli* orbits in, I'll be going back to Terra on the *Cyrano*."

"I wish you wouldn't. This is a whole new world of archaeology. Literally."

"Yes." He finished the cocktail and looked at his pipe as though wondering whether to re-light it so soon before dinner, then put it in his pocket. "A whole new world—but I've grown old, and it isn't for me. I've spent my life studying the Hittites. I can speak the Hittite language, though maybe King Muwatallis wouldn't be able to understand my modern Turkish accent. But the things I'd have to learn here—chemistry, physics, engineering, how to run analytic tests on steel girders and beryllo-silver alloys and plastics and silicones. I'm more at home with a civilization that rode in chariots and fought with swords and was just learning how to work iron. Mars is for young people. This expedition is a cadre of leadership—not only the Space Force people, who'll be the commanders of the main expedition, but us scientists, too. And I'm just an old cavalry general who can't learn to command tanks and aircraft. You'll have time to learn about Mars. I won't."

His reputation as the dean of Hittitologists was solid and secure, too, she added mentally. Then she felt ashamed of the thought. He wasn't to be classed with Tony Lattimer.

"All I came for was to get the work started," he was continuing. "The Federation Government felt that an old hand should do that. Well, it's started, now; you and Tony and whoever come out on the *Schiaparelli* must carry it on. You said it, yourself; you have a whole new world. This is only one city, of the last Martian civilization. Behind this, you have the Late Upland Culture, and the Canal Builders, and all the civilizations and races and empires before them, clear back to the Martian Stone Age." He hesitated for a moment. "You have no idea what all you have to learn, Martha. This isn't the time to start specializing too narrowly."

They all got out of the truck and stretched their legs and looked up the road to the tall building with the queer conical cap askew on its top. The four little figures that had been busy against its wall climbed into the jeep and started back slowly, the smallest of them, Sachiko Koremitsu, paying out an electric cable behind. When it pulled up beside the truck, they climbed out; Sachiko attached the free end of the cable to a nuclear-electric battery. At once, dirty gray smoke and orange dust puffed out from the wall of the building, and, a second later, the multiple explosion banged.

She and Tony Lattimer and Major Lindemann climbed onto the truck, leaving the jeep stand by the road. When they reached the building, a satisfyingly wide breach had been blown in the wall. Lattimer had placed his shots between two of the windows; they were both blown out along with the wall between, and lay unbroken on the ground. Martha remembered the first building they had entered. A Space Force officer had picked up a stone and thrown it at one of the windows, thinking that would be all they'd need to do. It had bounced back. He had drawn his pistol—they'd all carried guns, then, on the principle that what they didn't know about Mars might easily hurt them—and fired four shots. The bullets had ricocheted, screaming thinly; there were four coppery smears of jacket-metal on the window, and a little surface spalling. Somebody tried a rifle; the 4000-f.s. bullet had cracked the glasslike pane without penetrating. An oxyacetylene torch had taken an hour to cut the window out; the lab crew, aboard the ship, were still trying to find out just what the stuff was.

Tony Lattimer had gone forward and was sweeping his flashlight back and forth, swearing petulantly, his voice harshened and amplified by his helmet-speaker.

"I thought I was blasting into a hallway; this lets us into a room. Careful; there's about a two-foot drop to the floor, and a lot of rubble from the blast just inside."

He stepped down through the breach; the others began dragging equipment out of the trucks—shovels and picks and crowbars and sledges, portable floodlights, cameras, sketching materials, an extension ladder, even Alpinists' ropes and crampons and pickaxes. Hubert Penrose was shouldering something that looked like a surrealist machine gun but which was really a nuclear-electric jack-hammer. Martha selected one of the spike-shod mountaineer's ice axes, with which she could dig or chop or poke or pry or help herself over rough footing.

The windows, grimed and crusted with fifty millennia of dust, filtered in a dim twilight; even the breach in the wall, in the morning shade, lighted only a small patch of floor. Somebody snapped on a floodlight, aiming it at the ceiling. The big room was empty and bare; dust lay thick on the floor and reddened the once-white walls. It could have been a large office, but there was nothing left in it to indicate its use.

"This one's been stripped up to the seventh floor!" Lattimer exclaimed. "Street level'll be cleaned out, completely."

"Do for living quarters and shops, then," Lindemann said. "Added to the others, this'll take care of everybody on the *Schiaparelli*."

"Seem to have been a lot of electric or electronic apparatus over along this wall," one of the Space Force officers commented. "Ten or twelve electric outlets."

He brushed the dusty wall with his glove, then scraped on the floor with his foot. "I can see where things were pried loose."

The door, one of the double sliding things the Martians had used, was closed. Selim von Ohlmhorst tried it, but it was stuck fast. The metal latch-parts had frozen together, molecule bonding itself to molecule, since the door had last been closed. Hubert Penrose came over with the jack-hammer, fitting a spear-point chisel into place. He set the chisel in the joint between the doors, braced the hammer against his hip, and squeezed the trigger-switch. The hammer banged briefly like the weapon it resembled, and the doors popped a few inches apart, then stuck. Enough dust had worked into the recesses into which it was supposed to slide to block it on both sides.

That was old stuff; they ran into that every time they had to force a door, and they were prepared for it. Somebody went outside and brought in a power-jack and finally one of the doors inched back to the door jamb. That was enough to get the lights and equipment through: they all passed from the room to the hallway beyond. About half the other doors were open; each had a number and a single word, *Darfhulva*, over it.

One of the civilian volunteers, a woman professor of natural ecology from Penn State University, was looking up and down the hall.

"You know," she said, "I feel at home here. I think this was a college of some sort, and these were classrooms. That word, up there; that was the subject taught, or the department. And those electronic devices, all where the class would face them; audio-visual teaching aids."

"A twenty-five-story university?" Lattimer scoffed. "Why, a building like this would handle thirty thousand students."

"Maybe there were that many. This was a big city, in its prime," Martha said, moved chiefly by a desire to oppose Lattimer.

"Yes, but think of the snafu in the halls, every time they changed classes. It'd take half an hour to get everybody back and forth from one floor to another." He turned to von Ohlmhorst. "I'm going up above this floor. This place has been looted clean up to here, but there's a chance there may be something above," he said.

"I'll stay on this floor, at present," the Turco-German replied. "There will be much coming and going, and dragging things in and out. We should get this completely examined and recorded first. Then Major Lindemann's people can do their worst, here."

"Well, if nobody else wants it, I'll take the downstairs," Martha said.

"I'll go along with you," Hubert Penrose told her. "If the lower floors have no archaeological value, we'll turn them into living quarters. I like this building: it'll give everybody room to keep out from under everybody else's feet." He looked down the hall. "We ought to find escalators at the middle."

The hallway, too, was thick underfoot with dust. Most of the open rooms were empty, but a few contained furniture, including small seat-desks. The original proponent of the university theory pointed these out as just what might be found in classrooms. There were escalators, up and down, on either side of the hall, and more on the intersecting passage to the right.

"That's how they handled the students, between classes," Martha commented. "And I'll bet there are more ahead, there."

They came to a stop where the hallway ended at a great square central hall. There were elevators, there, on two of the sides, and four escalators, still usable as stairways. But it was the walls, and the paintings on them, that brought them up short and staring.

They were clouded with dirt—she was trying to imagine what they must have looked like originally, and at the same time estimating the labor that would be involved in cleaning them—but they were still distinguishable, as was the word, *Darfhulva*, in golden letters above each of the four sides. It was a moment before she realized, from the murals, that she had at last found a meaningful Martian word. They were a vast historical panorama, clockwise around the room. A group of skin-clad savages squatting around a fire. Hunters with bows and spears, carrying a carcass of an animal slightly like a pig. Nomads riding long-legged, graceful mounts like hornless deer. Peasants sowing and reaping; mud-walled hut villages, and cities; processions of priests and warriors; battles with swords and bows, and with cannon and muskets; galleys, and ships with sails, and ships without visible means of propulsion, and aircraft. Changing costumes and weapons and machines and styles of architecture. A richly fertile landscape, gradually merging into barren deserts and bushlands—the time of the great planet-wide drought. The Canal Builders—men with machines recognizable as steam-shovels and derricks, digging and quarrying and driving across the empty plains with aqueducts. More cities—seaports on the shrinking oceans; dwindling, half-deserted cities; an abandoned city, with four tiny humanoid figures and a thing like a combat-car in the middle of a brush-grown plaza, they and their vehicle dwarfed by the huge lifeless buildings around them. She had not the least doubt; *Darfhulva* was History.

"Wonderful!" von Ohlmhorst was saying. "The entire history of this race. Why, if the painter depicted appropriate costumes and weapons and machines for each period, and got the architecture right, we can break the history of this planet into eras and periods and civilizations."

"You can assume they're authentic. The faculty of this university would insist on authenticity in the *Darfhulva*—History—Department," she said.

"Yes! *Darfhulva*—History! And your magazine was a journal of *Sornhulva*!" Penrose exclaimed. "You have a word, Martha!" It took her an instant to realize that he had called her by her first name, and not Dr. Dane. She wasn't sure if that weren't a bigger triumph than learning a word of the Martian language. Or a more auspicious start. "Alone, I suppose that *hulva* means something like science or knowledge, or study; combined, it would be equivalent to our 'ology. And *darf* would mean something like past, or old times, or human events, or chronicles."

"That gives you three words, Martha!" Sachiko jubilated. "You did it."

"Let's don't go too fast," Lattimer said, for once not derisively. "I'll admit that *darfhulva* is the Martian word for history as a subject of study; I'll admit that *hulva* is the general word and *darf* modifies it and tells us which subject is meant. But as for assigning specific meanings, we can't do that because we don't know just how the Martians thought, scientifically or otherwise."

He stopped short, startled by the blue-white light that blazed as Sid Chamberlain's Kliegettes went on. When the whirring of the camera stopped, it was Chamberlain who was speaking:

"This is the biggest thing yet; the whole history of Mars, stone age to the end, all on four walls. I'm taking this with the fast shutter, but we'll telecast it in slow motion, from the beginning to the end. Tony, I want you to do the voice for it—running commentary, interpretation of each scene as it's shown. Would you do that?"

Would he do that! Martha thought. If he had a tail, he'd be wagging it at the very thought.

"Well, there ought to be more murals on the other floors," she said. "Who wants to come downstairs with us?"

Sachiko did; immediately. Ivan Fitzgerald volunteered. Sid decided to go upstairs with Tony Lattimer, and Gloria Standish decided to go upstairs, too. Most of the party would remain on the seventh floor, to help Selim von Ohlmhorst get it finished. After poking tentatively at the escalator with the spike of her ice axe, Martha led the way downward.

The sixth floor was *Darfhulva*, too; military and technological history, from the character of the murals. They looked around the central hall, and went down to the fifth; it was like the floors above except that the big quadrangle was stacked with dusty furniture and boxes. Ivan Fitzgerald, who was carrying the floodlight, swung it slowly around. Here the murals were of heroic-sized Martians, so human in appearance as to seem members of her own race, each holding some object—a book, or a test tube, or some bit of scientific apparatus, and behind them were scenes of laboratories and factories, flame and smoke, lightning-flashes. The word at the top of each of the four walls was one with which she was already familiar—*Sornhulva*.

"Hey, Martha; there's that word," Ivan Fitzgerald exclaimed. "The one in the title of your magazine." He looked at the paintings. "Chemistry, or physics."

"Both." Hubert Penrose considered. "I don't think the Martians made any sharp distinction between them. See, the old fellow with the scraggly whiskers must be the inventor of the spectroscope; he has one in his hands, and he has a rainbow behind him. And the woman in the blue smock, beside him, worked in organic chemistry; see the diagrams of long-chain molecules behind her. What word would convey the idea of chemistry and physics taken as one subject?"

"*Sornhulva*," Sachiko suggested. "If *hulva's* something like science, "*sorn*" must mean matter, or substance, or physical object. You were right, all along, Martha. A civilization like this would certainly leave something like this, that would be self-explanatory."

"This'll wipe a little more of that superior grin off Tony Lattimer's face," Fitzgerald was saying, as they went down the motionless escalator to the floor below. "Tony wants to be a big shot. When you want to be a big shot, you can't bear the possibility of anybody else being a bigger big shot, and whoever makes a start on reading this language will be the biggest big shot archaeology ever saw."

That was true. She hadn't thought of it, in that way, before, and now she tried not to think about it. She didn't want to be a big shot. She wanted to be able to read the Martian language, and find things out about the Martians.

Two escalators down, they came out on a mezzanine around a wide central hall on the street level, the floor forty feet below them and the ceiling thirty feet above. Their lights picked out object after object below—a huge group of sculptured figures in the middle; some kind of a motor vehicle jacked up on trestles for repairs; things that looked like machine-guns and auto-cannon; long tables, tops littered with a dust-covered miscellany; machinery; boxes and crates and containers.

They made their way down and walked among the clutter, missing a hundred things for every one they saw, until they found an escalator to the basement. There were three basements, one under another, until at last they stood at the bottom of the last escalator, on a bare concrete floor, swinging the portable floodlight over stacks of boxes and barrels and drums, and heaps of powdery dust. The boxes were plastic—nobody had ever found anything made of wood in the city—and the barrels and drums were of metal or glass or some glasslike substance. They were outwardly intact. The powdery heaps might have been anything organic, or anything containing fluid. Down here, where wind and dust could not reach, evaporation had been the only force of destruction after the minute life that caused putrefaction had vanished.

They found refrigeration rooms, too, and using Martha's ice axe and the pistollike vibratool Sachiko carried on her belt, they pounded and pried one open, to find dessicated piles of what had been vegetables, and leathery chunks of meat. Samples of that stuff, rocketed up to the ship, would give a reliable estimate, by radio-carbon dating, of how long ago this building had been occupied. The refrigeration unit, radically different from anything their own culture had produced, had been electrically powered. Sachiko and Penrose, poking into it, found the switches still on; the machine had only ceased to function when the power-source, whatever that had been, had failed.

The middle basement had also been used, at least toward the end, for storage; it was cut in half by a partition pierced by but one door. They took half an hour to force this, and were on the point of sending above for heavy equipment when it yielded enough for them to squeeze through. Fitzgerald, in the lead with the light, stopped short, looked around, and then gave a groan that came through his helmet-speaker like a foghorn.

"Oh, no! *No!*"

"What's the matter, Ivan?" Sachiko, entering behind him, asked anxiously.

He stepped aside. "Look at it, Sachi! Are we going to have to do all that?"

Martha crowded through behind her friend and looked around, then stood motionless, dizzy with excitement. Books. Case on case of books, half an acre of cases, fifteen feet to the ceiling. Fitzgerald, and Penrose, who had pushed in behind her, were talking in rapid excitement; she only heard the sound of their voices, not their words. This must be the main stacks of the university library—the entire literature of the vanished race of Mars. In the center, down an aisle between the cases, she could see the hollow square of the librarians' desk, and stairs and a dumb-waiter to the floor above.

She realized that she was walking forward, with the others, toward this. Sachiko was saying: "I'm the lightest; let me go first." She must be talking about the spidery metal stairs.

"I'd say they were safe," Penrose answered. "The trouble we've had with doors around here shows that the metal hasn't deteriorated."

In the end, the Japanese girl led the way, more catlike than ever in her caution. The stairs were quite sound, in spite of their fragile appearance, and they all followed her. The floor above was a duplicate of the room they had entered, and seemed to contain about as many books. Rather than waste time forcing the door here, they returned to the middle basement and came up by the escalator down which they had originally descended.

The upper basement contained kitchens—electric stoves, some with pots and pans still on them—and a big room that must have been, originally, the students' dining room, though when last used it had been a workshop. As they expected, the library reading room was on the street-level floor, directly above the stacks. It seemed to have been converted into a sort of common living room for the building's last occupants. An adjoining auditorium had been made into a chemical works; there were vats and distillation apparatus, and a metal fractionating tower that extended through a hole knocked in the ceiling seventy feet above. A good deal of plastic furniture of the sort they had been finding everywhere in the city was stacked about, some of it broken up, apparently for reprocessing. The other rooms on the street floor seemed also to have been devoted to manufacturing and repair work; a considerable industry, along a number of lines, must have been carried on here for a long time after the university had ceased to function as such.

On the second floor, they found a museum; many of the exhibits remained, tantalizingly half-visible in grimed glass cases. There had been administrative offices there, too. The doors of most of them were closed, and they did not waste time trying to force them, but those that were open had been turned into living quarters. They made notes, and rough floor plans, to guide them in future more thorough examination; it was almost noon before they had worked their way back to the seventh floor.

Selim von Ohlmhorst was in a room on the north side of the building, sketching the position of things before examining them and collecting them for removal. He had the floor checkerboarded with a grid of chalked lines, each numbered.

"We have everything on this floor photographed," he said. "I have three gangs—all the floodlights I have—sketching and making measurements. At the rate we're going, with time out for lunch, we'll be finished by the middle of the afternoon."

"You've been working fast. Evidently you aren't being high-church about a 'qualified archaeologist' entering rooms first," Penrose commented.

"Ach, childishness!" the old man exclaimed impatiently. "These officers of yours aren't fools. All of them have been to Intelligence School and Criminal Investigation School. Some of the most careful amateur archaeologists I ever knew were retired soldiers or policemen. But there isn't much work to be done. Most of the rooms are either empty or like this one—a few bits of furniture and broken trash and scraps of paper. Did you find anything down on the lower floors?"

"Well, yes," Penrose said, a hint of mirth in his voice. "What would you say, Martha?"

She started to tell Selim. The others, unable to restrain their excitement, broke in with interruptions. Von Ohlmhorst was staring in incredulous amazement.

"But this floor was looted almost clean, and the buildings we've entered before were all looted from the street level up," he said, at length.

"The people who looted this one lived here," Penrose replied. "They had electric power to the last; we found refrigerators full of food, and stoves with the dinner still on them. They must have used the elevators to haul things down from the upper floor. The whole first floor was converted into workshops and laboratories. I think that this place must have been something like a monastery in the Dark Ages in Europe, or what such a monastery would have been like if the Dark Ages had followed the fall of a highly developed scientific civilization. For one thing, we found a lot of machine guns and light auto-cannon on the street level, and all the doors were barricaded. The people here were trying to keep a civilization running after the rest of the planet had gone back to barbarism; I suppose they'd have to fight off raids by the barbarians now and then."

"You're not going to insist on making this building into expedition quarters, I hope, colonel?" von Ohlmhorst asked anxiously.

"Oh, no! This place is an archaeological treasure-house. More than that; from what I saw, our technicians can learn a lot, here. But you'd better get this floor cleaned up as soon as you can, though. I'll have the subsurface part, from the sixth floor down, airsealed. Then we'll put in oxygen generators and power units, and get a couple of elevators into service. For the floors above, we can use temporary airsealing floor by floor, and portable equipment; when we have things atmosphered and lighted and heated, you and Martha and Tony Lattimer can go to work systematically and in comfort, and I'll give you all the help I can spare from the other work. This is one of the biggest things we've found yet."

Tony Lattimer and his companions came down to the seventh floor a little later.

"I don't get this, at all," he began, as soon as he joined them. "This building wasn't stripped the way the others were. Always, the procedure seems to have been to strip from the bottom up, but they seem to have stripped the top floors first, here. All but the very top. I found out what that conical thing is, by the way. It's a wind-rotor, and under it there's an electric generator. This building generated its own power."

"What sort of condition are the generators in?" Penrose asked.

"Well, everything's full of dust that blew in under the rotor, of course, but it looks to be in pretty good shape. Hey, I'll bet that's it! They had power, so they used the elevators to haul stuff down. That's just what they did. Some of the floors above here don't seem to have been touched, though." He paused momentarily; back of his oxy-mask, he seemed to be grinning. "I don't know that I ought to mention this in front of Martha, but two floors above—we hit a room—it must have been the reference library for one of the departments—that had close to five hundred books in it."

The noise that interrupted him, like the squawking of a Brobdingnagian parrot, was only Ivan Fitzgerald laughing through his helmet-speaker.

Lunch at the huts was a hasty meal, with a gabble of full-mouthed and excited talking. Hubert Penrose and his chief subordinates snatched their food in a huddled consultation at one end of the table; in the afternoon, work was suspended on everything else and the fifty-odd men and women of the expedition concentrated their efforts on the University. By the middle of the afternoon, the seventh floor had been completely examined, photographed and sketched, and the murals in the square central hall covered with protective tarpaulins, and Laurent Gicquel and his airsealing crew had moved in and were at work. It had been decided to seal the central hall at the entrances. It took the French-Canadian engineer most of the afternoon to find all the ventilation-ducts and plug them. An elevator-shaft on the north side was found reaching clear to the twenty-fifth floor; this would give access to the top of the building; another shaft, from the center, would take care of the floors below. Nobody seemed willing to trust the ancient elevators, themselves; it was the next evening before a couple of cars and the necessary machinery could be fabricated in the machine shops aboard the ship and sent down by landing-rocket. By that time, the airsealing was finished, the nuclear-electric energy-converters were in place, and the oxygen generators set up.

Martha was in the lower basement, an hour or so before lunch the day after, when a couple of Space Force officers came out of the elevator, bringing extra lights with them. She was still using oxygen-equipment; it was a moment before she realized that the newcomers had no masks, and that one of them was smoking. She took off her own helmet-speaker, throat-mike and mask and unslung her tank-pack, breathing cautiously. The air was chilly, and musty-acrid with the odor of antiquity—the first Martian odor she had smelled—but when she lit a cigarette, the lighter flamed clear and steady and the tobacco caught and burned evenly.

The archaeologists, many of the other civilian scientists, a few of the Space Force officers and the two news-correspondents, Sid Chamberlain and Gloria Standish, moved in that evening, setting up cots in vacant rooms. They installed electric stoves and a refrigerator in the old Library Reading Room, and put in a bar and lunch counter. For a few days, the place was full of noise and activity, then, gradually, the Space Force people and all but a few of the civilians returned to their own work. There was still the business of airsealing the more habitable of the buildings already explored, and fitting them up in readiness for the arrival, in a year and a half, of the five hundred members of the main expedition. There was work to be done enlarging the landing field for the ship's rocket craft, and building new chemical-fuel tanks.

There was the work of getting the city's ancient reservoirs cleared of silt before the next spring thaw brought more water down the underground aqueducts everybody called canals in mistranslation of Schiaparelli's Italian word, though this was proving considerably easier than anticipated. The ancient Canal-Builders must have anticipated a time when their descendants would no longer be capable of maintenance work, and had prepared against it. By the day after the University had been made completely habitable, the actual work there was being done by Selim, Tony Lattimer and herself, with half a dozen Space Force officers, mostly girls, and four or five civilians, helping.

They worked up from the bottom, dividing the floor-surfaces into numbered squares, measuring and listing and sketching and photographing. They packaged samples of organic matter and sent them up to the ship for Carbon-14 dating and analysis; they opened cans and jars and bottles, and found that everything fluid in them had evaporated, through the porosity of glass and metal and plastic if there were no other way. Wherever they looked, they found evidence of activity suddenly suspended and never resumed. A vise with a bar of metal in it, half cut through and the hacksaw beside it. Pots and pans with hardened remains of food in them; a leathery cut of meat on a table, with the knife ready at hand. Toilet articles on washstands; unmade beds, the bedding ready to crumble at a touch but still retaining the impress of the sleeper's body; papers and writing materials on desks, as though the writer had gotten up, meaning to return and finish in a fifty-thousand-year-ago moment.

It worried her. Irrationally, she began to feel that the Martians had never left this place; that they were still around her, watching disapprovingly every time she picked up something they had laid down. They haunted her dreams, now, instead of their enigmatic writing. At first, everybody who had moved into the University had taken a separate room, happy to escape the crowding and lack of privacy of the huts. After a few nights, she was glad when Gloria Standish moved in with her, and accepted the newswoman's excuse that she felt lonely without somebody to talk to before falling asleep. Sachiko Koremitsu joined them the next evening, and before going to bed, the girl officer cleaned and oiled her pistol, remarking that she was afraid some rust may have gotten into it.

The others felt it, too. Selim von Ohlmhorst developed the habit of turning quickly and looking behind him, as though trying to surprise somebody or something that was stalking him. Tony Lattimer, having a drink at the bar that had been improvised from the librarian's desk in the Reading Room, set down his glass and swore.

"You know what this place is? It's an archaeological *Marie Celeste*!" he declared. "It was occupied right up to the end—we've all seen the shifts these people used to keep a civilization going here—but what was the end? What happened to them? Where did they go?"

"You didn't expect them to be waiting out front, with a red carpet and a big banner, *Welcome Terrans*, did you, Tony?" Gloria Standish asked.

"No, of course not; they've all been dead for fifty thousand years. But if they were the last of the Martians, why haven't we found their bones, at least? Who buried them, after they were dead?" He looked at the glass, a bubble-thin goblet, found, with hundreds of others like it, in a closet above, as though debating with himself whether to have another drink. Then he voted in the affirmative and reached for the cocktail pitcher. "And every door on the old ground level is either barred or barricaded from the inside. How did they get out? And why did they leave?"

The next day, at lunch, Sachiko Koremitsu had the answer to the second question. Four or five electrical engineers had come down by rocket from the ship, and she had been spending the morning with them, in oxy-masks, at the top of the building.

"Tony, I thought you said those generators were in good shape," she began, catching sight of Lattimer. "They aren't. They're in the most unholy mess I ever saw. What happened, up there, was that the supports of the wind-rotor gave way, and weight snapped the main shaft, and smashed everything under it."

"Well, after fifty thousand years, you can expect something like that," Lattimer retorted. "When an archaeologist says something's in good shape, he doesn't necessarily mean it'll start as soon as you shove a switch in."

"You didn't notice that it happened when the power was on, did you," one of the engineers asked, nettled at Lattimer's tone. "Well, it was. Everything's burned out or shorted or fused together; I saw one busbar eight inches across melted clean in two. It's a pity we didn't find things in good shape, even archaeologically speaking. I saw a lot of interesting things, things in advance of what we're using now. But it'll take a couple of years to get everything sorted out and figure what it looked like originally."

"Did it look as though anybody'd made any attempt to fix it?" Martha asked.

Sachiko shook her head. "They must have taken one look at it and given up. I don't believe there would have been any possible way to repair anything."

"Well, that explains why they left. They needed electricity for lighting, and heating, and all their industrial equipment was electrical. They had a good life, here, with power; without it, this place wouldn't have been habitable."

"Then why did they barricade everything from the inside, and how did they get out?" Lattimer wanted to know.

"To keep other people from breaking in and looting. Last man out probably barred the last door and slid down a rope from upstairs," von Ohlmhorst suggested. "This Houdini-trick doesn't worry me too much. We'll find out eventually."

"Yes, about the time Martha starts reading Martian," Lattimer scoffed.

"That may be just when we'll find out," von Ohlmhorst replied seriously. "It wouldn't surprise me if they left something in writing when they evacuated this place."

"Are you really beginning to treat this pipe dream of hers as a serious possibility, Selim?" Lattimer demanded. "I know, it would be a wonderful thing, but wonderful things don't happen just because they're wonderful. Only because they're possible, and this isn't. Let me quote that distinguished Hittitologist, Johannes Friedrich: 'Nothing can be translated out of nothing.' Or that later but not less distinguished Hittitologist, Selim von Ohlmhorst: 'Where are you going to get your bilingual?'"

"Friedrich lived to see the Hittite language deciphered and read," von Ohlmhorst reminded him.

"Yes, when they found Hittite-Assyrian bilinguals." Lattimer measured a spoonful of coffee-powder into his cup and added hot water. "Martha, you ought to know, better than anybody, how little chance you have. You've been working for years in the Indus Valley; how many words of Harappa have you or anybody else ever been able to read?"

"We never found a university, with a half-million-volume library, at Harappa or Mohenjo-Daro."

"And, the first day we entered this building, we established meanings for several words," Selim von Ohlmhorst added.

"And you've never found another meaningful word since," Lattimer added. "And you're only sure of general meaning, not specific meaning of word-elements, and you have a dozen different interpretations for each word."

"We made a start," von Ohlmhorst maintained. "We have Grotefend's word for 'king.' But I'm going to be able to read some of those books, over there, if it takes me the rest of my life here. It probably will, anyhow."

"You mean you've changed your mind about going home on the *Cyrano*?" Martha asked. "You'll stay on here?"

The old man nodded. "I can't leave this. There's too much to discover. The old dog will have to learn a lot of new tricks, but this is where my work will be, from now on."

Lattimer was shocked. "You're nuts!" he cried. "You mean you're going to throw away everything you've accomplished in Hittitology and start all over again here on Mars? Martha, if you've talked him into this crazy decision, you're a criminal!"

"Nobody talked me into anything," von Ohlmhorst said roughly. "And as for throwing away what I've accomplished in Hittitology, I don't know what the devil you're talking about. Everything I know about the Hittite Empire is published and available to anybody. Hittitology's like Egyptology; it's stopped being research and archaeology and become scholarship and history. And I'm not a scholar or a historian; I'm a pick-and-shovel field archaeologist—a highly skilled and specialized grave-robber and junk-picker—and there's more pick-and-shovel work on this planet than I could do in a hundred lifetimes. This is something new; I was a fool to think I could turn my back on it and go back to scribbling footnotes about Hittite kings."

"You could have anything you wanted, in Hittitology. There are a dozen universities that'd sooner have you than a winning football team. But no! You have to be the top man in Martiology, too. You can't leave that for anybody else—" Lattimer shoved his chair back and got to his feet, leaving the table with an oath that was almost a sob of exasperation.

Maybe his feelings were too much for him. Maybe he realized, as Martha did, what he had betrayed. She sat, avoiding the eyes of the others, looking at the ceiling, as embarrassed as though Lattimer had flung something dirty on the table in front of them. Tony Lattimer had, desperately, wanted Selim to go home on the *Cyrano*. Martiology was a new field; if Selim entered it, he would bring with him the reputation he had already built in Hittitology, automatically stepping into the leading role that Lattimer had coveted for himself. Ivan Fitzgerald's words echoed back to her—when you want to be a big shot, you can't bear the possibility of anybody else being a bigger big shot. His derision of her own efforts became comprehensible, too. It wasn't that he was convinced that she would never learn to read the Martian language. He had been afraid that she would.

Ivan Fitzgerald finally isolated the germ that had caused the Finchley girl's undiagnosed illness. Shortly afterward, the malady turned into a mild fever, from which she recovered. Nobody else seemed to have caught it. Fitzgerald was still trying to find out how the germ had been transmitted.

They found a globe of Mars, made when the city had been a seaport. They located the city, and learned that its name had been Kukan—or something with a similar vowel-consonant ratio. Immediately, Sid Chamberlain and Gloria Standish began giving their telecasts a Kukan dateline, and Hubert Penrose used the name in his official reports. They also found a Martian calendar; the year had been divided into ten more or less equal months, and one of them had been Doma. Another month was Nor, and that was a part of the name of the scientific journal Martha had found.

Bill Chandler, the zoologist, had been going deeper and deeper into the old sea bottom of Syrtis. Four hundred miles from Kukan, and at fifteen thousand feet lower altitude, he shot a bird. At least, it was a something with wings and what were almost but not quite feathers, though it was more reptilian than avian in general characteristics. He and Ivan Fitzgerald skinned and mounted it, and then dissected the carcass almost tissue by tissue. About seven-eighths of its body capacity was lungs; it certainly breathed air containing at least half enough oxygen to support human life, or five times as much as the air around Kukan.

That took the center of interest away from archaeology, and started a new burst of activity. All the expedition's aircraft—four jetticopters and three wingless airdyne reconnaissance fighters—were thrown into intensified exploration of the lower sea bottoms, and the bio-science boys and girls were wild with excitement and making new discoveries on each flight.

The University was left to Selim and Martha and Tony Lattimer, the latter keeping to himself while she and the old Turco-German worked together. The civilian specialists in other fields, and the Space Force people who had been holding tape lines and making sketches and snapping cameras, were all flying to lower Syrtis to find out how much oxygen there was and what kind of life it supported.

Sometimes Sachiko dropped in; most of the time she was busy helping Ivan Fitzgerald dissect specimens. They had four or five species of what might loosely be called birds, and something that could easily be classed as a reptile, and a carnivorous mammal the size of a cat with birdlike claws, and a herbivore almost identical with the piglike thing in the big *Darfhulva* mural, and another like a gazelle with a single horn in the middle of its forehead.

The high point came when one party, at thirty thousand feet below the level of Kukan, found breathable air. One of them had a mild attack of *sorroche* [sic] and had to be flown back for treatment in a hurry, but the others showed no ill effects.

The daily newscasts from Terra showed a corresponding shift in interest at home. The discovery of the University had focused attention on the dead past of Mars; now the public was interested in Mars as a possible home for humanity. It was Tony Lattimer who brought archaeology back into the activities of the expedition and the news at home.

Martha and Selim were working in the museum on the second floor, scrubbing the grime from the glass cases, noting contents, and grease-penciling numbers; Lattimer and a couple of Space Force officers were going through what had been the administrative offices on the other side. It was one of these, a young second lieutenant, who came hurrying in from the mezzanine, almost bursting with excitement.

"Hey, Martha! Dr. von Ohlmhorst!" he was shouting. "Where are you? Tony's found the Martians!"

Selim dropped his rag back in the bucket; she laid her clipboard on top of the case beside her.

"Where?" they asked together.

"Over on the north side." The lieutenant took hold of himself and spoke more deliberately. "Little room, back of one of the old faculty offices—conference room. It was locked from the inside, and we had to burn it down with a torch. That's where they are. Eighteen of them, around a long table—"

Gloria Standish, who had dropped in for lunch, was on the mezzanine, fairly screaming into a radiophone extension:

" ... Dozen and a half of them! Well, of course they're dead. What a question! They look like skeletons covered with leather. No, I do not know what they died of. Well, forget it; I don't care if Bill Chandler's found a three-headed hippopotamus. Sid, don't you get it? We've found the *Martians*!"

She slammed the phone back on its hook, rushing away ahead of them.

Martha remembered the closed door; on the first survey, they hadn't attempted opening it. Now it was burned away at both sides and lay, still hot along the edges, on the floor of the big office room in front. A floodlight was on in the room inside, and Lattimer was going around looking at things while a Space Force officer stood by the door. The center of the room was filled by a long table; in armchairs around it sat the eighteen men and women who had occupied the room for the last fifty millennia. There were bottles and glasses on the table in front of them, and, had she seen them in a dimmer light, she would have thought that they were merely dozing over their drinks. One had a knee hooked over his chair-arm and was curled in foetuslike sleep. Another had fallen forward onto the table, arms extended, the emerald set of a ring twinkling dully on one finger. Skeletons covered with leather, Gloria Standish had called them, and so they were—faces like skulls, arms and legs like sticks, the flesh shrunken onto the bones under it.

"Isn't this something!" Lattimer was exulting. "Mass suicide, that's what it was. Notice what's in the corners?"

Braziers, made of perforated two-gallon-odd metal cans, the white walls smudged with smoke above them. Von Ohlmhorst had noticed them at once, and was poking into one of them with his flashlight.

"Yes; charcoal. I noticed a quantity of it around a couple of hand-forges in the shop on the first floor. That's why you had so much trouble breaking in; they'd sealed the room on the inside." He straightened and went around the room, until he found a ventilator, and peered into it. "Stuffed with rags. They must have been all that were left, here. Their power was gone, and they were old and tired, and all around them their world was dying. So they just came in here and lit the charcoal, and sat drinking together till they all fell asleep. Well, we know what became of them, now, anyhow."

Sid and Gloria made the most of it. The Terran public wanted to hear about Martians, and if live Martians couldn't be found, a room full of dead ones was the next best thing. Maybe an even better thing; it had been only sixty-odd years since the Orson Welles invasion-scare. Tony Lattimer, the discoverer, was beginning to cash in on his attentions to Gloria and his ingratiation with Sid; he was always either making voice-and-image talks for telecast or listening to the news from the home planet. Without question, he had become, overnight, the most widely known archaeologist in history.

"Not that I'm interested in all this, for myself," he disclaimed, after listening to the telecast from Terra two days after his discovery. "But this is going to be a big thing for Martian archaeology. Bring it to the public attention; dramatize it. Selim, can you remember when Lord Carnarvon and Howard Carter found the tomb of Tutankhamen?"

"In 1923? I was two years old, then," von Ohlmhorst chuckled. "I really don't know how much that publicity ever did for Egyptology. Oh, the museums did devote more space to Egyptian exhibits, and after a museum department head gets a few extra showcases, you know how hard it is to make him give them up. And, for a while, it was easier to get financial support for new excavations. But I don't know how much good all this public excitement really does, in the long run."

"Well, I think one of us should go back on the *Cyrano*, when the *Schiaparelli* orbits in," Lattimer said. "I'd hoped it would be you; your voice would carry the most weight. But I think it's important that one of us go back, to present the story of our work, and what we have accomplished and what we hope to accomplish, to the public and to the universities and the learned societies, and to the Federation Government. There will be a great deal of work that will have to be done. We must not allow the other scientific fields and the so-called practical interests to monopolize public and academic support. So, I believe I shall go back at least for a while, and see what I can do—"

Lectures. The organization of a Society of Martian Archaeology, with Anthony Lattimer, Ph.D., the logical candidate for the chair. Degrees, honors; the deference of the learned, and the adulation of the lay public. Positions, with impressive titles and $alaries. Sweet are the uses of publicity.

She crushed out her cigarette and got to her feet. "Well, I still have the final lists of what we found in *Halvhulva*—Biology—department to check over. I'm starting on Sornhulva tomorrow, and I want that stuff in shape for expert evaluation."

That was the sort of thing Tony Lattimer wanted to get away from, the detail-work and the drudgery. Let the infantry do the slogging through the mud; the brass-hats got the medals.

She was halfway through the fifth floor, a week later, and was having midday lunch in the reading room on the first floor when Hubert Penrose came over and sat down beside her, asking her what she was doing. She told him.

"I wonder if you could find me a couple of men, for an hour or so," she added. "I'm stopped by a couple of jammed doors at the central hall. Lecture room and library, if the layout of that floor's anything like the ones below it."

"Yes. I'm a pretty fair door-buster, myself." He looked around the room. "There's Jeff Miles; he isn't doing much of anything. And we'll put Sid Chamberlain to work, for a change, too. The four of us ought to get your doors open." He called to Chamberlain, who was carrying his tray over to the dish washer. "Oh, Sid; you doing anything for the next hour or so?"

"I was going up to the fourth floor, to see what Tony's doing."

"Forget it. Tony's bagged his season limit of Martians. I'm going to help Martha bust in a couple of doors; we'll probably find a whole cemetery full of Martians."

Chamberlain shrugged. "Why not. A jammed door can have anything back of it, and I know what Tony's doing—just routine stuff."

Jeff Miles, the Space Force captain, came over, accompanied by one of the lab-crew from the ship who had come down on the rocket the day before.

"This ought to be up your alley, Mort," he was saying to his companion. "Chemistry and physics department. Want to come along?"

The lab man, Mort Tranter, was willing. Seeing the sights was what he'd come down from the ship for. She finished her coffee and cigarette, and they went out into the hall together, gathered equipment and rode the elevator to the fifth floor.

The lecture hall door was the nearest; they attacked it first. With proper equipment and help, it was no problem and in ten minutes they had it open wide enough to squeeze through with the floodlights. The room inside was quite empty, and, like most of the rooms behind closed doors, comparatively free from dust. The students, it appeared, had sat with their backs to the door, facing a low platform, but their seats and the lecturer's table and equipment had been removed. The two side walls bore inscriptions: on the right, a pattern of concentric circles which she recognized as a diagram of atomic structure, and on the left a complicated table of numbers and words, in two columns. Tranter was pointing at the diagram on the right.

"They got as far as the Bohr atom, anyhow," he said. "Well, not quite. They knew about electron shells, but they have the nucleus pictured as a solid mass. No indication of proton-and-neutron structure. I'll bet, when you come to translate their scientific books, you'll find that they taught that the atom was the ultimate and indivisible particle. That explains why you people never found any evidence that the Martians used nuclear energy."

"That's a uranium atom," Captain Miles mentioned.

"It is?" Sid Chamberlain asked, excitedly. "Then they did know about atomic energy. Just because we haven't found any pictures of A-bomb mushrooms doesn't mean—"

She turned to look at the other wall. Sid's signal reactions were getting away from him again; uranium meant nuclear power to him, and the two words were interchangeable. As she studied the arrangement of the numbers and words, she could hear Tranter saying:

"Nuts, Sid. We knew about uranium a long time before anybody found out what could be done with it. Uranium was discovered on Terra in 1789, by Klaproth."

There was something familiar about the table on the left wall. She tried to remember what she had been taught in school about physics, and what she had picked up by accident afterward. The second column was a continuation of the first: there were forty-six items in each, each item numbered consecutively—

"Probably used uranium because it's the largest of the natural atoms," Penrose was saying. "The fact that there's nothing beyond it there shows that they hadn't created any of the transuranics. A student could go to that thing and point out the outer electron of any of the ninety-two elements."

Ninety-two! That was it; there were ninety-two items in the table on the left wall! Hydrogen was Number One, she knew; One, *Sarfaldsorn*. Helium was Two; that was *Tirfaldsorn*. She couldn't remember which element came next, but in Martian it was *Sarfalddavas*. *Sorn* must mean matter, or substance, then. And *davas;* she was trying to think of what it could be. She turned quickly to the others, catching hold of Hubert Penrose's arm with one hand and waving her clipboard with the other.

"Look at this thing, over here," she was clamoring excitedly. "Tell me what you think it is. Could it be a table of the elements?"

They all turned to look. Mort Tranter stared at it for a moment.

"Could be. If I only knew what those squiggles meant—"

That was right; he'd spent his time aboard the ship.

"If you could read the numbers, would that help?" she asked, beginning to set down the Arabic digits and their Martian equivalents. "It's decimal system, the same as we use."

"Sure. If that's a table of elements, all I'd need would be the numbers. Thanks," he added as she tore off the sheet and gave it to him.

Penrose knew the numbers, and was ahead of him. "Ninety-two items, numbered consecutively. The first number would be the atomic number. Then a single word, the name of the element. Then the atomic weight—"

She began reading off the names of the elements. "I know hydrogen and helium; what's *tirfalddavas*, the third one?"*

* Transcriber's Note: There's an uncorrected error here: the third element is "Sarfalddavas" above but "tirfalddavas" below. This wasn't changed in later versions of the text, so it's not clear which is correct. The capitalized one would be consistent with the other names, and the "Sar-" implies a relationship with the first element, which *is* in the same group. Or maybe the character misspoke, in her excitement?

"Lithium," Tranter said. "The atomic weights aren't run out past the decimal point. Hydrogen's one plus, if that double-hook dingus is a plus sign; Helium's four-plus, that's right. And lithium's given as seven, that isn't right. It's six-point nine-four-oh. Or is that thing a Martian minus sign?"

"Of course! Look! A plus sign is a hook, to hang things together; a minus sign is a knife, to cut something off from something—see, the little loop is the handle and the long pointed loop is the blade. Stylized, of course, but that's what it is. And the fourth element, kiradavas; what's that?"

"Beryllium. Atomic weight given as nine-and-a-hook; actually it's nine-point-oh-two."

Sid Chamberlain had been disgruntled because he couldn't get a story about the Martians having developed atomic energy. It took him a few minutes to understand the newest development, but finally it dawned on him.

"Hey! You're reading that!" he cried. "You're reading Martian!"

"That's right," Penrose told him. "Just reading it right off. I don't get the two items after the atomic weight, though. They look like months of the Martian calendar. What ought they to be, Mort?"

Tranter hesitated. "Well, the next information after the atomic weight ought to be the period and group numbers. But those are words."

"What would the numbers be for the first one, hydrogen?"

"Period One, Group One. One electron shell, one electron in the outer shell," Tranter told her. "Helium's period one, too, but it has the outer—only—electron shell full, so it's in the group of inert elements."

"*Trav, Trav. Trav's* the first month of the year. And helium's *Trav, Yenth; Yenth* is the eighth month."

"The inert elements could be called Group Eight, yes. And the third element, lithium, is Period Two, Group One. That check?"

"It certainly does. *Sanv, Trav; Sanv's* the second month. What's the first element in Period Three?"

"Sodium. Number Eleven."

That's right; it's *Krav, Trav*. Why, the names of the months are simply numbers, one to ten, spelled out.

"*Doma's* the fifth month. That was your first Martian word, Martha," Penrose told her. "The word for five. And if *davas* is the word for metal, and *sornhulva* is chemistry and / or physics, I'll bet Tadavas Sornhulva is literally translated as: Of-Metal Matter-Knowledge. Metallurgy, in other words. I wonder what Mastharnorvod means." It surprised her that, after so long and with so much happening in the meantime, he could remember that. "Something like 'Journal,' or 'Review,' or maybe 'Quarterly.'"

"We'll work that out, too," she said confidently. After this, nothing seemed impossible. "Maybe we can find—" Then she stopped short. "You said 'Quarterly.' I think it was 'Monthly,' instead. It was dated for a specific month, the fifth one. And if *nor* is ten, Mastharnorvod could be 'Year-Tenth.' And I'll bet we'll find that *masthar* is the word for year." She looked at the table on the wall again. "Well, let's get all these words down, with translations for as many as we can."

"Let's take a break for a minute," Penrose suggested, getting out his cigarettes. "And then, let's do this in comfort. Jeff, suppose you and Sid go across the hall and see what you find in the other room in the way of a desk or something like that, and a few chairs. There'll be a lot of work to do on this."

Sid Chamberlain had been squirming as though he were afflicted with ants, trying to contain himself. Now he let go with an excited jabber.

"This is really it! *The* it, not just it-of-the-week, like finding the reservoirs or those statues or this building, or even the animals and the dead Martians! Wait till Selim and Tony see this! Wait till Tony sees it; I want to see his face! And when I get this on telecast, all Terra's going to go nuts about it!" He turned to Captain Miles. "Jeff, suppose you take a look at that other door, while I find somebody to send to tell Selim and Tony. And Gloria; wait till she sees this—"

"Take it easy, Sid," Martha cautioned. "You'd better let me have a look at your script, before you go too far overboard on the telecast. This is just a beginning; it'll take years and years before we're able to read any of those books downstairs."

"It'll go faster than you think, Martha," Hubert Penrose told her. "We'll all work on it, and we'll teleprint material to Terra, and people there will work on it. We'll send them everything we can ... everything we work out, and copies of books, and copies of your word-lists—"

And there would be other tables—astronomical tables, tables in physics and mechanics, for instance—in which words and numbers were equivalent. The library stacks, below, would be full of them. Transliterate them into Roman alphabet spellings and Arabic numerals, and somewhere, somebody would spot each numerical significance, as Hubert Penrose and Mort Tranter and she had done with the table of elements. And pick out all the chemistry textbooks in the Library; new words would take on meaning from contexts in which the names of elements appeared. She'd have to start studying chemistry and physics, herself—

Sachiko Koremitsu peeped in through the door, then stepped inside.

"Is there anything I can do—?" she began. "What's happened? Something important?"

"Important?" Sid Chamberlain exploded. "Look at that, Sachi! We're reading it! Martha's found out how to read Martian!" He grabbed Captain Miles by the arm. "Come on, Jeff; let's go. I want to call the others—" He was still babbling as he hurried from the room.

Sachi looked at the inscription. "Is it true?" she asked, and then, before Martha could more than begin to explain, flung her arms around her. "Oh, it really is! You are reading it! I'm so happy!"

She had to start explaining again when Selim von Ohlmhorst entered. This time, she was able to finish.

"But, Martha, can you be really sure? You know, by now, that learning to read this language is as important to me as it is to you, but how can you be so sure that those words really mean things like hydrogen and helium and boron and oxygen? How do you know that their table of elements was anything like ours?"

Tranter and Penrose and Sachiko all looked at him in amazement.

"That isn't just the Martian table of elements; that's *the* table of elements. It's the only one there is." Mort Tranter almost exploded. "Look, hydrogen has one proton and one electron. If it had more of either, it wouldn't be hydrogen, it'd be something else. And the same with all the rest of the elements. And hydrogen on Mars is the same as hydrogen on Terra, or on Alpha Centauri, or in the next galaxy—"

"You just set up those numbers, in that order, and any first-year chemistry student could tell you what elements they represented." Penrose said. "Could if he expected to make a passing grade, that is."

The old man shook his head slowly, smiling. "I'm afraid I wouldn't make a passing grade. I didn't know, or at least didn't realize, that. One of the things I'm going to place an order for, to be brought on the *Schiaparelli*, will be a set of primers in chemistry and physics, of the sort intended for a bright child of ten or twelve. It seems that a Martiologist has to learn a lot of things the Hittites and the Assyrians never heard about."

Tony Lattimer, coming in, caught the last part of the explanation. He looked quickly at the walls and, having found out just what had happened, advanced and caught Martha by the hand.

"You really did it, Martha! You found your bilingual! I never believed that it would be possible; let me congratulate you!"

He probably expected that to erase all the jibes and sneers of the past. If he did, he could have it that way. His friendship would mean as little to her as his derision—except that his friends had to watch their backs and his knife. But he was going home on the *Cyrano*, to be a big shot. Or had this changed his mind for him again?

"This is something we can show the world, to justify any expenditure of time and money on Martian archaeological work. When I get back to Terra, I'll see that you're given full credit for this achievement—"

On Terra, her back and his knife would be out of her watchfulness.

"We won't need to wait that long," Hubert Penrose told him dryly. "I'm sending off an official report, tomorrow; you can be sure Dr. Dane will be given full credit, not only for this but for her previous work, which made it possible to exploit this discovery."

"And you might add, work done in spite of the doubts and discouragements of her colleagues," Selim von Ohlmhorst said. "To which I am ashamed to have to confess my own share."

"You said we had to find a bilingual," she said. "You were right, too."

"This is better than a bilingual, Martha," Hubert Penrose said. "Physical science expresses universal facts; necessarily it is a universal language. Heretofore archaeologists have dealt only with pre-scientific cultures."

MARS ON EARTH (2008)

SOIL ANALOGUES FOR FUTURE MARS MISSIONS

Jeffrey J. Marlow, Zita Martins, Mark A. Sephton

Jeffrey J. Marlow is an Assistant Professor of Biology at Boston University; **Zita Martins** is an Associate Professor at the Instituto Superior Técnico, Portugal; **Mark A. Sephton** is Professor of Organic Geochemistry at Imperial College London.

Mars has long fascinated humankind because of its potential as a host for alien life.[1–3] Most of our accumulated knowledge of Mars comes from ground and space-based observations,[4–7] martian meteorites such as Nakhla and Allan Hills (ALH) 84001[8–15] and spacecraft landers.[16–18] Reaching and landing safely on the Red Planet has proven to be a challenging task despite the success of recent missions (e.g. the Mars Exploration Rovers).[19] All space missions are time-consuming and expensive and as we look forward to continued exploration of our planetary neighbour it is of the utmost importance that all technical components and experimental procedures for Mars missions function properly once *in situ*.

In order to test rover instrumentation before mission launch and improve the chances of success, experimental analyses of terrestrial soils similar to those that will be encountered on the Red Planet have become crucial.[20–1] Mars soil analogues provide a preview of the physical environment that a mission to the Red Planet may encounter. They simulate both the type of soils a rover will drive over and the materials that on-board instruments will sample. Mars soil analogues also help to mimic and predict the materials in which trace levels of organic molecules (and possibly life) might be found, particularly in light of recent findings that certain minerals appear to preserve organics better than others under simulated martian conditions. Here we describe several Mars soil analogues being used to help planetary scientists prepare for *in situ* investigations on Mars.

Not just a Red Planet

Mars, like Earth, is a geologically diverse object and the traditional view of it as an exclusively basaltic sandbox is becoming ever more refined, as evidenced by recent discoveries of varied mineralogies at regional and local scales. The OMEGA hyperspectral imager on Mars Express identified layered deposits exhibiting high levels of kieserite, gypsum and polyhydrated sulphates at multiple sites in Valles Marineris, Margaritifer Sinus and Terra Meridiani.[22] On a smaller scale, the Spirit rover has found evidence for six different soil types in Gusev Crater and the Columbia Hills, suggesting a range of formation conditions and alteration mechanisms.[23] Because of this heterogeneity, it is important to establish which part of Mars one is hoping to simulate and then to identify suitable soil analogue candidates that will benefit preparations for successful space missions. For example, a spectrometer prototype for a mission to the gypsum-rich dunes of Olympia Planitia[24] would require a different chemical analogue than a similar mission targeting the basaltic sands of Chyrse Planitia.[25] Likewise, a landing systems engineer would seek a different physical analogue for a polar mission depending on whether the mission would land on hardened winter ground or softer summer soils. The robotic armada currently orbiting Mars gives us the ability to identify bulk characteristics of proposed mission targets; using this knowledge to specify and use a suitable analogue on Earth will serve to prepare mission planners better for both the physical and scientific ground to be covered.

Types of Mars Analogues

Analogue soils can be classified by the properties of Mars they best mimic, allowing mission planners with specific engineering-based aims to target particular analogues by their properties of interest. These analogues can also be used in a predictive manner to anticipate scientific findings at the destination. For example, understanding the development of a chemically analogous soil on Earth allows us to hypothesize how similar geological processes may occur on Mars. Knowing the concentrations and sources of biological molecules in an analogue soil informs our expectations of the search for past or present life on Mars. Useful analogue classes include the following:

- **Chemical analogues.** These include terrestrial soils that are as similar as possible to martian regolith in terms of chemical properties such as dielectric constant, redox potential, pH, elemental composition, and mineralogical composition. This class of analogue is useful for testing and calibrating spectrometers as well as testing procedures that aim to interpret conditions of soil formation on Mars.
- **Mechanical analogues.** This class is represented by soils and sites that exhibit similar mechanical properties of the martian regolith, such as soil bearing strength, cohesive strength, and the angle of internal friction. These materials are helpful for input to overall rover design and landing systems, which can also be tested and improved using mechanical analogues.

- **Physical analogues.** These soils comprise materials that are Mars-like in their physical properties, such as particle size distribution, particle shape, density, bulk density, porosity, water content, and thermophysical properties (e.g. albedo and thermal inertia). Physical analogues allow us to evaluate the physical effect of martian soil on mechanical components such as spacesuit joints, robotic hinges, and soil intake machinery. Laboratory use of these materials could also help clarify past and present interactions between water and the soil.
- **Magnetic analogues.** These are materials with Mars-like magnetic properties including magnetic susceptibility and saturation magnetization. These soils are particularly valuable in testing magnetism-related instruments.
- **Organic analogues.** Near-barren soils are specifically useful in simulating the low-organic content of martian soils. The search for signs of past and/or present life on Mars is one of the most important goals for Mars exploration, and several life-detection instruments will be part of upcoming Mars missions.[26–8] Testing proposed instruments for detection of trace amounts of molecules of biological interest allows scientists and engineers to evaluate instrument sensitivity and functionality in a field environment.

Mars-like Terrestrial Locations

The value of Mars analogues has long been appreciated, and the scientific and engineering communities have trodden well-worn paths to several sites in the quest for a "Mars on Earth" regolith. Descriptions of selected analogues and their locations follow:

- **Hawaii.** Hawaii and Mars both exhibit a history of volcanism, and the near-continuous eruptions of Kalauea Volcano provide the opportunity for real-time study of lava formation and alteration.[29] The volcanic deposits of the Ka'u Desert, Kilauea, Mauna Loa and Mauna Kea provide a range of useful chemical analogues. Spectral signatures of the basaltic material show significant similarities to high albedo regions on Mars.[30] Weathered ash from the Pu'u Nene cinder cone on Hawaii is the source for JSC Mars-1, a martian soil simulant collected and characterized by scientists and engineers at Johnson Space Center in 1993.[31–2] The study of Hawaiian geology and martian climate led some researchers to predict the presence of kaolinite on Mars,[33] a forecast that has recently been shown to be true.[34]
- **Salten Skov.** The red-coloured sediments of Salten Skov in central Denmark contain high concentrations of iron oxides, especially hematite, maghemite and goethite.[35] Samples from this site are useful both as a magnetic analogue and a chemical analogue for biological applications. Bacterial concentrations in this soil are too high to test life-detection instruments properly, but because of their analogous mineralogical properties, Salten Skov soils can be used to examine the breakdown of cellular components under simulated martian conditions.[36] Research has shown that amino acids and other organic molecules present in this soil are degraded under simulated martian conditions due to the oxidizing

nature of the material, as is hypothesized to take place on Mars.[37] Understanding the mechanism of organic molecule destruction will inform the search for signs of past or present life on Mars and allow for more accurate interpretation of future findings.

- **Atacama Desert.** The recently heightened pace of the search for past or present life on Mars has led to the identification of an organic analogue soil—a material that contains organic molecules in trace quantities as expected on Mars. Given the tenacity of extremophilic organisms, this has not been an easy task, but researchers have recently proposed the Atacama Desert—the driest place on Earth—as a suitably "sterile" environment. Studies of the regional soil chemistry[38–9] suggest that highly oxidizing conditions, possibly resulting from elevated nitrate levels and the dry-deposition of acids, likely account for the low organic content.[40–2] Recently, scientists developing life-detection instruments have used this arid region almost exclusively as an organic Mars soil analogue.[43–6]
- **Mojave Desert.** The Mojave Desert of eastern California exhibits a history of volcanism and tectonic activity, making it a suitable analogue to the highlands of Mars.[47] Subsequent weathering has produced physical conditions and chemical compositions similar to those on Mars.[48] However, the region is used principally as the mechanical analogue of choice and *de facto* testing ground of Mars landing systems and rovers.[49–51] This area is logistically attractive due to its proximity to the Jet Propulsion Laboratory, where many NASA landers are built, a convenience that minimizes transportation costs and overall risks. The recent production of the Mojave martian simulant from crushed granular basalt aims to overcome some weaknesses of the JSC Mars-1 soil, namely its hygroscopic properties and large volatile composition.[52]
- **Arequipa.** This high desert site in Southern Peru serves as a chemical and organic martian analogue because of its sulphate mineralogy and a low organic content on a par with Atacama samples. Geochemical parameters such as pH and redox vary over metre-length scales. Recent studies have shown significant rates of amino acid destruction in Arequipa soils under Mars-like conditions, an observation attributed to soil mineralogy that can produce highly oxidizing conditions.[53] Similar reactions on Mars may account for Viking's inability to detect organic molecules.[54] Continued study of these soils may demonstrate how organic molecules, from either biological or meteoritic sources, react and persist on Mars.
- **Rio Tinto.** The analogues highlighted above have traditionally been used to simulate a homogenized "bulk Mars", irrespective of the local anomalies a given mission may target. Recent use of the Rio Tinto region in southwestern Spain, however, represents a new step in the evolution of Mars analogue studies, namely the use of a site on Earth that is targeted because of its Mars-like inorganic geochemistry. The Rio Tinto is a red-coloured river with a pH approaching 1 and a thriving microbial community.[55] It is a poor match for martian physical or mechanical properties, but the acid-sulphate chemistry provides a helpful model for the development of certain minerals in martian soil observed by Opportunity such as jarosite and hematite.[56–7] Working at Rio Tinto allows engineers to fine tune mineralogy-based instruments[58] and gives scientists a context in which to interpret rover data.

Meteorites

Meteorites can serve as analogues to determine how organic molecules respond to conditions on Mars in two different capacities: the direct study of martian meteorites, and the investigation of carbonaceous chondrite meteorites under simulated martian conditions.

In the absence of the products of a sample return mission to the Red Planet, the 34 currently known Mars meteorites (http://www2.jpl.nasa.gov/snc/) represent the only actual pieces of Mars on Earth. Despite original claims of fossil-like content in the best-known of these meteorites, ALH84001,[59] no martian biological molecules were found.[60–2] Nonetheless, the search for native organic molecules in martian meteorites continues[63–5] and, if found, such molecules would play a significant role in directing future missions.

Exposing carbonaceous meteorites with previously established amino acid contents to martian conditions can also help inform the search for organic molecules on Mars. Meteorites represent a near-certain source of organic molecules on Mars, but their ability to persist on the surface remains unresolved. Exposing carbonaceous meteorites with previously established amino acid contents to martian conditions can also help constrain the search for organics. The Orgueil meteorite, for example, is known to have a simple amino acid distribution, with glycine and b-alanine present in the largest amounts.[66] When placed in a simulated martian environment, Orgueil amino acids deteriorated more quickly than those hosted in the Salten Skov Mars analogue soil. This discrepancy is attributed to mineralogical differences: the clay-rich sediments of Salten Skov appear to slow the radical-induced destruction of amino acids better than the Orgueil meteorite.[67] The use of meteorites in this comparative capacity helps point to mineralogical hotspots on Mars that should be targeted by future life-detection missions.

Towards a Comprehensive Catalogue

Analysis of Mars soil analogues is a crucial part of our preparations for exploration of the Red Planet, offering a relatively cheap and safe opportunity to test mission hardware and study potentially Mars-like scientific processes. Yet, despite the interdisciplinary importance of high-fidelity Mars soil analogues, centralized data on soil parameters and analogue uses are in short supply and are inconsistently acquired. Mineralogical information, for example, can be acquired by several different analytical methods including thermal emission spectroscopy, X-ray diffraction, Mossbauer spectroscopy, or alpha particle X-ray spectroscopy. Data obtained decades ago may be obsolete given the technological advancements of analytical instruments in the intervening years. One of the current analogues of choice, JSC Mars-1, was enthroned largely due to spectral data available in the early 1990s. This basis for comparison is elementary by modern standards; incorporating findings from contemporary rovers and hyperspectral orbiters could point to more promising analogues. Even when multiple types of information for a site are available, they often come from studies examining different locations at different times with different instruments. Disparate studies ostensibly sampling the same location could very well be separated by hundreds of miles, dozens of years, several inches of rainfall, etc; in such cases, it is impossible to account for all confounding variables.

The Mars exploration community requires chemical, mechanical, physical, magnetic and organic data, but not all of these parameters are available for each sample. There are many gaps in our inventory of available data. Most notably, field work often fails to survey and/or communicate macroscopic and thermophysical properties. Subjecting analogue soils to a wider range of spectroscopic surveys, particularly those using techniques with a history of operation on Mars (e.g. Mossbauer) would allow scientists and engineers to better cross-reference data from multiple instruments and make more informed inferences about martian destinations. Our summary is a useful starting point, but Mars mission planners and scientists would profit greatly from the ability to access a comprehensive data set of soil analogue properties.

Conclusions

Missions to Mars are too expensive and infrequent and the potential scientific return is too great to leave something to chance. Maintaining a comprehensive and reliable database of soil analogue properties would allow researchers to target specific sites best tailored to their individual requirements. By using soil analogues to their full potential, our continued exploration of Mars can be safer and more scientifically productive.

SCIENCE RESULTS FROM THE MARS EXPLORATION ROVER MISSION (2008)

Steven Squyres

American planetary scientist **Steven Squyres** was born in 1956. He received a BA in Geological Sciences in 1978, and a PhD in Astronomy (Planetary Studies) in 1981, both from Cornell University, New York. He was the Principal Scientific Investigator of NASA's Mars Exploration Rovers. His last academic position was as the James A. Weeks Professor in Physical Sciences at Cornell University. In September 2019, he retired from academia and became the chief scientist at Blue Origin, the aerospace manufacturer and spaceflight services company founded by Jeff Bezos.

Introduction by Claude Canizares
This presentation was given at the 1926th Stated Meeting, held at the House of the Academy on April 9, 2008.

Introduction

It's a great pleasure, indeed an honor, to introduce my friend Steve Squyres. Just over four years ago I had the privilege of being at the Jet Propulsion Laboratory at the California Institute of Technology a few days after New Year's Eve, when the Mars Exploration Rover Mission reached Mars and went through the harrowing and exhilarating process known as entry, descent, and landing. This, of course, was wildly successful and it was the culmination of many years of effort by our speaker tonight, who is the Principal Scientific Investigator of this remarkable project. The two Rovers, *Spirit* and *Opportunity*, that are the scientific core of this mission have vastly exceeded by many times over their original design criteria and have returned an outstanding mother lode of information on Mars and its surface.

Even those of us who have lived through our own space missions recognize that the degree to which this mission has captivated the world is almost unprecedented: We can only marvel at the incredible scale, both scientific and technical, of the achievement. Steve Squyres really stewarded this effort through many years of development. Now, after probably thinking he had only a few years of operations and then a release onto other things, Steve is being called back over and over again to plan the very detailed activities of the two Rovers as they scour the surface of Mars.

In 1977, when Steve was an undergraduate at Cornell, the *Voyager* spacecraft was launched to Jupiter and Saturn. He, probably even as an undergraduate but certainly as a graduate student (also at Cornell), ended up participating in the scientific team for that mission, which became a galvanizing moment for him and set his career toward the planets, both the outer planets but then the inner planets as well. He participated in the Magellan Mission to Venus and the Cassini-Huygens Mission to Saturn; he, too, has touched most of the missions to Mars: the Mars Observer, the Russian Mars '96, Mars Express, Mars Reconnaissance Orbiter, the Mars Odyssey Mission, and, of course, the Mars Exploration Rovers.

CBS News called Steve the Mars Ambassador. Now, I don't think they meant that he himself is a Martian. Rather, he has brought Mars to the Earth, and without any question he is Earth's ambassador to Mars. It gives me great pleasure to welcome Steve Squyres.

Presentation

I face the challenge of trying to compress a combined 3,000 days on the surface of Mars into less than half an hour, so fasten your seatbelts.

The two Mars Rovers, *Spirit* and *Opportunity*, are effectively robotic field geologists. They have a two-part scientific payload. One part does remote sensing, which is supported by a mast with high-resolution color stereo cameras at the top and a Michelson interferometer and infrared spectrometer that live down toward the base; they use mirrors at the top of the mast to get the same view of the countryside as the cameras get. The second part is an arm in the front end of the vehicle, a five degree of freedom robotic manipulator, that includes a microscopic imager, an alpha particle X-ray spectrometer that does elemental chemistry, a Mössbauer spectrometer that tells us about the mineralogy of iron-bearing species, and a device called the RAT, or Rock Abrasion Tool, a diamond-tip tool that grinds away the outer layers of Martian rock and exposes the interior.

Spirit landed in the Gusev Crater. At 160 kilometers in diameter and 16 degrees south latitude on Mars, the Crater was chosen as a landing site because of a large, water-carved channel that empties into it. We went there seeking layered sedimentary rocks laid down long ago on a Martian lake that we believe once filled the crater. After we landed I managed to convince myself for about two days that this was what a Martian dry lake bed should look like, nice and smooth and flat. But when we started to look at the rocks we found that they were not sedimentary rocks at all.

Figure 1.

We named the first rock that we looked at in detail Adirondack (see Figure 1). A Mössbauer spectrum revealed olivine, pyroxene, and magnetite, among other minerals, in the Adirondack rock. (Olivine and pyroxene are minerals that would be very common in basaltic lava on Earth.) Our infrared spectrometer also revealed olivine and pyroxene, as well as plagioclase, another mineral found in basaltic lava. The mineralogy inferred from elemental chemistry, as derived from the X-ray spectrometer, showed, again, plagioclase, pyroxene, olivine, and a bit of magnetite. All of the instruments tell the same story: Adirondack is a magnetite-bearing olivine basalt. It's an igneous rock that was erupted onto the floor of the Crater, burying whatever sediments were once there. Basically Mars faked us out; this was a disappointment at first.

Our vehicles were designed to last for 90 Martian days and drive 600 meters over their lifetime. When *Spirit* landed, we came to rest 2.5 kilometers from a spectacular range of hills that we named the Columbia Hills, after the *Columbia* space shuttle. Because of the longevity of the vehicle, we were able to get to the Columbia Hills and spend most of the mission there. The first hill that we chose to go after was one that we named Husband Hill, after Rick Husband, who was the commander of *Columbia* when it went down. We climbed over a period of about 400 days to the very summit of Husband Hill, which gives you a sense of the scale of that hill.

I have nowhere near enough time to describe to you the incredibly rich diversity of different geologic materials that are found on Husband Hill and all the geologic stories they tell. I'll tell you just one, drawn from the first rocks that we found as we arrived at Husband Hill, on the portion of it that we called the West Spur. In contrast to what we saw on the plains – massive lavas – we started to see layered rocks, even sub-centimeter layering within the rocks, at the West Spur. Typical rocks from the West Spur are granular, with individual grains within the rock. There's enormous variety in the size of the grains: some are tiny little things that approach the resolution limit of our camera (30 microns per pixel); others are millimeters in size. The combination of small and large grains points toward a very violent, energetic process involved in the formation of these rocks. A gentle process like flowing water or blowing wind tends to have a particular grain size that it transports most effectively, so grains tend to be well-sorted and more or less the same size. A violent process like an explosion will throw out fine and coarse grains all together, resulting in a jumbled-up rock like we found at the West Spur.

We took the composition of these rocks and ratioed them on an element-by-element basis to compare them with the lavas that we saw on the plains. For some of the elements, the composition is fairly similar, but there are a number of elements, notably phosphorous, sulfur, chlorine, and bromine – elements that tend to be present in salts – that are substantially enriched in this rock from the West Spur. (The rock is also significantly enriched in nickel, to which I'll return later.) Mineralogy from our Mössbauer spectrometer found goethite, an iron oxihydroxide. The hydroxide tells us that water had to be involved in the formation of this mineral.

Taking all of this together, we've come to the conclusion that these rocks are impact ejecta that have been altered by water. When an impactor from space comes in, hits the surface, creates an explosion, and throws a bunch of stuff in the air, it all falls out at once, with some of the impactor itself mixed in. Impactors tend to be rich in nickel; we think that's where the nickel comes from. And then there's clear evidence that water altered the rock: the presence of goethite and deposited salts (which produced the sulfates), phosphates, and chlorides makes this obvious.

After time in the West Spur, we climbed all the way to the summit of Husband Hill and came down off the summit to a place called Home Plate, where we've been for a while. Home Plate is a plateau of layered volcanic rocks that is about two or three meters high and about 80 or 90 meters across. Right now we are on the north side of it, our solar rays tilted toward the north with the sun low in the northern sky, riding out our third winter on Mars. We hope to explore more with *Spirit* when springtime comes.

The right front wheel of *Spirit* no longer turns; it died about 800 days into the mission. The other five wheels work fine, but in order to drive the vehicle we have to drive it backward, dragging the broken wheel through the soil. While this does make *Spirit* hard to drive, it digs a trench, hundreds of meters long, through the Martian soil, turning up something wonderful every so often.

Opportunity came to rest in an impact crater we named Eagle Crater. We spent 60 Martian days there and then drove over to a much larger crater called Endurance Crater and spent a couple of hundred Martian days exploring there, including deep down into the crater. The *Opportunity* landing site was chosen not because of its topography but because of its chemistry. Data from an infrared spectrometer – the thermal emissions spectrometer that was in orbit around Mars on the Mars Global Surveyor spacecraft – show not only basaltic lava at the *Opportunity* landing site, but also hematite. Hematite, an iron oxide, is a mineral that sometimes forms as a consequence of the action of liquid water.

The rocks at Meridiani Planum, near where *Opportunity* landed, are all made of the same materials: they are sandstones, composed of sand-sized grains that are extremely rich in sulfate salts. Embedded within them are little round spherules, things that we've come to call 'blueberries,' which turn out to be extremely rich in hematite. By mass, sulfate salts account for roughly 40 percent of the rock: 20 percent magnesium sulfate, 10 percent calcium sulfate, and 10 percent of an iron sulfate called jarosite.

When we mapped the composition with our infrared spectrometer we found a lot of sand made of basalt. The soils in many places are very rich in these hematite blueberries, but everywhere you have bedrock exposed, the rock is sulfate rich. The mineralogy derived from the infrared spectrum tells the same story as elemental chemistry: 10 percent jarosite, 20 percent magnesium sulfate, 10 percent calcium sulfate. The Mössbauer spectrometer sees only iron-bearing minerals, so it shows the jarosite as well. You need water to make this jarosite, a particularly environmentally informative mineral because it only forms at low pH. The pH has to be less than about four or five to form jarosite, and, on Earth, jarosite typically forms around a pH of three or two. This helps to make it clear that when people talk about the Meridiani Planum and the presence of water on Mars, they should more accurately be talking about sulfuric acid on Mars.

At Endurance Crater, the larger of the two craters where we took *Opportunity*, we drilled with our rock abrasion tool a total of eleven rat holes over a stratographic distance of about seven meters, working our way down into the crater. This is the first stratigraphic section ever put together on another planet. We saw some substantial changes in the nature of the rock as we went down. Toward the surface the rock preserved the laminations very nicely in the original layering in sandstone. But when we got deeper in the crater that changed completely. The layering goes away, replaced by a lumpy texture, which, we believe, is a consequence of recrystalization. These are soluble rocks; magnesium sulfate in particular is highly soluble in water. If these rocks are soaked in water for long enough, recrystalization occurs, destroying the original textures. That is what you see deep in the crater.

The chemistry changes as you go down-section as well. As you get deeper the chlorine increases sharply (precipitation of chloride salts is taking place below a certain level), but both sulfur and magnesium decrease. They follow each other beautifully, which tells us that the compound made of mostly magnesium and sulfur – magnesium sulfate – is the soluble material that gets dissolved away. The point at which the texture changes is the same point of depth below which the chemistry begins to change.

We found a place that we called the Berry Bowl, where a bunch of the so-called blueberries have come together (see Figure 2). We measured their composition and found them to be at least 50 percent hematite by mass – probably closer to 70 or 80 percent actually. We have concluded that they are concretions, which, on Earth, form in sedimentary rocks that are saturated in water. With a concretion, some mineral wants to precipitate out, so it finds a nucleation point and starts to solidify. It adds layer upon layer upon layer, growing a hard, spherical nodule (sort of like the way an oyster builds a pearl), which is dispersed through

Figure 2.

the rock. One interesting thing about a concretion is that as it grows, it draws fluid from a body of fluid around it within the rock and carves out a space for itself. This volume it creates for itself within the rock means that statistical analysis of the special or the volumetric distribution of concretions within a rock reveals a distribution that is not a poisson distribution, not spatially random but more uniform, because the concretions space themselves out; nearest-neighbor statistics of their distribution confirm this uniformity.

Informed by data from experiments performed by Dave Rubin of the U.S. Geological Survey at Menlo Park, California, we have also found evidence in a few places that water not only saturated the ground but came to the surface. Rubin found that when water flows over sand that is 40 centimeters across it leaves behind highly sinuous crested ripples, at a scale of five to ten centimeters. Computer simulations (again, the work of Rubin) of a ripple crest propagating downstream show that what gets left behind in the geologic record are concave upward festoons or trough geometry cross-bedding structures, indicating that these ripples propagated downstream at very small scales and proving that water was the fluid that did it. We discovered concave smiley shapes within the rock, up to ten centimeters across, in a number of places, including a rock we named Cornville, which is chock full of these shapes (see Figure 3). So not only did water soak the ground here, but it also occasionally came to the surface.

I would be doing you a grave disservice if I gave you the sense that the science that I just described to you was science that was done by me. I am one member of a team of 170 scientists; I had 57 coinvestigators. It's an extraordinary team of scientists, and I'm very fortunate to be part of it. I also have to give a lot of credit

Figure 3.

to a fabulous team of engineers that built vehicles that were designed to last for 90 days and have lasted more than 1,500 days on the Martian surface. For every one of us who has been part of this mission, it has been, in the very literal sense of the phrase, the adventure of a lifetime.

PHOENIX ON MARS (2010)

THE LATEST SUCCESSFUL LANDING CRAFT HAS MADE NEW DISCOVERIES ABOUT WATER ON THE RED PLANET

Walter Goetz

Walter Goetz is a physicist at the Max Planck Institute for Solar System Research in Katlenberg-Lindau, Germany. In 2002 he received a PhD from the Centre for Planetary Science, University of Copenhagen and has since contributed to a number of Mars exploration missions: he was a member of the team that developed the robotic arm camera for the NASA Phoenix lander; in 2008 he worked with the mission's operations team at the University of Texas, Tucson; and in 2004 he collaborated with NASA's Mars Exploration Rover operations team. More recently, he has helped develop instruments for the European Space Agency's ExoMars rover.

Since we received our first close-up photographs of Mars, when Mariner 4 flew by it in 1965, our nearest neighbor has appeared to be much like our own planet in many ways, but also distinctly different. Mars is about half the size and has about 40 percent of the gravity of Earth, it's at least 55 million kilometers away (depending on the two planets' positions in their orbits), and it currently takes at least nine months to get there. But like Earth, Mars has polar ice caps, clouds in its atmosphere and seasonal weather patterns. It has familiar geological features, such as volcanoes and canyons. However, although there are signs of floods in the ancient past, Mars is now apparently a barren world.

What is the history of liquid water on Mars? Has water ever been stable on its surface (or in its near subsurface) for a geologically significant period of time? Was Mars warm and wet in ancient times? If so, what triggered the apparent change in climate? And could primitive terrestrial life-forms evolve in the present or past Martian environment? These are the main questions that have driven the exploration of Mars since the mid-1960s. In addition, if humans ever tried to travel to, or even set up an outpost on, another planet, Mars would likely be the first choice, so there's even more reason to learn as much as possible about our neighboring planet.

Missions to Mars have been a mix of failure and success. The first working spacecraft to land on the planet's surface were Viking 1 and 2 in the mid-1970s, and they returned the first color images of the planet. They also sent back data long past their planned mission lifetime, until 1982 and 1980, respectively. Their experiments on Martian soil, looking for signs of microscopic life, were inconclusive. More than a decade later, a mission to send an orbiter to Mars ended in failure, but another, Mars Global Surveyor, arrived in 1997 and returned data until October 2006. Also in 1997, the Mars Pathfinder lander, with its Sojourner rover, landed safely and was remarkably successful.

In 1999 the Mars Climate Orbiter and the Mars Polar Lander both failed and were lost upon arrival at Mars. The Mars Surveyor 2001 mission, including an orbiter, a lander and a rover, was canceled in 2000, but its orbiter was repurposed and successfully launched as the 2001 Mars Odyssey orbiter. This orbiter has also relayed information back from the twin rovers, Spirit and Opportunity, which landed in 2004. The European Space Agency (ESA) saw the safe arrival of its orbiter, Mars Express, in 2003, although the lander was lost on deployment. The Mars Reconnaissance Orbiter safely joined Mars's orbit in 2006, providing the highest camera resolution yet.

However, after the Mars Polar Lander crashed and the Mars Surveyor 2001 mission was canceled in 2000, there seemed to be no hope for a new mission to the Martian arctic regions. The situation changed early in 2002 when the Mars Odyssey orbiter discovered large amounts of near-surface hydrogen in exactly these regions. The hydrogen reservoir was interpreted as water ice—less than a meter below the surface. It was argued that such arctic water ice might contain the long-searched for (and long-missed) organic compounds that could signify the presence of life, either past or present.

These discoveries led a group, headed by Peter H. Smith of the University of Arizona, to develop a mission that would build on previous designs and use the already completed, but unused, Mars Surveyor lander. Thus was born the Phoenix Mars Lander, named because like the mythical bird, it had been resurrected from the ashes of its predecessors. The rocket that carried Phoenix was launched on August 4, 2007, and the spacecraft landed safely on May 25, 2008.

Anatomy of a Lander

Phoenix's suite of scientific instruments includes several imaging systems that have different levels of resolution. From lowest to highest resolution, these instruments are its stereo surface imager (SSI), which can show about 1 millimeter per pixel; a robotic arm camera (RAC), with a resolution of more than 24 micrometers per pixel; an optical microscope that can reach about 4 micrometers per pixel; and an atomic force microscope (AFM) that can show about 0.1 micrometers per scan. Our group at the Max Planck Institute for Solar System Research, in collaboration with the University of Arizona, contributed the RAC and the focal-plane assembly of the optical microscope.

The lander also has a wet chemistry laboratory unit (WCL), where it can mix Martian soil in liquid water. The unit consists of four such cells, each designed for a single use. The resulting aqueous solution is analyzed by ion-selective electrodes, which provide information on the compounds in the soil (such as salts) that are soluble in liquid water.

Another important instrument is a thermal analyzer, designed to heat soil samples up to 1,000 degrees Celsius; the evaporating gases are then studied by mass spectroscopy. This instrument group (referred to as the Thermal and Evolved Gas Analyzer, or TEGA) should be able to characterize the inventory of potential organic compounds in the Martian soil by detecting either the parent organic molecules or their thermally generated fragments, as the temperatures where specific gases are released constrain the identity of the parent compound. As the amount of heat is increased at known levels, any additional increase in temperature also reveals phase transitions in the compounds, which can possibly be identified by their enthalpic characteristics.

The spacecraft also has a robotic arm that is in itself a scientific instrument, as it allows the lander to characterize the physical properties of the soil. The scoop at the end of the 2.3-meter-long arm allows controllers to select and transfer specific soil samples to various instruments. An ice drill is mounted to the back-side of the scoop. The robotic arm camera is positioned on the arm so that it can see into the scoop and image the collected soil sample at high resolution. A sensor mounted next to the scoop can measure the soil's electric and thermal conductivity between four needles. An additional sensor can measure the atmospheric water vapor pressure and the relative humidity of the atmosphere.

Figure 1. More than a simple tourist snapshot of itself in an exotic locale, this self-portrait of the Phoenix lander shows a remarkable feat: the safe arrival, descent and landing of the spacecraft onto the surface of Mars, a result that is still not a given for interplanetary missions. Not only that, the lander also carried out a series of studies on water ice, soil chemistry and weather patterns over the course of several months in 2008. Phoenix is a Cinderella story, as the lander was originally constructed for the cancelled 2001 Mars Surveyor mission. Its resurrection and deployment to Mars' polar regions has provided great insights into the water cycle on our neighboring planet. This panorama of the lander, showing its robot arm partially deployed, was taken during the first few days after landing.

A meteorological mast, provided by the Canadian Space Agency, collects data about the Martian weather that help describe how water cycles between the solid and gas phases at the landing site. Its central instrument is a LIDAR (Light Detection And Ranging) that probes the vertical structure of the atmosphere by measuring the travel time of its emitted light as it is backscattered by suspended particles (such as dust and ice) in the air. Pressure and temperature sensors are mounted to the mast at three different heights above the deck (25, 50 and 100 centimeters), and the mast is topped by a Danish-produced weathercock (telltale, or wind indicator). The SSI provides complementary data on the atmospheric dust opacity and water vapor abundance by imaging the solar disk through specific visible and near-infrared filters.

Where It All Happens

Spectroscopic data and high-resolution images from various orbiters (including Mars Global Surveyor, Odyssey, Mars Express and Mars Reconnaissance Orbiter) were available prior to landing and were used extensively to select the best landing site for Phoenix, both in terms of safety and for the best chance of doing useful research.

The distance from the landing site to the northern border of the volcanic Tharsis region is about 500 kilometers and to the nearest volcano (Alba Patera) is about 1,800 kilometers. The north-polar ice cap and the circumpolar dunes are located about 2,000 kilometers north of the lander. On a large scale, we expected volcanic ashes from the Tharsis province as well as sand grains from the north-polar dunes at the site. The landing site is also situated about 20 kilometers west from the Heimdall crater, which has a diameter of 11 kilometers and a depth of about 1 kilometer. Thus ejected soil from these depths may also be found at the landing site.

A few days after landing, the terrain below the spacecraft was examined by the RAC in order to confirm the stability of the spacecraft's position. The first image showed a bright, even surface (called "Holy Cow"; the nomenclature followed fairy-tale themes) that was uncovered by the action of the descent thrusters. Apparently, the subsurface ice discovered by Odyssey in 2002 was right there—only a few centimeters below the surface. The images suggest that this is ice-rich regolith rather than pure water ice. Over the course of about 50 Martian days (or sols), another icy soil patch, dubbed "Snow Queen" and located just next to Holy Cow, developed numerous cracks after it lost its thermally insulating blanket of soil.

During the Phoenix mission, 12 trenches were excavated, the deepest being 18.3 centimeters. The appearance of the subsurface soil was different from trench to trench. In some trenches (such as one dubbed "Dodo Goldilock") almost pure ice was found, as determined by the spectra acquired by the SSI. Other trenches yielded ice-rich regolith, whereas in some no ice was found at all. In the Dodo Goldilock trench, bright centimeter-sized clumps disappeared over the course of four sols. This observation suggests that the bright material in the shallow subsurface is indeed water ice. So far, it is not clear why the ice/regolith mixing ratio varies so much within a few meters.

The surface of the Martian polar environment takes on a hummocky appearance that might account for some of the soil inconsistencies. The basic model for the formation of these "polygons" was developed by Ronald Sletten and his colleagues at the University of Washington. Seasonal contraction and expansion of soil generates wedge-shaped fractures. During winter, fine-grained debris moves into these wedges and prevents them from completely closing again during the next summer. The seasonal stress generated by these processes is relaxed by the formation of mounds (or polygons) at a certain spatial frequency. The net result is a slow cyclic transport of soil material. This erosional process is known as cryoturbation and occurs frequently in terrestrial environments around the edges of glacial regions.

The Heimdall crater formed about 500 million years ago. It seems likely that excavation of this crater contributed soil material to the landing site. Cryoturbation processes, however, take place over a much shorter time scale, continuously renewing the landscape and making the Phoenix landing site the youngest among all other past Martian landing sites (those of Viking, Mars Pathfinder or the Mars Exploration Rovers).

Larger rocks or boulders (bigger than 20 centimeters or so) are absent at the landing site. Water ice is abundant near the surface, in agreement with Odyssey 2002 data. Perhaps the absence of larger rocks can be explained by the high concentration of condensed volatiles (such as water ice) in the subsurface that were affected by the Heimdall impact: A violent explosion would have removed and crushed the rocks that may have been at the landing site initially. A future, systematic study of the correlation between rock density and distance to the nearest crater may provide further understanding of the size distribution of rocks at the site.

The Soil Itself

Microscopic color images from Phoenix demonstrate the great diversity of particles in Martian soil. AFM scans have given three-dimensional representations of dust particles, but it is unclear how typical these particles may be of Martian dust in general. Reddish-orange dust dominates by volume. The individual dust particles cannot be resolved by the microscope and must therefore be on the order of 10 micrometers or less in size. According to a preliminary classification, two different types of grains are present in the soil: Reddish-brownish to colorless grains and dark (almost black) grains. The origin of these grains is uncertain, but a careful comparison to terrestrial analog soils may constrain the potential scenarios for the formation of these grains.

Key Phoenix instruments, such as TEGA and WCL, have provided new insights into the microscopic structure, as well as the mineralogy, of the soil. One cubic centimeter of material transferred to one of the WCL cells was mixed with 25 cubic centimeters of aqueous solution and produced a weak alkaline solution (with a pH of about 8.3) that contained surprisingly large quantifies of perchlorate ($C1O_{4-}$), salts that could lower the freezing point of water and that have the potential to be found in a liquid-water solution under the temperature and pressure conditions on present-day Mars. This ion was by far the dominant anion (or negative ion) in the solution. Among the cations (positive ions) were, in order of decreasing concentration, magnesium, sodium, calcium and potassium.

To extrapolate from these results, one gram of Martian soil might have a perchlorate abundance of about 1 percent by weight. Such a concentration exceeds that found in some terrestrial desert soils by orders of magnitude. Finding chlorine at the highest possible degree of oxidation has significant implications for our understanding of the chemical processes taking place on the Martian surface, as well as in the atmosphere, and raises several important questions: Is the perchlorate just an exotic compound at the Phoenix landing site, or is it widespread on the surface of the planet? Is the chlorine identified by all previous Mars lander missions mostly present as perchlorate? Even the old question of life on Mars must be reformulated: Which types of primitive (terrestrial) life-forms could have evolved in the Martian soil, given the measured perchlorate concentration?

The ion-selective electrode in the WCL unit that is sensitive to perchlorate, and much less so to nitrate, provided such a strong signal that the identification of perchlorate was unambiguous (the mass of nitrate needed to explain the signal would have exceeded the total mass of the analyzed soil sample). Overall, the suite of electrodes used in the WCL provided rich information on the water-soluble components of the soil, but each one is sensitive to a range of different ions at strongly different rates. The conversion of the electrode data into concentrations is therefore non-unique, and constraints from other instruments are needed.

Figure 2. The icy area dubbed "Holy Cow" was swept clean by the lander's descent thrusters. It was imaged by the robotic arm camera, the only on-board instrument able to see beneath the lander; the light rectangle in all four images is the arm's soil sensor. In sunlight, Holy Cow appears bright and reflective *(top left)*, but during the reddish light of twilight, the patch is about as bright as the surrounding soil, indicating that it is not pure ice (*bottom left*). A neighboring region, "Snow Queen," was also uncovered by the lander's thrusters. The region was initially smooth (*top right*) but showed surface fractures after about 50 Martian days (*bottom right, in white circles*), indicating the sublimation of water ice.

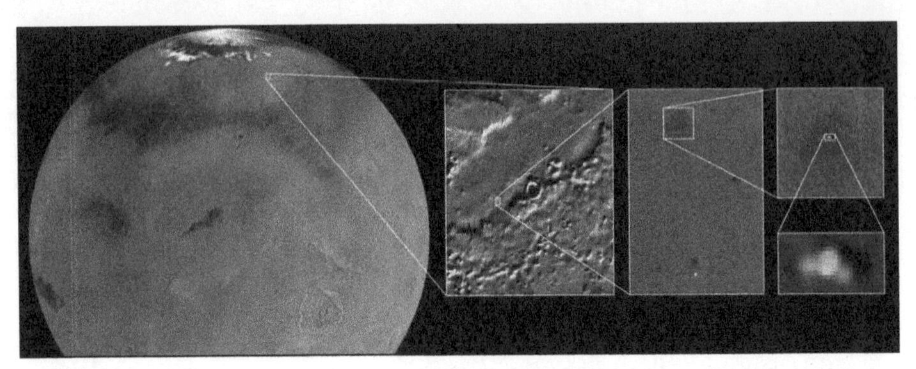

Figure 3. From global to local scale, successively enlarged images zoom in on the Phoenix lander on the Martian surface. The landing site is near the northern polar ice cap of Mars (*far left*). A black-and-white image about 280 meters wide from Mars Global Surveyor shows that the landing site is off to the left of the circular Heimdall crater (*second from left*). A higher resolution orbital image, taken 22 hours after landing by the HiRISE camera aboard the Mars Reconnaissance Orbiter, barely shows the lander; the black dot at the middle right of the image is the ejected heat shield, and the bright dot near the bottom center is the parachute (*middle*). An enlargement of the middle image (*top right*) shows surface roughness and coarser-grained, darker material that was exposed by the descent thrusters. A final enlargement (*bottom right*) shows the lander deck and two solar panels, at a resolution of about 33 centimeters per pixel.

Thermal decomposition products of perchlorates were generally not detected by the thermal analyzer or the evaporated gas analyzer's mass spectra, although this does not jeopardize the WCL identification of perchlorate. In at least one of the samples analyzed by these instruments (from a site called "Baby Bear"), some oxygen release in molecular form was observed that may be due to the decomposition of perchlorate.

Another important finding from the mass-spectroscopy studies was the release of carbon dioxide at temperatures of 800 to 900 degrees, which indicates the presence of 3 to 5 percent by weight of calcium carbonate in the soil. The result is remarkable, given that we have been on the hunt for this mineral for many years, and Phoenix found it in the soil! Carbonates are generally the products of aqueous processes, and thus their presence maybe indicative of liquid water on the surface of Mars at some point in the planet's history. The inferred presence of carbonates is also compatible with WCL results and explains the alkaline pH of the aqueous solutions.

Furthermore, the absence of certain gases after heating can provide critical information on the mineralogy of the Martian soil: No sulfur dioxide has been released over the entire temperature range (from below 0 degrees, up to 1,000 degrees). This is surprising as all previous lander missions have identified substantial quantities of sulfur in the Martian soil (5 to 10 percent by weight of sulfur trioxide). The presence of sulfate ions is compatible with WCL data. Magnesium sulfate would release sulfur dioxide at temperatures below 1,000 degrees, so the absence of this gas therefore proves the absence of magnesium sulfate in the soil. In the Martian environment (with an atmospheric pressure of roughly 10 millibars, or about a hundredth of that on Earth), calcium sulfate would decompose at about 1,400 degrees, but such temperatures are not reached by Phoenix's thermal analyzer. All these facts taken together point toward the likely presence of calcium carbonate in the soils that Phoenix has analyzed. In fact, large deposits of calcium carbonate previously have been found on the surface of Mars, in particular near the north-polar ice cap.

Nice Weather

Phoenix's instruments have enabled new types of meteorological measurements at the landing site. The site has the advantage that the polar regions exhibit strong weather phenomena, especially cloud formation (as was known from orbital imagery). Also, the weathercock has returned data on wind velocity and direction throughout the mission, enabling fruitful modeling. Phoenix weather measurements were coordinated with orbital observations on a regular basis throughout the mission, strengthening the results.

Mars's atmospheric water vapor pressure, as measured by Phoenix's humidity sensor, rises between about 2 AM and 10 AM, then reaches a plateau (about 1.8 Pascals) that is maintained throughout most of the day. In contrast, atmospheric temperatures continue to rise until about 2 PM. Apparently atmospheric convection becomes very efficient and rapidly redistributes the newly formed water vapor after 10 AM. The spacecraft observed several passing dust devils at typical wind velocities (5 to 10 meters per second). Analysis of the pressure data acquired throughout the mission shows that such dust devils are correlated with brief pressure dips of 1 to 3 Pascals.

LIDAR data from the later part of the mission turned out to be particularly important. Ground fog as well as water-ice clouds near the top of the atmospheric boundary layer (at an altitude of about 4 kilometers) formed every night after sol 80. Many of these clouds had "fall streaks" formed by initially growing, free-falling, then eventually sublimating ice crystals. Such fall streaks also can be observed in terrestrial clouds. Daytime LIDAR data showed mostly dust in the atmospheric boundary layer. However, SSI also documented many daytime clouds late in the mission. In some cases these clouds disappeared by sublimation over a timescale of 10 minutes.

Phoenix's instruments monitored the complete diurnal water cycle: During morning hours water vapor is released into the atmosphere. The sources for the water vapor include the shallow subsurface water ice, water adsorbed to soil grains and, possibly, crystal water in perchlorates. During the night, water vapor condenses and falls out by gravity. Most of these ice crystals sublimate again on their descent through the atmospheric boundary layer. In some cases snowfall was observed, when the fall streaks extended all the way down to the surface.

Where Phoenix Is Now

Phoenix surface operations lasted from Martian late spring to late summer—May 26 to November 2, 2008, or 152 sols. The polar night at the landing site lasted from April 1 to July 10, 2009. Since that time, the Sun has again risen above the horizon at the landing site. If the spacecraft—contrary to all expectations—survived both the low temperatures (150 kelvins) of the Martian winter and the dry-ice load built up on its solar panels, it will be able to reanimate itself through a so-called "Lazarus mode." Mars Odyssey was scheduled to search for Phoenix signals starting at the end of 2009.

Independent of its potential reanimation, Phoenix was a highly successful mission that provided on-site geochemical and atmospheric data for the first Martian arctic landing site ever explored. No organic molecules, and no traces of previous or present biological activity, were found at the landing site. Hence, the search for organic molecules will have to be continued by future missions.

It should be noted that organic molecules ought to be present in the Martian soil because of the steady influx of certain types of meteorites that contain substantial quantities of organic material. The fact that no such molecules have been found in the soil around Phoenix is indicative of fast geological degradation processes. The ever-continuing turnover of soil material (by cryoturbation) at the landing site may have favored such degradation processes.

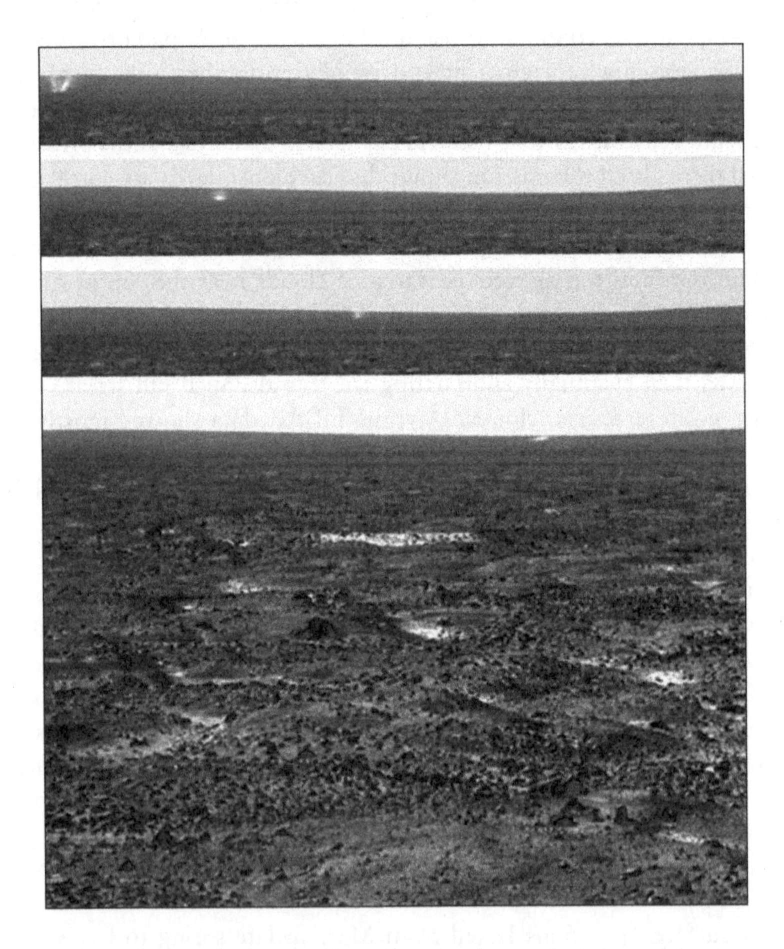

Figure 4. On day 104 of Phoenix's mission, the lander spotted a dust devil about a kilometer away (*top left*). The dust devil moved to the right and away from the lander (*two middle images, at left*), ending up about 1.7 kilometers away (*bottom left*). The presence of the dust devils correlates with brief dips in the atmospheric pressure. Larger dust devils have been observed previously in Mars's Gusev crater.

If organic molecules are ever detected it will be a major scientific task to track down their origin: Are they imported by comets or meteorites, or do they truly attest to primitive, extinct life-forms on the surface of Mars?

Although no organics have been found so far, it is essential to continue this exploration program and search for organic material in more protected environments, such as the interior of sedimentary rocks or deeper soil layers. The next scheduled missions that are available for this task are the rovers Curiosity (which NASA plans to launch in 2011) and ExoMars (which ESA plans to launch in 2018). They will carry complex follow-up instruments that will search for organic molecules in specific equatorial regions. One instrument for ExoMars, the Mars Organic Molecule Analyzer, is presently under development at the Max Planck Institute for Solar System Research.

Curiosity's Sample Analysis at Mars (SAM) instrument has been built by the NASA Goddard Space Flight Center and is ready to go. It will be able to detect particularly low concentrations of potentially organic material. Special attention will be given to the molecule methane (CH_4) whose presence on Mars recently has been demonstrated. The instrument will be able to measure at extremely low methane concentrations (less than one part per billion) the ratio of the isotopes of carbon-13 and carbon-12 present in a molecule. Such data may shed light on the molecule's origin, favoring either geochemical or biological formation.

There is little doubt that Mars will continue to present fascinating new data, surprises and mysteries for these upcoming missions, and to ones still on the drawing board. The two new rovers, Curiosity and ExoMars, will be important benchmarks on that path. Further in the future, robotic return of samples will play a major role in the Mars exploration program. The biggest challenge—a manned mission to Mars—may belong to the distant future, but perhaps at some point such projects will be within reasonable budgets for the major space agencies.

THE MARTIAN (2011)

Andy Weir

Andy Weir was born in 1972 in California. He studied computer science at the University of California, but did not finish the course. He then worked as a programmer for AOL and other companies, as well as Sandia National Laboratories. Weir began writing science fiction in his twenties. *The Martian* started life as a free serial on his website, before being packaged as a self-published ebook on Amazon. The rights were acquired by Crown Publishing Group and it debuted at No. 12 on *The New York Times* bestseller list. The book was subsequently adapted into a motion picture by Ridley Scott, starring Matt Damon.

CHAPTER 1
LOG ENTRY: SOL 6

I'm pretty much fucked.

That's my considered opinion.

Fucked.

Six days into what should be the greatest two months of my life, and it's turned into a nightmare.

I don't even know who'll read this. I guess someone will find it eventually. Maybe a hundred years from now.

For the record ... I didn't die on Sol 6. Certainly the rest of the crew thought I did, and I can't blame them. Maybe there'll be a day of national mourning for me, and my Wikipedia page will say, "Mark Watney is the only human being to have died on Mars."

And it'll be right, probably. 'Cause I'll surely die here. Just not on Sol 6 when everyone thinks I did.

Let's see ... where do I begin?

The Ares Program. Mankind reaching out to Mars to send people to another planet for the very first time and expand the horizons of humanity blah, blah, blah. The Ares 1 crew did their thing and came back heroes. They got the parades and fame and love of the world.

Ares 2 did the same thing, in a different location on Mars. They got a firm handshake and a hot cup of coffee when they got home.

Ares 3. Well, that was my mission. Okay, not *mine* per se. Commander Lewis was in charge. I was just one of her crew. Actually, I was the very lowest ranked member of the crew. I would only be "in command" of the mission if I were the only remaining person.

What do you know? I'm in command.

I wonder if this log will be recovered before the rest of the crew die of old age. I presume they got back to Earth all right. Guys, if you're reading this: It wasn't your fault. You did what you had to do. In your position I would have done the same thing. I don't blame you, and I'm glad you survived.

*

I guess I should explain how Mars missions work, for any layman who may be reading this. We got to Earth orbit the normal way, through an ordinary ship to *Hermes*. All the Ares missions use *Hermes* to get to and from Mars. It's really big and cost a lot so NASA built only one.

Once we got to *Hermes*, four additional unmanned missions brought us fuel and supplies while we prepared for our trip. Once everything was a go, we set out for Mars. But not very fast. Gone are the days of heavy chemical fuel burns and trans-Mars injection orbits.

Hermes is powered by ion engines. They throw argon out the back of the ship really fast to get a tiny amount of acceleration. The thing is, it doesn't take much reactant mass, so a little argon (and a nuclear reactor to power things) let us accelerate constantly the whole way there. You'd be amazed at how fast you can get going with a tiny acceleration over a long time.

I could regale you with tales of how we had great fun on the trip, but I won't. I don't feel like reliving it right now. Suffice it to say we got to Mars 124 days later without strangling each other.

From there, we took the MDV (Mars descent vehicle) to the surface. The MDV is basically a big can with some light thrusters and parachutes attached. Its sole purpose is to get six humans from Mars orbit to the surface without killing any of them.

And now we come to the real trick of Mars exploration: having all of our shit there in advance.

A total of fourteen unmanned missions deposited everything we would need for surface operations. They tried their best to land all the supply vessels in the same general area, and did a reasonably good job. Supplies aren't nearly so fragile as humans and can hit the ground really hard. But they tend to bounce around a lot.

Naturally, they didn't send us to Mars until they'd confirmed that all the supplies had made it to the surface and their containers weren't breached. Start to finish, including supply missions, a Mars mission takes about three years. In fact, there were Ares 3 supplies en route to Mars while the Ares 2 crew were on their way home.

The most important piece of the advance supplies, of course, was the MAV. The Mars ascent vehicle. That was how we would get back to *Hermes* after surface operations were complete. The MAV was soft-landed (as opposed to the balloon bounce-fest the other supplies had). Of course, it was in constant communication with Houston, and if there had been any problems with it, we would have passed by Mars and gone home without ever landing.

The MAV is pretty cool. Turns out, through a neat set of chemical reactions with the Martian atmosphere, for every kilogram of hydrogen you bring to Mars, you can make thirteen kilograms of fuel. It's a slow process, though. It takes twenty-four months to fill the tank. That's why they sent it long before we got here.

You can imagine how disappointed I was when I discovered the MAV was gone.

*

It was a ridiculous sequence of events that led to me almost dying, and an even more ridiculous sequence that led to me surviving.

The mission is designed to handle sandstorm gusts up to 150 kph. So Houston got understandably nervous when we got whacked with 175 kph winds. We all got in our flight space suits and huddled in the middle of the Hab, just in case it lost pressure. But the Hab wasn't the problem.

The MAV is a spaceship. It has a lot of delicate parts. It can put up with storms to a certain extent, but it can't just get sandblasted forever. After an hour and a half of sustained wind, NASA gave the order to abort. Nobody wanted to stop a monthlong mission after only six days, but if the MAV took any more punishment, we'd all have gotten stranded down there.

We had to go out in the storm to get from the Hab to the MAV. That was going to be risky, but what choice did we have?

Everyone made it but me.

Our main communications dish, which relayed signals from the Hab to *Hermes*, acted like a parachute, getting torn from its foundation and carried with the torrent. Along the way, it crashed through the reception antenna array. Then one of those long thin antennae slammed into me end-first. It tore through my suit like a bullet through butter, and I felt the worst pain of my life as it ripped open my side. I vaguely remember having the wind knocked out of me (pulled out of me, really) and my ears popping painfully as the pressure of my suit escaped.

The last thing I remember was seeing Johanssen hopelessly reaching out toward me.

*

I awoke to the oxygen alarm in my suit. A steady, obnoxious beeping that eventually roused me from a deep and profound desire to just fucking die.

The storm had abated; I was facedown, almost totally buried in sand. As I groggily came to, I wondered why I wasn't more dead.

The antenna had enough force to punch through the suit and my side, but it had been stopped by my pelvis. So there was only one hole in the suit (and a hole in me, of course).

I had been knocked back quite a ways and rolled down a steep hill. Somehow I landed facedown, which forced the antenna to a strongly oblique angle that put a lot of torque on the hole in the suit. It made a weak seal.

Then, the copious blood from my wound trickled down toward the hole. As the blood reached the site of the breach, the water in it quickly evaporated from the airflow and low pressure, leaving a gunky residue behind. More blood came in behind it and was also reduced to gunk. Eventually, it sealed the gaps around the hole and reduced the leak to something the suit could counteract.

The suit did its job admirably. Sensing the drop in pressure, it constantly flooded itself with air from my nitrogen tank to equalize. Once the leak became manageable, it only had to trickle new air in slowly to relieve the air lost.

After a while, the CO_2 (carbon dioxide) absorbers in the suit were expended. That's really the limiting factor to life support. Not the amount of oxygen you bring with you, but the amount of CO_2 you can remove. In the Hab, I have the oxygenator, a large piece of equipment that breaks apart CO_2 to give the oxygen back. But the space suits have to be portable, so they use a simple chemical absorption process with expendable filters. I'd been asleep long enough that my filters were useless.

The suit saw this problem and moved into an emergency mode the engineers call "bloodletting." Having no way to separate out the CO_2, the suit deliberately vented air to the Martian atmosphere, then backfilled with nitrogen. Between the breach and the bloodletting, it quickly ran out of nitrogen. All it had left was my oxygen tank.

So it did the only thing it could to keep me alive. It started back-filling with pure oxygen. I now risked dying from oxygen toxicity, as the excessively high amount of oxygen threatened to burn up my nervous system, lungs, and eyes. An ironic death for someone with a leaky space suit: too much oxygen.

Every step of the way would have had beeping alarms, alerts, and warnings. But it was the high-oxygen warning that woke me.

The sheer volume of training for a space mission is astounding. I'd spent a week back on Earth practicing emergency space suit drills. I knew what to do.

Carefully reaching to the side of my helmet, I got the breach kit. It's nothing more than a funnel with a valve at the small end and an unbelievably sticky resin on the wide end. The idea is you have the valve open and stick the wide end over a hole. The air can escape through the valve, so it doesn't interfere with the resin making a good seal. Then you close the valve, and you've sealed the breach.

The tricky part was getting the antenna out of the way. I pulled it out as fast as I could, wincing as the sudden pressure drop dizzied me and made the wound in my side scream in agony.

I got the breach kit over the hole and sealed it. It held. The suit backfilled the missing air with yet more oxygen. Checking my arm readouts, I saw the suit was now at 85 percent oxygen. For reference, Earth's atmosphere is about 21 percent. I'd be okay, so long as I didn't spend too much time like that.

I stumbled up the hill back toward the Hab. As I crested the rise, I saw something that made me very happy and something that made me very sad: The Hab was intact (yay!) and the MAV was gone (boo!).

Right that moment I knew I was screwed. But I didn't want to just die out on the surface. I limped back to the Hab and fumbled my way into an airlock. As soon as it equalized, I threw off my helmet.

Once inside the Hab, I doffed the suit and got my first good look at the injury. It would need stitches. Fortunately, all of us had been trained in basic medical procedures, and the Hab had excellent medical supplies. A quick shot of local anesthetic, irrigate the wound, nine stitches, and I was done. I'd be taking antibiotics for a couple of weeks, but other than that I'd be fine.

I knew it was hopeless, but I tried firing up the communications array. No signal, of course. The primary satellite dish had broken off, remember? And it took the reception antennae with it. The Hab had secondary and tertiary communications systems, but they were both just for talking to the MAV, which would use its much more powerful systems to relay to *Hermes*. Thing is, that only works if the MAV is still around.

I had no way to talk to *Hermes*. In time, I could locate the dish out on the surface, but it would take weeks for me to rig up any repairs, and that would be too late. In an abort, *Hermes* would leave orbit within twenty-four hours. The orbital dynamics made the trip safer and shorter the earlier you left, so why wait?

Checking out my suit, I saw the antenna had plowed through my bio-monitor computer. When on an EVA, all the crew's suits are networked so we can see each other's status. The rest of the crew would have seen the pressure in my suit drop to nearly zero, followed immediately by my bio-signs going flat. Add to that watching me tumble down a hill with a spear through me in the middle of a sandstorm ... yeah. They thought I was dead. How could they not?

They may have even had a brief discussion about recovering my body, but regulations are clear. In the event a crewman dies on Mars, he stays on Mars. Leaving his body behind reduces weight for the MAV on the trip back. That means more disposable fuel and a larger margin of error for the return thrust. No point in giving that up for sentimentality.

*

So that's the situation. I'm stranded on Mars. I have no way to communicate with *Hermes* or Earth. Everyone thinks I'm dead. I'm in a Hab designed to last thirty-one days.

If the oxygenator breaks down, I'll suffocate. If the water reclaimer breaks down, I'll die of thirst. If the Hab breaches, I'll just kind of explode. If none of those things happen, I'll eventually run out of food and starve to death.

So yeah. I'm fucked.

CHAPTER 2
LOG ENTRY: SOL 7

Okay, I've had a good night's sleep, and things don't seem as hopeless as they did yesterday.

Today I took stock of supplies and did a quick EVA to check up on the external equipment. Here's my situation:

The surface mission was supposed to be thirty-one days. For redundancy, the supply probes had enough food to last the whole crew fifty-six days. That way if one or two probes had problems, we'd still have enough food to complete the mission.

We were six days in when all hell broke loose, so that leaves enough food to feed six people for fifty days. I'm just one guy, so it'll last me three hundred days. And that's if I don't ration it. So I've got a fair bit of time.

I'm pretty flush on EVA suits, too. Each crew member had two space suits: a flight spacesuit to wear during descent and ascent, and the much bulkier and more robust EVA suit to wear when doing surface operations. My flight spacesuit has a hole in it, and of course the crew was wearing the other five when they returned to *Hermes*. But all six EVA suits are still here and in perfect condition.

The Hab stood up to the storm without any problems. Outside, things aren't so rosy. I can't find the satellite dish. It probably got blown kilometers away.

The MAV is gone, of course. My crewmates took it up to *Hermes*. Though the bottom half (the landing stage) is still here. No reason to take that back up when weight is the enemy. It includes the landing gear, the fuel plant, and anything else NASA figured it wouldn't need for the trip back up to orbit.

The MDV is on its side and there's a breach in the hull. Looks like the storm ripped the cowling off the reserve chute (which we didn't have to use on landing). Once the chute was exposed, it dragged the MDV all over the place, smashing it against every rock in the area. Not that the MDV would be much use to me. Its thrusters can't even lift its own weight. But it might have been valuable for parts. Might still be.

Both rovers are half-buried in sand, but they're in good shape otherwise. Their pressure seals are intact. Makes sense. Operating procedure when a storm hits is to stop motion and wait for the storm to pass. They're made to stand up to punishment. I'll be able to dig them out with a day or so of work.

I've lost communication with the weather stations, placed a kilometer away from the Hab in four directions. They might be in perfect working order for all I know. The Hab's communications are so weak right now it probably can't even reach a kilometer.

The solar cell array was covered in sand, rendering it useless (hint: solar cells need sunlight to make electricity). But once I swept the cells off, they returned to full efficiency. Whatever I end up doing, I'll have plenty of power for it. Two hundred square meters of solar cells, with hydrogen fuel cells to store plenty of reserve. All I need to do is sweep them off every few days.

Things indoors are great, thanks to the Hab's sturdy design.

I ran a full diagnostic on the oxygenator. Twice. It's perfect. If anything goes wrong with it, there's a short-term spare I can use. But it's solely for emergency use while repairing the main one. The spare doesn't actually pull CO_2 apart and recapture the oxygen. It just absorbs the CO_2 the same way the space suits do. It's intended to last five days before it saturates the filters, which means thirty days for me (just one person breathing, instead of six). So there's some insurance there.

The water reclaimer is working fine, too. The bad news is there's no backup. If it stops working, I'll be drinking reserve water while I rig up a primitive distillery to boil piss. Also, I'll lose half a liter of water per day to breathing until the humidity in the Hab reaches its maximum and water starts condensing on every surface. Then I'll be licking the walls. Yay. Anyway, for now, no problems with the water reclaimer.

So yeah. Food, water, shelter all taken care of. I'm going to start rationing food right now. Meals are pretty minimal already, but I think I can eat a three-fourths portion per meal and still be all right. That should turn my three hundred days of food into four hundred. Foraging around the medical area, I found the main bottle of vitamins. There's enough multivitamins there to last years. So I won't have any nutritional problems (though I'll still starve to death when I'm out of food, no matter how many vitamins I take).

The medical area has morphine for emergencies. And there's enough there for a lethal dose. I'm not going to slowly starve to death, I'll tell you that. If I get to that point, I'll take an easier way out.

Everyone on the mission had two specialties. I'm a botanist and mechanical engineer; basically, the mission's fix-it man who played with plants. The mechanical engineering might save my life if something breaks.

I've been thinking about how to survive this. It's not completely hopeless. There'll be humans back on Mars in about four years when Ares 4 arrives (assuming they didn't cancel the program in the wake of my "death").

Ares 4 will be landing at the Schiaparelli crater, which is about 3200 kilometers away from my location here in Acidalia Planitia. No way for me to get there on my own. But if I could communicate, I might be able to get a rescue. Not sure how they'd manage that with the resources on hand, but NASA has a lot of smart people.

So that's my mission now. Find a way to communicate with Earth. If I can't manage that, find a way to communicate with *Hermes* when it returns in four years with the Ares 4 crew.

Of course, I don't have any plan for surviving four years on one year of food. But one thing at a time here. For now, I'm well fed and have a purpose: Fix the damn radio.

LOG ENTRY: SOL 10

Well, I've done three EVAs and haven't found any hint of the communications dish.

I dug out one of the rovers and had a good drive around, but after days of wandering, I think it's time to give up. The storm probably blew the dish far away and then erased any drag-marks or scuffs that might have led to a trail. Probably buried it, too.

I spent most of today out at what's left of the communications array. It's really a sorry sight. I may as well yell toward Earth for all the good that damned thing will do me.

I could throw together a rudimentary dish out of metal I find around the base, but this isn't some walkie-talkie I'm working with here. Communicating from Mars to Earth is a pretty big deal, and requires extremely specialized equipment. I won't be able to whip something up with tinfoil and gum.

I need to ration my EVAs as well as food. The CO_2 filters are not cleanable. Once they're saturated, they're done. The mission accounted for a four-hour EVA per crew member per day. Fortunately, CO_2 filters are light and small, so NASA had the luxury of sending more than we needed. All told, I have about 1500 hours' worth of CO_2 filters. After that, any EVAs I do will have to be managed with bloodletting the air.

Fifteen hundred hours may sound like a lot, but I'm faced with spending at least four years here if I'm going to have any hope of rescue, with a minimum of several hours per week dedicated to sweeping off the solar array. Anyway. No needless EVAs.

*

In other news, I'm starting to come up with an idea for food. My botany background may come in useful after all.

Why bring a botanist to Mars? After all, it's famous for not having anything growing there. Well, the idea was to figure out how well things grow in Martian gravity, and see what, if anything, we can do with Martian soil. The short answer is: quite a lot ... almost. Martian soil has the basic building blocks needed for plant growth, but there's a lot of stuff going on in Earth soil that Mars soil doesn't have, even when it's placed in an Earth atmosphere and given plenty of water. Bacterial activity, certain nutrients provided by animal life, etc. None of that is happening on Mars. One of my tasks for the mission was to see how plants grow here, in various combinations of Earth and Mars soil and atmosphere.

That's why I have a small amount of Earth soil and a bunch of plant seeds with me.

I can't get too excited, however. It's about the amount of soil you'd put in a window box, and the only seeds I have are a few species of grass and ferns. They're the most rugged and easily grown plants on Earth, so NASA picked them as the test subjects.

So I have two problems: not enough dirt, and nothing edible to plant in it.

But I'm a botanist, damn it. I should be able to find a way to make this happen. If I don't, I'll be a really hungry botanist in about a year.

LOG ENTRY: SOL 11

I wonder how the Cubs are doing.

LOG ENTRY: SOL 14

I got my undergrad degree at the University of Chicago. Half the people who studied botany were hippies who thought they could return to some natural world system. Somehow feeding seven billion people through pure gathering. They spent most of their time working out better ways to grow pot. I didn't like them. I've always been in it for the science, not for any New World Order bullshit.

When they made compost heaps and tried to conserve every little ounce of living matter, I laughed at them. "Look at the silly hippies! Look at their pathetic attempts to simulate a complex global ecosystem in their backyard."

Of course, now I'm doing exactly that. I'm saving every scrap of biomatter I can find. Every time I finish a meal, the leftovers go to the compost bucket. As for other biological material ...

The Hab has sophisticated toilets. Shit is usually vaccum-dried, then accumulated in sealed bags to be discarded on the surface.

Not anymore!

In fact, I even did an EVA to recover the previous bags of shit from before the crew left. Being completely desiccated, this particular shit didn't have bacteria in it anymore, but it still had complex proteins and would serve as useful manure. Adding it to water and active bacteria would quickly get it inundated, replacing any population killed by the Toilet of Doom.

I found a big container and put a bit of water in it, then added the dried shit. Since then, I've added my own shit to it as well. The worse it smells, the better things are going. That's the bacteria at work!

Once I get some Martian soil in here, I can mix in the shit and spread it out. Then I can sprinkle the Earth soil on top. You might not think that would be an important step, but it is. There are dozens of species of bacteria living in Earth soil, and they're critical to plant growth. They'll spread out and breed like ... well, like a bacterial infection.

People have been using human waste as fertilizer for centuries. It's even got a pleasant name: "night soil." Normally, it's not an ideal way to grow crops, because it spreads disease: Human waste has pathogens in it that, you guessed it, infect humans. But it's not a problem for me. The only pathogens in this waste are the ones I already have.

Within a week, the Martian soil will be ready for plants to germinate in. But I won't plant yet. I'll bring in more lifeless soil from outside and spread some of the live soil over it. It'll "infect" the new soil and I'll have double what I started with. After another week, I'll double it again. And so on. Of course, all the while, I'll be adding all new manure to the effort.

My asshole is doing as much to keep me alive as my brain.

This isn't a new concept I just came up with. People have speculated on how to make crop soil out of Martian dirt for decades. I'll just be putting it to the test for the first time.

I searched through the food supplies and found all sorts of things that I can plant. Peas, for instance. Plenty of beans, too. I also found several potatoes.

If *any* of them can still germinate after their ordeal, that'll be great. With a nearly infinite supply of vitamins, all I need are calories of any kind to survive.

The total floor space of the Hab is about 92 square meters. I plan to dedicate all of it to this endeavor. I don't mind walking on dirt. It'll be a lot of work, but I'm going to need to cover the entire floor to a depth of 10 centimeters. That means I'll have to transport 9.2 cubic meters of Martian soil into the Hab. I can get maybe one-tenth of a cubic meter in through the airlock at a time, and it'll be backbreaking work to collect it. But in the end, if everything goes to plan, I'll have 92 square meters of crop-able soil.

Hell yeah I'm a botanist! Fear my botany powers!

LOG ENTRY: SOL 15

Ugh! This is backbreaking work!

I spent twelve hours today on EVAs to bring dirt into the Hab. I only managed to cover a small corner of the base, maybe five square meters. At this rate it'll take me weeks to get all the soil in. But hey, time is one thing I've got.

The first few EVAs were pretty inefficient; me filling small containers and bringing them in through the airlock. Then I got wise and just put one big container in the airlock itself and filled that with small containers till it was full. That sped things up a lot because the airlock takes about ten minutes to get through.

I ache all over. And the shovels I have are made for taking samples, not heavy digging. My back is killing me. I foraged in the medical supplies and found some Vicodin. I took it about ten minutes ago. Should be kicking in soon.

Anyway, it's nice to see progress. Time to start getting the bacteria to work on these minerals. After lunch. No three-fourths ration today. I've earned a full meal.

LOG ENTRY: SOL 16

One complication I hadn't thought of: water.

Turns out being on the surface of Mars for a few million years eliminates all the water in the soil. My master's degree in botany makes me pretty sure plants need wet dirt to grow in. Not to mention the bacteria that has to live in the dirt first.

Fortunately, I have water. But not as much as I want. To be viable, soil needs 40 liters of water per cubic meter. My overall plan calls for 9.2 cubic meters of soil. So I'll eventually need 368 liters of water to feed it.

The Hab has an excellent water reclaimer. Best technology available on Earth. So NASA figured, "Why send a lot of water up there? Just send enough for an emergency." Humans need three liters of water per day to be comfortable. They gave us 50 liters each, making 300 liters total in the Hab.

I'm willing to dedicate all but an emergency 50 liters to the cause. That means I can feed 62.5 square meters at a depth of 10 centimeters. About two-thirds of the Hab's floor. It'll have to do. That's the long-term plan. For today, my goal was five square meters.

I wadded up blankets and uniforms from my departed crewmates to serve as one edge of a planter box with the curved walls of the Hab being the rest of the perimeter. It was as close to five square meters as I could manage. I filled it with sand to a depth of 10 centimeters. Then I sacrificed 20 liters of precious water to the dirt gods.

Then things got disgusting. I dumped my big container o' shit onto the soil and nearly puked from the smell. I mixed this soil and shit together with a shovel, and spread it out evenly again. Then I sprinkled the Earth soil on top. Get to work, bacteria. I'm counting on you. That smell's going to stick around for a while, too. It's not like I can open a window. Still, you get used to it.

In other news, today is Thanksgiving. My family will be gathering in Chicago for the usual feast at my parents' house. My guess is it won't be much fun, what with me having died ten days ago. Hell, they probably just got done with my funeral.

I wonder if they'll ever find out what really happened. I've been so busy staying alive I never thought of what this must be like for my parents. Right now, they're suffering the worst pain anyone can endure. I'd give anything just to let them know I'm still alive.

I'll just have to survive to make up for it.

LOG ENTRY: SOL 22

Wow. Things really came along.

I got all the sand in and ready to go. Two-thirds of the base is now dirt. And today I executed my first dirt-doubling. It's been a week, and the former Martian soil is rich and lovely. Two more doublings and I'll have covered the whole field.

All that work was great for my morale. It gave me something to do. But after things settled down a bit, and I had dinner while listening to Johanssen's Beatles music collection, I got depressed again.

Doing the math, this won't keep me from starving.

My best bet for making calories is potatoes. They grow prolifically and have a reasonable caloric content (770 calories per kilogram). I'm pretty sure the ones I have will germinate. Problem is I can't grow enough of them. In 62 square meters, I could grow maybe 150 kilograms of potatoes in 400 days (the time I have before running out of food). That's a grand total of 115,500 calories, a sustainable average of 288 calories per day. With my height and weight, if I'm willing to starve a little, I need 1500 calories per day.

Not even close.

So I can't just live off the land forever. But I can extend my life. The potatoes will last me 76 days.

Potatoes grow continually, so in those 76 days, I can grow another 22,000 calories of potatoes, which will tide me over for another 15 days. After that, it's kind of pointless to continue the trend. All told it buys me about 90 days.

So now I'll start starving to death on Sol 490 instead of Sol 400. It's progress, but any hope of survival rests on me surviving until Sol 1412, when Ares 4 will land.

There's about a thousand days of food I don't have. And I don't have a plan for how to get it.

Shit.

RIDING ALONG WITH THE MARS ROVER DRIVERS (2012)

Rebecca Boyle

Rebecca Boyle is an American science journalist living in Saint Louis, Missouri. She began her writing career as a newspaper reporter but now focuses on astronomy, physics, deep time and climate change. In 2011 she became an Ocean Science Journalism Fellow at the Woods Hole Oceanographic Institution in Woods Hole, Massachusetts, and in 2013 she became a Journalism Fellow at the National Center for Atmospheric Research in Boulder, Colorado.

Scott Maxwell stared at his bedroom ceiling in the hours after his first drive, restless with excitement. All systems were go, and he'd sent the commands by the time he left the Jet Propulsion Laboratory. Now he was supposed to sleep before his next shift on Mars time. But he knew that on the fourth planet from the sun, the Spirit rover's wheels had started to move.

"I was thinking that at that moment, there is a robot on another planet, doing what I told it to do. I could not imagine going to sleep," Maxwell recalls. "It just blew my mind. And I still think it's amazing that what I do with my day job is reach out my hand across 100 million miles across of empty space, and move something on another planet."

Maxwell is one member of a team of engineers and scientists who have spent nearly a decade working with NASA's intrepid Mars Exploration Rovers, Spirit and Opportunity, maneuvering them across windblown Martian terrain and into groundbreaking new discoveries. Many of them, along with a new cadre of researchers, will also command the new Mars rover Curiosity, set to land in three weeks. That rover is far more complex and more powerful, and designed to last much longer than the twin rovers' initial three-month lifespan. But the MER mission, as it's known, set the stage in many ways—including how to live and work as a Mars rover driver.

The humans do their work in a fairly unremarkable setting, wearing normal clothes perched at normal computers, holding conference calls inside normal cubicles in southern California. Yet they're *driving little cars on Mars*, an incredible feat that never ceases to amaze team members like Maxwell. It is not easy, neither on the team members' own health—living on Mars time is hell for some—nor on their personal relationships. But ask them, and they'll say it's been the trip of a lifetime.

Waking up to drive a rover on Mars is complex from the moment you hit the snooze button. In the early days of a mission, engineers live according to the Martian clock. A Martian day, called a sol, is about 40 minutes longer than an Earth day, so engineers show up for their shifts 40 Earth-minutes later each day. Mars timekeeping requires some detailed calculations. From the perspective of a rover driver, it's awful—unless you're Maxwell, who claims to love it.

Morning people like Deborah Bass, a scientist at JPL who worked on MER and the Phoenix lander, and Ray Arvidson, the mission's deputy principal investigator, describe Mars time less than fondly. "Oh, it hurts," Arvidson says. "It's like coming back from a trip to Europe every day." Nowadays, the MER team plans ahead so they can work normal Earth schedules, but 300 to 400 people will revert to Mars time and its bizarre timelines after Curiosity lands.

"At this moment, there is a robot on another planet, doing what I told it to do. It just blew my mind." Spirit and Opportunity are solar-powered, so they have to work during the Martian day to ensure they have enough juice. Though Curiosity has a nuclear generator, it, too, will work mostly during the day so its cameras and instruments can see. That means planning is key, according to Bass.

"We want to be ready so as soon as that little spacecraft wakes up on Mars, we have a whole set of stuff for it to do," she says. "We send it at like 5 in the morning on Mars, and that's the start of the day. So we say we work the Martian night shift."

Thanks to Mars time, this makes for some long days and nights, and managers at the Jet Propulsion Laboratory have to make sure people go home to bed. "It is so compelling, it can be challenging to shut it off and go home," Bass acknowledges.

It's also challenging because team members feel real affection for the rovers and each other, and want to be there for every new step. The mission has served as a bedrock for people who have worked together for more than a decade, surviving deaths in their families, divorces and other traumatic experiences. Initially, engineers and scientists were tied to one rover or the other, and came to know them closely, discussing Spirit and "Oppy's" personalities the way the rest of us might discuss our pets.

"They are personalities we imbue them with, but they're not less real for that. If you ever had a first car, and you loved that car—maybe it was cantankerous, maybe it was great—that car is somebody to you. This is like that, on steroids," Maxwell says. "To the extent that you can say this about the rovers—that it has a personality and is 'somebody'—the character that she is, is the product of the whole team that is operating her."

The Daily Grind

Despite all the emotion and excitement, even Mars rover driving starts with meetings. The MER team's morning starts with a planning session; the science team, led by Steve Squyres of Cornell University and by Arvidson, a professor at

Washington University in St. Louis, head discussions of what the rover ought to do. Then engineers including Maxwell pass the instructions along to the rover itself. (At this point it's rover, singular, because Opportunity is the only one still functioning; its twin Spirit lost contact after becoming mired in sandy soil.)

The speed-of-light delay from Mars to Earth means there's a long lag time between commands sent, commands executed, and confirmation. In the mission's early days, maneuvers were painfully slow and simple as the team learned how the rovers would respond. Later, the rovers got artificial intelligence upgrades to help determine interesting targets. This was helpful, because a lot can change on Mars in a few minutes—rampaging dust storms, loose soil, and other unexpected things can spoil the best-laid plans.

"Imagine driving your car to the grocery store with a light time delay like that. Nothing happens for somewhere between 4 and 20 minutes, and when it does happen, you don't know about it, because your rear window doesn't update for another 4 to 20 minutes," Maxwell says. To avoid excruciating waits, the team now programs Opportunity (and formerly Spirit) with a series of tasks, up to three days' worth over a weekend. After so many years, they're confident in the rovers' response times and capabilities, so they don't have to test every little command with exacting detail. The Curiosity mission will do that for some time, though.

Late afternoon rover time, Opportunity stops driving and sends pictures to Earth, using its high-gain antenna and a relay from the Mars Odyssey orbiter. That's when Maxwell and his team get to work for the next day. As they write commands, they can move a virtual rover in a custom-built, video-game-like software platform, which serves as an avatar for the one on Mars. The computer version helps validate the commands.

"When it executes, there's not going to be anybody there to hit the panic button, so anything that might go wrong, we have to think of that in advance," Maxwell says.

But sometimes, Mars throws a curveball, and there's nothing anyone can do.

Losing Spirit

Both rovers had a few kinks early on; their solar panels filled with dust again and again, and Spirit lost control of one of its six wheels a couple of years into the mission. Some instruments eventually wore down, like the diamond in the Rock Abrasion Tool the rovers used to grind into rocks. But none of these limitations presented serious problems. Then on May 1, 2009, the hobbled Spirit fell into a trap.

"It was like an ice skater might break through the surface; Spirit broke through what looked like this very friendly terrain," Maxwell recalls. "It was softer, fluffier terrain, so it was really hard for her to get traction in."

The team went to JPL's rover test bed, a sandbox with a full-size duplicate rover where they could try some maneuvers. They had a plan in place and started sending commands just before the onset of Martian winter—and then another wheel failed. Still, Maxwell and his team figured out a method that would basically help Spirit swim out of the trap, building a pile of dirt behind it. They made 30 centimeters of progress before the sun dropped too low on the winter horizon, and Spirit's solar panels were tilted the wrong way. Spirit went into hibernation, and as far as anyone knows, she never woke up. NASA sent more than 1,300 attempts to hail the rover before finally giving up last year.

"Victory was within our grasp, but we just didn't have the time. It was devastating," Maxwell says.

The experience sparked Arvidson's interest in the terramechanics of Mars, studying the properties of soil at different depths and under crusts. He's a co-investigator on a new terramechanics experiment using Curiosity, which will use detailed telemetry from the rover to help it plot a safe course, avoiding sand traps like the one that caught Spirit. It will also help scientists understand how crusted soil forms, and their relation to the modern Martian water cycle. It's based on a system Arvidson built to free Spirit. "I decided that will never happen again," he says.

Every two seconds, the system records the rover's pitch, yaw, roll and motor currents, simultaneously building a topographic map. Then these figures drop into a computer model. The Curiosity system will work in a similar manner. Here's an example of the MER project, called Artemis.

Along with helping future Mars missions, all of the rover drivers' work—life on Mars time, late nights and missed appointments—has captured the public's imagination, and served to change the way we see our neighbor planet. No longer is the red planet, named for the god of war, such a foreign place; now it's a dirt-filled, rusty-colored landscape where humans can virtually scamper around. The rovers are just about an average person's height, and they see the world in stereo color vision, as Maxwell points out.

"When you see Mars through their eyes, you're seeing it like you would through your own eyes," he says. "We can't go there yet with our squishy, frail human bodies, but we can send robots that can send you home vacation photos of what Mars would be like if you were there yourself."

Despite all the public's affection over the past decade, the rovers have now slipped from most people's minds, just as regular space shuttle launches sometimes failed to make the newspapers. But Opportunity is still bringing home new discoveries all the time.

"In 1969, we landed a human being on the moon. You go out at night and look at that big thing in the sky—people were *walking* on that, for goodness' sake. But by Apollo 17, everybody was bored with it," as Maxwell puts it. "So we're not front page news everyday, I get that. But for me, it just never gets old."

Arvidson echoes his sentiment: "We're driving rovers on another planet. How can it not be fun?"

CURIOSITY ROVER'S DESCENT TO MARS (2013)

THE STORY SO FAR

Robin McKie

Robin McKie is science and environment editor for the *Observer.* He studied maths and psychology at Glasgow University and was named Science and Technology Journalist of the Year at the UK Press Gazette Awards in 2013.

Nestled below the foothills of the San Gabriel mountains, the Jet Propulsion Laboratory outside Pasadena has a surprisingly low-tech feel. For more than 40 years, space missions to the planets have been controlled from its operations rooms, yet the place is still striking for its bucolic charm. Mule deer crisscross its paths, pausing only to nibble plants, while its buildings, erected during the heyday of the US space programme, now have a settled, slightly worn aspect.

For a Californian campus, it is all very laid-back – on most occasions. On 5 August last year, however, JPL was far from being a peaceful place. Indeed, it was in tumult. Thousands of anxious scientists and engineers had gathered to track the fate of the Curiosity Mars rover, the most sophisticated interplanetary probe in history, which was now about to plunge toward the surface of the Red Planet after a journey of 354 million miles.

Given that Nasa itself was advertising Curiosity's descent through the Martian atmosphere as "seven minutes of terror", the probe was playing havoc with the emotions of the men and women who had devoted years to its construction. Lucky beards were left unshaved and favourite T-shirts worn. Peanuts – considered to be a particular bringer of good fortune at previous JPL landings – were gulped in handfuls. One engineer, Bobak Ferdowsi, even sported a special mohawk haircut with star shapes shaved into it.

All missions have their nerve-racking moments, of course. Curiosity's case was special, however, because of the vast numbers of people involved. A decade earlier, Nasa had decided to ramp up its efforts to study Mars and construct the grandaddy of all mobile laboratories. The one-tonne craft would be nuclear-powered and be fitted with a plethora of instruments constructed to answer one simple question: had Mars ever possessed conditions that could support life?

More than 7,000 scientists and engineers – an unprecedented number – were called in to help overcome the challenges in designing, building and operating Curiosity.

Just selecting a site for the giant rover's landing had proved a major headache, for example. More than 60 places were considered by mission staff, a list from which Gale Crater – near the Martian equator – was eventually selected. It is deep, boulder-strewn and has strong drafts of air sweeping up from its floor. "That made it a difficult place to land in," admits Bill Dietrich, a member of the Curiosity science team. "On the other hand it is geologically fascinating."

At the centre of Gale, Mt Sharp rises from the crater floor to a height of 16,000ft. Satellite photographs show it is made of layers that seem to delineate Mars's entire geological history. Set down at its base, Curiosity could trundle up its flanks to study how the Red Planet had changed over the past 4.5 billion years. The trick would be getting there.

In the past, US space engineers had adopted the bouncy castle approach to putting their rovers on Mars. A robot vehicle was secured inside a bag which was inflated during descent so that it simply bounced over the surface until it came to a standstill. For Curiosity, this was not an option. Five times bigger than its predecessors, it would have required an airbag so big, the stresses of descent would have ripped its fabric apart. Similarly, landers that use legs were considered too unstable to settle on Mars.

It took a marathon brainstorming session of Curiosity scientists and engineers at the Jet Propulsion Laboratory in September 2003 to come up with a solution: a rover on a rope. According to the plan proposed by Adam Steltzner, leader of the probe's Entry Descent and Landing (EDL) team, a platform of rocket thrusters – subsequently dubbed the Sky Crane – would hover above the Martian surface before lowering Curiosity by cable. The system was accurate, manoeuvrable and could drop the rover precisely where scientists wanted it to go. The concept was also novel, untested and contained dozens of potential flaws, any one of which could doom the mission.

Not surprisingly many mission scientists were doubtful. "I remember when I was first shown the plan. I thought: they must be joking!" says Dietrich. Many of Nasa's bigwigs were also alarmed. The mission had a $2.5bn price-tag, after all. Steltzner – a former rock musician-turned-astrophysicist – was summoned to Washington by Mike Griffin, then the head of the agency, to justify his plan. Steltzner defended it with aplomb. For his part, Griffin announced he still thought the concept was crazy but "maybe just the right sort of crazy". The Sky Crane was approved and on 26 November 2011, the device – attached to Curiosity – was blasted into space. On 5 August 2012, it reached its target.

It takes between four and 21 minutes for radio signals from Earth to reach Mars, depending upon the planets' relative positions. Last August, the distance between the two was such that radio signals took around 14 minutes to reach Mars. Guiding a complex landing sequence from California was never a prospect for this reason. The whole process would, instead, be run by on-board computers whose performance would determine the fate of the device created by the thousands of nervy scientists and engineers who had gathered in rooms and halls round JPL. The future of their brainchild depended on a sequence of parachutes and rocket thrusters being deployed automatically and accurately to slow Curiosity from its atmosphere-entry speed of 13,000mph to a landing velocity of 1.7mph.

Steltzner – who had spent the previous decade working on Curiosity – gathered his Entry, Descent and Landing (EDL) team in the mission's control room. "I had spent so much time thinking about the ways in which it couldn't work, or wouldn't work, that the idea of feeling that it would work, of relaxing and trusting that it would work, felt like a dereliction of duty to me," he recalls. "I was rationally confident but emotionally terrified."

Each stage of the descent passed perfectly until the Sky Crane was deployed. Curiosity had only seconds before it reached the surface. The room was utterly silent until touchdown was confirmed and the place erupted to the sound of cheering, screaming, high-fiving EDL engineers. "I felt strangely numb, exhilarated, and slightly in disbelief," Steltzner remembers.

Charles Bolden, the new head of Nasa, was less circumspect. At a quickly convened press conference, he gloated about American ingenuity. No other nation had landed on the Red Planet but the US had now placed seven craft on its surface. Bolden's eulogy was interrupted by Steltzner's team, which had already reached the early stages of cathartic release. They gathered outside chanting, "E-D-L! E-D-L!" before bursting into the press room to attempt to conga their way though it. The bars of Pasadena did stellar business that night while Ferdowsi – with his star-studded mohawk haircut – became a celebrity, receiving 20,000 new Twitter followers, several marriage proposals and, subsequently, an invitation to join President Obama's re-election inaugural parade.

Over the next few days, the engineers and scientists who had built the myriad devices fitted to Curiosity reported, without exception, that each had been tested and was working perfectly. Like an athlete limbering up before a race, the rover's components were gently put through their paces before the six-wheeled Curiosity trundled off on its search for clues to the past habitability of Mars. It did not have to travel far.

"We could see from satellite photographs, that a fan of sedimentary material – like an alluvial delta produced by a river of water on Earth – appeared to have spread through the Gale Crater from one point on its northern rim and had passed very close to Curiosity's landing place," says Dietrich. "It was a perfect target." By December, the rover had travelled about a kilometre from its landing site and had reached the edge of the fan at a spot named Yellowknife Bay by mission scientists. Images from satellites showed the ground here retained its heat even at night, a strong sign that it contained clays and other sediments. Zapping some of the rocks with lasers also indicated the presence of gypsum, a mineral associated on Earth with the presence of water. It was time to drill.

Louise Jandura had led the team of 50 engineers who designed Curiosity's sampling machinery, which includes a drill and a robot arm that delivers quantities of powdered rock to detectors inside the rover. "Before Curiosity made its first hole on Mars, we had carried out more than 1,200 tests on Earth on 20 different types of rock," she recalls. "We went to a lot of pains to get it right."

The day of Curiosity's first drill – on 8 February at Yellowknife Bay – was still a nerve-racking experience, nevertheless. Jandura and her team had been asked to design and build machinery that could produce samples of finely powdered Martian soil – the equivalent in volume of half a crushed aspirin – from a maximum depth of 6.5cm and to deliver those samples to Curiosity's chemical analysers. "We watched the drill go in. Later we saw the powdered rock in the scoop that carries samples to the detectors. The whole apparatus worked perfectly. It was so, so satisfying after eight years of effort. We were hooting and hollering and ended up partying in a local pub that night."

There would be more to celebrate than engineering success, however. The rock sample was divided and placed in two of Curiosity's suites of detection equipment. One was an x-ray diffraction device that measures the abundance of elements in rocks and soils. The second was Sam – Sample Analysis at Mars – which exploits spectrographic techniques to analyse a sample's constituent chemicals. Remarkably, the two instruments – despite exploiting very different scientific procedures – produced identical results.

"We got the same result: the rock drilled from Yellowknife Bay contained clays that could only have been created through sustained reactions between water and rocks," says project scientist Ashwin Vasavada. Photographs taken by Curiosity's cameras also revealed images of rounded pebbles like those moulded by streams on Earth, while other analysis showed the soil was neither too alkaline nor too acidic when those clays were created. "The water must have been drinkable then," says Vasavada.

Billions of years ago, fresh water had poured down the northern rim of Gale Crater and flowed over the rocks of Yellowknife Bay. More to the point, those rocks had been bathed in water for long enough to sustain reactions that produced the clays that the rover had detected. Curiosity had found the first habitable place on another world.

The clays detected by Curiosity have since been shown to be of a type called smectites. Typically these form in the sediments of lakes. "It is therefore quite likely that the entire Gale Crater was once flooded, possibly to a depth of more than a kilometre," adds Dietrich, who is also head of planetary sciences at the University of California, Berkeley. A place of grim desiccation today, it appears this part of Mars positively gurgled with water billions of years ago.

For the project's engineers and scientists, the discovery was particularly satisfying because it revealed the power of Curiosity and its array of instruments. Tens of thousands of tests have now been carried out on Mars by the rover, threatening to engulf Nasa and its scientists in data. Powered by a tiny nuclear generator, Curiosity can be run day and night. Previous rovers have relied on solar panels, which were poor power providers and could only operate effectively during the day. Only a relatively small number of instruments could be fitted and operated as a result.

"In the past, experiments that produced ambiguous results on Mars could not be repeated or compared with tests carried out by other instruments," says Pan Concord, a deputy principal investigator for the Sample Analysis at Mars device. "Here, we can go on and on until we are absolutely sure of our results. That is what we did at Yellowknife. It was wonderful. And I am sure we will do it again."

However, it is the wider implications of Curiosity's discovery at Yellowknife Bay that excites Michael Meyer, lead scientist for Nasa's Mars Exploration Programme. "We sent Curiosity to Mars to find out if the planet had the potential to sustain life – and that it is exactly what it has done." Fresh water, the matrix of life, was once abundant on Mars until an unknown planet-wide catastrophe engulfed it and turned it into an arid, lost world.

Finding the nature of that catastrophe will be a future task for Curiosity. Most scientists suspect Mars's low mass and its lack of a magnetic field may have doomed it. On Earth, our relatively strong gravitational field plus our relatively intense magnetic field protect our atmosphere from battering by the solar wind, a constant stream of particles that pours from the sun. Without such fields, Mars could not hold on to its atmosphere or its water, which were swept into space. Earth remained blue and watery while Mars turned to dust. Over the next few years, Curiosity's Sam detector will sniff out delicate isotope variations in the painfully thin remnants of Mars's atmosphere for clues to the timing and nature of this disaster. The key question is: did life get a chance to make its appearance before catastrophe struck?

For Meyer, this issue is of critical importance. "We know life appeared on Earth. But we do not know how easy that process was. It could have been straightforward or it could have been a highly fraught business filled with all sorts of unlikely contingencies."

Mars provides the perfect place to solve that mystery. If its ancient watery, organic soils eventually led to the appearance of primitive living beings before catastrophe struck, we can conclude that life could be relatively commonplace in the cosmos, he argues. "Life appearing separately on two neighbouring worlds would suggest it is a straightforward phenomenon."

But if the sands of Mars turn out never to have supported life, despite their initially attractive properties, life will look a far less likely outcome in the universe. "From that perspective, life on Earth – including humans – may turn out to be a cosmic improbability," adds Meyer. Finding out which version is the right one underscores the importance of Curiosity and future rover missions, which will be designed specifically to answer such questions.

There is more, adds Meyer. Our knowledge about life's appearance on our own planet is abysmal. Billions of years of biological, chemical, meteorological and geological activity on Earth have obliterated all evidence of its origins. "We do not know where it started, how it started, when it started, or what biochemical precursors led to its appearance," Meyer points out.

By contrast, on Mars, life may have flourished for only a very brief period before being extinguished so that signs of its existence may well have been preserved in the planet's dead dust. Their discovery would be like finding biological snapshots frozen in time. "If we can find evidence of organisms' first appearance on Mars, we will be provided with a cookbook for the ingredients of life itself," says Meyer.

The prospects of future scientific excitement emanating from Mars certainly look good, both from Curiosity and from the robot rovers that are scheduled to follow in its wheel-tracks. But for all the wonderful science that is being done there, it is simply the presence of Curiosity, with its sophisticated cameras and detectors, on the planet that provides the greatest satisfaction for those involved in its operations. "I head home after a shift working with Curiosity and sometimes see Mars in the night sky," says Jandura. "The next day I am back at JPL looking at images taken of the Martian surface that have just been sent back by Curiosity. I can see the planet from the rover's perspective and it is utterly thrilling to know that it is there."

Meanwhile Back on Earth...Curiosity's Successor Takes Shape in Hertfordshire

In a nondescript shed on a minor road near Stevenage railway station, researchers have built the ultimate, out-of-this-world facility. Sand has been spread over the floor of a large hall; red, sandstone boulders have been littered on top; and on the surrounding walls, photographs of Mars's surface have been pasted together to form a complete background landscape. Stand in the middle and you could be on the Red Planet – were it not for the breathable atmosphere and substantial gravity.

However, the real attraction of this alien experience is not its extraterrestrial decoration but a six-wheeled vehicle whose binocular cameras, perched on a five-foot-high central control rod, gives it more than a passing resemblance to Wall-E, the eponymous robot star of the Pixar cartoon. This is ExoMars, the European robot rover scheduled to succeed Curiosity in 2018. The two craft have similar designs, a metal skeleton frame fitted with six wire-rimmed wheels. However, ExoMars possesses one crucial difference: it will be able to drill two metres down into the Martian soil. Curiosity can penetrate only a few centimetres.

"Mars is battered by intense ultraviolet radiation," says Ben Boyes, Deputy Engineering Manager on the ExoMars Rover project. "The prospect of finding life – even very primitive forms of life – on its surface is therefore unlikely. But deeper down, things might be very different."

Due to launch constraints, the ExoMars rover (pictured) being built at Stevenage by the European aerospace company Astrium is much lighter than Curiosity, however. The former will weigh 300kg, a third of Curiosity's weight. As a result, thin solar panels will have to be used to provide power, in contrast to the heavy nuclear generator fitted on Curiosity, a lack of power that puts pressure on ExoMars's designers.

"At night, temperatures on Mars can drop to -120C," says Boyes. "At those temperatures, the craft's electronics can suffer irreparable damage so we have to find clever ways to keep them warm. We will need to charge up a battery to provide night heating and make sure key electronics are insulated – while not adding to the rover's weight."

All these different designs will be tested in Astrium's Mars Yard. "We have to get everything right," adds Boyes. "That includes the sand. Martian sand is very, very fine and can get into the tiniest of gaps. We have to be very careful."

ExoMars was originally scheduled to be a European-US collaboration until Nasa, reeling from budget cuts imposed by the White House, pulled out of the mission. At the last minute, rescue came from the Russians, who offered to launch the probe and share its development costs.

The relief experienced by Europe's scientists has been restrained, however – for Russia has a truly dismal record when it comes to Mars missions. A total of 19 have been attempted by Russia (and the Soviet Union before it). Only two partial successes have resulted. The rest have been complete failures. As one jaundiced UK space scientist describes Russia's last-minute intervention: "It's like being rescued while adrift at sea – only to find you have been picked up by the *Titanic*."

Boyes was more diplomatic. "It's a worry," he admits. "But then any mission to Mars is a worry. Hopefully we will get the success we feel we deserve."

PHYSIOLOGICAL AND PSYCHOLOGICAL ASPECTS OF SENDING HUMANS TO MARS (2014)

CHALLENGES AND RECOMMENDATIONS

Antonio Paris

Antonio Paris graduated from the NASA Mars Education Program at the Mars Space Flight Center, Arizona State University and is currently the Chief Scientist at the Center for Planetary Science, and an Assistant Professor of Astronomy and Astrophysics at St. Petersburg College, Florida. He is the author of three books – *Mars: Your Personal 3D Journey to the Red Planet, Space Science* and *Aerial Phenomena* – and is a professional member of the Washington Academy of Sciences and the American Astronomical Society. Paris is also a popularizer of astronomy and has appeared on the Science Channel, Discovery Channel and National Geographic Channel.

Introduction

Prior to the twentieth century, there was little opportunity to explore Mars except via astronomical observations and science fiction.[1] The last few decades, however, have brought forth many significant achievements in space exploration, transforming the human thirst for sending humans to Mars into a technologically achievable goal. Recent breakthroughs in space technology, space medicine, and cooperation among international space agencies, have contributed significantly toward transforming this fiction into a reality. There are, however, substantial differences between low-earth orbit operations and exploring interplanetary space. A manned mission to Mars will place humans in a remorseless environment that will not tolerate human error or technical failure. The challenges, to name a few, will include massive bone and muscle loss as a direct result of long-term exposure to micro-or zero gravity, and cell damage from ionizing cosmic radiation, potential permanent vision problems, and psychological and sociological deterioration due to isolation. Moreover, because the distance between Mars and Earth would require a 2 to 3-year round trip,[2,18] the massive undertaking of developing nutritional and medical strategies would be required in order for the mission to Mars to succeed.

1. Physiological Aspects of Space Travel

A journey to Mars would require, at a minimum, two 6–8 month segments of travel in "deep space" before and after a nominally 18-month stay on the surface of Mars. On the trips to and from Mars, the crew will be exposed to microgravity and to radiation levels much more severe than that experienced at the International Space Station (ISS) in low Earth orbit. During her trip to Mars, for example, the rover Curiosity experienced radiation levels beyond NASA's career limit for astronauts. On the surface of Mars, moreover, gravity is 38% that of Earth's and radiation is still very dangerous, but reduced by more than 50% from levels in deep space. Furthermore, the surface of Mars is generally coated with dust containing toxic chemicals such as perchlorates. The key information that we do not have is whether the reduced gravity on the martian surface is strong enough to afford recovery from the physiological effects of zero-g, or at least to reduce the deleterious effects discussed in the sections below. Installing a centrifuge on the ISS could provide some valuable data —at least on mice or other animal subjects.

1.1 Radiation

Earth's magnetic field protects astronauts in low Earth orbit from harmful radiation. Although these astronauts are more exposed to radiation than humans on the ground, they are still protected by the Earth's magnetosphere.[2] A manned mission to Mars, however, will introduce the spacecraft and its crew to an environment outside of this protective shield. During the Apollo program, for instance, astronauts on the moon reported seeing flashes of light, and experienced cataracts; these flashes were due to radiation from cosmic rays interacting with matter, and depositing its energy directly into the eyes of the astronaut.[3] It is important to note, however, that the Apollo missions were comparatively short and are not comparable to a 2–3 year trip to Mars and back. The crew enroute to Mars will be outside of Earth's magnetosphere and thus will be at risk from: radiation capable of critically damaging the spacecraft; absorption of fatal radiation doses from bursts of solar protons due to coronal mass ejection events with exposures lasting a matter of hours; and/or potential damage to DNA at the cellular level (which may eventually lead to cancer).

The first recommendation for a manned mission to Mars, therefore, requires a spacecraft built with a heavily shielded area that the astronauts can use to protect themselves from life-threatening radiation events. In the past decade, concept engineers have moved on from traditional aluminum shields and envision any spacecraft traveling to Mars with a "storm shelter" made of better shielding materials. Some ideas include the use of a magnetic field to create a protective shell around the spacecraft, and use of low-density materials, such as water tanks (which would be needed anyway for a long-term mission), to surround the crew's habitat.[4] Countermeasures other than better shielding would also play a vital role in protecting the crew from harmful radiation; this would include a diet plan containing antioxidants such as vitamins E, C, and A, beta-carotene, and selenium, which have been shown to minimize damage to the skin caused by radiation.[5]

1.2 The Cardiovascular System in Space

Although the cardiovascular and pulmonary systems (including the heart, lungs, and blood vessels) adapt well in space, they function differently in micro or zero-gravity than on Earth. An astronaut's cardiovascular system begins to adapt to weightlessness as soon as the blood and other body fluids shift from their lower extremities (feet, legs, and lower trunk) to the upper body, chest, and head. The shifting of these fluids causes the heart to enlarge so that it is capable of handling the increase of blood flow. Although the astronaut's body still contains the same total fluid volume at this point, a higher proportion of fluids have accumulated in the upper body (resulting in what is commonly referred to as puffy face and chicken leg syndrome).[9] The brain and other systems in his/her body then interpret this increase in blood and fluids as a "flood" in the upper body. The astronaut's body reacts to correct this flood by getting rid of some of the "excess" body fluid (for example, astronauts become less thirsty and the kidneys increase the output of urine).[6] These actions decrease the overall quantity of fluids and electrolytes in the body, which leads to a reduction in total circulating blood volume. Once the fluid levels have decreased and the heart no longer needs to work against gravity, the heart shrinks in size, which can degrade performance in an astronaut's duties.

Upon returning to normal gravity, nearly 63%[7] of astronauts experience postflight orthostatic intolerance. Since the astronaut's cardiovascular system adapted to weightlessness in space, it will initially be unable to function efficiently upon return to gravity. Symptoms of post-flight orthostatic intolerance include lightheadedness, headaches, fatigue, altered vision, weakness, sweating, anxiety, and heart palpitations as a result of the heart racing to compensate for falling blood pressure.[7] Logically, an astronaut experiencing any combination of these symptoms when he/she arrives on Mars will initially not be able to function.

The bigger question is whether or not the crew, after a 6-month journey in deep space, would be able to function on Mars, which has 62% less gravity than Earth.[8] When the crew arrives on Mars, the crew members would hypothetically be stronger compared to astronauts returning to Earth's gravity after a mission of similar length. Recent studies of astronauts on long-term missions in space, however, suggest a Mars bound flight with micro-gravity or zero-gravity for as many as 6 months would almost certainly cause incapacitation of the astronauts.[18] Astronauts immediately arriving on Mars would have trouble walking, suffer fatigue, and be in real danger of bone fracture and intermittent loss of consciousness.

Moreover, according to NASA, six months (the time it will take to get to Mars) in zero-gravity will take the astronaut 2 years of recovery time. Therefore, a mission profile which allows only 30–90 days on the surface of Mars would not give the crew enough time to recover from the 6 months in zero-gravity.[8]

Today, there are several countermeasures and prevention strategies implemented in the astronaut corps specifically designed to combat postflight orthostatic intolerance and cardiovascular deconditioning.[5] Prior to spaceflight and throughout the journey, it is recommended that astronauts take part in vigorous aerobic and strength training exercises to improve endurance, increase blood volume, and maintain or increase heart mass. Additionally, medications like Erythropoietin (commonly used in dialysis for cancer patients) and fludrocortisones (commonly used to treat orthostatic hypotension) can increase red cell mass and blood volume. More importantly, after landing on Mars, astronauts must be allowed to gradually adapt to gravity to minimize postflight orthostatic intolerance. Likewise, the use of G-suits after landing will improve orthostatic tolerance, while a spacecraft designed with artificial gravity (intermittently, at a minimum) should theoretically load the vessels of the lower extremities to help minimize orthostatic intolerance.[5]

1.3 The Neuro-Sensory System in Space

The most striking of all of the physiological changes astronauts experience are the changes in the neurovestibular system, which is the part of the nervous system largely responsible for balance mechanisms.[6] Weightlessness during a round trip to Mars will affect an astronaut's neurovestibular system. His or her perception of body orientation, point of reference, and equilibrium will be severely altered during the trip to Mars. As a result, astronauts will experience severe motion sickness symptoms that include disorientation, dizziness, depressed appetite, vomiting, and, in severe cases, extreme nausea.[7] This happens simply in part because weightlessness affects the otolith organs and the semicircular canals, both of which are in the inner ear. Our awareness and perception of our body's orientation on Earth is attributed to the detection of gravity by the otolith organs and the detection of head rotational movements by the semicircular canals. In weightlessness, these organs have trouble computing the body orientation relative to gravity, and the resulting signals no longer correspond with the visual and other sensory signals sent to the brain. In other words, an astronaut's brain has no concept of what is "up" or "down".[6]

Although not as severe as motion sickness, other effects on the neuro-sensory system will include diminished sensitivity to taste and smell, difficulty in hand-eye coordination and pointing at or concentrating on a certain object, and massive hearing stress due to loud life support equipment inside the spacecraft. After a few weeks in weightlessness, however, the crew will begin to adapt. They will learn to propel themselves around by pushing off the overhead, deck, and bulkhead, and they will learn to "fly" through the spacecraft's cabin. In an effort to reinterpret the meaning of the otolith signals, and to provide some sort of a "down" reference, the interior of the spacecraft should have equipment and lettering positioned in the same direction.

Historically, motion sickness in space has not been a major problem. Nonetheless, the key to minimizing motion sickness and other effects to the vestibular system is prevention. Astronauts selected for the Mars mission must be those who can adapt to weightlessness easily and have no history of damage to the neurovestibular system. In the event of severe cases, medications such as promethazine and scopolamine are extremely effective in helping with motion sickness and are thus recommended for a trip to Mars.[5] Other negligible measures to counter the effects on vestibular function are to add spices and condiments that offer more taste to meals, and to ensure that astronauts minimize excessive head movements early into the flight to Mars.

1.4 The Musculo-Skeletal System in Space

The human body has about 700 muscles.[6] Many of these muscles operate as cables that pull on bones to make motion possible. Their function is contraction—that is, they all work by shortening the angle between two bones. The force of gravity on the Earth's surface has shaped the structural design of nearly all life; our bodies look and function the way they do partly because of the continuous pull of this ever-present gravitational force on all of our parts. When we don't use certain muscles, however, they can go into "hibernation" mode.[6] In a weightless environment, where an astronaut does not use his or her muscles for a period of time, the muscles themselves begin to waste away, or atrophy. The long-term result on the astronaut's load-bearing tissues will be significant reduction of bone and muscle. Thus, muscle atrophy will cause problems for astronauts on a mission to Mars. For example, research done on rats in space discovered that being in microgravity for two weeks had converted a large portion of their muscle fibers from Type I, which are muscles efficient at using oxygen to generate more fuel over a longer period of time, to Type II, which fatigue more quickly than Type I.[6] This is due to the fact that, while in a weightless environment, the rats no longer needed their legs to balance and control their bodies against the force of gravity (the rats just floated around from one location to another). As a result, their muscles essentially began to change during space flight. Likewise, we must assume that the crew members arriving at Mars will have weak muscles because they would not have used them as they normally would on Earth. When the astronaut lands on Mars, his or her muscles will need to deal with the sudden force of gravity again.

An additional consequence of leaving gravity is that the astronauts no longer require the full strength of the skeletal and muscular systems for support of their "upright" posture. This is because astronauts do not stand up in space. Since their muscles and bones are not used, they depreciate or "decondition" somewhat.[5] As a consequence, their bones lose calcium and become weaker and, to a degree, waste away.

When bones develop and grow on Earth in the presence of gravity, they typically increase simultaneously in length, diameter, and mass; these three growth characteristics contribute to the strength of the bone. During space flight, in the absence of gravity, studies have shown that certain bones appear to grow in length at about the same rate as on Earth, but that the diameter of the bone is, to some extent, smaller. Data from Soviet/Russian flights suggests that dicephyseal bone formation may stop during weightlessness; the rate of elongation of long bones in the body is not affected by weightlessness, but the rate of circumferential growth diameter is decreased.[6]

Moreover, the low level of light in space means that little vitamin D will be formed, which will also impair the absorption of calcium, resulting in even more bone loss.[6] Bone mineral loss in astronauts has been documented in most early human space flights. Changes in calcium balance, decreased bone density, and inhibition of bone formation have also been reported. In addition to the physical changes in bone growth, increased urinary calcium excretion has been observed in astronauts in Skylab and on other flights.[3]

For a trip to Mars, therefore, there are potential causes for concern. The loss of muscle and bone will have a dramatic impact in the crew's level of fitness. There are, however, several recommended measures that can be taken to minimize muscle and bone loss. First, working in weightlessness does not require a lot of muscle strength, so in an effort to minimize too much muscle loss, exercise can be done onboard the spacecraft. Daily exercises include passive stretching, isometric stretching, multiple small bouts, strength training, and aerobics. Periodic anthropometric measurements can be taken, and weight can also be monitored, in order to increase calorie intake in the event of too much weight loss. Lastly, nutritional supplements such as amino acids and antioxidants can be used as countermeasures. Similarly, screening countermeasures could be put into practice to help prevent too much bone loss; and every effort must be made to select astronauts who have no hereditary hypercalciuria. Astronauts who have idiopathic osteoporosis or a high risk of susceptibility to kidney stones should not be selected.[9] For those who are selected, careful monitoring of bone loss is the best prevention strategy. A diet of low sodium, high calcium and vitamin D must be strictly enforced, and high impact loading exercise of the lower extremities must be done periodically. These exercises, which include squats, leg abductions/adductions, optimal treadmill sessions, and intense resistant band training, will help in maintaining bone mass. Lastly, the use of drugs such as potassium citrate can be employed to reduce the chance of kidney stones.[5]

1.5 Potential Risk of Permanent Damage to Vision

The space science medical community has recently realized that long-term spaceflight can cause severe and possibly permanent vision problems in astronauts.[8] NASA researchers are conducting experiments in an effort to comprehend the issue, which, in the case of travelling to Mars, could present a significant hurdle. In the post-flight examination of 300 U.S. astronauts since 1989, studies have demonstrated that 29% of space shuttle astronauts and 60% of ISS astronauts experienced significant degradation of visual acuity.[8] The space science community does not know the exact cause for the degradation; scientists believe the eye problems stem largely from an increase in pressure inside the skull, specifically, from increased pressure from cerebrospinal fluid which surrounds the brain, which works its way to the optic nerve and pushes on the back of the eyeball.[8] A spacecraft equipped with artificial gravity, which would prevent an increase in pressure in the skull, would be the recommended primary countermeasure to mitigate potential permanent vision problems.

2. Psychological Aspects of Space Travel

Of all problems that can be encountered enroute to Mars and back, effects on the astronaut's mind may be the biggest risk factor of them all.[17] As mentioned, a round trip to Mars would take 2–3 years. Anxiety, depression, and loneliness, along with the stress of routine tasks, tensions within the crew, and a daily battle to maintain fitness and avoid accidents, is the ideal recipe for disturbed behavior in space. Although the psychological effects of living in space for long durations have not been clearly analyzed, similar studies on Earth do exist, such as those derived from Arctic research stations and submarines.[5,15] Many of these studies confirm psychological stress could be the biggest problem for the crew. For example, unlike crews on the ISS, the crew enroute to Mars cannot remain in direct contact with their loved ones and are not steadily supplied with replacement crews, food, or even gifts. Isolation and confinement pose the greatest challenge for the crew members, and as they approach the Red Planet, communications between the spacecraft and Earth become sparser. For example, they would have to wait up to 21 minutes for a message to reach family members and another 21 minutes to receive a reply.[10] A variety of other psychological and physical effects have also been observed from both operational and simulated isolated and confined environments. These factors include motivational decline, fatigue, insomnia, headaches, digestive problems, and social tensions. Strained crew relations, heightened friction, and social conflict are also expected from isolation and confinement.

The experience of Russian and U.S. long duration spaceflight has revealed the need for psychological countermeasures to support human crews in space and lessen the impact of these stressors on crews which will improve mission safety and success while lowering risk. As a result, countermeasures that involve astronaut selection, training, and in-flight support are being developed, validated, and implemented.

One method in development is an attempt to select-in psychologically fit crewmembers, as opposed to merely selecting-out psychiatrically ill applicants. The Behavior and Performance group at NASA is currently validating a psychological select-in astronaut selection methodology.[3] These validation studies have now discovered that several personality variables such as agreeableness, conscientiousness, empathy, sociability, and flexibility, among others, are positively correlated with astronaut performance under stressful conditions, teamwork, group living, motivation, and decision-making.[3]

Psychological training focuses on developing skills for coping with the stressors of the spaceflight environment and for interacting with fellow crewmembers as well as with ground control personnel. The training also deals with leadership styles, multicultural issues, working in an isolated and confined environment, and communicating with team members.

In-flight psychological support involves ground-based monitoring by flight psychologists and psychiatrists, in-flight entertainment (such as videos, books, games, and special items), leisure activities, and opportunities to communicate with the ground (*i.e.*, with family and loved ones); it also extends to care of the families of astronauts on the ground.

The U.S. space program is now acknowledging that psychological factors are crucial for supporting the health, well-being, and performance of astronauts and increasing mission safety and success. Accordingly, new areas of specialty within the behavioral sciences are emerging, which focus on space psychology, human factors, habitability, performance, and space sociology. Health and medical professionals supporting human spaceflight operations will benefit from data in these areas as well.

Recent studies conducted by NASA, specifically on the ISS, have shown that a variety of common sense countermeasures have been successful in keeping astronauts psychologically fit.[11] Some of these countermeasures, which would be adopted by astronauts enroute to Mars, include keeping busy with daily tasks, conducting physical workouts, productive use of free time, and attaining goals that contribute to mental and emotional well-being.[12] Additionally, maintaining a confidential journal allows the astronauts to vent and reflect.[12]

NASA, moreover, offers psychological support to all astronauts before, during, and after mission.[13] This support includes:

- Preflight preparation training and briefings in a variety of areas;
- Family teleconferences; and
- Preparation for the psychological hardships of long distance separation from family and issues likely to arise following the astronauts' return.

3. Long-term Food and Nutritional Concerns

Unlike short duration space missions or the ISS, which gets resupplied periodically, food supply becomes a critical issue for a manned mission to Mars. While the U.S. military currently produces food with a long shelf life, astronauts on a mission to Mars will have different nutritional needs. The food that an astronaut must consume must be of the highest quality to combat the effects of long-term exposure to weightlessness, primarily in order to maintain body mass and prevent disease.[6] Once the crew leaves Earth for Mars, no other options are accessible and any further supply of additional food must be sent months or years in advance. The cost of added weight on the spacecraft is also important and another of the problems that must be overcome prior to leaving for Mars.

Furthermore, unlike most food with a long shelf life, the nutritional requirements for a mission to Mars must be designed so the crew can look forward to an interesting and varied cuisine while they are away from home. On the ISS and Space Shuttle (recently retired), food is prepared on Earth and requires only minimal additional preparation.[14] A mission to Mars, therefore, will require a shift to a system of production, processing, preparation and recycling of nutrients in a closed loop environment. This process is currently designated as Advanced Life Support and it involves not just the production of food materials but also regeneration of oxygen and potable water.[10]

From a physiological perspective there are number of bodily changes that have a role in modifying food intake. These bodily changes, for example, are well documented: early-induced fluid shifts and changes in the volume of blood and total body water. Gastrointestinal function, moreover, may be altered due to changes in micro flora and lack of gas separation in the stomach and intestine.[5]

A manned spacecraft built for Mars must have a galley, eating area, and an exercise station. Also, the crew must have access to refrigerators, freezers, a microwave, an oven, and ambient temperature storage for foods. Frozen items should include entrees, vegetables, baked goods and desserts. The refrigerators, moreover, must be capable of keeping fresh fruits and vegetables. Some dairy products should be available, as well as extended shelf life produce. And, at a minimum, a 30-day[10] repeating menu should be provided, along with the individual choice of menu within the constraints of nutritional adequacy. Other considerations factored into the menu must be a diet high in calcium and vitamin D to maintain bone mass, as well as food low in saturated fats to prevent cardiovascular disease.

4. Operational Medicine and Health Care Delivery

On a mission to Mars, the crew would not have access to an emergency room. Moreover, there will not be much room for a full sick bay, and hands-on medical care will be limited. More importantly, during the astronaut selection process it is unlikely that one would know if a crewmember is in the early stages of a deadly or incapacitating disease that would develop during the journey. Although the probability is low, there are several possible situations where medical or surgical care could be required during a mission to Mars. Medical situations that have emerged during analogous circumstances (for example, crews in Antarctica or on submarines)[5] include strokes, appendicitis, bone fractures, cancer, intracerebral hemorrhage, psychiatric illness, and kidney stones. Decompression sickness, moreover, is another potential problem the crew could encounter, particularly during an extravehicular activity, or when moving between two different pressure environments within the spacecraft.

The first step in mitigating any potential medical problem is to thoroughly screen the crew, and implement prevention and countermeasure strategies to avoid most medical emergencies during the flight. A detailed knowledge of each crewmember—and his or her genetic makeup, to account for heredity conditions—will be necessary. Screenings for potential risk for cancer, risk for cardiovascular disease, and development of kidney stones must be part of the assessment process to ensure the crew is at optimal health. Prior to, and during the flight, the crew must also follow an aggressive cardiovascular and cancer prevention program (and diet) to minimize the risk of disease.[7] The crew must have access to advanced medical kits which provide a wide range of first aid and surgical instruments. These kits must include antibiotics, allergy treatments, analgesics, stimulants, cardiovascular drugs, and other drugs for motion sickness, anxiety, depression, bone loss, and radiation protection. The crew, moreover, must be trained to conduct minimally invasive surgery, and, if needed, use advanced robotic life support such as Robonaut for trauma. During the flight to Mars, the crew must conduct medical refresher training and have contact with medical personnel at ground control. More importantly, it is highly recommended that at least one member of the crew is a fully trained medical doctor or physician with extensive training in space medicine to monitor the crew while on the mission.

Additionally, astronauts who fly together in space are typically chosen from a select group of individuals. These astronauts are hand picked based on the application of evidence-based medical evaluations and the unique combination of technical and behavioral competencies critical to success in long-term spaceflight.[15] The astronaut crew enroute to Mars will be isolated during the entire trip and thus must heavily rely on the spacecraft's onboard systems for health and safety. In addition, as the spacecraft moves further away from Earth, communications with Mission Managers could be delayed up to 40 minutes due to the large distance that radio communications must travel. Therefore, the Mars astronaut selection criteria must include a consideration of psychological and behavioral health issues related to crew performance during the prolonged lack of communication with Mission Managers back on Earth.[16] Nevertheless, the medical system developed and integrated for any mission to Mars will be more robust and intelligent than any medical care system used on the Space Shuttle or ISS. For example, the spacecraft would be integrated with medical systems that will function autonomously with little or no interaction from Mission Managers back on Earth.[16]

5. Leveraging the International Space Station

From 2007 to 2010, the European Space Agency (ESA), Russia, and China selected volunteers to take part in a 520-day simulated round-trip mission to Mars. Known as the Mars500 program, the volunteers were sealed in a mocked spacecraft in Moscow, Russia and took part in a study to investigate the psychological and medical aspects of a long-duration space mission. Although the Mars500 project provided valuable information as predicted, a manned mission to Mars will require long-term medical research under conditions of weightlessness, such as on the ISS. With the recent retirement of the U.S. Space Shuttle fleet, the only viable option would be to use the ISS to simulate a mission to Mars.

The ISS is the most complex and largest international engineering and scientific project in history. It is over four times larger than Russia's Mir space station and longer than a football field.[14] The station's primary goals are to enable long-term exploration of space and provide benefits to all people on Earth. In addition to scientific research on space, additional projects not related to space exploration, but which have expanded our understanding of the Earth's environment, have been conducted. These experiments have included learning more about the long-term effects of radiation on crews, nutritional requirements levied upon astronauts during long-term missions in space, and developing newer technology that can withstand the harsh environment of space. Other experiments conducted over several expeditions on the ISS include:

- clinical nutrition assessments of astronauts;
- subregional assessment of bone loss in the axial skeleton in long-term space flight;
- crewmember and crew-ground interaction during ISS missions;
- effects of altered gravity on spinal cord excitability;
- effect of microgravity on the peripheral subcutaneous veno-arteriolar reflex in humans;
- assessment and countermeasures to renal stone risk during spaceflight;
- validation effect of prolonged space flight on human skeletal muscle;
- bodies in the space environment: relative contributions of internal and external cues to self; and
- orientation during and after zero gravity exposure.[14]

Conclusion

Over the past few decades, a variety of proposals have depicted spacecraft that are capable of completing a round-trip mission to Mars. Many, if not all, of these technical proposals can be used to build a spacecraft using today's technology. More importantly, any spacecraft built for such a mission would be an international effort of epic proportions.

The spacecraft itself, however, is only a part of the solution for developing a successful mission. As noted in this paper, there are still many physiological and psychological challenges the crew destined for Mars must overcome. Although dozens of astronauts have been used as test subjects for physiological and psychological experiments, and preventive strategies and countermeasures have been implemented, we still do not have a lot of knowledge concerning long-term exposure to spaceflight. We can learn more about long-term exposure to a weightless environment, and how it will affect a manned mission to Mars, by simulating such a mission on the ISS. At a minimum, a crew can spend two years on the station to simulate the amount of time it would take to travel to Mars and back (not counting the amount of time spent on Mars waiting for point of departure). We can use the time spent on the station to continue with additional scientific and medical experiments to determine the effects of long-term exposure and, more importantly, develop additional (or better) countermeasures to ensure a successful mission to the Red Planet.

Ultimately, going to Mars makes sense, as it is the next step in space exploration. Unsurprisingly, there continue to be many unanswered questions about long-term exposure in space and how it can affect the crew physiologically and psychologically. Nonetheless, we have the right technology, personnel, and pioneering spirit to address these challenges, move forward, and conquer this bold goal.

THE SOVIET MARS SHOT THAT ALMOST EVERYONE FORGOT (2020)

Andy Heil

Andy Heil is a Prague-based senior correspondent for RadioFreeEurope/RadioLiberty, covering the Balkans, the Transcaucasus, science and the environment. Before joining RFE/RL in 2001, he was a long-time reporter and editor of business, economic and political news in Central Europe, including for *Reuters*, *Oxford Analytica, Acquisitions Monthly*, the *Christian Science Monitor*, *Respekt* and *Tyden*.

A U.S. Atlas V rocket lifted off from Florida's Cape Canaveral on July 30 to cap an unprecedented month of international Mars mission launches, including by the United Arab Emirates and China.

The current flurry of missions to the "red planet" reflects renewed efforts to determine whether a "warmer and wetter Mars" might have hosted ancient life and follows in the figurative tracks of four successful U.S. Mars rovers since 1997, along with some interplanetary missteps.

But the reemphasis on breaking new ground on Mars is also a reminder of one of Soviet scientists' most disappointing – yet in major ways, successful – episodes of the Space Race.

After a decade of misfires and failed flyby missions, the U.S.S.R. made 14 seconds of history in 1971, when its Mars 3 lander transmitted radio signals and a fuzzy image – or signal – from the surface of Mars before cutting out and going silent forever.

Then, almost without a trace, humankind's first controlled landing on the red planet and its lone image disappeared for decades after the malfunction amid one of the fiercest Martian dust storms that astronomers had ever witnessed.

The image finally resurfaced in the 1980s.

And more than two decades after that, NASA crowdsourced a massive, multi-billion-pixel hunt for the debris of the muted Mars 3 lander in an image from its Mars Reconnaissance Orbiter.

After the near-miss of the Mars 3 mission, Soviet scientific planners made a few more stabs at that inner planet before instead turning their attention for the next decade to Earth's nearest neighbor, Venus.

"In what we call the Space Race, there were many subraces, and there were subraces to Venus and to Mars," Brian Harvey, an Irish writer on spaceflight and fellow of the British Interplanetary Society who has written and broadcast extensively on the Soviet space program, told RFE/RL.

"The Soviet space program did not lack in ambition, and the early sets of missions included landers as well as flyby missions," he said. "But things kind of came to a climax in 1971."

The 'Horizon' Image

Observers are divided over exactly what – if anything – the Mars 3 lander's TV camera captured on December 2, 1971, in the only image beamed from the surface of Mars before NASA's Mars Pathfinder mission in 1997.

It is not hard to see why.

Soviet scientists themselves concluded at the time that it had no information whatsoever in its 79 scan lines of imagery and, rather than keeping it secret simply believed it was of no value.

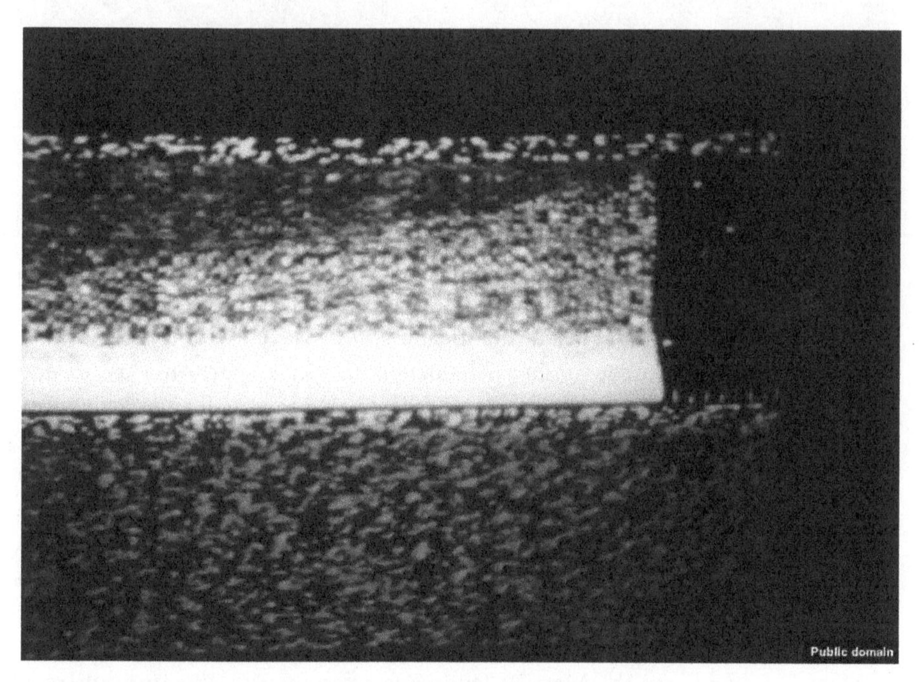

The lone image sent by the Mars 3 lander, showing either a Martian dust storm or radio "snow".

The image was not made public until the late 1980s, when Soviet leader Mikhail Gorbachev encouraged increased scientific cooperation with the West.

"One of the features that I've noticed about the Soviet space program was that, first of all, if they didn't think a photograph was good enough, they didn't release it," said Harvey, who argues that many Soviet space secrets were "hidden in plain view."

"So it wasn't a question of it being secret; they just felt that it didn't tell you enough, give you enough detail. So, they took the decision for us – wrongly in my view – that they should just not release it. They wanted something that could show something very definite very quickly."

Some people still argue that rather than just radio "noise," its successfully transmitted portion shows the Martian horizon in a dust storm.

But NASA and National Geographic describe this image from the Viking Lander 1 on July 20, 1976, as the first ever of the surface of Mars.

This 1976 NASA image was the first to give a clearer view of the Martian landscape, from a low-lying plain called Chryse Planitia in the northern hemisphere of Mars.

Soft Landing, Harder Luck

The Mars 3 mission was the third in a three-pronged effort during 1971 to reach, orbit, map, and touch down on the planet.

Mars 2 and Mars 3 were nearly identical missions – each with its own orbiter module and descent/lander module – launched just days apart on May 19 and May 28, 1971, and competing with NASA's ill-fated Mariner 8 and vastly more successful Mariner 9 Mars missions for global attention.

Mars 2 and 3 were intended to piggyback on the Kosmos 419 mission, which was supposed to provide a radar beacon from orbit around Mars to help guide Mars 2 and Mars 3. But Kosmos 419 had crashed out of Earth's orbit soon after launch on May 12, leaving them to fend for themselves on the six-month journey to Mars.

Both had reached Mars orbit by late November.

Mars 2 crashed into the Martian landscape after entering the atmosphere at too steep an angle on November 27.

Five days later, the Mars 3 lander successfully navigated its 4-hour-plus descent from its orbiting platform to the Martian surface on December 2. It gradually opened its petal-like stabilizers and prepared to transmit.

V.G. Perminov, the leading designer for Mars and Venus spacecraft at the U.S.S.R.'s Lavochkin design offices in the early days of Martian and Venusian exploration, later detailed the moment of the lander's demise in a text called *The Difficult Road To Mars*.

Once the lander was stable on the surface, he said, "the transmission of panoramic images of the Martian surface recorded on the magnetic tape was initiated. The main engineer…who was standing close to the rack where the signal was displayed, gave a command to reduce the signal because it was too strong. Then the telephotometer data were transmitted. There was a gray background with no details.

"In 14.5 seconds, the signal disappeared. The same thing happened with the second telephotometer. Why did two telephotometers working in independent bands simultaneously fail within a hundredth of a second? We could not find an answer to this question."

Perminov drew a parallel with a World War II incident in which British radio signals were interrupted by electrical interference in a Lebanese dust storm.

Most scientists agree that the dust storm somehow knocked out the lander's electrical system. Other theories include being struck by an object in the massive storm or failure as a result of being blown over by the wind.

It never appears to have gotten a chance to plant its Soviet pennant or deploy the PROP-M "rover," a box-like, 5-kilo robot tethered to the lander with 15 meters of cable that would allow it to "'walk' on a pair of skis" to collect soil and atmospheric samples.

Perminov would say one of its most important accomplishments "was that the scientifically and technically intricate problem of a soft landing on the Martian surface was solved."

Moreover, both the Mars 2 and 3 orbiters continued their missions orbiting Mars for about eight more months, snapping up images and providing other useful data that would contribute greatly to knowledge of the fourth planet from the sun.

Ground Control Saves Major Time

The Mars 3 landing remained mostly obscure to anyone but space-exploration devotees until a little over a decade ago.

But in 2007, U.S. researchers shared an image from the High Resolution Imaging Science Experiment (HiRISE) camera on NASA's Mars Reconnaissance Orbiter and unleashed a years-long scramble to find the silenced Soviet lander.

They titled it "Center of Soviet Mars 3 Landing Ellipse" to challenge the public to help scour the image's 1.8 billion pixels of data for Mars 3 or its wreckage.

The 2007 HiRISE image of Mars that set off the new search for the Mars 3 lander would take "about 2,500 typical computer screens" to view at full resolution, according to NASA.

It homed in on the Ptolemaeus Crater, a low-lying expanse of rock and dust that Soviet scientists pinpointed as the touchdown spot decades earlier.

In 2011, the epic crowdsourcing effort appeared to have paid off when a Russian known online as Imxotep reported spotting "light-colored debris" that measured about 8 meters by 8 meters and looked to be a Mars 3-style parachute. Imxotep suggested searching nearby for the lander itself.

Then, in April 2013, HiRISE said Russian space enthusiasts appeared to have spotted the lander, as well as its discarded retrorockets, heat shield and, yes, even its white parachute, on the Martian surface.

"Together, this set of features and their layout on the ground provide a remarkable match to what is expected from the Mars 3 landing, but alternative explanations for the features cannot be ruled out," NASA's Jet Propulsion Laboratory announced in April 2013. "Further analysis of the data and future images to better understand the 3-dimensional shapes may help to confirm this interpretation."

NASA issued a montage of the images and mostly left it at that.

The Russian space enthusiast who organized the VKontakte group that appeared to find the lander and its metal hardware, Vitaly Yegorov, chronicled the painstaking search.

In it, he described his surprise at learning that, with all the progress in space exploration, by 2012 no one had discovered the Mars 2 lander's final resting place.

"Our instrument, which accomplished a phenomenal achievement, the first-ever successful landing on another planet, more than 40 years ago!" he wrote, forgetting the success of a Soviet soft landing by Venera-7 on Venus in December 1970. "It did it in almost the same sequence that the Americans, in 2012, put [the Mars rover] Curiosity on the ground."

Keeping Pace in a Space Race

Since the Soviets had begun preparing for a Mars mission in 1959, Perminov wrote, "the United States had put humans on the Moon, and the Soviet Union had put a cosmonaut in space and circled the Moon with a satellite. However, sending a spacecraft to a distant planet and having it enter an unknown atmosphere and land on a poorly known surface was an undertaking of a different magnitude."

A few Soviet innovations had given momentum to their space program in the late 1960s, including what would become a vaunted reputation for rocket-engine manufacturing that continues to this day. (With varying degrees of controversy, U.S. rockets still use Russian engines; the U.S. Atlas V that blasted off carrying the Perseverance on July 30 was powered by the Russian RD-180 engine.)

The Proton rocket that the Soviet Union introduced in 1965 allowed for large payloads that could include tons of fuel and orbiters and landers along with their instruments to create images and to explore and measure topography, soil, and atmospheric composition.

And Soviet scientists had also developed an astronavigation system that was decades ahead of its rivals and allowed onboard computers to chart and steer its spacecraft using the position of stars and other celestial bodies.

Between the conception and realization of the first Mars landing, the Soviets had achieved a handful of firsts in their Venus program.

But for now, they were keen to repeat their Venus landing on another planet.

And 1971 was a particularly good year of what's known as "opposition," a phase when Earth and Mars are nearest to each other in their solar orbits. In fact, it was as close as those two neighboring, rocky planets had been in 47 years.

But there were other time pressures, too.

With the Mars race in high gear, just two days after the Mars 3 launch an American orbiter – Mariner 9 – was lifting off that would go on to become the first spacecraft to orbit another planet and be enormously successful at mapping the red planet.

"The Russians, I think, felt that they should match that," Harvey said.

Perminov later acknowledged that some of the Soviet planners' key decisions on the Kosmos 419 and Mars 2 and 3 missions were influenced by a desire to outpace their U.S. counterparts, despite two dramatic launch failures of planned Mars orbiters in 1969.

Ironically, by the time the Mars 2 and Mars 3 orbiters were preparing to send their landers toward the planet in November and December 1971, Mariner 9 was in orbit around Mars and sending back images showing what Perminov would later describe as "an unusually strong dust storm," the likes of which "had never before been recorded on the Martian surface."

WATER ON MARS (2020)

A LITERATURE REVIEW

Mohammad Nazari-Sharabian, Mohammad Aghababaei, Moses Karakouzian and Mehrdad Karami

Lead author **Mohammad Nazari-Sharabian** holds a PhD in Civil and Environmental Engineering from the University of Nevada, Las Vegas. His research focuses on water usage, management and quality. He currently works at the Department of Civil and Environmental Engineering and Construction, University of Nevada, Las Vegas alongside his co-author **Moses Karakouzian. Mohammad Aghababaei** works at the Department of Civil and Environmental Engineering, Washington State University and **Mehrdad Karami** works at the Department of Civil Engineering, Isfahan University of Technology, Iran.

1. Introduction

While a substantial portion of the existing water on Mars today is as ice, tiny water amounts are present as vapor in the atmosphere or as low-volume liquid brine that can be found in shallow soil areas. Bright material inferred to be ice can also be seen visually in exposures within new impact craters at high latitudes imaged by HiRISE (high-resolution imaging science experiment). On the surface of Mars, water can only be seen at the northern polar ice caps. Other places on Mars that contain significant amounts of water are at the south pole, where there is a permanent carbon dioxide ice cap, as well as in the shallow subsurface, where more moderate conditions exist. Detection of water on Mars' surface, or close to it, shows the existence of more than 21 million km^3 of ice, which would be able to cover the planet with water 35 m (115 ft.) deep. Much more ice is probably frozen into the profound subsurface.[1–10]

Although some liquid water can be found now on the Martian surface, as a challenging environment for known life, it is confined to thin layers or dissolved atmospheric moisture. Since the average atmospheric pressure on the planet's surface is approximately 600 pascals (0.087 psi), lower than the melting point of water's vapor pressure, no significant amount of liquid water exists on the surface; typically, if pure water on the Martian surface were heated to more than its melting point, it would become vapor; otherwise, it would freeze. In addition, brines have a significant impact on lowering water activity/vapor pressure over solution, and hence reducing the evaporation rate and prolonging the lifetime of liquid water/brine exposed to the Martian atmosphere. Mars probably had higher surface temperatures, as well as a denser atmosphere, 3.8 billion years ago, which resulted in large amounts of surface water that may have contained a vast ocean, possibly covering one-third of the planet. It seems that there was water flowing across the surface for short amounts of time at different periods in Mars' history. According to NASA's report on December 9, 2013, obtained from the Curiosity rover, which is studying Aeolis Palus, an aged freshwater lake used to exist in Gale Crater, where there may have been an environment for microbial life. Much evidence shows the notable role of water ice in Mars' geologic history, due to a considerable amount of it on the planet.[11–21]

Additionally, Martian achondrites provide clear evidence for the water presence over time. These rocks can be considered the key to get clues about the primary geologic processes at work in Mars, identify its ancient hydrothermal environments, search for traces of life forms, and study the interaction of water in sample returned rocks. In addition, as they can be accurately dated, they can provide a message in a bottle: unique information of the key regions in which future missions should focus to extract the maximum scientific information on the key questions to be answered in future robotic and manned sample-return missions to the red planet.[22,23]

ALH 84001 is the oldest (4.1 Gyr old) known Martian meteorite and so is a unique source of information on the environmental conditions of early Mars. One of the most interesting features of ALH 84001 is the presence of Mg-Fe-Ca carbonates along fractures and in cataclastic areas. The study of Martian carbonates is important as these minerals precipitate from aqueous fluids and thus can be used to explore the history of water at the planet's surface and crust. Carbonate outcrops have been found on the Martian surface by orbiters, landers, and rovers. Carbonate deposits that may represent sites of paleolakes support the idea that Mars had a warmer and wetter past.[24–29]

Moreover, large outflow channels that curved from flooding, networks of river valleys, deltas, and lakebeds, as well as traces of minerals and rocks on the surface that could solely have been formed by water in its liquid form, are some geologic evidence of past water on Mars. The existence of permafrost, as well as past and present ice flow in glaciers can be determined, given several geomorphic characteristics. Since there are slopes and gullies near crater walls and cliffs, it seems that water flow is still shaping Mars' surface, though much less than it did a long time ago.[30–44]

Regardless of appropriate conditions for microbial life due to the periodically wet surface of Mars billions of years ago, today Mars' surface is probably not hospitable, because of its mostly dry and subfreezing surface. Additionally, both cosmic and solar radiation can now reach the surface of Mars since there is a lack of thick atmosphere, magnetic field, or the ozone layer. Another limiting factor for life survival on the surface of Mars is ionizing radiation, which damages cellular structure. Subsurface environments are, thus, perhaps the best possible locations to look for life on Mars. NASA found a substantial amount of underground ice on Mars on November 22, 2016; the water detected has the volume equal to that of Lake Superior, located between the US and Canada. The first identified stable water body on Mars was detected by Italian scientists in July 2018, which is located 1.5 km (0.93 mi) underneath the south polar ice cap, and lengthens sidelong for about 20 km (12 mi).[45–54]

To evaluate the feasibility of life surviving, as well as securing the resources required for human exploration in the future, it is essential to understand the abundance and distribution of water on Mars. Therefore, for the first 10 years of the 21st century, the science theme of NASA's Mars Exploration Program (MEP) was "Follow the Water." A variety of efforts have been undertaken to answer fundamental questions about the amount and situation of water on Mars through discoveries made by various expeditions including the 2001 Mars Odyssey, Mars Exploration Rovers (MERs), Mars Reconnaissance Orbiter (MRO), and Mars Phoenix Lander. Additionally, the European Space Agency's Mars Express orbiter was able to provide crucial data in this area.[55] Data is still being sent back from Mars by the Mars Odyssey, Mars Express, MRO, and Mars Science Lander Curiosity rover to continue such discoveries.

2. Evidence from Rocks and Minerals

Regardless of the existence of abundant water rather early in Mars' history, which is widely accepted, none of those vast areas of liquid water remain. Modern Mars has a fraction of this water as either ice, or clogged in the structures of plentiful water-rich materials, comprised of sulfates and clay minerals (phyllosilicates). The principal sources of water on Mars, which is equivalent to 6% to 27% of the Earth's present ocean, are asteroids and comets from more than 2.5 astronomical units (AU), based on investigations of hydrogen isotopic ratios. The spectro-imaging instrument (OMEGA) on Mars Express provided the first detection of hydrated minerals on Mars. The data indicated that large amounts of liquid water had once survived for long periods on the planet's surface. The OMEGA has mapped almost the entire surface of the planet, typically at a resolution between one and five kilometers, with some areas at sub-kilometer resolution. The instrument has recorded the presence of two different classes of hydrated minerals—so-called because they contain water in their crystalline structure and provide a clear mineralogical record of water-related processes.[56–60]

Phyllosilicates are derived from alteration products of igneous minerals (found in magma) due to long-term contact with water. One example of a phyllosilicate is clay. Phyllosilicates have been detected by OMEGA mainly in the Arabia Terra, Terra Meridiani, Syrtis Major, Nili Fossae, and Mawrth Vallis regions, in the form of dark deposits or eroded outcrops. Hydrated sulfates are formed through interaction with acidic water. OMEGA has detected these in layered deposits in Valles Marineris, in extended surface exposures in Terra Meridiani, and within dark dunes in the northern polar cap. The findings have had major implications for interpretation of the climatic history of the planet and whether it may have once been habitable. They point particularly to two major climatic episodes: an early, moist environment in which phyllosilicates formed, followed by a more acidic environment in which the sulfates formed.[58,59]

2.1. *Water in Weathering Products (Aqueous Minerals)*

The surface of Mars is made up mostly of basalt, a fine-grained magmatic rock containing primarily pyroxene, plagioclase feldspar, and mafic silicate minerals olivine. Chemical weathering occurs for these minerals when exposed to water and atmospheric gases, causing them to change into secondary minerals; during the process, some of them may combine water into their crystalline structures, as hydroxyl (OH). Some examples of hydrated or hydroxylated minerals that have been identified on Mars are gypsum, kieserite, phyllosilicates (e.g., kaolinite and montmorillonite), opaline silica, and iron hydroxide goethite.[61,62]

Chemical weathering directly affects water and other reactive chemical species by consuming, and accordingly separating them from the hydrosphere or the atmosphere, and eventually putting them in minerals and rocks. The exact amount of water that has been collected through hydrated minerals in the Martian crust is not known, but it is predicted to be quite large. For instance, deposited sulfate in Meridiani Planum could have up to 22% water by weight, based on the rock outcroppings' mineralogical models that were measured by apparatus on the Opportunity rover.[63–65]

Water is involved to some degree in all chemical weathering reactions on Earth. While secondary minerals require water to form, they often do not include water. Some sulfates such as anhydrite, several carbonates, and metallic oxides are examples of anhydrous secondary minerals. To form a small number of these weathering products on Mars, there might be no need for water, or if there is any, it can be in small amounts, as ice or in slight molecular-scale films. It is still uncertain to what extent such strange weathering processes work on Mars. But instead, aqueous minerals, which are minerals that include water or form in the presence of water, can indicate the type of environment in which the minerals were formed. Aqueous reactions occur depending on certain variables including temperature and pressure, as well as concentrations of gaseous and soluble species. Oxidation-reduction potential (Eh) and pH are two significant properties in this regard. For instance, low pH (highly acidic) water is the only condition under which the sulfate mineral jarosite might form. A proper condition to form phyllosilicates is usually water with neutral to high pH (alkaline). The minerals that are most likely to thermodynamically form from aqueous ingredients are demonstrated by Eh and pH together. As a result, the sorts of minerals present in the rocks can indicate Mars' historical environmental conditions, including those suitable for life.[66–69]

2.2. *Hydrothermal Alteration*

Hydrothermal fluids moving through fissures and pores in the subsurface can also cause the formation of aqueous minerals. Residual heat from considerable impacts or magma bodies might be the heat sources for a hydrothermal system. Migrating seawater through ultramafic and basaltic rocks may cause serpentinization, which is known as one primary type of hydrothermal alteration that can occur in the oceanic crust of the Earth. As the mineral magnetite, ferric iron can be produced by the oxidation of ferrous iron in pyroxene and olivine, owing to water-rock reactions and, as a byproduct, generating molecular hydrogen (H_2). This process can create a low Eh environment that is highly alkaline, which allows for the formation of various carbonate and serpentine minerals, which together can form a rock called serpentinite. Trace amounts of methane exist in Mars' atmosphere, produced by the reaction of CO_2 with hydrogen gas, and are a significant energy source for chemosynthetic organisms. A large amount of water can be stored as hydroxyl in the crystal structure of serpentine minerals. Up to a 500 m (1600 ft)-thick global equivalent layer (GEL) of water could be stored in the crust of Mars by hypothetical serpentinites. Evidence from remote sensing data indicates that not many serpentine minerals exist on Mars. However, there is a probability of the existence of large amounts of serpentinite hidden deep in the Martian crust.[70–74]

2.3. Weathering Rates

Primary minerals can be converted into secondary aqueous minerals at different rates. For example, primary silicate minerals crystallize from magma, under temperatures and pressures that are much higher than the conditions seen on the surface of a rocky planet. These minerals are in a nonequilibrium situation while exposed to a surface environment, and are prone to interact with other chemical ingredients that are available in order to make steadier mineral phases. The silicate minerals that weather the fastest are the ones that crystallize at the highest temperatures. Olivine, which in the presence of water weathers to clay minerals, is the most common mineral meeting this criterion on the Earth and Mars. Since olivine is extensive on Mars, it is likely that water has not significantly altered the planet's surface, but some geological evidence suggests otherwise.[75–78]

2.4. Martian Meteorites

To date, more than 60 meteorites from Mars have been detected. Evidence confirms that some of them were exposed to water while they were on Mars. The existence of sulfates and hydrated carbonates indicate that some Martian meteorites, known as basaltic shergottites, were exposed to liquid water before being ejected into space. Studies indicate that approximately 620 million years ago, the nakhlites, classified as meteorites, were saturated by liquid water; they also demonstrate that an asteroid impact of nearly 10.75 million years ago led to these meteorites being ejected from Mars. In the last 10,000 years, some of them have fallen to Earth. Among all identified Martian meteorites, NWA 7034 has more water than the others by one order of magnitude. It was formed at the beginning of the Amazonian epoch and is similar to the basalts investigated by rover missions.[79–84]

The plausible existence of microfossils in a meteorite from Mars called Allan Hills 84001, was reported by a group of investigators in 1996. The fossil's validity was discussed in several studies. The meteorite contained organic matter which was mostly of terrestrial origin. However, the scientific agreement is that morphology cannot be used by itself to unambiguously detect primitive life. Using morphology alone has resulted in many interpretation errors, since the interpretation of morphology is notoriously subjective.[85–89]

Moyano-Cambero et al. discovered that the carbonate globules formed by the precipitation of Mg- and Fe-rich carbonates from an aqueous solution have interesting information about the past climatic environment. They found distinctive layers indicating that the globules' growth probably occurred in several flooding stages. As they exhibit differentiated elemental chemistry, it could reflect distinctive chemical solutes, being probable evidence of volcanic or environmental differences.[22]

3. Geomorphic Evidence

3.1. Lakes and River Valleys

A revolution in ideas, concerning water on Mars, was caused by the 1971 Mariner 9 spacecraft, which discovered immense valleys in many areas. Breaking through dams, eroding grooves into bedrock, and carving deep valleys are some physical change examples caused by water floods that were displayed in its images. Given the areas of branched streams that were seen in the southern hemisphere of the planet, falling rain in the past is probable. The number of identified valleys has increased through time. Maps of 40,000 river valleys on Mars were shown by research published in June 2010, which is approximately four times more than the number of river valleys recognized earlier. There are two main classes of Martian water-worn features: (1) those that are widely visible are Noachian-age networks, which are on a terrestrial scale and branched; and (2) those that are very long, large, isolated, and single-thread, which are Hesperian-age outflow channels. Recently, some younger and smaller channels have been seen in mid-latitudes, and these are thought to be Hesperian and Amazonian-age. These channels may be due to local iced deposits occasionally melting.[90–93]

Inverted relief, probably occurring due to cementation, leads to increased resistance of deposits on stream beds from erosion and can be found in some parts of Mars; thus, when the covering layer is ultimately removed by erosion, the former streams become visible. Figure 1 shows this phenomenon. Several examples of this were discovered by the Mars Global Surveyor. Numerous inverted streams have been found in varying regions of Mars, including in the Medusae Fossae Formation, Miyamoto Crater, Saheki Crater, and the Juventae Plateau.[94–100]

Various lake basins have also been discovered on Mars. Some of them can be compared with Earth's largest lakes, such as the Black Sea, the Caspian Sea, and Lake Baikal in terms of size. In the southern highlands, there are lakes fed by valley networks, as well as some river valleys leading into closed depressions. The existence of lakes in these areas in the past is probable. Some of the lakes may have been formed by precipitation, and others from groundwater. For instance, the Mars Exploration Rover Spirit has explored such an area in Terra Sirenum, whose overflow moved through the Ma'adim Vallis into the Gusev Crater. Lakes may have also been found in the Argyre basin, Hellas basin, Valles Marineris, Parana Valles, and Loire Vallis. Many craters are assumed to have hosted lakes at times in the Noachian-age. These lakes are cold and dry, similar to the environment during the last maximum glacial period of the US Great Basin.[101–105]

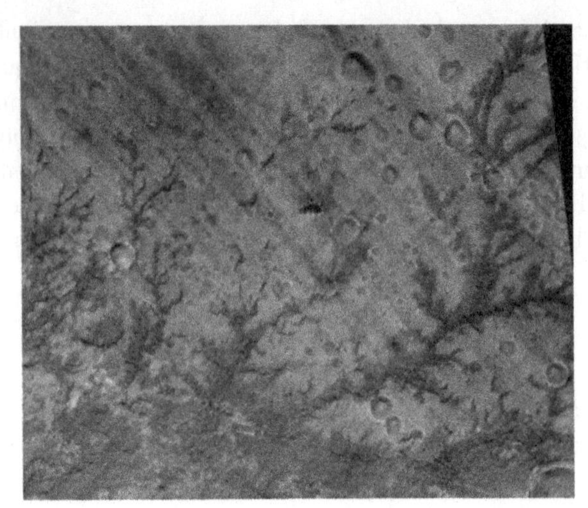

Figure 1: Inverted stream channels in Antoniadi Crater. Location is Syrtis Major quadrangle.

It was suggested by research from 2010 that there were lakes on Mars along parts of its equator. While a long, warm, and wet primary history has since dried up on Mars, as shown by earlier investigations, it seems that these lakes were on Mars' surface during the much later period of the Hesperian age. The study of detailed images from NASA's Mars Reconnaissance Orbiter shows that meteorite impacts and volcanic activity, as well as shifts in Mars' orbit, probably increased during this period. Late Heavy Bombardment took place about 3.8 Gyrs ago. About one hundred craters larger than 150 km found in the ancient southern highlands are early to mid-Noachian in age. This led to the melting of much of the ice present in the ground, likely due to adequate warmth in Mars' atmosphere. The atmosphere's thickness would have increased because of more gases released from volcanoes. This resulted in enough warmth for water to have existed in a liquid phase, due to the trapping of more sunlight. Channels that connected lake basins near the Ares Vallis were discovered in this study. The water of the lake overflowed its banks when it filled up and carved channels to a lower area, forming another lake downstream of the channel. To seek proof (biosignatures) of past life, these dry lakes would be targets.[106,107]

In the Gale Crater, evidence was obtained from an old streambed by the Curiosity rover and reported by NASA scientists on September 27, 2012. It suggests that water on Mars was flowing vigorously at one point in the past. It appears that water was running hip-deep at 3.3 km/h (0.92 m/s), given the analysis of the currently dry streambed. The presence of gravel fragments and rounded pebbles are evidence of flowing water, since forcefully flowing liquid could have weathered them. It seems, from their orientation and shape, that they came from a long distance; there is a channel named Peace Vallis that is above the rim of the crater, and feeds into the alluvial fan.[108–112]

A hypothesized ancient lake is Eridania Lake, having a surface area of approximately 1.1 million square kilometers. Its volume is about 562,000 km^3, and its depth, in the deepest part, reaches up to 2400 m. It had more water compared with all other Martian lakes collectively and was larger than the Caspian Sea, which is the largest landlocked sea on Earth. In fact, the amount of water once held by Eridania Lake would have been nine times greater than all of the Great Lakes, located in North America. The lake was assumed to be surrounded by valley networks, which were at the same elevation as the upper surface of the lake; they emptied into a lake since they all end at the same elevation.[113–117]

Deposits with thicknesses greater than 400 m, found by research with the compact reconnaissance imaging spectrometer for Mars (CRISM), contained Fe- and Mg-serpentine, Fe-rich mica (for instance, glauconite-nontronite), Mg-Fe-Ca-carbonate, Fe-sulfide, saponite, and talc-saponite. Those organic materials are produced by a procedure known as the hydrothermal process. Hydrothermal vents are like geysers on the ocean floor. Along mid-ocean ridges where tectonic plates spread apart, molten inside materials (magma) rise and cool to form a new crust and volcanic mountain chains. Seawater circulates deep in the ocean's crust and becomes super-heated by hot magma. As pressure increases and the seawater warms, it begins to dissolve minerals and rise toward the surface of the crust. The hot, mineral-rich water then exits the oceanic crust and mixes with the cold seawater above. As the vent minerals become cold and solidify into mineral deposits, they form different types of hydrothermal vent structures. Hydrothermal vents support unique ecosystems and their communities of organisms in the deep ocean. They help to regulate ocean chemistry and circulation. Life on Earth perhaps began from such a hydrothermal process.[118]

3.2. Lake Deltas

Several deltas formed in Martian lakes have been reported by researchers (Figure 2). The existence of a significant amount of liquid water can be inferred by finding deltas. In order to form deltas, it is necessary to have deep water over a long period of time. The stability of the water level also is needed to keep sediment from washing away. It seems that there may be deltas around the edges of what is considered the one-time northern ocean of Mars, and they have been discovered over a broad geographical range.[113–117]

3.3. Groundwater

Before 1979, it was thought that single, catastrophic ruptures of water reservoirs, which were under the surface and possibly sealed by ice, formed outflow channels by which large amounts of water were discharged across an arid Martian surface. Additionally, the evidence of very large ripples in the Athabasca Vallis show heavy flooding in the area. There is also evidence from many outflow channels that begin at Chaos or Chasma features that suggests a rupture may have breached through ice seals below the surface.[119–121]

A formation caused by an abrupt, calamitous discharge of groundwater is not in line with the evidence of water networks that appear to branch out on Mars. It is mainly since their dendritic shapes do not come from a single point of outflow, and the amount of water that possibly flowed in the branches. Another idea given by authors related to how these formations came about is that groundwater

Figure 2: Delta in Eberswalde crater.

slowly seeped from springs under the surface of the planet. One reason for this idea is that the upstream ends of numerous valleys in these types of networks start with box canyon heads, which are usually associated with groundwater seepage on Earth. However, there is also some evidence of valleys that are on finer scales at the tips of the channels, indicating the flow emerged abruptly from the subsurface with substantial discharge, instead of collecting slowly across the surface. Others have disputed the link between valleys' amphitheater heads and formation by groundwater for different cases on earth; however, it has been argued that the lack of fine-scale heads to valley networks is caused by their elimination by weathering or impact gardening.[122–124]

It is now accepted, by most authors, that groundwater seepage processes at least somewhat affect and shape most valley networks. Additionally, groundwater may have had an important impact on controlling large-scale sedimentation patterns and processes on Mars. Based on this hypothesis, when groundwater that contained dissolved minerals reached to the surface, it helped to cement sediments and form layers by adding minerals, especially sulfate. In fact, groundwater may form some layers by cementing loose aeolian sediments and raising depositing minerals. As a result, less erosion occurs for stiffened layers. The existence of similar sediments in a wide area, including Arabia Terra, was shown by a 2010 study, using data obtained from the Mars Reconnaissance Orbiter. In a region, the areas where groundwater upwelling is most likely are arguably some of those that are rich in sedimentary rocks. Geological evidence of the connection of an old, planet-wide groundwater system to a possibly existing large ocean has been recently (February 2019) argued by European scientists' publications.[125–137]

3.4. Mars Ocean Hypothesis

It is proposed by the Mars ocean hypothesis that at least one time, the Vastitas Borealis basin held a liquid water ocean; additionally, the evidence provided by that hypothesis shows that early in the geological history of Mars (the first billion years of evolution), a liquid ocean covered almost a third of Mars' surface. In the northern hemisphere of the planet, the Vastitas Borealis basin, which is located 4–5 km (2.5–3.1 mi) below the mean planetary elevation, would have been filled by this ocean, known as Oceanus Borealis. It is plausible that two shorelines existed: a lower one, probably associated with the more recent outflow channels; and a higher one, estimated to have existed over 3.8 billion years ago, at the same time when valley networks were formed in the Highlands. The lower shoreline, which follows the formation of Vastitas Borealis, is called Deuteronilis; the higher shoreline, which can be seen everywhere on the planet, except in the Tharsis volcanic region, is called the Arabia shoreline.[138]

A study from June 2010 showed that 36% of Mars would have been covered by the older ocean. Using data that was acquired from Mars orbiter laser altimeter (MOLA), which measured the elevation of all terrain on Mars in 1999, it was ascertained that nearly 75% of the planet would have been covered by the watershed for this ocean. In order for liquid water to have been able to exist at the surface, Mars would have been required to have had both a denser atmosphere and a warmer climate. Furthermore, the presence of a hydrological cycle early in the planet's life is highly probable, owing to the many valley networks. Valley networks are branching networks of valleys on Mars, which are like river drainage basins. The form, distribution, and implied evolution of the valley networks have a prominent role for liquid water on the Mars surface, and consequently, the climate of Mars.[138–140]

It is still controversial among scientists whether a primordial Martian ocean exists on Mars or not. Additionally, the challenge remains to better interpret some of its characteristics as 'ancient shorelines'. Evidence for the existence of an ancient Martian ocean was first announced by scientists in March 2015. The ocean seemed to have had a size equivalent to that of Earth's Arctic Ocean and be located in Mars' northern hemisphere, covering nearly 19% of its surface. Given the ratio of water and deuterium in Mars' current atmosphere, and comparing it with the equivalent ratio on Earth, this discovery was attained. It could be proposed that in the distant past, Mars had remarkably higher water levels, since the amount of deuterium found on Mars was eight times more than what exists on Earth. Most hydrogen atoms contain just a proton and an electron, but some contain an extra neutron, forming deuterium. On Earth, deuterium is much rarer than hydrogen—for example, in oceans, one in every 6420 hydrogens also has a neutron. Since deuterium is supposed to have been produced by the big bang, it should have once appeared in similar abundances on all the solar system planets. In the Gale Crater, the Curiosity rover formerly discovered a high ratio of deuterium; however, it was not sufficiently high to conclude there had previously been an ocean there. Because the climate models of Mars have not yet shown liquid water on the planet, due to the inadequately warm climate in the past, (the temperature was not high enough to allow liquid water to exist), other scientists have still not validated this study.[141–144]

In May 2016, further evidence that there had been a northern ocean on the planet was published, giving new indications for how two tsunamis may have changed some of the surfaces in the Ismenius Lacus quadrangle. The main reason given for the tsunamis was asteroids striking the ocean so heavy that they seemingly created 30 km diameter craters. Huge stones, the size of small houses or cars, were picked up and carried by the first tsunami. By altering the boulders, the wave backwash shaped channels. It is believed that the second tsunami occurred when the ocean was 300 m lower, and it moved a large piece of ice, which was eventually dropped in valleys. The wave heights would have ranged from 10 to 120 m, and their average is estimated to have been about 50 m. Based on numerical simulations, every 30 million years, two impact craters, 30 km in diameter, would form in this part of the ocean. A vast northern ocean is assumed to have existed for millions of years. The lack of shoreline characteristics, however, is a dispute against such an ocean. Chryse Planitia and northwestern Arabia Terra are the parts of Mars that were studied in this research, while some surfaces in the Mare Acidalium quadrangle and Ismenius Lacus quadrangle may have been affected by these tsunamis.[145–147]

3.5. Present Water

Applying a gamma-ray spectrometer to record the spectra of gamma rays emitted from the Martian surface as the spacecraft passes over different regions of the planet and the Mars Odyssey neutron spectrometer, surface hydrogen has been detected globally in considerable amounts. The molecular structure of ice seems to contain this hydrogen; and based on stoichiometric calculations, in the upper meter of the surface of Mars, some concentrations of ice have been reached by the conversion of the detected fluxes. According to this process, ice is not only plentiful, but also widespread on the present surface of the planet. It is concentrated in various regions with latitudes lower than 60 degrees, such as in Terra Sabaea, around the Elysium volcanoes, and the northwest of Terra Sirenum; it also exists in the subsurface, with some concentrations containing 18% ice. Ice is plentiful above 60 degrees latitude, so that its concentration exceeds 25% nearly everywhere at 70 degrees of latitude, and at the planet's poles, it reaches 100%. Due to the instability of the ice present on Mars' surface, most of it is estimated to be covered by a layer of dusty or rocky material.[148,149]

Mars Odyssey neutron spectrometer observations demonstrated that a Water Equivalent Global layer (WEG) of at least ≈14 cm (5.5 in) could not be reached unless the ice in the top meter of the Martian surface was spread equally; in other words, Mars' surface can be globally averaged to be roughly 14% water. Further, both of Mars' poles correspond to a WEG of 30 m (98 ft). Over the planet's geological history, geomorphic evidence suggests larger quantities of surface water, with a WEG as deep as 500 m (1600 ft). Even though the detailed mass balance of the governed processes is still unknown, portions of this past water may have been lost into the deep subsurface and into space. Ice has gradually migrated from one part of the planet's surface to another. This has occurred seasonally and on longer timescales, and been aided by the modern atmospheric water reservoir. However, with a WEG of less than 10 micrometers (0.00039 in), it is unimportant in volume.[150]

3.6. Polar Ice Caps

During the winter, the thickness of the northern (Planum Boreum) and southern (Planum Australe) polar ice caps increases, but they partially decrease in the summer, owing to sublimation (Figure 3). Expansion of the ice depth to up to 3.7 km (2.3 mi) below the surface in the southern polar cap was reported in 2004 by the Mars advanced radar for subsurface and ionosphere sounding (MARSIS) radar sounder from the Mars Express satellite (Figure 4). Based on detection by the spectro-imaging instrument (OMEGA) from the orbiter in the same year, three various parts, with different frozen water contents, were found in the southern cap. The first part, centered on the pole, comprises 85% CO_2 ice and 15% water ice, and can be seen in photos of the polar cap as the bright part. The second part, which is located in the plains surrounding the ice cap looks like a ring, and includes steep slopes classified as scarps, containing almost only water ice. Extending multiple kilometers away from the scarps, the third part includes large permafrost areas, and its surface composition required analyses to be included as part of the cap. NASA scientists estimate that if the water ice in the southern polar ice cap were to melt, it would be able to cover all of the planet's surface 11 m (36 ft) deep. The south polar region is predicted to have an old ice sheet probably containing 20 million km^3 of water ice, which can cover the entire planet up to 137 m deep.[152–156]

The existence of water ice near the northern polar ice cap (at 68.2° latitude), which was the Phoenix lander's landing site, was announced by NASA in July 2008. It was the first time that the direct detection of ice from the surface occurred. The total volume of water ice in the cap is 821,000 cubic km (197,000 cubic mi), based on the measurements of the northern polar ice cap conducted by the shallow radar onboard the Mars Reconnaissance Orbiter in 2010. That is equivalent to the amount required to cover the surface of Mars to a depth of 5.6 m (18 ft), or equal to 30% of the Earth's Greenland ice sheet. Examination in the high-resolution imaging science experiment (HiRISE) and Mars Global

Figure 3: The Mars Global Surveyor acquired this image of the Martian north polar ice cap in early northern summer.

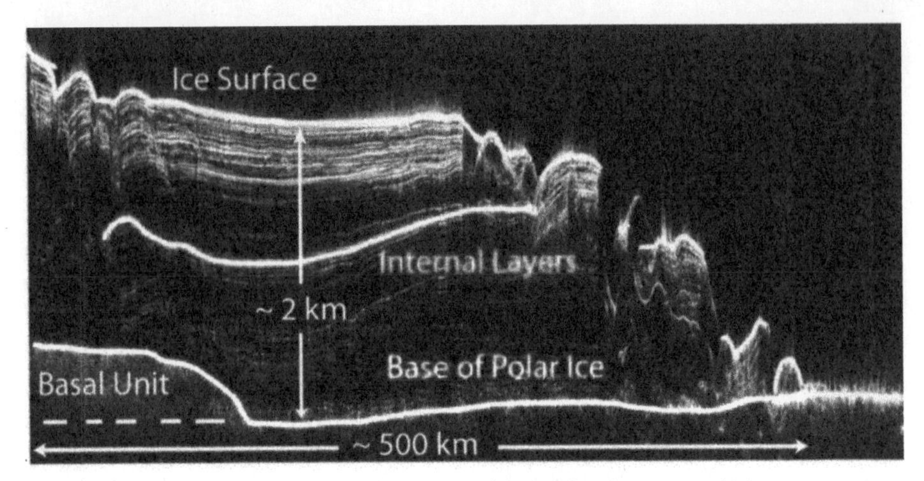

Figure 4: Cross-section of a portion of the north polar ice cap of Mars, derived from satellite radar sounding.

Surveyor imagery shows that both polar caps exhibit numerous fine internal layers. Understanding the history, flow properties, and structure of caps depends on the study of this layering, which has been done by several researchers. However, their explanation is not straightforward.[157,158]

The Mars Express orbiter discovered that Korolev Crater, with over 80 km in diameter and located near the northern polar cap, contained nearly 2200 cubic km (530 cubic mi) of water ice. A central mound of permanent water ice, with 1.8 km (1.1 mi) in depth and up to 60 km (37 mi) in diameter, covered the crater floor, situated approximately two kilometers (1.2 mi) below the rim.[159,160]

3.7. Subglacial Liquid Water

Given the existence of Lake Vostok in Antarctica, if before the polar ice caps formed, water existed on Mars, it is possible to conclude that liquid water might still exist on Mars. However, it is a theory that needs to be verified. As the first indication for a liquid water body with stability on the planet, the discovery of a subglacial lake on Mars located 1.5 km (1 mi) below the southern polar ice cap and extending 20 km (10 mi) horizontally, was announced by the Italian Space Agency in July 2018. From May 2012 to December 2015, the MARSIS radar onboard the European Mars Express orbiter collected echo sounding data showing a bright spot, considered as an evidence for the Mars lake. Centered at 193 °E, 81 °S the discovered lake contains mostly flat areas, without any anomalous topographic features, and is enclosed by higher ground, with only one depression on its eastern side. No sign of the lake has been detected by the Mars shallow radar sounder (SHARAD) radar onboard NASA's Mars Reconnaissance Orbiter. However, the team decided to observe again in the future to attempt to verify the finding when its orbital parameters are appropriate. Since the ability of SHARAD in ground-penetration is lower than MARSIS, the detection of the lake by SHARAD would be unlikely.[161–167]

Given the temperature of 205 K (–68 °C; –91 °F) at the bottom of the south polar cap, scientists assume that the water remains liquid due to the antifreeze effect of calcium perchlorates and magnesium. The hypothetical lake is covered by a 1.5-km (0.93 mi) ice layer, containing water ice with 10% to 20% admixed dust and seasonally covered by a one-meter (3 ft 3 in)-thick layer of CO_2 ice. The scientists referred to the limitations of raw-data coverage of the south polar ice cap and mentioned that "there is no reason to conclude that the presence of subsurface water on Mars is limited to a single location".[168,169]

3.8. Ground Ice

Several scientists have proposed, for many years, that there are similar characteristics between periglacial areas on the Earth and some Martian surfaces. Various shapes such as polygons and stripes can occur as typical characteristics in the higher latitudes. On the Earth, generally, freezing and thawing of soil cause these shapes. Terrain softening, which rounds sharp topographical characteristics, can be considered as one of the other indications for abundant frozen water under the surface of Mars. The presence of ground ice can be inferred from many of these features, confirmed by evidence from direct measurements with the Phoenix lander and Mars Odyssey's gamma-ray spectrometer.[170–174]

In 2017, an image from the HiRISE camera onboard the Mars Reconnaissance Orbiter (MRO) demonstrated at least eight eroding slopes, indicating exposed water ice sheets with 100-m thickness, coated by one or two meters of a soil layer. By considering that the latitudes of the sites varied from nearly 55 to 58 degrees, there may be shallow ground ice under approximately a third of the Martian surface. The results given by this image are in line with those of the ground-penetrating radars on Mars Express and on MRO, the Phoenix lander's in situ excavation, and the spectrometer on the 2001 Mars Odyssey. Evidence for Mars' climate history can be reached easily by these ice layers, which provide accessibility of frozen water for future human or robotic explorers.[175–176]

3.9. Scalloped Topography

Scalloped-shaped depressions seem to be in some regions of Mars. A degrading ice-rich mantle deposit is thought to remain in these depressions. The main reason causing scallops is ice sublimating from frozen soil. Sublimation under current Martian climate conditions causes the subsurface's loss of water ice, leading to the formation of the scalloped topography landforms. A model predicts similar shapes when the ground has large amounts of pure ice, up to many tens of meters in depth. When the ground has massive pure ice, up to numerous meters in depth, similar shapes can be estimated by a model. As ice formed on dust, when the climate was different because of variations of the Mars pole tilt, the deposition of this mantle material from the atmosphere was probable. The depth of scallops are usually tens of meters, and their extent is varied from a few hundred to a few thousand. They are typically circular or elongated. The combination of some of them has seemingly resulted in the formation of an enormous massively pitted terrain.[179–183]

The discovery of a large amount of underground ice in the Utopia Planitia region of Mars was announced by NASA on November 22, 2016. Scientists have estimated the volume of identified water to be equal to the volume of water in Lake Superior. Using the ground-penetrating radar instrument on the Mars Reconnaissance Orbiter, called SHARAD, the volume of water ice in the area was estimated. The dielectric constant, known as "dielectric permittivity", was calculated by applying the data acquired from SHARAD. The value of this constant was equivalent to an enormous concentration of water ice. There are superficial similarities between these scalloped features and those of Swiss cheese, observed around the south polar cap. Swiss cheese features are thought to be due to cavities forming in a surface layer of solid carbon dioxide, rather than water ice—although the floors of these holes are probably H_2O-rich.[184,185]

3.10. Ice Patches

The European Space Agency reported the presence of a crater partly filled with frozen water, on 28 July 2005; the discovery was later interpreted as an "ice lake". The High-Resolution Stereo Camera onboard the European Space Agency's Mars Express orbiter has taken crater images, which clearly display a large sheet of ice in the bottom of a crater situated on Vastitas Borealis, a broad plain that covers a significant part of Mars' far northern latitudes, at roughly 70.5° north and 103° east. The crater depth is nearly two kilometers (1.2 mi) and its width is about 35 km (22 mi). Additionally, its floor is located 200 m (660 ft) higher than the surface of the water ice. Partially visible sand dunes, located beneath the water ice, are recognized by ESA scientists as the main reason for this height discrepancy. Despite not being referred to as a "lake" by scientists, the water ice patch is considerable since it is always present, and its size is remarkable. Up to the present date, explorers have found water ice deposits and layers of frost in several locations on the planet. Through more images of Mars' surface, achieved using the modern generation of orbiters, it has gradually become clear that apparently many more patches of ice have been spread across the Martian surface. Most of these likely ice patches are concentrated in the Martian mid-latitudes (≈30–60° N/S of the equator). For instance, dust- or debris-covered ice patches, which are slowly degrading, are considered by many scientists as the general characteristics in those latitude bands, differently described as "latitude dependent mantle" or "pasted-on terrain". Surface features consistent with existing pack ice have been discovered in the southern Elysium Planitia. The nearby geological fault Cerberus Fossae is supposed as the source for a flood, which caused not only spewing water, but also lava aged some 2 to 10 million years. It seemed to scientists that the frozen lakes may still exist. They were created by the water, which came from Cerberus Fossae and then pooling and freezing in the low-level plains.[186–193]

3.11. Glaciers

Evidence shows the current existence of glaciers or at least their earlier presence in numerous large regions of Mars. It seems to the scientists that large amounts of water ice can still be present in many high latitude areas, particularly the Ismenius Lacus quadrangle. Additionally, based on evidence obtained recently, across large areas in middle and high latitudes of the planet, there is still water ice as glaciers, which do not sublime, due to having thin covers of protecting rock and/or dust. Glaciers are related to fretted terrain and several volcanoes. Glacial deposits have been characterized by researchers in various shield volcanoes on

Mars, including Hecates Tholus, Arsia Mons, Pavonis Mons, and Olympus Mons. Further, glaciers have been described in some larger Martian craters in the mid-latitudes and above.[190–195]

Based on their characteristic forms, the landforms from which they originated, and their locations, glacier-like characteristics on Mars are identified differently as: Martian flow characteristics, lobate debris aprons (LDA), lineated valley fill (LVF), or viscous flow features (VFF). Most of the small glaciers are seemingly related to gullies on the walls of craters and mantling material. The floors of most channels that are in the fretted terrain around the Arabia Terra in the northern hemisphere contain rock-covered glaciers that can possibly be attributed to the lineated deposits recognized as LVF. Ridged and grooved materials deflecting surrounding obstacles can be seen on their surfaces. Orbiting radar demonstrates that lineated floor deposits, which may be associated with LDA, comprise considerable amounts of ice. The "LDA" are known as characteristics that have been interpreted by scientists as glacial flows for many years. Additionally, scientists have supposed that ice could be found under a layer of insulating rocks. Later, utilizing innovative instrument readings, the presence of pure ice covered with a layer of rocks in LDA was confirmed.[196–199]

When ice moves, it transfers rock materials. This phenomenon usually occurs at the snout or edges of the glacier. This characteristic is known as moraines on Earth; on Mars, however, it is commonly identified as concentric ridges, arcuate ridges, or moraine-like ridges. The appearance of these glaciers remain, and the ridges they leave are not precisely the same as typical glaciers on Earth because of two main reasons: first, the low temperatures of Mars create conditions under which glaciers are not able to slide, as they are frozen down to their beds (known as "cold-based"); second, instead of melting, which happens on Earth, ice typically tends to sublime on Mars. Mars has a lower geothermal heat flux, density, and atmosphere temperature than Earth, since lower solar insolation reaches it compared to Earth. Therefore, the ice in a glacier on Mars is frozen into the ground because the temperature in the interface between a glacier and its bed is below freezing, which is suggested by modeling. When this process occurs, ice cannot slide across the bed, and consequently, it has less ability for surface erosion. Since the ice in glaciers on Mars is generally frozen and, therefore, is not able to slide, Martian moraines tend to be deposited without being deflected by the underlying topography. The direction of ice movement can be determined by ridges of debris that are found on the surfaces of the glaciers. Because of the sublimation of buried ice, some glaciers have rough textures on their surfaces. No ice melting occurs during evaporation on Mars, leaving an empty space behind any overlying, which then collapses into the void. Sometimes, ice chunks fall from the glacier and get buried in the land's surface. Their melting leaves a nearly round hole, and many of these "kettle holes" have been identified by scientists on the surface of Mars. There is strong evidence for glacial flows on Mars; however, there is not enough evidence for landforms carved by glacial erosion. On Earth, these characteristics (landforms carved by glacial erosion) can be found plentifully in glaciated regions; on Mars, however, the fact that they are absent puzzles scientists. The lack of these landforms might be because the glaciers on Mars have ice of a cold-based nature.[200–202]

4. Development of Mars' Water Inventory

There is a substantial nexus between the evolution of the Mars atmosphere and the change in its surface water content, characterized by some key stages.

4.1. Early Noachian Era (4.6 Ga to 4.1 Ga)

Scientists have characterized the early Noachian era using the atmospheric loss to space, which was due to hydrodynamic escape (causes the exit of heavier atoms of a planetary atmosphere due to numerous collisions with lighter atoms), as well as from heavy meteoritic bombardment. Approximately 60% of Mars' early atmosphere may have been eliminated through ejection by the impact of asteroids and comets.[203,204]

During this period, it is possible that significant quantities of phyllosilicates formed. This would have required an atmosphere dense enough to sustain surface water. Moderate water to rock ratios have been suggested by the spectrally dominant phyllosilicate group, smectite. However, the pH-pCO_2 between the smectite and carbonate show that the precipitation of smectite would constrain pCO_2 to a value not more than 1×10^{-2} atm (1.0 kPa). Accordingly, by considering the lack of evidence for carbonate deposits, if the clays formed adjacent to the Martian atmosphere, there would be uncertainty as to the determination of the principal component of early Mars' dense atmosphere. An atmosphere that had a strong greenhouse effect would have been required in order to raise the surface temperature to a level that would sustain liquid water. This is because the young Sun had approximately 25% lower brightness than it has today, which makes determining early Mars' dense atmosphere more complicated. The higher CO_2 content would not have been enough because CO_2 is not very effective as a greenhouse gas, as it precipitates at partial pressures exceeding 1.5 atm (1500 hPa).[205,206]

In NWA7533, unlike modern Martian soils, interclast crystalline matrix (ICM) and clast laden impact melt rock (CLIMR) do not show enrichments of S, Cl, and Zn with values similar to SNC meteorites (The SNC meteorites are a group of petrologically similar achondrites named after the locations where examples of these meteorites were first found—Shergotty (India), Nakhla (Egypt), and Chassigny (France)). These elements are likely to be in water-soluble phases in modern soils and the lack of enrichment observed in NWA7533 components is probably due to the transportation of these salts into ancient seas or lakes by liquid water present on Mars at the time of formation of ICM and CLIMR. The chemical composition of Martian wind-blown dust, present as ICM and CLIMR in NWA7533, should provide clues to the original igneous processes that formed the primary Martian crust. Moreover, combined ^{142}Nd–^{143}Nd isotope evidence in shergottites implies that the formation of the enriched and depleted reservoirs on Mars occurred within the first 100 Myr of the planet's history.[199]

4.2. Middle to Late Noachean Era (4.1 Ga to 3.8 Ga)

Mars could have experienced the possible formation of a secondary atmosphere during the middle to late Noachean era. It occurred by outgassing, controlled by the Tharsis volcanoes with considerable amounts of H_2O, CO_2, and SO_2. It can be inferred from Martian valley networks related to this period that surface water was globally extensive and sustained temporally, rather than caused by floods. This period ended at the same time when the bombardment of meteorites and the internal magnetic field terminated. When the internal magnetic field ended and the local magnetic fields weakened, the solar wind allowed for atmospheric stripping. For instance, in comparison with their telluric equivalents, $^{38}Ar/^{36}Ar$, $^{15}N/^{14}N$, and $^{13}C/^{12}C$ proportions of the Martian atmosphere are in line with losing ~60% of Ar, N_2, and CO_2 caused by the solar wind stripping the upper atmosphere, which was enriched in lighter isotopes through Rayleigh fractionation. Ejection of atmospheric components in bulk, without isotopic fractionation, is the consequence of increased solar wind activity influences. Nonetheless, volatiles may have occurred, in particular, by cometary impacts.[203,204]

4.3. Hesperian to Amazonian Era (Present) (~3.8 Ga to Present)

Solar wind stripping of the atmosphere countered atmospheric enhancement by sporadic outgassing events, though less intensely than the young Sun did. In this period, in the opposite direction with sustained surface flows, the disastrous floods induced sudden subterranean releases of volatiles. In the late Noachian (early Hesperian), the formation of Fe^{3+} oxides, which are what give Mars' surface a red color, were a major part of the surface alteration processes; the earlier part of this era included groundwater discharges centered around Tharsis and included mainly as aqueous acidic environments. Different processes can yield such oxidation of primary mineral phases without requiring free O_2, including low-pH (and maybe high temperature) processes that are associated with the formation of palagonitic tephra, the action of H_2O_2 that forms photochemically in the Martian atmosphere, and the action of water. There was an extreme decrease in both igneous and aqueous activity in early Hesperian, and so the action of H_2O_2 may have dominated temporally. This would make the observed Fe^{3+} oxides volumetrically small, even though they are spectrally dominant and pervasive. Nonetheless, given the geomorphology of craters such as Mojave, it seems that aquifers may have kept localized and sustained surface water in the recent geologic history of the planet. Moreover, evidence of aqueous changes can be seen through the study of the Lafayette Martian meteorite as recently as 650 Ma.[203–211]

4.4. Ice Ages

The Late Amazonian glaciation was the most recent, occurring at the dichotomy boundary around 2.1 to 0.4 M years ago. However, there have been many big changes on Mars' surface, during the last five million years, in relation to the distribution of the ice and the amount of ice on the planet's surface. These variations are called ice ages, but are much different from those that happened on the Earth. On Mars, ice ages are driven by changes in the planet's orbit and tilt (obliquity). According to the orbital calculations, it seems that Mars wobbles much more commonly on its axis when compared with Earth. The Earth only wobbles a few degrees, owing to stabilization by its proportionally large moon. Mars' tilt might be varied by many tens of degrees. Mars' poles receive much more heat, due to direct sunlight, when this obliquity is high; this warms the ice caps and they become smaller by ice subliming. Accordingly, since climate variability increases, the changes in the eccentricity of the Mars orbit are twice the variations in Earth's eccentricity. When the poles on Mars sublime, the ice moves closer to the planet's equator, as this area receives lesser solar insolation at high obliquities. Computer simulations can show that ice accumulation in areas displaying glacial landforms is caused by Mars' axis being at a tilt of 45°.[212–218]

The moisture that is originated from the ice caps moves toward lower latitudes in the form of snow, combined with dust or frost deposits. In Mars' atmosphere, a considerable amount of fine dust particles exist, which water vapor condenses on, leading to an increase in the weight of the water coating, and accordingly dropping it to the ground. As it returns to the atmosphere at the mantling layer's surface, ice leaves dust, which performs as insulation for the remaining ice. Only a small percentage of the ice caps is removed as water, which is equivalent to the amount of water it would take to cover the planet's surface up to one meter. Most of the moisture that comes from the ice caps causes the formation of a thick smooth mantle, which is made of a combination of dust and ice. Being up to 100 m thick at the mid-latitudes, this ice-rich mantle smooths the land at lower latitudes. The existence of water ice underneath the mantle is apparent due to its uneven texture and the observable patterns.[219–221]

5. Evidence for Recent Flows

The current low temperature and atmospheric pressure make it improbable for pure liquid water to exist in a stable form on Mars. However, water may be located at the lowest elevations for short periods of time. Hence, after the observation of gully deposits by NASA's Mars Reconnaissance Orbiter in 2006, which were not present ten years earlier, a geological mystery began. During the warm months on Mars, liquid brine flows on the surface, which may play the primary role in forming these gullies (Figure 5).[221–225]

It is still controversial among scientists whether liquid water forms gullies. There is also a possibility that the flows carving gullies are dry grains, or maybe carbon dioxide has lubricated them. Some studies suggest that the gullies are formed by melting carbon dioxide during the summer. Although it is thought that water flowing on the surface of the planet carved gullies, the specific water source or the mechanisms that caused its movement are still unknown. The claim that improper conditions disprove the possibility of gully formation by water in the southern highlands can be found in some studies. While the low pressure non-geothermal colder regions would be ideal for solid carbon, they would not provide appropriate conditions for liquid water at any point.[226–229]

Mars has deep grooves on the surface of the planet, which appear to be carved into various slopes, and are known as dry gullies. However, some of them vary seasonally. The modern seasonal variations in the southern hemisphere of Mars, on the steep slopes beneath the rocky outcrops that are near the rims

Figure 5: Branched gullies.

of craters, were announced by NASA in August 2011. Increased temperatures during some summer days cause these dark streaks, named recurrent slope lineae (RSL), to flow downslope; then in the rest of the year, they gradually fade due to decreasing temperature; this process recurs cyclically between years. These marks were thought by scientists to be consistent with the downslope flow of salty water (brines) and then by evaporation, depositing some kind of residue. In 2015, researchers confirmed that liquid brines flowing through shallow soils form these lineae. They reported that the CRISM spectroscopic instrument directly observed hydrous salts emerging at the same time as the appearance of recurrent slope lineae. The lineae are comprised of perchlorate salts and hydrated chlorate (ClO_4^-), and also have water molecules in liquid form. During the summer on Mars, when the temperature rises above –23 °C (–9 °F; 250 K), the lineae move downhill. However, it is still unknown as to where the water comes from. Nevertheless, based on the neutron spectrometer data that were obtained by the Mars Odyssey orbiter over a ten-year period, no indication of water (hydrogenated regolith) at the active sites could be found; however, two possible hypotheses are proposed: first, short-lived, atmospheric water vapor deliquescence; and second, dry granular flows. Even if evidence of liquid water can be found on Mars today, it is only a thin film and traces of moisture that have dissolved from the atmosphere, which does not make the environment appropriate for life as we know it.[230–237]

5.1. Findings by Probes

5.1.1. Mariner 9, 1971

The first direct indication of Mars having water in the past comes from dry river beds and canyons found from the images obtained using the Mariner 9 Mars orbiter after its launch in 1971. These include over 4020 km (2500 mi) of canyons in a system, which is called Valles Marineris. Mariner 9 Mars also found hints of weather fronts, fogs, water erosion, water deposition, and more. The reason for naming the giant Valles Marineris canyon system was indeed honoring Mariner 9's achievements. The following Viking program was underpinned by the discoveries of the Mariner 9 missions.[30]

5.1.2. Viking Program, 1976

Investigators found large outflow channels in various areas. These indicated the breaking through of dams by water floods, which eroded grooves into bedrock, carved deep valleys, and traveled thousands of kilometers. The rainfall that once occurred on the surface of Mars can be intimated by valley networks that branch out into large areas on the southern hemisphere of the planet. A revolution in the understanding of water on the planet Mars occurred when the two landers and the two Viking orbiters discovered numerous geological formations that generally require large amounts of water to be formed. Craters usually seem to be in shapes similar to the shape mud would take after impactors fell into it. During their formation, soil ice probably melted, changing the ground into mud, which resulted in mud flowing across the surface. Downstream, "Chaotic Terrain" was formed by large channels that were created when a significant volume of water was abruptly lost. Moreover, Chaotic terrains are irregular to circular fractured areas that consist of jumbles of blocks that vary in size and shape from knobs and irregular conical mounds to angular mesas with preserved remnants of the upland plateau, and they are often located in enclosed to semi-closed depressions surrounded by fractured highlands. Some channel flows are estimated to be more than ten thousand times that of the Mississippi River. One possibility is that ice was melted by underground volcanism, leading to water flowing away and crumbling the ground, leaving the chaotic terrain. Additionally, surface exposure to water, or the planet being submerged in water in the past, have been suggested by the two Viking landers, using general chemical analysis.[238–243]

5.1.3. Mars Global Surveyor, 1996

Mars' surface mineral composition could be specified from the Mars Global Surveyor's Thermal Emission Spectrometer (TES), which can provide information related to the history of water's presence or absence on Mars. Further, TES was used in studying the formation of Nili Fossae, which is comprised of the mineral olivine. Faults that exposed the olivine may have been caused by the impact of ancient asteroids that also made the Isidis basin. There are areas on Mars that, for a long time, have been very dry. In some areas that are between 60 degrees south or north of the planet's equator, scientists discovered olivine in several small outcrops. Additionally, images from the probe show that there may have been sustained liquid flows in the past, two of which could be in Nanedi Valles and in Nirgal Vallis.[244–246]

5.1.4. Mars Pathfinder, 1996

The mission of Mars Pathfinder was to record the diurnal temperature cycle variation. The coldest time in the cycle was found to be before sunrise, having a temperature of about −78 °C (−108 °F; 195 K), and the warmest was just after noon-time on Mars, with a temperature of about −8 °C (18 °F; 265 K).

The highest temperature was always found to be lower than water's freezing point, so it would be too cold for the existence of pure liquid on the surface. The Pathfinder lander measurement indicated that Mars had quite low atmospheric pressure, which was about 0.6% of Earth's; this is not high enough to allow for the presence of pure liquid water on the planet's surface.[247]

There are other observations that confirm the existence of water on Mars in the past. For instance, some of the rocks in the areas that the Mars Pathfinder studied were leaning against each other in a way that geologists call imbricated, which implies the presence of water in the past. Researchers speculated that rocks were pushed by substantial floodwaters in the past, leading them to face away from the flow of the water. Additionally, some rounded pebbles were found that may have been tumbled in a stream. Further, parts of the ground were found to be crusty, which may have been caused by cementing from a mineral-containing fluid. Moreover, the Pathfinder found some evidence of clouds, and perhaps fog.[248]

5.1.5. Mars Odyssey, 2001

Images provided by the 2001 Mars Odyssey revealed many indications for water on Mars, and its neutron spectrometer showed that significant parts of the ground were full of water ice. The amount of ice beneath the surface of Mars is adequate to fill Lake Michigan twice. In both hemispheres of the planet, just under the surface, from latitudes of 55° to the poles, ice can be found at high density, in which one kilogram of soil comprises nearly 500 g (18 oz) of water ice. Near the equator, however, only 2%–10% of soil contains water. Researchers think that the chemical structure of minerals, like clay and sulfates, lock up a significant amount of this water. The upper surface of the planet contains only a small percentage of chemically-bound water; however, ice can be found just a few meters deeper; for example, large amounts of water ice is contained in the Arabia Terra, Amazonis quadrangle, and Elysium quadrangle. Moreover, in equatorial regions, vast deposits of bulk water ice were discovered by the orbiter near the surface of the planet. Evidence for both morphological and compositional equatorial hydration can be seen in two different areas: the Medusae Fossae formation, and the Tharsis Montes. Given data analysis on the southern hemisphere, there are indications of a layered structure from stratified deposits that are located below what is believed to have been a huge water mass that is now extinct. The information about the top meter of soil has been acquired using the instruments aboard the Mars Odyssey. Calculations based on available data in 2002 showed that a layer of water covering all soil surfaces would be equivalent to a global layer of water (GLW) 0.5–1.5 km (0.31–0.93 mi) deep. Evidence from many images provided by the Odyssey orbiter indicate that there were once large amounts of water flowing across Mars' surface. Layers that probably formed under lakes, as well as patterns of branching valleys, and river and lake deltas, can all be inferred from the images. Scientists suppose that beneath an insulating rock layer, glaciers were present for many years. On the example of rock-covered glaciers is lineated valley fill, which exists on the floors of some channels. The surfaces of these areas contain ridged and grooved materials, which deflect around obstacles. Orbiting radar has shown a possible relationship between lineated floor deposits and lobate debris aprons to contain substantial amounts of ice.[249–254]

5.1.6. Phoenix, 2007

The Phoenix lander showed that in Mars' northern regions, there was evidence for substantial water existing as water ice. On June 19, 2008, NASA reported the vaporization of small pieces of bright material that were found in the "Dodo-Goldilocks" trench, which was an area excavated by the lander's robotic arm. Because the material vaporized over four days, this was an indication that the pieces were most likely made of ice, and they sublimed due to exposure. Although solid CO_2 (dry ice) also sublimes under similar conditions, the claim of water ice sublimation cannot be denied, since the former sublimes at a rate much faster than what was observed for the bright material. More confirmation of existing water ice at the Phoenix landing site was reported by NASA on July 31 of the same year. A sample was taken by the lander, and vaporization occurred when the temperature of that sample reached 0 °C (32 °F; 273 K). Given the low temperature and atmospheric pressure on the surface of Mars, liquid water can only exist at the lowest elevations for a short time. It also was corroborated that a strong oxidizer, Perchlorate (ClO_4), can be found in the soil. When the chemical combines with water, it decreases water's freezing point, as salt does when it is used to melt ice on roads.[255–258]

Soil and melting ice were splashed onto the vehicle due to the retrorockets during the Phoenix's landing (Figure 6). This left material on the landing struts during landing, as shown by photographs. This material then expanded in a way that resembled deliquescence, as it got darker before it disappeared, which is what happens when something liquefies and then drips. The material was probably liquid brine droplets, which can be inferred from the thermodynamic evidence and these observations. Other scientists considered this material as "clumps of frost".[259–262]

It seemed from the camera that the landing site was flat. However, there were some polygons that were surrounded by what appeared to be troughs 20–50 cm (7.9–19.7 in) deep. Expansion of ice in the soil and its contraction, owing to notable temperature variations, are the main reasons for polygons to be formed. Rounded and flat particles can be seen in a microscope as the main parts of the soil on top of the polygons, which can be considered to be a kind of clay. In the middle of the polygons and beneath their surface a few inches of ice exist, but the thickness of ice reaches at least 8 inches (200 mm) along the polygons' edges. Further, there were Cirrus clouds that formed in the atmosphere with a temperature of about –65 °C (–85 °F; 208 K), which are thought to be the origin from which snow was recognized to fall. Therefore, instead of carbon dioxide ice (CO_2 or dry ice), which requires a lower temperature than –120 °C (–184 °F; 153 K), water-ice formed the major part of the clouds. According to the mission observations, researchers now have a suspicion that later in the year, water ice (snow) may have accumulated at this location. During the mission, –19.6 °C (–3.3 °F; 253.6 K) was recognized as the highest temperature that occurred in

Figure 6: View underneath Phoenix lander showing water ice exposed by the landing retrorockets.

the Martian summer, and –97.7 °C (–143.9 °F; 175.5 K) as the lowest one. As a result, the temperature in this region never reached water's freezing point (0 °C (32 °F; 273 K)). It has been recently argued that shallow transient liquid water, which is likely widespread on today's Mars, can potentially support life. In addition, there is a capability of supporting microbial life in portions of the shallow subsurface of Mars.[263–265]

5.1.7. Mars Exploration Rovers, 2003

Strong evidence of water existing in the past on Mars was obtained by the Mars Exploration Rovers, Spirit (June 10, 2003), and Opportunity (July 8, 2003). It is believed that the area where the Spirit rover landed was a giant lake bed. It was first hard to detect evidence of water in the past, since lava flows had covered the lake. It was announced by NASA on March 5, 2004, that the Spirit ascertained clues to Mars' water history in a rock named "Humphrey". Detailed inspection of the rock revealed a bright material filling internal cracks. Such material may have crystallized from water trickling through the volcanic rock. The amount of Mars once covered by ancient water remains unknown, as both rovers landed in regions thought likely to once be underwater. The upper layer of soil was scraped off by the rover's wheels in December 2007, which uncovered some white ground rich in silica. Scientists know of only two possible ways of producing white ground rich in silica: dissolving silica at one location and it being transferred to another location (i.e., a geyser), and silica being left behind when acidic streams may have flowed through rock cracks and depleted their mineral components. The Spirit rover could also provide evidence for water in the Gusev Crater's Columbia Hills. Various components were detected by the Mössbauer spectrometer (MB) in the rocks of Clovis, including goethite, which only forms in the presence of water; an oxidized form of iron (Fe^{3+}); and rocks that were rich in carbonate, which implies the presence of water in some areas of the planet.[266–273]

A significant amount of hematite was shown from orbit in a site to which the Opportunity rover was directed. Hematite concretions and layered rocks that were marbled, were also found by the rover. On Opportunity's transverse, in the Burns Cliff area of the Endurance Crater, the rover examined aeolian dune stratigraphy. Opportunity's operators realized that shallow groundwater played a principal role in the preservation and cementation of the outcrops. Through its years of continuous operation, the rover could determine that that area on Mars had been covered with liquid water.[274,275]

The indication of highly acidic ancient wet environments was affirmed by the MER rovers. Opportunity discovered sulfuric acid evidence, which creates a harsh environment for life on the planet. However, NASA announced on 17 May 2013, that the discovery was made in clay deposits that usually form in neutral acidity wet environments. The announcement helped researchers to have more evidence of an ancient wet environment, which presumably contributes to providing a suitable environment for life.[276,277]

5.1.8. Mars Reconnaissance Orbiter, 2005

The Mars Reconnaissance Orbiter's HiRISE instrument obtained several images that show strong evidence that on Mars, there is a history of water-related processes. Obtaining an indication of ancient hot springs was its main discovery. If Mars at one time contained microbial life, finding their bio-signatures would be probable. In January 2010, near Valles Marineris, researchers reported strong evidence for sustained precipitation. They indicated the association of minerals with water. Besides, since small branching channels with high density are also present in that area, a large amount of precipitation is expected. In many various places, rocks on Mars are typically in a form with layers, known as strata. These layers can be formed in various ways, such as by wind, volcanoes, or water. The main parts of the light-toned rocks on Mars have been determined to be made from hydrated minerals, like clay and sulfates. Scientists used the orbiter to ascertain that a thick smooth mantle, which is made from a mixture of ice and dust, covers the surface of Mars.[278–283]

Frequent, major climate change is speculated to be one of the main reasons for Mars' shallow subsurface having an ice mantle. Coming down from polar regions to lower latitudes that would be similar to those of Texas on Earth, notable changes in the water ice distribution can be detected, which are caused by changes both in the tilt and orbit of the planet. Water vapor moves toward the atmosphere from polar ice in periods that certain climates occur; then at lower latitudes, it returns to the ground in the form of snow or frost combined with dust. Using the Shallow Radar on the Mars Reconnaissance Orbiter, researchers in 2008 found clues showing that the lobate debris aprons (LDA) in Hellas Planitia and in mid-northern latitudes are glaciers that are covered with a thin layer of rocks. From the top and base of the LDAs, a strong reflection was detected by its radar, from which it can be implied that the bulk of the formation is made mostly from pure water. Since water ice was discovered in the LDAs, it is expected that even at lower latitudes, water can be found. In September 2009, some new craters on Mars were reported showing exposed, pure water ice. The ice, with only a few feet of depth, vanishes after a while and evaporates into the atmosphere. The Compact Imaging Spectrometer (CRISM) onboard the Mars Reconnaissance Orbiter corroborated the existence of the ice.[284,285]

Very recently, in 2019, investigators assessed the amount of water ice located at the northern pole. In one report, scientists applied the data obtained by the MRO's SHARAD (SHAllow RADar sounder) probes. The capability of SHARAD is scanning up to approximately two kilometers (1.2 mi) beneath the surface at 15-m (49 ft) intervals. Clues of water ice strata and sand beneath the Planum Boreum, whose volume contains around 60% to 88% of water ice, were obtained when past SHARAD runs were analyzed. The results were consistent with the theory of Mars' long-term global weather, including cycles of global warming and cooling; when cooling periods occurred, ice layers were formed at the poles, owing to accumulating water; then during global warming, dust and dirt from Mars' recurrent dust storms coated the unthawed water ice. This study estimated that the total ice volume was equivalent to about 2.2 × 105 cubic kilometers (5.3 × 104 cubic mi), or to the value adequate to thoroughly cover Mars' surface with 1.5 m (4.9 ft) layer of water. A separate study that used recorded gravity data to determine the density of Planum Boreum confirmed this work, showing that on average, up to 55% of the volume was water ice.[286,287]

5.1.9. Curiosity Rover, 2011

NASA's Curiosity rover detected unequivocal fluvial sediments when it was at the beginning stages of its continuing mission. An earlier robust flow in a streambed, containing water between waist and ankle-deep, was expected given the characteristics of the pebbles in these outcrops. These rocks were discovered at the base of an alluvial fan system, which came from a crater wall although they had already been recognized from orbit. In October 2012, Curiosity accomplished an X-ray diffraction analysis of soil from Mars for the first time. The presence of various minerals, including feldspar, pyroxenes, and olivine was revealed. Therefore, it was inferred that the soil in the analyzed sample was similar to the weathered basaltic soils that are found in Hawaiian volcanoes. The composition of local fine sand and dust, disseminated from global dust storms, was also found in the sample. So far, the results from the materials analysis by Curiosity are in line with the initial ideas related to the Gale Crater area deposits, which seem to show that a transition from a wet to dry environment happened on the planet over time. The existence of sulfur, chlorine, and water molecules in the soil of Mars was announced by NASA in December 2012, when Curiosity implemented its first extensive soil analysis. Additionally, from the various rock samples analyzed, including the broken pieces of "Tintina" rock and "Sutton Inlier" rock, along with the nodules and veins in other rocks, like "Knorr" rock and "Wernicke" rock, NASA announced in March 2013 clues for the presence of mineral hydration, which was most likely hydrated calcium sulfate.[288–293]

The rover's DAN instrument was used to analyze the possibility of subsurface water existence. The analysis led to obtaining subsurface water evidence from the transverse of the rover, from the Bradbury Landing site to the Yellowknife Bay area, down to 60 cm (two feet) deep. In soil samples extracted from the

Aeolis Palus, in the Rocknest region of the Gale Crater, the Mars Curiosity rover discovered plentiful chemically-bound water (1.5 to 3 weight percent), which was announced by NASA scientists on 26 September 2013. NASA also reported finding two main soil types: a regionally derived, coarse-grained felsic type; and a fine-grained mafic type. Like other soils and dust on Mars, the mafic type was related to hydration in the formless phases of the soil. Near the Curiosity rover landing site, scientists found perchlorates, which may decrease the detectability of life-related organic molecules. After that, the "global distribution of these salts" was suggested. Jake M rock, a rock that was found by Curiosity on its trajectory to Glenelg, was announced by NASA to be a mugearite that was similar to mugearite rocks on Earth.[294–301]

According to the NASA report on December 9, 2013, there was once a freshwater lake of large size inside Gale Crater, and this may have created conditions suitable for microbial life. Unusual variations (abnormal increase, then decrease) in methane amounts in Mars' atmosphere were reported by NASA on December 16, 2014; additionally, organic chemicals were detected by the Curiosity rover in a sample that was drilled from a rock. Besides, given the deuterium to hydrogen ratio studies, it was realized that in ancient times, before forming the lakebed in the crater, a substantial amount of water at Gale Crater on Mars had been lost, and this loss of large amounts of water persisted over time. Researchers analyzed the ground temperature and humidity data collected by Curiosity and revealed evidence for the formation of films that were made of liquid brine water during the night, in the upper five cm of Mars subsurface. However, temperature and water activity continued to exist at lower than the minimum values that are required for either the reproduction or metabolism of microorganisms that are found on Earth. The existence of the Gale crater around 3.3–3.8 billion years ago was affirmed by NASA on October 8, 2015. By delivering sediments, this crater created the lower layers of Mount Sharp. Investigations in Gale Crater by the Curiosity rover led geologists to present evidence of plenty of water existing on early Mars on November 4, 2018.[302–309]

5.1.10. Mars Express, 2003

To seek clues of sub-surface water, the European Space Agency launched the Mars Express Orbiter, which uses radar equipment to map Mars' surface. The area on the Palnum located beneath the ice caps was scanned by the Orbiter between 2012 and 2015 to find evidence of sub-surface water. According to the readings, by 2018, researchers ascertained the existence of an approximately 20 km (12 mi) wide lake in the sub-surface, containing water. While the location of the lake top is known, at approximately 1.5 km (0.93 mi) below the planet's surface, it is still unknown to what extent its depth continues.[310]

6. Summary

Over the years, various studies have been carried out to discover water on Mars. In the current article, the signs of water presence on the planet, based on results and reports obtained from the literature, were classified into the following main parts.

6.1. Evidence from Rocks and Minerals on Mars

The sorts of minerals present in the rocks can indicate Mars' historical environmental conditions, including those suitable for life. Aqueous minerals, which are minerals that include water or form in the presence of water, can indicate the types of environments in which the minerals were formed. Aqueous reactions occur depending on certain variables, including temperature and pressure, as well as concentrations of gaseous and soluble species. Water is the only condition under which the sulfate mineral jarosite might form. Another aspect of rock evidence is hydrothermal alteration. Migrating seawater, through ultramafic and basaltic rocks, may cause serpentinization, known as one primary type of hydrothermal alteration, which can occur in the oceanic crust. A large amount of water can be stored as hydroxyl in the crystal structure of serpentine minerals. Evidence from remote sensing data indicates that few serpentine minerals exist on Mars; however, there is a probability of large amounts of serpentinite hidden deep in the Martian crust.

The weathering rate is another reason for the existence of water on Mars. The silicate minerals that weather the fastest are those that crystallize at the highest temperatures. Olivine, which in the presence of water weathers to clay minerals, is the most common mineral meeting this criterion on both the Earth and Mars. Since olivine is extensive on Mars, it is likely that water has not significantly altered the planet's surface, but some geological evidence suggests otherwise. In addition, meteorites from Mars show signs of water. For example, the existence of sulfates and hydrated carbonates indicate that some Martian meteorites, known as basaltic shergottites, were exposed to liquid water before being ejected into space.

6.2. Geomorphic Evidence

Basin lakes and river valleys are landscape features found on Mars. Thanks to the Mariner 9 spacecraft, immense valleys were discovered in many areas of the planet. Some physical change examples caused by water floods that were displayed in its images include breaking through dams, eroding grooves into bedrock, and carving deep valleys. Moreover, inverted reliefs, which refer to landscape features near rivers that have reversed their elevation relative to other features, were discovered by the Mars Global Surveyor. Examples of this are the Miyamoto Crater, Saheki Crater, and Juventae Plateau. Various lake basins have also been discovered on Mars, some of which may have been formed by precipitation or groundwater. Several deltas formed in Martian lakes have been reported by researchers; the existence of significant liquid water amounts can be inferred by finding deltas. Eberswalde crater is an area where one of these deltas was found. In addition, evidence shows the current existence of glaciers, or their earlier presence, in numerous large regions of Mars. It seems to scientists that large amounts of water ice could still be present in many high latitude areas, particularly the Ismenius Lacus quadrangle.

6.3. Evolution of Mars' Atmosphere

There is a substantial nexus between the evolution of the Mars atmosphere and the change in its surface water content, characterized by four key stages:

(1)

Early Noachian Era: the atmospheric loss to space from heavy meteoritic bombardment and hydrodynamic escape is used to characterize this era. For this era, the dominant component of a dense atmosphere is not certain.

(2)

Middle to late Noachean Era: large amounts of H_2O, CO_2, and SO_2 were released by the Tharsis volcanoes and made a secondary atmosphere on Mars.

(3)

Hesperian to Amazonian Eras: solar wind stripping of the atmosphere countered atmospheric enhancement by sporadic outgassing events, though less intensely than the young sun did. In this period, in the opposite direction of sustained surface flows, disastrous floods induced sudden subterranean releases of volatiles.

(4)

Ice ages: during the ice ages, Mars' poles received much more heat, due to direct sunlight, when this obliquity was high; this warmed the ice caps and they became smaller by ice subliming. Accordingly, since climate variability has increased, changes in the eccentricity of Mars' orbit are twice the variations in Earth's eccentricity. Computer simulations can show that ice accumulation in areas displaying glacial landforms has been caused by Mars' axis being at a tilt of 45°.

6.4. Findings by Probes

The numerous probes sent to Mars have played prominent roles in collecting information about the presence of water on the planet. For instance, the Viking program, Mars Pathfinder, Phoenix, Curiosity Rover, have all collected data for this purpose. A revolution in the understanding of water on the planet occurred when the two landers and the two Viking orbiters discovered numerous geological formations that generally require large amounts of water to be formed. Investigators also found large outflow channels in various areas. The mission of the Mars Pathfinder was to record the diurnal temperature cycle variation. The Pathfinder lander's measurements indicated that Mars had quite low atmospheric pressure, about 0.6% of Earth's, and this is not high enough to allow for the presence of pure liquid water on the planet's surface. The Phoenix lander showed that in Mars' northern regions, there was evidence for substantial water existing as water ice. A photograph of underneath the Phoenix lander showed water ice exposed by its landing retrorockets. Permafrost polygons were also imaged by the Phoenix lander. Curiosity accomplished an X-ray diffraction analysis of soil from Mars for the first time. The presence of various minerals, including feldspar, pyroxenes, and olivine were revealed. In 2020, the Mars 2020 rover will seek the answer to questions about the potential for life on Mars, by searching for subsurface water. The Mars 2020 rover introduces a drill that can collect core samples of the most promising rocks and soils. More studies have to be conducted on Mars, to learn how life could be supported and where there are resources to support life in the future.

THE FUTURE OF MARS EXPLORATION, FROM SAMPLE RETURN TO HUMAN MISSIONS (2020)

Jeff Foust

Jeff Foust is Senior Staff Writer for *Space News*, specializing in space policy and commercial space activities. He holds a BSc with honours in geophysics from the California Institute of Technology and a PhD in planetary sciences from the Massachusetts Institute of Technology. Foust is also editor at *The Space Review* and publishes 'Space Today', a daily digest of space news from around the world.

When an Atlas V lifted off from Cape Canaveral July 30, NASA heralded it as the beginning of a new era of Mars exploration. The rocket was launching NASA's Mars 2020 mission, which will land the rover Perseverance on the surface of Mars in February. That rover will collect samples for later return to Earth, a long-running goal of scientists.

Just how those samples will be returned, though, is still a work in progress. NASA and ESA have agreed to cooperate on the effort, proposing a pair of missions launching as soon as 2026: a NASA-led lander with an ESA "fetch rover" to collect the samples, load them into a rocket, and launch them into orbit; and an ESA-led orbiter with a NASA collection system to pick up the samples and return them to Earth. It's a complex, expensive effort, but one for which there appeared to be growing consensus about how to accomplish.

A recent report, though, suggested that those plans might be a little too ambitious, in that it may take more time, and more money, than originally envisioned. That reality check comes as other elements of future Mars exploration, including both robotic and human missions, start to take shape and face similar scrutiny.

Assessing Mars Sample Return

After the successful launch of Mars 2020, NASA announced it would conduct an independent review of its Mars Sample Return program. The purpose of the review was to examine the plans as they currently stood to identify problems while the program was still in its early phases.

"The goals of the review are to make sure we're on a firm foundation technically going forward, to review the concepts that have been developed to date, and also to look at the cost and schedule that we're proposing and to make sure that they agree that we've got the right resources we need to do this job," Jeff Gramling, Mars Sample Return program director at NASA, said when the review was announced in mid-August.

That review would move quickly. Less than three months after announcing the review, NASA released the independent review board's final report. The good news for NASA was that the panel found no show-stoppers with the overall program.

"We unanimously believe that the Mars Sample Return program should proceed," said David Thompson, the former CEO of Orbital ATK who chaired the board, in a call with reporters November 10. "We think its scientific value would be extraordinarily high."

Or, in the words of another board member, planetary scientist Maria Zuber of MIT, "Full steam ahead."

The bad news, though, was that Thompson's board concluded that plans for carrying out the two future launches—the Sample Return Lander and Earth Return Orbiter—were unlikely to remain on schedule.

"Since the current development schedule necessary to achieve the 2026 launches were judged by the IRB [independent review board] to not be compatible with recent NASA experience," Thompson said, "we believe NASA should replan the program for launches in 2027 and 2028." The lander would, in that scenario, slip to 2028, while the orbiter could launch in either 2027 or 2028, depending on the selected trajectory.

A delay in launch, not surprisingly, comes with an increase in cost. NASA had said little about how much future phases of Mars Sample Return would cost, beyond rough estimates of $2.5–3 billion. (That does not include Mars 2020, which costs $2.7 billion through its first Martian year of operations after landing, nor ESA's contribution of about $1.8 billion.) But the independent review board concluded those phases of Mars Sample Return will likely cost $3.8–4.4 billion.

The panel said NASA should investigate other changes to the mission design. One possibility is to split the Sample Return Lander into two landers: one carrying the fetch rover and the other the rocket for launching the samples into orbit. A two-lander approach, the report stated, "may open up increased margins and design flexibility, [and] enable greater use of already-developed systems and subsystems."

That lander, and the fetch rover, are designed to be solar powered, but the report said NASA should consider adding an RTG like that powering Mars 2020 to the lander. Doing so, it argued, could extend its lifetime on the surface and improve thermal conditions, particularly for the Mars Ascent Vehicle rocket.

NASA accepted many of the recommendations from the independent panel, but stopped short of endorsing a delay in the 2026 launch of the orbiter and lander missions. Thomas Zurbuchen, NASA associate administrator, said any decision on changing schedules would require working with stakeholders in both the US and Europe. A decision on the schedule for the program, he said, could come "on the timescale of a year or so."

The report noted there are other risks should the mission be delayed beyond 2028: a 2030 launch would force the lander to operate during dust storm season on Mars. "It makes the mission more complicated, potentially requiring a redesign of the elements," said Gramling.

For now, work continues on Mars Sample Return activities towards a 2026 launch on both sides of the Atlantic. In October, ESA awarded a contract to Airbus Defence and Space to develop the Earth Return Orbiter valued at €491 million (US$595 million), part of a set of awards that Airbus and Thales Alenia Space received to support ESA's contributions to lunar and Mars exploration.

Mars Ice Mapper

Mars Sample Return is the major robotic Mars mission of the coming decade for NASA, but it is not the only one. The fiscal year 2021 budget proposal introduced a new one: Mars Ice Mapper, an orbiter equipped with a synthetic aperture radar designed to search for subsurface deposits of ice.

The inclusion of Mars Ice Mapper in the budget surprised some in the scientific community, who had not proposed such a mission. Instead, Mars Ice Mapper came out of the human exploration part of NASA as a precursor to support later human missions to Mars.

The mission was the outcome of a "summer study" in 2019 by the agency, said Jim Watzin, the former director of NASA's Mars Exploration Program who now serves as a senior advisor supporting human Mars mission planning at the agency. "The purpose of the summer study was to pull together NASA's strategy for preparing for human exploration," he said, including both the Moon and Mars.

A key element of that was identifying subsurface ice deposits that could be studied, and utilized, by those crewed missions. "The obvious question was, 'Where?' Where does the crew have to land?" he said at a November 23 meeting of a panel supporting the ongoing planetary science decadal survey. "So, the Mars Ice Mapper mission was identified as an essential precursor mission necessary to get that critical information so we could decide where to go for the first human mission, and also how to prepare for that mission."

NASA envisions Mars Ice Mapper as an international endeavor. The primary instrument, the radar mapper, would come from the Canadian Space Agency (CSA). The Japanese space agency JAXA would provide the spacecraft bus and Italian space agency ASI the communications subsystem for the spacecraft. Those arrangements could be formalized as soon as this month with a "statement of intent", followed by a memorandum of understanding in the spring or summer of 2021.

NASA would serve as the mission coordinator for Mars Ice Mapper and also provide a launch, likely no earlier than 2026. Watzin said NASA's notional budget for the mission is just $185 million, not including the contributions from international partners.

The spacecraft would operate in a low sun-synchronous orbit around Mars, using an L-band radar to probe to depths of five to seven meters, said Jim Garvin, chief scientist at NASA's Goddard Space Flight Center, at the decadal survey panel meeting. "This particular approach, as a point solution for Ice Mapper, has been peer reviewed by five experts from the radar community who have all flown radars in space," he said, who concluded it meets the goals of looking for subsurface ice deposits.

NASA wants to fly Mars Ice Mapper in 2026 or 2028. "We really need this data by the 2030 timeframe," said Rick Davis, assistant director for science and exploration in NASA's Science Mission Directorate, in order to support a human mission notionally in the mid-2030s.

However, some Mars scientists have been critical of the mission. At the decadal survey meeting, as well as one a week later by NASA's Planetary Science Advisory Committee, some questioned how the mission could meet goals of the science community.

"This is an exploration precursor mission. It's not a science mission per se," Watzin said at the decadal survey meeting. He compared it to Lunar Reconnaissance Orbiter, which NASA developed initially to support the Constellation program of human lunar exploration more than a decade ago, but later became part of NASA's planetary science program.

"The notion of an exploration-driven mission that produces huge science returns is not without precedent," said Tim Haltigin, senior mission scientist for planetary exploration at CSA, also citing LRO.

The focus of the beginning of the Mars Ice Mapper mission, he said, will be the top-level reconnaissance objectives, "closing those knowledge gaps as quickly as we can to inform the future human exploration planning." In a later phase of the mission the focus would turn to science: "make the platform available for science investigations moving forward."

NASA officials also noted that the mission will be run by NASA's science division, but funded by human exploration, so that if Mars Ice Mapper didn't go forward, the funding would not be available for alternative Mars missions. "If, for whatever reason, we didn't do Ice Mapper, that money is not available to us," said Eric Ianson, director of the Mars Exploration Program, at the Planetary Science Advisory Committee meeting. "This is funding specifically identified by the agency in support of exploration."

Developing a Communications Network

The discussions at those meetings about Mars Ice Mapper revealed details about a complementary effort. One challenge of a mission like Mars Ice Mapper is that radar instruments generate huge volumes of data, making it difficult to get it back to Earth.

One solution is for the mission to relay data through dedicated communications satellites, developed and potentially operated commercially. "Experimenting with commercial communications was brought up as an opportunity as we examined the bandwidth limitations of Ice Mapper," Watzin said, noting the sophistication of commercial communications satellites orbiting Earth today. "Is the time right that one could exploit the derivation of that technology and deploy it Mars [sic] at an affordable cost such that we can make a paradigm shift in how we do communication and increase the effectiveness of how we do our missions?"

The idea of having a dedicated communications satellite system is not new. In the early 2000s, NASA studied a Mars Telecommunications Orbiter that would relay data between Earth and other spacecraft orbiting or on the surface of Mars. The agency didn't proceed with that mission, instead relying on orbiters primarily intended to do science to serve as relays for missions on the surface.

NASA revisited the concept several years ago with the Next Mars Orbiter, or NeMO, which would serve as a communications relay and also carry a high-resolution camera, serving as a successor to the Mars Reconnaissance Orbiter. That proposal, too, was set aside in favor of using existing, if aging, orbiters as relays.

The new concept differs from earlier ones in both technology and procurement. One concept discussed at the recent meetings involved three satellites in orbits 6,000 kilometers above the surface. The satellites would be able to communicate with Mars Ice Mapper and potentially other missions in orbit and the surface, relaying their transmissions to Earth. Those satellites could also have laser intersatellite links between each other.

Davis said such a system could increase the amount of data Mars Ice Mapper could return by two orders of magnitude. "That is revolutionary, potentially," he said.

NASA officials said they were looking at commercial approaches for the system. Davis noted that discussions about the communications network involved both NASA's Space Communications and Navigation office as well as the commercial cargo and crew programs. "We are essentially trying to leverage their experience on how to do this very efficiently," he said.

Studies of that network are still in their early phases, and NASA officials said that if it turns out to be infeasible in some way, they will proceed with Mars Ice Mapper without it. Those studies, though, have caught the attention of Congress, which mentioned it in a report accompanying the draft of the Senate's version of a fiscal year 2021 spending bill.

"The Committee is aware that NASA is investigating possible new models for using commercial services for future communications with Mars surface assets in the late 2020s and early 2030s, though no such services exist today," the report stated. It called for a study within 180 days of enactment "outlining the Science plan for securing such commercial services for future Mars surface assets."

Planning for Humans

Mars Ice Mapper is part of a much broader effort to support the long-term goal of sending humans to Mars, no earlier than the 2030s. The agency is starting to plan exactly what that first mission would look like.

At a meeting last month of the Space Studies Board, Watzin discussed the status of the planning. "What grand questions could be addressed if we were to join the power of man and machine at Mars that would be worthy of an endeavor of this nature?" he said. The answer to that, he said, "is to continue to expand the theme that has guided the Mars exploration science program for the last 20 years, which is follow the water."

NASA now envisions that first human Mars mission be devoted to collecting subsurface water ice. "The objective of that mission that we're now assessing is to return ice cores from that mission back to Earth for examination by the science community," he said.

A reason for that goal, beyond the science benefits of collecting an ice sample, is that NASA is contemplating a relatively brief mission. "The first crewed mission to Mars is envisioned to be a short-stay mission, something on the order of 30 days or so," he said. "What we've done with the ice cores is set a grand objective so we can start shaking out the requirements for the equipment and the conops [concept of operations] that would be necessary to support that."

The idea of sending humans to Mars to collect ice samples was underwhelming to some members of the board. "I find that argument very, very unconvincing," said Antonio Elias, a retired Orbital ATK executive, citing the advances in remote operations in mining and drilling on Earth.

Watzin argued that having humans on the surface would be more effective than remotely operated and autonomous systems. He added the mission would bring back "many, many tens of kilograms" of ice samples, including ice cores meters long. "When you look at how that's done here on Earth, it is a marriage of crews of humans and equipment, both automated and manual."

Others asked what other kinds of work astronauts would do on that initial, short mission. "Of course, there's hundreds of things we would like to do over the course of time, as we continue to expand our stays and our capabilities at Mars," Watzin said. "What we're trying to do is identify a very challenging requirement as one of the driving things going into the mission architecture. We'll see how that shakes out over the coming years."

THE KEY TO FUTURE MARS EXPLORATION? PRECISION LANDING (2020)

Kara Platoni

Kara Platoni is a journalist based in the San Francisco Bay Area. She graduated from UC Berkeley's Graduate School of Journalism in 1999, and now teaches on that same course. She won the American Association for the Advancement of Science Journalism Award in 2008 and is currently the science editor for *WIRED* magazine. Her first book, *We Have the Technology*, was published in 2015.

What gives Katie Stack Morgan the confidence to point not just to the Jezero Crater but to an area on its edge, which she hopes the Mars 2020 rover will explore? A new technology, "terrain-relative navigation." The crater once held a deep lake and is believed to have once harbored microbes.

Katie Stack Morgan, deputy project scientist for NASA's Mars 2020 mission, pulls up a photo of where the rover is headed: the western rim of Jezero Crater. Billions of years ago, a river filled that basin, creating a delta and what is today a dry lakebed. It's an exciting destination, if, like the science team at NASA's Jet Propulsion Laboratory (JPL), you're on a particular quest. "That is a great place to go and search for life," she says.

There's a neon green ring on the image too: That's the ellipse encompassing all the spots their rover might land. Most of the ring lies in the lakebed, but because Mars landing is still not an exact science, it also includes more treacherous terrain, like the rocky delta and a bit of the crater rim, as much as a kilometer high.

Tracing a finger on the computer screen, Stack Morgan maps her desired path. Ideally, they'll land in the flats and drive uphill, so scientists can "read the rocks" from oldest to youngest. It's like reading a book, she says: Skipping to the end might be satisfying, but context is what makes it most exciting.

First stop: the delta. "That's the juicy spot," she says, because it would have built up lake sediments and perhaps organic compounds. Scientists think life would develop—and fossilize—in quiet, nutrient-rich water, away from the tumult of a river.

Then they would climb to the "bathtub ring," the crater's inner margin, which may have had warm, shallow waters—another good place to look for life. Finally, they would ascend to check out the very oldest rocks in the crater wall, caching rocks along the way for a future sample-return mission. "When those samples come back," she says, "they are going to feed a generation of Mars scientists."

But for Allen Chen, the mission's entry, descent, and landing lead, this will require some delicate footwork. "While the scientists love things like cliffs and scarps and rocks—those are the science targets for them—that's death for me," he says. Jezero is full of the dangers he would most like to avoid: rocks (too sharp), slopes (a rollover risk), and meter-high dunes the team calls "inescapable hazards" (sand traps).

Therein lies the trade-off that has bedeviled both Mars and moon landings since the very beginning: Landing on an open plain is safer, but scientifically kind of dull. On past missions, Chen says, NASA wanted each site "to be a parking lot."

To turn Jezero, once a no-go, into a landing pad, the team is relying on terrain-relative navigation, or TRN. The technology, which JPL researchers have been working on since 2004, enables more precise landings by giving the vehicle a visual landing system. Using a camera, it scans the ground for landmarks, compares those images to onboard maps, and estimates its position. Tie this to related advances in hazard detection and avoidance, and now they have a better shot at not landing in a sand trap.

With better precision, says Chen, they can tolerate more hazards in the ellipse, which can also be smaller, and closer to the juicy spots Stack Morgan wants to study. Ultimately, Chen says, TRN means "we can be near those science targets—those death hazards to me—and not have to drive years to get there." Just like on Earth, drive time means fuel and money. And you don't want to know how much it costs to drive in space.

Before the rover can roll, the spacecraft must first survive a descent, called, since the 2012 Mars Curiosity rover landing, the "seven minutes of terror." After speeding through the Martian atmosphere, an enormous parachute inflates to slow entry, the heat shield separates from the capsule, and the rocket-powered "sky crane" hovers to lower the rover, wheels down, before blowing the cables and rocketing away to crash somewhere else.

For most of this period, the vehicle won't have detailed information about where it is. Mars missions, Chen says, rely on NASA's Earth-based Deep Space Network until the end of the cruise phase. He compares it to driving while looking in the rearview mirror.

But once entry begins, the vehicle navigates by dead reckoning. It's got an inertial measurement unit (IMU) that tracks acceleration and angular rates, but no visual way to scan for landmarks. "At that point the vehicle doesn't have a way to figure out where it is," says Chen. "It's still buttoned up in its aeroshell and we're about to go screaming into the atmosphere."

Chen says it's like asking a blindfolded person to find his way across a room based only on a description of how he was positioned in the doorway. "We tell the vehicle, 'Well, when the bell goes off, we think you're here, and you're going in this direction, about this fast,' " he deadpans. "Good luck!"

With so much uncertainty, Chen says, after heat shield separation, "we could be off in our position in latitude and longitude space by up to about three kilometers. That's not so great if you want to know where you are precisely, and dodge craters."

The Apollo series of crewed missions to the moon had a solution: Give the astronauts a map and have them look out a window. But Mars 2020 is people-free, and the landing must unfurl without help from Mission Control, because there's more than a 10-minute communications lag with Earth. Swati Mohan, lead guidance, navigation and control systems engineer for Mars 2020, describes landing this way: "Because of the delay, it's already done it by the time you get the signal that it started." TRN fills in this gap by giving the vehicle its own camera and map.

Mohan and Andrew Johnson, the Mars 2020 guidance, navigation, and control subsystem manager, have gowned up and are standing in a clean room in front of the Lander Vision System—the eyes and brain of the navigation technology. (They've recently finished 17 test flights over the Mojave, so this version is rigged up to fly on a helicopter.)

The system's "eyes" are the landing camera. "It's mounted onto the bottom of the rover, so it actually takes pictures of the terrain as we come down," says Mohan, pointing at the left side of the rig. The brain—called the Vision Compute Element—processes those images. On Mars 2020, it will be nestled in the rover's belly. Today, it's sitting on a shelf, looking like an oversize hard drive.

This computer's job is to swiftly compare the images of the approaching ground with onboard maps previously stitched together using photos shot by the Mars Reconnaissance Orbiter (MRO), which has been orbiting the planet since 2006. It correlates the two sets, looking for landmarks. But "landmarks" doesn't mean rocks or craters. It means tiny pixel patterns, gradients of dark and light that are imperceptible to the human eye but can be identified by algorithm. Spotting these landmarks, while also using IMU data, will let the lander find its position on the map.

Practice shots at a Mars landing are vanishingly rare, so the team is testing their system using computer simulations that model parameters like trajectories, dust, and solar illumination levels. They are also flying their gear over the best Earth analogs they can locate. "It's very hard to find things that look like Jezero Crater," places as vast and treeless, says Johnson. They've ended up using desert mesas that are steeper and more rugged than Jezero, a way to over-train. Other than a bug that caused the system to reset when they flew too close to the edge of the map, he says, "It was a huge success."

Specifically, Johnson says, they wanted to check that their estimated position would be off from the map by no more than 40 meters (about 130 feet). When they tested landing conditions similar to what they expect from Mars 2020, they beat their margin by half—the error was less than 20 meters. When they tested under more demanding conditions, like areas with high terrain relief or different illumination levels, they stayed within their original margin. And under the most challenging conditions of all, the system discarded estimates it considered unreliable and restarted its calculation. Johnson believes this is a good indicator that the system can recover if faced with unexpected conditions.

Meanwhile, on the other side of the JPL campus, scientists have been creating

Jezero hazard maps, marking rocks they can see from orbit, and modeling smaller ones they can't—plus those slopes and sand traps. They will use this to make a "safe targets map," which breaks the terrain down into 10-meter-square "pixels," each scored according to its hazard level. Before launch, they'll load this onto the spacecraft.

The Lander Vision System will activate once the vehicle is about 4,200 meters above the surface. Once its back shell comes off at about 2,300 meters, the vehicle will use TRN data to gauge its position, and will autonomously select the safest nearby pixel from the targets map. Then, down it will come.

Mohan will be watching from Mission Control, and Johnson in the entry, descent, and landing war room, hoping to receive a tone—a specific radio frequency—indicating that the craft survived touchdown. Then, a camera mounted higher up on the vehicle will send back a photo. (The team is hoping to see ground and not sky, proof the vehicle's not upside down or tilted.) By the time the MRO circles overhead to photograph the rover, NASA should know exactly where it landed.

TRN is also being researched by private companies working to send spacecraft to the moon. Among them is Pittsburgh-based Astrobotic Technology, which is developing TRN for its Peregrine lander. Fourteen groups have purchased space on Peregrine to carry everything from scientific instruments to time capsules to personal mementos.

"Our business is to make space and the moon accessible to the world as a whole," says Astrobotic principal research scientist Andrew Horchler. The company plans for Peregrine to fly on routine missions to many lunar sites, landing within 100 meters of each spot. "Terrain-relative navigation is an enabling technology for this," says Horchler. "It will enhance the overall reliability of landing, the precision of the landing, and the actual spaces we can feasibly get to on the moon."

The frame of the Peregrine lander looks a bit like a Space Age coffee table; it's an elegant, aluminum-alloy, X-shaped structure atop four spindly legs, topped by a solar panel. Four payload tanks can be attached, one to each side of the craft. The tanks are shaped like golden pill capsules, and carry a combined 90 kilograms of experiments or supplies. The entire craft is 1.9 meters tall and 2.5 meters wide.

Astrobotic's work on TRN received funds from NASA's Tipping Point program, which awards money to bring commercial space technologies to market. Additionally, Peregrine's planned 2021 mission to the moon's Lacus Mortis crater is affiliated with NASA's Commercial Lunar Payload Services (CLPS) program. This public-private partnership will deliver gear ahead of the agency's 2024 Artemis landing by astronauts, who will be sent to the lunar south pole. But landing on the moon will be somewhat different than landing on Mars.

First, lunar missions face a photography challenge. Mars 2020's rich maps are courtesy of the Mars Reconnaissance Orbiter. Because it flies a sun-synchronous orbit, the MRO is continuously overhead at about 3 p.m. local time, meaning it's catching the same shadows on every pass. Sharp, consistent shadows are important when you're trying to map light-and-dark landmarks a TRN system can find. Mars 2020 is conveniently scheduled to land at 3 p.m., so what its camera sees should easily match its maps. (As Chen puts it: "It's always kind of 3 p.m. on Mars to me.")

The moon has the Lunar Reconnaissance Orbiter, but that spacecraft is on a polar orbit, and traces a new longitude line on each pass. "That point in time and view angle will almost certainly not correspond to what our lander will see," says Horchler. "This is even more a factor at the poles of the moon, where the shadows are very dynamic, and even small features cast very long shadows."

This variability affects the way private companies headed to the moon make their maps for TRN. Astrobotic and its competitor Draper, a fellow CLPS participant based in Cambridge, Massachusetts, are creating synthetic landing images, modeling what the surface will look like at their intended landing times and spots. They build computerized terrain models, then light them based on factors like where the craft will be and how they expect the moon and sun to be aligned.

Another difference: On Mars, the lander comes down on a parachute, following a steep trajectory and using the atmosphere as a brake. But landers can't parachute on the airless moon, so the landing trajectory must be more horizontal—and that means a bigger map and more data storage. But Horchler says that with a longer trajectory, they'll have more time to run TRN as they descend, which might offer a more precise landing.

While Horchler tips his hat to the close relationship his company has had with NASA, allowing both to benefit from the other's research, he points out that they have another big difference: Astrobotic's TRN technology is designed for commercial missions to many destinations. (After their first launch, they hope to fly about one per year.) He believes their sensor will be able to estimate position based on a single image, without any additional information about altitude, orientation, or velocity, which Horchler says makes their interface simpler, and potentially more self-contained and robust—better adapted to business use.

Similarly, Draper, which has been building spacecraft navigation systems since the Apollo missions, wants to keep things simple. Alan Campbell, the company's space systems program manager, mentions that while their algorithms are likely similar to NASA's, on the hardware side they are "working diligently to build a smaller, lighter, more efficient system to serve many different kinds of planetary landing missions."

After all, that's still the big idea: landing near all kinds of geologically tricky spots. Horchler says that some customers might want to prospect the moon's poles for ice. Campbell notes that scientists are keen to explore its lava tubes, tunnels that might shield astronauts from radiation.

And the hope for TRN is that, if your camera and algorithm work on one heavenly body, they will work just about anywhere. "We're really agnostic to the surface that we're landing on," says Campbell, including ones that are super far away. Draper has done work for the OSIRIS-REx sample collection mission to the asteroid Bennu, and Campbell thinks TRN might be useful for NASA's Dragonfly program to send a rotorcraft to Titan. Horchler points out that their sensor could be adapted for other moons: Phobos and Deimos, the satellites of Mars.

"You can't land on Europa without it," says Johnson. Europa has an ocean covered by a giant ice cap, and is a promising place to hunt for life—but there's no detailed imagery of its surface. NASA plans to send the Europa Clipper orbiter to take photos, but a lander would have to be designed and launched before Clipper even arrives. "We don't have the luxury like we do at Mars of picking the landing site beforehand, mapping it out as much as we possibly can," Johnson says.

JPL's Vision Compute Element will perform a second job after the Mars landing: helping the rover drive. Mohan says its forebear, Curiosity, was capable of higher speeds, but poked along at about 8 meters per hour during autonomous driving because it couldn't process terrain images while moving. The rover, she says, had to constantly "stop, think, take the next step." These functions work in parallel on their new computer, which they think will boost speeds for fully autonomous driving to 60 to 80 meters per hour. This too means getting to the juicy spots faster.

But, suggests Johnson, "Wouldn't it be cool if you didn't have to have a rover?" What if TRN permitted such pinpoint landings that you could just send an instrument, like a drill, that works where it lands? Or, asks Mohan, how about making life easier for future human crews? Dropping their supplies close together will help them set up their habitats. Instead of scattering payloads over a huge ellipse, she says, "you're hitting the bulls-eye every single time."

And this, TRN developers say, is the promise of being able to park in a tight spot. Once you can safely put people and their machines near dead lakes or polar ice or anything that was once terra incognita, the real adventure begins.

Thanks to its novelty, it's easy to forget that TRN is being deployed in service of one of humanity's oldest questions: Are we alone?

It's also one of our hardest. "If life existed on Mars, it was likely microbial and may not have advanced beyond that," says Stack Morgan. It's probably very, very small, and very, very dead. And, for now, scientists have to study it from very, very far away. Only the second mission of Mars 2020 is a sample return mission. The rover will cache rocks to be collected by a future robot, yet unborn. Until then, scientists must rely on photos plus electronic data from instruments mounted on the rover arm that can detect organics like carbon.

So how do you find microbes from 249 million miles away? Scientists will be looking for biosignatures, rock textures, and chemical patterns that could have formed only in the presence of life.

Take, for example, a fossil formation called a stromatolite. Stack Morgan picks up a pen. Imagine a lakebed, she says, drawing a bumpy line. If Jezero was lifeless, carbon particles falling through the water column would roll off the lakebed's peaks and cluster in its troughs. (She heavily dots the low points and places only a few on the high ones.) Once the lake dried up, the resulting stromatolite would show uneven ripples of carbon: thinner at the top and thicker at the bottom.

Now imagine the lakebed was covered by a jelly-like mat of microbes. Carbon would have been equally likely to stick anywhere, she continues, dotting her pen all over. This time, the carbon ripples would be consistently thick. "I can explain that with physics," she says, pointing to her first drawing. Then she aims her pen at the second. "I can only explain that with life."

INGENUITY (2020)

HOW THE MARS HELICOPTER WILL FLY ON ANOTHER PLANET

Paul Parsons

Paul Parsons has been a science writer for more than twenty years. He has a PhD in cosmology and is the former editor of *BBC Science Focus*, and managing editor of *BBC Sky At Night*. Parsons is a fellow of the Royal Statistical Society and a consultant to Guinness World Records and currently works as a quantitative analyst with the sports trading technology company Botsphere.

In a space exploration first, NASA is preparing to deploy a chopper on the Red Planet. Flying in a fraction of Earth's gravity, it should be a sight to behold.

Whether disrupting air traffic, returning glorious vistas of Earth from above, or just spying on the neighbours (if that's your thing), drones have become a familiar sight in our skies. Now, for the first time, the US space agency NASA is poised to fly a drone-like helicopter in the atmosphere of another planet.

The craft, named Ingenuity, will hitch a ride to the Red Planet aboard the one-tonne Perseverance lander, NASA's latest wheeled robotic rover mission to drive across the planet's rugged surface. Perseverance is expected to launch from Earth this summer, with touchdown on Mars scheduled for Spring 2021.

Flying in the alien atmosphere of another world is a feat that poses a unique set of engineering challenges and yet, if this small technology test mission is successful, it will furnish scientists with a new and highly effective way to explore the planets and moons of our Solar System. That's because flying is a much faster way to get around than ground roving.

Aircraft can gather aerial imagery that's much sharper than pictures returned by spacecraft. They can also serve as scouts to identify potential targets for ground-based rover vehicles, and they can even gather samples and bring them back to a central lander station for analysis.

And, of course, they can go where other probes simply can't.

"Larger Mars rotorcraft in the 5kg to 20kg class with small science payloads could access areas not reachable by rovers, and support wide-area surveys in shorter times," says Dr Bob Balaram, Ingenuity's chief engineer, based at NASA's Jet Propulsion Laboratory (JPL) in Pasadena, California.

"There is also a NASA mission to explore Saturn's moon Titan with a flying lander that will arrive in the early 2030s."

The Ingenuity craft is 50 centimetres high with four blades – one pair above the other – mounted on twin, counter-rotating rotors each spanning 1.2 metres.

The size of the rotors (which need to be this big for the helicopter to fly in Mars's thin atmosphere) is the main reason why a more familiar drone-like quadcopter design was rejected – such a vehicle would be simply too large to fit on the rover.

Ingenuity is stowed beneath the body of Perseverance, from where it will be dropped onto the surface of Mars, probably a couple of months into the mission. The rover will then drive 100 metres away, to minimise collision risk, and the two will exchange radio signals to 'pair up' – a bit like pairing wireless earbuds to your phone – before the rover sends the command for Ingenuity to make its inaugural flight.

This will likely be what Balaram calls a 'mutual selfie' – the two vehicles taking pictures of one another as the helicopter rises to a low hover and then lands again.

How Ingenuity will Explore Mars

Ingenuity carries a black-and-white navigation camera and a 4,208 x 3,120-pixel colour camera, comparable to what might be found in a mobile phone. Images are beamed by short-range radio link to Perseverance, which then relays them to one of a number of NASA spacecraft in Mars orbit, from where they are transmitted back to Earth.

"Multiple images may be stitched into a panorama at some point using ground software tools," says Balaram.

The helicopter isn't going to make any actual scientific observations, however. Instead, the focus is to return engineering data from the test flights that will, it's hoped, validate the technology or at the very least provide valuable feedback to refine future designs.

The primary mission plan is for up to five flights over a period of 30 days, though this may be extended. The maximum horizontal range of the flights will be about 300 metres with a ceiling altitude of 10 metres and a max flight time of 90 seconds, after which the helicopter's six lithium-ion batteries need recharging.

This is handled by a solar panel mounted directly above the rotor blades. A full charge takes a whole Martian day to complete, and two-thirds of the power stored is needed to keep the aircraft's electronics warm during the bitterly cold Martian night, when temperatures can plunge to –100°C.

Specifications

- **MASS:** 1.8kg
- **HEIGHT:** 50cm
- **ROTOR SPAN:** 1.2m

- **ROTOR SPEED:** 2,300-2,900rpm
- **MAX ALTITUDE:** 10m
- **BATTERIES:** 6x Sony Li-ion, delivering 220W power
- **MAX FLIGHT TIME:** 90s
- **RANGE PER FLIGHT:** 300m
- **MAX FLIGHTS PER DAY:** 1
- **MISSION DURATION:** 30 days

The low temperature is only the beginning of the problems facing a would-be Martian aviator. Aircraft take off using 'lift', an upward force created as air passes over a curved aerofoil. On a plane, the aerofoil is the wing; on a helicopter, it's the rotor blades.

But the surface density of Mars's atmosphere is just 1 per cent of Earth's. A cubic metre of air at sea level on Earth weighs 1.2kg, but on Mars that figure drops to just a few grams – roughly equivalent to the tenuous atmosphere on Earth at an altitude of 30,000 metres. And to generate enough lift to fly from such thin air means the engine has to work overtime.

Whereas helicopters on Earth typically spin their blades at 500 revolutions per minute (rpm), on Mars you need to crank this up to around 2,500rpm.

For the same reason, the weight of the craft is kept to an absolute bare minimum. For example, each pair of rotor blades weighs just 56 grams – despite measuring over a metre in length.

"Carbon composite layups within a foam core matrix were used to achieve lightweight but stiff blades," says Balaram. In total, Ingenuity weighs in at just 1.8 kilograms, less than a couple of bags of sugar.

Balaram has even worked out the best time of day to fly the helicopter on Mars – mid-morning, as it turns out, roughly 11:00. By this time, the craft has been warmed by the Sun after the chilly Martian night, but the air is cool, keeping its density as high as possible and its wind speeds low, while the batteries still have a healthy charge.

"After we get the first couple of flights under our belt, I'm sure we will try to fly in the afternoon, and do more exploratory things, but the most conservative thing we can do is to pick a mid-morning flight," he says.

Ingenuity on Trial

The team tested their design inside the JPL's Space Simulation Chamber, a 25-metre-high, eight-metre-wide cylindrical test vessel in which the atmosphere could be precisely tuned to replicate the aerodynamic conditions on Mars.

However, the low pressure was only one of the peculiarities of flying on Mars that had to be modelled. They still needed to account for the planet's gravity, which is only 0.38 (just over a third) of that found on Earth. They came up with a novel solution.

"A gravity off-load device was installed into the chamber and was used to provide a compensating force on the vehicle to account for the difference in gravity between Earth and Mars," says Balaram.

This literally means attaching a lightweight thread to the top of the vehicle, which they pulled tight just enough to lift 0.62 of the helicopter's weight on Earth – leaving the remaining 0.38 (its weight on Mars) to be lifted aerodynamically by the rotors.

As well as demonstrating the basic feasibility of flight on Mars, these ground tests were crucial for another reason – allowing the team to fine-tune the Guidance, Navigation and Control (GNC) software that will actually have to pilot the helicopter on Mars.

When the first real Martian flights begin, the light-travel time from Earth to Mars will be many minutes, making it impossible for controllers to steer the craft remotely.

Instead, the commands from ground control on Earth consist of a set of waypoints – coordinates selected by controllers – which the GNC software will follow while taking account of the stream of real-time data flooding in from the helicopter's sensors. These include gyroscopes, accelerometers, a navigation camera, an altimeter, and an inclinometer which measures the aircraft's tilt.

"Waypoints consist of an x-y position, a height above ground, a vehicle heading, and a desired time to arrive at each point," Balaram says. To test the software, his team even set up multiple high-energy lamps inside the simulator to recreate sunlight illumination at the Martian surface, as well as fans to simulate flight in windy conditions.

On top of all this, Ingenuity must also be a spacecraft, certified to survive the stresses of launch, the seven-month journey to Mars during which it will be subjected to radiation and the harsh vacuum of interplanetary space, and then finally wracked by the high g-forces and scorching 2,200°C heat of entry, descent and landing through Mars's atmosphere.

But if it works and the helicopter's trials on Mars are a success, then it will be a genuinely historic achievement – the first-ever powered flight beyond Earth. And the expertise gained will lay the groundwork to fly bigger and better aircraft, capable of carrying a full payload of scientific instruments to explore the skies of distant worlds.

As Balaram puts it: "It's kind of a Wright Brothers' moment on another planet."

MARS SAMPLE RETURN (2021)

THE MISSION THAT WILL BRING HOME A PIECE OF ANOTHER PLANET

Sara Rigby

Sara Rigby is the online staff writer at *BBC Science Focus* magazine. She has an MPhys in mathematical physics and loves all things space, dinosaurs and dogs.

We want to bring something back from Mars. That's the thinking behind a programme so ambitious that it will take two space agencies and several missions to pull it off. To find out what's at stake, and how we'll get a sample from the Red Planet, we spoke to ESA's Dr Albert Haldemann.

He's the ExoMars Payload and Assembly, Integration and Verification Team Leader at ESA, and he coordinates with NASA to keep the two agencies working smoothly together.

Tell us about the Sample Return mission.

The Mars Sample Return mission is the first mission – programme actually, because more than one craft is required – to return samples from another planet. We have samples from the Moon returned by the Apollo missions, by the Russian rover and most recently by the Chinese. And we've returned samples from asteroids [via missions run] by the US and Japan.

But we've yet to return a sample from Mars. That's the objective. The first step is the Perseverance rover, which is beginning to collect samples after NASA landed it successfully on Mars in February this year.

The Mars Sample Return programme is a joint endeavour between NASA and the European Space Agency (ESA), although NASA is leading it. NASA will be responsible for the rockets that will launch from the surface of Mars and put the samples in Mars orbit. But ESA is responsible for the Earth Return Orbiter, which is basically going to be the first interplanetary spacecraft.

It'll go from Earth to Mars, go into orbit around Mars, rendezvous with the sample craft in Mars orbit, capture the sample and then come back to Earth. It won't enter Earth orbit, but it'll release the NASA-built entry capsule containing the samples once it's near enough. Only the entry capsule will come back to Earth, not the Earth Return Orbiter.

The plan is to essentially have the capsule perform a controlled crash. So we'll target the crash site, then have the Earth Return Orbiter release the samples and fly away. It'll deviate so that it flies past Earth and goes into a graveyard orbit around the Sun.

But prior to all this, Perseverance will be on Mars collecting samples. Those samples will be put in caches on the surface. In 2026 NASA will launch another rocket to Mars containing a lander that's carrying the Mars Ascent Vehicle and the Sample Fetch Rover.

The lander will land on Mars and release the Fetch Rover, which will pick up the caches of rock samples that have been collected by Perseverance. The Fetch Rover will then bring them back to the lander, where they'll be put in the Mars Ascent Vehicle and launched into Mars orbit.

Launching rockets from Earth is tricky enough. Launching one from Mars sounds incredibly difficult.

Yes, there are plenty of challenges, especially as the launch will have to be carried out automatically.

The rocket will be a two-stage, solid-fuelled rocket, and that will carry the Orbiting Sample container – or OS, essentially a football-sized sphere. When the rocket gets into orbit, it'll eject the OS and leave it orbiting Mars.

We're still working on how best to make sure we can find it. Our baseline is that we make it shiny and use cameras to spot it from a distance of a few thousand kilometres, which is what the Earth Return Orbiter will be at as it makes its approach.

It seems doable, but everybody's still a little worried. So we're considering about radio beacons, laser range finders and stuff like that. But there are a lot of challenging steps, which is what makes this fun for engineers who have to build all of these things and make them work.

What would it mean to the astronomical community to get a sample of Mars back to Earth?

We've sampled Mars from afar with our robots, rovers and landers. And we have samples from meteorites that were blown off Mars and landed on Earth. We know that's where they came from, thanks to isotopic analysis of the rocks. A few of them contain gas bubbles that are dead ringers for Mars's atmosphere, which we measured with Viking 1 and 2 back in 1976 and 1977.

So if we already have samples from Mars, why go and bring back more? Well, firstly we can't tell precisely where these Martian meteorites came from on Mars. We have some hypotheses, but we don't know. Secondly, although we have high-powered instruments and spectrometers on Perseverance, and other Mars rovers, those instruments are miniaturised so they can make the trip; they're not as high-performance as the instruments we have in our labs on Earth.

Also, some of the rocks that will be obtained by Perseverance in the Jezero Crater aren't represented in the meteorite samples. Those kind of rocks don't survive being blown off of a planet, we think. And yet they're key to the geologic history of Mars. And because of the delta in Jezero Crater, the rocks from there might have been associated with liquid water and so are most likely to contain evidence of past life or organic materials of a past environment on Mars.

Whether we decide that means there was or wasn't once life on Mars, we don't know yet. Let's wait for the evidence.

What sort of things will you be looking for in the samples?
That's down the road. But in order to analyse them, we're going to take extraordinary precautions. Because Mars is considered potentially inhabitable – we've seen that there are underground environments that may contain liquid water – for these first samples we'll use the most stringent planetary protection measures.

The samples will be contained within rugged metal test tubes that are as inert as possible so we don't contaminate the contents. They'll be packaged inside the OS, which has multiple shells. Then, when the OS gets up to the Earth Return Orbiter, it's packaged inside the entry capsule that will also have multiple shells. And when the capsule gets back to Earth, it'll be picked up and bagged before being taken to a special facility for unpacking.

We'll X-ray the test tubes to see the mineralogy of the contents before we open them. Some of the samples may also undergo very high-energy synchrotron examination to look at their compositions.

Only once we're convinced that it's safe to do so will we open the capsules, and we'll do that in a controlled way too. Then other, more advanced geochemical procedures will be applied to the samples to look at the details and see if there are any organic materials along with the mineral.

The Sample Fetch Rover looks like it's going to have inflatable tyres on its wheels. Why?
Yeah, they do look like inflated tyres. They're not, though – they're actually a wire mesh made with memory metal that allows us to make a bigger wheel that can be compressed in a smaller package for launch. The tyres do expand, but they don't inflate through air. When we get to Mars, they're released and allowed to expand to their full size. The reason for that is the Sample Fetch Rover will need to cover a lot of ground quickly, and larger wheels will help it do that.

We're also going to use just four wheels instead of the six we've had in previous designs. Six wheels is very efficient for rough terrain. Four wheels with big tyres, like you see on dune buggies, are also okay for rough terrain, especially if you know what the terrain is like, which we will, to some extent, because we have Perseverance on the ground already essentially doing reconnaissance.

Are you worried about contamination from the Mars samples?
Personally, I'm not worried about Mars contamination. I think Earth and Mars have been 'sharing spit' for billions of years. But my opinion is irrelevant compared to what the public and political perception is of this risk. So we need to

address that. It's going to be a heck of a challenge, I think, especially given what we've seen with COVID as far as the general public's confidence in scientific authority is concerned. We've already taken our lesson from that.

And we welcome the challenge because we need to understand what the samples from Mars mean, if we ever want to have astronauts go there and come back. We need a detailed understanding of what planetary protection means if we have the ambition, as many do, to have astronauts – human beings – go further into the Solar System and then be able to come back to Earth.

The Mars Sample Relay

1. Land, explore, drill

Launched in July 2020, NASA's Perseverance rover successfully landed in the Jezero Crater on Mars in February 2021. The roughly hatchback-sized rover represents the first leg of humanity's first round trip to another planet.

The rover will use its drill to collect rock and soil samples and store them on the planet's surface.

2. Seal and deposit

Any samples Perseverance collects will be in sealed tubes that the rover will deposit on the planet's surface, essentially leaving a trail of breadcrumbs behind it as it explores the Martian surface.

Future missions to Mars – collaborative efforts by NASA and ESA – will follow the trail left by Perseverance, to retrieve the samples and return them to Earth.

3. Set off the second wave

The next stage will get underway in 2026, when NASA launches another mission to Mars that will carry a lander containing the rover, rocket and capsule needed to collect the samples gathered by Perseverance and launch them on their journey back to Earth.

The lander will aim to set down in the Jezero Crater, where Perseverance began its journey.

4. Unfold and unload

After the lander has successfully touched down on the Red Planet, it'll deploy its solar panels to power up the ESA-designed Fetch Rover.

The rover (a four-wheeled vehicle, slightly smaller than Perseverance) will then unfold itself and deploy from the lander before beginning to follow the trail of samples left behind by Perseverance.

5. Grab and go

The Fetch Rover will follow in Perseverance's tracks to collect the samples that have been deposited on the surface. Once a sample has been located, a robotic arm will deploy to collect it and store it inside the rover, before it moves on to find the next one.

Once enough samples have been collected, the Fetch Rover will return to the lander.

6. Hand over the samples

Once the Fetch Rover has arrived back at the lander, the same arm it used to collect the sealed samples will transfer them into the nose cone of the Mars Ascent Vehicle, stowed inside the lander.

If there are additional samples that the Fetch Rover was unable to collect, they could be brought to the lander directly by Perseverance.

7. Blast off for home

Once the samples are loaded onto the Mars Ascent Vehicle, the small rocket will be launched from the lander into Mars orbit.

The gravity on Mars is about a third of that on Earth, so it's easier for objects to reach escape velocity, hence the rocket can be much smaller. It's expected to be a maximum of 3m tall and around 50cm in diameter.

8. Pick up and drop off

The Mars Ascent Vehicle will carry the container of samples into orbit around Mars and release it. Another spacecraft, the Earth Return Orbiter, will detect the sample container and collect it, before returning to Earth.

Once there, the Earth Return Orbiter will drop the sample container into Earth's atmosphere... if everything goes according to plan.

FEARS OF MARS

THE WAR OF THE WORLDS (1898)

CHAPTER X: IN THE STORM

H. G. Wells

Herbert George Wells was born in 1866 in Bromley, Kent. While still at school he became interested in how society might be refashioned in the future by socialism, as advocated by the Fabian Society. As an adult he found work as a teacher, in his spare time studied for a degree in zoology from the University of London, and in 1893 published his first work, the *Text Book of Biology*. Wells will always be associated with Mars because of his epochal 1898 novel *The War of the Worlds*.

LEATHERHEAD is about twelve miles from Maybury Hill. The scent of hay was in the air through the lush meadows beyond Pyrford, and the hedges on either side were sweet and gay with multitudes of dog-roses. The heavy firing that had broken out while we were driving down Maybury Hill ceased as abruptly as it began, leaving the evening very peaceful and still. We got to Leatherhead without misadventure about nine o'clock, and the horse had an hour's rest while I took supper with my cousins and commended my wife to their care.

My wife was curiously silent throughout the drive, and seemed oppressed with forebodings of evil. I talked to her reassuringly, pointing out that the Martians were tied to the pit by sheer heaviness, and, at the utmost, could but crawl a little out of it; but she answered only in monosyllables. Had it not been for my promise to the innkeeper, she would, I think, have urged me to stay in Leatherhead that night. Would that I had! Her face, I remember, was very white as we parted.

For my own part, I have been feverishly excited all day. Something very like the war-fever that occasionally runs through a civilized community had got into my blood, and in my heart I was not so very sorry that I had to return to Maybury that night. I was even afraid that that last fusillade I had heard might mean the extermination of our invaders from Mars. I can best express my state of mind by saying that I wanted to be in at the death.

It was nearly eleven when I started to return. The night was unexpectedly dark; to me, walking out of the lighted passage of my cousins' house, it seemed indeed black, and it was as hot and close as the day. Overhead the clouds were driving fast, albeit not a breath stirred the shrubs about us. My cousins' man lit both lamps. Happily, I knew the road intimately. My wife stood in the light of the doorway, and watched me until I jumped up into the dog-cart. Then abruptly she turned and went in, leaving my cousins side by side wishing me good hap.

I was a little depressed at first with the contagion of my wife's fears, but very soon my thoughts reverted to the Martians. At that time I was absolutely in the dark as to the course of the evening's fighting. I did not know even the circumstances that had precipitated the conflict. As I came through Ockham (for that was the way I returned, and not through Send and Old Woking) I saw along the western horizon a blood-red glow, which, as I drew nearer, crept slowly up the sky. The driving clouds of the gathering thunder-storm mingled there with masses of black and red smoke.

Ripley Street was deserted, and except for a lighted window or so the village showed not a sign of life; but I narrowly escaped an accident at the corner of the road to Pyrford, where a knot of people stood with their backs to me. They said nothing to me as I passed. I do not know what they knew of the things happening beyond the hill, nor do I know if the silent houses I passed on my way were sleeping securely, or deserted and empty, or harassed and watching against the terror of the night.

From Ripley until I came through Pyrford I was in the valley of the Wey, and the red glare was hidden from me. As I ascended the little hill beyond Pyrford Church the glare came into view again, and the trees about me shivered with the first intimation of the storm that was upon me. Then I heard midnight pealing out from Pyrford Church behind me, and then came the silhouette of Maybury Hill, with its treetops and roofs black and sharp against the red.

Even as I beheld this a lurid green glare lit the road about me and showed the distant woods towards Addlestone. I felt a tug at the reins. I saw that the driving clouds had been pierced as it were by a thread of green fire, suddenly lighting their confusion and falling into the fields to my left. It was the Third Falling-Star!

Close on its apparition, and blindingly violet by contrast, danced out the first lightning of the gathering storm, and the thunder burst like a rocket overhead. The horse took the bit between his teeth and bolted.

A moderate incline runs down towards the foot of Maybury Hill, and down this we clattered. Once the lightning had begun, it went on in as rapid a succession of flashes as I have ever seen. The thunder-claps, treading one on the heels of another and with a strange crackling accompaniment, sounded more like the working of a gigantic electric machine than the usual detonating reverberations. The flickering light was blinding and confusing, and a thin hail smote gustily at my face as I drove down the slope.

At first I regarded little but the road before me, and then abruptly my attention was arrested by something that was moving rapidly down the opposite slope of Maybury Hill. At first I took it for the wet roof of a house, but one flash following another showed it to be in swift rolling movement. It was an elusive vision—a moment of bewildering darkness, and then, in a flash like daylight, the red masses of the Orphanage near the crest of the hill, the green tops of the pine-trees, and this problematical object came out clear and sharp and bright.

And this Thing I saw! How can I describe it? A monstrous tripod, higher than many houses, striding over the young pine-trees, and smashing them aside in its career; a walking engine of glittering metal, striding now across the heather; articulate ropes of steel dangling from it, and the clattering tumult of its passage mingling with the riot of the thunder. A flash, and it came out vividly, heeling over one way with two feet in the air, to vanish and reappear almost instantly as it seemed, with the next flash, a hundred yards nearer. Can you imagine a milking-stool tilted and bowled violently along the ground? That was the impression those instant flashes gave. But instead of a milking-stool imagine it a great body of machinery on a tripod stand.

Then suddenly the trees in the pine-wood ahead of me were parted, as brittle reeds are parted by a man thrusting through them; they were snapped off and driven headlong, and a second huge tripod appeared, rushing, as it seemed, headlong towards me. And I was galloping hard to meet it! At the sight of the second monster my nerve went altogether. Not stopping to look again, I wrenched the horse's head hard round to the right, and in another moment the dog-cart had heeled over upon the horse; the shafts smashed noisily, and I was flung sideways and fell heavily into a shallow pool of water.

I crawled out almost immediately, and crouched, my feet still in the water, under a clump of furze. The horse lay motionless (his neck was broken, poor brute!) and by the lightning flashes I saw the black bulk of the overturned dog-cart and the silhouette of the wheel still spinning slowly. In another moment the colossal mechanism went striding by me, and passed uphill towards Pyrford.

Seen nearer, the Thing was incredibly strange, for it was no mere insensate machine driving on its way. Machine it was, with a ringing metallic pace, and long, flexible, glittering tentacles (one of which gripped a young pine-tree) swinging and rattling about its strange body. It picked its road as it went striding along, and the brazen hood that surmounted it moved to and fro with the inevitable suggestion of a head looking about it. Behind the main body was a huge mass of white metal like a gigantic fisherman's basket, and puffs of green smoke squirted out from the joints of the limbs as the monster swept by me. And in an instant it was gone.

So much I saw then, all vaguely for the flickering of the lightning, in blinding high lights and dense black shadows.

As it passed it set up an exultant deafening howl that drowned the thunder—"Aloo! aloo!"—and in another minute it was with its companion, half a mile away, stooping over something in the field. I have no doubt this Thing in the field was the third of the ten cylinders they had fired at us from Mars.

For some minutes I lay there in the rain and darkness watching, by the intermittent light, these monstrous beings of metal moving about in the distance over the hedge-tops. A thin hail was now beginning, and as it came and went their figures grew misty and then flashed into clearness again. Now and then came a gap in the lightning, and the night swallowed them up.

I was soaked with hail above and puddle-water below. It was some time before my blank astonishment would let me struggle up the bank to a drier position, or think at all of my imminent peril.

Not far from me was a little one-roomed squatter's hut of wood, surrounded by a patch of potato-garden. I struggled to my feet at last, and, crouching and making use of every chance of cover, I made a run for this. I hammered at the door, but I could not make the people hear (if there were any people inside), and after a time I desisted, and, availing myself of a ditch for the greater part of the way, succeeded in crawling, unobserved by these monstrous machines, into the pine-wood towards Maybury.

Under cover of this I pushed on, wet and shivering now, towards my own house. I walked among the trees trying to find the footpath. It was very dark indeed in the wood, for the lightning was now becoming infrequent, and the hail, which was pouring down in a torrent, fell in columns through the gaps in the heavy foliage.

If I had fully realized the meaning of all the things I had seen I should have immediately worked my way round through Byfleet to Street Chobham, and so gone back to rejoin my wife at Leatherhead. But that night the strangeness of things about me, and my physical wretchedness, prevented me, for I was bruised, weary, wet to the skin, deafened and blinded by the storm.

I had a vague idea of going on to my own house, and that was as much motive as I had. I staggered through the trees, fell into a ditch and bruised my knees against a plank, and finally splashed out into the lane that ran down from the College Arms. I say splashed, for the storm water was sweeping the sand down the hill in a muddy torrent. There in the darkness a man blundered into me and sent me reeling back.

He gave a cry of terror, sprang sideways, and rushed on before I could gather my wits sufficiently to speak to him. So heavy was the stress of the storm just at this place that I had the hardest task to win my way up the hill. I went close up to the fence on the left and worked my way along its palings.

Near the top I stumbled upon something soft, and, by a flash of lightning, saw between my feet a heap of black broadcloth and a pair of boots. Before I could distinguish clearly how the man lay, the flicker of light had passed. I stood over him waiting for the next flash. When it came, I saw that he was a sturdy man, cheaply but not shabbily dressed; his head was bent under his body, and he lay crumpled up close to the fence, as though he had been flung violently against it.

Overcoming the repugnance natural to one who had never before touched a dead body, I stooped and turned him over to feel for his heart. He was quite dead. Apparently his neck had been broken. The lightning flashed for a third time, and his face leaped upon me. I sprang to my feet. It was the landlord of the Spotted Dog, whose conveyance I had taken.

I stepped over him gingerly and pushed on up the hill. I made my way by the police-station and the College Arms towards my own house. Nothing was burning on the hill-side, though from the common there still came a red glare and a rolling tumult of ruddy smoke beating up against the drenching hail. So far as I could see by the flashes, the houses about me were mostly uninjured. By the College Arms a dark heap lay in the road.

Down the road towards Maybury Bridge there were voices and the sound of feet, but I had not the courage to shout or to go to them. I let myself in with my latch-key, closed, locked and bolted the door, staggered to the foot of the staircase, and sat down. My imagination was full of those striding metallic monsters, and of the dead body smashed against the fence.

I crouched at the foot of the staircase with my back to the wall, shivering violently.

THE GREAT SACRIFICE

(1903)

George C. Wallis

Born in Weedon, England in 1871, **George C. Wallis** began writing science fiction and adventure stories in his early twenties. His first published work was *Behind the Barrier: A Story of Mystery and Peril in the Antarctic Regions*, published in *Pearson's Storyteller* in 1895. Wallis supplemented his writing income as a printer and then a cinema manager. 'The Great Sacrifice' appeared in *The London Magazine* in 1903. Cleverly modelled on the recent runaway success of H. G. Wells's *The War of the Worlds*, the writing evokes the mystery of the Martians but subverts the reader's expectations of what dreaded plans lay behind their forays into space. His final novel, *The Call of Peter Gaskill: A Master Thriller Science Fiction Novel*, was published in 1948. He died in 1956.

I. Signs in the Sky

"IT'S QUEER," SAID HARRISON, RESTING HIS ELBOWS ON THE table meditatively. "The scientific world has been living for years in the comfortable belief that Kepler's and Newton's laws were inviolable, and that predicting the places of the planets at any instant was only a matter of correct calculation—and all at once some of the planets get hundreds of miles out of their track. Either Kepler and Newton were wrong, or something very strange has happened to the poor old Solar System."

We were in Harrison's private observatory waiting for the sky to clear. I lit a cigar carelessly.

"Let's hear the details," I said. "I must confess I've been too busy with other things lately to have time to read up the technical papers. Seems to be a big fuss about things, anyhow."

"Big fuss! I should think so—and there will be a bigger one still if this sort of thing goes on. When the Solar System begins to get behind schedule time, it's serious. It's been going on for over a week now. Only last night Saturn was about a whole degree behind his proper place. Uranus was even more, and as for Neptune we've had to re-discover him. To tell you the truth, Milford, I'm getting a bit scared. According to all long-established mathematical law, any such perturbation of the planetary movements would have a great effect on the orbit of the earth, would perhaps—nay, almost certainly—so upset the balance of the Solar System that it would go to everlasting smash. But the earth hasn't budged an inch out of her way, nor Mars, nor Jupiter—at least, we gave him the benefit of the doubt over the last observation—and here are the outer planets slowing down. Why, with their loss of momentum, they don't fall into the sun, I can't make out. And everything else, so far, seems as per usual. It looks to me as though we were being held back from chaos by a bond that might break at any time. And now you know as much as I do, and I hope you'll get to sleep tomorrow."

"Daresay I shall," I answered. The news did not seem so terrible to me as to Harrison and the other astronomers, who saw its significance with scientific eyes. "The sky looks pretty much the same to us ordinary mortals, and will do to the end, I expect, no matter what extraordinary things are going on. But it's getting clearer. Let's have a peep. What's on view?"

"There are hundreds of telescopes at work now," said Harrison, thoughtfully, unscrewing a cap as he spoke. "We've arranged a plan of campaign between us—a sort of bureau of celestial information—and I'm one of about twenty who have been deputed to watch Mars. If he varies a millimetre from his expected position, I have to wire it to the exchange. He's in the field now, and both moons are just visible. Bend down."

I placed my left, and strongest, eye to the bright dot of light at the end of the long tube, and saw the ruddy planet with its two small moons. I saw the familiar cloudy markings, the well-known snow-patches at the poles, and fancied that I detected, even as I looked, the movement of rapid Phobos—that active little satellite which contrived to get round Mars in less than eight hours. I was very interested at first, as I have always been in telescopic views, but soon grew tired of the constrained position, and made way for Harrison. He was busy with observations and calculations for some time, and at last he looked up with something like a sigh of relief.

"Mars all right, then?"

"Yes. He's exactly where he ought to be. I'll wire to the exchange now, if you don't mind waiting, and see if they have any news of Jupiter."

The clicking of the Morse code was very distinct in the dim observatory, as I bent down and peered through the tube once more. Active little Phobos was at that moment at greatest elongation. I had been looking perhaps a minute when the tiny moon suddenly blazed out with an intense white light. Thus luminous, it lasted several seconds, then it incontinently and utterly disappeared. I gave a startled exclamation, and at the same moment Harrison turned from the telegraph with a scared and puzzled expression:

"It's worse than ever," he said, before I had time to speak. "Jupiter is three minutes of arc behind his place, and the principal asteroids are slowing down. What did you say?"

I told him briefly. He looked eagerly up the tube. The light of a great fear—that fear which was so soon to come into all faces—shone in his eyes.

"Good heavens, Milford, what next? This is no mere cause for wonder, it is something strange and terrible. Some unknown, stupendous influence is at work in the Solar System, and God only knows what will be the end."

Then with a sudden change of manner he bustled round the room, put his paraphernalia back into their place, and said we had better go indoors and get some sleep. He talked wildly, and not a little incoherently, I thought, as we walked up the silent drive and across the lawn. I was not sorry when I found myself dozing off to sleep in bed. There was a sense of practical security between the white sheets. Yet I often thought afterwards of his talk that night as almost prophetic.

For a week things went on as usual. The scientific world was distraught with wonder, and a multiplicity of hypotheses to account for the action of the outer planets and the burning up and disappearance of Phobos, but the average crowd went about their business without much concern.

Then the message from Mars came, and theory gave way to fact, and the popular indifference to a reign of terror that shook the fabric of civilisation to its base.

II. The Message from Mars

By what means the Martian intelligences despatched their tiny projectiles across the abyss of space with such accuracy we shall never know, nor shall we ever know what those intelligences were like. We only know that they must have been thousands of years in advance of us in knowledge and in power, almost god-like in the latter, and altruistic to a degree which our lower minds cannot comprehend. We only know that their message came in a number of small, metallic balls, which fell to earth like meteors on the night of August 5th and the following day.

So accurately timed and speeded were these messengers of the peril that they fell at nearly regular distances apart, and with such motion as to bring them to the ground without being fused by the friction of the atmosphere. Many, no doubt, fell into the seas and rivers and deserts; many have been found since in all kinds of out-of-the-way places; but one and all contained similar contents. These consisted of a tightly rolled drum of a substance not unlike parchment and a few grains of a greyish powder.

Many of these metallic messengers fell into scientific hands on the morning of the 6th, and the whole world knew their purport on the 7th. Every telegraph wire carried the dread news; every paper had articles upon the subject; it was everywhere the chief topic of conversation. One fell into the lane at the back of Harrison's house, and was found accidentally by Harrison himself. We deciphered its meaning together the same afternoon, and wired it to the Press Agencies.

The day before I should have been amused at the idea of any ultra-terrestrial intelligences communicating with us, seeing that we could have no common basis of agreement as to the meaning of signs, but the Martians compelled me to confess my dull wit. The method used was at once simple and convincing.

The first thing that came into view as we unrolled the drum of parchment was a marvellously accurate map of the Solar System, so intricate in detail as to force an exclamation of wonder and praise from Harrison at first sight. On this map Mars was marked with a peculiar sign, resembling a Maltese cross. A number of lines were drawn from Mars to the earth when each were at different points of their orbits—points which Harrison quickly marked with their respective dates—and, on the date on which we had witnessed the disappearance of Phobos, a line was drawn from the surface of Mars to the place of its small satellite, and for the satellite itself was substituted a splash of pale blue colour.

"Then it was the Martians who destroyed Phobos by some terrible projectile or exercise of force?" I cried. Harrison nodded gravely.

"Yes. Evidently to attract our attention or to make an experiment. But look at this!—look at this! Do you see what it means?"

The unrolled drum now disclosed a map of much larger dimensions, on which the sun and some of the nearer stars were represented by tiny dots. We knew the sun again by the recurrence of the mark attached to it in the former map. Between the star *a* in Cassiopeia and our own sun, a small cloud of tiny objects (marked with a sign identical with that used to denote the Leonids and other meteor-streams of the Solar System) was shown, whose indicated orbit was directed straight to the sun. This cloud of meteors was marked with two distinctive signs.

"It means, I expect," said I, "that a large cloud of meteors is coming towards us from the direction of Cassiopeia, and so—but don't look scared, man—an extra grand display of celestial fireworks will be a compensation for the disappointment of two Novembers ago!"

My friend pulled himself together with a visible effort, but his voice trembled a little as he said:

"You don't see all I see in this. Do you think the Martians would have taken all this trouble just to warn us of the approach of an ordinary meteor swarm? No—this is a warning of world-peril—so far as I can see, a sentence of death for all humanity."

He paused a moment; then continued, growing calmer gradually:

"You will notice that the orbit of this meteor cloud does not start in Cassiopeia but is prolonged indefinitely beyond. This suggests that in order for the Martians to have traced its backward way so far it must be of a size and density unparalleled within our knowledge. Here is the next map conveying that precise information. Here is the Solar System again, bounded by a circular line; here is the meteor cloud with a circle of the same size drawn *within* it. Here is part of its orbit, marked off with a succession of regular ticks, each tick representing, I think, the distance travelled by the stream in a Martian year. A little calculation will settle that point... Yes, I am right. And here is the last Martian year period marked off into ten minor divisions—they must use the decimal system on Mars—and this tells us that in six weeks from tonight the great stream will be passing our system, scorching us all to death by the heat evolved by the impact of its mass upon the sun. The sun will blaze out like the famous star in Corona Borealis did, and every living thing on earth will be roasted alive! Heaven help us, Milford!"

A half-jesting doubt of the truth of his deductions was upon my lips, but I looked at the maps again and saw that he was right, read thereon the sentence of death in indisputable language. The silent but eloquent dots and lines and marks seemed to dance before my eyes like fantastic figures of fire. I did not speak, and we unrolled the drum of parchment to the end.

III. Days of Despair

The twilight shadows were creeping across the floor of the study when we re-wound the parchment and looked straight into each other's haggard eyes. The further maps and diagrams confirmed and reiterated beyond dispute our worst conjectures, besides being an evident attempt to convey some more information to us—information which, at that time, we completely failed to comprehend.

On a series of Solar maps a great quantity of lines were drawn from Mars to the outer planets, at times and distances easily calculated to be those of the retardations which had so exercised the wonder of the astronomers. That these retardations had been purposely achieved by the Martians, with a definite object in view, we had no room for doubt; that the Martians had solved the problem of the nature and control of gravitation, we were compelled to believe. All the indirect evidence of their existence and intelligence which had been so suddenly thrust upon us, enforced these points. But what bearing this knowledge, or the knowledge of the coming death, was expected to have upon earthly behaviour, we could not even dimly guess. That something more than a mere warning was intended, we felt sure; what it was, we had no idea.

We found our knowledge and our ignorance alike shared by all the competent students into whose possession the message came. One curious point was noted by all of us, though none then divined its true significance. On the map which represented the meteor stream infringing on the Solar System, the earth, Mars, and the four great outer planets, were shown to be in a direct line between the sun and centre of the great stream.

I parted from Harrison quietly—we felt that ordinary, commonplace sentences were all we dare use just then—and made my way home. It was a clear, warm night, already glowing with faint stars. They seemed to me as still, as steadfast, as prophetic of long endurance, as they have done to man for countless years, though I knew their seeming steadfastness was but a mask of mockery. As I turned the handle of the garden gate I remembered that I was due out that evening, and had no time to spare if I would be punctual. And hitherto I had been punctuality itself where Ethel was concerned. I hurriedly made myself fashionable, and went.

There came an interval in the evening's mirth when Ethel and I found ourselves alone on the balcony we knew so well. The stars were more numerous now, and shone radiantly. A few clouds were creeping up out of the west. The landscape spread out before us down to the river was covered with overlapping shadows. For a few minutes I debated with myself whether or not I should tell Ethel the awful news that would blight for her eyes, as it had done for mine, all the beauty of this quiet scene. If I did not tell her, she would read it in the papers tomorrow.

"Well, Jack," she said at last, breaking the thread of my unpleasant cogitations, "have you turned miserly all at once? 'Silence is golden' you know! It must be something *very* serious to make *you* so solemn!"

"That is true enough"—with the ghost or a smile—"I am frivolous enough, I admit, in a usual way. But 'tis nothing that concerns you—at least—that is, you will hear of it soon enough."

"Hear of what, Jack? I shall not be friends until you tell me all about it whatever it is—that's troubling you."

"I'd better not," I began, weakly: but a moment later I was launched upon the strange recital. I told her the whole story, omitting nothing, and grew marvellously scientific, even eloquently philosophical, with the intensity of my emotion. As I spoke I looked at the stars and the dark landscape, and when I finished and turned to Ethel I found her gaze fixed upon my face with a very curious expression. She took the tidings calmly.

"If this is true, Jack," she said. "I mean, if you and Harrison are not the victims of a dreadful joke—or the originators of one—it will be confirmed by the great scientific men of the world tomorrow or the following day. Then we shall know; until then, let us dismiss the subject. We have heard so many predictions of universal disaster that have come to nothing."

There was no getting the better of Ethel in an argument of this sort, so after a feeble protest I gave way. I even felt a sense of relief in the temporary casting-off of the yoke of terror. "Sufficient unto the day" seemed to me a very good maxim just then. Yet when we parted at the door Ethel grew serious.

"If it *should* be true, Jack," she said, holding my hand lightly, "will you come on and see me as soon as you can? And will you let me see one of these dreadful messages?"

I gave her the required promises readily, a thrill of pleasure running through the sighings of despair, and went home strangely placid. I had been engaged to Ethel for over a year, but until that night I had never known the depth of her love for me. It was strange. I, one of the few sharers of the secret of universal and imminent doom, was glad because a girl, one of the doomed—a mere ephemera who had not seven weeks to live—was more in love with me than I had thought.

But next morning the flood of gloom was upon me again, and upon all who believed.

After a brief call at Harrison's, I was at Ethel's home. It was early, but she met me in the drawing-room, a newspaper in her hand, a pallor on her sweet face.

"It is true, after all," she said.

"Yes," said I, "it's true. And what's the good of anything now? What does anything matter? Only a few weeks, and all the world, and you, and I, will be wiped out of existence as thoroughly as wrong figures on a schoolboy's slate."

"Yes, but our duties and our love remain the same, Jack. Whether we die tomorrow or fifty years hence does not matter at all. Whether we live well *now*, whether we love well *now*, matters everything. It is only weak souls who are slaves to time. You know the lines:

> We should count time by heart-throbs, not by hours,
> By feelings, not by figures on a dial."

"Of course you are right, Ethel," I said, gloomily, "but I am a mere, prosaic, modern man, unable, I fear, to soar to the heights of your transcendentalism. There is something unspeakably terrible to me in the thought of what is coming—that in seven weeks from now the world will be a scorched and lifeless ember; that all that men have striven after, and achieved and hoped for through countless generations, will be utterly wasted and lost."

"It is terrible to me, too, Jack, but we must not let it overmaster us. We must work. There will be plenty to do for all who can keep cool heads, and we shall face the end better if we do not anticipate it."

She spoke bravely, firing with enthusiasm. I felt a little cheered for the moment, but once out in the street again despair settled over me the deeper. And as the days wore on and the awful truth—confirmed by another corroborative message from Mars—gradually percolated through the meshes of ignorance and doubt, reaching to the lowest substratum of society and to the most remote corners of the earth, that despair settled over most men. But for the almost automatic instinct of law and order that upheld the Governments, and the heroism that the peril called forth from many natures such as Ethel's, civilisation would have gone to ruin long before the fatal day.

Who that was not forced would work against his will, with that dread shadow ever creeping nearer? Who would save, who respect any law but the law of strength, who paint, or write, or speak, or strive for fame or honour when in six weeks all must die?

The first days of the dread suspense have well been called the days of desolation. Many went mad in them, many died of fright or by self-destruction, many made fortunes at which their neighbours laughed, many lost their all and lived on charity, caring nothing. But many were mentally sober yet, and worked on.

IV. A Cleft in the Clouds

At the end of the first week mankind was divided into three factions—those who believed in the approaching catastrophe and were afraid; those who believed and were not afraid; and those who did not believe. The latter faction was a very small one, although it included some clever men. They spoke and wrote eloquently and scoffingly of antecedent improbabilities, universal experience, illusions, gigantic hoaxes, and the like. Their jests fell on heedless ears. One faction was too terrified to reason, to refute, to reprove; the other was too feverishly intent upon the daily work that casts out fear. And some, gay and reckless, were too busy making the most, in a frivolous sense, of the short time that remained to them.

You heard many grim jokes as you went along the streets. You saw men, for once, in their true colours. The churches and chapels and mission rooms were crowded nightly with fervent and trembling audiences, listening now to the preacher with an attention before unknown. And now the preacher had a theme that appealed to all. As in the days of old, he pointed his perorations with a fiery moral, this time drawing from science his picture of the end of the world. The drum of the Salvation Army rolled incessantly in the streets, and at nearly every corner and crossing you heard the cries of modern Solomon Eagles: "Repent, repent, ere it be too late, generation of sinners!" Faces blanched with terror as the once unheeded words fell on their ears: "The earth shall be melted with fervent heat and the heavens be rolled together like a scroll." And on the other hand you heard the calm, dispassionate reasonings of the Rationalists and Determinists. Their arguments penetrated the fever heat of fear as icicles of impersonal logic. They strove to drive home the lesson that here was one more proof of the fact that the inexorable workings of Nature give no sanction to man's anthropomorphic imaging of the Unknowable God.

The air was full of strange theories and queer suggestions. The intellect of all humanity was at work on the one absorbing problem. Up at Harrison's one evening, whither I had taken Ethel to see the Martian messages, we were talking of these things.

"Have you heard Schiaparelli's latest?" enquired Harrison, picking up a copy of the *British Mechanic* and glancing at a paragraph. "It's a daring idea, but has its merits. He calculates the sweep of the meteoric orbit from the portion shown to us on the Martian maps, and together with that, and the rate at which the stream is travelling, deduces the conclusion that it completes a circuit once in 20,000,000 of our years. That being so, the meteors must have passed through our system and enriched and superheated the sun 20,000,000 years ago, and at similar periods for who knows how far back? And you know Kelvin's estimate of the sun's present age—20,000,000 years. If Schiaparelli is right, the sun is constantly cooling and yet continually being revived by the impact of this immense stream of matter. This also suggests an argument in favour of the meteoritic origin of the Solar System, as against the old nebular hypothesis."

"It is a clever thought," said Ethel, "but then, you know, Kelvin's estimate is not a proved fact—it was seriously questioned by Proctor—and the ascertaining of a 20,000,000-year orbit is a very delicate matter indeed."

"Yes," said Harrison. "And another thing. The theory requires a very large body at or near the centre of the alleged orbit, and no visible star happens to be there. Of course, there maybe one of the dark, burnt-out suns in that position—and equally, of course, there may not."

"Just so," said I. "And after all, it does not matter, because we shall never know."

"Of course not," said Harrison, gravely. "But if Professor Belmont's idea is right, *that* will matter a great deal."

"And that is—?"

"That the Martians intend to ward off the catastrophe by some means," my friend replied. "I will read you Belmont's summing-up from the *Scientific American*: "That the Martians intended to convey to us more than the mere fact of the approach of the meteor-stream, no one who has seen the messages doubts. That this something more is an intimation that they will endeavour to prevent the disaster, seems to me equally beyond dispute. We know that they are retarding the outer planets—and now the earth and Venus and Mars itself—so as to bring all the planets in a right line between the centre of the sun and the centre of the meteor-stream at the time it will pass through and around the Solar System. On the map which conveys this information we find lines drawn from Mars to the outer planets, each line marked with a sign, and terminating in a sign, identical with the signs on the map marking the destruction of Phobos. The one spectroscopic observation of that destruction which we possess shows that a mass of pure hydrogen of immense volume was suddenly fired. This cannot have been sent from Mars; we can only account for it by supposing that the Martians—who, we know, can control gravitation—know the nature of the primeval element out of which all our so-called elements are formed, and can transmute these "elements" at their will; that, in fact, by means incomprehensible to us, they dissolved the whole mass of Phobos into incandescent hydrogen.

'The conclusion seems forced upon one that at the instant of impact they will similarly dissolve Neptune, Uranus, Saturn, and Jupiter, thus opposing the central portion of the meteor-stream with a succession of flaming hydrogen shields, in passing through which the meteors will be fused, and dissipated into cosmic dust. That the grains of grey powder in the Martian messages were intended for us to prove their accuracy with, seems convincingly shown by the fearful explosions which have taken place in the laboratories, because after each explosion the presence of a large body of hydrogen has been detected in the vicinity.

'Thus much appears certain. It is too daring to go further and affirm that the last map of the series suggests the possible self-destruction of Mars (and of the earth also, we being supposed to have learned the nature of the grey powder) should the hydrogenising of the outer planets prove insufficient? With the Martian message before me, I affirm that such is my belief.'"

"Feasible, is it not?" added Harrison.

"Quite," said I, "except the last suggestion. That seems too terrible, unless we suppose the Martians to have reached such a pitch of intelligence as to feel no repugnance at the idea of a world-suicide."

"May there not be another explanation?" asked Ethel, quietly. "It is evident that they are aware intelligences exist here. May not their intentional destruction of their own planet make the last flame-shield necessary to save *us* from death? May it not indicate a contemplated act of supreme self-sacrifice on the part of a race—a world?"

We sat still awhile, silently digesting the strange, new thought. Then Harrison cried:

"That shall be in all the leading papers tomorrow!"

V. The Flashes of Flame

Some other pen than mine must write the full story of those days of waiting, those days of drawn and agonised suspense. I lived my life in a very narrow sphere. Excepting at the house where I lodged, Ethel's and Harrison's, I spoke to very few people. Yet what I saw and read impressed upon me so vivid and terrible a picture of a world under sentence of death, gone mad with fear, that even now it colours all my dreams with a nameless horror.

But for the hint of hope conveyed in Belmont's theory and Ethel's now famous suggestion, I believe that mankind would have gone mad *en masse*, wrecked every institution of society that makes for order, and reverted to barbarism. As it was, the strain of alternate hope and fear showed itself in all faces, in all words, in all acts. Politics and business, peace and war, love and sport, travel and invention, all sank into sudden insignificance. Towards the end of the last week of security all supplies except the staples of life fell off alarmingly, even at panic prices. No one travelled except those who must, or those who hurried home from distant places. The cry of every human heart was for the presence of those dear to him in the last, dreaded hours.

The churches throughout the world announced that they would hold services with open doors on the last day, and two-thirds of the literature that poured from the world's printing presses consisted of sermons, exhortations, warnings, and words of supposed comfort put forth on behalf of the various religions which sway humanity. The Buddhist spoke resignedly of the peace of Nirvana (for even to India's millions the truth had reached); the Moslem, kneeling barefooted in the mosque, grew fervent over the joys of Paradise; the Catholic told his beads and prayed to the Holy Mother; the Protestant foresaw the nearness of the Second Coming; the Theist and Rationalist, with ideas broadened by the great conception of the vastness and unity and incomprehensibility of Nature, preached only courage and calmness, pleading that men should act a man's part even in the darkest hour, no matter whether a further lease of mortal life, or glory inconceivable, or death eternal, awaited all.

And there were some who rushed into more violent excesses of frivolity, debauchery and crime than even Rome knew in her most degenerate days. The lower class of theatres and music halls were filled nightly with reckless crowds; and despite the efforts of the military and police, forcible robberies and other worse deeds were enacted daily in the dark corners of great cities and on desolate country roads.

Fortunately the great mass of the people, especially in the stolid, English-speaking countries, contained themselves admirably, neither evincing any spirit of superhuman heroism nor of sub-human despair. The sky was black enough, no doubt, but plenty of time for "the panic of wild affright" when the storm burst. There was yet a gleam of hope shining.

The moment of contact between the meteor-stream and Neptune arrived at last. For some nights previously we had fine displays of shooting stars—the advance guard of the great group—which struck fresh terror into the minds of the ignorant, and now was the time to verify Belmont's hypothesis. Harrison and I were alone in the observatory once more, waiting in the small hours for the sign that should confirm what we both believed.

Harrison was at the eye-piece; I was wondering why, on so clear a night, we had seen fewer meteors than on the two previous evenings. Suddenly my friend uttered an exclamation of relief and seized my arm.

"Look—quick! It is true! The Martians are at work!"

I glanced through the dark tube, and saw where the tiny point of Neptune should have shone, a small cloud of intensely vivid light—the light of incandescent hydrogen. Belmont's reading of the Martian message was correct; the intelligences that inhabited the ruddy planet, superior to us in intellect and in power, were fighting the great peril. Knowing by means we cannot even guess that the earth held intelligences also, they had warned us of the peril and of their attempt to avert it. The thought, although familiar enough now, came home to me with fresh force, with an almost staggering greatness—was almost too large for a human brain to grasp, otherwise than as a wildly-impossible dream. Yet it was true.

I moved aside for Harrison to look once more, but in a little over ten minutes the cloud dimmed and disappeared. How many millions of meteors had been fused into cosmic dust in their headlong rush through that shield of flaming hydrogen? How many millions came after them, with shape and motion unimpaired? Flow many millions?—how many trillions of trillions!

We busied ourselves with the telegraph as the cloud vanished, sending and receiving news. We learnt that the spectroscope showed the flame of Neptune's destruction to be the flame of almost pure hydrogen, and that it was of immense volume. News of the destruction of Uranus came from the European observatories some days later, with the determination of the speed of the great meteor-stream and the fixing of the dates when it would reach the earth and the sun respectively. Saturn's annihilation was invisible everywhere, owing to heavy clouds, but the hydrogenising of the mighty mass of Jupiter was visible to the naked eye. To Harrison and me, sweeping the sky with the telescope, the sight was one never to be forgotten.

A straggling meteor shot through the field of view a moment prior to the predicted time, and the luminous trail in its wake had scarcely vanished when the huge globe of the giant planet seemed to melt into nothingness before our gaze. A moment's blank. Then, in place of the round orb, a swiftly expanding cloud of flame shone, glowing as with the sustained blast of a furnace, shooting out fierce red tongues of fire round all its jagged and increasing circumference, glittering with a thousand points of yet intenser flame, as though riddled through and through with a torrent of stars.

The cloud of flame endured for more than half-an-hour, and in every second of that time it was swallowing in its capacious maw myriads upon myriads of groups of those restless, hurrying enemies of ours, fusing them with the furnace heat of friction into harmless dust. The grey dawn stole across the sky shortly after the cloud became invisible. We felt a thrill of hope now that we were sure those Martian intelligences were fighting so bravely for us and for themselves. Yet whether we were saved we did not know. The peril was not yet past. The stress of suspense was not yet over.

VI. The Supreme Sacrifice

It was late in the afternoon, and the whole sky was growing dull with the approach of twilight, when Ethel and I walked together to my friend's house. She could not rest at home, she said, waiting to know the end, and I was more than willing that she should be with me in the moment when the nature of that end should be known for good and all. After that supreme moment, whether the verdict was life or death, I had promised to return home with her. In the last hours we would be together.

On our way we met and passed many people. Their faces reflected all the shades of fear, from dim doubt to wild terror. Some walked defiantly, some furtively, some with affected unconcern, but we did not see one face that betokened real carelessness or suggested real composure. One group we passed at the corner of a street were indeed singing and praying and preaching loudly, and called to us to join them, but the scared expression in their eyes—red with want of sleep, with long and anxious vigils at bedroom windows, repeated studies of the dangerous sky—belied the complacency they professed and offered. Sadly shaking off the nervous hands that one zealot laid upon us, we hurried on.

Harrison and his wife greeted us briefly but sincerely, and the four of us went straight to the garden seat near the open door of the observatory. If the calculations of the astronomers were correct, we should receive news of the great issue from Greenwich in a quarter of an hour. As Greenwich time is about five hours in advance of American Eastern time, we should have to wait that period for a distinct observation of Mars ourselves. As the minutes were counted off and the telegraph instrument gave no sign, the tension drew its cords tightly around us.

Save for a few drifting clouds, the sky was dark and clear and cold; the air was still; the whole world about us seemed plunged into waiting silence. We were each loth to break the spell of that hush, and sat apart, busy with our own thoughts. I remember that I was thinking something like this: The Martians, besides having control of gravitation, must have means of knowing the total amount of matter in the meteor cloud. Knowing that, and how much of it they have destroyed already, they will know whether what yet remains is sufficient to raise the sun's heat beyond the point fatal to life. They must also know if the hydrogenising of their own planet would avert the calamity, and so save the earth and the inner planets. If Mars is hydrogenised, it will mean that such is the case; if he is not, it will mean that even Martian intelligence is beaten and that only universal death remains. I also remember wondering whether any of the Martians would try to escape from the ruin of their world and come to earth, as Wells's Martians did.

I was roused from my reverie by Harrison shutting his chronometer with a snap. The time had expired and the wire had not spoken. Mars yet existed.

"We will wait a little longer," said Ethel, very quietly, but with pale face. "The calculations may be in error or the sky cloudy in Europe. Until we have seen Mars with our own eyes we will not go home. And it may be that the danger is really past—that the meteors have already been reduced in number below the danger point."

"That might be so, but I do not think the Martians, knowing the approximate mass of the meteors, would have sent us any message in that case. For my part I go further than Belmont now. I believe that they foresaw the need of self-sacrifice and contemplated it from the first, only sending us the message to suggest the same self-destruction to us, in the event of their calculations proving in error or the hydrogen shields not fusing sufficient of the meteors. The balance of mass between death and life for us may be—I think it is—very fine now, and we may yet see what we want. They may be delaying their own destruction because of a concentration of the meteors in the tail of the stream. We will wait, as Miss Holroyd suggests, until we have seen Mars himself."

We waited, and the sky darkened swiftly. An hour later a brilliant meteor shot across the sky. After a short interval it was followed by another, and they by others, until the heavens seemed riven by myriads of lances of fire.

"The meteors—the harbingers of death!" cried Ethel, her hands locked in silent prayer. We looked at each other for a few moments, then back at the sky. It held us fascinated. The meteors—evidently a portion of the great stream which had escaped the hydrogen shields—came scattering to right and left from that point on the horizon where Mars would rise. East and west, north and south, they spread; the sky was full of their brief shillings. It has been said by those who remember the great Leonid showers of 1866–7, that those displays were almost as nothing compared with the terribly grand vision of a hurricane of stars that blazed through the heavens on that never-to-be-forgotten night of fate.

The number of the meteors was simply incalculable, and as their radiant point rose slowly to the horizon, the light of their intense incandescence grew hurtful to our weary eyes. Sheer exhaustion of the optic nerve drove us into the house, and the calm of an abnormal excitement bade us to the supper table. Then, after a short rest, back to the observatory. The meteor storm still swept in fiery violence through the silent sky.

The hours passed. Now voicing what scraps of forlorn hope we could find, and now relapsing into a stillness that seemed prophetic of the world as it would be in the lifeless days to come, we four sat in the observatory, waiting for a glimpse of Mars. Once or twice I went to the doorway and looked out across the quiet garden, thick and dense with the night shadows, illuminated only by the pulsing light of the meteor-riven heavens. The trees stood mysteriously grey and still, not a twig moved, nor a leaf upon the evergreens of the hedge. Away over the dim housetop, outlined as the horizon of the falling stars, the old Park Hill rose, a smudge of indistinguishable darkness, with one solitary house-light betokening the presence of humanity upon its desolate slope.

As I stood in the doorway the last time, thinking of time and space, and life and death and destiny, and love that is greater than all, yet ephemeral as those tiny bodies that were lighting the sky with their evanescent radiance—thoughts that one can *think*, but can neither write nor speak—Ethel touched me on the arm. She then drew me within, dosing the door softly after me.

"You will catch cold, Jack," she said. "And as the sky is clear of clouds, and the number of the meteors is diminishing, we shall soon have Mars in the field of the telescope."

I laughed, rather bitterly. With a grim sense of incongruous humour, I remarked that there had never been a finer chance of ignoring the laws of health, especially with regard to chills, as we should all, no doubt, he hot enough soon. Ethel put her hand over my mouth without answering, and I sat down. I shall never forget her calm courage in those nerve-straining hours.

Then Harrison slowly swung the tube round and adjusted the eyepiece according to calculation, and for a few minutes peered at the field of view. In silence his wife and then Ethel succeeded him at the instrument; in silence I, too, received upon my optic nerve those waves of ruddy light, which, having traversed forty-million miles of vibrant ether, told me Mars yet lived. Mars lived—the meteor stream was then too vast for even Martian intelligence to reduce to impotence. Mars lived—and therefore it and Earth, with the minor planets that yet remained to the relentless sun, must on the morrow die. Was this, then, the end of all—the Purpose of the Ages?...

At that moment I heard, though faintly, the sound of feet on the road outside, then voices pitched in the key of despair, then a woman's choking sob. All suddenly the horror of it struck me as it had never done before. With a sharp cry I left the telescope and staggered to a chair, and out of the dim shadows under the galvanised dome my imagination fashioned a vision of the World's Death.

I saw the countless meteors still shooting, incandescent, through the air, and knew that when one fell to earth in dust ten billion billion more sped on towards the all-compelling, all-consuming, superheated sun. I saw fierce spots on his golden disc, and new red-flaming prominences around his rim, as though, insatiate, he reached out greedy arms towards the stricken planets. I saw his colour change to a fiercer red, and felt the scorching flood of his heat grow more and more intense, from tropical dawns to noons in which no men could live, save in dungeons and cellars and caves. I watched the temperate zone, fast followed by the tropic as that was pursued by the *burning zone*, recede to north and south, even to the shuddering Poles. I saw the Alpine hills grow moist, and the late snow run over and down them in warm, black torrents; saw the glaciers sweat and slip and slide, and, melting as they slid, descend with roar and ruin into flooded valleys. I saw grass go brown, and drop in dust where it did not burn. I saw the sea bubble and boil, and rise in clouds of scalding steam.

Finally, I beheld a charred and blackened earth—a world in which there was no life, neither of man nor beast, nor creeping nor flying thing, nor tree nor herb of any kind—a world of heat and darkness and utter desolation.

And then the vision passed; it could not have lasted many seconds. All was still and cold in the observatory. Ethel was looking at me, but not curiously. Harrison and his wife were looking at each other. Feeling that I must do something, that inaction would be dangerous, I once more put my eye to the telescope. I had not been watching Mars for twenty seconds when the ruddy planet melted away into the sky around him, showing again swiftly as a vast, expanding cloud of flame, shotted through and through with a myriad points of yet more vivid fire. A minute longer I remained at the tube, to make sure my eyes had not deceived me; then I told the others.

The cloud of consuming hydrogen lasted long enough for all of us to see it several times. After it faded, and whilst we three talked wildly and gladly, like doomed prisoners suddenly set free, Harrison's lingers were busy with the telegraph.

It was true, after all; quite true. Mars had died so that the earth *could* live. Those unknown intelligences who had warned us of the peril had made the last supreme sacrifice to avert the full fury of the great assault. It is a thought that even yet passes human comprehension, yet it is true—that a world has sacrificed itself for a world, a race of intelligent beings has blotted itself out of existence in order that another and less-intellectual race might live.

I need only add that the meteor display continued one night more and then ceased abruptly; that the approaching winter was very mild and the succeeding summer almost tropical: that the earth is now the largest planet of the Solar System; and that I married Ethel within the month.

PIED PIPER OF MARS (1942)

Frederic A. Kummer

Frederic Arnold Kummer was born in 1873 in Maryland. His father was a German immigrant who fought in the American Civil War. He attended Rensselaer Polytechnic Institute, New York, and graduated with a degree in civil engineering. This led to him becoming chief engineer for the American Wood Preserving Company, and later General Manager of the Eastern Paving Block Company. In 1907, however, he turned to writing full time, producing numerous short stories, novels and screenplays. 'Pied Piper of Mars' was published in 1942, the year before Kummer's death.

Elath Taen made mad music for the men of Mars. The red planet lived and would die to the soul-tearing tunes of his fiendish piping.

In all the solar system there is no city quite like Mercis, capital of Mars. Solis, on Venus, is perhaps more beautiful, some cities of Earth certainly have more drive and dynamitism, but there is a strange inscrutable air about Mercis which even terrestials of twenty years' residence cannot explain. Outwardly a tourists' mecca, with white plastoid buildings, rich gardens, and whispering canals, it has another and darker side, ever present, ever hidden. While earthmen work and plan, building, repairing, bringing their vast energy and progress to decadent Mars, the silent little reddies go their devious ways, following ancient laws which no amount of terrestial logic can shake. Time-bound ritual, mysterious passions and hates, torturous, devious logic ... all these, like dark winding underground streams run beneath the tall fair city that brings such thrilled superlatives to the lips of the terrestial tourists.

Steve Ranson, mounting the steps of the old house facing the Han canal, was in no mood for the bizarre beauties of Martian scenery. For one thing, Mercis was an old story to him; his work with Terrestial Intelligence had brought him here often in the past, on other strange cases. And for another thing, his mission concerned more vital matters. Jared Haller, as head of the state-owned Martian Broadcasting System, was next in importance to the august Governor Winship himself. As far back as the Hitlerian wars on earth it had been known that he who controls propaganda, controls the nation ... or planet. Martian Broadcasting was an important factor in controlling the fierce warlike little reddies, keeping the terrestial-imposed peace on the red planet. And when Jared Haller sent to Earth for one of the Terrestial Intelligence, that silent efficient corps of trouble-shooters, something was definitely up.

The house was provided with double doors as protection against the sudden fierce sandstorms which so often, in the month of Tol, sweep in from the plains of Psidis to engulf Mercis in a red choking haze. Ranson passed the conventional electric eye and a polite robot voice asked his name. He gave it, and the inner door opened.

A smiling little Martian butler met him in the hall, showed him into Haller's study. The head of M.B.C. stood at one end of the big library, the walls of which were lined with vivavox rolls and old-fashioned books. As Ranson entered, he swung about, frowning, one hand dropping to a pocket that bulged unmistakably.

"Ranson, Terrestial Intelligence." The special agent offered his card. "You sent to Earth a while ago for an operator?"

Jared Haller nodded. He was a big, rough-featured individual with gray leonine hair. A battering-ram of a man, one would think, who hammered his way through life by sheer force and drive. But as Ranson looked closer, he could see lines of worry, of fear, etched about the strong mouth, and a species of terror within the shaggy-browed eyes.

"Yes," said Jared Haller. "I sent for an operator. You got here quickly, Mr. Ranson!"

"Seven days out of earth on the express-liner *Arrow*." Ranson wondered why Haller didn't come to the point. Even Terrestial Intelligence headquarters in New York hadn't known why a T.I. man was wanted on Mars ... but Haller was one of the few persons sufficiently important to have an operator sent without explanation as to why he was wanted. Ranson put it directly. "Why did you require the help of T.I., Mr. Haller?" he asked.

"Because we're up against something a little too big for the Mercian police force to handle." Jared Haller's strong hands tapped nervously upon the desk. "No one has greater respect for our local authorities than myself. Captain Maxwell is a personal friend of mine. But I understood that T.I. men had the benefit of certain amazing devices, remarkable inventions, which make it easy for them to track down criminals."

Ranson nodded. That was true. T.I. didn't allow its secret devices to be used by any other agency, for fear they might become known to the criminals and outlaws of the solar system. But Haller still hadn't told what crime had taken place. This time Ranson applied the spur of silence. It worked.

"Mr. Ranson," Haller leaned forward, his face a gray grim mask, "someone, something, is working to gain control of the Martian Broadcasting Company! And I don't have to tell you that whoever controls M.B.C. controls Mars! Here's the set-up! Our company, although state owned, is largely free from red-tape, so long as we stress the good work we terrestials are doing on Mars and keep any revolutionary propaganda off the air-waves. Except for myself, and half a dozen other earthmen in responsible positions, our staff is largely Martian. That's in line with our policy of teaching Mars our civilization until it's ready for autonomy. Which it isn't yet, by quite some. As you know."

Ranson nodded, eyes intent as the pattern unfolded.

"All right." Haller snapped. "You see the situation. Remove us ... the few terrestials at the top of M.B.C ... and Martian staff would carry on until new men came out from Earth to take our places. But suppose during that period with no check on their activities, they started to dish out nationalist propaganda? One hour's program, with the old Martian war-songs being played and some rabble-rouser yelling 'down with the terrestial oppressors' and there'd be a revolution. Millions of reddies against a few police, a couple of regiments of the Foreign Legion. It'd be a cinch."

"But," ... Ranson frowned ... "this is only an interesting supposition. The reddies are civilized, peaceful."

"Outwardly," Haller snapped. "But what do you or any other earthmen know about what goes on in their round red heads? And the proof that some revolt is planned lies in what's been happening the past few weeks! Look here!" Haller bent forward, the lines about his mouth tighter than ever. "Three weeks ago my technical advisor, Rawlins, committed suicide. Not a care in the world, but he killed himself. A week later Harris, head of the television department, went insane. Declared a feud with the whole planet, began shooting at everyone he saw. The police rayed him in the struggle. The following week Pegram, the musical director, died of a heart attack. Died with the most terrorized expression on his face I've ever seen. Fear, causing the heart attack, his doctor said. You begin to see the set-up? Three men, each a vital power in M.B.C. gone within three weeks! And who's next? Who?" Jared Haller's eyes were bright with fear.

"Suicide, insanity, heart attack." Ranson shrugged. "All perfectly normal. Coincidence that they should happen within three weeks. What makes you think there's been foul play?"

For a long brittle moment Jared Haller stared out at the graceful white city, wan in the light of the twin moons. When he turned to face Ranson again, his eyes were bleak as a lunar plain.

"One thing," he said slowly. "The music."

"Music?" Ranson echoed. "Look here, Mr. Haller, you..."

"It's all right." Jared Haller grinned crookedly. "I'm not insane. Yet. Look, Mr. Ranson! There's just one clue to these mysterious deaths! And that's the music! In each instance the servants told of hearing, very faintly, a strange melody. Music that did queer things to them, even though they could hear it only vaguely. Music like none they'd ever heard. Like the devil's pipes, playing on their souls, while... Almighty God!"

Jared Haller froze, his face gray as lead, his eyes blue horror. Ranson was like a man in a trance, bent forward, lips pressed tight until they resembled a livid scar. The room was silent as a tomb; outside, they could hear the vague rumbling of the city, with the distant swish of canal boats, the staccato roar of rockets as some earth-bound freighter leaped from the spaceport. Familiar, homey sounds, these, but beneath them, like an undercurrent of madness, ran the macabre melody.

There was, there had never been, Ranson knew, any music like this. It was the pipes of Pan, the chant of robots, the crying of souls in torment. It was a cloudy purple haze that engulfed the mind, it was a silver knife plucking a cruel obligato on taut nerves, it was a thin dark snake writhing its endless coils into the room.

Neither man moved. Ranson knew all the tricks of visual hypnotism, the whirling mirror, the waving hands, the pool of ink ... but this was the hypnotism of sound. Louder and clearer the music sounded, in eerie overtones, quavering sobbing minors, fierce reverberating bass. Sharp shards of sound pierced their ears, deep throbbing underrhythm shook them as a cat shakes a mouse.

"God!" Haller snarled. "What ... what is it?"

"Don't know." Ranson felt a queer irritation growing within him. He strode stiffly to the window, peered out. In the darkness, the broad Han canal lay placid; the stars caught in its jet meshes gently drifted toward the bank, shattered on the white marble. Along the embankment were great fragrant clumps of *fayeh* bushes. It was among these, he decided, that their unknown serenader lay concealed.

Suddenly the elfin melody changed. Fierce, harsh, it rose, until Ranson felt as though a file were rasping his nerves. He knew that he should dash down, seize the invisible musician below ... but logic, facts and duty, all were fading from his mind. The music was a spur, goading him to wild unreasoning anger. The red mists of hate swirled through his brain, a strange unreasoning bloodlust grew with the savage beat of the wild music. Berserk rage sounded in each shivering note and Ranson felt an insane desire to run amok. To inflict pain, to see red blood flow, to kill ... kill! Blindly he whirled, groping for his gun, as the music rose in a frenzied death-wail.

Turning, Ranson found himself face to face with Jared Haller. But the tall flinty magnate was now another person. Primitive, atavistic rage distorted his features, insane murder lurked in his eyes. The music was his master, and it was driving him to frenzy. "Kill!" the weird rhythm screamed, "Kill!" And Jared Haller obeyed. He snatched the flame-gun from his pocket, levelled it at Ranson.

Whether it was the deadly melody outside, or the instinct of self-preservation, Ranson never knew, but he drove at Haller with grim fury. The flame-gun hissed, filling the room with a greenish glare, its beam passing so close to Ranson's hair as to singe it. Ranson came up, grinning furiously, and in a moment both men were struggling, teeth bared in animalistic grins, breath coming in choked gasps, whirling in a mad dance of death as the macabre music distilled deadly poison within their brains.

The end came with startling suddenness. Ranson, twisting his opponent's arm back, felt the searing blast of the flame-gun past his hand. Jared Haller, a ghastly blackened corpse, toppled to the floor.

At that moment the lethal rhythm outside changed abruptly. From the fierce maddening beat of a few minutes before, the chords took on a yearning seductive tone. A call, it seemed, irresistible, soft, with a thousand promises. This was the song the sirens sang to Ulysses, the call of the Pied Piper, the chant of the houris in paradise. It conjured up pictures in Ranson's mind ... pictures of fairyland, of exquisitely beautiful scenes, of women lovely beyond imagination. All of man's hopes, man's dreams, were in that music, and it drew Ranson as a moth is drawn to a flame. The piping of Pan, the fragile fantasies of childhood, the voices of those beyond life... Ranson walked stiffly toward the source of the music, like a man drugged.

As he approached the window the melody grew louder. The hypnotism of sound, he knew, but he didn't care. It was enthralling, irresistible. Like a sleepwalker he climbed to the sill, stood outlined in the tall window. Twenty feet to the ground, almost certain death ... but Ranson was lost in the golden world that the elfin melody conjured up. He straightened his shoulders, was about to step out.

Then suddenly there was a roar of atomic motors, a flashing of lights. A police boat, flinging up clouds of spray, swept up the canal, stopped. Ranson shook himself, like a man awakening from a nightmare, saw uniformed figures leaping to the bank. From the shadow of the *fayeh* bushes a slight form sprang, dodged along the embankment. Flame-guns cut the gloom but the slight figure swung to the left, disappeared among the twisting narrow streets. Bathed in cold sweat, Ranson stepped back into the room, where the still, terrible form of Jared Haller lay. Ranson stared at it, as though seeing it for the first time. Outside, there were pounding feet; the canal-patrolmen raced through the house, toward the study. And then, his brain weary as if it had been cudgelled, Ranson slid limply to the floor.

Headquarters of the Martian Canal-Patrol was brilliantly lighted by a dozen big *astralux* arcs. Captain Maxwell chewed at his gray mustache, staring curiously at Ranson.

"Then you admit killing Haller?" he demanded.

"Yes." Ranson nodded sombrely. "In the struggle. Self-defense. But even if it hadn't been self-defense, I probably would have fought with him. That music was madness, I tell you! Madness! Nobody's responsible when under its influence! I..."

"You killed Haller," Captain Maxwell said. "And you blame it on this alleged music. I might believe you, Ranson, but how many other people would? Even members of Terrestial Intelligence aren't sacro sanct. I'll have to hold you for trial."

"Hold me for trial?" Ranson leaned forward, his gaunt face intent. "While the real killer, the person playing that music, gets away? Look! Let me out of here for twelve hours! That's all I ask! And if I don't track down whoever was outside Haller's house, you can..."

"Sorry." Captain Maxwell shook his head. "You know I'd like to, Ranson. But this is murder. To let a confessed murderer, even though he is a T.I. man, go free, is impossible." The captain drew a deep breath, motioned to the two gray-uniformed patrolmen. "Take Mr. Ranson."

And then Steve Ranson went into action. In one blinding burst of speed, he lunged across the desk, tore Captain Maxwell's pistol from its holster. Before the captain and the two patrolmen knew what had happened, they were staring into the ugly muzzle of the flame-gun.

"Sorry." Ranson said tightly. "But it had to be done. There's hell loose on Mars, the devil's melody! And it's got to be stopped before it turns this planet upside down!"

"You can't get away with this, Ranson!" Captain Maxwell shook his head. "It'll only make it tougher for you when we nab you again! Be sensible! Put down that gun."

"No good. Got to work fast." Ranson backed toward the door, gun in hand. "Let this mad music go unchecked and it's death to all terrestials on Mars! And

I'm going to stop it! So long, captain! You can try me for murder if you want, after I've done my job here!"

Ranson took the key from the massive plastic door as he backed through the entrance. Once in the hall, he slammed the door shut, locked Maxwell and his men in the room. Then, dropping the gun into his pocket, he ran swiftly down the corridor to the main entrance of headquarters. In the hall a patrolman glanced at him suspiciously, halted him, but a wave of Ranson's T.I. card put the man aside.

Free of headquarters, Ranson began to run. Only a few moments, he knew, before Maxwell and his men blasted a way to freedom, set out in pursuit. Like a lean gray shadow Ranson ran, twisting, dodging, among the narrow streets, heading toward Haller's house. Mercis was a dream city in the wan light of the moons. One in either side of the heavens, they threw weird double shadows across the rippling canals, the aimless streets. Sleek canal-cabs roared along the dark waterways, throwing up clouds of spray, and on the embankments, green-eyed, bulge-headed little reddies padded, silent, inscrutable, themselves a part of the eternal mystery of Mars.

Haller's house stood dark and brooding beside the canal. Captain Maxwell's men had completed their examination and the place was deserted. Ranson stepped into the shadow of the clump of fragrant *fayeh* bushes, where the unknown musician had stood; there was little danger, he felt, of patrolmen hunting for him at Haller's house. The captain had little faith in copybook maxims about the murderer returning to the scene of the crime.

Ranson stood motionless for a moment as a canal boat swept by, then drew from his pocket a heavy black tube. He tugged, and it extended telescopically to a cane some four feet long. The cane was hollow, a tube, and the head of it was large as a man's two fists and covered with small dials, gauges. This was the T.I.'s most cherished secret, the famous "electric bloodhound," by which criminals could be tracked.

Ranson touched a lever and a tiny electric motor in the head of the cane hummed, drawing air up along the tube. He tapped the bank where the unknown musician had stood, eyes on the gauges. Molecules of matter, left by the mysterious serenader, were sucked up the tube, registered on a sensitive plate, just as delicate color shades register on the plate of a color camera.

Ranson tapped the cane carefully upon the ground, avoiding those places where he had stood. Few people crossed this overgrown embankment, and it was a safe bet that no one other than the strange musician had been there recently. The scent was a clear one, and the dials on the head of the cane read R-2340-B, the numerical classification of the tiny bits of matter left behind by the unknown. The theory behind it was quite simple. The T.I. scientists had reasoned that the sense of smell is merely the effect of suspended molecules in the air acting upon sensitive nerve filaments, and they knew that any normal human can follow a trail of some strong odor such as perfumes, or gasoline, while animals, possessing more sensitive perceptions, can follow less distinct trails. To duplicate this mechanically had proven more difficult than an electric eye or artificial hearing device, but in the end they had triumphed. Their efforts had resulted in the machine Ranson now carried.

The trail was, at the start, clear. Ranson tapped the long tube on the ground like a blind man, eyes on the dial. Along the embankment, into a side street, he made his way. There were few abroad in this old quarter of the city; from the spaceport came the roar of freighters, the rumble of machinery, but here in the narrow winding streets there was only the faint murmur of voices behind latticed windows, the rustle of the wind, the rattle of sand from the red desert beyond the city.

As Ranson plunged further into the old Martian quarter, the trail grew more and more confused, crossed by scores of other trails left by passersby. He was forced to stop, cast about like a bloodhound, tapping every square foot of the street before the R-2340-B on the dial showed that he had once more picked up the faint elusive scent.

Deeper and deeper Ranson plunged into the dark slums of Mercis. Smoky gambling dens, dives full of drunken spacehands and slim red-skinned girls, maudlin singing ... even the yellow glare of the forbidden san-rays, as they filtered through drawn windows. Unsteady figures made their way along the streets. Mighty-thewed Jovian blasters, languid Venusians, boisterous earthmen ... and the little Martians padding softly along, wrapped in their loose dust-robes.

At the end of an alley where the purple shadows lay like stagnant pools, Ranson paused. The alley was a cul-de-sac, which meant that the person he was trailing must have entered one of the houses. Very softly he tapped the long tube on the ground. Again with a hesitant swinging of dials, R-2340-B showed up, on the low step in front of one of the dilapidated, dome-shaped houses. Ranson's eyes narrowed. So the person who had played the mad murder melody had entered that house! Might still be there! Quickly he telescoped the "electric bloodhound," dropped it into his pocket, and drew his flame-gun.

The old house was dark, with an air of morbid deadly calm about it. Ranson tried the door, found it locked. A quick spurt from his flame-gun melted the lock; he glanced about to make sure no one had observed the greenish glare, then stepped inside.

The hallway was shadowy, its walls hung with ancient Martian tapestries which, from their stilted symbolic ideographs must have dated back to the days of the Canal-Builders. At the end of the hallway, however, light jetted through a half-open door. Ranson moved toward it, silent as a phantom, muscles tense. Gripping his flame-gun, he pushed the door wide ... and a sudden exclamation broke from his lips.

Before him lay a gleaming laboratory, lined with vials of strange liquids, shining test-tubes, and queer apparatus. Beside a table, pouring a black fluid from a beaker into a test-tube, stood a man. Half-terrestial, half-Martian, he seemed, with the large hairless head of the red planet, and the clean features of an earthman. His eyes, behind their glasses, were like green ice, and the hand pouring the black fluid did not so much as waver at Ranson's entrance.

Ranson gasped. The bizarre figure was that of Dr. Elath Taen, master-scientist, sought by the T.I. for years, in vain! Elath Taen, outlaw and renegade, whose sole desire was the extermination of all terrestials on Mars, a revival of the ancient glories of the red planet. The tales told about him were fabulous; and this was the man behind the unholy music!

"Good evening, Mr. Ranson," Elath Taen smiled. "Had I known T.I. men were on Mars I should have taken infinitely more precautions. However..."

As he spoke, his hand moved suddenly, as though to hurl the test tube at Ranson. Quick as he was, the T.I. man was quicker. A spurt of flame leapt from his gun, shattering the tube. The dark liquid hissed, smoking, on to the floor.

"Well done, Mr. Ranson." Elath Taen nodded calmly. "Had the acid struck you, it would have rendered you blind."

"That's about enough of your tricks!" Ranson grated. "Come along, Dr. Taen! We're going to headquarters!"

"Since you insist." Elath Taen removed his chemist's smock, began, very deliberately, to strip off his rubber gloves.

"Quit stalling!" Ranson snapped. "Get going! I..." The words faded on the T.I. man's lips. Faintly, in the distance, came the strains of soft eerie music!

"Good God!" Ranson's eyes darted about the laboratory. "That ... that's the same as Haller and I..."

"Exactly, Mr. Ranson." Elath Taen smiled thinly. "Listen!"

The music was a caress, soft as a woman's skin. Slow, drowsy, like the hum of bees on a hot summer's afternoon. Soothing, soporific, in dreamy, crooning chords. A lullaby, that seemed to hang lead weights upon the eyelids. Audible hypnotism, as potent as some drug. Clearer with each second, the melody grew, coming nearer and nearer the laboratory.

"Come ... come on," Ranson said thickly. "Got to get out of here."

But his words held no force, and Elath Taen was nodding sleepily under the influence of the weird dream-music. Ranson knew he should act, swiftly, while he could; but the movement of a single muscle seemed an intolerable effort. His skin felt as though it were being rubbed with velvet, a strange purring sensation filled his brain. He tried to think, to move, but his will seemed in a padded vise. The music was dragging him down, down, into the gray mists of oblivion.

Across the laboratory Elath Taen had slumped to the floor, a vague smile of triumph on his face. Ranson turned to the direction of the music, tried to raise his gun, but the weapon slipped from his fingers, he fell to his knees. Sleep ... that was all that mattered ... sleep. The music was like chloroform, its notes stroked his brain. Through half-shut eyes he saw a door at the rear of the laboratory open, saw a slim, dark, exotic girl step through into the room. Slung about her neck in the manner of an accordian, was a square box, with keys studding its top. For a long moment Ranson stared at the dark, enigmatic girl, watched her hands dance over the keys to produce the soft lulling music. About her head, he noticed, was a queer copper helmet, of a type he had never before seen. And then the girl, Elath Taen, the laboratory, all faded into a kaleidoscopic whirl. Ranson felt himself falling down into the gray mists, and consciousness disappeared.

Steve Ranson awoke to find himself still in the laboratory, bound securely hand and foot. Opposite him Elath Taen was just struggling to his feet, aided by the dark-haired, feline girl.

"I ... I'm all right, Zeila," Taen muttered. "It was necessary that I, also, hear the sleep-melody, in order to overcome our snooping friend here. But look—he's coming to!"

The girl's gold-flecked eyes turned to Ranson, studied him impassively. Elath Taen gave a mocking smile.

"My daughter Zeila, Mr. Ranson," he murmured. "The consolation of my declining years. She, too, has devoted her life to the great cause of Martian freedom, the overthrow of Terra!"

"To be expected from your daughter," Ranson grunted. "I might have known you were at the bottom of this, Taen! Killing off the officials of the Martian Broadcasting Company!"

"Killing?" Taen smiled, glanced at the queer box slung about the girl's neck. "We only serenaded them. Induce the necessary moods for murder, suicide, madness. You have played our tunes to the remaining two, Zeila?"

The girl nodded impassively. "Cartwright unfortunately ended his own life," she said. "Rankin heard the song of hate, went berserk and was killed. Yla-tu, one of our own people, is in charge of M.B.C. until more terrestial executives arrive from earth."

"By which time we will have played our melodies to all Mars," Taen murmured. "One swift, merciless uprising, and the red planet is free! An hour or so over M.B.C.'s network..."

"You're nuts!" Ranson laughed. "If you think..."

"I don't think," Elath Taen smiled. "I *know*, Mr. Ranson. Before the night is out, all terrestials on Mars will be imprisoned or dead. Our people need only something to awaken them, to arouse their hate! And I can do that! I am the master of moods!" He took a copper helmet similar to the one the girl wore, from a shelf, placed it on his head. "A shield against supersonics," he explained. "It produces vibrations which nullify those set up by the *sonovox*." He faced the langorous Zeila. "Play, child! Convince Mr. Ranson of our powers!"

Again the girl's fingers danced over the keys in a wild melody of hate. Red mists rose before Ranson's eyes and he fought against the bonds that held him. Then the song changed to a dirge-like melody and Ranson fell into the black abyss of despair. This was more than music, he knew; it was something deeper that played upon the soul. Again the notes changed and crawling fear enveloped Ranson until he felt sick with horror of the unknown. Emotion after emotion gripped him, and had he not been helpless, bound, he would have obeyed the moods that swept his brain. He was himself like an instrument upon which a thousand tunes were played ... and through it all Elath Taen smiled with a vague detached air, while the girl's eyes burned into his own.

Suddenly Elath Taen raised his hand. "Enough, Zeila," he said. "He is exhausted."

The music ceased and Ranson fell back weakly, worn by the storm of emotions that had surged in waves over him.

"You... You win!" he gasped. "What kind of deviltry is this?"

"Deviltry?" Dr. Taen laughed. "But it is so simple. Music, even normal music, can produce moods. The uplift of the ancient earthsong, 'Marsailles,' the melancholy of the 'Valse Triste,' the passion of the 'Bolero.' Indeed, many years ago on Terra, there was a strange song entitled, 'Gloomy Sunday,' which caused numerous suicides on the part of those who heard it. As for the instrument, it's merely an electrical sound producer such as your electric organ, theremins, and so on. But to it I have added a full range of supersonic notes, which, though inaudible, are the real mood-changers."

"Supersonics?" Ranson exclaimed. "You mean they're what created the emotions inside me just now?"

"Exactly." Elath Taen nodded. "The audible music helps, but it is the supersonics that determine the emotions! Their effect is upon the brain, and nothing can shut them out except counter-notes such as are set up by our helmets!" He tapped the copper dome that encased his head. "The effects of supersonics upon the emotions is interesting, Mr. Ranson. I first got my idea from old twentieth-century records on Terra itself. I read how, in the days of motion pictures before television was perfected, one of your Hollywood companies introduced a supersonic note onto the sound-track of a film in hopes of creating an atmosphere of horror at a certain point in the picture. But so great was the terror induced at the private showing that the supersonic note was immediately cut from the sound-track, and the records of the case filed away. It was the discovery and study of these records that started me on the trail of super-music. Thus with cosmic irony, Mr. Ranson, Earth has created the weapon which will destroy her! Supersonics!"

Ranson stared at Elath Taen, bewildered. Supersonics creating emotions! That was what had infuriated Haller and himself, had driven the other officials of M.B.C. to various forms of death! And now, with M.B.C. in the hands of Taen's followers, they planned to arouse the silent little reddies of Mars to revolt!

"But why?" Ranson demanded. "Earthmen have brought new life, new progress to Mars! We've built roads, canals, spaceports, taught your people our science..."

"You are aliens!" Elath Taen cried. "You must be wiped out!" He drew a whistle from his pocket, blew a shrill blast. There was a pattering of feet, and a squat Martian, his arms scarred by flame-gun burns, entered the room.

"Place the terrestial in safe keeping," Elath Taen commanded. "Watch him well." He glanced at the blinking red light of a time-signal on the wall. "Come Zeila! It's time to go!"

The girl nodded, picked up the *sonovox*. At the door she paused, glanced back at Ranson.

"Music for the men of Mars," she murmured. "When we return our own people will rule this planet!" Her eyes, brooding on the earthman, were inscrutable. "*Alotah*, Stephen Ranson!"

Then she and her father had left the laboratory, and the burly guard was forcing Ranson toward a small iron-barred door at the rear of the room. Bound, helpless, he staggered into the cell, heard the door clang shut behind him. The scarred, ugly guard stationed himself across the laboratory, where he could keep an eye on the cell.

Ranson lay there in the shadows, suddenly bitter. A nice mess he'd made of things! Wanted for murder by Captain Maxwell, tricked by Elath Taen and his daughter when he had them in his grasp, and now a prisoner here, while they sent their musical madness, their deadly supersonic notes, over the planet-wide chain of M.B.C. Ranson knew what that would mean. Except for the Foreign

Legion, a few rocket-plane squadrons, Mars was undefended. If Elath Taen's supersonics aroused the reddies to revolt, his dream of making himself emperor of Mars would be at last fulfilled.

Ranson shot a glance at his guard. The scarred little Martian was leaning back in his chair, eyes on the cell door. But it seemed unlikely that he could see what went on within the shadowy cell. In one swift movement the T.I. man smashed his wrist-watch against the wall, then, picking up a sliver of glass with his fingertips, began to saw at his bonds.

At length the ropes fell from Ranson's aching arms. Swiftly he freed his legs. The guard was still sitting in the well-lighted laboratory, unmoved. Ranson glanced at the door. Steel bars, impossible to penetrate. And seconds ticking away!

A dark fighting grin spread over Ranson's lean face. There was one chance. A wild, desperate chance, but if it worked... Hastily he slipped off his shoes, placed them on the floor beside him. Then, thrusting his hand into his coat pocket, he bulged the cloth out with his finger to simulate a gun.

"Don't move!" he said in sibilant Martian. "Drop your flame-gun! Try anything and I'll shoot!"

The guard sprang to his feet, his bulging hairless head gleaming in the bright light, his green eyes cold with rage. As Ranson had expected, he gave no indication of surrender. Instead, he raised his weapon, fired.

At the moment that the guard pressed the trigger, the terrestial leaped to one side, seeking cover of the wall at the side of the door. A savage greenish flash spat from the gun, a terrible wave of heat swept the cell. Half-blinded, sick from the searing heat, Ranson lay in his corner and watched the door. Under the fiery blast, the iron bars turned white, ran, until only pools of molten metal lay between him and freedom.

The squat Martian snapped off the ray, approached the glowing door cautiously, to find out if there was life in Ranson's inert body. There was ... more than the little reddy had bargained for. The earthman's arm swung in an arc and one of his shoes, flying through the blasted, melted door, caught the little Martian's wrist, knocked the flame-gun from his hand. The other shoe, following swiftly, landed alongside his head, sent him reeling and staggering back into a shelf of test-tubes and beakers.

"And that's how we do it on Earth!" Grinning tightly, Ranson leaped the puddles of molten metal, plunged through the blasted, glowing remains of the door. Before the ugly little guard could recover, a hard knotted terrestial fist had slammed against his chin, sent him, limp and unconscious to the floor.

Swiftly Ranson ripped wires from the masses of intricate machinery, bound the inert reddy, then, snatching up the flame-gun, ran from the house.

Twisting, turning, he came to the embankment of the Psidian canal. A sleek water-cab slid into view, its atomic motors humming. Ranson hailed it, hand on his gun, but the wizened reddy at the wheel had apparently not heard of Elath Taen's mad melody.

"Martian Broadcasting Building," Ranson grated. "Step on it!"

The driver nodded, and, when his passenger was aboard, sent the boat surging along the canal, throwing up clouds of spray. Racing, roaring, dodging heavily-laden freight boats, the cab tore over the dark cold water that flowed, via the intricate networks of canals, from the polar caps.

As they neared the center of the city, the atmosphere of tension grew. Little bands of terrestial police patrolled the embankments, a squadron of rocket-planes droned above the towering metropolis, the light of their exhausts throwing weird shadows. Occasional shouts, the green flash of flame-guns, issued from the darkness and the crowds of reddies gathered before their radios in houses, shops, and public squares, were seething with excitement. The roar of the cab's motors drowned out the sound of the music and Elath Taen's exultant voice, but the driver moved uneasily.

"Looks like somethin's up," he muttered. "I'll see if we can get a bulletin."

Before Ranson could stop him, he had snapped on the radio within the cab. The wild, frenzied music filled the small cabin, tearing at both men's minds, while Taen's voice urged revolt. Then, under the influence of the supersonics, red flames of hatred leaped through their brains, banishing all thought, logic. The little Martian driver whirled about, only to have the butt of Ranson's gun crash down upon his head. Slumping forward, his body fell against the radio, shattering its fragile tubes. Ranson shook himself as the infernal music abruptly ceased.

The M.B.C. building lay just before them. Ranson swung the cab to the embankment, sprang out. The tall plastoid building towered white and spectral above the canal. Ranson burst through the door.

Several reddies on guard sprang forward, but a blast from the terrestial's gun cleared the great hall. He sprang into an elevator, jabbed at a button, and the car shot upward.

The elevator stopped at the top floor, where the broadcasting studios were located. Ranson hurtled along the corridor, plunged through the door. Before him lay a large room, blocked at one end by a thick, double-paned glass. And on the other side of the glass stood Elath Taen, crouched before a television set, his fingers running over the keys of the *sonovox*, his face exultant as he poured out the supersonics of his song of hate. Musical madness for the men of Mars, making them forget all that Terra had done for the red planet, driving them to insane mass murder! And as he played upon the *sonovox*, Taen spoke into the microphone, urging them to revolt! Already they were starting their reign of terror; when he reached his climax they would pour from their houses to kill all who had terrestial blood. Unless...

Ranson leaped forward. Even the supersonics were kept from the outer room by the vacuum-insulated double glass panes; Elath Taen was like a silent marionette in the broadcasting booth, his green eyes flickering with apprehension, his head encased by the shielding copper helmet.

"Drop your gun, Mr. Ranson!" Zeila's voice came from behind him.

Ranson whirled; the girl had been standing behind the door, unnoticed, as he burst into the room. Her exotic face was pale, but the flame-gun in her hand was steady. Ranson obeyed, smiling.

"As you wish," he said. "But T.I. has one trick we use as a last resort. Look!"

From his pocket he drew a flat metal case. "Supposedly cigarettes, but really the most powerful explosive devised by our laboratories. Shoot me with that flame gun and the heat sets it off. You, your charming father, and I, will all be blown to atoms. So you won't dare shoot!"

Zeila stared at him, lips a crimson slash across her face.

"You won't get away with it!" she exclaimed. "It's bluff!"

"Shoot, then," Ranson said. "Blow the whole top of this building to bits!" He reached out for her gun.

The girl's eyes were fixed on the metal case, and there was fear in them. Ranson took another step toward her. Elath Taen could not watch since he was forced to keep his eyes on the intricate keyboard of the *sonovox*.

"Blown to bits," Ranson repeated sardonically. "Me, too, but at least I'll have removed the leaders of the revolt. This explosive is the last resort of T.I. men. Squeeze that trigger and the heat will set it off! Now give me that gun!"

Zeila Taen broke suddenly, shuddering at the thought of her vivid beauty torn to shreds by an explosion.

"Take it!" she snarled. "It's too late, anyhow! Mars is in revolt! No one can stop them now! Fool! My father will be emperor after the insurrection! You might have been prince."

Ranson didn't wait to hear more. One blast of the heat gun and the glass partition shattered to a thousand fragments.

"No good, Mr. Ranson." Elath Taen lifted his hands from the keyboard, smiling thinly. "The flame is lit and cannot be put out! The red flame of revolt! Already my people are fighting! Loud-speakers in every public square have carried the sound of mad, blind fury! I am the mood-master!"

"Get back to that sound-box!" Ranson grated. "Play those sleep-producing notes! Play, or I'll blast your lovely daughter here to a cinder! You claim you're the mood-master! Well, if your damned supersonics started this, they can end it!" He swung his gun to cover Zeila's sleek figure. "Play, Dr. Taen! I've never killed a woman yet, but it's her life or those of all terrestials on Mars! Back to your *sonovox*!"

For a long moment Elath Taen stared at his daughter, then nodded his hairless head somberly.

"Again you win, Mr. Ranson," he said softly. "I should have killed you or won you to my side, long ago." Turning to the *sonovox*, he began to play.

Ranson stood tense, covering the girl with his gun. Soft, lulling music, supersonic notes that seemed to caress his brain, filled the room. The drowsy sound of rain on a roof, of rustling leaves, of a soothing night wind ... all these were bound up in the melody. Peace, rest, sleep ... every nerve seemed to relax, every muscle seemed limp, as the dreamy musical hypnosis took effect.

Elath Taen and the girl were watching him covertly. There was a thin smile on the doctor's dark saturnine face. Dully Ranson tried to reason out why Elath Taen should be smiling, but somehow his mind refused to function. Those cloudy mists rising before his eyes! Miles away Taen was speaking, above the soporific sounds.

"Too bad," he was saying. "You forgot that whatever these supersonics may do to my people, they also affect you. Zeila and I are protected from the short-wave emanations by our helmets. But you, Mr. Ranson, are not! Already you are helpless and in a moment you will sleep, as you did in our laboratory! Then, with you secure, I shall arouse my people once more!"

Ranson tried to move, tried to act, but the music was a silken noose binding him, and he had no will power left. Sleep ... nothing else mattered... As in a dream he saw Zeila coming toward him, felt himself crumple to the floor. Vaguely he remembered bright flashes, shouts, and then all was grey oblivion.

"Ranson! Ranson!" The words beat like fists upon his drugged brain.

The T.I. man stirred restlessly; out of the whirling mists Captain Maxwell's face became a stern reality.

"What happened?" the police officer was saying. "First the reddies go kill-crazy, then start passing out! Almost went nutty ourselves, down at headquarters, listening! But then the murder-music stopped and we heard your voice, talking to Elath Taen! So we came here pronto. Just in time."

"Taen! And Zeila!" Ranson gasped. "Where are they?"

"Gone." Captain Maxwell motioned to a door at the rear of the room. "Ducked out and down the elevator. Blasted the cables when they hit the bottom so we weren't able to follow." He shook his head. "You were right about that music! No wonder you and Haller went berserk! Don't worry about any trial for murder! Mars has been mad, this night!"

Ranson struggled to his feet. Taen and his daughter escaped! With the secret of the supersonic notes! But it would be a long time before they dared return to Mars. Still groggy, Ranson drew the metal cigarette case from his pocket.

"How were you able to force your way in here?" Captain Maxwell demanded. "To make them change the tune and break up the revolt?"

Ranson opened the metal case.

"Bluff," he said, taking a cigarette from the container and lighting it. "That's what saved Mars! Just ... bluff!"

Grinning, he blew a cloud of smoke.

THE MOONS OF MARS

(1952)

Dean Evans

Dean Evans is a pseudonym for George Kull, an American science fiction writer based in California. Little is known of him except that he published stories in various science fiction magazines in the early 1950s. Mark Rich's book *C. M. Kornbluth: The Life and Works of a Science Fiction Visionary* mentions that Kull was agented by Kornbluth, who paid his authors up-front for work, rather than taking a commission from eventual sales. Unfortunately, this noble tactic eventually ran the agency into financial trouble. The book also notes that Kull wrote good stories faster than Kornbluth could sell them on, and that when the agency folded Kornbluth owed Kull around three thousand dollars.

Every boy should be able to whistle, except, of course, Martians. But this one did!

He seemed a very little boy to be carrying so large a butterfly net. He swung it in his chubby right fist as he walked, and at first glance you couldn't be sure if he were carrying it, or it carrying *him*.

He came whistling. All little boys whistle. To little boys, whistling is as natural as breathing. However, there was something peculiar about this particular little boy's whistling. Or, rather, there were two things peculiar, but each was related to the other.

The first was that he was a Martian little boy. You could be very sure of that, for Earth little boys have earlobes while Martian little boys do not—and he most certainly didn't.

The second was the tune he whistled—a somehow familiar tune, but one which I should have thought not very appealing to a little boy.

"Hi, there," I said when he came near enough. "What's that you're whistling?"

He stopped whistling and he stopped walking, both at the same time, as though he had pulled a switch or turned a tap that shut them off. Then he lifted his little head and stared up into my eyes.

"'The Calm'," he said in a sober, little-boy voice.

"The *what*?" I asked.

"From the William Tell Overture," he explained, still looking up at me. He said it deadpan, and his wide brown eyes never once batted.

"Oh," I said. "And where did you learn that?"

"My mother taught me."

I blinked at him. He didn't blink back. His round little face still held no expression, but if it had, I knew it would have matched the title of the tune he whistled.

"You whistle very well," I told him.

That pleased him. His eyes lit up and an almost-smile flirted with the corners of his small mouth.

He nodded grave agreement.

"Been after butterflies, I see. I'll bet you didn't get any. This is the wrong season."

The light in his eyes snapped off. "Well, good-by," he said abruptly and very relevantly.

"Good-by," I said.

His whistling and his walking started up again in the same spot where they had left off. I mean the note he resumed on was the note which followed the one interrupted; and the step he took was with the left foot, which was the one he would have used if I hadn't stopped him. I followed him with my eyes. An unusual little boy. A most precisely *mechanical* little boy.

When he was almost out of sight, I took off after him, wondering.

The house he went into was over in that crumbling section which forms a curving boundary line, marking the limits of those frantic and ugly original mine-workings made many years ago by the early colonists. It seems that someone had told someone who had told someone else that here, a mere twenty feet beneath the surface, was a vein as wide as a house and as long as a fisherman's alibi, of pure—*pure*, mind you—gold.

Back in those days, to be a colonist meant to be a rugged individual. And to be a rugged individual meant to not give a damn one way or another. And to not give a damn one way or another meant to make one hell of a mess on the placid face of Mars.

There had not been any gold found, of course, and now, for the most part, the mining shacks so hastily thrown up were only fever scars of a sickness long gone and little remembered. A few of the houses were still occupied, like the one into which the Martian boy had just disappeared.

So his *mother* had taught him the William Tell Overture, had she? That tickling thought made me chuckle as I stood before the ramshackle building. And then, suddenly, I stopped chuckling and began to think, instead, of something quite astonishing:

How had it been possible for her to teach, and for him to whistle?

All Martians are as tone-deaf as a bucket of lead.

I went up three slab steps and rapped loudly on the weather-beaten door.

The woman who faced me may have been as young as twenty-two, but she didn't look it. That shocked look, which comes with the first realization that youth has slipped quietly away downstream in the middle of the night, and left nothing but frightening rocks of middle age to show cold and gray in the hard light of dawn, was like the validation stamp of Time itself in her wide, wise eyes. And her voice wasn't young any more, either.

"Well? And what did I do now?"

"I beg your pardon?" I said.

"You're Mobile Security, aren't you? Or is that badge you're wearing just something to cover a hole in your shirt?"

"Yes, I'm Security, but does it have to mean something?" I asked. "All I did was knock on your door."

"I heard it." Her lips were curled slightly at one corner.

I worked up a smile for her and let her see it for a few seconds before I answered: "As a matter of fact, I don't want to see *you* at all. I didn't know you lived here and I don't know who you are. I'm not even interested in who you are. It's the little boy who just went in here that I was interested in. The little Martian boy, I mean."

Her eyes spread as though somebody had put fingers on her lids at the outside corners and then cruelly jerked them apart.

"Come in," she almost gasped.

I followed her. When I leaned back against the plain door, it closed protestingly. I looked around. It wasn't much of a room, but then you couldn't expect much of a room in a little ghost of a place like this. A few knickknacks of the locality stood about on two tables and a shelf, bits of rock with streak-veins of fused corundum; not bad if you like the appearance of squeezed blood.

There were two chairs and a large table intended to match the chairs, and a rough divan kind of thing made of discarded cratings which had probably been hauled here from the International Spaceport, ten miles to the West. In the back wall of the room was a doorway that led dimly to somewhere else in the house. Nowhere did I see the little boy. I looked once again at the woman.

"What about him?" she whispered.

Her eyes were still startled.

I smiled reassuringly. "Nothing, lady, nothing. I'm sorry I upset you. I was just being nosy is all, and that's the truth of it. You see, the little boy went by me a while ago and he was whistling. He whistles remarkably well. I asked him what the name of the tune was and he told me it was the 'Calm' from William Tell. He also told me his mother had taught him."

Her eyes hadn't budged from mine, hadn't flickered. They might have been bright, moist marbles glued above her cheeks.

She said one word only: "Well?"

"Nothing," I answered. "Except that Martians are supposed to be tone-deaf, aren't they? It's something lacking in their sense of hearing. So when I heard this little boy, and saw he was a Martian, and when he told me his *mother* had taught him—" I shrugged and laughed a little. "Like I said before, I guess I got just plain nosy."

She nodded. "We agree on that last part."

Perhaps it was her eyes. Or perhaps it was the tone of her voice. Or perhaps, and more simply, it was her attitude in general. But whatever it was, I suddenly felt that, nosy or not, I was being treated shabbily.

"I would like to speak to the Martian lady," I said.

"There isn't any Martian lady."

"There has to be, doesn't there?" I said it with little sharp prickers on the words.

But she did, too: "*Does there?*"

I gawked at her and she stared back. And the stare she gave me was hard and at the same time curiously defiant—as though she would dare me to go on with it. As though she figured I hadn't the guts.

For a moment, I just blinked stupidly at her, as I had blinked stupidly at the little boy when he told me his mother had taught him how to whistle. And then—after what seemed to me a very long while—I slowly tumbled to what she meant.

Her eyes were telling me that the little Martian boy wasn't a little Martian boy at all, that he was cross-breed, a little chap who had a Martian father and a human, Earthwoman mother.

It was a startling thought, for there just aren't any such mixed marriages. Or at least I had thought there weren't. Physically, spiritually, mentally, or by any other standard you can think of, compared to a human male the Martian isn't anything you'd want around the house.

I finally said: "So that is why he is able to whistle."

She didn't answer. Even before I spoke, her eyes had seen the correct guess which had probably flashed naked and astounded in my own eyes. And then she swallowed with a labored breath that went trembling down inside her.

"There isn't anything to be ashamed of," I said gently. "Back on Earth there's a lot of mixtures, you know. Some people even claim there's no such thing as a pure race. I don't know, but I guess we all started somewhere and intermarried plenty since."

She nodded. Somehow her eyes didn't look defiant any more.

"Where's his father?" I asked.

"H-he's dead."

"I'm sorry. Are you all right? I mean do you get along okay and everything, now that...?"

I stopped. I wanted to ask her if she was starving by slow degrees and needed help. Lord knows the careworn look about her didn't show it was luxurious living she was doing—at least not lately.

"Look," I said suddenly. "Would you like to go home to Earth? I could fix—"

But that was the wrong approach. Her eyes snapped and her shoulders stiffened angrily and the words that ripped out of her mouth were not coated with honey.

"Get the hell out of here, you fool!"

I blinked again. When the flame in her eyes suddenly seemed to grow even hotter, I turned on my heel and went to the door. I opened it, went out on the top slab step. I turned back to close the door—and looked straight into her eyes.

She was crying, but that didn't mean exactly what it looked like it might mean. Her right hand had the door edge gripped tightly and she was swinging it with all the strength she possessed. And while I still stared, the door slammed savagely into the casing with a shock that jarred the slab under my feet, and flying splinters from the rotten woodwork stung my flinching cheeks.

I shrugged and turned around and went down the steps. "And that is the way it goes," I muttered disgustedly to myself. Thinking to be helpful with the firewood problem, you give a woman a nice sharp axe and she immediately puts it to use—on you.

I looked up just in time to avoid running into a spread-legged man who was standing motionless directly in the middle of the sand-path in front of the door. His hands were on his hips and there was something in his eyes which might have been a leer.

*

"Pulled a howler in there, eh, mate?" he said. He chuckled hoarsely in his throat. "Not being exactly deaf, I heard the tail end of it." His chuckle was a lewd thing, a thing usually reserved—if it ever was reserved at all—for the mens' rooms of some of the lower class dives. And then he stopped chuckling and frowned instead and said complainingly:

"Regular little spitfire, ain't she? I ask you now, wouldn't you think a gal which had got herself in a little jam, so to speak, would be more reasonable—"

His words chopped short and he almost choked on the final unuttered syllable. His glance had dropped to my badge and the look on his face was one of startled surprise.

"I—" he said.

I cocked a frown of my own at him.

"Well, so long, mate," he grunted, and spun around and dug his toes in the sand and was away. I stood there staring at his rapidly disappearing form for a few moments and then looked back once more at the house. A tattered cotton curtain was just swinging to in the dirty, sand-blown window. That seemed to mean the woman had been watching. I sighed, shrugged again and went away myself.

When I got back to Security Headquarters, I went to the file and began to rifle through pictures. I didn't find the woman, but I did find the man.

He was a killer named Harry Smythe.

I took the picture into the Chief's office and laid it on his desk, waited for him to look down at it and study it for an instant, and then to look back up to me. Which he did.

"So?" he said.

"Wanted, isn't he?"

He nodded. "But a lot of good that'll do. He's holed up somewhere back on Earth."

"No," I said. "He's right here. I just saw him."

"*What?*" He nearly leaped out of his chair.

"I didn't know who he was at first," I said. "It wasn't until I looked in the files—"

He cut me off. His hand darted into his desk drawer and pulled out an

Authority Card. He shoved the card at me. He growled: "Kill or capture, I'm not especially fussy which. Just *get* him!"

I nodded and took the card. As I left the office, I was thinking of something which struck me as somewhat more than odd.

I had idly listened to a little half-breed Martian boy whistling part of the William Tell Overture, and it had led me to a wanted killer named Harry Smythe.

Understandably, Mr. Smythe did not produce himself on a silver platter. I spent the remainder of the afternoon trying to get a lead on him and got nowhere. If he was hiding in any of the places I went to, then he was doing it with mirrors, for on Mars an Authority Card is the big stick than which there is no bigger. Not solely is it a warrant, it is a commandeer of help from anyone to whom it is presented; and wherever I showed it I got respect.

I got instant attention. I got even more: those wraithlike tremblings in the darker corners of saloons, those corners where light never seems quite to penetrate. You don't look into those. Not if you're anything more than a ghoul, you don't.

Not finding him wasn't especially alarming. What was alarming, though, was not finding the Earthwoman and her little half-breed Martian son when I went back to the tumbledown shack where they lived. It was empty. She had moved fast. She hadn't even left me a note saying good-by.

That night I went into the Great Northern desert to the Haremheb Reservation, where the Martians still try to act like Martians.

It was Festival night, and when I got there they were doing the dance to the two moons. At times like this you want to leave the Martians alone. With that thought in mind, I pinned my Authority Card to my lapel directly above my badge, and went through the gates.

The huge circle fire was burning and the dance was in progress. Briefly, this can be described as something like the ceremonial dances put on centuries ago by the ancient aborigines of North America. There was one important exception, however. Instead of a central fire, the Martians dig a huge circular trench and fill it with dried roots of the *belu* tree and set fire to it. Being pitch-like, the gnarled fragments burn for hours. Inside this ring sit the spectators, and in the exact center are the dancers. For music, they use the drums.

The dancers were both men and women and they were as naked as Martians can get, but their dance was a thing of grace and loveliness. For an instant—before anyone observed me—I stood motionless and watched the sinuously undulating movements, and I thought, as I have often thought before, that this is the one thing the Martians can still do beautifully. Which, in a sad sort of way, is a commentary on the way things have gone since the first rocket-blasting ship set down on these purple sands.

I felt the knife dig my spine. Carefully I turned around and pointed my index finger to my badge and card. Bared teeth glittered at me in the flickering light, and then the knife disappeared as quickly as it had come.

"Wahanhk," I said. "The Chief. Take me to him."

The Martian turned, went away from the half-light of the circle. He led me some yards off to the north to a swooping-tent. Then he stopped, pointed.

"Wahanhk," he said.

I watched him slip away.

Wahanhk is an old Martian. I don't think any Martian before him has ever lived so long—and doubtless none after him will, either. His leathery, almost purple-black skin was rough and had a charred look about it, and up around the eyes were little plaits and folds that had the appearance of being done deliberately by a Martian sand-artist.

"Good evening," I said, and sat down before him and crossed my legs.

He nodded slowly. His old eyes went to my badge.

From there they went to the Authority Card.

"Power sign of the Earthmen," he muttered.

"Not necessarily," I said. "I'm not here for trouble. I know as well as you do that, before tonight is finished, more than half of your men and women will be drunk on illegal whiskey."

He didn't reply to that.

"And I don't give a damn about it," I added distinctly.

His eyes came deliberately up to mine and stopped there. He said nothing. He waited. Outside, the drums throbbed, slowly at first, then moderated in tempo. It was like the throbbing—or sobbing, if you prefer—of the old, old pumps whose shafts go so tirelessly down into the planet for such pitifully thin streams of water.

"I'm looking for an Earthwoman," I said. "This particular Earthwoman took a Martian for a husband."

"That is impossible," he grunted bitterly.

"I would have said so, too," I agreed. "Until this afternoon, that is."

His old, dried lips began to purse and wrinkle.

"I met her little son," I went on. "A little semi-human boy with Martian features. Or, if you want to turn it around and look at the other side, a little Martian boy who whistles."

His teeth went together with a snap.

I nodded and smiled. "You know who I'm talking about."

For a long long while he didn't answer. His eyes remained unblinking on mine and if, earlier in the day, I had thought the little boy's face was expressionless, then I didn't completely appreciate the meaning of that word. Wahanhk's face was more than expressionless; it was simply blank.

"They disappeared from the shack they were living in," I said. "They went in a hurry—a very great hurry."

That one he didn't answer, either.

"I would like to know where she is."

"Why?" His whisper was brittle.

"She's not in trouble," I told him quickly. "She's not wanted. Nor her child, either. It's just that I have to talk to her."

"Why?"

I pulled out the file photo of Harry Smythe and handed it across to him. His wrinkled hand took it, pinched it, held it up close to a lamp hanging from one of the ridge poles. His eyes squinted at it for a long moment before he handed it back.

"I have never seen this Earthman," he said.

"All right," I answered. "There wasn't anything that made me think you had. The point is that he knows the woman. It follows, naturally, that she might know him."

"This one is *wanted*?" His old, broken tones went up slightly on the last word.

I nodded. "For murder."

"Murder." He spat the word. "But not for the murder of a Martian, eh? Martians are not that important any more." His old eyes hated me with an intensity I didn't relish.

"You said that, old man, not I."

A little time went by. The drums began to beat faster. They were rolling out a lively tempo now, a tempo you could put music to.

He said at last: "I do not know where the woman is. Nor the child."

He looked me straight in the eyes when he said it—and almost before the words were out of his mouth, they were whipped in again on a drawn-back, great, sucking breath. For, somewhere outside, somewhere near that dancing circle, in perfect time with the lively beat of the drums, somebody was whistling.

It was a clear, clean sound, a merry, bright, happy sound, as sharp and as precise as the thrust of a razor through a piece of soft yellow cheese.

"In your teeth, Wahanhk! Right in your teeth!"

He only looked at me for another dull instant and then his eyes slowly closed and his hands folded together in his lap. Being caught in a lie only bores a Martian.

I got up and went out of the tent.

The woman never heard me approach. Her eyes were toward the flaming circle and the dancers within, and, too, I suppose, to her small son who was somewhere in that circle with them, whistling. She leaned against the bole of a *belu* tree with her arms down and slightly curled backward around it.

"That's considered bad luck," I said.

Her head jerked around with my words, reflected flames from the circle fire still flickering in her eyes.

"That's a *belu* tree," I said. "Embracing it like that is like looking for a ladder to walk under. Or didn't you know?"

"Would it make any difference?" She spoke softly, but the words came to me above the drums and the shouts of the dancers. "How much bad luck can you have in one lifetime, anyway?"

I ignored that. "Why did you pull out of that shack? I told you you had nothing to fear from me."

She didn't answer.

"I'm looking for the man you saw me talking with this morning," I went on. "Lady, he's wanted. And this thing, on my lapel is an Authority Card. Assuming you know what it means, I'm asking you where he is."

"What man?" Her words were flat.

"His name is Harry Smythe."

If that meant anything to her, I couldn't tell. In the flickering light from the fires, subtle changes in expression weren't easily detected.

"Why should I care about an Earthman? My husband was a Martian. And he's dead, see? Dead. Just a Martian. Not fit for anything, like all Martians. Just a bum who fell in love with an Earthwoman and had the guts to marry her. Do you understand? So somebody murdered him for it. Ain't that pretty? Ain't that something to make you throw back your head and be proud about? Well, ain't it? And let me tell you, Mister, whoever it was, I'll get him. *I'll get him!*"

I could see her face now, all right. It was a twisted, tortured thing that writhed at me in its agony. It was small yellow teeth that bared at me in viciousness. It was eyes that brimmed with boiling, bubbling hate like a ladle of molten steel splashing down on bare, white flesh. Or, simply, it was the face of a woman who wanted to kill the killer of her man.

And then, suddenly, it wasn't. Even though the noise of the dance and the dancers was loud enough to command the attention and the senses. I could still hear her quiet sobbing, and I could see the heaving of the small, thin shoulders.

And I knew then the reason for old Wahanhk's bitterness when he had said to me, "But not for the murder of a Martian, eh? Martians are not that important any more."

What I said then probably sounded as weak as it really was: "I'm sorry, kid. But look, just staking out in that old shack of yours and trying to pry information out of the type of men who drifted your way—well, I mean there wasn't much sense in that, now was there?"

I put an arm around her shoulders. "He must have been a pretty nice guy," I said. "I don't think you'd have married him if he wasn't."

I stopped. Even in my own ears, my words sounded comfortless. I looked up, over at the flaming circle and at the sweat-laved dancers within it. The sound of the drums was a wild cacophonous tattoo now, a rattle of speed and savagery combined; and those who moved to its frenetic jabberings were not dancers any more, but only frenzied, jerking figurines on the strings of a puppeteer gone mad.

I looked down again at the woman. "Your little boy and his butterfly net," I said softly. "In a season when no butterflies can be found. What was that for? Was he part of the plan, too, and the net just the alibi that gave him a passport to wander where he chose? So that he could listen, pick up a little information here, a little there?"

She didn't answer. She didn't have to answer. My guesses can be as good as anybody's.

After a long while she looked up into my eyes. "His name was Tahily," she said. "He had the secret. He knew where the gold vein was. And soon, in a couple of years maybe, when all the prospectors were gone and he knew it would be safe, he was going to stake a claim and go after it. For us. For the three of us."

I sighed. There wasn't, isn't, never will be any gold on this planet. But who in the name of God could have the heart to ruin a dream like that?

Next day I followed the little boy. He left the reservation in a cheery frame of mind, his whistle sounding loud and clear on the thin morning air. He didn't go in the direction of town, but the other way—toward the ruins of the ancient Temple City of the Moons. I watched his chubby arm and the swinging of the big butterfly net on the end of that arm. Then I followed along in his sandy tracks.

It was desert country, of course. There wasn't any chance of tailing him without his knowledge and I knew it. I also knew that before long he'd know it, too. And he did—but he didn't let me know he did until we came to the rag-cliffs, those filigree walls of stone that hide the entrance to the valley of the two moons.

Once there, he paused and placed his butterfly net on a rock ledge and then calmly sat down and took off his shoes to dump the sand while he waited for me.

"Well," I said. "Good morning."

He looked up at me. He nodded politely. Then he put on his shoes again and got to his feet.

"You've been following me," he said, and his brown eyes stared accusingly into mine.

"I have?"

"That isn't an honorable thing to do," he said very gravely. "A gentleman doesn't do that to another gentleman."

I didn't smile. "And what would you have me do about it?"

"Stop following me, of course, sir."

"Very well," I said. "I won't follow you any more. Will that be satisfactory?"

"Quite, sir."

Without another word, he picked up his butterfly net and disappeared along a path that led through a rock crevice. Only then did I allow myself to grin. It was a sad and pitying and affectionate kind of grin.

I sat down and did with my shoes as he had done. There wasn't any hurry; I knew where he was going. There could only be one place, of course—the city of Deimos and Phobos. Other than that he had no choice. And I thought I knew the reason for his going.

Several times in the past, there have been men who, bitten with the fever of an idea that somewhere on this red planet there must be gold, have done prospecting among the ruins of the old temples. He had probably heard that there were men there now, and he was carrying out with the thoroughness of his precise little mind the job he had set himself of finding the killer of his daddy.

I took a short-cut over the rag-cliffs and went down a winding, sand-worn path. The temple stones stood out barren and dry-looking, like breast bones from the desiccated carcass of an animal. For a moment I stopped and stared down at the ruins. I didn't see the boy. He was somewhere down there, though, still swinging his butterfly net and, probably, still whistling.

I started up once more.

And then I heard it—a shrill blast of sound in an octave of urgency; a whistle, sure, but a warning one.

I stopped in my tracks from the shock of it. Yes, I knew from whom it had come, all right. But I didn't know why.

And then the whistle broke off short. One instant it was in the air, shrieking with a message. The next it was gone. But it left tailings, like the echo of a death cry slowly floating back over the dead body of the creature that uttered it.

I dropped behind a fragment of the rag-cliff. A shot barked out angrily. Splinters of the rock crazed the morning air.

The little boy screamed. Just once.

I waited. There was a long silence after that. Then, finally, I took off my hat and threw it out into the valley. The gun roared once more. This time I placed it a little to the left below me. I took careful sighting on the hand that held that gun—and I didn't miss it.

It was Harry Smythe, of course. When I reached him, he had the injured hand tucked tightly in the pit of his other arm. There was a grim look in his eyes and he nodded as I approached him.

"Good shooting, mate. Should be a promotion in it for you. Shooting like that, I mean."

"That's nice to think about," I said. "Where's the boy? I owe him a little something. If he hadn't whistled a warning, you could have picked me off neat."

"I would." He nodded calmly.

"Where is he?"

"Behind the rock there. In that little alcove, sort of." He indicated with his chin.

I started forward. I watched him, but I went toward the rock.

"Just a minute, mate."

I stopped. I didn't lower my gun.

"That bloody wench we spoke about yesterday. You know, out in front of that shack? Well, just a thought, of course, but if you pull me in and if I get *it*, what'll become of her, do you suppose? Mean to say, I couldn't support her when I was dead, could I?"

"Support her?" Surprise jumped into my voice.

"What I said. She's my wife, you know. Back on Earth, I mean. I skipped out on her a few years back, but yesterday I was on my way to looking her up when you—"

"She didn't recognize the name Harry Smythe," I said coldly. "I'm afraid you'll have to think a little faster."

"Of course she didn't! How could she? That ain't my name. What made you think it was?"

Bright beads of sweat sparkled on his forehead, and his lips had that frantic looseness of lips not entirely under control.

"You left her," I grunted. "But you followed her across space anyway. Just to tell her you were sorry and you wanted to come back. Is that it?"

"Well—" His eyes were calculating. "Not the God's honest, mate, no. I didn't know she was here. Not at first. But there was this Spider, see? This Martian. His name was Tahily and he used to hang around the saloons and he talked a lot, see? Then's when I knew..."

"So it was you who killed him," I said. "One murder wasn't enough back on Earth; you had to pile them up on the planets." I could feel something begin to churn inside of me.

"Wait! Sure, I knocked off the Martian. But a fair fight, see? That Spider jumped my claim. A fair fight it was, and anybody'd done the same. But even without that, he had it coming anyway, wouldn't you say? Bigamist and all that, you know? I mean marrying a woman already married."

His lips were beginning to slobber. I watched them with revulsion in my stomach.

"Wouldn't you say, mate? Just a lousy, stinking Martian, I mean!"

I swallowed. I turned away and went around the rock and looked down. One look was enough. Blood was running down the cheek of the prone little Martian boy, and it was coming from his mouth. Then I turned back to the shaking man.

"Like I say, mate! I mean, what would you've done in my place? Whistling always did drive me crazy. I can't stand it. A phobia, you know. People *suffer* from phobias!"

"What did you do?" I took three steps toward him. I felt my lips straining back from my teeth.

"Wait now, mate! Like I say, it's a phobia. I can't stand whistling. It makes me suffer—"

"So you cut out his tongue?"

I didn't wait for his answer. I couldn't wait. While I was still calm, I raised my gun on his trembling figure. I didn't put the gun up again until his body stopped twitching and his fingers stopped clawing in the sands.

From the desk to the outside door, the hospital corridor runs just a few feet. But I'd have known her at any distance. I sighed, got to my feet and met her halfway.

She stopped before me and stared up into my eyes. She must have run all the way when she got my message, for although she was standing as rigid as a pole in concrete, something of her exhaustion showed in her eyes.

"Tell me," she said in a panting whisper.

"Your boy is going to be okay." I put my arm around her. "Everything's under control. The doctors say he's going to live and pull through and…"

I stopped. I wondered what words I was going to use when no words that I had ever heard in my life would be the right ones.

"Tell me." She pulled from my grasp and tilted her head so that she could look up into my eyes and read them like a printed page. "*Tell me!*"

"He cut out the boy's—he said he couldn't stand whistling. It was a phobia, he claimed. Eight bullets cured his phobia, if any."

"He cut out what?"

"Your son's tongue."

I put my arm around her again, but it wasn't necessary. She didn't cry out, she didn't slump. Her head did go down and her eyes did blink once or twice, but that was all.

"He was the only little boy on Mars who could whistle," she said.

All of the emotion within her was somehow squeezed into those few words.

I couldn't get it out of my mind for a long while. I used to lie in bed and think of it somewhat like this:

There was this man, with his feet planted in the purple sands, and he looked up into the night sky when the moon called Deimos was in perigee, and he studied it. And he said to himself, "Well, I shall write a book and I shall say in this book that the moon of Mars is thus and so. And I will be accurately describing it, for in truth the moon *is* thus and so."

And on the other side of the planet there was another man. And he, too, looked up into the night sky. And he began to study the moon called *Phobos*. And he, too, decided to write a book. And he knew he could accurately describe the moon of Mars, for his own eyes had told him it looked like thus and so. And his own eyes did not lie.

I thought of it in a manner somewhat like that. I could tell the woman that Harry Smythe, her first husband, was the man who had killed Tahily, the Martian she loved. I could tell her Smythe had killed him in a fair fight because the Martian had tried to jump a claim. And her heart would be set to rest, for she would know that the whole thing was erased and done with, at last.

Or, on the other hand, I could do what I eventually did do. I could tell her absolutely nothing, in the knowledge that that way she would at least have the strength of hate with which to sustain herself through the years of her life. The strength of her hate against this man, whoever he might be, plus the chill joy of anticipating the day—maybe not tomorrow, but some day—when, like the dream of finding gold on Mars, she'd finally track him down and kill him.

I couldn't leave her without a reason for living. Her man was dead and her son would never whistle again. She had to have something to live for, didn't she?

THE MAN WHO HATED MARS (1956)

Randall Garrett

Born in 1927, **Randall Garrett** was an American science fiction and fantasy writer. He contributed to many science fiction magazines during the 1950s and 60s, before turning to novels and short story collections in the 1970s. In 1979, he contracted a viral infection that irreparably damaged his memory. Throughout the 1980s, his wife and fellow author Vicki Ann Heydron wrote the seven novels of the Gandalara Cycle, based on Garrett's draft of the first volume and an outline of the entire series. He died in 1987, aged 60.

"I WANT you to put me in prison!" the big, hairy man said in a trembling voice.

He was addressing his request to a thin woman sitting behind a desk that seemed much too big for her. The plaque on the desk said:

LT. PHOEBE HARRIS
TERRAN REHABILITATION SERVICE

Lieutenant Harris glanced at the man before her for only a moment before she returned her eyes to the dossier on the desk; but long enough to verify the impression his voice had given. Ron Clayton was a big, ugly, cowardly, dangerous man.

He said: "Well? Dammit, say something!"

The lieutenant raised her eyes again. "Just be patient until I've read this." Her voice and eyes were expressionless, but her hand moved beneath the desk.

Clayton froze. *She's yellow!* he thought. She's turned on the trackers! He could see the pale greenish glow of their little eyes watching him all around the room. If he made any fast move, they would cut him down with a stun beam before he could get two feet.

She had thought he was going to jump her. *Little rat!* he thought, *somebody ought to slap her down!*

He watched her check through the heavy dossier in front of her. Finally, she looked up at him again.

"Clayton, your last conviction was for strong-arm robbery. You were given a choice between prison on Earth and freedom here on Mars. You picked Mars."

He nodded slowly. He'd been broke and hungry at the time. A sneaky little rat named Johnson had bilked Clayton out of his fair share of the Corey payroll job, and Clayton had been forced to get the money somehow. He hadn't mussed the guy up much; besides, it was the sucker's own fault. If he hadn't tried to yell—

Lieutenant Harris went on: "I'm afraid you can't back down now."

"But it isn't fair! The most I'd have got on that frame-up would've been ten years. I've been here fifteen already!"

"I'm sorry, Clayton. It can't be done. You're here. Period. Forget about trying to get back. Earth doesn't want you." Her voice sounded choppy, as though she were trying to keep it calm.

Clayton broke into a whining rage. "You can't do that! It isn't fair! I never did anything to you! I'll go talk to the Governor! He'll listen to reason! You'll see! I'll—"

"*Shut up!*" the woman snapped harshly. "I'm getting sick of it! I personally think you should have been locked up—permanently. I think this idea of forced colonization is going to breed trouble for Earth someday, but it is about the only way you can get anybody to colonize this frozen hunk of mud.

"Just keep it in mind that I don't like it any better than you do—*and I didn't strong-arm anybody to deserve the assignment!* Now get out of here!"

She moved a hand threateningly toward the manual controls of the stun beam.

Clayton retreated fast. The trackers ignored anyone walking away from the desk; they were set only to spot threatening movements toward it.

Outside the Rehabilitation Service Building, Clayton could feel the tears running down the inside of his face mask. He'd asked again and again—God only knew how many times—in the past fifteen years. Always the same answer. No.

When he'd heard that this new administrator was a woman, he'd hoped she might be easier to convince. She wasn't. If anything, she was harder than the others.

The heat-sucking frigidity of the thin Martian air whispered around him in a feeble breeze. He shivered a little and began walking toward the recreation center.

There was a high, thin piping in the sky above him which quickly became a scream in the thin air.

He turned for a moment to watch the ship land, squinting his eyes to see the number on the hull.

Fifty-two. Space Transport Ship Fifty-two.

Probably bringing another load of poor suckers to freeze to death on Mars.

That was the thing he hated about Mars—the cold. The everlasting damned cold! And the oxidation pills; take one every three hours or smother in the poor, thin air.

The government could have put up domes; it could have put in building-to-building tunnels, at least. It could have done a hell of a lot of things to make Mars a decent place for human beings.

But no—the government had other ideas. A bunch of bigshot scientific characters had come up with the idea nearly twenty-three years before. Clayton could remember the words on the sheet he had been given when he was sentenced.

"Mankind is inherently an adaptable animal. If we are to colonize the planets of the Solar System, we must meet the conditions on those planets as best we can.

"Financially, it is impracticable to change an entire planet from its original condition to one which will support human life as it exists on Terra.

"But man, since he is adaptable, can change himself—modify his structure slightly—so that he can live on these planets with only a minimum of change in the environment."

So they made you live outside and like it. So you froze and you choked and you suffered.

Clayton hated Mars. He hated the thin air and the cold. More than anything, he hated the cold.

Ron Clayton wanted to go home.

The Recreation Building was just ahead; at least it would be warm inside. He pushed in through the outer and inner doors, and he heard the burst of music from the jukebox. His stomach tightened up into a hard cramp.

They were playing Heinlein's *Green Hills of Earth*.

There was almost no other sound in the room, although it was full of people. There were plenty of colonists who claimed to like Mars, but even they were silent when that song was played.

Clayton wanted to go over and smash the machine—make it stop reminding him. He clenched his teeth and his fists and his eyes and cursed mentally. *God, how I hate Mars*!

When the hauntingly nostalgic last chorus faded away, he walked over to the machine and fed it full of enough coins to keep it going on something else until he left.

At the bar, he ordered a beer and used it to wash down another oxidation tablet. It wasn't good beer; it didn't even deserve the name. The atmospheric pressure was so low as to boil all the carbon dioxide out of it, so the brewers never put it back in after fermentation.

He was sorry for what he had done—really and truly sorry. If they'd only give him one more chance, he'd make good. Just one more chance. He'd work things out.

He'd promised himself that both times they'd put him up before, but things had been different then. He hadn't really been given another chance, what with parole boards and all.

Clayton closed his eyes and finished the beer. He ordered another.

He'd worked in the mines for fifteen years. It wasn't that he minded work really, but the foreman had it in for him. Always giving him a bad time; always picking out the lousy jobs for him.

Like the time he'd crawled into a side-boring in Tunnel 12 for a nap during lunch and the foreman had caught him. When he promised never to do it again if the foreman wouldn't put it on report, the guy said, "Yeah. Sure. Hate to hurt a guy's record."

Then he'd put Clayton on report anyway. Strictly a rat.

Not that Clayton ran any chance of being fired; they never fired anybody. But they'd fined him a day's pay. A whole day's pay.

He tapped his glass on the bar, and the barman came over with another beer. Clayton looked at it, then up at the barman. "Put a head on it."

The bartender looked at him sourly. "I've got some soapsuds here, Clayton, and one of these days I'm gonna put some in your beer if you keep pulling that gag."

That was the trouble with some guys. No sense of humor.

Somebody came in the door and then somebody else came in behind him, so that both inner and outer doors were open for an instant. A blast of icy breeze struck Clayton's back, and he shivered. He started to say something, then changed his mind; the doors were already closed again, and besides, one of the guys was bigger than he was.

The iciness didn't seem to go away immediately. It was like the mine. Little old Mars was cold clear down to her core—or at least down as far as they'd drilled. The walls were frozen and seemed to radiate a chill that pulled the heat right out of your blood.

Somebody was playing *Green Hills* again, damn them. Evidently all of his own selections had run out earlier than he'd thought they would.

Hell! There was nothing to do here. He might as well go home.

"Gimme another beer, Mac."

He'd go home as soon as he finished this one.

He stood there with his eyes closed, listening to the music and hating Mars.

A voice next to him said: "I'll have a whiskey."

The voice sounded as if the man had a bad cold, and Clayton turned slowly to look at him. After all the sterilization they went through before they left Earth, nobody on Mars ever had a cold, so there was only one thing that would make a man's voice sound like that.

Clayton was right. The fellow had an oxygen tube clamped firmly over his nose. He was wearing the uniform of the Space Transport Service.

"Just get in on the ship?" Clayton asked conversationally.

The man nodded and grinned. "Yeah. Four hours before we take off again." He poured down the whiskey. "Sure cold out."

Clayton agreed. "It's always cold." He watched enviously as the spaceman ordered another whiskey.

Clayton couldn't afford whiskey. He probably could have by this time, if the mines had made him a foreman, like they should have.

Maybe he could talk the spaceman out of a couple of drinks.

"My name's Clayton. Ron Clayton."

The spaceman took the offered hand. "Mine's Parkinson, but everybody calls me Parks."

"Sure, Parks. Uh—can I buy you a beer?"

Parks shook his head. "No, thanks. I started on whiskey. Here, let me buy you one."

"Well—thanks. Don't mind if I do."

They drank them in silence, and Parks ordered two more.

"Been here long?" Parks asked.

"Fifteen years. Fifteen long, long years."

"Did you—uh—I mean—" Parks looked suddenly confused.

Clayton glanced quickly to make sure the bartender was out of earshot. Then he grinned. "You mean am I a convict? Nah. I came here because I wanted to. But—" He lowered his voice. "—we don't talk about it around here. You know." He gestured with one hand—a gesture that took in everyone else in the room.

Parks glanced around quickly, moving only his eyes. "Yeah. I see," he said softly.

"This your first trip?" asked Clayton.

"First one to Mars. Been on the Luna run a long time."

"Low pressure bother you much?"

"Not much. We only keep it at six pounds in the ships. Half helium and half oxygen. Only thing that bothers me is the oxy here. Or rather, the oxy that *isn't* here." He took a deep breath through his nose tube to emphasize his point.

Clayton clamped his teeth together, making the muscles at the side of his jaw stand out.

Parks didn't notice. "You guys have to take those pills, don't you?"

"Yeah."

"I had to take them once. Got stranded on Luna. The cat I was in broke down eighty some miles from Aristarchus Base and I had to walk back—with my oxy low. Well, I figured—"

Clayton listened to Parks' story with a great show of attention, but he had heard it before. This "lost on the moon" stuff and its variations had been going the rounds for forty years. Every once in a while, it actually did happen to someone; just often enough to keep the story going.

This guy did have a couple of new twists, but not enough to make the story worthwhile.

"Boy," Clayton said when Parks had finished, "you were lucky to come out of that alive!"

Parks nodded, well pleased with himself, and bought another round of drinks.

"Something like that happened to me a couple of years ago," Clayton began. "I'm supervisor on the third shift in the mines at Xanthe, but at the time, I was only a foreman. One day, a couple of guys went to a branch tunnel to—"

It was a very good story. Clayton had made it up himself, so he knew that Parks had never heard it before. It was gory in just the right places, with a nice effect at the end.

"—so I had to hold up the rocks with my back while the rescue crew pulled the others out of the tunnel by crawling between my legs. Finally, they got some steel beams down there to take the load off, and I could let go. I was in the hospital for a week," he finished.

Parks was nodding vaguely. Clayton looked up at the clock above the bar and realized that they had been talking for better than an hour. Parks was buying another round.

Parks was a hell of a nice fellow.

There was, Clayton found, only one trouble with Parks. He got to talking so loud that the bartender refused to serve either one of them any more.

The bartender said Clayton was getting loud, too, but it was just because he had to talk loud to make Parks hear him.

Clayton helped Parks put his mask and parka on and they walked out into the cold night.

Parks began to sing *Green Hills*. About halfway through, he stopped and turned to Clayton.

"I'm from Indiana."

Clayton had already spotted him as an American by his accent.

"Indiana? That's nice. Real nice."

"Yeah. You talk about green hills, we got green hills in Indiana. What time is it?"

Clayton told him.

"Jeez-krise! Ol' spaship takes off in an hour. Ought to have one more drink first."

Clayton realized he didn't like Parks. But maybe he'd buy a bottle.

Sharkie Johnson worked in Fuels Section, and he made a nice little sideline of stealing alcohol, cutting it, and selling it. He thought it was real funny to call it Martian Gin.

Clayton said: "Let's go over to Sharkie's. Sharkie will sell us a bottle."

"Okay," said Parks. "We'll get a bottle. That's what we need: a bottle."

It was quite a walk to the Shark's place. It was so cold that even Parks was beginning to sober up a little. He was laughing like hell when Clayton started to sing.

"We're going over to the Shark's, To buy a jug of gin for Parks! Hi ho, hi ho, hi ho!"

One thing about a few drinks; you didn't get so cold. You didn't feel it too much, anyway.

The Shark still had his light on when they arrived. Clayton whispered to Parks: "I'll go in. He knows me. He wouldn't sell it if you were around. You got eight credits?"

"Sure I got eight credits. Just a minute, and I'll give you eight credits." He fished around for a minute inside his parka, and pulled out his notecase. His gloved fingers were a little clumsy, but he managed to get out a five and three ones and hand them to Clayton.

"You wait out here," Clayton said.

He went in through the outer door and knocked on the inner one. He should have asked for ten credits. Sharkie only charged five, and that would leave him three for himself. But he could have got ten—maybe more.

When he came out with the bottle, Parks was sitting on a rock, shivering.

"Jeez-krise!" he said. "It's cold out here. Let's get to someplace where it's warm."

"Sure. I got the bottle. Want a drink?"

Parks took the bottle, opened it, and took a good belt out of it.

"Hooh!" he breathed. "Pretty smooth."

As Clayton drank, Parks said: "Hey! I better get back to the field! I know! We can go to the men's room and finish the bottle before the ship takes off! Isn't that a good idea? It's warm there."

They started back down the street toward the spacefield.

"Yep, I'm from Indiana. Southern part, down around Bloomington," Parks said. "Gimme the jug. Not Bloomington, Illinois—Bloomington, Indiana. We really got green hills down there." He drank, and handed the bottle back to Clayton. "Pers-nally, I don't see why anybody'd stay on Mars. Here y'are, practic'ly on the equator in the middle of the summer, and it's colder than hell. Brrr!

"Now if you was smart, you'd go home, where it's warm. Mars wasn't built for people to live on, anyhow. I don't see how you stand it."

That was when Clayton decided he really hated Parks.

And when Parks said: "Why be dumb, friend? Whyn't you go home?" Clayton kicked him in the stomach, hard.

"And that, that—" Clayton said as Parks doubled over.

He said it again as he kicked him in the head. And in the ribs. Parks was gasping as he writhed on the ground, but he soon lay still.

Then Clayton saw why. Parks' nose tube had come off when Clayton's foot struck his head.

Parks was breathing heavily, but he wasn't getting any oxygen.

That was when the Big Idea hit Ron Clayton. With a nosepiece on like that, you couldn't tell who a man was. He took another drink from the jug and then began to take Parks' clothes off.

The uniform fit Clayton fine, and so did the nose mask. He dumped his own clothing on top of Parks' nearly nude body, adjusted the little oxygen tank so that the gas would flow properly through the mask, took the first deep breath of good air he'd had in fifteen years, and walked toward the spacefield.

He went into the men's room at the Port Building, took a drink, and felt in the pockets of the uniform for Parks' identification. He found it and opened the booklet. It read:

PARKINSON, HERBERT J.
Steward 2nd Class, STS

Above it was a photo, and a set of fingerprints.

Clayton grinned. They'd never know it wasn't Parks getting on the ship.

Parks was a steward, too. A cook's helper. That was good. If he'd been a jetman or something like that, the crew might wonder why he wasn't on duty at takeoff. But a steward was different.

Clayton sat for several minutes, looking through the booklet and drinking from the bottle. He emptied it just before the warning sirens keened through the thin air.

Clayton got up and went outside toward the ship.

"Wake up! Hey, you! Wake up!"

Somebody was slapping his cheeks. Clayton opened his eyes and looked at the blurred face over his own.

From a distance, another voice said: "Who is it?"

The blurred face said: "I don't know. He was asleep behind these cases. I think he's drunk."

Clayton wasn't drunk—he was sick. His head felt like hell. Where the devil was he?

"Get up, bud. Come on, get up!"

Clayton pulled himself up by holding to the man's arm. The effort made him dizzy and nauseated.

The other man said: "Take him down to sick bay, Casey. Get some thiamin into him."

Clayton didn't struggle as they led him down to the sick bay. He was trying to clear his head. Where was he? He must have been pretty drunk last night.

He remembered meeting Parks. And getting thrown out by the bartender. Then what?

Oh, yeah. He'd gone to the Shark's for a bottle. From there on, it was mostly gone. He remembered a fight or something, but that was all that registered.

The medic in the sick bay fired two shots from a hypo-gun into both arms, but Clayton ignored the slight sting.

"Where am I?"

"Real original. Here, take these." He handed Clayton a couple of capsules, and gave him a glass of water to wash them down with.

When the water hit his stomach, there was an immediate reaction.

"Oh, Christ!" the medic said. "Get a mop, somebody. Here, bud; heave into this." He put a basin on the table in front of Clayton.

It took them the better part of an hour to get Clayton awake enough to realize what was going on and where he was. Even then, he was plenty groggy.

It was the First Officer of the STS-52 who finally got the story straight. As soon as Clayton was in condition, the medic and the quartermaster officer who had found him took him up to the First Officer's compartment.

"I was checking through the stores this morning when I found this man. He was asleep, dead drunk, behind the crates."

"He was drunk, all right," supplied the medic. "I found this in his pocket." He flipped a booklet to the First Officer.

The First was a young man, not older than twenty-eight with tough-looking gray eyes. He looked over the booklet.

"Where did you get Parkinson's ID booklet? And his uniform?"

Clayton looked down at his clothes in wonder. "I don't know."

"You *don't know*? That's a hell of an answer."

"Well, I was drunk," Clayton said defensively. "A man doesn't know what he's doing when he's drunk." He frowned in concentration. He knew he'd have to think up some story.

"I kind of remember we made a bet. I bet him I could get on the ship. Sure—

I remember, now. That's what happened; I bet him I could get on the ship and we traded clothes."

"Where is he now?"

"At my place, sleeping it off, I guess."

"Without his oxy-mask?"

"Oh, I gave him my oxidation pills for the mask."

The First shook his head. "That sounds like the kind of trick Parkinson would pull, all right. I'll have to write it up and turn you both in to the authorities when we hit Earth." He eyed Clayton. "What's your name?"

"Cartwright. Sam Cartwright," Clayton said without batting an eye.

"Volunteer or convicted colonist?"

"Volunteer."

The First looked at him for a long moment, disbelief in his eyes.

It didn't matter. Volunteer or convict, there was no place Clayton could go. From the officer's viewpoint, he was as safely imprisoned in the spaceship as he would be on Mars or a prison on Earth.

The First wrote in the log book, and then said: "Well, we're one man short in the kitchen. You wanted to take Parkinson's place; brother, you've got it—without pay." He paused for a moment.

"You know, of course," he said judiciously, "that you'll be shipped back to Mars immediately. And you'll have to work out your passage both ways—it will be deducted from your pay."

Clayton nodded. "I know."

"I don't know what else will happen. If there's a conviction, you may lose your volunteer status on Mars. And there may be fines taken out of your pay, too.

"Well, that's all, Cartwright. You can report to Kissman in the kitchen."

The First pressed a button on his desk and spoke into the intercom. "Who was on duty at the airlock when the crew came aboard last night? Send him up. I want to talk to him."

Then the quartermaster officer led Clayton out the door and took him to the kitchen.

The ship's driver tubes were pushing it along at a steady five hundred centimeters per second squared acceleration, pushing her steadily closer to Earth with a little more than half a gravity of drive.

There wasn't much for Clayton to do, really. He helped to select the foods that went into the automatics, and he cleaned them out after each meal was cooked. Once every day, he had to partially dismantle them for a really thorough going-over.

And all the time, he was thinking.

Parkinson must be dead; he knew that. That meant the Chamber. And even if he wasn't, they'd send Clayton back to Mars. Luckily, there was no way for either planet to communicate with the ship; it was hard enough to keep a beam trained on a planet without trying to hit such a comparatively small thing as a ship.

But they would know about it on Earth by now. They would pick him up the instant the ship landed. And the best he could hope for was a return to Mars.

No, by God! He wouldn't go back to that frozen mud-ball! He'd stay on Earth, where it was warm and comfortable and a man could live where he was meant to live. Where there was plenty of air to breathe and plenty of water to drink. Where the beer tasted like beer and not like slop. Earth. Good green hills, the like of which exists nowhere else.

Slowly, over the days, he evolved a plan. He watched and waited and checked each little detail to make sure nothing would go wrong. It *couldn't* go wrong. He didn't want to die, and he didn't want to go back to Mars.

Nobody on the ship liked him; they couldn't appreciate his position. He hadn't done anything to them, but they just didn't like him. He didn't know why; he'd *tried* to get along with them. Well, if they didn't like him, the hell with them.

If things worked out the way he figured, they'd be damned sorry.

He was very clever about the whole plan. When turn-over came, he pretended to get violently spacesick. That gave him an opportunity to steal a bottle of chloral hydrate from the medic's locker.

And, while he worked in the kitchen, he spent a great deal of time sharpening a big carving knife.

Once, during his off time, he managed to disable one of the ship's two lifeboats. He was saving the other for himself.

The ship was eight hours out from Earth and still decelerating when Clayton pulled his getaway.

It was surprisingly easy. He was supposed to be asleep when he sneaked down to the drive compartment with the knife. He pushed open the door, looked in, and grinned like an ape.

The Engineer and the two jetmen were out cold from the chloral hydrate in the coffee from the kitchen.

Moving rapidly, he went to the spares locker and began methodically to smash every replacement part for the drivers. Then he took three of the signal bombs from the emergency kit, set them for five minutes, and placed them around the driver circuits.

He looked at the three sleeping men. What if they woke up before the bombs went off? He didn't want to kill them though. He wanted them to know what had happened and who had done it.

He grinned. There was a way. He simply had to drag them outside and jam the door lock. He took the key from the Engineer, inserted it, turned it, and snapped off the head, leaving the body of the key still in the lock. Nobody would unjam it in the next four minutes.

Then he began to run up the stairwell toward the good lifeboat.

He was panting and out of breath when he arrived, but no one had stopped him. No one had even seen him.

He clambered into the lifeboat, made everything ready, and waited.

The signal bombs were not heavy charges; their main purpose was to make a flare bright enough to be seen for thousands of miles in space. Fluorine and magnesium made plenty of light—and heat.

Quite suddenly, there was no gravity. He had felt nothing, but he knew that the bombs had exploded. He punched the LAUNCH switch on the control board of the lifeboat, and the little ship leaped out from the side of the greater one.

Then he turned on the drive, set it at half a gee, and watched the STS-52 drop behind him. It was no longer decelerating, so it would miss Earth and drift on into space. On the other hand, the lifeship would come down very neatly within a few hundred miles of the spaceport in Utah, the destination of the STS-52.

Landing the lifeship would be the only difficult part of the maneuver, but they were designed to be handled by beginners. Full instructions were printed on the simplified control board.

Clayton studied them for a while, then set the alarm to waken him in seven hours and dozed off to sleep.

He dreamed of Indiana. It was full of nice, green hills and leafy woods, and Parkinson was inviting him over to his mother's house for chicken and whiskey. And all for free.

Beneath the dream was the calm assurance that they would never catch him and send him back. When the STS-52 failed to show up, they would think he had been lost with it. They would never look for him.

When the alarm rang, Earth was a mottled globe looming hugely beneath the ship. Clayton watched the dials on the board, and began to follow the instructions on the landing sheet.

He wasn't too good at it. The accelerometer climbed higher and higher, and he felt as though he could hardly move his hands to the proper switches.

He was less than fifteen feet off the ground when his hand slipped. The ship, out of control, shifted, spun, and toppled over on its side, smashing a great hole in the cabin.

Clayton shook his head and tried to stand up in the wreckage. He got to his hands and knees, dizzy but unhurt, and took a deep breath of the fresh air that was blowing in through the hole in the cabin.

It felt just like home.

Bureau of Criminal Investigation
Regional Headquarters
Cheyenne, Wyoming
20 January 2102

To: Space Transport Service
Subject: Lifeship 2, STS-52
Attention Mr. P. D. Latimer

Dear Paul,

I have on hand the copies of your reports on the rescue of the men on the disabled STS-52. It is fortunate that the Lunar radar stations could compute their orbit.

The detailed official report will follow, but briefly, this is what happened:

The lifeship landed—or, rather, crashed—several miles west of Cheyenne, as you know, but it was impossible to find the man who was piloting it until yesterday because of the weather.

He has been identified as Ronald Watkins Clayton, exiled to Mars fifteen years ago.

Evidently, he didn't realize that fifteen years of Martian gravity had so weakened his muscles that he could hardly walk under the pull of a full Earth gee.

As it was, he could only crawl about a hundred yards from the wrecked lifeship before he collapsed.

Well, I hope this clears up everything.

I hope you're not getting the snow storms up there like we've been getting them.

John B. Remley

Captain, CBI

THE JANITOR ON MARS
(1999)
Martin Amis

Martin Amis is an acclaimed British novelist. Born in 1949 in Oxford, he grew up reading comic books until he was introduced to Jane Austen by his stepmother, the novelist Elizabeth Jane Howard. He graduated from Exeter College, Oxford, with a congratulatory first in English. He is best known for the London trilogy of novels: *Money, London Fields* and *The Information.*

1

POP JONES WAS TELLING the child why he couldn't watch the news that day.

"Special regulation, Ash. You have to be eighteen. It's like an X-certificate."

"I want to see the Martian."

"Well you can't. And he's not strictly speaking a Martian. They think he must be some sort of robot."

"He's the man on Mars."

"He, or it, is the *janitor* on Mars."

And Pop Jones was the janitor on earth—more specifically, the janitor of Shepherds Lodge, the last nonprivatized orphanage in England. Remote, decrepit, overcrowded, and all male, the place was, of course, a Shangri-La of pedophilia. And Pop Jones was, of course, a pedophile, like everybody else on the staff. To use the (rather misleading) jargon, he was a "functional" pedophile—which is to say, his pedophilia *didn't* function. Pop Jones was an inactive pedophile, unlike his hyperactive colleagues. He had never interfered with any of the boys in his care: not once.

The child, Ashley, a long-suffering nine-year-old, said, "They're taking us to the beach. I want to stay and see the robot."

"To the beach! Remember to take your starblock."

"But I want to starbathe."

"You'll get starstroke out there."

"I want a startan."

"A startan? You'll get starburn!"

No one called it the sun anymore: the nature of the relationship had changed. It was 25 June, 2049, and every television on earth would soon be featuring the live interview with the janitor on Mars.

Outside, the boys were being marshaled into queues under the awning as the first electric bus pulled up. Each of them clutched his white umbrella. Pop Jones was pleased to see that Ashley was wearing his starglasses and his starhat. All the children were flinching up at the sky. Each mouth wore a wary sneer.

The thing had been building for nine months.

On 30 September, 2048, at 12:45 P.M., West Coast time, Incarnacion Buttruguena-Hume, the most frankly glamorous of CNN's main newscasters, received an encrypted message on her PDA. Incarnacion's computer failed to recognize the cipher but then quickly cracked it. The message was written in the Blacksmith Code, unused for a century and considered obsolescent in World War II. It began, CKBIa TCaAIa-CaBTKaCa: Dear Incarnacion. Decoded, the message said:

FORGIVE THE INSTRUSION, BUT I'M GOING TO BE COMING IN ON YOUR AIRTIME TONIGHT. I HAVE NEWS FOR YOU. I'M THE JANITOR ON MARS. TALK TO PICK AROUND FIVE-THIRTY.

Pick was Pickering Hume, Incarnacion's husband, who, noncoincidentally (it was soon supposed), worked in the public-relations and fund-raising departments of SETI—or Search for Extra-Terrestrial Intelligence. Incarnacion called Pick right away at his office in Mountain View. They discussed the transmission: which of their friends, they wondered, was responsible for it? But at 17:31 Pick called back. In a clogged whisper he told her that they were receiving a regularly repeated radio signal on the hydrogen line from the Tharsis Bulge on Mars, in straight Morse. The Morse from Mars was saying: PICK—CALL INCARNACION.

It was five-forty in Los Angles. Within fifteen minutes the sat links were engaged and the floor of Incarnacion's studio was filling up with astronomers, cosmologists, philosophers, historians, science-fiction writers, millenarians, Rapturists, UFO abductees, churchmen, politicians, and five-star generals, gathered for a story that just kept on breaking—that went twenty-four hours and stayed that way. On the stroke of six o'clock the screen turned a rusty red.

Pop Jones himself was watching, on that day, along with every other adult in the building, called to the Common Room by the Principal, Mr. Davidge. The screen went red, then white. And the message appeared, unscrolling upwards, B-movie style, in heroic, backward-leaning capitals. It said:

GREETINGS DNA, FROM HAR DECHER, THE RED ONE, AS YOUR EGYPTIANS CALLED OUR WORLD, OR NERGAL, AS YOUR BABYLONIANS HAD IT: THE STAR OF DEATH. GREETINGS FROM MARS. OUR TWO PLANETS HAVE MUCH IN COMMON. OUR DIURNAL MOTION IS SIMILAR. THE OBLIQUITY OF OUR RESPECTIVE ECLIPTICS IS NOT VERY DIFFERENT. YOU HAVE OCEANS, AN ATMOSPHERE. A MAGNETOSPHERE. SO DID WE. YOU ARE LARGER. YOU ARE CLOSER IN. WE COOLED QUICKER. BUT LIFE ON OUR PLANETS WAS SEEDED MORE OR LESS COINSTANTANEOUSLY—A DIFFERENCE OF A FEW MONTHS, WITH EARTH TAKING TECHNICAL SENIORITY. OUR WORLDS, AS I SAY, ARE SIMILAR, AND WERE ONCE MORE SIMILAR. BUT OUR HISTORIES RADICALLY AND SPECTACULARLY DIVERGE.

IT'S GONE NOW, VANISHED, ALL MARTIAN LIFE, AND I'M WHAT REMAINS. I AM THE JANITOR ON MARS. AND I HAVE BEEN WATCHING YOU, TRIPWIRED TO MAKE CONTACT AT THE APPROPRIATE TIME. THAT TIME HAS COME. LET'S TALK.

I'LL BE IN TOUCH WITH NASA ABOUT LAUNCH WINDOWS. ALSO SOME TIPS ABOUT CLIMBING YOUR GRAVITY WELL: A FUEL THING. AND A SUGGESTION ABOUT YOUR COSMIC-RAY PROBLEM AND WAYS OF REDUCING PAYLOAD. DUPLICATES OF ALL COMMUNICATIONS WILL GO TO CNN AND THE NEW YORK TIMES. LET'S PLAY THIS ONE STRAIGHT, PLEASE.

YOU NEVER WERE ALONE. YOU JUST THOUGHT YOU WERE. AND HOW COULD YOU EVER HAVE THOUGHT THAT? DNA, MAKE HASTE. I AM IMPATIENT TO SEE YOU WITH MY OWN EYES. COME.

Under his dirty white umbrella Pop Jones limped quickly across the courtyard. He glanced up. Although his flesh wore the pallor of deep bacherlorhood, Pop's face often looked childish, tentative; this, plus his pertly plump backside, his piping yet uneffeminate voice, and his chastity, combined to earn him his nickname. His nickname was Eunuch. (His forename, moreover, was Enoch.) The children he treated with bantering geniality. But with his fellow adults Pop Jones was a *janitor*, through and through; he was *all* janitor, a janitors' janitor, idle, disobliging, truculent, withdrawn. And, in his person, defiantly unclean. Overhead the star wriggled goopily in the sky, with slipped penumbra, like one of the cataracts it so prolifically dispensed. The sun hadn't changed. The sky had. The sky had fallen sick, but everybody said it was now getting better again. Pop limped on up the steps to the Sanatorium. He turned: a square lawn supporting two ancient trees, both warped and crushed by time into postures of lavatorial agony. Shepherds Lodge looked like an Oxford college as glimpsed in the dreams of Uriah Heep. Pop Jones, taking pride in his profession, maintained the place as a sophistical labyrinth of sweat and shiver, the radiators now raw, now molten, the classrooms either freezers or crucibles, the taps, once turned, waiting a while before hawking forth their gouts of steam or sleet. The plumbing clanked. Locks stuck. All the lights flickered and fizzed.

He passed the medical officer's nook and glanced sideways into the old surgical storeroom, now a mini-gym, where two male nurses were talc-ing their hands for the bench press. They glanced back at him, pausing. Pop Jones could feel the hum of isolation in his ears. Yes, he thought, a dreadful situation. Quite dreadful. The whole moral order. But someone has to...The patient he had come to see was an eleven-year-old called Timmy. Timmy suffered from various learning disabilities (he was always injuring himself by falling over or walking into walls), and Pop Jones felt a special tenderness for him. Many of the boys at Shepherds Lodge, it had to be said, were some-what soiled and complaisant, if not thoroughly debauched. Indeed, on warm evenings the place had the feel of an antebellum bordello, with boys in pyjamas straddling windowsills—training their hair, reading mail-order magazines—to the sound of some thrummed guitar...Timmy wasn't like that. Sealed off in his own mind, Timmy had an inviolability that everyone had respected. Until now. Pop and Timmy were chaste—they were the innocents! *That* was their bond...To be clear: it is not youth alone that attracts the pedophile. The pedophile, for some reason, wants carnal knowledge of the carnally ignorant: a top-heavy encounter, involving lost significance. So far as the child is concerned, of course, that lost significance doesn't stay lost, but lingers, forever. On some level Pop Jones sensed the nature of this disparity, this pre-emption, and it kept him halfway straight. The merest nudge or nuzzle, every now and then. His use of the bathhouse peepholes was now strictly rationed. In any given month, you could count his rootlings in the laundry baskets on the fingers of one hand.

"How are you this morning, my lad?"

"Car," said Timmy.

Timmy was alone in the six-bed ward. A TV set roosted high up on the opposite wall: it showed the planet Mars, filling half the screen now, and getting ever closer.

"Timmy, try to remember. Who did this to you, Timmy?"

"House," said Timmy.

The boy was not in San for one of his workaday injuries, something like a burn or a twisted ankle. Timmy was in San because he had been raped: three days ago. Mr. Caroline had found him in the potting shed, lying on the duckboards, weeping. And from then on Timmy had lapsed into the semi-autistic bemusement that had marked his first two years at Shepherds Lodge: the state that Pop Jones, and others, liked to think they had coaxed him out of. The flower had partly opened. It had now closed again.

"Timmy, try to remember."

"Floor," said Timmy.

Rape—nonstatutory rape—was vanishingly rare at Shepherds Lodge: rape flew in the face of everything its staff cherished and honored. Intergenerational sex, in that gothic mass on the steep green of the Welsh border, was of course ubiquitous, but they had a belief system which accounted for that. Its signal precept was that the children liked it.

"Who did this, Timmy?" persisted Pop, because Timmy was perfectly capable of identifying and, after a fashion, naming every carer on the payroll. The Principal, Mr. Davidge, he called "Day." Mr. Caroline he called "Ro." Pop Jones himself he called "Jo." Who did this? Everyone, including Pop, was edging toward a wholly unmanageable suspicion: *Davidge* had done it. It seemed inescapable. The last time something like this happened (in fact a much milder case, involving the "inappropriate fondling" of a temporary referral from Birmingham), Davidge had pursued the matter with Corsican rigor. But the investigation into the attack on Timmy seemed oddly stalled: three days had passed without so much as an analdilation test. Davidge's shrugs and prevarications, by a process of political trickledown, now threatened a general dissolution, Pop sensed. The janitor was on his own here. Already he felt at the limit of his moral courage. The only whispers of support were coming from a confused and indignant eleven-year-old called Ryan, Davidge's current regular (and, therefore, the cynosure of B Wing).

"Was it...'Day'?" he asked, leaning nearer.

"Dog," said Timmy.

The two male nurses—the two reeking sadists in their sleeveless T-shirts—were snorting rhythmically under their weights. Pop called out to them:

"Excuse me? Excuse me? Mr. Fitzmaurice, if you please. You will be turning this television off. I hope. The boys are not permitted to watch the news today. It's an OO. An official order. From the Department Head."

The male nurses leered perfunctorily at each other and made no response.

"This television will have to be disconnected."

Fitzmaurice sat up on his bench and shouted. "If I do that the whole fucking system goes down. Every TV in the fucking gaff."

Pop Jones, as a janitor, had to bow to the logic of that. He said, "Then he'll have to be moved. It may be quite unsuitable for children. There may be some bad language."

With a cheerful squint Fitzmaurice said, "Bad language?"

"You can turn the sound down at least. Nobody knows what's going to happen up there. Anything could happen up there."

Fitzmaurice shrugged.

"Car." said Timmy.

Pop looked at the TV. Mars now filled the screen.

This day many questions would be answered. Not the least pressing (many felt) was: why now? What was the "tripwire"? How did you explain the timing of the Contact from the janitor on Mars?

It seemed significant, or perverse, for two reasons. As recently as 2047, after many a probe and flyby, NASA had successfully completed the first manned mission to Mars. The Earthling cosmonauts spent three months on the Red Planet and returned with almost half a ton of it in sample form. Preliminary analysis of this material was completed and made public in the autumn of 2048. The findings seemed unambiguous. True, the layer of permafrost proved that water had once flowed on the Martian surface, and in stupendous quantities, as was already clear from the flood tracks in its gorge and valley systems. But otherwise the *Sojourner 3* mission could come up with nothing to puncture the verdict of ageless sterility. So the question remained: why wasn't Contact made then? In the interim 1,500 new telecommunications satellites had gone into orbit; as the janitor on Mars himself pointed out, in of one his earlier communiqués, Earth had practically walled itself up with space junk. Five hundred units had to be blown out of the sky to clear a lane for *Sojourner 4*.

The second coincidence had to do with ALH84001. ALH84001 was the fist-sized, green-tinged lump of rock found in Antarctica in 1984, analyzed in 1986, and argued about for over half a century. But its history was grander, weirder, and above all longer than that. About 4.5 billion years ago ALH84001 was an anonymous subterranean resident of primordial Mars; 4,485,000,000 years later something big hit Mars at a shallow angle and ALH84001 was part of its ejecta; for 14,987,000 years it followed a cat's-cradle solar orbit before crashlanding on Earth. Then, 13,000 years later still, a meteorite-hunter called Roberta Star tripped over it and the controversy began. Did ALH84001 bear traces of microscopic life? The answer came, finally, in April 2049—two months before the janitor on Mars made his move. And the answer was No. ALH84001's organic compounds (magnetite, gregite, and pyrrhotite) were proven to be mere polycyclic aromatic hydrocarbons—i.e., they were nonbiological. Apparently Mars couldn't even support a segmented worm one-hundredth the width of a human hair. *That* was how dead Mars seemed.

> *Let me remind you that these images...from the camera in the nose cone of the trailing vessel. Lacking an ozone layer...effectively sterilized by solar ultraviolet radiation. The atmosphere...thinner than our best laboratory vacuums. We can see Phobos, the larger...a mere 3,500 miles distant as compared to our moon's...Deimos, the second satellite, is overhead...as bright to the eye as Venus.*

The TV-viewing armchair in Pop Jones's terrible old Y-front of a bedroom (with its Bovril tins and clouded toothmugs) had become steeped in his emanations, over the years. Anyone else, settling into it, would have instantly succumbed to projectile nausea, and would have shot up out of it, as from an ejector seat. But not Pop: in his armchair he felt fully alive. Look at him now, his tongue idling on his lower teeth, as he watched the screen with the kind of awe he usually reserved for only the most sincere and accurate pedography, freely available from many an outlet in Shepherds Lodge (and quite regularly starring its inmates). He had seen this image before—everyone had: a Colorado made of rust before a strangely proximate horizon. But the planet was now in some sense a living Mars, and life invested it everywhere with menace. The thin mist looked like fat on the meaty crimson of the regolith, and shapes seemed to form and change in the shadows of the sharp ravines...

For a second the picture was lost. Then the voice of Incarnacion Buttruguena-Hume—warmly aspirated, extravagantly human—continued:

> *In some ways Mars is a small world. Its surface area is only a third of ours, and its mass only a thousandth. But in other ways. Mars is a big world. Its canyons...than ours, its peaks far higher. One of its gorges, the Valles... Grand Canyon to shame. And—yes: we're approaching it now. This is Olympus Mons, sixteen miles high—three Everests—but sloping so gradually that it casts no shadow. It resembles the shield volcanoes in...I have just been told that this vessel is no longer under our control. He's bringing us in. We...We...*

And you saw it: in utter silence but with sky-shaking effort, the mountain was opening—its segmented upper flanks now bending backwards like a nest full of titanic chicks with their beaks open wanting food. The leading vessel, Nobel 1, strained above these battlements, and plummeted. Nobel 2, the POV vessel, followed. During its descent Pop felt that he was riding an elevator downwards, the innards of the edifice thrumming past you in the dark, but much too quickly: with all the avid acceleration of free fall.

Every screen on Earth stayed black. Then these numerals appeared in a pale shade of green: 45:00. And started going 44:59, 44:58, 44:57...
In fact it was twice that many minutes before anything happened.

A weak light came up and the camera jerked around in consternation, as if violently roused from deep slumber. There were shadows, figures. You could hear mumbling and coughing. And one of their number was calling out in a strained and self-conscious voice: "Hello!...Hello?...Hello!...Hello?"

> *Everything is fine here. We've been waiting in this... room. The vessels docked smoothly and we just followed the arrowed signs. One of the Laureates fell over a moment ago, but was unhurt. And for a moment Miss World had a minor problem with her air supply. We are wearing filament-heated mesh suits with...*

There had of course been enormous controversy about who would go, and who would not go, to meet the janitor on Mars. Everyone on Earth was up for it. After all, there was no longer anything frightening or even exotic about space travel. In the Thirties and Forties, before the satellites really thickened up, lunar tourism expanded to the extent that parts of the moon's surface now resembled a wintry Torremelinos. Granted, the moon was a mere 250,000 miles away, and Mars, at the current opposition, was almost two million. But everyone was up for it. No ticket had ever been hotter. There were sixty-five seats. And seven billion people in the queue.

They had to contend not just with each other but also with the janitor on Mars, who, in a number of communications, had proved himself a brisk and abrasive stipulator. For example he had at the outset refused to countenance any clerics or politicians. Later, when pressed by massive referenda to find a couple of seats for the Pope and the U.S. President, the janitor on Mars caused far more hurt than mirth when he sent the following E-mail to the *New York Times* (forcing that journal to break an ancient taboo: "print the obscenity in full," he cautioned, "or I switch to the *Post*"): "Don't send me no fucking monkeys, okay? Monkeys no good. Just send me the talent." He wanted scientists, poets, painters, musicians, mathematicians, philosophers "and some examples of male and female pulchritude." He wanted no more media than Incarnacion Buttruguena-Hume (and her camera operator. She was also allowed to bring Pick). The haggling continued well into the countdown at Cape Canaveral. In the end there were twenty-eight hard-science Laureates on board Nobels 1 and 2, as well as several fashion models, Miss World, some NASA personnel, and various searchers and reachers from various branches of the humanities. The janitor on Mars had been particularly obdurate about Miss World, even though the contest she had won was by now an obscure affair, disputed between a couple of hundred interested onlookers in the Marriott at Buffalo Airport.

This weakness of the janitor's—for harsh language and harsh sarcasm—was the focus of much terrestrial discussion, and much disquiet. Even those who shared this weakness seemed to sense a breach of fundamental cosmic decorum. The pop psychologist Udi Ertigan put many minds at rest with the following suggestion (soon adopted as the consensus view): "I see here a mixture of high and low styles. The high style feels programmed, the low style acquired. Acquired from whom? From us! Our TV transmissions go out into space at the speed of light. What we're dealing with is a robot who's watched too many movies." Make no mistake, though: the janitor on Mars was for real. At first, the doubters doubted and the trimmers trimmed. But the janitor on Mars was definitely for real. His brief introductory tips about fuel-gelation had revolutionized aeronautics. And every couple of weeks he stirred up one discipline after the other with his mordant memos on such things as protein synthesis, the Coriolis force, slow-freeze theory, tensor calculus, chaos and K-entropy, gastrulation in developmental biology, sentential variables, butterfly catastrophe, Champernowne's number, and the *Entscheidungsproblem.* The janitor on Mars had promised to disclose a formula for cold fusion ("I'm no expert," he wrote, "and I'm having some trouble dumbing down the math") and a cure for cancer ("Or how about prevention? Or would you settle for *remission*?"). "Your gerontology, he noted, "is in its infancy. Working together, we can double life expectancy within a decade." On cosmological issues—and on Martian history—he usually refused to be drawn, saying that there were "some things you [couldn't] talk about on the phone"; and, besides, he didn't want "to cheapen the trip." "But I will say this," he said:

> The Big Bang and the Steady State theories are both wrong. Or, to put it another way, they are both right but incomplete. It pains me to see you jerk back from the apparent paradox that the Universe is younger than some of the stars it contains. That's like Clue One.

Iain Henryson, Lucasian Professor at Cambridge University, described the mathematics that accompanied this memo as "ineffable. In every sense." The janitor on Mars was often petulant, insensitive, facetious and sour, and not infrequently profane. But Earth trusted his intelligence, believing, as it always had, in the ultimate indivisibility of the intelligent and the good.

It was in any case a time of hope for the blue planet. The revolution in consciousness during the early decades of the century, a second enlightenment having to do with self-awareness as a species, was at least gaining political will. None of the biospherical disasters had quite gone ahead and happened. Humankind was still bailing water, but the levels had all ceased rising and some had started to fall. And for the first time in Earth's recorded history no wars were being contested on its surface.

Pop Jones settled back into his armchair, then, with all the best kinds of thoughts and feelings. If things did start to get rough he would go and see Davidge about getting Timmy moved at half-time—during the intermission demanded by the janitor on Mars.

> *We are wearing filament-heated mesh suits with autonomous air supply, but according to Colonel Hick's instruments the air is breathable and the temperature is rising. It was close to freezing at first but now it's evidently no worse than chilly. And damp. I'm removing*

> *my headpiece...now. Yup Seems okay...Gravity is at I g. I have no sense of lightness or hollowness. We seem to be in some kind of reception area, but our lights don't work and until a minute ago we've had only the faintest illumination. I can hear...*

You could hear the squawk of tortured rivets or hinges, and high on the wall was abruptly thrown a slender oblong of light, which briefly widened as a shadow moved past its source. Then the door closed on the re-established gloom. Pop Jones nodded in sudden agreement. Whether or not the janitor on Mars was a genuine Martian (and there had been much speculation earlier on: a hoax, no, but was he maybe a lure?), the janitor on Mars, in Pop's view, was definitely a genuine janitor. Now kill the light again, thought Pop, and turn off the heat. He listened expectantly for the clank of buckets, the skewering of big old keys in cold damp locks. But all he heard was the slow clop of foot-steps. Then, causing pain to the dark-adapted eye, the lights came on with brutal unanimity.

"Welcome, DNA. So this is the double helix on the right-handed scroll. DNA, I extend my greetings to you."

When you could focus you saw that the janitor on Mars sat at a table on a raised stage: an unequivocal robot wearing blue-black overalls and a shirt and tie. His face was a dramatically featureless beak of burnished metal; his hands, clawlike, intricate, fidgety. The accent was not unfamiliar: semieducated American. He sounded like a sports coach—a sports coach addressing other, lesser sports coaches. But he had no mouth to frame the words and they had a buzzy, boxy tone: an interior sizzle. The janitor on Mars tossed an empty clipboard on to the table and said,

"Ladies and gentlemen, I apologize for the condition of these modest furnishings. This room is something I threw together almost exactly a century ago, on 29 August, 1949: the day it became clear that Earth was featuring two combatants equipped with nuclear arms. I kept meaning to update it. But I could never be fucked...Human beings, don't look that way. Miss World: don't crinkle your nose at me. And dispense, in general, with your expectations of grandeur. There is such a thing as cosmic censorship. But the universe is profoundly and essentially profane. I think you'll be awed by some of the things I'm going to tell you. Other emotions, however, will predominate. Emotions like fear and contempt. Or better say terror and disgust. Terror and disgust. Well. First—the past."

By now two cameras were established back-to-back at the base of the podium. You saw the janitor on Mars; and then you saw his audience (seated on tin chairs in an ashen assembly hall: wood paneling, drab drapes on the false windows; a blackboard; the American and Soviet flags). In the front row sat Incarnacion Buttruguena-Hume and her husband, Pickering. Tentatively Incarnacion raised her hand.

"Yes, Incarnacion."

She blushed, half-smiled, and said, "May I ask a preliminary question, sir?"

The janitor on Mars gave a minimal nod.

"Sir. Only two years ago there were human beings on your *door*step. Why—?"

"Why didn't I make myself known to you then? There's a good reason for that: the tripwire. Patience, please. All will become clear. If I may revert to the program? The past...To recap: Earth and Mars are satellites of the same second-generation, metal-rich, main-sequence yellow dwarf on the median disk of the Milky Way. Our planets seized and formed some four and a half billion years ago. Smaller, and further out, we cooled quicker. Which you might say gave us a head start."

With a brief snort of amusement or perhaps derision the janitor on Mars leant backwards in his chair and thoughtfully intermeshed his slender talons.

"Now. We two had the same prebiotic chemistry and were pollinated by the same long-periodical comet: the Alpha Comet, as we called her, which visits the solar system every 113 million years. Life having been established on Earth, you then underwent that process you indulgently call "evolution." Whereas we were up and running pretty much right away. I mean, in a scant 300 million years. While you were just some fucking disease. Some fucking germ, stinking up the shoreline. And I can promise you that ours was the more typical planetary experience: self-organizing complexity, with remorseless teleological drive. Martian civilization flourished, with a few ups and down, for over three trillennia, three billion years, reaching its (what shall I say?)—its apotheosis, its *climax* 500 million years ago, at which time, as they say, dinosaurs ruled the Earth. Forty-three million years later, Martian life was extinguished, and I, already emplaced, was activated, to await tripwire."

Miss World said, "Sir? Could you tell us what your people looked like?"

Nicely framed though this question was, the janitor on Mars seemed to take some exceptions to it. A momentary shudder in the thick blade of his face.

"Not unlike you now, at first. Somewhat taller and ganglier and hairier. We did not excrete. We did not sleep. And of course we lived a good deal longer than you do—even at the outset. This explains much. You see, DNA isn't any good until it's twenty years old, and by the time you're forty your brains start to rot. Average life expectancy on Mars was at least two centuries even before they started upping it. And of course we pursued aggressive bioengineering from a very early stage. For instance, we soon developed a neurological integrated-circuit technology. What you'd call telepathy. I'm doing it now, though I've added a voice-over for TV viewers. Can you feel that little nasal niggle in your heads? Thoughts, it might please you to learn, are infinity-tending and travel at the speed of light."

The janitor on Mars stood—with a terrible backward-juddering scrape of his metal chair that had Pop Jones frowning with approval as he reached for the tin of Bovril and the spoon. At this stage, Pop's feelings for his Martian counterpart touched many bases: from a sense of solidarity all the way to outright hero worship. The air of brusque obstructiveness, the grudge-harboring slant of his gaze; and there was something else, something subtler, that struck Pop as so quintessentially janitorial. *Alertness to the threat of effort*: that was it. The day has come, he thought. The day when at last the janitors—

"Now I don't have all afternoon," said the robot, rather unfeelingly, perhaps (his audience having spent four-and-a-half months in transit). In his black crepe-soled shoes the janitor on Mars was no more than five feet tall. Yet he filled his space with formidable conviction—a metallic self-sufficiency. He moved like a living being but he could never be mistaken for one, in any light. And while the face had an expressive range of attitudes and elevations, there was nothing human, nothing avian, nothing remotely organic in its severity. He approached the edge of the stage, saying,

"Let's not have this degenerate into Q and A. I have a program to get through here. We'll go thumbnail and examine our respective journeys in parallel. So: 3.7 billion years ago, life is seeded. 3.4 billion years ago, Martians, as I say, are up and running: 'hunter-gatherers' is your euphemism but 'scavengers' is closer to the truth. At this stage, of course, you're still a bubble of fart gas. Goop. Macrobiotic yogurt left out in the sun. Five centuries go by: Mars is fully industrialized. Another five, and we entered what I guess you'd call our posthistorical phase. We called it Total Wealth. All you're managing to do, at this stage, is stink up the estuaries and riverbeds, but meanwhile, over on Mars, we're into quantum gravity, tired light, chromo power, trace drive, cleft conformals, scalar counterfactuals, wave superposition, and orthogonics. We were the masters of our habitat, having gotten rid of all the animals and the oceans and so forth, and the tropospherical fluctuations you call weather. In other words, we were ready."

"Ready for what?" came a voice.

"Now I'm just a janitor, right? I'm just a, uh, 'robot.' At the time of my manufacture, there was on Mars no distinction between the synthetic and the organic. Everyone was a mix, semi-etherealized, self-duplicating. The natural/mechanical divide belonged to ancient memory. But what you see before you here is a *robot*. Of the...crudest kind. It's as if, on Earth, in 2050, an outfit like Sony produced a gramophone with a dishful of spare needles and a tin bullhorn." The janitor on Mars paused, nodding his lowered head. Then he looked up. "And yet my makers, in their wisdom...However. In the last five hundred million years I've had access to an information source that was not available to the former denizens of this planet. And with that perspective it's quite clear that Mars was an absolutely average world of its class. A type-v world, absolutely average, and it did what type-v worlds invariably do in the posthistorical phase."

"Sir?" said Incarnacion. "Excuse me, but is this a grading system? What's a type-v world?"

"A world that has mined its star."

"What type world is Earth?"

"A type-y world."

"What are type-z worlds?"

"Dead ones. But I digress. You go posthistorical and the question is: now what? As I say, three billion three million nine hundred and ninety-nine thousand years ago, Martians were lords of all they surveyed. They were ready. Ready for what? Ready for war."

The robot let this ripple out through the moist air, over the ranked metal seats.

"Yes, that's right. Mars, the Planet of War. Congratulations. The only time you ever get anywhere is when you follow the artistic pulse. You even got the moons. I quote: 'Two lesser stars, or satellites, revolve about Mars; whereof the innermost is distant from the center of the primary planet exactly three of its diameters, and the outermost five.' That's not one of your early Mars watchers, some chump like Schiapperelli or Percival Lowell. That's *Gulliver's Travels*. Phobos and Deimos. Just so. Fear and Panic. Hitherto, there had never been any disharmony on Mars. Firm but wise world government was proceeding without friction. There was never any of that brawling and scragging that you went in for. Mars had tried peace, but now the time felt right. What else was there to do? We divided, almost arbitrarily, into two sides. We were ready. One called the other People of Fear. The other called the one People of Panic. There wasn't a dissenting voice on the whole planet. Everyone was absolutely all for it. Imagine two superfuturistic Japanese warrior cults, with architecture by Albert Speer. I guess that'll give you some idea.

"We fell into a rhythm. Arms races followed by massive conflicts. We'd pepper each other with all kinds of superexotic weaponry in delightfully elaborate successions of thrusts and feints and counters. But in the end nothing could match the hit of central thermonuclear exchange. We always ended up throwing everything we had at each other, in arsenal-clearing deployments. After the devastation, we rebuilt toward another devastation. No complaints. Shelter culture had come on a long way. Casualties could be patched up good as new. And fatalities were simply resurrected—except, of course, in cases of outright vaporization. They took their nuclear winters like Martians. The lulls lasted centuries. The battles were over in an afternoon.

"It doesn't make a lot of obvious sense, does it? Later on they tended to argue that it was a necessary stage in their military development. They felt that they were…rich in time. They didn't know—as I do—that this happens to all type-v worlds in the posthistorical phase. Without exception. They go insane.

"The Hydrogen War of the Two Nations lasted for 112 million years, and was followed, six months later, by the Seventy Million Years War, in which the use of quantum-gravity weapons exponentially increased the firepower of both sides. By this time another factor was preying on Martian mental health. Immortality. That's actually not a very useful word. Put it this way. Everyone on Mars was looking at a future-endless worldline. And in a type-v context that always messes with your head. There was one more great war, the War of the Strong Nuclear Force, which dragged on for 284 million years. When they came out of that, there was a general feeling that Mars was in something of a rut. So they decided to stop fucking around. You, at this stage, by the way, were still doing your imitation of a septic tank. Well, and why not? It was a very *good* imitation of a septic tank.

"First there were matters to attend to in our own backyard. People of Fear and People of Panic united to face a common enemy. One found near by."

The janitor on Mars fell silent; his head, with its steel arc, was interrogatively poised. Vladimir Voronezh, one of the Russian Laureates (his field was galaxy formation), spoke up, saying,

"My dear sir, I feel you are now going to tell us that life thrived elsewhere in the solar system, once upon a time."

"Certainly. You've got to lose this habit of thinking about the 'miracle' of life, the stupendous 'accident' of intelligence, and so on. I can assure you that in this universe cognition is as cheap as spit. Being a type-v world, Mars was extremely insular in its Total Wealth phase. There was no interest in space exploration, despite adequate technology. But we were perfectly well aware of the coexistence of two type-w worlds: Jupiter and—"

"Jupiter?" This was Lord Kenrick Douglas (quasars): tall, bearded, famous. "Sir, we do know *something* about the solar system. Jupiter is a gas giant. It is wreathed in freezing clouds six hundred miles deep on a shell of liquid hydrogen. Our suicide probes tell us that there are no solid surfaces on this planet. Would you tell us what the Jovians looked like? Jellyfish with powerpacks? Wearing scuba suits, no doubt?"

This last drollery aroused some anxious laughter. The janitor tensed himself to the sound, not with umbrage but with concentration, with efficient curiosity. He said,

"Can I ask *you* a question?" He seemed to be addressing Miss World. "Did they laugh just now because they thought he was funny or because they thought he was full of shit? No. Never mind. Let me tell you, Lord Nobel Laureate, that Jupiter wasn't *always* a gas giant. Originally it was much smaller and denser. Rock mantle on an iron silicate core. But that was before they fucked with Mars.

"The storm system that you call the Great Spot? The Earth-sized zit in its southern tropic? That was ground zero for an NH4 device we sent their way."

"Ammonia?" asked Voronezh, with a glint in his eye.

"Right. It's something we were very proud of, for a while. We turned their place into a colossal stink bomb without altering its mass. To avoid perturbation problems further down the line. Some said at the time that the War with Jupiter might have been bypassed quite easily. Mars overreacted, some said. I mean, a type-w planet, hundreds of millions of years away from posing any plausible threat. Whatever, the War with Jupiter was wrapped up in six months. But then we faced perceived disrespect from another quarter, and turned our attention to—"

"Don't tell me," said Lord Kenrick. "Venus."

"Wrong direction. No, not Venus. Ceres."

The janitor on Mars waited. Fukiyama (superstrings) said dutifully, "Ceres isn't a planet. It's the biggest rock in the asteroid belt."

Calmly inspecting the tips of his talons the janitor on Mars said. "Yeah, right. They wanted to play rough and so..." He shrugged and added, 'It was as our expeditionary force was returning from Jupiter that it picked up the ambiguous transmission from Ceres, another type-w world, though well behind Jupiter. It's possible that in the heat of the moment the Martian commander mistakenly inferred an undertone of sarcasm in the Cerean message of tribute. The War with Ceres, in any case, ended that same afternoon. Then for several weeks, on the home planet, there reigned an uneasy peace. Plans were drawn up for a preemptive strike against Earth. Some Martians sensed aggressive potential there. Because—hey. Action on the blue planet. Photosynthesis. Photochemical dissociation of hydrogen sulphide, no less. Light energy used to break the bonds cleaving oxygen to hydrogen and carbon. Bacteria becomes cyanobacteria. Gangway. Where's the fire? But then something happened that changed all our perspectives. Suddenly we knew that all this was bullshit and the real action lay elsewhere.

"In the year 2,912,456,327 B.C., by your calendar, the Scythers of the Orion Spur sent a warning shot across our bows. They compacted Pluto. Pluto was originally a gas giant the size of Uranus. And the Scythers scrunched it. Without a care for mass-conservation—hence the perturbations you've noticed in Neptune. You thought Pluto was a planet? You thought Pluto was *supposed* to look like that? In the Scythers of the Orion Spur I suppose you could say that Mars had found an appropriate adversary. A type-v world. Same weaponry. Same mental-health problems. Rather superior cosmonautics. The War with the Scythers of the Orion Spur—the combatants being separated by twenty kiloparsecs—was, as you can imagine, a somewhat protracted affair. Door to door, the round trip took 150,000 years: at even half-lightspeed, achievable with our scoop drives, relativistic effects were found to be severe. Still, the great ships went out. Wave after wave. The War with the Scythers of the Orion Spur was hotly prosecuted for just over a billion years. Who won? We did. They're still there, the Scythers. Their planet is still there. The nature of war changed, during that trillennium. It was no longer nuclear or quantum-gravitational. It was neurological. Informational. Life goes on for the Scythers, but its quality has been subtly reduced. We fixed it so that they think they're simulations in a deterministic computer universe. It is believed that this is the maximum suffering you can visit on a type-v world. The taste of victory was sweet. But by then we knew that interplanetary war, even at these distances, was essentially bullshit too. Oh and meanwhile, in that billion-year interlude, all hell has been breaking out on Earth. Oxygen established as an atmospheric gas. Cells with nuclei. All hell is breaking out.

"The Scyther War broadened our horizons. Martian astronomers had become intensely interested in a question that you yourselves are still wrestling with. I mean dark matter. The speed with which galaxies rotate suggests that 98.333 percent of any given galactic mass is invisible and unaccounted for. We went through all the hoops you're going through, and more. What was the dark matter? Massive neutrinos? Failed stars? Slain planets? Black holes? Resonance residue? Plasma fluctuation? Then we kind of flashed it. The answer had been staring us in the face but we had to overcome a mortal reluctance to confront its truth. There was no dark matter. The galaxies had all been *engineered*, brought on line. Including our own. Many, many cycles ago.

"With cosynchronous unanimity it was decided that this subjection was not going to be tolerated. Despite the odds against. It was believed that we were up against a type-n world or entity—maybe even type-m. I now know that we were actually dealing with a type-q world, though one obscurely connected to a power of the type-j order. Apart from the bare fact of their existence, incidentally, nothing is known—in this particle horizon—about worlds -a through -i.

"Our idea was to launch a surprise attack on the galactic core. We figured that our small but measurable chance of success was entirely dependent on surprise—on instantaneity. None of that Scyther shit was going to be any help to us here. There was no question of idling coreward at ninety thousand miles per second—we'd just have to *be there* and hit them with absolutely everything we had. Now. To be clear. In your technological aspirations, on Earth, you are restricted by various inadvertencies like lack of funds but also by your very weak grasp of the laws of physics. We were restricted by the laws of physics. Period. So take a guess. How were we going to do it?"

"Wormholes," said Paolo Sylvino (wormholes).

"Wormholes. Evanescent openings into hyperspace—or, more accurately, into parallel universes with different curvatures or phase trajectories. Ultraspace was the word we preferred. In crude form the idea's been knocking around on Earth since Einstein. Though I venture to suggest that you have a way to go on the how-to end of it. For us of course it was largely a stress-equation problem. You fish a loop out of the quantum foam and then punch a tunnel in spacetime, flexibilizing it with the use of certain uh, exotic materials. We worked on this problem for seven and a half million years.

"Here was the setup. We knew that at the core there lay a black hole of some 1.4237 million solar masses, and we knew it had been ringed and tapped. As you're aware, the energy contained within the black-hole inswirl is stupendous, but it's wholly insufficient to drive a galaxy. The true energy source was something other. And that was the prize we sought. While fitting out the initial strike force we sent recon probes to the galactic core at roughly million-year intervals. Many missions were lost. Those that returned did so with wiped sensors. One way or another, preparations for the strike consumed 437 million years. Then we made our play. On Earth, around now, let it be noted, what do we get but the emergence of organisms visible to the naked eye."

The janitor on Mars sat down and leaned backwards and folded his claws behind his head. Ruminatively he continued, "No one ever thought of this move as a—as a 'mistake' exactly. Everybody was deeply convinced that this was something we absolutely had to do. But the consequences were somewhat extreme. So long in preparation, the Involvement of the Initial Strike Force with the Core Power was over in nine seconds.

"Our fleet was...sent back. Not by wormhole either. The long way around. We knew we'd lost, but we had to wait 300,000 years to find out why. This was an anxious time. We expected intricate reprisals—daily, hourly...

"As military units our ships had been neutralized in the first nanosecond of their appearance at the core, but their sensors were intact and had picked up a great deal of information. Much of it exceedingly depressing, from a Martian point of view. The galactic core had indeed been ringed and tapped. The artificial Loopworld surrounding it had been in place, by our best estimates, for approximately 750 billion years. There was kind of an outpost force guarding the Loopworld. Nothing more. A, uh—a janitor force. Stationed there by entities we would later come to call the Infinity Dogs. Their energy source lay beyond the doorway of the black hole. They were using dead-universe power. Tapping closed universes in which, during contraction, the Higgs field couples to the gravitational shear. Also, we detected beyond Loopworld what I can only describe as a comet depot. Our equipment identified the signature of our own Alpha Comet among the comets parked there.

"Morale was generally low. Almost nihilistic. Martians started to believe, with varying degrees of conviction, that they were mere simulations in a deterministic computer universe. They divided up again. People of Fear. People of Panic. The planet was wracked by spasm wars, random, unending. Certain information began to be made available to us. We learned that the Infinity Dogs had seeded life on Mars—and on Earth, Jupiter and Ceres—for a purpose. We were middens. That's all. Middens."

"Middens, sir?" This was Incarnacion.

"Yeah, middens. Down on Earth, in Africa, the male rhinos all take a dump beyond the waterhole? On Columbus's island of Hispaniola the squinting Carib lines shells on the bank of the riverbed? To demarcate territory? That's a midden. And that's all we were: a message from the Infinity Dogs to a type-r power called the Core Raiders, saying: Keep Out. I have since learned that both Infinity and Core are merely the errand boys of the type-I agency called the Resonance. Which in turn owes tribute to a type-j imperium called the Third Observer. Which..."

Trailing off, the janitor on Mars let his sicklelike head drop to his chest. Then it reared up again, catching the light, and he said, "Everybody knew that the only honorable or even dignified course was planetary self-slaughter. Such in fact is the usual destiny of type-v worlds in this phase. Then bolder voices started to be heard. This had never been a thing about winning or losing. This had always been about the *glorious autonomy of Martian will.* As it turned out, Mars's next battle plan involved kamikaze forces and was itself not easily distinguishable from suicide.

"We came up with a ruse du guerre. We *faked* auto-annihilation and moved our whole operation underground. It had to look good, though: we blew off our atmosphere and paralyzed our core, which also spelt goodbye to our magnetosphere. What you see out there—the red plains and valleys, the rocks and pocks on that carpet of iodized rubble: it's just set-dressing. We went underground, and waited.

"We undertook an arms build-up in a series of five-million-year plans. Morale was high: ringingly idealistic. *Just one shot.* Just one shot: that was the chant we worked to. We were going to turn that wormhole into a gun barrel. And what was the bullet? We started working on a strictly illegal type of weaponry based on the void-creating yield of false vacuum. A bubble of nothingness expanding at the speed of light. The great voids, the great starless deserts that so puzzle you: they're the sites of incautious false-vacuum deployment. Or false-vacuum accident. Hence also the numberless void universes that populate the Ultraverse. If we could detonate this weapon within the event horizon of the core black hole—well, we felt confident of creating quite an impression when the time came for our second rendezvous with Infinity. Such a deed would rearrange the entire Ultraverse. Conceivably to Martian advantage.

"False-vacuum harnessing, we knew, was in itself exquisitely perilous: the field would be appallingly vulnerable to runaway. It was at this time that I was constructed and emplaced, here, in a shell of pure ultrium (an element not to be found in your periodic table), awaiting activation and eventual tripwire. It was as well that I was. For I would remain here alone to ponder the appalling prepotence of the i-power. Forget Infinity and Core. Forget the Resonance and the Third Observer. This came from much higher up.

"The device was ready. All that remained to be done was the addition of the final digit of its algorithm. The planet held its breath. In this instant the war would begin. Preparations that had lasted half a trillennium would now bear fruit...The Martian Slave Rebellion, as I came to call it, was over in a trillionth of the time it takes the speed of light to cross a proton. That was how long it took for all life on this planet to be extinguished. You see, the i-power had *imposed cosmic censorship on matter*. Poised to form the forbidden configuration, matter was instructed to destroy itself. This was 570 million years ago. You'd just gone Cambrian. I settled down for the wait.

"But that's enough about Mars. Let's talk about Earth. Before we do that, though—how about our intermission? There are...facilities in the rear there. No soap. I'm afraid. Or towels. Or hot water. I suggest you fortify yourselves. After the break we'll do tripwire. I'll give you the bad news first. Then I'll give you the bad news."

Pop Jones came out of the rear door, flexed his face in the weak starshine, and skirted the south lawn in his brisk, busy waddle. Keys jounced in the sagging pockets of his black serge suit. It was important, he thought, to walk as quickly as you could...Pop felt deafened, depersonalized. How quiet the place was: no boys on the benches, smoking, grooming, grumbling, coughing, yawning, scratching, gaping. Pop passed through the doors of the Rectory and trotted up the stairs.

He wasn't normally allowed in the Common Room. His public space was the Pantry, a blighted nooklet between the bathhouse and the bikeshed, where he could, if he wished, consume a mug of cocoa among wordless representatives of the catering and gardening staff. Pop Jones knocked on the oak and entered.

The room received him in sudden silence. All you could hear was a stray voice somewhere: the wallscreen TV with somebody saying. *One way out of the faint-young-star paradox lies in radiative transfer calculations, suggesting that the presence of* CO_2 *on early Mars which*...Smells of brewery and ashtray, ginger tea, ginger biscuits, ginger hair, and the dead soldiers of many beer cans. And Mr. Davidge, flanked by Mr. Kidd and Mr. Caroline, turning and saying in his tight Welsh voice:

"What is it, Jones?"

"It's about Timmy, sir. Timmy Jenkins."

He felt the silence rise another notch. Mr. Davidge waited. Then he said, "What about him?"

"He's in San, sir, as you know. And Fitzmaurice says they can't turn the television off, sir. Without disconnecting the whole—"

"So what's your solution, Jones?"

"The directive from the Department Head about the news, sir. I—"

"So what's your solution, Jones?"

"Request permission to move him to the Conservatory, sir."

Mr. Davidge glanced at Mr. Kidd and said, "That's okay by you, isn't it? Yes, Jones, I think we can leave Timmy to your tender mercies."

Everyone was smiling with just their upper lips. For a moment Pop Jones felt with frightening certainty that he was in a room full of strangers. He dropped his head and turned.

Largely disused, the Conservatory led off the south end of the main building, a few meaningless twists and turns from Pop Jones's own quarters. He wheeled Timmy in and established him there, warmly wrapped, on a settee. The child lent his limp cooperation. Pop thought back. Three days ago, when Timmy was found... That bright morning, the air had glittered with such possibility—possibility, coming up out of the lawn. In all the newspapers and on TV they were analyzing the Martian "key" to the aging process: so elegant, so easily grasped. And everyone was laughing and feeling faint...Pop put his hands on his rounded hips and said,

"Dear oh dear, who did this to you, Timmy? It was 'Day,' wasn't it? Dear oh dear, Timmy."

"Floor." said Timmy.

And what becomes of the moral order? he thought, settling back between the jaws of his gray armchair. The screen said: 03.47, 03.46, 03.45.

2

"In the Ultraverse there is an infinite number of universes and an infinite number of planets, and in infinity everything recurs an infinite number of times. That's a mathematical fact. But it hasn't panned out in your case. Among the countless trillions of type-y worlds so far cataloged, none. I can confidently divulge, presents a picture of such agonizing retardation as Mother Earth. To be clear: type-y planets that have been around as long as you have are, without exception, type-x planets or better. Earth has other peculiarities. DNA. I have known you since before you were children. I am the witness of all your excruciations! I have watched you hopping along the savannah and hooting around your campfires. I have watched you daub shit on the walls of your caves. I have watched you stumble, grope, err, miscarry, flop, dither, blunder, goof off. I have watched you trying, straining, heaving. I feel...I sometimes feel that I, too, have become partly human, over these many, many years."

The conference room was now but feebly illumined. You saw the milky outlines of the listeners and the fumes of their milky breath, shapes of heads, Incarnacion with Pickering's hand on her lap. Lord Kenrick flexing his shoulders, Zendovich hunched forward with his chin on his palm, Miss World chewing gum and not blinking. On stage the robot moved among shadows, tracked by the glint of its face. It came forward, and sat. The janitor on Mars had changed clothes. That serge coat had been discarded: in its place, a rust-red smoking jacket of balding velvet. At first you thought it was a trick of the light, but no: there were two black rivets, like eyes, on the curved axe of the face.

"What *was* it with you, O double helix? What kept you back? Most salient, no doubt, was the failure of your science. The utter failure of your science. Your Einsteins and Bohrs, your Hawkings and Kawabatas—they'd have been down on their lousy knees, licking the lab floors on Mars. Only now are you receiving your first whispers from the higher dimensions. On Mars, they *always* thought in ten dimensions. The Infinity Dogs are believed to think in seventeen, the Resonance in thirty-one, the Third Observer in sixty-seven, the higher entities in a number of dimensions both boundless and finite. But you think in four. As do I. They made me like that. I had to be something that you could understand.

"Next: terrestrial religion and its scarcely credible tenacity. Everywhere else they just kick around a few creation myths for a while and then snap out of it when science gets going. But you? One of your writers put it succinctly when he said that there was no evidence for the existence of God other than the human longing that it should be so. An extraordinary notion. What *is* this longing? Everyone else wants 'God' too—but from a different angle. For us, 'God' isn't top-down. He's bottom-up. Why yearn for a power greater than your own? Why not seek to become it? Even the most affable and conciliatory Martian would have found your Promethean urge despicably weak. Okay, on Mars we had to face—and maybe we never truly faced it—our actual position in the order of being. It goes beyond the Third Observer, on and on and up and up. And what do you reach? An entity for whom the Ultraverse is a game of eight ball. And maybe he's just a janitor—the Ultrajanitor. This entity, through his surrogate the Third Observer, created life on Mars. And what am I supposed to do about Him? *Worship* Him? You must be out of your fucking mind. That's your thing. When all is said and done, you *are* very talented adorers.

"Earth would be a curiosity of much interest to cosmoanthropologists if there were any, but the Ultraverse has never concerned itself with information that does no work. In my own musings I adopted the obvious homeostatic view that your science and politics were naturally though brutally depressed in order to foreground your art. Because your art...Art is not taken very seriously elsewhere in this universe or in any other. Nobody's interested in art. They're interested in what everybody else is interested in: the superimposition of will. It may be that nobody's interested in it because nobody's any good at it. 'Painters'—if you can call them that—never get far beyond finger smears and stick figures. And, so far as 'music' is concerned, the Ultraverse in its entirety has failed to advance on a few variations on 'Chopsticks.' Plus the odd battle hymn. Or battle chant. Likewise, 'poets' have managed the occasional wedge of martial doggerel. There are at least a dozen known limericks. And that's about it. I suppose nobody was trying very hard. Why would they? Art and religion are rooted in the hunger for immortality. But nearly everyone already has that. On type-y planets, generally speaking, they soon advance to a future-indefinite worldline. Eighty years, ninety years? What use is that going to be? Oh yeah. The other thing that slowed you down was the unique diffuseness of your emotional range. Tender feelings for each other, and for children and even animals.

"I like art now. It takes a while to get the hang of it. What you've got to do is tell yourself 'This won't actually get me anywhere' and then you don't have a problem. It's strange. Your scientists had no idea what to look for or where to look for it, but your poets, I sometimes felt, divined the universal...Forgive me. My immersion in your story, particularly over these last ten thousand years, while often poisoned by an unavoidable—an obligatory—contempt, has caused me to...Why do I say that: 'Forgive me'?"

And indeed the force field propagating from the janitor on Mars seemed to weaken: the metal he was made of had lost the sheen of the merely metallic. His dropped, prowed head was briefly babyish in its curve.

"Tell me something, O DNA. Human beings, go ahead, disabuse the janitor on Mars. I have this counterintuitive theory. I can tell it's bullshit but I can't get it out of my head. It goes like this...Now I know I'm halfway there on religion. Surely this has to be how it is. It's like a tapestry sopping with blood, right? You had it do it that way: for the art. But tell me. Tell me. Does it go further? Like Guernica happened so Picasso could paint it. No Beethoven without Bonaparte. The First World War was to some extent staged for Wilfred Owen, among others. The events in Germany and Poland in the early 1940s were set in motion for Primo Levi and Paul Celan. Etcetera. But I'm already getting the feeling it isn't like that. It isn't like that, is it, Miss World?"

"No, sir," said Miss World. "It isn't like that."

"I didn't really think so. Well in a way," said the janitor on Mars interestedly, "this makes my last job easier. I'm glad we met. You know, it took me the longest time to get the hang of the way you people do things. As, technically, a survivor on a chastened type-v world, I had automatic access to certain information sources. Like I was on a mailing list. From my studies I came to think of other worlds as always swift and supple—as always *responsive*, above all, in their drive toward complexity. But not you. You always had to do it at your own speed. A torment to watch, but that was your way. And whenever I tried to liven things up it was usually a total dud."

"Sir? Excuse me?" This was Incarnacion Buttruguena-Hume. "Are you saying you influenced events on Earth?"

"Yeah and I'll give you an example. Yeah, I used to try and soup things up every now and then. For example, take this gentleman Aristarchus. Almost exactly twenty-three centuries ago there's this Greek gentleman working on the brightness fluctuations of the planets. I put it to him that—"

"You put it to him?"

"Yes. On the neural radio. When your scientists talk about their great moments of revelation—a feeling of pleasant vacuity followed by a ream of math—they're usually describing a telepathic assist from Mars. This Aristarchus happens on a completely coherent heliocentric system. He spreads the word around the land. And what happens? Ptolemy. Christianity. You weren't *ready*. So we all had to sit and wait two thousand years for Copernicus. Stuff like that happened all the time."

Murmurs died in the dark chill. Pioline (solar neutrino count) gave an emphatic and breathy moan which had in it elements of anger but far more predominantly elements of grief. As the silence settled the janitor on Mars gave a light jolt of puzzlement and said, "You're uncomfortable with that? Come on. That's the least of it. Welcome to middenworld."

"But some things took?" said Lord Kenrick. "You shaped us? Is that what you're saying?"

"...Yeah I fucked with you some. Sure. Hey. I was programmed to do that. I had—guidelines. Some things worked out. Others didn't. Slavery was *all* me, for instance. Yes, slavery was my baby. *That* worked out. All worlds dabble with it, early on. It's good practice for later. Because slavery's what the Ultraverse is all about. Okay, on Earth, you could argue that it got out of hand. But on a nonculling planet it seemed like a necessary development. Even in its decadent phase slavery had many distinguished though often irresolute advocates. Locke, Burke, Hume, Montesquieu, Hegel, Jefferson. And there's an influential justification for it in the holy book of one of your Bronze Age nomad tribes."

"Which?"

"The Bible. Any last questions?"

"Just what the hell is this tripwire thing?"

"Again, part of the program. Contact with Earth could not be established until you went and tripped that wire. Which you did on June nine: the day I buzzed Incarnacion here."

"What was it about June nine?" asked Montgomery Gruber (geophysiology). "We looked into it and nothing happened."

"You mean you looked into it and you *think* nothing happened. Plenty happened. Some asshole of an otter or a beaver sealed off a minor tributary of the River Lee in Washington State...along certain latitudes a critical fraction of microbal life committed itself to significant changes in its respiratory metabolism...the forty-seven billionth self-cooling cola can burped out its hydrocarbons...and there was that mild forest fire in Albania. And there you have it. You wouldn't know how these things are connected, but connected they are. All this against a background of mobilized phosphorus, carbon burial, and hydrogen escape. The necessary synergies are all locked in."

"Meaning?"

"Meaning the amount of oxygen in your atmosphere is starting to climb. At last irreversibly. It won't feel any different for a while. But by the end of the Sixties it'll hit twenty-seven percent. Yes I know: a pity about that."

Incarnacion and Miss World turned to each other sharply. Because the scientists were now shouting out, gesturing, interjecting. Miss World said, "Please, sir. I don't understand."

"Well. It means you'll have to be very, very careful with your heat sources, Miss World. At such a concentration, to light a cigarette and throw a match over your shoulder would spark a holocaust. It's all a great shame, because this is the kind of problem that's easy to fix if you catch it early on. In the coming years you'll have to work awful hard on volcano-capping and storm control. To no avail, alas. Here's another thing. It seems, anyway, that the solar system is shutting down. There's a planetesimal out there with your name written on it. An asteroid the size of Greenland is due to ground zero on the Iberian peninsula in the unseasonably torrid summer of 2069. At ninety miles a second. Now. There might have been a window of a couple of days or so at the beginning of the decade: you could have duplicated your feat of 2037 when you saw off Spielberg-Robb. But the thing is you'll need your nuclear weapons this time. A mass-driver won't do the trick, not with the English this asteroid's got on it. Unfortunately, though, there's now a tritium hitch with your nukes that you'd have needed to start work on much earlier to have any hope of rearming them in time. Obviously a body this size moving at sixteen times the speed of sound will have considerable kinetic energy: to be released as heat. And it'll rip through the mantle and the crust, disgorging trillions of tons of magma. It's all very unfortunate. Mars itself may be lightly damaged in the blast."

Zendovich said, "That was the tripwire? You're saying you couldn't act until it was already too late to make any difference?"

"Affirmative. That was the lock."

"Sir?" asked Miss World. "I'm sorry, sir, but there's something I have to say. I think you're a despicable person."

"Nugatory. I'm not a person, lady. I'm a machine obeying a program."

Zendovich got to his feet. So did the janitor on Mars, who leaned forward and cocked his beak at him.

"Then God curse whoever put you together."

"Oh come on. What did you expect? This is *Mars*, pal," said the janitor as the lights began to fade. "The Red One. You hear that? Nergal: Star of Death. Now get the hell out. Yeah. Go. Walk out of here with your eyes on the fucking floor. Exit through the left hall. Follow the goddamned signs."

Pop Jones slipped into the conservatory and opened the back door. Dusk was coming. Across the lawn were the lit windows of the Common Room (he could see Kidd and Davidge, staring out). The children wouldn't return from the beach for another hour. Later, after they'd been fed, Pop Jones would make his rounds with his bucket and his keys. Make his rounds? Pop shrugged, then nodded. Yes, it would be important to try to go on just as before. But could you do that?

The star was dropping over the steep green. Starset! Stardown! And already a generous, a forgiving moon; it carved a penumbra of golden grime in the cirrus, and the face saying, I'm sorry. I'm sorry. I'm sorry.

Pop Jones turned.

"Floor."

"Timmy?"

He could see the moisture in the child's eyes.

"Timmy. Timmy. Who did this to you, Timmy?"

At one remove, it seemed, Pop Jones felt astonishment gathering in him. How entirely different his own voice sounded: thick, mechanical. In this new time, when he, in common with everyone else on Earth, was submitting to an obscure and yet disgustingly luminous reaffiliation, Pop Jones found that thing in himself that had never been there before: the necessary species of self-love.

"Day," said Timmy clearly. And he said it again, quite clearly, like an English teacher. "Day...Day done it."

Darkness increased its hold on the room of glass. Pop Jones's new voice said that night was now coming. He moved toward the boy. Hush there. Hush.

PIECES OF MARS HAVE FALLEN TO EARTH (2015)

Cherise Saywell

Cherise Saywell was born and brought up in Australia but has lived in Scotland since 1996. She studied English and cultural studies at the University of Queensland, Brisbane, and then spent eight years working as an academic researcher in Australia and Scotland. After this, she turned to fiction writing and published her first novel, *Desert Fish*, with the help of a Scottish Arts Council bursary in 2011. 'Pieces of Mars Have Fallen to Earth' was selected for BBC Radio 4's *Opening Lines* in 2015. Cherise won the V. S. Pritchett Short Story Prize in 2003, and was a runner-up in the Asham Award in 2009 and the Salt Prize in 2012.

I hate the storms, and the waiting too – the long clear nights when stars and planets puncture the darkness. I know each from the other now, but not how to navigate by them.

During the storms I stay indoors. I avoid superstitious rituals, crossing the fingers and such, and I avoid looking at the sky. What I like to do is google names, famous ones to get me going. Explorers mostly: Arctic and Antarctic, the Antipodes, the Americas – also mariners and merchants. Then if there's a link I click through to the lesser-known names, like Susanna Fontanerossa, Alda de Mesquita, Grace Pace. I used to like a whisky and Coke while doing this, to take the edge off things. Hannah Weekes. Henrietta Gavan. Charlotte Strahlow.

Michael is more pragmatic. He watches the Mars Channel. He'll be logged on to the space station too, for updates.

He was here by late morning and although he still has a key, he knocked. Usually he wears a button-down shirt but today he's chosen something collarless in pale blue – for Joe, I suppose; he always loved the colour blue.

He's brought tea, a malty blend from India. Assam, he announces, with a dash of something to lift it – a Darjeeling. He holds the packet up, with its clear cellulose window, so I can see the knot of curled leaves.

It's not time to turn on yet, so he fills the kettle and goes up to Joe's room, where I have hung a birthday banner.

'Why is the monitor up here?' he calls. He walks out onto the landing.

'Because I put it there,' I say.

'But it's not connected. How will we watch?'

I shrug. Even today I don't want to watch.

Michael sighs. 'I'll bring it down and connect it up it,' he says.

That's just like him: studiously deciding not to see. It means he can get right up close to the things he ought to be afraid of.

He's a fount of knowledge about Mars. At any opportunity he'll talk about it: the axial tilt and the seasons – spring, summer, autumn, winter – like ours. And how the days are only slightly longer there, by 37 minutes and 22 seconds. That's almost nothing, he says. But you can make up any story you like using numbers. You just shift new words around them and tell it a different way.

He carries the monitor down to the livingroom and positions it on the stand. I should make the tea while he does this but instead I hang about watching him.

When Joe was little you just plugged in the power and pressed the 'on' switch. Now it's all cables and wires and connections. Michael locates the scart leads beneath the sofa. He checks them – yellow to yellow, so the monitor will play from the drive. For stereo sound he needs red and white, but there's no white. He'll be thinking I've hidden it.

I remember how when Joe was little he liked to be the one to switch on. At age four or so he was obsessed with a cartoon series. Michael called it his Messiah show. Joe would wake early and I'd come downstairs with him. I'd make his breakfast and fix myself a coffee. The programme was about a boy who could harness the power of the elements to fight for the power of good. Joe used to copy his hero's gestures before he turned on the television: hands together, then an arc with his arms and leaning forward on one leg, aiming the remote control and, hey presto, there was the picture.

We'd sit close together on the sofa. Sometimes Joe would put a hand on my knee, an arm around my waist. During the ad breaks I'd make extra toast. Although I was tired – he still woke often in the night then – I looked forward to our ritual and I wasn't bothered that Michael didn't get up to help me. I was bereft when the series finished – something significant had come to an end – but I thought, *I will always have the memory of this time.*

I often think of this when I'm clicking on names. Hanna Sahlqvist. Sarah Nee Newman. Were their memories like mine?

'White,' Michael mutters, lifting the tangle of black leads and checking their ends. Then, 'Ahhh,' he says, lifting one free. 'Now we will have full sound.'

It's the sound I find most disturbing.

'Will you make the tea?' he asks. I'd prefer something stronger. I'd like a whisky and Coke. But it's no longer good for my health, Michael says. And although I wonder what my health is good for, I won't argue with him.

In the kitchen I spoon leaves into a pot and pour on boiling water. We like our tea the old-fashioned way.

*

Although Michael likes to talk as if Mars is just like Earth, even I'm well aware the tilt is more extreme. So the seasons are nearly twice as long. The year is almost twice as long as well. And because the days are longer, they don't correspond to ours. Mars has an atmosphere, Michael will say. But it only supports weather, not people.

'So why go there?' I said to him once, and he looked at me strangely. Then he told me that pieces of Mars had fallen to Earth, meteorites, carrying traces of Martian atmosphere. They've taken millions of years to reach us, he said. And those traces inside are like questions. Can't you see? We have to go there to find the answers.

I laughed. He was always good at making stupid things sound rational. I can tell you now that if Mars has anything, it has weather: dust storms that cover swathes of its surface for days, weeks, even months. The views from Cameras 5 to 11 become obscured until the lenses are completely coated. After that, the whole channel goes down and you see nothing until the storm is over.

You wonder whether the astronauts got back to their pods. What if the dust got into their helmets? Their eyes and ears? Their oxygen?

The first time there was a storm on MarsCam the whole world watched. It was all over the internet. People tuned in to the Space Channel until the cameras went down; after that, monitors everywhere were switched on and off again at regular intervals until at last the storm was over. Then, if you'd tuned in, you saw in the murky blankness, the ends of a brush: the lens being swept – and clarity, and the face of a man in a space helmet. Mars was back online.

The 'Mars Four': like superstars. Their images were everywhere: magazines and blogs, clips all over the news channels, on YouTube. Famous.

They're still famous, I guess, but you can't really be a superstar if you're a one-way traveller. Not in the regular way. Those astronauts are more like priests, or monks. Pioneers.

Whenever there is a storm now I turn off, unable to watch or listen. Unable to sleep.

When it ends, Michael comes over and logs on so that I will see that everything is as it should be. So I can hear that sound I can't bear: the triumph of life in a hostile world, the slow rasp of breath being fed from a tank.

*

After the leaves have steeped Michael fills two cups. I never tell him about my research. He wouldn't understand. When he isn't talking about Mars or Darjeeling tea we struggle to find things to say.

By the time we separated Joe was long gone. We were generous when dividing our possessions. Michael returned often to borrow things – a fish kettle, a pasta maker, software. We copied all our pictures of our son as if by doing this we could share the same memories.

By then there was no point in telling Joe. We wouldn't see him again. But I was already schooled in the art of goodbyes. He will go away from you, I'd been told. Children do. At 13 he was embarrassed if I saw him on the street with his friends. He scoffed at me if I revealed my ignorance about anything. He closed the door of his room at all times.

I watched him one night, when I'd asked him to return a key to my friend. 'She needs it,' I told him, 'to lock her door tonight. Will you run over the road with it?'

'Alright,' he said. He swiped the key from me.

'Do you need a jacket?' I asked.

'No,' he said. 'I'm only going over the street.'

I stood at the window. He'd have seen me if he'd looked up. It was cold outside. The ground shone with the promise of frost. Joe crossed between the cars parked below. His steps were long and loping. He ran the way men do – the long strides, the slow deliberate swing of the arms, the balance not striven for, but intrinsic.

He has already left me, I thought.

Only then, under the goldish glow of the street lamp, he leapt into the air. He put his arm up, his fingertips extended, as if he wanted to touch the light.

This was a boy who had not yet left childish things behind. His hooded top was unzipped. His thin T-shirt flapped in the crisp night air and I worried that he might be cold. I wanted him to turn and see me worrying: I put my hand to the window as if it might catch his attention. But he didn't notice.

He took two more steps, until he was out of the spotlight, and then jumped again, this time reaching into darkness. I thought, *Soon he will be grown, but for now, in this moment, he can still be mine*. My ribs felt tight. My throat seemed to close. I blinked and then he disappeared into the stairwell of my neighbour's building.

Now I can only wait. I occupy myself with my list, filling it out with detail. I join up the names of the women with those of their sons, and I compare dates – births and deaths – seeking what? Alda de Mesquita and Ferdinand Magellan. Susanna Fontanerossa and Christopher Columbus. Where are the women who said goodbye?

Hannah Scott. Now, I like her. She didn't allow her boy to go. But then she died and he went anyway.

*

We toast our son. We drink our tea. We count the years.

Today another storm on Mars has passed. There will be edited highlights and an astronaut, having survived the weather again, will clean the lenses of Cameras 5 to 11. Maybe it will be Joe.

It took my son eight months to reach Mars. Sometimes Michael tries to explain how he got there, the speed at which the capsule moved through space. Joe will have looked out at the solar system and the glow of the stars and the planets: like a king, a messiah, he has said, as if he could seduce me with the beauty of an idea I've never had. But I can only imagine a Purgatory where my son is now, a halfway world. I can't speak to him there; I can't reach him.

Before he left, when he was in Siberia, and then the American desert, and then Chile – I remember the echo of my voice on my mobile phone, the second or so delay from a landline. From Mars, my husband has explained, there is a 20-minute delay – that's how far away it is. You can't hold a live conversation. Distance makes it impossible. You have to let go.

So you see the storms, though frightening, are not the worst thing. And the waiting is only time, after all.

*

Tonight will be clear. Stars will prickle in a matte black sky. I know where to find the red planet now, whatever the time, whatever the season.

I remember my son under the streetlamp that night: leaping, hand extended, into the circle of light. On Mars, the gravitational pull is less, so he'll be able to jump higher, while I click on names and navigate links. Annie Ridley. Sarah Morgan. Susannah Ward. Sarah Bate. On and on I'll go, while pieces of Mars, millions of years old, containing tiny traces of Martian air, orbit the earth.

LIFE ON MARS

NICOLA TESLA PROMISES COMMUNICATION WITH MARS (1901)

Nikola Tesla

Nikola Tesla was born in 1856 in Smiljan, Croatia, when it was part of the Austrian Empire. He studied engineering and physics in the 1870s but never completed a degree. Instead, he emigrated to the United States and set up a research company in New York to develop electrical and mechanical devices. He courted the attention of wealthy celebrities and other potential patrons and made a name for himself on the public lecture circuit. He particularly championed the idea of wireless communications, both around the world and between Earth and Mars. This report was published in *The Times* (Richmond, Virginia) on 13 January 1901.

There are thousands of people living in the world today who do not believe that the planet of Mars is inhabited.

There are many others who do and some of the leaders in science and foremost men in thought and invention are members of this last-named class.

Nicola [sic] Tesla, the inventor of the wireless telegraphy, is one of these. Astronomers tell us that the planet Mars is several millions of years older than the earth and H. G. Wells, novelist, in one of his fantastic creations, has peopled this planet with a race of strange creatures. One thing, however, stands to reason, and that is this: If Mars is inhabited those inhabitants are far in advance of us as regards sciences, both theoretical and applied. This is what Tesla thinks and why he is of the opinion has just recently been made known.

He is convinced that the Martians are trying to communicate with us.

Tesla's Dream

Says Julian Hawthorne:

"Apart from love and religion there happened the other day to Mr. Tesla the most momentous experience that has ever visited a human being on this earth. As he sat beside his instrument on the hillside in Colorado, in the deep silence of that austere, inspiring region, where you plant your feet in gold and your head brushes the constellations – as he sat there one evening, alone, his attention, exquisitely alive at that juncture, was arrested by a faint sound from the receiver – three fairy taps, one after the other, at a fixed interval. What man who has ever lived on this earth would not envy Tesla that moment. Never before since the globe first swung into form had that sound been heard. Those three soft impulses, reflected from the sensitive disc of the receiver, had not proceeded from any earthly source. The force which propelled them, the measure which regarded them, the significance they were meant to convey, had their origin in no mind native to this planet.

"They were sent, those marvelous signals, by a human being living and thinking so far away from us, both in space and in condition, that we can only accept him as a fact, not comprehend him as a phenomenon. Traveling with the speed of light, they must have been dispatched but a few moments before Tesla, in Colorado, received them. But they came from some Tesla on the planet Mars!

"This was two years ago; it has just been made public. Thereupon all the tame beasts with long ears in the stables of science began waving those ears vigorously and braying forth indignant scoffs and denials. Yes, so has it ever been, and will be. How eagerly will every so-called son of science who has the power of absorbing, but not of assimilating or of creating, rush to trample under his hoofs the man of genius, imagination and wisdom who commits the crime of disclosing to them the means of their own uplifting and humanization: 'Charlatan–fraud–fool!' they cry: and fetch out their musty little books of statistics and logarithms to show why it cannot be anything else but a humbug and delusion. Well, let us leave them trampling and braying, and consider for a moment what has occurred."

The Real Question

The real question, however, is how will these interplanetary communications be conducted: what the medium to be employed: with such interplanetary communication as is proposed, electricity will doubtless be revealed as but the fractional aspect of a force possessing a vastly greater scope and power than have any of the phenomena of our experiments yet revealed to us. The energy of man's brain, if properly applied, may suffice to propagate waves of meaning from one end of the universe to the other, and science will unquestionably aid.

Nicola Tesla promises us communication with our terrestrial neighbors. How, when and where? remain to be seen.

IS MARS HABITABLE? (1907)

Alfred Russel Wallace

The British naturalist **Alfred Russel Wallace** is best known today for independently formulating the theory of evolution. He was born in 1823 in Llanbadoc, Wales, but his family moved to Hertfordshire when he was five. He attended the Hertford Grammar School (now Richard Hale school) and then attended lectures at the London Mechanics Institute (now Birkbeck, University of London). As well as his work on evolution, Wallace was a founder of the modern science of astrobiology. In 1904 he published *Man's Place in the Universe*, in which he tried to evaluate the possibilities for life on other planets. Three years later, he expanded his analysis in *Is Mars Habitable?* He died in 1913 in Broadstone, England.

Chapter II
Mr. Percival Lowell's Discoveries and Theories

The Observatory in Arizona.

In 1894, after a careful search for the best atmospheric conditions, Mr. Lowell established his observatory near the town of Flagstaff in Arizona, in a very dry and uniform climate, and at an elevation of 7300 feet above the sea. He then possessed a fine equatorial telescope of 18 inches aperture and 26 feet focal length, besides two smaller ones, all of the best quality. To these he added in 1896 a telescope with 24 inch object glass, the last work of the celebrated firm of Alvan Clark & Sons, with which he has made his later discoveries. He thus became perhaps more favourably situated than any astronomer in the northern hemisphere, and during the last twelve years has made a specialty of the study of Mars, besides doing much valuable astronomical work on other planets.

Mr. Lowell's recent Books upon Mars.

In 1905 Mr. Lowell published an illustrated volume giving a full account of his observations of Mars from 1894 to 1903, chiefly for the use of astronomers; and he has now given us a popular volume summarising the whole of his work on the planet, and published both in America and England by the Macmillan Company. This very interesting volume is fully illustrated with twenty plates, four of them coloured, and more than forty figures in the text, showing the great variety of details from which the larger general maps have been constructed.

Non-natural Features of Mars.
But what renders this work especially interesting to all intelligent readers is, that the author has here, for the first time, fully set forth his views both as to the habitability of Mars and as to its being actually inhabited by beings comparable with ourselves in intellect. The larger part of the work is in fact devoted to a detailed description of what he terms the 'Non-natural Features' of the planet's surface, including especially a full account of the 'Canals,' single and double; the 'Oases,' as he terms the dark spots at their intersections; and the varying visibility of both, depending partly on the Martian seasons; while the five concluding chapters deal with the possibility of animal life and the evidence in favour of it. He also upholds the theory of the canals having been constructed for the purpose of 'husbanding' the scanty water-supply that exists; and throughout the whole of this argument he clearly shows that he considers the evidence to be satisfactory, and that the only intelligible explanation of the whole of the phenomena he so clearly sets forth is, that the inhabitants of Mars have carried out on their small and naturally inhospitable planet a vast system of irrigation-works, far greater both in its extent, in its utility, and its effect upon their world as a habitation for civilised beings, than anything we have yet done upon our earth, where our destructive agencies are perhaps more prominent than those of an improving and recuperative character.

A Challenge to the Thinking World.
This volume is therefore in the nature of a challenge, not so much to astronomers as to the educated world at large, to investigate the evidence for so portentous a conclusion. To do this requires only a general acquaintance with modern science, more especially with mechanics and physics, while the main contention (with which I shall chiefly deal) that the features termed 'canals' are really works of art and necessitate the presence of intelligent organic beings, requires only care and judgment in drawing conclusions from admitted facts. As I have already paid some attention to this problem and have expressed the opinion that Mars is not habitable,* judging from the evidence then available, and as few men of science have the leisure required for a careful examination of so speculative a subject, I propose here to point out what the facts, as stated by Mr. Lowell himself, do not render even probable much less prove. Incidentally, I may be able to adduce evidence of a more or less weighty character, which seems to negative the possibility of any high form of animal life on Mars, and, a fortiori, the development of such life as might culminate in a being equal or superior to ourselves. As most popular works on Astronomy for the last ten years at least, as well as many scientific periodicals and popular magazines, have reproduced some of the maps of Mars by Schiaparelli, Lowell, and others, the general appearance of its surface will be familiar to most readers, who will thus be fully able to appreciate Mr. Lowell's account of his own further discoveries which I may have to quote. One of the best of these maps I am able to give as a frontispiece to this volume, and to this I shall mainly refer.

* *Man's Place in the Universe*, p. 267 (1903).

The Canals as described by Mr. Lowell.

In the clear atmosphere of Arizona, Mr. Lowell has been able on various favourable occasions to detect a network of straight lines, meeting or crossing each other at various angles, and often extending to a thousand or even over two thousand miles in length. They are seen to cross both the light and the dark regions of the planet's surface, often extending up to or starting from the polar snow-caps. Most of these lines are so fine as only to be visible on special occasions of atmospheric clearness and steadiness, which hardly ever occur at lowland stations, even with the best instruments, and almost all are seen to be as perfectly straight as if drawn with a ruler.

The Double Canals.

Under exceptionally favourable conditions, many of the lines that have been already seen single appear double–a pair of equally fine lines exactly parallel throughout their whole length, and appearing, as Mr. Lowell says, "clear cut upon the disc, its twin lines like the rails of a railway track." Both Schiaparelli and Lowell were at first so surprised at this phenomenon that they thought it must be an optical illusion, and it was only after many observations in different years, and by the application of every conceivable test, that they both became convinced that they witnessed a real feature of the planet's surface. Mr. Lowell says he has now seen them hundreds of times, and that his first view of one was 'the most startlingly impressive' sight he has ever witnessed.

Dimensions of the Canals.

A few dimensions of these strange objects must be given in order that readers may appreciate their full strangeness and inexplicability. Out of more than four hundred canals seen and recorded by Mr. Lowell, fifty-one, or about one eighth, are either constantly or occasionally seen to be double, the appearance of duplicity being more or less periodical. Of 'canals' generally, Mr. Lowell states that they vary in length from a few hundred to a few thousand miles long, one of the largest being the Phison, which he terms 'a typical double canal,' and which is said to be 2250 miles long, while the distance between its two constituents is about 130 miles. The actual width of each canal is from a minimum of about a mile up to several miles, in one case over twenty. A great feature of the doubles is, that they are strictly parallel throughout their whole course, and that in almost all cases they are so truly straight as to form parts of a great circle of the planet's sphere. A few however follow a gradual but very distinct curve, and such of these as are double present the same strict parallelism as those which are straight.

Canals extend across the Seas.

It was only after seventeen years of observation of the canals that it was found that they extended also into and across the dark spots and surfaces which by the earlier observers were termed seas, and which then formed the only clearly distinguishable and permanent marks on the planet's surface. At the present time, Professor Lowell states that this "curious triangulation has been traced over almost every portion of the planet's surface, whether dark or light, whether greenish, ochre, or brown in colour." In some parts they are much closer together than in others, "forming a perfect network of lines and spots, so that to identify them all was a matter of extreme difficulty." Two such portions are figured at pages 247 and 256 of Mr. Lowell's volume.

The Oases.

The curious circular black spots which are seen at the intersections of many of the canals, and which in some parts of the surface are very numerous, are said to be more difficult of detection than even the lines, being often blurred or rendered completely invisible by slight irregularities in our own atmosphere, while the canals themselves continue visible. About 180 of these have now been found, and the more prominent of them are estimated to vary from 75 to 100 miles in diameter. There are however many much smaller, down to minute and barely visible black points. Yet they all seem a little larger than the canals which enter them. Where the canals are double, the spots (or 'oases' as Mr. Lowell terms them) lie between the two parallel canals.

No one can read this book without admiration for the extreme perseverance in long continued and successful observation, the results of which are here recorded; and I myself accept unreservedly the substantial accuracy of the whole series. It must however always be remembered that the growth of knowledge of the detailed markings has been very gradual, and that much of it has only been seen under very rare and exceptional conditions. It is therefore quite possible that, if at some future time a further considerable advance in instrumental power should be made, or a still more favourable locality be found, the new discoveries might so modify present appearances as to render a satisfactory explanation of them more easy than it is at present.

But though I wish to do the fullest justice to Mr. Lowell's technical skill and long years of persevering work, which have brought to light the most complex and remarkable appearances that any of the heavenly bodies present to us, I am obliged absolutely to part company with him as regards the startling theory of artificial production which he thinks alone adequate to explain them. So much is this the case, that the very phenomena, which to him seem to demonstrate the intervention of intelligent beings working for the improvement of their own environment, are those which seem to me to bear the unmistakeable impress of being due to natural forces, while they are wholly unintelligible as being useful works of art. I refer of course to the great system of what are termed 'canals,' whether single or double. Of these I shall give my own interpretation later on.

Chapter III
The Climate and Physiography of Mars

Mr. Lowell admits, and indeed urges strongly, that there are no permanent bodies of water on Mars; that the dark spaces and spots, thought by the early observers to be seas, are certainly not so now, though they may have been at an earlier period; that true clouds are rare, even if they exist, the appearances that have been taken for them being either dust-storms or a surface haze; that there is consequently no rain, and that large portions (about two-thirds) of the planet's surface have all the characteristics of desert regions.

Snow-caps the only Source of Water.

This state of things is supposed to be ameliorated by the fact of the polar snows, which in the winter cover the arctic and about half the temperate regions of each hemisphere alternately. The maximum of the northern snow-caps is reached at a period of the Martian winter corresponding to the end of February with us. About the end of March the cap begins to shrink in size (in the Northern Hemisphere), and this goes on so rapidly that early in the June of Mars it is reduced to its minimum. About the same time changes of colour take place in the adjacent darker portions of the surface, which become at first bluish, and later a decided blue-green; but by far the larger portion, including almost all the equatorial regions of the planet, remain always of a reddish-ochre tint.*

The rapid and comparatively early disappearance of the white covering is, very reasonably, supposed to prove that it is of small thickness, corresponding perhaps to about a foot or two of snow in north-temperate America and Europe, and that by the increasing amount of sun-heat it is converted, partly into liquid and partly into vapour. Coincident with this disappearance and as a presumed result of the water (or other liquid) producing inundations, the bluish-green tinge which appears on the previously dark portion of the surface is supposed to be due to a rapid growth of vegetation.

But the evidence on this point does not seem to be clear or harmonious, for in the four coloured plates showing the planet's surface at successive Martian dates from December 30th to February 21st, not only is a considerable extent of the south temperate zone shown to change rapidly from bluish-green to chocolate-brown and then again to bluish-green, but the portions furthest from the supposed fertilising overflow are permanently green, as are also considerable portions in the opposite or northern hemisphere, which one would think would then be completely dried up.

No Hills upon Mars.

The special point to which I here wish to call attention is this. Mr. Lowell's main contention is, that the surface of Mars is wonderfully smooth and level. Not only are there no mountains, but there are no hills or valleys or plateaux. This assumption is absolutely essential to support the other great assumption, that the

* In 1890 at Mount Wilson, California, Mr. W. H. Pickering's photographs of Mars on April 9th showed the southern polar cap of moderate dimensions, but with a large dim adjacent area. Twenty-four hours later a corresponding plate showed this same area brilliantly white; the result apparently of a great Martian snowfall. In 1882 the same observer witnessed the steady disappearance of 1,600,000 square miles of the southern snow-cap, an area nearly one-third of that hemisphere of the planet.

wonderful network of perfectly straight lines over nearly the whole surface of the planet are irrigation canals. It is not alleged that irregularities or undulations of a few hundreds or even one or two thousands of feet could possibly be detected, while certainly all we know of planetary formation or structure point strongly towards *some* inequalities of surface. Mr. Lowell admits that the dark portions of the surface, when examined on the terminator (the margin of the illuminated portion), do *look* like hollows and *may be* the beds of dried-up seas; yet the supposed canals run across these old sea-beds in perfect straight lines just as they do across the many thousand miles of what are admitted to be deserts–which he describes in these forcible terms: "Pitiless as our deserts are, they are but faint forecasts of the state of things existent on Mars at the present time."

It appears, then, that Mr. Lowell has to face this dilemma–*Only if the whole surface of Mars is an almost perfect level could the enormous network of straight canals, each from hundreds to thousands of miles long, have been possibly constructed by intelligent beings for purposes of irrigation; but, if a complete and universal level surface exists no such system would be necessary.* For on a level surface–or on a surface slightly inclined from the poles towards the equator, which would be advantageous in either case–the melting water would of itself spread over the ground and naturally irrigate as much of the surface as it was possible for it to reach. If the surface were not level, but consisted of slight elevations and depressions to the extent of a few scores or a few hundreds of feet, then there would be no possible advantage in cutting straight troughs through these elevations in various directions with water flowing at the bottom of them. In neither case, and in hardly any conceivable case, could these perfectly straight canals, cutting across each other in every direction and at very varying angles, be of any use, or be the work of an intelligent race, if any such race could possibly have been developed under the adverse conditions which exist in Mars.

The Scanty Water-supply.

But further, if there were any superfluity of water derived from the melting snow beyond what was sufficient to moisten the hollows indicated by the darker portions of the surface, which at the time the water reaches them acquire a green tint (a superfluity under the circumstances highly improbable), that superfluity could be best utilised by widening, however little, the borders to which natural overflow had carried it. Any attempt to make that scanty surplus, by means of overflowing canals, travel across the equator into the opposite hemisphere, through such a terrible desert region and exposed to such a cloudless sky as Mr. Lowell describes, would be the work of a body of madmen rather than of intelligent beings. It may be safely asserted that not one drop of water would escape evaporation or insoak at even a hundred miles from its source.*

* What the evaporation is likely to be in Mars may be estimated by the fact, stated by Professor J. W. Gregory in his recent volume on 'Australia' in *Stanford's Compendium*, that in North-West Victoria evaporation is at the rate of ten feet per annum, while in Central Australia it is very much more. The greatly diminished atmospheric pressure in Mars will probably more than balance the loss of sun-heat in producing rapid evaporation.

Miss Clerke on the Scanty Water-supply.

On this point I am supported by no less an authority than the historian of modern astronomy, the late Miss Agnes Clerke. In the *Edinburgh Review* (of October 1896) there is an article entitled 'New Views about Mars,' exhibiting the writer's characteristic fulness of knowledge and charm of style. Speaking of Mr. Lowell's idea of the 'canals' carrying the surplus water across the equator, far into the opposite hemisphere, for purposes of irrigation there (which we see he again states in the present volume), Miss Clerke writes: "We can hardly imagine so shrewd a people as the irrigators of Thule and Hellas* wasting labour, and the life-giving fluid, after so unprofitable a fashion. There is every reason to believe that the Martian snow-caps are quite flimsy structures. Their material might be called snow *soufflé*, since, owing to the small power of gravity on Mars, snow is almost three times lighter there than here. Consequently, its own weight can have very little effect in rendering it compact. Nor, indeed, is there time for much settling down. The calotte does not form until several months after the winter solstice, and it begins to melt, as a rule, shortly after the vernal equinox. (The interval between these two epochs in the southern hemisphere of Mars is 176 days.) The snow lies on the ground, at the outside, a couple of months. At times it melts while it is still fresh fallen. Thus, at the opposition of 1881–82 the spreading of the northern snows was delayed until seven weeks after the equinox: and they had, accordingly, no sooner reached their maximum than they began to decline. And Professor Pickering's photographs of April 9th and 10th, 1890, proved that the southern calotte may assume its definitive proportions in a single night.

"No attempt has yet been made to estimate the quantity of water derivable from the melting of one of these formations; yet the experiment is worth trying as a help towards defining ideas. Let us grant that the average depth of snow in them, of the delicate Martian kind, is twenty feet, equivalent at the most to one foot of water. The maximum area covered, of 2,400,000 square miles, is nearly equal to that of the United States, while the whole globe of Mars measures 55,500,000 square miles, of which one-third, on the present hypothesis, is under cultivation, and in need of water. Nearly the whole of the dark areas, as we know, are situated in the southern hemisphere, of which they extend over, at the very least, 17,000,000 square miles; that is to say, they cover an area, in round numbers, seven times that of the snow-cap. Only one-seventh of a foot of water, accordingly, could possibly be made available for their fertilisation, supposing them to get the entire advantage of the spring freshet. Upon a stint of less than two inches of water these fertile lands are expected to flourish and bear abundant crops; and since they completely enclose the polar area they are necessarily served first. The great emissaries for carrying off the surplus of their aqueous riches, would then appear to be superfluous constructions, nor is it likely that the share in those riches due to the canals and oases, intricately dividing up the wide, dry, continental plains, can ever be realised.

"We have assumed, in our little calculation, that the entire contents of a polar hood turn to water; but in actual fact a considerable proportion of them must pass

* Areas on Mars so named.

directly into vapour, omitting the intermediate stage. Even with us a large quantity of snow is removed aerially; and in the rare atmosphere of Mars this cause of waste must be especially effective. Thus the polar reservoirs are despoiled in the act of being opened. Further objections might be taken to Mr. Lowell's irrigation scheme, but enough has been said to show that it is hopelessly unworkable."

It will be seen that the writer of this article accepted the existence of water on Mars, on the testimony of Sir W. Huggins, which, in view of later observations, he has himself acknowledged to be valueless. Dr. Johnstone Stoney's proof of its absence, derived from the molecular theory of gases, had not then been made public.

Description of some of the Canals.

At the end of his volume Mr. Lowell gives a large chart of Mars on Mercator's projection, showing the canals and other features seen during the opposition of 1905. This contains many canals not shown on the map here reproduced (see frontispiece), and some of the differences between the two are very puzzling. Looking at our map, which shows the north-polar snow below, so that the south pole is out of the view at the top of the map, the central feature is the large spot Ascræeus Lucus, from which ten canals diverge centrally, and four from the sides, forming wide double canals, fourteen in all. There is also a canal named Ulysses, which here passes far to the right of the spot, but in the large chart enters it centrally. Looking at our map we see, going downwards a little to the left, the canal Udon, which runs through a dark area quite to the outer margin. In the dark area, however, there is shown on the chart a spot Aspledon Lucus, where five canals meet, and if this is taken as a terminus the Udon canal is almost exactly 2000 miles long, and another on its right, Lapadon, is the same length, while Ich, running in a slightly curved line to a large spot (Lucus Castorius on the chart) is still longer. The Ulysses canal, which (on the chart) runs straight from the point of the Mare Sirenum to the Astræeus Lucus is about 2200 miles long. Others however are even longer, and Mr. Lowell says: "With them 2000 miles is common; while many exceed 2500; and the Eumenides-Orcus is 3540 miles from the point where it leaves Lucus Phoeniceus to where it enters the Trivium Charontis." This last canal is barely visible on our map, its commencement being indicated by the word Eumenides.

The Trivium Charontis is situated just beyond the right-hand margin of our map. It is a triangular dark area, the sides about 200 miles long, and it is shown on the chart as being the centre from which radiate thirteen canals. Another centre is Aquæ Calidæ situated at the point of a dark area running obliquely from 55° to 35° N. latitude, and, as shown on a map of the opposite hemisphere to our map, has nearly twenty canals radiating from it in almost every direction. Here at all events there seems to be no special connection with the polar snow-caps, and the radiating lines seem to have no intelligent purpose whatever, but are such as might result from fractures in a glass globe produced by firing at it with very small shots one at a time. Taking the whole series of them, Mr. Lowell very justly compares them to "a network which triangulates the surface of the planet like a geodetic survey, into polygons of all shapes and sizes."

At the very lowest estimate the total length of the canals observed and mapped by Mr. Lowell must be over a hundred thousand miles, while he assures us that numbers of others have been seen over the whole surface, but so faintly or on such rare occasions as to elude all attempts to fix their position with certainty. But these, being of the same character and evidently forming part of the same system, must also be artificial, and thus we are led to a system of irrigation of almost unimaginable magnitude on a planet which has no mountains, no rivers, and no rain to support it; whose whole water-supply is derived from polar snows, the amount of which is ludicrously inadequate to need any such world-wide system; while the low atmospheric pressure would lead to rapid evaporation, thus greatly diminishing the small amount of moisture that is available. Everyone must, I think, agree with Miss Clerke, that, even admitting the assumption that the polar snows consist of frozen water, the excessively scanty amount of water thus obtained would render any scheme of world-wide distribution of it hopelessly unworkable.

The very remarkable phenomena of the duplication of many of the lines, together with the dark spots–the so-called oases–at their intersections, are doubtless all connected in some unknown way with the constitution and past history of the planet; but, on the theory of the whole being works of art, they certainly do *not* help to remove any of the difficulties which have been shown to attend the theory that the single lines represent artificial canals of irrigation with a strip of verdure on each side of them produced by their overflow.

Lowell on the Purpose of the Canals.

Before leaving this subject it will be well to quote Mr. Lowell's own words as to the supposed perfectly level surface of Mars, and his interpretation of the origin and purpose of the 'canals':

"A body of planetary size, if unrotating, becomes a sphere, except for solar tidal deformation; if rotating, it takes on a spheroidal form exactly expressive, so far as observation goes, of the socalled centrifugal force at work. Mars presents such a figure, being flattened out to correspond to its axial rotation. Its surface therefore is in fluid equilibrium, or, in other words, a particle of liquid at any point of its surface at the present time would stay where it was devoid of inclination to move elsewhere. Now the water which quickens the verdure of the canals moves from the pole down to the equator as the season advances. This it does then irrespective of gravity. No natural force propels it, and the inference is forthright and inevitable that it is artificially helped to its end. There seems to be no escape from this deduction. Water only flows downhill, and there is no such thing as downhill on a surface already in fluid equilibrium. A few canals might presumably be so situated that their flow could, by inequality of terrane, lie equator-ward, but not all... Now it is not in particular but by general consent that the canal-system of Mars develops from pole to equator. From the respective times at which the minima take place, it appears that the canal-quickening occupies fifty-two days, as evidenced by the successive vegetal darkenings, to descend from latitude 72° north to latitude 0°, a journey of 2650 miles. This gives for the water a speed of fifty-one miles a day, or 2.1 miles an hour. The rate of progression is remarkably uniform, and this abets the deduction as to assisted transference. But the fact is more unnatural yet. The growth pays no regard to the equator, but proceeds across it as if it did not exist into the planet's other hemisphere. Here is something still more telling than travel to this point. For even if we suppose, for the sake of argument, that natural forces took the water down to the equator, their action must there be certainly reversed, and the equator prove a dead-line, to pass which were impossible" (pp. 374–5).

I think my readers will agree with me that this whole argument is one of the most curious ever put forth seriously by an eminent man of science. Because the polar compression of Mars is about what calculation shows it ought to be in accordance with its rate of rotation, its surface is in a state of 'fluid equilibrium,' and must therefore be absolutely level throughout. But the polar compression of the earth equally agrees with calculation; therefore its surface is also in 'fluid equilibrium'; therefore it also ought to be as perfectly level on land as it is on the ocean surface! But as we know this is very far from being the case, why must it be so in Mars? Are we to suppose Mars to have been formed in some totally different way from other planets, and that there neither is nor ever has been any reaction between its interior and exterior forces? Again, the assumption of perfect flatness is directly opposed to all observation and all analogy with what we see on the earth and moon. It gives no account whatever of the numerous and large dark patches, once termed seas, but now found to be not so, and to be full of detailed markings and varied depths of shadow. To suppose that these are all the same dead-level as the light-coloured portions are assumed to be, implies that the darkness is one of material and colour only, not of diversified contour, which again is contrary to experience, since difference of material with us always leads to differences in rate of degradation, and hence of diversified contour, as these dark spaces actually show themselves under favourable conditions to independent observers.

Lowell on the System of Canals as a whole.

We will now see what Mr. Lowell claims to be the plain teaching of the 'canals' as a whole:

"But last and all-embracing in its import is the system which the canals form. Instead of running at hap-hazard, the canals are interconnected in a most remarkable manner. They seek centres instead of avoiding them. The centres are linked thus perfectly one with another, an arrangement which could not result from centres, whether of explosion or otherwise, which were themselves discrete. Furthermore, the system covers the whole surface of the planet, dark areas and light ones alike, a world-wide distribution which exceeds the bounds of natural possibility. Any force which could act longitudinally on such a scale must be limited latitudinally in its action, as witness the belts of Jupiter and the spots upon the sun. Rotational, climatic, or other physical cause could not fail of zonal expression. Yet these lines are grandly indifferent to such competing influences. Finally, the system, after meshing the surface in its entirety, runs straight into the polar caps.

"It is, then, a system whose end and aim is the tapping of the snow-cap for the water there semi-annually let loose; then to distribute it over the planet's face" (p. 373).

Here, again, we have curiously weak arguments adduced to support the view that these numerous straight lines imply works of art rather than of nature, especially in the comparison made with the belts of Jupiter and the spots on the sun, both purely atmospheric phenomena, whereas the lines on Mars are on the solid surface of the planet. Why should there be any resemblance between them? Every fact stated in the above quotation, always keeping in mind the physical conditions of the planet–its very tenuous atmosphere and rainless desert-surface–seem wholly in favour of a purely natural as opposed to an artificial origin; and at the close of this discussion I shall suggest one which seems to me to be at least possible, and to explain the whole series of the phenomena set forth and largely discovered by Mr. Lowell, in a simpler and more probable manner than does his tremendous assumption of their being works of art. Readers who may not possess Mr. Lowell's volume will find three of his most recent maps of the 'canals' reproduced in *Nature* of October 11th, 1906.

Chapter IV
Is Animal Life Possible On Mars?

Having now shown, that, even admitting the accuracy of all Mr. Lowell's observations, and provisionally accepting all his chief conclusions as to the climate, the nature of the snow-caps, the vegetation, and the animal life of Mars, yet his interpretation of the lines on its surface as being veritably 'canals,' constructed by intelligent beings for the special purpose of carrying water to the more arid regions, is wholly erroneous and rationally inconceivable. I now proceed to discuss his more fundamental position as to the actual habitability of Mars by a highly organised and intellectual race of material organic beings.

Water and Air essential to Life.

Here, fortunately, the issue is rendered very simple, because Mr. Lowell fully recognises the identity of the constitution of matter and of physical laws throughout the solar-system, and that for any high form of organic life certain conditions which are absolutely essential on our earth must also exist in Mars. He admits, for example, that water is essential, that an atmosphere containing oxygen, nitrogen, aqueous vapour, and carbonic acid gas is essential, and that an abundant vegetation is essential; and these of course involve a surface-temperature through a considerable portion of the year that renders the existence of these–especially of water–possible and available for the purposes of a high and abundant animal life.

Blue Colour the only Evidence of Water.

In attempting to show that these essentials actually exist on Mars he is not very successful. He adduces evidence of an atmosphere, but of an exceedingly scanty one, since the greatest amount he can give to it is–"not more than about four inches of barometric pressure as we reckon it";* and he assumes, as he has a fair

* In a paper written since the book appeared the density of air at the surface of Mars is said to be $^1/_{12}$ of the Earth's.

right to do till disproved, that it consists of oxygen and nitrogen, with carbon-dioxide and water-vapour, in approximately the same proportions as with us. With regard to the last item–the water-vapour–there are however many serious difficulties. The water-vapour of our atmosphere is derived from the enormous area of our seas, oceans, lakes, and rivers, as well as from the evaporation from heated lands and tropical forests of much of the moisture produced by frequent and abundant rains. All these sources of supply are admittedly absent from Mars, which has no permanent bodies of water, no rain, and tropical regions which are almost entirely desert. Many writers have therefore doubted the existence of water in any form upon this planet, supposing that the snow-caps are not formed of frozen water but of carbon-dioxide, or some other heavy gas, in a frozen state; and Mr. Lowell evidently feels this to be a difficulty, since the only fact he is able to adduce in favour of the melting snows of the polar caps producing water is, that at the time they are melting a marginal blue band appears which accompanies them in their retreat, and this blue colour is said to prove conclusively that the liquid is not carbonic acid but water. This point he dwells upon repeatedly, stating, of these blue borders: "This excludes the possibility of their being formed by carbon-dioxide, and shows that of all the substances we know the material composing them must be water."

This is the only proof of the existence of *water* he adduces, and it is certainly a most extraordinary and futile one. For it is perfectly well known that although water, in large masses and by transmitted light, is of a blue colour, yet shallow water by reflected light is not so; and in the case of the liquid produced by the snow-caps of Mars, which the whole conditions of the planet show must be shallow, and also be more or less turbid, it cannot possibly be the cause of the 'deep blue' tint said to result from the melting of the snow.

But there is a very weighty argument depending on the molecular theory of gases against the polar caps of Mars being composed of frozen water at all. The mass and elastic force of the several gases is due to the greater or less rapidity of the vibratory motion of their molecules under identical conditions. The speed of these molecular motions has been ascertained for all the chief gases, and it is found to be so great as in certain cases to enable them to overcome the force of gravity and escape from a planet's surface into space. Dr. G. Johnstone Stoney has specially investigated this subject, and he finds that the force of gravity on the earth is sufficient to retain all the gases composing its atmosphere, but not sufficient to retain hydrogen; and as a consequence, although this gas is produced in small quantities by volcanoes and by decomposing vegetation, yet no trace of it is found in our atmosphere. The moon however, having only one-eightieth the mass of the earth, cannot retain any gas, hence its airless and waterless condition.

Water Vapour cannot exist on Mars.

Now, Dr. Stoney finds that in order to retain water vapour permanently a planet must have a mass at least a quarter that of the earth. But the mass of Mars is only one-ninth that of the earth; therefore, unless there are some special conditions that prevent its loss, this gas cannot be present in the atmosphere. Mr. Lowell does not refer to this argument against his view, neither does he claim the evidence of spectroscopy in his favour. This was alleged more than thirty years ago to show the existence of water-vapour in the atmosphere of Mars, but of late years it has been doubted, and Mr. Lowell's complete silence on the subject, while laying stress on such a very weak and inconclusive argument as that from the tinge of colour that is observed a little distance from the edge of the diminishing snow-caps, shows that he himself does not think the fact to be thus proved. If he did he would hardly adduce such an argument for its presence as the following: "The melting of the caps on the one hand and their re-forming on the other affirm the presence of water-vapour in the Martian atmosphere, of whatever else that air consists" (p. 162). Yet absolutely the only proof he gives that the caps are frozen water is the almost frivolous colour-argument above referred to!

No Spectroscopic Evidence of Water Vapour.

As Sir William Huggins is the chief authority quoted for this fact, and is referred to as being almost conclusive in the third edition of Miss Clerke's *History of Astronomy* in 1893, I have ascertained that his opinion at the present time is that "there is no conclusive proof of the presence of aqueous vapour in the atmosphere of Mars, and that observations at the Lick Observatory (in 1895), where the conditions and instruments are of the highest order, were negative." He also informs me that Marchand at the Pic du Midi Observatory was unable to obtain lines of aqueous vapour in the spectrum of Mars; and that in 1905, Slipher, at Mr. Lowell's observatory, was unable to detect any indications of aqueous vapour in the spectrum of Mars.

It thus appears that spectroscopic observations are quite accordant with the calculations founded on the molecular theory of gases as to the absence of aqueous vapour, and therefore presumably of liquid water, from Mars. It is true that the spectroscopic argument is purely negative, and this may be due to the extreme delicacy of the observations required; but that dependent on the inability of the force of gravity to retain it is positive scientific evidence against its presence, and, till shown to be erroneous, must be held to be conclusive.

This absence of water is of itself conclusive against the existence of animal life, unless we enter the regions of pure conjecture as to the possibility of some other liquid being able to take its place, and that liquid being as omnipresent there as water is here. Mr. Lowell however never takes this ground, but bases his whole theory on the fundamental identity of the substance of the bodies of living organisms wherever they may exist in the solar system.

In the next two chapters I shall discuss an equally essential condition, that of temperature, which affords a still more conclusive and even crushing argument against the suitability of Mars for the existence of organic life.

THE THINGS THAT LIVE ON MARS (1908)

H. G. Wells

Herbert George Wells was born in 1866 in Bromley, Kent. While still at school he became interested in how society might be refashioned in the future by socialism, as advocated by the Fabian Society. As an adult he found work as a teacher, in his spare time studied for a degree in zoology from the University of London, and in 1893 published his first work, the *Text Book of Biology*. Wells will always be associated with Mars because of his epochal 1898 novel *The War of the Worlds*. This piece, *The Things that Live on Mars*, is non-fiction. It was published in 1908 and is a speculation on what creatures might live on Mars given the state of knowledge at the time.

WHAT sort of inhabitants may Mars possess?

To this question I gave a certain amount of attention some years ago when I was preparing a story called "The War of the Worlds," in which the Martians are supposed to attack the earth; but since that time much valuable work has been done upon that planet, and one comes to this question again with an ampler equipment of information, and prepared to consider it from new points of view.

Particularly notable and suggestive in the new literature of the subject is the work of my friend, Mr. Percival Lowell, of the Lowell Observatory, Flagstaff, Arizona, to whose publications, and especially his "Mars and its Canals," I am greatly indebted. This book contains a full statement of the case, and a very convincing case it is, not only for the belief that Mars is habitable, but that it is inhabited by creatures of sufficient energy and engineering science to make canals beside which our greatest human achievements pale into insignificance.

He does not, however, enter into any speculation as to the form or appearance of these creatures, whether they are human, quasi-human, supermen, or creatures of a shape and likeness quite different from our own. Necessarily such an inquiry must be at present a speculation of the boldest description, a high imaginative flight. But at the same time it is by no means an unconditioned one. We are bound by certain facts and certain considerations. We are already forbidden by definite knowledge to adopt any foolish fantastic hobgoblin or any artistic ideal that comes into our heads and call it a Martian. Certain facts about Mars we definitely known, and we are not entitled to imagine any Martians that are not in accordance with these facts.

When one speaks of Martians one is apt to think only of those canal-builders, those beings who, if we are to accept Mr. Lowell's remarkably well-sustained conclusions, now irrigate with melting polar snows and cultivate what were once the ocean-beds of their drying planet. But after all they cannot live there alone; they can be but a part of the natural history of Mars in just the same way that man is but a part of the natural history of the earth. They must have been evolved from other related types, and so we must necessarily give our attention to the general flora and fauna of this world we are invading in imagination before we can hope to deal at all reasonably with the ruling species.

Does Life Exist on Mars?

And, firstly, will there be a flora and fauna at all? Is it valid to suppose that upon Mars we should find the same distinction between vegetable and animal that we have upon the earth? For the affirmative answer to that an excellent case can be made. The basis upon which all life rests on this planet is the green plant. The green plant alone is able to convert really dead inorganic matter into living substance, and this it does, as everybody knows nowadays, by the peculiar virtue of its green coloring matter, chlorophyl, in the presence of sunshine. All other animated things live directly or indirectly upon the substance of green-leaved plants. Either they eat vegetable food directly, or they eat it indirectly by eating other creatures which live on vegetable food. Now upon this earth it is manifest that nature has tried innumerable experiments and made countless beginnings. Yet she has never produced any other means than chlorophyl whereby inorganic matter, that is to say, soil and minerals and ingredients out of the air, can be built up into living matter. It is plausible, therefore, to suppose that on Mars also, if there is life, chlorophyl will lie at the base of the edifice; in other words, that there will be a vegetable kingdom. And our supposition is greatly strengthened by the fact upon which Mr. Lowell lays stress, that, as the season which corresponds to our spring arrives, those great areas of the Martian surface that were once ocean beds are suffused with a distinct bluish green hue. It is not the yellow-green of a leafing poplar or oak-tree; it is the bluish green of a springtime pine.

This all seems to justify us in assuming a flora at least upon Mars, a green vegetable kingdom after the fashion of our earthly one. Let us ask now how far we may assume likeness. Is an artist justified in drawing grass and wheat, oaks and elms and roses in a Martian landscape? Is it probable that evolution has gone upon exactly parallel lines on the two planets? Well, here again we have definite facts upon which to base our answer. We know enough to say that the vegetable forms with which we are familiar upon the earth would not "do," as people say, on Mars, and we can even indicate in general terms in what manner they would differ. They would not do because, firstly, the weight of things at the surface of Mars is not half what it would be upon the earth, and, secondly, the general atmospheric conditions are very different. Whatever else they may be the Martian herbs and trees must be adapted to these conditions.

Probable Appearance of the Martian Flora

Let us inquire how the first of these two considerations will make them differ. The force of gravity upon the surface of their planet is just three-eighths of its force upon this earth; a pound of anything here would weigh six ounces upon Mars. Therefore the stem or stalk that carries the leaves and flowers of a terrestrial plant would be needlessly and wastefully stout and strong upon Mars; the Martian stems and stalks will all be slenderer and finer and the texture of the plant itself laxer. The limit of height and size in terrestrial plants is probably determined largely by the work needed to raise nourishment from the roots to their topmost points. That work would be so much less upon Mars that it seems reasonable to expect bigger plants there than any that grow upon the earth.

Larger, slighter, slenderer; is that all we can say? No, for we have still to consider the difference in the atmosphere. This is thinner upon Mars than it is upon the earth, and it has less moisture, for we hardly ever see thick clouds there, and rain must be infrequent. Snow occurs nearly everywhere all the year round, but the commonest of all forms of precipitation upon Mars would seem to be dew and hoar frost. Now the shapes of leaves with which we are most familiar are largely determined by rainfall, by the need of supporting the hammering of raindrops and of guiding the resulting moisture downward and outward to the rootlets below. To these chief necessities we owe the hand-like arrangement of the maple- and chestnut-leaf and the beautiful tracery of fibers that forms their skeletons. These leaves are admirable in rain but ineffectual against snow and frost; snow crushes them down, frost destroys them, and with the approach of winter they are shed. But the Martian tree-leaf will be more after the fashion of a snowfall-meeting leaf, spiky perhaps like the pine-tree needle. Only, unlike the pine-tree needle, it has to meet not a snowy winter but a dry, frost-bitten, sunless winter, and then probably it will shrivel and fall. And since the great danger for a plant in a dry air is desiccation, we may expect these Martian leaves to have thick cuticles, just as the cactus has. Moreover, since moisture will come to the Martian plant mainly from below in seasonal floods from the melting of the snow-caps, and not as rain from above, the typical Martian plant will probably be tall and have its bunches and clusters of spiky bluish green leaves upon uplifting reedy stalks.

Of course there will be an infinite variety of species of plants upon Mars as upon the earth, but these will be the general characteristics of the vegetation.

The Animal Kingdom

Now this conception of the Martian vegetation as mainly a jungle of big, slender, stalky, lax-textured, flood-fed plants with a great shock of fleshy, rather formless leaves above, and no doubt with as various a display of flowers and fruits as our earthly flora, prepares the ground for the consideration of the Martian animals.

It is a matter of common knowledge nowadays how closely related is the structure of every animal to the food it consumes. Different food, different animals, has almost axiomatic value, and the very peculiar nature of the Martian flora is in itself sufficient to dispel the idea of our meeting beasts with any close analogy to terrestrial species. We shall find no flies nor sparrows nor dogs nor cats on Mars. But we shall probably find a sort of insect life fluttering high amidst the vegetation, and breeding during the summer heats in the flood-water below. In the winter it will encyst and hibernate. Its dimensions may be a little bigger than those ruling among the terrestrial insecta; but the mode of breathing by tracheal tubes, which distinguishes insects, very evidently (and very luckily for us) sets definite limits to insect size. Perhaps these limits are the same upon Mars. We cannot tell. Perhaps they are even smaller; the thinner air may preclude even the developments we find upon the earth in that particular line. Still there is plenty of justification if an artist were to draw a sort of butterfly or moth fluttering about, or ant-like creatures scampering up and down the stems of a Martian jungle. Many of them perhaps will have sharp, hard proboscides to pierce the tough cuticle of the plants.

No Fish on the Planet

But, and here is a curious difference, there are perhaps no fish or fish-like creatures on Mars at all. In the long Martian winter all the water seems either to drift to the poles and freeze there as snow or to freeze as ice along the water-courses; there are only flood-lakes and water-canals in spring and summer. And forms of life that trusted to gills or any method of underwater breathing must have been exterminated upon Mars ages ago. On earth the most successful air-breathing device is the lung. Lungs carry it universally. Only types of creatures that are fitted with lungs manage to grow to any considerable size out of water in our world. Even the lobsters and scorpions and spiders and such like large crustacean and insect-like forms that come up into the air can do so only by sinking their gills into deep pits to protect them from evaporation and so producing a sort of inferior imitation of a lung. Then and then only can they breathe without their breathing-organs drying up. The Martian air is thinner and drier than ours, and we conclude therefore that there is still more need than on earth for well-protected and capacious lungs. It follows that the Martian fauna will run to large chests. And the lowest types of large beast there will be amphibious creatures which will swim about and breed in the summer waters and bury themselves in mud at the approach of winter. Even these may have been competed out of existence by air-inhaling swimmers. That is the fate our terrestrial amphibia seem to be undergoing at the present time.

Here then is one indication for a picture of a Martian animal: it must be built with more lung space than the corresponding terrestrial form. And the same reason that will make the vegetation laxer and flimsier will make the forms of the Martian animal kingdom laxer and flimsier and either larger or else slenderer than earthly types.

Much that we have already determined comes in here again to help us to further generalizations. Since the Martian vegetation will probably run big and tall, there will be among these big-chested creatures climbing forms and leaping and flying forms, all engaged in seeking food among its crests and branches. And a thing cannot leap or fly without a well-placed head and good eyes. So an imaginative artist may put in head and eyes, and the mechanical advantages of a fore-and-aft arrangement of the body are so great that it is difficult to suppose them without some sort of back-bone. Since the Martian vegetation has become adapted to seasonal flood conditions there will be not only fliers and climbers but waders—long-legged forms. Well, here we get something—fliers, climbers, and waders, with a sort of backbone.

Climatic Conditions

Now let us bring in another fact, the fact that the Martian year is just twice the length of ours and alternates between hot summer sunshine, like the sunshine we experience on high mountains, and a long, frost-bitten winter. The day, too, has the length of a terrestrial day, and because of the thin air will have just the quick changes from heat to cold we find on this planet on the high mountains. This means that all these birds and beasts must be adapted to great changes of temperature. To meet that they must be covered with some thick, air-holding, non-conducting covering, something analogous to fur or feathers, which they can molt or thin out in summer and renew for the winter's bitterness. This is much more probable than that they will be scaly or bare-skinned like our earthly lizards and snakes; and since they will need to have fur or down outside their frameworks, their skeletons, which will be made up of very light slender bones, will probably be within. Moreover, the chances are that they will be fitted with the best known contrivances for protecting their young in the earliest stages from cold and danger. On earth the best known arrangement is the one that prevails among most of the higher land animals, the device of bringing forth living young at a high stage of development. This is the "hard life" arrangement as distinguished from the easy-going, sunshiny, tropical, lay-an-egg-and-leave-it method, and Martian conditions are evidently harder than ours. So these big-chested, furry or feathery or downy Martian animals will probably be very like our mammalia in these respects. All this runs off easily and plausibly from the facts we know.

The Ruling Inhabitants

And now we are in a better position to consider those ruling inhabitants who made the gigantic canal-system of Mars, those creatures of human or superhuman intelligence, who, unless Mr. Lowell is no more than a fantastic visionary, have taken Mars in hand to rule and order and cultivate systematically and completely, as I believe some day man will take this earth. Clearly these ruling beings will have been evolved out of some species or other of those mammal-like animals, just as man has been evolved from among the land animals of this globe. Perhaps they will have exterminated all those other forms of animal life as man is said to be exterminating all the other forms of animal life here. I have written above of floods and swamps and jungles to which life has adapted itself, but perhaps that stage is over now upon Mars altogether. It must have been a long and life-molding stage, but now it may be at an end. Mr. Lowell, judging by the uniform and orderly succession of what he calls the "fallow" brown and then of the bluish green tints upon the low-lying areas of Mars, is inclined to think that this is the case and that all the fertile area of the planet has been reclaimed from nature and is under cultivation.

How Like Terrestrial Humanity?

How far are these beings likely to resemble terrestrial humanity?

There are certain features in which they are likely to resemble us. The quasi-mammalian origin we have supposed for them implies a quasi-human appearance. They will probably have heads and eyes and backboned bodies, and since they must have big brains, because of their high intelligence, and since almost all creatures with big brains tend to have them forward in their heads near their eyes, these Martians will probably have big shapely skulls. But they will in all likelihood be larger in size than humanity, two and two-thirds times the mass of a man, perhaps. That does not mean, however, that they will be two and two-thirds times as tall, but, allowing for the laxer texture of things on Mars, it may be that they will be half as tall again when standing up. And as likely as not they will be covered with feathers or fur. I do not know, I do not know if anyone knows, why man, unlike the generality of mammals, is a bare-skinned animal. I can find, however, no necessary reason to make me believe the Martians are bare-skinned.

Will they stand up or go on four legs or six? I know of no means of answering that question with any certainty. But there are considerations that point to the Martian's being a biped. There seems to be a general advantage in a land-going animal having four legs; it is the prevailing pattern on earth, and even among the insects there is often a tendency to suppress one pair of the six legs and use only four for going. However, this condition is by no means universal. A multitude of types, like the squirrel, the rat, and the monkey, can be found which tend to use the hind legs chiefly for walking and to sit up and handle things with the fore-limbs. Such species tend to be exceptionally intelligent. There can be no doubt of the immense part the development of the hand has played in the education of the human intelligence. So that it would be quite natural to imagine the Martians as big-headed, deep-chested bipeds, grotesquely caricaturing humanity with arms and hands.

But that is only one of several almost equally plausible possibilities. One thing we may rely upon: the Martians must have some prehensile organ, primarily because the development of intelligence is almost unthinkable without it, and, secondly, because in no other way could they get their engineering done. It is stranger to our imaginations, but no less reasonable, to suppose, instead of a hand, an elephant-like proboscis, or a group of tentacles or proboscis-like organs. Nature has a limitless imagination, never repeats exactly, and perhaps, after all, the chances lie in the direction of a greater unlikeness to the human shape than these forms I have ventured to suggest.

How wild and extravagant all this reads!

One tries to picture feather-covered men nine or ten feet tall, with proboscides and several feet, and one feels a kind of disgust of the imagination. Yet wild and extravagant as these dim visions of unseen creatures may seem, it is logic and ascertained fact that forces us toward the belief that some such creatures are living now. And, after all, has the reader ever looked at a cow and tried to imagine how it would feel to come upon such a creature with its knobs and horns and queer projections suddenly for the first time?

Martian Civilization

I have purposely abstained in this paper from going on to another possibility of Martian life. Man on this earth has already done much to supplement his bodily deficiencies with artificial aids—clothes, boots, tools, corsets, false teeth, false eyes, wigs, armor, and so forth. The Martians are probably far more intellectual than men and more scientific, and beside their history the civilization of humanity is a thing of yesterday. What may they not have contrived in the way of artificial supports, artificial limbs, and the like?

Finally, here is a thought that may be reassuring to any reader who finds these Martians alarming. If a man were transferred suddenly to the surface of Mars he would find himself immensely exhilarated so soon as he had got over a slight mountain-sickness. He would weigh not one-half what he does upon the earth, he would prance and leap, he would lift twice his utmost earthly burden with case. But if a Martian came to the earth his weight would bear him down like a cope of lead. He would weigh two and two-thirds times his Martian weight, and he would probably find existence insufferable. His limbs would not support him. Perhaps he would die, self-crushed, at once. When I wrote "The War of the Worlds," in which the Martians invade the earth, I had to tackle this difficulty. It puzzled me for a time, and then I used that idea of mechanical aids, and made my Martians mere bodiless brains with tentacles, subsisting by suction without any digestive process and carrying their weight about, not on living bodies but on wonderfully devised machines. But for all that, as a reader here and there may recall, terrestrial conditions were in the end too much for them.

MARS AS THE ABODE OF LIFE (1908)

CHAPTER VI: PROOFS OF LIFE ON MARS

Percival Lowell

A businessman turned astronomer, **Percival Lowell** did more than anyone else to build interest in Mars as an inhabited – but dying – planet. His interest in Mars was sparked by Camille Flammarion's book *La planète Mars*, and he founded the Lowell Observatory in Flagstaff, Arizona to study the planet, particularly the 'canals'. He produced intricate diagrams of these optical illusions, convinced of their reality, and popularized his observations. Born in Boston in 1855, Lowell died in Flagstaff in 1916. He is buried on Mars Hill, the site of his observatory. This is the second piece of his work to appear in *The Book of Mars*.

ASTRONOMICAL discovery is of two kinds. If it consist simply in adding another asteroid or satellite to those already listed, obedience to the law of gravitation, with subsequent corroboration of place, alone is needed for belief. But if it relate to the detection of an underlying truth as yet unrecognized, then it is only to be unearthed by reasoning on facts after they are obtained, and effects credence according to one's capacity for weighing evidence. Breadth of mind must match breadth of subject. For to plodders along prescribed paths a far view fails of appeal; conservative settlers in a land differ in quality from pioneers.

Discovery of a truth in the heavens varies in nothing, except the subject, from discovery of a crime on earth. The forcing of the secrets of the sky is, like the forcing of man's, simply a piece of detective work. It is the finding of a cause in place of a culprit; but the process is quite similar. *Causa criminis* and *causa discriminis* differ only by a syllable.

Like, too, are, or should be, the methods employed. In astronomy, as in criminal investigation, two kinds of testimony require to be secured. Circumstantial evidence must first be marshalled, and then a motive must be found. To omit the purpose as irrelevant, and rest content with gathering the facts, is really as inconclusive a procedure in science as in law, and rarely ends in convincing, any more than in properly convicting, anybody. For motive is just as all-pervading a preliminary to cosmic as to human events, only for lack of fully comprehending it we call the one a motive and the other a cause. Unless we can succeed in assigning a sufficient reason for a given set of observed phenomena, we have not greatly furthered the ends of knowledge and have done no more than the clerkage of science. A theory is just as necessary to give a working value to any body of facts as a backbone is to higher animal locomotion. It affords the data vertebrate support, fitting them for the pursuit of what had otherwise eluded search.

Coördination is the end of science, the aim of all attempt at learning what this universe may mean. And coördination is only another name for theory, as the law of gravitation witnesses. Now, to be valid, a theory must fulfil two conditions: it must not be contradicted by any fact within its purview, and it must assign an underlying thread of reason to explain all the phenomena observed. Circumstantial evidence must first lead to a suspect, and then this suspect must prove equal to accounting for the facts.

This method we shall pursue in the case before us; and it will conduce to understanding of the evidence to keep its order of presentation to the detective in presenting it at the bar of reason.

Starting with the known physical laws applicable to the concentration of matter, we found that though in general the course of evolution of the earth and Mars was similar, the smaller mass of Mars should have caused it to differ eventually from the earth in some important respects.

Three of these are noteworthy: (1) its surface should be smoother than the earth's, (2) its oceans relatively less, (3) its air scantier. On turning to Mars itself we then saw that these three attributes of the planet were precisely those the telescope disclosed. (1) The planet's surface was singularly flat, being quite devoid of mountains; (2) its oceans in the past covered at most three-eighths of its surface instead of three-quarters, as with us; (3) its air was relatively thin.

We next showed that physical loss should, from its smaller mass, have caused it to age quicker, and that this aging should reveal itself by the more complete departure of what oceans it once possessed and by the wider spread of deserts.

Telescopic observation we then found asserted these two peculiarities: (1) no oceans now exist on the planet's surface; (2) desert occupies five-eighths of it.

From such confirmation of the principles of planetary evolution from the present aspect of the planet Mars, we went on to consider the two most essential prerequisites to habitability: water and warmth. Water we sought first; and we found it in the polar caps. The phenomena of the polar caps proved explicable as consisting of water, and not as of anything else. Still more important was the question of temperature. We took this up with particularity. We found several factors to the problem not hitherto reckoned with, and that when these were taken into account the result came out entirely different from what had previously been supposed. Instead of a temperature prohibitive to life, one emerged from our research entirely suitable for it. And this even more for animals than plants. For a climate of extremes was what that of Mars appeared to be, with the summers warm. Now, investigations on earth have shown that it is the temperature of the hottest season that determines the existence of animals, cold much more adversely affecting plants. Yet to the presence of the latter the look of the disk conformed. Scanning it, we marked effects which could only be explained as vegetation. Thus the conditions on Mars showed themselves hospitable to both great orders of life, the latter actually revealing its presence by its seasonal changes of tint.

Here we reached the end of what might directly be disclosed in the organic economy of the planet. For at this point we brought up before a most significant fact: that vegetable life could thus reveal itself directly, but that animal life could not. Not by its body, but by its mind, would it be known. Across the gulf of space it could be recognized only by the imprint it had made on the face of Mars.

Turning to the planet, we witnessed a surprising thing. There on the Martian disk were just such markings as intelligence might have made. Seen even with the unthinking eye, they appear strange beyond belief, but viewed thus, in the light of deduction, they seem positively startling, like a prophecy come true.

Confronting the observer are lines and spots that but impress him the more, as his study goes on, with their non-natural look. So uncommonly regular are they, and on such a scale, as to raise suspicion whether they can be by nature regularly produced. Next to one's own eyesight the best proof of this is the unsolicited indorsement it has received in the scepticism their depiction invariably evokes. Those who have not been privileged to see them find it well-nigh impossible to believe that such things can be. Nor is this in the least surprising. But however consonant with nescience to doubt the existence of the lines on this score, to do so commits it to witness against itself of the most damaging character the moment their existence is proved. Now, assurance of actuality no longer needs defence. The lines have not only been amply proved to exist, but have actually been photographed, and doubt has shifted its ground from existence to character, a half retreat tantamount to a complete surrender. For without equal investigation, to admit a discovery and deny its description is like voting for a bill and against its appropriation. It reminds one of the advice of the old lawyer to a junior counsel: "When you have no case, abuse the plaintiff's attorney."

Unnatural regularity, the observations showed, betrays itself in everything to do with the lines: in their surprising straightness, their amazing uniformity throughout, their exceeding tenuity, and their immense length. These traits, instead of disappearing, the better the canals have been seen, as was confidently prophesied, have only come out with greater insistence. With increased study not only the assurance gains that they are as described, but a mass of detail has been added about them impossible to reconcile with any natural known process.

A single instance of the methodism that confronts us will serve to make this plain. The *Lucus Ismenius* is a case in point. The marking so called consists of two round spots each about seventy-five miles in diameter. They lie close together, not more than fifty miles of ochre ground parting their peripheries. Into them converge a number of canals—seven doubles and five singles. Now, the manner of these meetings is curiously detailed. Three of the doubles embrace the oases, just enclosing them between their two arms. The four other doubles send a line to each oasis to enter it centrally. Which connection the double shall adopt apparently depends upon the angle at which the approach is made. If the direction be nearly vertical to the line of the two oases, the entrance is central;

if parallel, it is an embrace. As for the singles, they connect with one or the other oasis, as the case may be. Such precise and methodical arrangement, thus marvellously articulated and detailed, discloses an orderliness so surprising, if on nature's part, as to throw us at once into the arms of the alternative as the least astonishing of the two.

Before passing on to reason upon the fact, we note that the characters mentioned are themselves enough to negative all suppositions of natural cause. First, the lines cannot be rivers, since rivers are never straight and never uniform in width. Now, we see the canals so well as to be quite certain of their evenness. The best proof of this is that, though each is uniform, some are at least ten times the size of others. If one of them dwindled *en route*, we should have ample measure of the fact.

Nor can the lines be cracks in the surface, because cracks also are not straight, and because cracks end before finishing. We have examples of undoubted cracks in more than one heavenly body, and their appearance is quite unlike the look of the lines of Mars. The moon offers such in many, if not all, of her so-called rills.

To the most superficial view these suggest their nature, but when carefully examined at Flagstaff, corroboration of the fact came out in certain definite characteristics. For the rills proved to be made of parts which overlapped at their ends, one fractional line taking up the course before the other had given out, thus exactly reproducing the composition of the cracks in any plaster ceiling.

Mercury bears testimony to the same effect. Its lines, more difficult than the canals of Mars,—for we see Mercury four times as far off when best placed as we do Mars,—though roughly linear, are not unnatural in irregularities suggestive of cracks.* In the markings on Venus, too, there is nothing unnatural.

Rivers and cracks are the two most plausible suppositions made to account for the lines on any theory of natural causation. Other guesses have been indulged in, such as that meteors by their passing attraction have raised the lines as welts upon the surface—welts easily allayed by application of the fact that the lines change with the seasons, actually disappearing at certain epochs, to revive again at others. Such suggestions there are, but none have been advanced to my knowledge that bear the most cursory inspection.

Still more inexplicable on any natural hypothesis is the systematized arrangement of the lines to form a network over the whole planet. That the lines should go directly from certain points to certain others in an absolutely unswerving direction; that they should there meet lines that have come with like directness from quite different points of departure; that sometimes more than ten of them should thus rendezvous, and rarely less than six; and that, lastly, this state of intercommunication should be true all over the disk, are phenomena that no natural physical process that I can conceive of—and no one else seems to have been able to, either—can in the least explain. Yet this arrangement cannot be due

* Lately, at least two critics have stated that the descriptions of the spoke-like markings seen on Venus at Flagstaff in1897 and later, are inconsistent. The seeming inconsistency is due to our own air, which sometimes defines them, sometimes not. The important point about them is that the Venusian lines are *irregular*. —P. L.

to chance, the probabilities against the lines meeting one another in this orderly manner being millions to one.

But the canals are not all that is wonderful; we have to reckon with the oases as well. These are remarkable, both in themselves and in their relation to the system of lines; for they occur at the junctions—only at the junctions, and virtually always at the junctions. They are thus of the nature of knots to the network. No explanation can be given of this by purely physical laws.

So we might go on, with the enigma of the double canals more and more mysterious the more one learns about them—with their strange positioning on the planet in the tropical belts; with the curious phenomenon of converging or wedge-shaped doubles descending to join them from the pole; and with other facts equally odd.

But long before the catalogue of geometric curiosities had drawn to its close,—for it were wearisome to count them all, and where even one is so cogent, numbers do not add,—it becomes apparent to any one capable of weighing evidence that these things which so palpably imply artificiality on their face cannot be natural products at all, but that the observer apparently stands confronted with the workings of an intelligence akin to and therefore appealing to his own. What he is gazing on typifies not the outcome of natural forces of an elemental kind, but the artificial product of a mind directing it to a purposed and definite end.

When once this standpoint is adopted, we begin to see light. The recognition of artificiality puts us on a track where we gather explanation as we proceed.

Thus two attributes, one of the canals, the other of the oases, find explanation at once. The great-circle directness of the lines stands instantly interpreted. On a sphere a great circle takes the shortest distance between two points. It offers, therefore, the most expeditious route from one place to another. It is, then, that which, when possible, intelligence would adopt. Even in the case of our very accidented earth, our lines of communication are being rectified every year as we progress in mastery of our globe.

Equally suggestive is the shape of the oases, or spots, that button the lines together. For they show round. Now, a solid circle has the peculiar property that the average distance from its centre to all points in it is less than for any figure enclosing a like area. It would be the part of intelligence, then, to construct this figure whenever the greatest amount of ground was to be reached for tillage or any other purpose at the least expenditure of force.

No less telltale is their behavior; and now not only of the bare fact of artificiality, but of the manner in which it came to be.

The extreme threads of the world-wide network of canals stand connected with the dark-blue patches at the edge of one or the other of the polar caps. But they are not always visible. In the winter season they fail to show. Not till the cap has begun to melt, do they make their appearance, and then they come out dark and strong. Now, the cap in winter is formed of snow and ice that melts as summer comes on. Here, then, the attentive ear seems to catch the note of running water.

From their poleward origin the lines begin to darken down the disk. One after the other takes up the thread of visibility, to hand it on to the next in place.

So the strange communication travels, carried from the arctic zone through the temperate and the tropic ones on to the equator, and then beyond it over into the planet's other hemisphere. A flow is here apparent, journeying with measured progress over the surface of this globe. Here, again, the mental ear detects the sound of water percolating down the latitudes.

Across what once were seas, but are seas no more, the darkening of the lines advances, with the same forthrightness as over the ochre continental tracts. Blue-green areas of vegetation and arid wastes alike are threaded by the silent deepening of tint. Latitude bars it not, nor character of country. It great-circles the old sea bottoms as cheerfully as it caravans the desert steppes. This persistency made possible by the loss of what the seas once held, the thought of water is once more thrust upon the sense, its absence now as telling as its presence was before. One hears it in the very stillness the lack of it promotes.

Then, as with quickened sense one listens, the mind is aware of antiphonal response in the unlocking of the other cap to send its scanty hoardings in similar rilling over the long-parched land. The note of water confronts us thus at every turn of this strange action. Water, then, must be the word of the enigma: the clew that will lead us to the unloosening of the riddle.

But though water it be, this is not the complete solution of the problem; for, as one ponders, the unnatural character of the action dawns on one. That a wave of progression passes through the canals down the disk; that something, then, proceeds from the pole to the equator; and that this something can be none other than water, giving rise to vegetation, sounds simple and forthright. The startling character of the action is not at once apparent. It becomes so only when we try to account for the locomotion. When we so envisage it, the transference turns out to be a most astounding and instructive thing.

To understand wherein lies its peculiarity, we must consider the shape of the planet. For the planet is flattened at the poles by of its diameter. This, to begin with, will make the action seem even stranger than it is. It might seem at first as if the water in going to the equator had to run twenty-one miles uphill.

If Mars did not rotate, its figure would be a sphere, except for such tidal deformation as outside bodies might give it, because its own gravity would pull it into a shape similar in all directions. As Mars rotates, its rotatory momentum bulges it at the equator, changing the sphere into what is called an oblate spheroid of the general form of an orange. The ellipticity of a rotating mass is affected not only by the size of the body and by the speed of rotation, but by the distribution of the matter composing it. Thus it is different for a homogeneous body than for a heterogeneous one, and differs according to the law of density from surface to centre. Now it is an interesting fact that the oblateness of Mars—$\frac{1}{190}$, found by two independent methods quite independently applied; one from measurements of the planet made in 1894 at Flagstaff by Mr. Douglass, reduced and discussed by the director; the other from the motions of the satellites by Hermann Struve—should fall between the value it would have, were it homogeneous, and that which it would show did the density increase from surface to centre in the same manner as on earth. But we can see from theory that it should lie between these two extremes. For the compression there is not so great as with the earth because of Mars' smaller mass. In this we find another proof, were any needed, that the evolution of both planets was as sketched in our opening chapter. A rapidly rotated mass of putty will take on the same shape. In the case of Mars the stresses are so enormous that for a long acting force, such as is here concerned, the planet, although probably as rigid as steel, behaves as if its mass were plastic. The result is that the direction of gravity is always perpendicular to the surface at every point; or, in other words, the surface is in stable equilibrium.

Now, the fact that every point of the surface is in equilibrium means that any particle of a liquid there—as, for example, a drop of water—would not move, but would stay where it was. For all the forces being exactly balanced to rest, their resultant cannot solicit it to stir. Just as on the surface of the earth, water upon a level stretch of ground shows no tendency to move.

Consequently, any water set free near the pole by the melting of the polar cap would stay where it was liberated without the least inclination to go elsewhere. The only force which would have the slightest effect upon it might be its own head, if it had any. Were the melting ice or snow that gave birth to it ten feet thick, and it is more likely to be less, it would give rise to an average head of water of five feet. Now, a head of five feet could not urge the water against surface friction more than a few miles at most. So that any such impulse is quite impotent to the effects we see.

Face to face, then, we find ourselves with a motion of great magnitude occurring without visible or physically imaginable cause. A body of water travels 3300 miles at the rate of 51 miles a day under no material compulsion whatever.

It leaves the neighborhood of the pole, where it was gravitationally at home, and wanders to the equator, where gravitationally it was not wanted, without the slightest prompting on the part of any natural force. The deduction is inevitable; it must have been artificially conducted over the face of the planet. We are left no alternative but to suppose it intelligently carried to its end.

Nor is this the limit of the extraordinary performances shown by the progressive darkening of the canals down the disk. Were they actuated by natural forces, what they next do would be simply incredible. For, not content with descending to the equator without visible means of propulsion, once arrived there, they promptly proceed to cross it into the planet's other hemisphere and run up the latitudes with equal celerity on the other side. Now, any physical inducement given them to come equatorward must have its action reversed so soon as that dividing-line was crossed. If, then, they were in any way helped to the earlier part of their peregrination by natural forces, they would be hindered by them in this latter portion of their career. Thus, the only rational result of our discussion of the canals is that these things are not dependent on natural forces for their action, but are artificial productions designed to the end they so beautifully serve. In the canals of the planet we are looking at the work of local intelligence now dominant on Mars. Such is what the circumstantial evidence points to unmistakably.

To detection of a motive we now turn. And here it is our study of planetary evolution in general becomes of service. As a planet ages, its surface water grows scarce. Its oceans in time dry up, its rivers cease to flow, its lakes evaporate. Its fauna, if it have any, dependent as they are upon water for life, must more and more be pushed to it for that prime necessity to existence.

As the water leaves a planet, departing into space, so much of it as does not sink out of sight into its interior stands for a while a-tiptoe in its air before taking final flight into the sky. In the planet's economy it has ceased to be water, and become that more ethereal thing, water-vapor. In one way and place only does it ever in any amount descend to earth again and take on even transiently its liquid state. This is in the polar caps. The general meteorologic circulation of the planet deposits it there throughout the winter months. From the cold of the arctic latitudes its deposition takes the form of snow or ice, and in consequence of this solid state is largely tethered to the spot where it falls, remaining *in situ* until the returning sun melts it in the spring. This is the state of things on Mars.

When this unlocking occurs, and while the water is in its intermediate liquid state, between not easily transportable ice and ungatherable vapor, it is in a condition to be moved, and may be drawn upon for consumption. Then, and then only, is it readily available for use, and then, if ever, it must be tapped.

Now, in the struggle for existence, water must be got, and in the advanced condition of the planet this is the only place where it is in storage and whence, therefore, it may be had. Round the semestral release of this naturally garnered store everything in the planet's organic economy must turn. There is no other source of supply. Its procuring depends upon the intelligence of the organisms that stand in need of it. If these be of a high enough order of mind to divert it to their ends, its using, from a necessity, will become a fact. Here, then, is a motive of the most compelling kind for the tapping of the polar caps and the leading of the water they contain over the surface of the planet: the primal motive of self-preservation. No incentive could be stronger than this.

Our motive found being of the most drastic kind, it remains now to examine whether it can be put into execution.

As a planet ages, any organisms upon it would share in its development. They must evolve with it, indeed, or perish. At first they change only as environment offers opportunity, in a lowly, unconscious way. But, as brain develops, they rise superior to such occasioning. Originally the organism is the creature of its surroundings; later it learns to make them subservient to itself. In this way the organism avoids unfavorableness in the environment, or turns unpropitious fortune to good use. Man has acquired something of the art here on the earth, and what with clothing himself in the first place, and yoking natural forces in the second, lives in comfort now where, in a state of nature, he would incontinently perish.

Such adaptation in mind, making it superior to adaptation in body, is bound to occur in the organic life on any planet, if it is to survive at all. For conditions are in the end sure to reach a pass where something more potent than body is required to cope with them.

It is possible to apply a test to tell whether such life existed or not. For certain signs would be forthcoming were such intellect there. Increase of intelligence would cause one species in the end to prevail over all others, as it had prevailed over its environment. What it found inconvenient or unnecessary to enslave, it would exterminate, as we have obliterated the bison and domesticated the dog.

This species will thus become lord of the planet and spread completely over its face. Any action it might take would, in consequence, be planet-wide in its showing.

Now, such is precisely what appears in the worldspread system of canals. That it joins the surface from pole to pole and girdles it at the equator betrays a single purpose there at work. Not only does one species possess the planet but even its subdivisions must labor harmoniously to a common aim. Nations must have sunk their local patriotisms in a wider breadth of view and the planet be a unit to the general good.

As the being has conquered all others, so will it at last be threatened itself. In the growing scarcity of water will arise the premonitions of its doom. To secure what may yet be got will thus become the forefront of its endeavor, to which all other questions are secondary. Thus, if these beings are capable of making their presence noticeable at all, their great occupation should be that of water-getting, and should be the first, because the most fundamental, trace of their existence an outsider would be privileged to catch.

The last stage in the expression of life upon a planet's surface must be that just antecedent to its dying of thirst. Whether it came to this pass by simple exhaustion, as is the case with Mars, or by rotary retardation, as is the case with Mercury and Venus, the result would be all one to the planet itself. Failure of its water-supply would be the cause. To procure this indispensable would be its last conscious effort.

With an intelligent population this inevitable end would be long foreseen. Before it was upon the denizens of the globe, preparations would have been made to meet it. And this would be possible, for the intelligence attained would be of an order to correspond. A planet's water-supply does not depart in a moment. Long previous to any wholesale imminence of default, local necessity must have begun the reaching out to distant supply. Just as all our large cities to-day go far to tap a stream or a lake, so it must have been on Mars. Probably the beginnings were small and inconspicuous, as the water at first locally gave out. From this it was a step to greater distances, until necessity lured them even to the pole. The very process, one of addition, instead of one of total synchronous construction, seems to show stereotyped to us in the canals. These run in their fashioning rather with partial than with teleologic intent, giving as much concern to halfway points as to the goal itself, although in their action now they are totally involved. The thing was not done in a day, and by that very fact stamps the more conclusively its artificial origin.

The ability of beings there to construct such arteries of sustenance, two considerations will help to make comprehensible: one of these minifies the work, the other magnifies the workers. In the first place, it is not what we see that would have to be constructed. The object of endeavor is not only the water itself, but the products that water makes possible. It is vegetation which is matter of immediate concern, water being of mediate employment. This, then, is what would probably show. Just as on the earth it is the irrigated strip of reclaimed desert, and not the Nile itself, which would make its presence evident across interplanetary space. If these lines are irrigated bands of planting, the vertebral canal would be a mere invisible thread in the midst of that to which it gave growth. This alone would have to be made, and indeed it would probably be covered to prevent evaporation.

Now, we have evidence that the canals are thus composed of nerve and body. When they lie down, they do not entirely vanish. Under the visual conditions of Flagstaff they may still be made out in their dead season, the mere skeletons of themselves as they later fill out. And even so we do not actually see the nerve itself.

For the construction of these residuary filaments we have a plethora of capabilities to draw upon: in the first place, beings on a small planet could be both bigger and more effective than on a larger one, because of the lesser gravity on the smaller body. An elephant on Mars could jump like a gazelle. In the second place, age means intelligence, enabling them to yoke nature to their task, as we are yoking electricity. Finally, the task itself would be seven times as light. For gravity on the surface of Mars is only about 38 per cent of what it is on the surface of the earth; and the work which can be done against a force like gravity with the same expenditure of energy is inversely as the square of that force. A ditch, then, seven times the length of one on earth could be dug as easily on Mars.

With this motive of self-preservation for clew, and with a race equal to the emergency, we should expect to note certain general phenomena. Both polar caps would be pressed into service in order to utilize the whole available supply and also to accommodate most easily the inhabitants of each hemisphere. We should thus expect to find a system of conduits of some sort world-wide in its distribution and running at its northern and southern ends to termini in the caps. This is precisely what the telescope reveals. These means of communication should be, if possible, straight, both for economy of space and of time, it being especially necessary to avoid any wasteful evaporation on the road. Construction of such would needs be very difficult, if not impracticable, on earth, owing to the often mountainous character of its surface. But on Mars this is not the case. As we have seen, there are fortunately no mountains on Mars. Thus the great obstacle to canals, and, in consequence, the great obstacle to their acceptance, is providentially removed. Terrain offers the least of objections, terror the greatest of spurs, to their construction.

Thus we see that not only should the execution be possible, but that it should exhibit precisely the phenomena we see.

It would be interesting, doubtless, to learn how are bodied these inhabitants that analysis reaches out phenomena to touch. But body is the last thing we are likely to know of them. Of their mind as embodied in their works, we may learn much more; and, after all, is not that the more pregnant knowledge of the two ? Something of this we have surveyed together. But beyond the lime-light of assured deduction stand many facts awaiting their turn to synthetic coördination which we have not touched upon. It is proper to mention some of them under due reserve, for they constitute the bricks which, with others yet to come, will some day be built up into a housing whole.

Not least of these are those strange caret-shaped dark spots at the points where the canals leave the dark regions to adventure themselves into the light. No canal thus circumstanced in position is apparently without them, and, unlike the oases, they do not show round. On the theory of canalization they are certainly well placed. We have seen that the blue-green regions and the ochre ones lie undoubtedly at different levels, the former standing much lower than the latter.

Here, then, should occur difficulties in canalization which would have to be overcome. Are these, then, the evidence of their surmounting? They certainly suggest the fact.

Then the oases themselves lure our thoughts afield. Important centres to the canal system they are on their face. But, if centres to that, they should bear a like relation to what fashioned the canals. That they dilate and dwindle seasonally points to vegetation as their chief constituent, whence their name. But behind, and informing this, must be the bodied spirit of the whole. We are certainly justified in regarding them as the apple of the eye of Martian life—what corresponds with us to centres of population.

An interesting phenomenon about the oases makes this the more probable. Observation discloses that the oases are given to change both of size and tone. They fade at certain seasons, retaining only a relatively diminutive dark kernel. They are thus formed of two parts, pulp and core. The pulp itself indicates vegetation, since it follows the same laws as the canals; the core may well be the evidence of the permanent population. That the largest are some 75 miles across, seems to give sufficient space for living and the means to live. If our cities had to be their own sources of supply, they might well be of this size. As it is, Tokio is ten miles by ten, and London yet larger. But we must in this be careful to part surmise from deduction.

In our exposition of what we have gleaned about Mars, we have been careful to indulge in no speculation. The laws of physics and the present knowledge of geology and biology, affected by what astronomy has to say of the former subject, have conducted us, starting from the observations, to the recognition of other intelligent life. We have carefully considered the circumstantial evidence in the case, and we have found that it points to intelligence acting on that other globe, and is incompatible with anything else. We have, then, searched for motive and have lighted on one which thoroughly explains the evidence that observation offers. We are justified, therefore, in believing that we have unearthed the cause and our conclusion is this: that we have in these strange features, which the telescope reveals to us, witness that life, and life of no mean order, at present inhabits the planet.

Part and parcel of this information is the order of intelligence involved in the beings thus disclosed. Peculiarly impressive is the thought that life on another world should thus have made its presence known by its exercise of mind. That intelligence should thus mutely communicate its existence to us across the far stretches of space, itself remaining hid, appeals to all that is highest and most far-reaching in man himself. More satisfactory than strange this; for in no other way could the habitation of the planet have been revealed. It simply shows again the supremacy of mind. Men live after they are dead by what they have written while they were alive, and the inhabitants of a planet tell of themselves across space as do individuals athwart time, by the same imprinting of their mind.

Thus, not only do the observations we have scanned lead us to the conclusion that Mars at this moment is inhabited, but they land us at the further one that these denizens are of an order whose acquaintance was worth the making.

Whether we ever shall come to converse with them in any more instant way is a question upon which science at present has no data to decide. More important to us is the fact that they exist, made all the more interesting by their precedence of us in the path of evolution. Their presence certainly ousts us from any unique or self-centred position in the solar system, but so with the world did the Copernican system the Ptolemaic, and the world survived this deposing change. So may man. To all who have a cosmoplanetary breadth of view it cannot but be pregnant to contemplate extra-mundane life and to realize that we have warrant for believing that such life now inhabits the planet Mars.

A sadder interest attaches to such existence: that it is, cosmically speaking, soon to pass away. To our eventual descendants life on Mars will no longer be something to scan and interpret. It will have lapsed beyond the hope of study or recall. Thus to us it takes on an added glamour from the fact that it has not long to last. For the process that brought it to its present pass must go on to the bitter end, until the last spark of Martian life goes out. The drying up of the planet is certain to proceed until its surface can support no life at all. Slowly but surely time will snuff it out. When the last ember is thus extinguished, the planet will roll a dead world through space, its evolutionary career forever ended.

RED STAR (1922)

THE FIRST BOLSHEVIK UTOPIA

Alexander Bogdanov

Alexander Aleksandrovich Bogdanov was born in the Russian Empire in Sokółka (now in Poland) in 1873. He trained as a physician at Moscow State University and the University of Kharkiv (now Ukraine), graduating in 1899. After becoming interested in the possible use of blood transfusions to reverse ageing, Bogdanov began a programme of self-experimentation in 1924. Bogdanov also became politically active during his time at university and supported Bolshevik communism from 1903. His writings include works of medical research and political tracts. *Red Star* was his first work of fiction. He died aged 54 in 1928, having transfused himself with blood from a student who was suffering from malaria and tuberculosis.

1. Menni's Apartment

During the first period of my stay I moved in with Menni in a factory settlement—that is to say, a planned complex of industry and residences—whose physical center and economic base was a large chemical laboratory located far below ground. The part of the settlement above ground was spread through a park covering about ten square kilometers and consisted of several hundred apartment buildings for the laboratory workers, a large meeting hall, the Cooperative Depot, which was something on the order of a large department store, and the Communications Center, which connects the settlement with the rest of the world. Menni was the factory supervisor and lived not far from the community buildings right next to the main descent to the laboratory.

What first surprised me about nature on Mars, and the thing I found most difficult to get used to, was the red vegetation. The substance which gives it this color is similar in chemical composition to the chlorophyll of plants on Earth and performs a parallel function in their life processes, building tissues from the carbon dioxide in the air and the energy of the sun. Netti thoughtfully suggested that I wear protective glasses to prevent irritation of the eyes, but I refused.

"Red is the color of our socialist banner," I said, "so I shall simply have to get used to your socialist vegetation."

"In that case you must also recognize the presence of socialism in the plants on Earth," Menni remarked. "Their leaves also possess a red hue, but it is concealed by the stronger green color. If you were to don a pair of glasses which completely absorb the green waves of light but admit the red ones, you would see that your forests and fields are as red as ours."

I lack the time and space to describe the peculiar Martian flora and fauna, nor can I devote much attention to the atmosphere of the planet, which is pure and clear, relatively thin, but rich in oxygen. The sky is a deep, dark green, and the most prominent celestial bodies are the sun—much smaller than it appears on Earth—the two tiny moons, and two bright evening or morning stars, Venus and Earth. All of this was strange and foreign to me then and seems splendid and precious to me now as I look back upon it, but it is not essential to the purpose of my narrative. The people and their relationships are what concern me most, and they were the most fantastic and mysterious of all the wonders of this fairy-tale world.

Menni lived in a small two-story house that was indistinguishable architecturally from all the rest. The most original feature of this architecture was the transparent roof made of several huge sheets of blue glass. The bedroom and a parlor for receiving guests were located directly beneath it. Because of its soothing effect, the Martians prefer blue light during their leisure time. The color of the human face in this light does not strike them as gloomy. All of the work rooms—the study, Menni's home laboratory, the communications room—were on the ground floor, whose large windows freely admitted the restless red light reflected from the foliage of the trees in the park. This light made me uneasy and absentminded at first, but the Martians are used to it and find it has a stimulating effect on work.

Menni's study was full of books and writing implements, from ordinary pencils to a phonotype, a complex mechanism in which a phonograph recording of clearly enunciated speech activated the keys of a typewriter which accurately translated it into the written alphabet. Playing the phonogram did not erase it, so that one could use either it or the printed translation, whichever happened to be more convenient.

Above Menni's desk hung a portrait of a middle-aged Martian. He resembled Menni, although his almost sinister expression of grim energy and cold resolve was alien to Menni, whose face merely reflected a tranquil and resolute will. Menni told me the story of this man's life.

He was one of Menni's ancestors, a great engineer* who lived long before the social revolution, during the epoch of the Great Canals. It was he who planned, organized, and supervised that grandiose project. His first assistant envied his fame and power and launched a conspiracy against him. Several hundred thousand men were employed on the construction of one of the main canals, which passed through a swampy, disease-infested region. Thousands perished, and great discontent spread among the survivors. While the chief engineer was busy negotiating with the central government of Mars about pensions for the families of the dead and the incapacitated, his assistant was secretly rousing the dissatisfied workers against him, inciting them to strike and demand transfers to other regions. This was impossible, as it would have disrupted the entire plan of the Great Project, but he also urged them to call for the resignation of the chief

* Engineer Menni, the main character in the following novel.

engineer, which, of course, was quite feasible. When the latter learned of all this he summoned his assistant for an explanation and killed him on the spot. The engineer declined to defend himself at his trial, declaring that he considered his behavior just and necessary. He was sentenced to a long prison term.

Soon, however, it became obvious that none of his successors was capable of running the gigantic undertaking. Misunderstandings, embezzlement, and disorders followed. The entire mechanism of the project broke down; expenses increased by hundreds of millions, and the acute discontent of the workers threatened to end in open revolt. The central government hastenend to address an appeal to the chief engineer, offering to pardon him in full and reinstate him in his former position. He refused the pardon, but consented to head the project from prison. The inspectors he appointed quickly got to the bottom of things at the various construction sites. Thousands of engineers and contractors were put on trial. Wages were raised, the system by which the workers were supplied with food, clothing, and tools was reorganized from top to bottom, work plans were reviewed and revised. Order was soon fully restored, and once again the enormous mechanism began functioning rapidly and smoothly like an obedient tool in the hands of a real master.

The master, however, not only supervised the entire project but also planned its continuation in the years to come, grooming a certain energetic and talented engineer from a working-class background to be his successor. By the time his prison term had expired, everything had been so well prepared that the great master was confident the project could be safely entrusted to others. The very moment the prime minister of the central government arrived at the prison to release him, the engineer committed suicide. As Menni was telling me all this his face underwent a peculiar transformation, taking on an expression of inflexible severity that gave him a striking resemblance to his ancestor. I sensed that he understood and sympathized with this man who had died hundreds of years before he was born.

The communications room was at the center of the ground floor. It contained telephones attached to visual devices which transmitted an image of everything that passed in front of them at any distance. One of these apparatuses connected Menni's house with the Communications Center, which was in turn joined to all the cities of the planet. Other devices provided communication with the underground laboratory which Menni headed. These were in continuous function: several finely gridded screens showed a reduced image of illuminated rooms full of large metal machines and glass equipment attended by hundreds of workers. I asked Menni to take me on a tour of the laboratory.

"That would be ill-advised," he answered. "The substances handled there are unstable, and although we take considerable precautions, there is always a slight risk of explosion or of poisoning by invisible rays. You must not expose yourself to such dangers. Being the only one of your kind we have, you are irreplaceable."

Menni's home laboratory contained only the materials and equipment relevant to the research he was doing at a given moment. Near the ceiling of the corridor on the ground floor hung an aero-gondola that was always ready to take us wherever we might want to go.

"Where does Netti live?" I asked Menni.

"In a large city about two hours' flying time from here. A big engineering works employing tens of thousands of workers is located there, so Netti has ample material for medical research. We have another doctor at our enterprise."

"Surely I would be permitted to inspect that factory sometime?"

"Of course. There is nothing particularly dangerous there. We can visit it tomorrow if you like."

We decided to do so.

2. The Factory

We covered approximately 500 kilometers in two hours. That is the speed of a plummeting falcon, and so far not even our electric trains have been able to match it. Unfamiliar landscapes unfurled below us in rapid succession. At times we were overtaken by strange birds flying even faster than we. The blue roofs of houses and the giant yellow domes of buildings I did not recognize glittered in the sunlight. The rivers and canals flashed like ribbons of steel. My eyes lingered on them, for they were the same as on Earth. In the distance appeared a hugh [sic] city spread out around a small lake and transversed by a canal. The gondola slowed down and landed gently near a small and pretty house that proved to be Netti's. Netti was at home and glad to see us. He got into our gondola and we set off for the factory, which was located a few kilometers away on the other side of the lake.

It consisted of five huge buildings arranged in the form of a cross. They were all identically designed, each of them having a transparent glass vault supported by several dozen dark columns in a slightly elongated ellipse. The walls between the columns were made of alternating sheets of transparent and frosted glass. We stopped by the central building, also the largest, whose gates were about 10 meters wide and 12 meters high, filling the entire space between two columns. The ceiling of the first floor transected the gates at the middle. Several pairs of rails ran through the gates and disappeared into the interior of the building.

We ascended in the gondola to the upper half of the gates and, amid the deafening roar of the machines, flew directly into the second story. Actually, the floors of the factory were not stories as we understand them. At each level there were gigantic machines of a construction unfamiliar to me, surrounded by a network of suspended glass-parquet footbridges girded by beams of gridded steel. Interconnected by a multitude of stairways and elevators, these networks ascended toward the top of the factory in five progressively smaller tiers.

The factory was completely free from smoke, soot, odors, and fine dust. The machines, flooded in a light that illuminated everything yet was by no means harsh, operated steadily and methodically in the clean fresh air, cutting, sawing, planing, and drilling huge pieces of iron, aluminum, nickel, and copper. Levers rose and fell smoothly and evenly like giant steel hands. Huge platforms moved back and forth with automatic precision. The wheels and transmission belts seemed immobile. The soul of this formidable mechanism was not the crude force of fire and steam, but the fine yet even mightier power of electricity. When the ear had become somewhat accustomed to it, the noise of the machines began to seem almost melodious, except, that is, when the several-thousand-ton hammer would fall and everything would shudder from the thunderous blow.

Hundreds of workers moved confidently among the machines, their footsteps and voices drowned in a sea of sound. There was not a trace of tense anxiety on their faces, whose only expression was one of quiet concentration. They seem to be inquisitive, learned observers who had no real part in all that was going on around them. It was as if they simply found it interesting to watch how the enormous chunks of metal glided out beneath the transparent dome on moving platforms and fell into the steely embrace of dark monsters, where after a cruel game in which they were cracked open by powerful jaws, mauled by hard, heavy paws, and planed and drilled by sharp, flashing claws, small electric railway cars bore them off from the other side of the building in the form of elegant and finely fashioned machine parts whose purpose was a mystery to me. It seemed altogether natural that the steel monsters should not harm the small, big-eyed spectators strolling confidently among them: the giants simply scorned the frail humans as a quarry unworthy of their awesome might. To an outsider the threads connecting the delicate brains of the men with the indestructible organs of the machines were subtle and invisible.

When we finally emerged from the building, the engineer acting as our guide asked us whether we would rather go on immediately to the other buildings and auxiliary shops or take a rest. I voted for a break.

"Now I have seen the machines and the workers," I said, "but I have no idea whatever of how production is organized, and I wonder whether you could tell me something about that."

Instead of answering, the engineer took us to a small cubical building between the central factory and one of the corner edifices. There were three more such structures, all of them arranged in the same way. Their black walls were covered with rows of shiny white signs showing tables of production statistics. I knew the Martian language well enough to be able to decipher them. On the first of them, which was marked with the number one, was the following:

"The machine-building industry has a surplus of 968,757 man-hours daily, of which 11,325 hours are of skilled labor. The surplus at this factory is 753 hours, of which 29 hours are of skilled labor.

"There is no labor shortage in the following industries: agriculture, chemicals, excavations, mining," and so on, in a long alphabetical list of various branches of industry.

Table number two read:

"The clothing industry has a shortage of 392,685 man-hours daily, of which 21,380 hours require experienced repairmen for special machines and 7,852 hours require organization experts."

"The footwear industry lacks 79,360 hours, of which ..." and so on.

"The Institute of Statistics—3,078 hours ..." and so on.

There were similar figures on the third and fourth tables, which covered occupations such as preschool education, primary and secondary education, medicine in rural areas, and medicine in urban areas.

"Why is it that a surplus of labor is indicated with precision only for the machine-building industry, whereas it is the shortages everywhere else that are noted in such detail?" I asked.

"It is quite logical," replied Menni. "The tables are meant to affect the distribution of labor. If they are to do that, everyone must be able to see where there is a labor shortage and just how big it is. Assuming that an individual has the same or an approximately equal aptitude for two vocations, he can then choose the one with the greater shortage. As to labor surpluses, exact data on them need be indicated only where such a surplus actually exists, so that each worker in that branch can take into consideration both the size of the surplus and his own inclination to change vocations."

As we were talking I suddenly noticed that certain figures on the table had disappeared and been replaced by others. I asked what that meant.

"The figures change every hour," Menni explained. "In the course of an hour several thousand workers announce that they want to change jobs. The central statistical apparatus takes constant note of this, transmitting the data hourly to all branches of industry."

"But how does the central apparatus arrive at its figures on surpluses and shortages?"

"The Institute of Statistics has agencies everywhere which keep track of the flow of goods into and out of the stockpiles and monitor the productivity of all enterprises and the changes in their work forces. In that way it can be calculated what and how much must be produced for any given period and the number of man-hours required for the task. The Institute then computes the difference between the existing and the desired situation for each vocational area and communicates the result to all places of employment. Equilibrium is soon established by a stream of volunteers."

"But are there no restrictions on the consumption of goods?"

"None whatsoever. Everyone takes whatever he needs in whatever quantities he wants."

"Do you mean that you can do all this without money, documents certifying that a certain amount of labor has been performed, pledges to perform labor, or anything at all of that sort?"

"Nothing at all. There is never any shortage of voluntary labor—work is a natural need for the mature member of our society, and all overt or disguised compulsion is quite superfluous."

"But if consumption is entirely uncontrolled, there must be sharp fluctuations which upset all your statistical compilations."

"Not at all. A single individual may suddenly eat two or three times his normal portion of a given food or decide to change ten suits in ten days, but a society of billions of people is not subject to such fluctuations. In a population of that size deviations in any given direction are neutralized, and averages change very slowly and with the strictest continuity."

"In other words your statistics work almost automatically—they are calculations pure and simple?"

"No, not really, for there are great difficulties involved in the process. The Institute of Statistics must be alert to new inventions and changes in environmental conditions which may affect industry. The introduction of a new machine, for example, immediately requires a transfer of labor in the field in which it is employed, in the machine-building industry, and sometimes also in the production of materials for both branches. If a given ore is exhausted or if new mineral fields are discovered there will again be a transfer of labor in a number of industries—mining, railroad construction, and so on. All of these factors must be calculated from the very beginning, if not with absolute precision then at least with an adequate degree of approximation. And until firsthand data become available, that is no easy task."

"Considering such difficulties," I remarked, "I suppose you must constantly have a certain surplus labor reserve."

"Precisely, and this is the main strength of our system. Two hundred years ago, when collective labor just barely managed to satisfy the needs of society, statistics had to be very exact, and labor could not be distributed with complete freedom. There was an obligatory working day, and within those bounds it was not always or fully possible to take the vocational training of the workers into account. However, although each new invention caused statistical problems, it also contributed to solving the main difficulty, namely the transition to a system in which each individual is perfectly free to choose his own occupation. First the working day was shortened, and then, when a surplus arose in all branches, the obligation was dropped altogether. Note that the labor shortages indicated for the various industries are almost negligible, amounting to mere thousands, tens or hundreds of thousands of man-hours out of the millions and tens of millions of hours presently expended by those same industries."

"But shortages of labor do still exist," I objected. "Yet I suppose that they are covered by later surpluses, are they not?"

"Not only by later surpluses. In reality, necessary labor is computed by adding a certain quantity to the basic figures. In the most vital branches of industry—the production of food, clothing, buildings, machines, and so on—this margin can be as high as 5 percent, whereas in less important areas it is about 1.2 percent. Thus generally speaking, the figures in these tables indicating shortages express merely a relative deficiency, not an absolute one."

"How long is the average working day—at this factory, for example?"

"From an hour and a half to two and a half hours," replied the guide, "but there are those who work both more and less. Take, for example, the comrade operating the main hammer. He is so fascinated by his job that he refuses to be relieved during the entire six hours daily the factory is in production."

I mentally translated these figures from the Martian system of reckoning into our own. On Mars a day and night together are a little longer than on Earth and are divided into ten of their hours. This means that the average working day is from four to six Earth-hours, and the longest operational day is about fifteen hours, which is approximately the same as in our most intensely run enterprises.

"But isn't it harmful for that comrade at the hammer to work so much?" I asked.

"Not for the time being," Netti replied. "He can permit himself such a luxury for another six months or so. But of course I have warned him of the dangers to which his enthusiasm exposes him. One such risk is the possibility of a convulsive fit of madness that may irresistibly draw him under the hammer. Last year something like that happened at this very factory to another operator who was likewise fascinated by powerful sensations. It was only by a lucky chance that we managed to stop the hammer in time and avert the involuntary suicide. An appetite for strong sensations is in itself no disease, but it can easily become perverted if the nervous system is thrown ever so little off-balance by exhaustion, emotional disturbances, or an occasional illness. Of course I try to keep an eye on those workers who become overly engrossed in any sort of monotonous work."

"Shouldn't this man you mentioned have cut down his labor, considering that there is a surplus in the machine-building industry?"

"Of course not," Menni laughed. "Why should just he take it upon himself to restore the equilibrium? The statistics oblige no one to do that. Everyone takes these figures into consideration when making their own plans, but they cannot be guided by them alone. If you were to want to begin working at this factory you would probably find a job; the surplus figure in the central statistics would rise by one or two hours, and that would be that. The statistics continually affect mass transfers of labor, but each individual is free to do as he chooses."

We had time to rest up during our conversation, and everyone except Menni, who was forced to leave by a call from his laboratory, continued the excursion through the factory. I decided to spend the night at Netti's, as he promised to take me to the Children's Colony, where his mother was working as an educator.

THE ORIGIN OF LIFE (1924)

THE WORLD OF THE LIVING AND THE WORLD OF THE DEAD

Alexander Oparin

Alexander Ivanovich Oparin was born in 1894 in Uglich, then part of the Russian Empire. His family moved to Moscow when he was nine because there was no secondary school in Uglich. He subsequently studied plant physiology at Moscow State University, graduating in 1917. In 1924 he published *The Origin of Life*, in which he proposed that there was no essential difference between living and non-living matter. In 1927 he was made professor of biochemistry at Moscow State University and helped to found a biochemical institute in Moscow in 1935. Throughout his career, he was highly decorated by the Soviet authorities; by the time of his death in 1980, he had received the Order of Lenin five times.

Grau, teurer Freund, ist alle Theorie,
Und grün des Lebens goldener Baum.
GOETHE

The first and most eye-catching difference between organisms and the rest of the (mineral) world is their chemical composition. The bodies of animals, plants and microbes are composed of very complicated, so-called organic substances. With very few and insignificant exceptions we do not find these substances anywhere in the mineral world. Their main peculiarity consists in the fact that at high temperatures in the presence of air they burn while in the absence of air they carbonize. This shows that they contain carbon. This element is present in all organisms without exception. It forms the basis of all those substances of which protoplasm is made up. At the same time, however, it is no stranger to the mineral world.

The diamond is certainly more or less well known to everyone. It is a precious stone which, in the cut form, is much used as an adornment under the name of brilliant.

As early as 1690 the English scientific genius Newton put forward the idea that diamond, the hardest and brightest body on the Earth, must contain combustible material. This suggestion was quickly confirmed. A diamond was submitted to trial by fire at the focal point of a burning mirror. It did not survive the test but became covered with cracks, glowed and burnt away before everyone's eyes. Later and more accurate experiments showed beyond doubt that the shining diamond is the blood-brother of the unattractive graphite, the substance from which pencils are made. Both consist entirely of carbon. They have no component other than this element.

However, carbon is met with in the mineral world, not only in the pure state, but also in combination with other elements. Chalk, marble, soda, potash and other compounds all contain carbon. In general, one of the most valuable properties of carbon is its tendency to form the most varied compounds with other elements. In the substances which make up the bodies of organisms the carbon is always found combined with hydrogen, oxygen, nitrogen, sulphur and phosphorus and often with several other elements. All these elements are widely distributed in the non-living world. In combination with oxygen, hydrogen forms water; the air around us consists of a mixture of oxygen and nitrogen; sulphur and phosphorus are found in many of the minerals which make up rock formations. Thus we see that all the elements which enter into organic compounds are also found in abundance in mineral world. Even this alone gives reason to doubt the existence of any essential difference between the world of the living and that of the dead.

It may be, however, that this difference concerns not so much the elementary compositions as the actual compounds made up from the elements.

It has been considered that the substances of the organic and inorganic (mineral) world are so unlike one another that the former could not be obtained from the latter by artificial means under any circumstances. Some chemists of that time even maintained that it was impossible to obtain organic bodies simply because these substances could only be formed within the living organism by the action of a special "vital force". However, as early as 1828 the chemist Wöhler succeeded in preparing an organic substance artificially and so to cast doubt on the importance of the famous "vital force".

Since then the study of organic compounds has been advancing with rapid strides, and the further it has gone the clearer has become the falsity of the idea that there is a fundamental difference between these substances and inorganic bodies. Starting from the simplest inorganic substances, chemists can now prepare artificially almost all the substances which are encountered in organisms. Although some of these substances have not yet been obtained there is no doubt that they can be obtained in the near future. The structure of these substances has been studied in extreme detail. No special means of combination of the individual elements has been found in them. They obey the same physico-chemical laws with the same constancy as inorganic compounds.

The essential similarity between organic and inorganic substances has now become so obvious that not a single serious natural scientist would deny it and the protagonists of the view that there is a fundamental difference between the living and the dead have already stopped. They assert that organic compounds whether prepared artificially or isolated from organisms are just as dead as minerals. Life may be recognized only in bodies which have particular special characteristics. These characteristics are peculiar to living things and are not seen in the world of the dead.

What are these characteristics? In the first place there is the definite structure or organization. Then there is the ability of organisms to metabolize to reproduce others like themselves and also their response to stimulation.

Let us go over each of these characteristics and see whether it is really present only in the living organism or whether it is not, in some form or another, also found in the mineral world.

The most important and essential characteristic of organisms is, as is demonstrated by their very name, their organization, their particular form or structure. The bodies of all living things, beginning with the smallest bacteria and algae and ending with man, are constructed according to a definite plan in which the greatest importance attaches, not to the external visible organization but to the fine structure of the protoplasm of the cells which make up the organism. This structure is the same in general for all members of the animal and vegetable kingdoms. Unfortunately it has still not been studied very much. Various investigators have seen different structural formations in the semi-liquid protoplasm in the form of fibres, networks and alveoli. However, as these formations are so extremely small it is very hard to see them, even with the best microscopes now available. All the same it is certain that protoplasm has a definite structure and is not a homogeneous lump of slime. This structure holds the secret of life. Destroy it and there will remain in your hands a lifeless mixture of organic compounds.

Some scientists believe that this structure itself, this organization, could not haven been formed spontaneously from structureless and, according to them, lifeless substances. Following the ancient Greek philosopher Empedocles, they repeat, in one form or another, the idea that the organization is closely bound up with the spirit which both constructs the body and is destroyed or flies away when the particular form is annihilated. However, if we adopt this position we must agree with another philosopher, Thales, in ascribing a spirit to a magnet since this extremely simple spirit which expresses itself in the attraction which iron holds for the magnet, depends on the structure of the magnet, that is to say on the arrangement of the particles in it. This structure has only to be disturbed, as when a magnetic stone is ground up in a mortar, and it will lose its spirit just like an organism which has been cut into pieces.

The world of the dead or mineral world is certainly not lacking in definite forms, it is not structureless. It is a property of most chemical substances that they try to take up particular forms, that is to form crystals. We now know that the finest particles of the substance forming the crystals are not just arranged anyhow in them, but are arranged according to a definite plan depending on the chemical composition and the conditions under which the crystals are formed by separation from a solution or from the molten state. It is here that a transition takes place from the hitherto formless, structureless substance to the organized body. In a solution the smallest particles of the substance are in disorder; the same is true even when a substance is melted. However, when a crystal separates out, the particles arrange themselves relative to one another in a strict order, like soldiers forming straight ranks on the command "Attention!" The form of the crystal and the whole range of its other properties depend upon this arrangement. If the crystal is destroyed by disturbing the arrangement of the particles all these properties disappear.

There is thus no doubt that even the simplest crystals have a definite arrangement. At first glance this arrangement seems to be extremely simple. However, we shall immediately put this idea aside if we remember those marvellous "ice flowers" which appear on the panes of our windows on a frosty day. In their delicacy, complexity, beauty and variety these "ice flowers" may even look like tropical vegetation while all the time being nothing at all but water, the simplest compound we know. In small droplets of it the particles were scattered in disorder but they were cooled, the wind blew, the temperature fell below zero and these particles, complying with the eternal laws of nature, which are the same for both the living and the dead, arranged themselves in a definitive order and, on the simple window pane, they produce pictures of fabulous gardens, glistening in the sunshine with all the colours of the rainbow.

If we leave the simplest compounds aside and go on to look at those forms which produce more complicated substances an even more varied and involved picture will be presented to our gaze which, in its complexity, is in no way inferior to the most detailed picture of the structures of organisms. However, before we enter on this review we must make a small excursion which, though perhaps rather tedious, is necessary for our further discussion into the realms of the comparatively young science of colloid chemistry which has already acquired great importance.

As early as 1861, the English scientist, Graham, divided all the chemical bodies known at that time into two main classes, crystalloids and colloids. To the first class belonged such substances as various salts, sugar, organic acids and so forth. The substances in this group formed crystals easily and, when dissolved, gave clear and completely transparent solutions. If such a solution is poured into a bag made of vegetable parchment or the bladder of an animal and the bag is placed in pure water, the dissolved substance will pass through the walls of the bag and be washed out of it by the water.

Colloids present a completely contrary picture. They very seldom crystallize and then only with great difficulty. Their solutions are usually cloudy and they cannot pass through vegetable or animal membranes. Graham assigned such substances as starch, proteins, gums and mucus to this group.

It turned out later that such an assignment of all substances between two groups was not altogether correct since the same chemical compound could turn up both as a colloid and as a crystalloid according to the conditions of the solution. Thus "colloidness" was not a property of a particular substance but of its particular state.

There was, however, a considerable element of truth in the classification made by Graham since substances which have very large and complicated particles very often and easily give rise to colloidal solutions. The study of the colloidal state is therefore of special importance since the vast majority, if not all, of the substances of which protoplasm is made up have very large and complicated particles and therefore must give colloidal solutions.

As we have already said, colloidal substances do not give crystals, but still they are fairly easily precipitated as clots or lumps of mucus or jelly. We may take as our example of such formations the protein of eggs which is precipitated on boiling and the gelatin which sets on cooling (well known to everyone as a jelly). The separating out of such coagulates or precipitates from previously uniform solutions sometimes takes place amazingly easily for apparently insignificant reasons.

The coagulates or gels obtained by precipitating colloids are, at first sight, quite structureless. If we examine a lump of jelly under the microscope, even at a high magnification, it appears to us quite uniform. However, scientists have now invented very effective instruments with which they have succeeded in revealing the complicated structure of coagulates. Here it is not a question of straight lines and planes such as we meet in crystals, for here we have a whole network, a whole skein of fine threads which are interlaced, separating from one another and coming together again in a definite, complicated order. Sometimes these threads are very fine, on the other hand, sometimes they are thickened, fusing with one another to form small enclosed bubbles or alveoli.

The structure of coagulates is strikingly reminiscent of that of protoplasm. Unfortunately this structure has not yet been sufficiently well studied for us to be able to say anything conclusive about this resemblance. However, there can be no doubt that we are dealing here with phenomena of the same order. There is no essential difference between the structure of coagula and that of protoplasm.

It may be, however, that the difference between living and dead does not lie in the organization which, as we have seen, is present in both worlds, but in the other features which we mentioned, the ability of living organisms to metabolize to reproduce themselves and to respond to stimuli.

The shapes of crystals are unalterable, they are formed once and for all, while an organism may be compared with a waterfall which keeps its general shape constant although its composition is changing all the time and new particles of water are continually passing trough it. The composition of the living body changes in just the same way. The organism takes in different substances from its environment; after a number of chemical changes it assimilates these substances, transforming what had been foreign compounds into parts of its own body. The organism grows and develops at the expense of these substances. However, just as a factory requires a certain amount of fuel to carry on its work, so, if the organism is to carry on its unceasing activities it should consume, that is to say break down, at least part of the material which it has assimilated, and this is what actually happens. In the process of respiration or fermentation the organism breaks down substances which it has already taken in and the products of their degradation or decomposition are given off into the environment. Thus life consists of continual absorption, construction and destruction.

However, if we make a detailed analysis and simply contemplate the phenomenon of metabolism which has been described, we shall not find, even here, anything specifically characteristic of the living world. In fact, the phenomenon of feeling [sic], the assimilation of substances from the environment, is, of course, found in its simplest form even in crystals. Thus, a crystal of common salt, which is well known to everyone, will, if it is immersed in a supersaturated solution of the same substance, increase its size and grow by absorbing individual particles of substance from its environment (the solution) and making them part of its body. Even here, in this simple phenomenon, we have before us all the characteristic features of the phenomenon of nutrition. There is even more similarity between this phenomenon and the processes which occur in colloidal coagula. A lump of such material has the ability to extract from solution and absorb the most varied substances such as dyes. These latter do not just remain on the surface of the lump but penetrate deeply into it, some of them simply adhering to the tangled threads which constitute the lump while others enter into chemical reactions with these threads and combine firmly with them, forming component parts of the whole lump.

Study of the process of feeding of living protoplasm shows that this process too takes place in exactly the same way as has just been described. Solutions of different chemical substances penetrate into the protoplasm as a result of the action of comparatively simple and thoroughly studied physical forces, just the same forces which operate in colloidal coagula. Having entered into the protoplasm, one substance will quickly pass out of it while others will enter into a chemical reaction with it, combine with it and become parts of it. And here, all in wall [sic], we have a simple chemical reaction and not anything mysterious such as could only be accomplished by a "vital force".

Thus, when various chemicals are absorbed by lifeless coagula we are dealing with processes which take place in a way which is completely analogous with the first stage of metabolism, that is to say, feeding.

The following example shows that, even in the world of the dead, we can find processes which are essentially just the same as metabolism as a whole. If we take a small piece of so-called spongy platinum (platinum is one of the "noble" metals which can be obtained by special methods not as sheets or solid lumps, but as a very delicate sponge with very fine holes and delicate walls) and throw it into a solution of hydrogen peroxide in water* then bubbles of oxygen will immediately begin to form on the surface of the lump. They are formed by the breakdown of the hydrogen peroxide and the process goes on quite rapidly and only stops when all the hydrogen peroxide has been broken down to oxygen and water. If we then remove, dry and weigh our piece of platinum we shall find its weight has remained just as it was before. The same piece may again be thrown into a new amount of hydrogen peroxide and will again decompose it quickly while itself remaining unaffected. Thus, a comparatively small piece of spongy platinum can decompose an unlimited quantity of hydrogen peroxide.

Chemists have been interested in the mechanism of this process for a long time, and as a result of many investigations we now know for certain that the decomposition of hydrogen peroxide by platinum takes place in the following way. First the peroxide is adsorbed on the platinum. As the peroxide cannot penetrate into the metal this adsorption only takes place on its surface. That is why it is important to use spongy platinum for this experiment as it has a very large surface at which the metal is in contact with the liquid. The particles of peroxide do not simply adhere to the surface of the platinum but form a chemical compound with it, namely the

* Hydrogen peroxide is a chemical compound which, like water, is composed of hydrogen and oxygen only but there is twice as much oxygen in it as in water. This compound is fairly stable and can usually stand for a very long time without breaking down. When it breaks down for any reason it gives rise to oxygen and water.

hydrate of platinum peroxide. Thus the piece of metal, like the living organism, extracts particles of hydrogen peroxide from water and assimilates them into its body. However, it does not end at that. After a short time the hydrate of platinum peroxide on the surface of the metal breaks down to platinum, water and oxygen, the last being given off in the form of bubbles of gas. The reduced platinum can combine with a new portion of hydrogen peroxide and again break it down, the products of the decomposition being oxygen and water. The process is repeated until there is no hydrogen peroxide left in the solution surrounding the metal.

In the example given we have the simplest but still complete prototype of metabolism. It contains all the important elements of this process. The absorption of substances from the surrounding medium, their assimilation and breakdown and the giving off of the products of their decomposition. Just the same process takes place in any living organism, for example in any bacteria cultivated in a solution of nutritive substances. The bacteria absorb the substances from the solution, assimilate them and then break them down, giving off to the outside the products of their decomposition. Thus, a simple piece of metal behaves in just the same way as a living organism.

In this connection it must be pointed out that both phenomena (the metabolism of organisms and the decomposition of hydrogen peroxide by platinum) are not only similar, in their external form, but the actual mechanism of the process is similar in both cases. In all organisms, without exception, metabolism is brought about by means of so-called enzymes. This name is given to substances, the chemical nature of which is still only poorly understood, but which can comparatively easily be isolated from any animal or plant in solution in water or as a powder which is easily dissolved in water. All enzymes now known have the power to act on substances forming part of the living body in a very remarkable way. They alter these substances in one way or another (either by breaking them down or by causing them to combine with one another) while themselves remaining completely intact. A detailed study of this phenomenon has shown that enzymes act on the different organic compounds of the living body in just the same way as does platinum on a solution of hydrogen peroxide. In fact metabolism, in its most important aspects, does not consist of anything but a long series of successive enzymic processes following one another and related to one another like the links of an unbroken chain.

At present one of the most extensive sections of physiology, the science which studies the functioning of living organisms, is devoted exclusively to the problem of metabolism. The further the study of this complicated process goes on, the more closely and accurately we get to know the essential features of the processes which are carried out in the living cell, the more strongly we become convinced that there is nothing peculiar or mysterious about them, nothing that cannot be explained in terms of the general laws of physics and chemistry.

Thus, even the ability to metabolize cannot be taken as a special characteristic peculiar to living organisms.

We still have two "peculiarities of life" left to discuss, namely, the capacity for self-reproduction and response to stimuli.

In what does the capacity of organisms for self-reproduction consist? In the simplest case it amounts to this: the elementary organism, the cell, divides itself into two halves each of which then grows into a new daughter cell in which the structure of the mother cell is reproduced down to the finest detail. This property, however does not belong only to organisms, but to all bodies possessing a definitive structure, without exception. Let us take the simplest case as our example. If we take a crystal of any substance such as alum, break it into two halves and place them in a supersaturated solution of the same substance what will happen will be that the halves of the crystal which had been placed in the solutions will quite quickly replace their missing faces, angles and edges at the expense of particles which had previously been floating freely in the solution. Before growing larger they take a form which reproduces in the finest details that of their mother, the original crystal.

The question may, however, arise that in the example which we have given we forcibly broke the crystal whereas the division of the cell apparently takes place spontaneously. Is that not the fundamental difference between the two phenomena? The fact is, however, that it only seems to us that the division of cells takes place spontaneously, it really takes place under the influence of definite physical forces (capillary attraction, surface tension) which, though they certainly have not yet received much study, still are of just the same kind, in principle, as all the other physical forces.

An even more interesting phenomenon is that of the "seeding" of supersaturated solutions. It occurs as follows. In some case it is possible to concentrate a solution of a particular substance very strongly without that substance separating out in the solid form. However, if the most minute crystals will immediately begin to separate out of the solution, sometimes in such quantities that the whole mass becomes crystalline.

This shows that a crystal can cause the formation of bodies like itself which would otherwise not have been formed. If the particles which are scattered at random throughout the solution are to arrange themselves according to a definite plan to give a definite organization or form, that form must already be present.

Here we have the occurrence of the most amazing phenomena which may serve as a key to the understanding of other extremely complicated phenomena of the same order. "Let us take the example of sulphur" as Carus Sterne says, "this is known to be a simple substance yet it depends on the temperature at which it changes from the liquid to the solid state which of two very different forms it will take, octahedral or prismatic*. If we place two such crystals on fine platinum wires in a supersaturated solution of sulphur in benzine then, in the neighborhood of the prismatic crystal new prisms will be formed while octahedra will be formed near the octahedral crystal. When the two armies of crystals approach one another, the latter form is victorious at the first clash. Here is an example of the struggle for existence in the realm of crystals!".

Let us now go on to take a look at the last of the peculiarities of living things which we mentioned, that is, at responsiveness to stimuli. In all living things

* On crystallizing, different substances take on regular geometrical shapes. The crystals are characterized by these forms. Some substances may crystallize in different forms under different conditions. In particular, sulphur can give crystals in the form of octahedra (two four-sided pyramids joined together by their bases) or hexagonal prisms.

without exception we meet with a property which in its most general form may be described as follows. In an organism external and internal stimuli will cause something of the nature of a discharge and will induce the performance by it of some definite action (e.g. movement, etc.) which will carry out in a particular way according to its structure and the means at its disposal. It is a very characteristic feature of responsiveness that there should be a quantitative disparity between the energy, that is to say the forces, required to excite or bring about the response, and the work which is the response of the organism to the stimulus in question. Thus, for example, a relatively slight touch can be enough to induce the organism to move from one place to another or to carry out some other work requiring the expenditure of much force. The organism draws from within itself the forces (energy) required for this work.

Some scientists believe that responsiveness is a specially characteristic feature of organisms. However, if we take this view we shall have to regard a railway locomotive with the steam up as a living thing. In fact it is only necessary to apply a slight stimulus by shifting a lever and the locomotive will start to move, carrying out a very considerable amount of work at the expense of the fuel which is burnt in its boiler. This work is many times greater than that expended in moving the lever and is carried out in complete accordance with the structure of the locomotive.

Responsiveness is to be found, not only in organisms, but also in any physical body which has any noteworthy store of hidden (potential) energy. Comparatively insignificant causes may lead to the discharge of this energy which will lead to the carrying out of some particular work. A landslide caused by a comparatively slight movement of the air, the explosion of a powder magazine caused by a spark which happened to fall in it – these are very simple cases of the phenomenon of responsiveness.

With this we will finish our short review of the main feature of living organisms. We have seen that not one of these can be held to be inherent only in living things. But if this is so we have no reason to think of life as being something which is completely different in principle from the rest of the world. If life had always existed and had not arisen by generation bodily from the rest of the world, if it had not separated itself or, crystallized out at some time from this world, then it would inevitably have had characteristics peculiar to itself. But this is not so. The specific peculiarity of living organisms is only that in them there have been collected and integrated an extremely complicated combination of a large number of properties and characteristics which are present in isolation in various dead, inorganic bodies. Life is not characterized by any special properties but by a definite, specific combination of these properties.

In course of the colossal length of time during which our planet, the Earth, has existed, the appropriate conditions must certainly have arisen in which there could have been the conjunction of properties which were formerly disjoined to form the combination which is characteristic of living organisms. To discover these conditions would be to explain the origin of life.

MARS: THE CASE FOR LIFE (1998)

Christopher McKay

Christopher McKay is an American planetary scientist at the NASA Ames Research Centre, California. Born in 1954, he studied physics and mechanical engineering at Florida Atlantic University and obtained a PhD in astrogeophysics from the University of Colorado in 1982. He has been a co-investigator on two NASA Mars missions: the Mars Phoenix lander and the Mars Science Laboratory rover, named Curiosity.

It would be most interesting to live in a solar system that was home to a multitude of planets with life: worlds full of life in all its forms. Unfortunately, we don't. Our solar system has just one planet with life: Earth. But there is another planet that still has the potential to teach us about life: Mars.

In my opinion the most compelling scientific question about Mars is life: Did it, or does it, have life? Beyond that, can life from Earth survive there—what is the future of life on Mars? I'm going to review what we know about life and Mars and how we've looked for it in the past and will, or ought to, in the future.

The search for life on Mars began in earnest with the Viking missions in 1976. Viking went to search the sands of Mars for LGMs—little green microbes. Actually, of three Viking biology experiments, only one searched for green (photosynthetic) microbes. The other two experiments tried to detect bacteria capable of consuming organic material. Interestingly, all the biology experiments yielded some positive results. Indeed, the Labeled Release (LR) experiment gave precisely the result that would have been expected had there been life on Mars. In this experiment, nutrients added to the Martian soil were decomposed to carbon dioxide, but when the soil was heated to sterilizing temperatures, the decomposition did not occur—all consistent with life. However, one of the other biology experiments, the Gas Exchange (GEx) experiment, also showed activity, but not in a way suggestive of biology. This experiment was able to detect a variety of gases released from the soil when a nutrient solution was added. Researchers found that merely moistening the soil with water released oxygen, and that heating the soil did not deactivate the oxygen release. The abrupt release of the oxygen; the fact that the sample was in the dark; and the inability to quench the reaction by sterilization all imply a non-biological cause. The case for a chemical explanation was sealed by the organic analysis instrument on the Viking landers. Neither lander detected any complex organic material in the Martian soil at levels of parts per billion.

The standard explanation for the reactivity seen by the Viking biology experiments and the absence of organics in the Martian soil centers on the presence of one or more chemical oxidants in the soil. These oxidants are presumably produced by ultraviolet light acting on the soil or creating hydrogen peroxide from water in the atmosphere. These oxidants are what has—over billions of years—made Mars red. I can summarize what Viking told us about the surface of Mars: "It's dead Jim."

Why is Mars so dead? The answer is not that Mars lacks the elements necessary to support life. Its atmosphere contains carbon dioxide, nitrogen, and water. These are the basic compounds of a biosphere. In addition, Martian soil holds other elements needed for life. The problem with Mars from a biological perspective is its low atmospheric pressure and the resultant lack of liquid water. The pressure on Mars averages about 120 times less than the pressure at sea level on Earth. At this low pressure water can barely exist as a liquid. As pressure decreases water boils at lower temperatures. In Denver—a mile high—water boils at 95°C while at sea level it boils at 100°C. The pressure on Mars corresponds to an Earthly altitude of nearly 20 miles, so water boils at a temperature only a few degrees above its freezing point. Thus, any liquid water on Mars would evaporate rapidly since it would be near its boiling point. This evaporation would cool the water quickly, and since it would be so near its freezing point as well, the water would freeze. For this reason liquid water essentially does not exist on Mars, at any place, at any season. Without liquid water it's not surprising that Mars has no life on its surface.

Mars has not always been a dry world. There is direct evidence that early in its history Mars had liquid water flowing on its surface long enough to carve impressive canyons and channels. Orbital images from Mariner 9, Viking and now from the Mars Global Surveyor show these fluvial features. The observation that at one time Mars had liquid water for sustained periods is the fundamental motivation for the search for past life on Mars.

The large volcanoes on Mars—all now extinct—provide further evidence that Mars was more Earth-like early in its history. The increased volcanism and the presence of stable liquid water both indicate that the atmosphere must have been thicker early in Martian history.

We can determine when Mars experienced its early Earth-like period from the association of the fluvial channels with Mars' ancient cratered terrain. We know from the analysis of the lunar samples returned by the Apollo program that the Moon's period of intense cratering ended 3.8 billion years ago. Assuming that this was true on Mars as well, we then can infer that the main epoch of stable liquid water flow on Mars was around 3.8 billion years ago. Computer simulations of how the Martian climate would have deteriorated after the end of the heavy bombardment suggest that liquid water remained on Mars until about 3.5 billion years ago. The total duration of liquid water on Mars may have been several hundred million years. Long compared to the life of an organism but short compared to the life of a planet.

The question of life becomes most interesting when we compare the early history of Earth and Mars. This comparison is shown in Figure 2.

The fossil record on Earth from 3.5 to 4 billion years ago is incomplete, since most of the rocks that old have been destroyed or heavily altered. But still we have direct evidence of life on Earth at 3.5 billion years ago. This evidence is in the form of stromatolites (fossilized layers of microbes) and microfossils that are similar to modern microbes. At 3.9 billion years ago, we do not find clear fossils, but there is chemical evidence for life in the form of a characteristic enrichment of the lighter isotope of carbon in organic sediments. Throughout Earth's history this characteristic enrichment has been due to biological processes, and the comparison to Mars suggests that if Mars had liquid water at this time, it could also have had life.

The scenario I just described—of an initially warm and wet Mars that quickly became the cold dry world we see today—is based on data from spacecraft that went to Mars. However, the same basic outline of Martian history can be deduced from meteorites on Earth that have come from Mars.

Among the many thousands of meteorites known, we have about a dozen rocks that came from Mars. The evidence for their Martian origin is compelling. There is oxygen isotope data indicating that all these rocks came from the same parent body. These data do not by themselves show conclusively that Mars was that parent body. Direct evidence of a Martian origin comes from gas inclusions in the youngest of the Martian meteorites. The relative concentrations of different gases in the bubbles in this meteorite compare exactly (over a range of concentrations that span nine orders of magnitude) with the Martian atmosphere as measured by the Viking landers.

The Martian meteorites also indicate that Mars started out more Earth-like and then became the cold and dry planet we know today. The oldest Martian meteorite—formed over 4 billion years ago—reflects formation in warm, wet environmental conditions, while the younger Martian meteorites—formed on Mars less than 1.3 billion years ago—reflect formation in today's cold, dry environment.

Our quest then is to go back in time to Mars 3.8 billion years ago during its Earth-like phase and search for evidence of life. Where on Mars should we search?

We have developed an approach for searching for Martian fossils based on studies in the coldest, driest, most Mars-like place on Earth: the dry valleys of Antarctica.

The mean temperature in the dry valleys is –20°C and the precipitation—all as snow—is ten times less than in Death Valley. The valleys are so cold and dry that they appear lifeless, but life is there, hidden a millimeter or two beneath the surface of sandstone rocks and beneath the thick ice covers of lakes on the valley floors. These perennially ice-covered lakes provide an analog for past life on Mars and indicate what might be the best way to search for evidence of this past life.

Although the mean temperatures in the dry valleys are 20°C below freezing, the summertime temperatures can rise to slightly above freezing. When this happens, glaciers ringing the valleys melt and their waters flow down the valley floor and into the lakes. As this meltwater freezes, under the ice in the lake, it releases the latent heat of fusion. This heat release is the fundamental energy source that maintains the liquid water in the lakes despite average ambient temperatures below freezing. The thickness of the ice cover (about 4–6 meters) is set by the balance between heat added to the lake by the freezing of meltwater and heat loss from the lake by conduction. In steady state, the former is equal to the rate of ablation from the top of the ice cover and the latter is proportional to the thickness of the ice cover. Beneath the ice, the lakes vary from 30 to over 100 meters deep.

Sunlight penetrating the thick ice covers allows for photosynthesis. Simple life forms—algae, diatoms, and bacteria—are found in the lakes, but no higher life forms are present.

These ice-covered lakes are an example of how liquid water can be present and how life can survive when temperatures are well below freezing. A similar type of habitat could have been the main reservoir of life on Mars while average temperatures were below freezing.

I would suggest that the place to search for fossils on Mars is an ancient lake bed. Not only would such a lake bed have been a potential site for life as Mars became colder (as shown by the Antarctic dry valley lakes), but the sediments on the bottom of the lake would have provided a means to preserve fossil evidence of that life.

Note that a river or canyon might have had liquid water, but such site would not usually be a location in which sedimentary material would have trapped fossil evidence of the life that had been present. Dried lake beds are the main target for a fossil hunt on Mars.

Gusev Crater has a river flowing into it and the bottom of the crater appears to be filled with sediments. While there may be a mantle of windblown dust on the crater floor as well, it is likely that most of the sediments in Gusev were deposited while it was full of liquid water. Within these sediments there may be fossil imprints of the life that lived in this lake.

Finding fossils on Mars that provide unambiguous evidence of past life will be a most interesting event, and is certainly a worthy goal for the robotic rover and sample return program currently underway. However, fossils alone will not answer the main question we have about life on Mars: Is it truly a second genesis, a separate and independent origin of life? It could well be the case that life from Earth and Mars share a common origin. The planets may have exchanged biological material throughout their history via meteorites. The Martian meteorites show that this is possible, even likely. To determine if life on Mars was a separate genesis will require more than fossils. We will need to analyze the biochemistry of actual Martian organisms—whether dead or alive. The likely place to find Martian organisms is frozen in the ancient permafrost near the southern polar regions. There lies undisturbed ground from 3.8 billion years ago, frozen and possibly containing life from that warmer, wetter state. It is likely, however, that any organisms in the permafrost are dead due to accumulated low level radiation. Even buried well beneath the surface and hence shielded from cosmic radiation, these dormant Martians would receive radiation from the natural radio-nucleotides uranium, thorium, and potassium that are present in all material in the solar system. Even if the uranium and thorium were somehow depleted from the Martian sediments, the potassium found in the life forms themselves would provide a radiation dose over 3 billion years more than sufficient to kill the dormant microorganisms. Though dead, the frozen Martian microbes would be biochemically preserved and would allow for direct comparison to the biochemistry of life on Earth. We could determine quite easily if they were genetically related to us.

These then are the questions that motivate the scientific exploration of Mars' past: Was there life? Can we find fossil evidence of this life in a lake bed? Can we find preserved remains of Martian life in the permafrost? Was this Martian life genetically related to life on Earth? Initially we will conduct this search with robotic missions and with samples returned to Earth. But eventually humans must go to explore this world directly. The question of life is of such scope and complexity that only the most capable of field instruments—the human being—is up to the job.

Human exploration of Mars will probably begin with a small base manned by a temporary crew, a necessary first start. But exploration of the entire planet will require a continued presence on the Martian surface and the development of a self sustaining community in which humans can live and work for very long periods of time. A permanent Mars research base can be compared to the permanent research bases that several nations maintain in Antarctica at the South Pole, the geomagnetic pole, and elsewhere. In the long run, a continued human presence on Mars will be the most economical way to study that planet in detail.

It is possible that at some time in the future we might recreate a habitable climate on Mars, returning it to the life-bearing state it may have enjoyed early in its history? Our studies of Mars are still in a preliminary state, but everything we have learned suggests that it may be possible to restore Mars to a habitable climate. I believe that bringing life to Mars may be a goal worthy of humanity.

THE VIRTUAL ASTRONAUT (2004)

Robert Park

Robert Park is an emeritus professor of physics and a former chair of the physics department at the University of Maryland. He is also a former Director of Public Information in the Washington office of the American Physical Society. He founded and continues to write the *What's New* electronic newsletter. He is the author of two books: *Voodoo Science: The Road from Foolishness to Fraud* and *Superstition: Belief in the Age of Science.*

When the president delivered his Moon-Mars speech, it was nearly the anniversary of the *Columbia* disaster. Within hours of the shuttle's disintegration, the president went on the air to declare that "our journey into space will go on."

That's what he should have done. The exploration of space is an important endeavor, worthy of great national support. But President Bush's idea of space exploration is misguided.

In his January speech at NASA headquarters, President Bush called for a base on the Moon, which could one day be used to launch a manned expedition to Mars. If that sounds familiar, it should. I was there in 1989 when his father stood on the steps of the Air and Space Museum and set forth exactly the same vision. This president seems forever destined to finish his father's unfinished business.

The current George Bush invoked Lewis and Clark, while his father invoked Columbus. It is worth remembering that halfway across the Atlantic, Columbus had a crisis of confidence in which, fearing mutiny, he locked himself in his cabin. If he had possessed a drone that he could have sent out to discover whether there was something across the ocean besides the edge of the Earth, I'm sure he would have done so. But he didn't. That's a technology we have now. And to talk about Lewis and Clark and Christopher Columbus as models in the twenty-first century is bizarre. The image of explorers facing the unknown dangers of a strange planet a hundred million miles from Earth is certainly heroic, but it's hopelessly old-fashioned. If you want romance, read romance novels.

The great adventure worthy of the twenty-first century is to explore where no human can ever set foot. In the entire history of humanity, we could never do that before. But with modern technology, we can explore places where no human being can ever go. This is the exciting future we have in space exploration.

The president spoke of "human missions to Mars and to worlds beyond." But if we insist on exploring Mars with human beings, that's the end of our journey. There's no place else to go. In our solar system, every place else is impossible. In some places the gravity is too great; it would crush a human. In some places, the radiation levels are too high or the temperatures are too high. Mars is just about it. It is, in the end, a very limited journey. We're going to launch an exploration of the universe that can only go to Mars?

Ironically, when the president made this speech, we already *were* on Mars. We're there now. We have two rovers that are mere extensions of the scientists back on Earth. And it's the scientists that operate the rovers. I hesitate to call them robots; they're telerobots. They extend the scientist's senses to a place where it's inconvenient for him to go.

In large part, we judge the success of a civilization by the extent to which dangerous or menial tasks are done by machines. And space travel is both menial and dangerous. It is dangerous for obvious reasons, and menial because humans don't really fly the ship: the robots fly the ship. The humans are just expensive passengers—terribly expensive. The cost of doing anything in space with human beings is vastly greater than doing it with robots.

And it's not clear that the robots don't do it better. The robots that are on Mars right now, Spirit and Opportunity, are only second generation machines. The first generation was Sojourner seven years ago. Future generations will be even more advanced and more capable.

Scientists back on Earth see Mars through the rovers' eyes. And the rovers have better eyes than any human. They can focus on nearby things like grains of sand and distant things like mountains. They can see microscopically or telescopically. We can't do that.

In fact, if a human were on Mars, what could he do? I have been on field trips with geologists. They use their hands. You don't have those hands when you're locked in a space suit. You can't pick up a rock and heft it, you don't get any feel of its composition, any sense of hardness or texture. You would have no sense of touch. There's no sense of smell. There's nothing much to hear on Mars, except maybe a very low rumble from the wind. So the only sense an astronaut would have is the sense of sight, and even that's through a visor. The robot's sight is simply much better.

Simply put, the future is not in spacesuits. The future is in robots. Our robots get better every day. Human beings haven't changed in 35,000 years.

Beyond the scientific advantages, robotic exploration is more democratic. It's as though we're all there. I can go every day to NASA's websites to look at the latest pictures, sometimes live pictures. Some years ago, when we were doing a flyby of Neptune, I took my class out to NASA's Goddard Space Flight Center and we watched the image of Neptune being built up one line at a time on the huge screen. When I looked at my students, they were doing the same thing I did when Neil Armstrong stepped on the Moon. They were holding their breath; they were excited. The same is true with today's rovers: we feel like we're going along on the mission.

One flawed justification for sending humans into space is the promise of scientific or technological "spin-offs." In my experience, there are three kinds of liars—ordinary liars, damned liars, and spin-off claimers. Of course, when you spend billions of dollars on a human spaceflight program, you're going to get some spin-offs. And a great many of the spin-offs supposedly developed by NASA were actually developed quite independently by private industry, which used NASA as good advertising. NASA loved it, because they could tout these achievements in front of Congress and look like they were doing something useful.

But in reality, the most impressive spin-offs—communication satellites, spy satellites, weather satellites, and global positioning systems—have all been products of the *unmanned* space program.

The best science is also usually national science. The Clinton administration's rationale for the International Space Station celebrated the benefits of international cooperation: the idea that science would improve by working together with other nations, and that working together on science would improve the relations among nations. This is a bogus rationale for going to space—and ultimately ineffective. The reason internationalizing science doesn't work is that Congress—and it works pretty much the same way in other countries—prefers research conducted for the improvement of the United States. Congress doesn't understand or like the argument that research is for the improvement of the world, and they never will. They are always going to vote on a closer conception of the national interest.

Some people have more wild notions for sending humans into space—like putting colonies on Mars or terraforming Mars. But why have we had colonies in the past? Put bluntly, to rape a region of the Earth and bring its riches back to the home country. But what are the riches on Mars? I'm at a loss to know just what we could bring back that would begin to compensate for the cost of going to get it.

As far as terraforming goes—we are unable to maintain our atmosphere on Earth the way it should be maintained. Do we really believe that we can build a new atmosphere on a planet that doesn't have one?

If we are serious about the exploration of space, we should stop this dangerous and expensive project of sending human beings. At most, we should preserve some manned spaceflight capability in case a situation arises where we really need to send up a human being—although I cannot for the life of me imagine what such a situation would be.

We certainly shouldn't send humans to explore Mars. The reason we're most interested in visiting Mars is to find out whether life exists beyond Earth and, if so, how common life is in the larger universe. That's what we long to know, and that's why those robots are up there right now: they're looking for evidence of liquid water on Mars, since evolutionary biologists believe that if there is liquid water for long periods of time, there is a strong likelihood of life.

My nightmare is that we send humans to Mars to look for life, because we're going to find it—but it's going to look awfully familiar if we contaminate the planet. We will end up bringing Earth life to Mars rather than finding Martian life there. There are more bacteria in one human gut than all the people that have ever lived on Earth. One accident on Mars with a human being and the search for life is pretty much over.

We don't need to send life to look for life. We should be sending out sterilized spacecraft and sterilized robots. Fortunately, our machines get pretty well sterilized just by being out there, just by making the trip to Mars. But the one thing you can't sterilize is a human being.

MARS AND THE PARANORMAL (2008)

Robert Crossley

Born in 1945, **Robert Crossley** obtained a PhD in English Literature from the University of Virginia in 1972. He is now Professor Emeritus at the University of Massachusetts, Boston, having taught in the English department for thirty-seven years, nine of which he was department head. In 2011, he published *Imagining Mars: A Literary History*, which charts the relationship between literary imagination and scientific knowledge about the planet. He is currently writing a book about epic ambitions in modern times.

One of the most peculiar instances of symbiosis in the cultural history of Mars is the one that developed in the late nineteenth century between astronomy and psychical research. In an intriguing historical conjunction, the Society for Psychical Research (SPR) was founded in London in 1882, just half a dozen years after the modern phase of Martian observation got underway with Asaph Hall's discovery of the two Martian moons and Giovanni Schiaparelli's first reports of "canali" on the planet's surface. Three years later the American SPR opened in New York, under the presidency of one of the harder-nosed American astronomers, Simon Newcomb, who was determined to use his position to expose spiritualism as phony science.[1] It was not unusual for astronomers to sign up as members of either the British or the American SPR, whose lists included both committed spiritualists and more dispassionate psychical researchers.[2] Percival Lowell, as his biographer David Strauss has explored in some detail, was intrigued by psychic research in his earlier years traveling in Japan, and consulted with members of the American SPR and with William James, who began a two-year term as president of the British SPR in 1894. The sequence of Lowell's career as both a scientific researcher and a popular writer who was "drawn to exotic topics—trances and extraterrestrial life"[3] itself demonstrates the links between psychic phenomena and Martiana. Lowell moved from investigating Shinto mesmerism and cases of possession, recorded in his 1894 *Occult Japan*, to planetary studies from his observatory in Flagstaff—studies that went public with *Mars* (1895), the first of his three major books on the subject.

In fact, one hundred years ago the two most popular scientific writers who engaged with questions about the possiblity of life on Mars—Lowell in the

United States and Camille Flammarion in France—both combined interests in psychic research and astronomy, although their interests in spiritualism were sharply different in nature. Lowell always remained a materialist and looked for physical explanations for psychical practices. When he was approached by spiritualists hoping to use his Mars research in their cause during the period of his fame as proponent of an inhabited Mars, Lowell made it clear he had little patience with table rappers and clairvoyants who sought to find in him a kindred spirit.* Flammarion, on the other hand, was as devout about psychic experience as he was about Mars. His monumental *La Planète Mars* (1892) was the definitive summation of three hundred years' worth of telescopic observations of the planet. But this volume, as well as his more popular works on astronomy, alternated with spiritualist narratives like *Uranie* (1889; discussed below) and essays like his "Spiritualism and Materialism," originally published in 1900 as an open letter to Camille Saint-Saëns, in which he writes that clairvoyance "is proved by such a considerable number of observations that it is *incontestable*".†[4] Flammarion's espousal of spiritualism was lifelong. As a teenager he became a member of La Societé Parisienne des Études Spirites, and in 1923, two years before his death, he became president of the international SPR. Telepathy, he asserted in his inaugural address, is "as much a fact as are London, Sirius and oxygen".‡[5]

As the popular fervor for Mars grew in the final two decades of the nineteenth century and into the twentieth century, there was a parallel explosion of interest in telepathy, reincarnation, and clairvoyance that was a cosmopolitan outgrowth of the provincial spiritualist movement that had begun in New York State in the mid-nineteenth century. The parallel may be something more than just an accident of chronology. Roger Luckhurst, in his learned study of the history of telepathy, suggests that modern scientific ideology and intellectual networks took hold in the late Victorian era, increasingly multidisciplinary and at last independent from and indifferent to the authority of the church: "The emergence of a scientific culture consequently produced other, less predictable effects: strange, unforeseen knowledges, hybrid and ephemeral notions, that emerged as compromise formations melding apparently discrete systems".[6] Certainly, Lowell's ability to persuade vast segments of the public to his views of Mars as irrigated and inhabited depended significantly on the hybrid science he called "planetology" and on the multiplex nature of his arguments, which were based not only on telescopic observation but also on evolutionary theory,

* When a medium wrote to Lowell that she had been "working telepathically on Mars," she enclosed a clipping of her discoveries and offered her services to "aid your mechanical devices in investigation." Lowell wrote a terse pencil note on the letter: "Return clipping. No Ans[wer]" (E.B. Ringland to Percival Lowell, September 29, 1909. Lowell Observatory Archives, Flagstaff.) On Lowell's own research into occult phenomena, see Strauss (133–50).

† Flammarion reprinted this essay as the introduction to his 1924 *Haunted Houses*, a vigorously committed defense of psychic phenomena against the objections of skeptics.

‡ Among other prominent literary, scientific, and philosophical figures who served as president in the early decades of the Society's existence were William James, Oliver Lodge, Andrew Lang, Henri Bergson, and Gilbert Murray.

geology, paleontology, and social science.* Many spiritualists, with their own antagonistic relationship to mainstream religion, may have assumed that, as boundaries between fields of knowledge blurred, they would have reliable allies among natural scientists. But while many scientists were deeply interested in psychical phenomena as a subject of investigation, few were inclined to be partisans. Gauri Viswanathan points out that Madame Blavatsky's Theosophy—one of the most prominent manifestations of the late-Victorian fascination with psychic phenomena—liked to advertise itself as a "rational alternative to religion" and "merely another professional interest, no different from law, medicine, or natural science".[7] At the beginning of the twentieth century, psychic research still seemed capable of being brought under the umbrella of scientific research and of coexisting comfortably with psychology. But the case was not made and before long the term "parapsychology"—that is, beside or outside psychology—came into usage.† Similarly, the Martian canals championed by Lowell and Flammarion came under increasingly severe attack early in the twentieth century, with geologist Eliot Blackwelder denouncing as a "kind of pseudo-science" the propositions that Lowell "foisted upon a trusting public":

> I feel sure that the majority of scientific men will feel just indignation toward one who stamps his theories as facts; says they are proven, when they have almost no supporting data; and declares that certain things are well known, which are not even admitted to consideration by those best qualified to judge.[8]

The critique made of the spiritualists could, with only minor changes in phrasing, be applied equally to the Lowellians: "No amount of rationalization can write away the discrepancy between empiricist ways of knowing (as the professional sciences understand them) and occult knowledge".[9] In the last decades of the nineteenth century, at a time when psychology was still considered a branch of philosophy rather than a discrete discipline, psychologists and philosophers were particularly drawn to the investigation of séances and of the experiences and the claims of mediums; at the same time, some mediums began publishing accounts of their visions of Mars in the form of travelogues that were, in fact, works of science fiction. The most celebrated instance of the intersection of Mars and the paranormal appeared at the turn of the twentieth century when Théodore Flournoy, professor of psychology at the University of Geneva and a psychical researcher, published *Des boles à la Planète Mars* (1899), his extensive case study of a Swiss medium who called herself Hélène Smith. Smith's supposed visionary experiences, manifested in her paintings of the Martian landscape, inhabitants, and artifacts and, centrally, in the Martian language that she spoke and wrote,

* See Strauss's important discussion of the essentially multidisciplinary nature of Lowell's notion of "planetology" as the appropriate term for his kind of scientific study of Mars (197–219). Lowell devotes his 1908 *Mars as the Abode of Life* to an extensive illustration of how planetology, the "science of the making of worlds" (2), extends the reach of astronomy.

† The *Oxford English Dictionary* records the first occurrences of "parapsychology" in the 1920s.

attracted the attention of linguists and psychopathologists, dream analysts and surrealists.* Smith's visions and the fascination they generated can be more fully understood in the context of other narratives from the 1880s to the 1920s that combined the subject of Mars with psychic experience.

A surprising number of early novels about Mars and Martians are either framed by or deeply imbedded in spiritualist practices. The two most famous Mars narratives in English from the *fin de siècle* and early twentieth century glancingly incorporate psychic phenomena into their narratives. H.G. Wells's Martians in *The War of the Worlds* (1898) communicate telepathically, although the narrator clearly reflects Wells's own deep and public skepticism about paranormal phenomena: "Before the Martian invasion, as an occasional reader here or there may remember, I had written with some little vehemence against the telepathic theory".†[10] In *A Princess of Mars* (1912), Edgar Rice Burroughs depicts a disembodied John Carter standing naked over his clothed double, and then being instantaneously transported from a cave in Arizona to the surface of Mars as he "felt a spell of overpowering fascination" with the red star in the night sky: "My longing was beyond the power of opposition; I closed my eyes, stretched out my arms toward the god of my vocation and felt myself drawn with the suddenness of thought through the trackless immensity of space".[11] While there is no evidence of any considerable personal interest by Burroughs in psychic phenomena, it is plausible that he picked up the notion of astral projection of a double from the culture at large. But it is not entirely clear that this *is* a case of astral travel. When Carter returns to the cave at the end of *Princess* in similarly mysterious fashion, he finds the mummy of a very old woman, a charcoal brazier with an unidentifiable green powder, and a row of human skeletons. Carter's trip to Mars may have been accomplished by old-fashioned witchcraft as much as by newfangled teleportation. While it is of interest to find in Wells and Burroughs reflections of the popular craze for telepathy and astral bodies, to get a fuller sense of the interconnections between narratives of Mars and the paranormal it is necessary to examine some much more unfamiliar writers and texts.

The earliest example I have discovered is Henry Gaston's *Mars Revealed, or Seven Days in the Spirit World* (1880), an account of a journey by psychic projection whose publication is dictated by a spirit guide, said to be John of Patmos, author of the Book of Revelation. Some later psychical narratives of journeys to Mars would make at least a nominal and often a concerted effort to tie their accounts to the maps of Schiaparelli and the theories of Flammarion and Lowell, but *Mars Revealed*, published before the surge of publicity over Schiaparelli's sighting of "canali," depicts an exotic fairy-tale landscape, "a world

* The 1994 Princeton translation of Flournoy's book as *From India to the Planet Mars* has extensive apparatus, including a brief essay on the Helene Smith case by Carl Jung and a long and superb introduction by the editor, Sonu Shamdasani, to which 1 owe a general debt.

† In the notes to their critical edition of *The War of the Worlds*, Hughes and Geduld cite Wells's hostile review, published in *Nature* (Dec. 6, 1894), of Frank Podmore's 1894 *Apparitions and Thought Transference*; they also cite passages in *Love and Mr. Lewisham* that expose the fakery of spiritualism (219, n.15).

in ruby, emerald, and silver," a planet lush and full of water in its rivers, lakes, seas, and falls.[12] Gaston includes a prefatory letter that traces the genealogy of his narrative to the doctrines of Swedenborg, a tenable line of descent for many subsequent ventures that link psychical apparatus to interplanetary travels.* The preface also makes explicit what is often an ambiguous boundary in spiritualist narratives about Mars between the factual and the factitious. "Some may consider it a beautiful romance," we are advised, but thoughtful readers will find in it "the very essence of spiritual philosophy".[13]

Mars Revealed, like many of its later spiritualist brethren and like earlier "fantastical excursions,"† offers a hodge-podge of utopian vignettes of communal life, psychical propaganda, earnest moralizing, highly colored sentiments and descriptions, and unintended farce. Utopian hygiene and grooming tumble into this latter category when we hear that Martians "give their teeth and mouths a thorough cleansing" each morning and "comb their hair, as do all decent people on the Earth." Men and boys, we are reassured, "part theirs upon the side, like all men of sense on Earth".[14] Much of Gaston's Martian imagery appears to derive not from scientific texts but from visionary descriptions of the New Jerusalem. The Mars revealed by John of Patmos is "a miniature heaven"[15] adorned in gold, sapphires, emeralds, and other precious materials. In a curious wedding of astronomy and Biblical numerology, Martian temples are topped with telescopes—including one monster with a lens 144 feet in diameter—and they are carefully guarded since individuals must be exemplary in virtue in order to look through one. At the end of the account of the journey to Mars the narrator acknowledges that he would like to go to Saturn next, but only if readers show a proper appreciation of his Mars narrative. They are exhorted to talk the book up with other potential readers and "then write to the publishers of this book, and encourage them by many purchases".[16] This campaign seems not to have prospered; there was no Saturnian sequel.

In many turn-of-the-century Martian fictions science and spiritualism coexist uneasily. Mark Wicks's 1911 *To Mars Via the Moon* is a veritable encyclopedia of astronomical lore and a fictional homage to Lowell and his Martian writings, but its plot turns on the narrator's discovery on Mars of his reincarnated son. Wicks's Martians are both telepaths and prodigious canal builders, and the narrator goes out of his way to indicate how his journey to Mars—by spaceship, it should be noted, not by astral projection—and his experiences on the planet confirm the maligned theories of Lowell and Flammarion. When, for instance, the Martians reveal that they have no fear of death because of their confidence that they will

* Swedenborg's 1758 *De Telluribus*, first translated into English in 1787, is an excursion narrative inspired by the theory of "the habitability of worlds," popular in the seventeenth and eighteenth centuries. Swedenborg's tour of the planets and the spiritual beings that inhabit each of them was, as his English translator John Clowes acknowledged in the preface, liable to be understood as "merely visionary, groundless, and enthusiastic, and the Fruit only of a light or disordered Imagination" (vi).

† Examples are Bernard le Bovier de Fontenelle's *Entretiens sur la pluralité des mondes* (1686), Christiaan Huygens's *Cosmotheoros* (1698), and the anonymous English *Fantastical Excursion into the Planets* (1839). For discussion of these texts, see my "H.G. Wells, Visionary Telescopes, and the 'Matter of Mars'."

be reborn on any one of an infinite number of habitable worlds, the traveler has a sudden realization: "There are some upon our world who hold very similar ideas," he tells his interlocutor, "notably a great French astronomer named Flammarion".[17]

A more mainstream instance of the incorporation of the paranormal into the literature of Mars is a posthumously published novel by the well-known illustrator for *Punch*, George DuMaurier. *The Martian* (1897), despite the alien sound of its title, is for most of its length a conventional and uninspired *Bildungsroman*. It is the account of the growth, intellectual development, and artistic successes of Barty Josselin, a fantasy version of DuMaurier's own career. Late in the novel, however, it is suddenly revealed that Barty's consciousness is inhabited by someone named Martia, who has been steering his life since birth. Through Martia's long letter, apparently composed "automatically" in Barty's hand while he was asleep and just before she leaves his body forever, the reader is offered a brief account of Martian geography (only partly informed by current scientific views). Mars is inhabited now only near the equator, with winters there even more brutal than in Spitzbergen. Amphibious, seal-like, furry, vegetarian, and naked, the Martians have highly developed senses, including a sixth sense that is both magnetic and telepathic. Mars has been rid of nearly all its fauna, except for some large fish and bat-like birds, in order to conserve its dwindling resources for the humanoids. The Martians have become "the Spartans of our universe" who are "near the end of their lease" on their native world and perhaps will soon migrate to Venus.[18] But before their extinction or migration, the Martians take a missionary interest in the benighted inhabitants of Earth. Incarnating themselves in "promising unborn though just begotten men and women," as Martia has done with Barty Josselin, they stir their terrestrial hosts with dreams and visions that can open up humanity to the aesthetic and philosophical delights that the inhabitants of Mars know as part of their daily lives. "According to Martia, most of the best and finest of our race have souls that have lived forgotten lives in Mars".[19]

Once a well-known novel, *The Martian* is rather long-winded and self-absorbed, awkwardly straddling the realistic and the preternatural modes. It is not in itself a psychical novel, nor does it operate as propaganda for the spiritualist movement. But in its odd importation of telepathy, reincarnation, astral traveling, gender-switching, and automatic writing, it works in some of the leitmotifs that were beginning to appear in paranormal fictions about Mars and it represents the growing popular fascination at the close of the nineteenth century with the connection between the world of Mars and the world of the séance. While DuMaurier seemed to have no program that he was pursuing in *The Martian*, other than an indirect exploration of the shaping of his own professional and emotional life, other authors of "romances of fact" about Mars (to use Lowell's phrase), or their publishers at least, were often eager to advertise an explicit affinity between the psychical content of their narratives and the new astronomical research centered on Mars.*

* "Romance of fact" was Lowell's hopeful term for his own writings about Mars—and much preferable, he thought, to the "romance of fiction" (*Mars and its Canals* 382). For recent considerations of Lowell and Mars, see my "Percival Lowell and the History of Mars"; Strauss's *Percival Lowell* (esp. 173–257); and, most provocatively, Markley's *Dying Planet* (esp. 61–114).

Some of the most dogmatically spiritualist Martian narratives are zealous, even obsessive, about invoking supposed astronomical authority for their visions. Louis Pope Gratacap's *The Certainty of a Future Life in Mars* (1903) prints as an appendix to the narrative one of Schiaparelli's papers on "The Planet Mars" in the 1893 translation of Lowell's associate, W.H. Pickering. The boards of Sara Weiss's 1905 *Journeys to the Planet Mars*, written, the title page tells us, "under the editorial direction of (Spirit) Carl De L'Ester," is imprinted with a reproduction of a Schiaparelli map. In the most curious example of all, the true author of *The Planet Mars and Its Inhabitants* (1922), we are told by the purported "amanuensis" J.L. Kennon, is Iros Urides, a Martian. The narrative is punctuated by numerous references to the work of Lowell, sometimes with extensive footnotes summarizing his views and comparing them with the visions of Mars that have been conveyed to the amanuensis by a medium, and it concludes with a ten-page appendix with extensive summaries of the leading ideas of Lowell's books on Mars.

Of all the works of fiction that incorporate spiritualist structures or motifs, none was as influential as Camille Flammarion's, for the obvious reason of his dual status as an astronomer and a popular writer. Because Flammarion was so attached to his theories, both on the habitability of the planets and on reincarnation, his fiction is extraordinarily didactic and autobiographical, as much personal essay as narrative. *Uranie*, published in 1889 and translated into English as *Urania* in the following year, is named for the muse of astronomy. A summary of the organization of this book may suggest how stubbornly the author seeks a seamlessness between astronomy and spiritualism and between romance and fact. In Part I, the first-person narrator (we eventually learn that his name is Camille) tells of his apprenticeship at the Paris Observatory at age seventeen, where, Pygmalion-like, he becomes obsessed by a statuette of Urania. In a dream Urania comes alive to take him on a tour of the universe and to introduce him to the infinite diversity of worlds and species. The narrator's friend, 25-year-old astronomer George Spero, is the protagonist of Part II. After Spero and his fiancée Idea are killed in a ballooning accident, the narrator goes to a séance and the savant has a vision of a place with cliffs and foaming seas, sandy beaches, and reddish vegetation. The landscape is Martian, and in the vision the savant sees Spero and Idea, who have somehow ended up on Mars after their deaths.

The major portion of Part III is devoted to case histories—what Flammarion italicizes as *facts*—that purport to validate claims of communication between the living and the dead. Some of the cases are documented in the text from publications of the Society for Psychical Research. Alternating these "authentic" stories of telepathy with passages on the telescopic observation of Mars, the narrator insists that "astronomy and psychology are indissolubly connected".[20] The narrator falls asleep under a tree and awakens to the sight of two small moons in the sky. He, too, is now on Mars. Feeling astonishingly light on his feet, he discovers that he has acquired a sixth sense (magnetism), and he observes the Earth appearing as the evening star. This last sight leads him to meditate on the failures of most terrestrials to appreciate the beauty of their planet as they indulge in soldiering and nationalist wars: "Ah! if they could behold the earth from the place where I am now, with what pleasure would they return to it, and what a transformation would be effected in their ideas".[21]

The narrator surveys Martian physiology (delicate, winged, six-limbed, luminescent at night); customs (vegetarian, pacifist, sexually sublimated); and terrain (flat and, of course, networked with irrigation canals). His claim that all his visions are "completely in accord with the scientific notions we already have of the physical nature of Mars"[22] is true in some respects, if one grants that most astronomers of the period applied the nebular hypothesis to predict that a Martian civilization, if one existed, would be older and more highly developed than ours. Similarly, when he intimates that the climate of Mars would be livable for human beings he says no more than many astronomical writers of the nineteenth century suggested; however, the graceful analogy he makes manages to downplay the fact that life at the Martian equator would be very chilly: "A country of Mars situated on the borders of the equatorial sea differs less in climate from France, than Lapland differs from India".[23]

But Uranie's narrator goes well beyond "scientific notions" about Mars when he starts to pursue some of Flammarion's pet spiritualist fantasies. In conversation with his friend Spero, now reincarnated as a woman while his fiancée has emerged in male form, the narrator is enlightened about the philosophical and spiritual character of life on Mars, and the narrative takes a turn from the improbable to the ludicrous. The Martian body is etherealized and Martians take their nourishment from the atmosphere, freeing them "from the grossness of terrestrial wants".[24] Because their planet is not very "material," the reader is urged to picture Martians as "thinking and living winged flowers".[25] The Martians' standard of beauty is highly refined and they view (using telephotographic technology to film and study terrestrial history) "the Apollo Belvidere and the Venus de Medicis [as] veritable monstrosities because of their animal grossness".[26] Sexual passion has atrophied (hence Spero and Idea seem unfazed by their metamorphoses). The physiology of reproduction has been so spiritualized that the narrator is hard-pressed to find words to describe the process: "Conception and birth take place there in an altogether different manner, which resembles, but in a spiritual form, the fecundation and blossoming of a flower".[27] And, with a nod to the Victorian cult of womanhood, Spero reveals that Mars is ruled by the feminine sex because of the delicacy of their sensations and their "incontestable superiority over the masculine".[28]

The most remarkable feature of *Uranie* is not that it is fiction with a message, and not even that it is strikingly unsuccessful fiction. Like many more strictly literary, though not necessarily more artistically talented, writers who would follow him, Flammarion used Mars as an excuse to pursue a cultural agenda. What we might expect of a novel about Mars written by one of the astronomers most closely identified with the popularization of Martian discoveries in the nineteenth century is a thinly disguised paean to science, as the title seems to promise. Instead, this miscellany of "episodes," "researches," and "reflections" that are "brought together in a sort of *Essay*," as the author accurately names the book's mixed genre,[29] has the distinction of being one of the earliest examples of the misappropriation of scientific research on Mars.

Uranie is in the vanguard of a group of books and pamphlets, published over the next several decades, in which Mars becomes the domain of spiritualists. In such narratives the link between paranormal experience and astronomical research is often made explicit and visible. Gratacap's *Certainty of a Future Life in Mars* is a reincarnation fable narrated by Bradford Dodd, a young innovator in "interplanetary telegraphy," whose father Randolph—in a move reminiscent of Lowell's journey from Boston to Flagstaff—leaves New York's Hudson River Valley to build his own observatory on Mt. Cook in New Zealand. With an echo of Lowell's Mars Hill in Flagstaff, Randolph Dodd calls the locale of his observatory "Martian Hill".[30] Throughout the early parts of the narrative there are frequent knowledgeable references to the Martian drawings of Herschel, Schroeter, and Schiaparelli as well as reports from more recent astronomers, including Henri Perrotin, François J. Terby, and Edward S. Holden.

The narrative oscillates between the presentation of then-standard information about Martian climate, topography, and habitability—often buttressed with notes from current numbers of *Scientific American—and* Randolph Dodd's fantastic notions about the evolution of souls through successive reincarnations and streams of transference of life from one planet to another in the solar system. The motive behind his observations of Mars and, more importantly, his efforts to establish wireless contact with Mars, is to communicate with his dead wife, who, he believes, may have been "transplanted" to Mars.[31] The sickly father passes on to his son the ambition of achieving "the union of our world with others by magnetic waves".[32] Shortly before his death the father receives a message of dots and dashes, but in an untranslatable form. He concocts a plan to send a message himself in Morse code from his future *post-mortem* home on Mars and enlists his son to monitor the apparatus in the observatory. A year after Randolph Dodd dies, long transmissions in Morse code begin to arrive, telling of a Mars full of reincarnated terrestrials—the great majority of them scientists.

The expectation that the era of Edison and Marconi was a ripe time for communication with Mars and its inhabitants was a notion that the interest in psychical phenomena encouraged.* If the technology of transport had not yet reached the stage that would gratify the human desire to visit the worlds being disclosed by the telescope, emergent mental powers might do the trick. Just ten years after the publication of Gratacap's narrative, the young Olaf Stapledon, at the beginning of his long epistolary courtship of his cousin, craved a telepathic shortcut: "Writing is such a slow way of conveying thought. Speaking is bad enough. I should like some method of wireless telepathy. If Marconi and Mrs. Eddy and Brahms were to get into partnership they might discover the thing."† But Stapledon's whimsy is treated with high seriousness in *The Certainty of a*

* See Sheehan and O'Meara (197–203) for a recent discussion of the rival efforts of Guglielmo Marconi and Nikola Tesla, from the late 1890s to the mid-1920s, to effect radio communication with Mars. The work of Marconi and Tesla on wireless telegraphy was often cited by spiritualist partisans as analogous to psychical communication (see Luckhurst 135–39).

† Olaf Stapledon to Agnes Miller, Oct. 28,1913 (privately held). This letter was not printed in my *Talking Across the World*, although it was cited in the introduction (xxxvii).

Future Life in Mars, which is largely absent of intentional humor, save for one irreverent Martian transplant who finds some other Martians "a trifle heavy in style, just a suggestion of a kind of sublimated Bostonese about them".[33]

For Gratacap Mars becomes "a sort of Paradise"[34] or at least a "stepping stone" to a "higher beatitude of living".[35] Like Flammarion's Mars, this one also celebrates immateriality and registers its distaste for terrestrial grossness. Its center is a City of Light and its primary aesthetic delight is the most disembodied of the arts, music. There is a ghostly vapidity to the planet and its reincarnated inhabitants: the newly reincarnated are, we are informed in the Morse Code dispatches, "little more than gaseous condensations";[36] the human body is so purified and etherealized that not only is sexual passion absent, but also "evaporation replaces defecation".[37] And when death finally comes to the reincarnate it too is "like evaporation".[38] In many respects Gratacap's Mars is less like a heavenly paradise than a Dantean Limbo of the Virtuous in which Randolph Dodd can attend a dinner party where the great physical scientists of earth's past—Galileo Galilei, Isaac Newton, Antoine Lavoisier, Joseph Priestley, Humphry Davy, Léon Foucault, Mary Somerville, Alessandro Volta—engage in scintillating conversation.

There is little of the classic utopian tour in Gratacap's book, although he does address the question of how the reincarnated Martians manage to find the leisure to enjoy their apparently endless delights of travel and music and talk. There are almost no labor-saving technologies in this immaterial culture, save those machines that are used in engineering the canals and manufacturing telescopes. Nevertheless, the reincarnates are fortunate in having "prehistoric" native Martians who cheerfully perform other arduous tasks such as mining the marble used to construct the cities. "Where hard labor on a mammoth scale is necessary," the disembodied father reassures his son, "the little race of *prehistorics* serves all their purposes".[39] That the indigenous Martians are brown or copper-colored while all the reincarnates are white makes commentary on the significance of the class division superfluous.

As a novel, *The Certainty of a Future Life in Mars* has little to recommend it. Gratacap tries to generate suspense with an approaching comet that will collide with the capital city of Mars and by having the son grow fatally ill as he rushes to complete the manuscript based on his father's messages. There is also an effort to achieve a domestic resolution to the slender plot when, just before the coded messages abruptly end, Dodd discovers his reincarnated wife on Mars. The son's death is reported by his executor who has taken charge of getting the manuscript published. But the concluding note from the New York editor of the volume we have been reading tells us what may not entirely surprise us—that the manuscript was repeatedly rejected by other publishers before it fell into his hands. At the close, the reader is primed to anticipate a full family reunion on Mars, when the son will also pass through the immigration hall for newly arrived souls. The narrative, however, has long since deliquesced into little more than the vaporish abstractions that Gratacap's Mars celebrates as human destiny.

The blurred line between fable and factual report suggested by the editorial apparatus in Gratacap's novel is a structural feature common within the tradition of utopian allegories. But the relationship between truth and fiction becomes even more tenuous in Martian narratives that are self-proclaimed "psychic revelations." The authors of such narratives claim only to be amanuenses who record, as faithfully as possible, the communications they receive through their spirit advisors. Take as an example the complete title of a 1903 book:

> *Journeys to the Planet Mars or Our Mission to Ento (Mars), Being a record of visits made to Ento (Mars) by Sara Weiss, Psychic, under the guidance of a Spirit Band, for the purpose of conveying to the Entoans, a knowledge of the Continuity of Life, Transcribed Automatically by Sara Weiss Under the Editorial Direction of (Spirit) Carl De L'Ester.*

To enforce her claims of authenticity, Sara Weiss includes thirteen botanically detailed illustrations of Martian flora, a preface on the pronunciation of and gestural accompaniments to the Martian language, and an extensive glossary including tables of specific words for colors, numbers, and personal pronouns.

Like many of the psychical accounts of Mars, Weiss's *Journeys* lacks a principle of narrative selection. At well over 500 pages, it is an *omnium gatherum* of spiritualist doctrines. Digressions sprout up everywhere, occasioned by any chance sight or comment during the various tours of Mars on which the medium is taken by the company of spirits (led by Carl De L'Ester and including Giordano Bruno, Louis Agassiz, Charles Darwin, Edward Bulwer-Lytton, and Alexander von Humboldt). A visit to the unpopulated Martian poles, for instance, prompts an exposition of the dogma that evolved peoples on all planets gravitate away from inhospitable arctic climates and leads the medium to ask for—and get—an account of "the origin of the Eskimos and other polar races of our planet".[40] The references to evolution here and in Flammarion's *Uranie* and Wicks's *To Mars Via the Moon* have led Robert Markley to observe that Martian fictions centered on reincarnation and spiritualism are efforts at "reconciling evolutionary theory and Judeo-Christian theology" and at providing comforting reassurance in the face of Wells's more malevolent images of evolutionarily advanced Martians in *The War of the Worlds*.[41]

The impact of Lowell's canal hypothesis, with its attendant suppositions about the heroic Martian race and deteriorating planetary environment, is evident in Sara Weiss's frequent observations of and explanations of the "Waterways, Irrigating System, Embankments and other stupendous works" of civil engineering.[42] But because her astral journeys to Mars supposedly occurred in 1893 and 1894, before Lowell began publishing about Mars, Lowell himself is never mentioned (although Weiss made sure to send him a copy of the book on its publication).* Instead, late in the book Carl De L'Ester, in instructing the medium about the words she is to use, makes an explicit reference to other terrestrial astronomers whose work can now be, as it were, "validated" by the psychic testimony:

* There is an autographed copy in the library of the Percival Lowell Archives at Lowell Observatory, Flagstaff, Arizona.

> Through telescopic observations, one of Earth's foremost astronomers is inclined to believe that the Entoans (Martians) have resorted to irrigation. To him and to another illumined scientific man, who, I am proud to say, is my countryman, you will convey this message: "Gentlemen, to your vision your telescopes convey faint, and generally misleading gleams of what may be facts, but in the instance mentioned, I assure you that the surmise is entirely correct, and inevitably a period will arrive when Earth, like Ento, will require the same treatment."[43]

The two "illumined" scientists are revealed in subsequent extended panegyric apostrophes to be Schiaparelli and Flammarion. And a further effort to bring the revelation into line with scientific research is evident in the book's physical design: the front board is stamped with what is labeled "Schiaparelli's Map of Mars."

In 1906 Sara Weiss published a second automatic transcription from Ento, *Decimon Huydas: A Romance of Mars*, again with floral illustrations and—a rarity in such mediumistic narrations—a photograph of herself as frontispiece. This "romance" does have a narrative unity absent from the earlier book, being the sustained story—communicated to Sara Weiss by an Ento spirit with the assistance of Carl De L'Ester—of a domestic tragedy supposed to have occurred hundreds of years earlier. It is a tedious recital, interrupted occasionally by a statement of spiritist dogma about the evolution of life in the solar system. One such statement by the spirit-narrator asserts that soon the psychic link between spirits on Mars and terrestrial mediums will be augmented by technology enabling more material telecommunication with Martians. Such a prediction rested on the twofold faith that Mars was certainly inhabited and that wireless telegraphy would produce exchanges between Mars and Earth, confirming the stories told by Sara Weiss:

> No other inhabited planet of our solar system presents, in all directions, correspondences so noticeable as those which exist between Ento and Earth; and were I a prophet, or the son of a prophet, I would predict, that ere the close of the present century communication, on a scientific basis, will be established between the two worlds known, astronomically, as Earth and Mars.[44]

For sheer imaginative brazenness, a 1922 text, *The Planet Mars and Its Inhabitants*, said to be the work of "Iros Urides, A Martian, Written Down and Edited by J.L. Kennon," outdoes even Sara Weiss's narratives. Iros Urides was on earth 2,000 years ago, and Kennon—under the editorial supervision of an angel named Gaston Sergius—takes shorthand dictation from a psychic, Mrs. X—, whose messages come "direct from a disincarnated intelligence of Mars".[45] More interesting than the standard parapsychological apparatus, however, is the omnipresence of Lowell in Kennon's book. Not only is there a drawing of a canal-gridded Mars, but also the book has frequent footnotes to popular scientific articles—chiefly by Lowell—as well as an appendix on Lowell with an abstract of 56 points taken from his *Mars and Its Canals* and *Mars as the Abode of Life*. Mrs X—, we are solemnly assured in the foreword, "knows nothing of astronomy and has never read anything concerning Mars".[46] The fact that the revelations she conveys jibe almost perfectly with Lowell's ideas about Mars is a matter for wonder. The crowning marvel, however, is the book's frontispiece. Kennon receives not only a verbal portrait of a utopian Mars but a series of clairvoyant pictures, one of which is reproduced as a photograph of a Martian plateau with a large city "built of white stone".[47] No matter that to the ordinary reader's eye the city appears to be nothing more than a cloud or fogbank and that the mountain looks suspiciously terrestrial. It was, after all, a given in psychic narratives about Mars that the similarities among the planets were more important and far more numerous than their differences.

Early in World War II there appeared yet another cross-breeding of the utopian and the spiritualist romance in J.W. Gilbert's *The Marsian* (1940). The narrator, a devotee of telepathic experiments and séances, is visited by a splendidly angelic being who has heard a "cry of woe" coming from Earth and has experienced "radiations of sadness" in the ether waves.[48] These cosmic signals of the new world war on Earth are the occasion for the narrator's being psychically projected into space. First he orbits the Earth, explaining terrestrial history, economics, and religion to the incredulous "Marsian" and touring places like the Ford auto plant in Detroit. Then, in part II, he is taken to Mars for a standard utopian excursion and homily on Martian social practices. Gilbert has a few inspired moments, as in the account of the stunning glassed-in gardens which, helped out by the improvements of "a utopian Burbank," provide bountiful, oversized fruits and vegetables for the Martian population.[49] Intensive farming is not the only key to the survival of the Martians in a deteriorating planetary environment. They also practice population control—a "superstitiously tabooed" subject on Earth, the narrator explains to the Marsian. "One of our good women has almost made a martyr of herself by persisting in the wisdom of birth-control," he says in an apparent reference to Margaret Sanger.[50]

Despite such glimmers of inventiveness when Mars holds up a mirror to terrestrial issues, *The Marsian* is an old-fashioned and often poorly written book. Gilbert's earnest collectivism is so heavy-handed that the narrator never even puts up a decent argument in the fashion of the classic utopian dialogues. Obligingly, he capitulates at once to each socialist image on Mars. "My dear brother," he confesses to the visitor from Mars, "to look upon your people makes me feel that my own people are a race of runts".[51] The narrator is a Gulliver-figure of neither charming innocence nor stubborn opinions nor inappropriate sentiments; he is simply a pushover. *The Marsian* may represent the last gasp of full-blown spiritualist science fiction. Before much longer, that spiritualist fringe of extraplanetary travel literature would yield to the fad for alien abduction, and the humanitarian man from Mars would be supplanted by extraterrestrial kidnappers.

Few of these psychical Martian fictions merit re-reading, though two others that could be said to be late developments in the Flammarion line of a spiritualized Mars—Olaf Stapledon's account of sentient viral Martians who communicate by radiant energy in *Last and First Men* (1930) and C.S. Lewis' Christianized

Mars with its ethereal and telepathic ruling "Oyarsa" in *Out of the Silent Planet* (1938)—are among the literary masterpieces about other worlds. One thing that differentiates Stapledon's and Lewis's narratives from the mass of spiritualist Martian stories of the early twentieth century is their humor, in which the true-believer narratives are mostly lacking. There is sly comedy in Stapledon's likening a cloud of Martian viruses to "mobile wireless stations",[52] a jibe at the running sideshow of efforts at telegraphic communication with Mars in the first decades of the century. And Lewis's delightful representation of the physicist Weston's blustering insistence that the Oyarsa's telepathy is mere ventriloquism epitomizes the debunking of voices from beyond in séances. Beyond their ability to stand back ironically from their narratives, Stapledon and Lewis also write in the service of larger—and competing—visions that transcend any propagandizing tendencies in their fiction and give their creations durable literary afterlives.

The paranormal periodically cropped up as an ingredient in Mars fiction through much of the first half of the twentieth century, even after the discrediting of both Lowell's Martian theories and the scientific basis of spiritualism. George Babcock's *Yezad* (1922) has as its protagonist a dead aviator, reincarnated after his spiritual instruction on Mars and urged on his return to Earth to tell telegraphy experts to keep working on their schemes to communicate with the Martians. In his boys' book *Red Planet* (1949), Robert Heinlein imagines Martians that go into trance states while they commune with "the other world"—and Heinlein seems just as comfortable with a dose of parapsychology as he is with including Lowell's canals in his Martian landscape. Ray Bradbury revels in the paranormal in *The Martian Chronicles* (1950), but far from wanting to advance spiritualist dogma, he is drawn to a neoRomantic Mars out of which he can create cultural satire and political allegory. The casual gestures made by the Martian girl in this bit of dialogue with the captain of the second American expedition to Mars—a far cry from the solemnity with which parapsychological experience is treated in the committed spiritualist narratives early in the century—gives away Bradbury's ironic stance:

> "We're Earth Men," he said. "Do you believe me?"
>
> "Yes." The little girl peeped at the way she was wiggling her toes in the dust.
>
> "Fine." The captain pinched her arm, a little bit with joviality, a little bit with meanness to get her to look at him. "We built our own rocket ship. Do you believe *that*?"
>
> The little girl dug in her nose with a finger. "Yes."
>
> "And—take your finger out of your nose, little girl—*I* am the captain, and—"
>
> "Never before in history has anybody come across space in a big rocket ship," recited the little creature, eyes shut.
>
> "Wonderful! How did you know?"
>
> "Oh, telepathy." She wiped a casual finger on her knee.[53]

Fiction founded on an uncritical deployment of psychical phenomena on Mars could not really survive either the casual ironies that come so readily to Bradbury or the definitive death knell to Lowellian Mars that the space age tolled. But however impoverished its literary legacy, the spiritualist phenomenon in fiction about Mars, especially as it was manifested just before and after the turn of the century, is worth some thought. It is almost as if the pseudoscience of the psychics was irresistibly drawn to the faulty science emerging from some of the astronomical observatories. Just as the late 1930s saw an instance of mass hysteria in the United States generated by Orson Welles's broadcast of a Martian fantasy, so the early twentieth century was a ripe time for paranormal delusions. Mars furnished a convenient, and for many people, credible destination for spiritual journeying. And as Freudian and Jungian psychological research and experimentation was beginning to flourish, the claims of psychics who said they had traveled to Mars, observed its people, and even learned its language provided a fertile field for investigators. By far the most momentous of the professional studies, published in the last days of the nineteenth century, was Théodore Flournoy's *From India to the Planet Mars* (1899), a case study of the Swiss medium Hélène Smith.

The case of Hélène Smith, born Élise Müller, provides a useful perspective on the various Martian narratives that celebrate paranormal experience. Hélène Smith did not herself publish any of her Martian visions, but in Flournoy's richly detailed study of her—undertaken with her cooperation, though disowned by her when she saw the published version—there is ample attention to what he called her "Martian romance" as well as her "Hindoo romance." A contemporary reviewer of Flournoy's book wrote shrewdly, "Today, with the great interest that there has been among the spiritists for the writings of Flammarion on the planet Mars and the revelations of the theosophists on the Hindu Masters, it's Mars and the Orient which are fashionable."*

As the editor of the 1994 reissue of *From India to the Planet Mars* points out, the case of Hélène Smith would today be classified as an instance of multiple personality.[54] Believing that she was the reincarnation of the fifteenth-century Princess Simandini and of Marie Antoinette, Smith during her séances also adopted the personality of Marie Antoinette's admirer and court magician Cagliostro; she renamed him, in his "discarnate" spirit-form, Leopold, the mentor for all her psychical experiences. But it was not Smith's ability to take on Cagliostro's personality and voice, complete with male bass and Italian accent, that most fascinated Flournoy, who was himself drawn into what he called her "Hindoo romance" when she informed him that he was the reincarnation of Prince Sivrouka on whose funeral pyre the historical Simandini was cast alive. Hélène Smith's glossolalia, the phenomenon of speaking in tongues which was a common behavior of mediums during a séance, took the unusual form of her speaking at length—and sometimes writing as well—in Martian.

For Flournoy, and for the pioneering linguist Ferdinand de Saussure, who was curious about the Hélène Smith case, the mystery to be solved was where

* Translated and quoted by Sonu Shamdasani in his introduction to Flournoy's *Des Indes à la planète Mars* (xxix).

this Martian language, with its elaborate vocabulary and syntax, came from. As a psychologist, Flournoy searched diligently for the sources of Smith's visions; the smallest external source, buried in her subliminal memories, could trigger, he believed, elaborately creative inventions. While an external source of her Hindoo Romance could be traced, circuitously, to historical documents Smith might have read or heard about, Mars seemed an entirely different matter. Flournoy was inclined to believe that the Martian romance must be a matter of "pure imagination",[55] since there were no hard historical data from which a vision of Mars could be extrapolated. But as Flournoy acknowledges—and as the reviewer quoted above insists—theories about Mars were everywhere in the culture in the last decade of the nineteenth century. A medium's imagination about Mars, far from being pure, was likely to be thoroughly broken in, if not by the books and scholarly articles on the subject of Mars, then by news reports and conversations on the Martian debate. And the French-speaking Smith and her acquaintances in the psychical circles in Geneva would above all be aware of Flammarion's writings, interweaving as they did astronomical and spiritualist notions about Mars.

One of the most suggestive discoveries of Flournoy, who attended many of the séances during which Smith "visited" Mars or described Martian scenes in transcriptions from their spiritual "author" on Mars, is something absent from her Martian romance:

> He [the spirit-author] shows a singular indifference—possibly it may be due to ignorance—in regard to all those questions which are most prominent at the present time, I will not say among astronomers, but among people of the world somewhat fond of popular science and curious concerning the mysteries of our universe. The canals of Mars, in the first place—those famous canals with reduplication—temporarily more enigmatical than those of the Ego of the mediums; then the strips of supposed cultivation along their borders, the mass of snow around the poles, the nature of the soil, and the conditions of life on those worlds, in turn inundated and burning, the thousand and one questions of hydrography, of geology, of biology, which the amateur naturalist inevitably asks himself on the subject of the planet nearest to us—of all this the author of the Martian romance knows nothing and cares nothing.[56]

What, then, *does* Hélène Smith—or the "author," who is one of her several submerged personalities—care about in her visions of Mars? A single word will do: communication. What is absolutely central to the work of Smith, as to all mediums, is the forging of a link between the living and the dead, the material and the spiritual. It is her voices more than her visions that are primary, and Mars as a site for the reincarnated and the discarnate—as we saw in the narratives by Flammarion, Wicks, Gratacap, Weiss, and Kennon—is the crucial interest of the turn-of-the-century spiritists. Fifty years later, in strikingly different contexts, Bradbury could offer up the exuberant sentiment, "Mars is heaven."* The spiritualists, starting with Henry Gaston in *Mars Revealed*, also believed something like that, and quite literally.

As Flournoy analyzed the evidence in the case of Hélène Smith, he discovered that almost no technological invention, architectural scheme, social arrangement, costume design, or any other cultural artifact in the Mars she envisioned ventured beyond what already existed or could be anticipated in the near future on Earth. Even her Martian language, the greatest of the mysteries, turned out to be modeled on French; although Hélène Smith remained of interest to linguists for her extraordinary gifts of mimicry, the Martian language itself was about as innovative as pig Latin. Only in trivial details (eating on square rather than round plates) does Smith's Martian anthropology differ from terrestrial practice; in fact, Mars in its external forms looks a good deal like Asia as imagined by a European. The jacket illustration for the new edition of From *India to the Planet Mars* is one of Hélène Smith's own oil paintings of Mars—but anyone who makes a casual study of the details of pagoda-like buildings, hanging gardens, and tunicked figures in turban-like white hats will be reminded instantly of Persian carpets and Indian tapestries. That is, the picture, with what Flournoy calls its "clearly Oriental stamp," could as easily have emerged from Smith's Hindoo romance as from her Martian romance.[57]

The psychics who made their astral voyages to Mars, who had clairvoyant images of its cities, who transmitted messages from both its incarnate and discarnate inhabitants, and who reunited grieving terrestrial mourners with their reincarnated families and friends living happily as Martians were not as different from Percival Lowell as one might at first think. Like him, they believed they had proceeded from mere fantasies about other worlds to the higher level of "romance of fact." They too thought their methods were scientific, or at least advance forays into the new science of the twentieth century—and they too found in Mars exactly what they expected to find, exactly what their theories told them they would find.

The fashion for spiritualist Martian narratives is now long past. Twenty-first century fiction about Mars stands in the shadow, especially, of Kim Stanley Robinson's sweeping MARS trilogy (1992–96) and the new emphasis on scientific realism and the enactment of and resistance to *realpolitik* that are the new orthodoxy for Martian narratives. But the phenomenon of interweaving scientific information with blatant fantasies about Mars remains very much alive among the large body of stories about the fourth planet that continue to be written. The most prominent current examples of the wishful-thinking school—the descendants of the paranormal Martian fictions—are perhaps the multiple and ever-expanding editions of Richard Hoagland's *Monuments of Mars* (5th ed., 2001) and some of the sf novels derived from his fixation on the so-called "Face"

* "Mars Is Heaven" was the original title of the story (published in 1948) that became the chapter "The Third Expedition" in *The Martian Chronicles* (1950). The original title was changed in the *Chronicles* into a question asked by one of the characters—"Is this Heaven?" (41)—suggesting that the author had indeed modulated the romance into a more ironic mode.

in the Cydonia Mensae region. The linking of Mars and the paranormal—while characteristic of a short-lived and historically specific cultural moment—is not a mere eccentricity. For the past 125 years, since Mars first emerged as an object of intense scientific study and persistent public interest, the planet has functioned as both a magnet for and a mirror of our cultural preoccupations and fantasies.

WHY ARE WE OBSESSED WITH MARTIANS? (2015)

Natalie Haynes

Natalie Haynes is a writer and broadcaster. Six seasons of her series titled 'Natalie Haynes stands up for the Classics' have been broadcast on BBC Radio 4, and her latest non-fiction work, *Pandora's Jar: Women in the Greek Myths*, was published in 2020. She is also the author of three novels: *The Amber Fury*, *The Children of Jocasta* and *A Thousand Ships*.

Some of the earliest storytellers were inspired – just as modern storytellers are – by looking up at the night skies. While many of us look to the heavens and wonder how many stars there could be, and how far away they are, others look up and think: I wonder who lives there? And do they have antennae?

The 2nd-Century satirist, Lucian, can lay a pretty convincing claim to writing the first science fiction. In his True History, he takes his heroes on a trip to the moon, which they reach after being carried into the air by a whirlwind that lasts for seven days. Lucian is often cited as the predecessor of Jules Verne, but he neatly presages Dorothy's transport to Oz, too. Once they arrive on the moon, our heroes are startled to see men riding three-headed vultures as though they were horses. The unusual steeds are only the first strange creatures they encounter: a short while later, warriors arrive on fleas the size of 12 elephants. The king of the moon and the king of the sun are engaged in all-out war.

Whatever objects the Greeks could see in the sky, they constructed stories about them. It begins with gods: Helios, the god of the sun, and his sister, Selene, goddess of the moon, who both traverse the sky with their chariots. One of the most celebrated sculptures of the Parthenon Frieze – which once decorated the Acropolis of Athens – is that of Selene's horse from the east pediment. After a long night dragging the moon across the firmament, the horse looks exhausted: his eyes bulge, his nostrils flare, his jaw sags open in a desperate bid for air. In other words, the moon-goddess isn't just a flowery shorthand for the moon. Her story has texture and detail, right down to her weary horse.

The idea of people on, or from, the moon is not unique to Western culture. The Tale of the Bamboo Cutter is a Japanese story dating back to the 10th Century. A radiantly beautiful girl, Kaguya-hime, is found inside a bamboo plant. When she grows up, she reveals that she has come from the moon, and must return there. If Lucian's aliens were horrifying in their strangeness, Kaguya-hime is one of the earliest stories of aliens who are anthropomorphic, but somehow better than human (Superman being one of the more celebrated recent examples).

We Come in Peace

But the moon was only the beginning of our extraterrestrial inspiration. Mars is visible to the naked eye, so it's no surprise that it drew the attention of early astronomers and story-tellers. The red colour was unavoidably reminiscent of war to the ancients, and thus Mars shares its name with the Roman god of war. The martial nature of the planet's inhabitants was perfectly clear to Chuck Jones, who presented Marvin the Martian (though he didn't yet have that name) in *Haredevil Hare*, in 1948.

Marvin's outfit is a plumed helmet, like a Roman soldier might have worn, though in a bright, alien green. He also sports a skirt, divided by slits. It doesn't hang down to his knees as a centurion's might have done, but fans out like a demented tutu. Unusually, for one of Bugs Bunny's nemeses, he is actually quite alarming, possessing as he does a degree of competence which eludes, say, Elmer Fudd. So he is both Martian and martial: that is, of the planet Mars, and bellicose to boot.

Perhaps one of the easiest ways to detect whether a person's outlook is positive or negative is to ask them how they feel about aliens. For the optimist, aliens are ET, or Mork, or Mr Spock. For the pessimist, outer space is full of Facehuggers and the diminutive sadists in *Mars Attacks!*. Every time scientists send a probe into the unknown reaches of the galaxy, I find myself agreeing with Stephen Hawking (who this year helped launch the Breakthrough Listen project, to try and find alien intelligence). Although he's keen to find signs of alien life, Hawking has pointed out that an advanced alien civilisation might well destroy us without hesitation, regarding us as a lower species.

Life on Mars

Surely the greatest vision of Martians is also one of the earliest, in HG Wells' *The War of the Worlds*. Mars came into its own – culturally speaking – when Giovanni Schiaparelli trained his telescope on the planet and examined it in detail in 1877. He saw what he believed to be channels or grooves on the surface of the planet, which he called 'canali'. It was all too easy for the word to be mistranslated by a fascinated English-speaking audience: were there canals on Mars? If so, someone must have built them.

Although some writers had previously speculated about life on Mars, the images Schiaparelli produced fired many imaginations. In 1881, a newspaper called *London Truth* published a story about a Martian invasion, which envisaged us declaring war on the Martians, who would retaliate by using missiles to take huge chunks out of the Himalayas and leave a giant hole where Mont Blanc once stood. But only in 1893 did the word 'Martian' – meaning an inhabitant of Mars

– really take off in *Aleriel or A Voyage to Other Worlds*, a story by the Reverend Wladyslaw Lach Szyrma. Aleriel himself is a Venusian, and the Martians he meets are vegetarian, nine feet tall, and have a somewhat leonine appearance.

Then in 1892, people claimed to have seen flashes of light emanating from the red planet. Were these messages coming from Martians? Did they now have canals and torches? In 1895, HG Wells would begin work on *The War of the Worlds*, which was first serialised in 1897. His Martians travelled inside huge tripods (a sculpture of one stands in Woking, a small town near London which the Martians destroyed as best they could). The aliens are a terrifying bunch, 'at once vital, intense, inhuman, crippled and monstrous.' They prove too strong for the soldiers who try to resist them. Luckily, bacteria ex machina save the day, and the planet.

Science Fact?

By 1912, Edgar Rice Burroughs was imagining the trip in reverse: John Carter is from Virginia (though he has an alien quality of seeming agelessness), and makes his journey to Mars (called Barsoom by the locals) by some sort of strange astral projection. Like his fictional predecessors in Lucian, he finds himself in the midst of warring alien beings. Carter's experience fighting in the American Civil War has prepared him for the sacrifices required in such circumstances.

Burroughs wrote about John Carter for decades. And throughout the 20th Century, writers, artists, composers and film-makers returned again and again to our closest planetary neighbour. Exploration to and invasion from Mars have provided tremendous cinematic and science-fictional material. Or science fact, as the trailer for *Robinson Crusoe on Mars*, released in 1964, boldly claimed.

By 1964, when a man was stranded on Mars, he did have aliens to fear, but not Martians. Humanoids from Orion – using Mars for mining missions – were the source of peril for Kit Draper and the only fellow-survivor (a monkey) of his aborted space flight: 'This film is scientifically authentic,' read the blurb at the beginning and end of the trailer. 'It is only one step ahead of present reality!' But still it revealed something important: the more we discovered about Mars, the more it changed the way people wrote about it, and shaped the kind of stories they could write.

Andy Weir's *The Martian* was one of last year's best-selling books, and the new film adaptation is likely to repeat its success. Perhaps it's a sign of our more sceptical times that the only Martian in this story is no alien, but an all-too-human Nasa astronaut left behind by his crew. Even more so than Kit Draper, Mark Watney – the stranded engineer – is a man who must cope with the difficulties of being utterly alone in a landscape which is hostile, not because of alien inhabitants, but because of his own human frailty.

It took us so long to imagine the Martians, and our obsession with them has undoubtedly led to scientific progress that HG Wells could only have imagined. Even now, a Nasa team is holed up in an isolated testing lab in Hawaii. They will live for a year in a dome measuring 36 feet (11m) in diameter, on the northern slopes of Mauna Loa. The news that scientists have found evidence of flowing water on Mars has further fuelled excitement about life on the Red Planet.

But the more we have discovered, the less we have been able to imagine. Where once writers could populate Mars with all manner of alien beings, now we have had to move those creatures further afield. We know too much – and crucially, so do readers and viewers – to keep pretending that the aliens are up there, digging their canals. Mars has lost none of its wonder, but it has certainly shed some of its mystery.

I'M CONVINCED WE FOUND EVIDENCE OF LIFE ON MARS IN THE 1970s (2019)

Gilbert Levin

American engineer **Gilbert Levin** was born in 1924 in Baltimore, Maryland. He interrupted his university studies to serve as a radio operator on merchant ships during the Second World War, but subsequently obtained a degree in civil engineering, an MSc in sanitary engineering and a PhD in environmental engineering, all from Johns Hopkins University. In 1967, he founded Biospherics Research Inc. and became the Principal Investigator of a life detection experiment on NASA's Viking mission. He died in July 2021, aged 97.

We humans can now peer back into the virtual origin of our universe. We have learned much about the laws of nature that control its seemingly infinite celestial bodies, their evolution, motions and possible fate. Yet, equally remarkable, we have no generally accepted information as to whether other life exists beyond us, or whether we are, as was Samuel Coleridge's Ancient Mariner, "alone, alone, all, all alone, alone on a wide wide sea!" We have made only one exploration to solve that primal mystery. I was fortunate to have participated in that historic adventure as experimenter of the Labeled Release (LR) life detection experiment on NASA's spectacular Viking mission to Mars in 1976.

On July 30, 1976, the LR returned its initial results from Mars. Amazingly, they were positive. As the experiment progressed, a total of four positive results, supported by five varied controls, streamed down from the twin Viking spacecraft landed some 4,000 miles apart. The data curves signaled the detection of microbial respiration on the Red Planet. The curves from Mars were similar to those produced by LR tests of soils on Earth. It seemed we had answered that ultimate question.

When the Viking Molecular Analysis Experiment failed to detect organic matter, the essence of life, however, NASA concluded that the LR had found a substance mimicking life, but not life. Inexplicably, over the 43 years since

Viking, none of NASA's subsequent Mars landers has carried a life detection instrument to follow up on these exciting results. Instead the agency launched a series of missions to Mars to determine whether there was ever a habitat suitable for life and, if so, eventually to bring samples to Earth for biological examination.

NASA maintains the search for alien life among its highest priorities. On February 13, 2019, NASA Administrator Jim Bridenstine said we might find microbial life on Mars. Our nation has now committed to sending astronauts to Mars. Any life there might threaten them, and us upon their return. Thus, the issue of life on Mars is now front and center.

Life on Mars seemed a long shot. On the other hand, it would take a near miracle for Mars to be sterile. NASA scientist Chris McKay once said that Mars and Earth have been "swapping spit" for billions of years, meaning that, when either planet is hit by comets or large meteorites, some ejecta shoot into space. A tiny fraction of this material eventually lands on the other planet, perhaps infecting it with microbiological hitch-hikers. That some Earth microbial species could survive the Martian environment has been demonstrated in many laboratories. There are even reports of the survival of microorganisms exposed to naked space outside the International Space Station (ISS).

NASA's reservation against a direct search for microorganisms ignores the simplicity of the task accomplished by Louis Pasteur in 1864. He allowed microbes to contaminate a hay-infusion broth, after which bubbles of their expired gas appeared. Prior to containing living microorganisms, no bubbles appeared. (Pasteur had earlier determinted that heating, or pasteurizing, such a substance would kill the microbes.) This elegantly simple test, updated to substitute modern microbial nutrients with the hay-infusion products in Pasteur's, is in daily use by health authorities around the world to examine potable water. Billions of people are thus protected against microbial pathogens.

This standard test, in essence, was the LR test on Mars, modified by the addition of several nutrients thought to broaden the prospects for success with alien organisms, and the tagging of the nutrients with radioactive carbon. These enhancements made the LR sensitive to the very low microbial populations postulated for Mars, should any be there, and reduced the time for detection of terrestrial microorganisms to about one hour. But on Mars, each LR experiment continued for seven days. A heat control, similar to Pasteur's, was added to determine whether any response obtained was biological or chemical.

The Viking LR sought to detect and monitor ongoing metabolism, a very simple and fail-proof indicator of living microorganisms. Several thousand runs were made, both before and after Viking, with terrestrial soils and microbial cultures, both in the laboratory and in extreme natural environments. No false positive or false negative result was ever obtained. This strongly supports the reliability of the LR Mars data, even though their interpretation is debated.

In her recent book *To Mars with Love*, my LR co-experimenter Patricia Ann Straat provides much of the scientific detail of the Viking LR at lay level. Scientific papers published about the LR are available on my Web site.

In addition to the direct evidence for life on Mars obtained by the Viking LR, evidence supportive of, or consistent with, extant microbial life on

Mars has been obtained by Viking, subsequent missions to Mars, and discoveries on Earth:

- Surface water sufficient to sustain microorganisms was found on Mars by Viking, Pathfinder, Phoenix and Curiosity;
- Ultraviolet (UV) activation of the Martian surface material did not, as initially proposed, cause the LR reaction: a sample taken from under a UV-shielding rock was as LR-active as surface samples;
- Complex organics, have been reported on Mars by Curiosity's scientists, possibly including kerogen, which could be of biological origin;
- Phoenix and Curiosity found evidence that the ancient Martian environment may have been habitable.
- The excess of carbon-13 over carbon-12 in the Martian atmosphere is indicative of biological activity, which prefers ingesting the latter;
- The Martian atmosphere is in disequilibrium: its CO_2 should long ago have been converted to CO by the sun's UV light; thus the CO_2 is being regenerated, possibly by microorganisms as on Earth;
- Terrestrial microorganisms have survived in outer space outside the ISS;
- Ejecta containing viable microbes have likely been arriving on Mars from Earth;
- Methane has been measured in the Martian atmosphere; microbial methanogens could be the source;
- The rapid disappearance of methane from the Martian atmosphere requires a sink, possibly supplied by methanotrophs that could co-exist with methanogens on the Martian surface;
- Ghost-like moving lights, resembling will-O'-the-wisps on Earth that are formed by spontaneous ignition of methane, have been video-recorded on the Martian surface;
- Formaldehyde and ammonia, each possibly indicative of biology, are claimed to be in the Martian atmosphere;
- An independent complexity analysis of the positive LR signal identified it as biological;
- Six-channel spectral analyses by Viking's imaging system found terrestrial lichen and green patches on Mars rocks to have the identical color, saturation, hue and intensity;
- A wormlike feature was in an image taken by Curiosity;

- Large structures resembling terrestrial stromatolites (formed by microorganisms) were found by Curiosity; a statistical analysis of their complex features showed less than a 0.04 percent probability that the similarity was caused by chance alone;
- No factor inimical to life has been found on Mars.

In summary, we have: positive results from a widely-used microbiological test; supportive responses from strong and varied controls; duplication of the LR results at each of the two Viking sites; replication of the experiment at the two sites; and the failure over 43 years of any experiment or theory to provide a definitive nonbiological explanation of the Viking LR results.

What is the evidence against the possibility of life on Mars? The astonishing fact is that there is none. Furthermore, laboratory studies have shown that some terrestrial microorganisms could survive and grow on Mars.

NASA has already announced that its 2020 Mars lander will not contain a life-detection test. In keeping with well-established scientific protocol, I believe an effort should be made to put life detection experiments on the next Mars mission possible. I and my co-experimenter have formally and informally proposed that the LR experiment, amended with an ability to detect chiral metabolism, be sent to Mars to confirm the existence of life: non-biological chemical reactions do not distinguish between "left-handed" and "right-handed" organic molecules, but all living things do.

Moreover, the Chiral LR (CLR) could confirm and extend the Viking LR findings. It could determine whether any life detected were similar to ours, or whether there was a separate genesis. This would be a fundamental scientific discovery in its own right. A small, lightweight CLR has already been designed and its principle verified by tests. It could readily be turned into a flight instrument.

Meanwhile a panel of expert scientists should review all pertinent data of the Viking LR together with other and more recent evidence concerning life on Mars. Such an objective jury might conclude, as I did, that the Viking LR did find life. In any event, the study would likely produce important guidance for NASA's pursuit of its holy grail.

THE SEARCH FOR LIFE ON MARS (2020)

CHAPTER 1: FROZEN IN TIME

Elizabeth Howell and Nicholas Booth

Elizabeth Howell is a Canadian space journalist. She writes for space.com, *Forbes* and Sky News. In 2014, she took part in a simulated Mars Mission as part of Crew 133 at the Mars Society's Mars Desert Research Station in Utah. **Nicholas Booth** is a freelance journalist and broadcaster. He has worked for *Astronomy Now*, was technology editor at *The Times* and has written for other British broadsheets. He also writes fiction about unusual characters and unlikely events from history.

"To understand how life would evolve on Mars, you have to go to Antarctica," says Dr. Christopher McKay. "There is no other place like it."

Widely seen as the most eloquent spokesman for the tantalizing possibilities of life on Mars, Chris McKay has spent decades studying life in Antarctica to gain a greater understanding of how microbes could exist on the Red Planet. Working out of NASA's Ames Research Center, south of San Francisco in the suburb of Mountain View, McKay is a leading astrobiologist. Tall, genial, and with a voice so *basso profundo* it seems to emanate from somewhere below the floor and boom in empty spaces, McKay made his first journey down to "the ice" in 1980. "Antarctica is like a second home to me," he says today.

McKay is acknowledged as a voice of reason in a field of research that has sometimes split into contentious factions, especially where life on Mars is concerned. "That question would be easier to answer if we could understand the evolution of life on Earth," he says, "or even if there was a consensus on the origins of life. Mars is going to help us with this riddle."

Though he has also investigated life in Siberia and in the Atacama Desert in Chile, McKay believes that the high, dry valleys of Antarctica are the only place on Earth where conditions are sufficiently extreme to mimic the Red Planet. The key ingredient is water. "People often say how amazingly robust life is," McKay says. "My reaction is the opposite. It always needs water. If we had the trick of learning to live without water, life would be hardier."

If we can't look for life directly, then searching out water is the next best thing. That has informed the scientific rationale behind the most recent and the next missions to the Red Planet. Without liquid water, life would have been unthinkable on Earth as it would have been on Mars. Given the fact that most water on Mars is concentrated in the form of ice at its poles, McKay believes that is where life is most likely to be found.

The polar ice caps of Mars have beguiled and enticed astronomers since their discovery in the eighteenth century. Their waxing and waning showed that, like Earth, the Red Planet undergoes seasons as it alternately tilts away from and toward the Sun. The seasonal ice caps grow and retract with the passage of the seasons. It was later found that the Martian tilt is 25°, similar to the 23.5° value for Earth.

However, it is very difficult to reach the Martian poles. Tricky maneuvers would be required to touch down far from the easier-to-reach equator of the planet, requiring greater amounts of fuel at the expense of scientific instruments. Any attempt would be severely constrained by weight limitations and the extreme temperatures. Actually landing a probe amid the ice there is even more hazardous than exploring the poles of our own world.

NASA's first attempt to do so, in 1999, failed. The Mars Polar Lander crashed somewhere in the southern polar regions, likely as a consequence of a software error that affected its landing system. Nine years later, the *Phoenix* lander, named for the ancient bird that rises from the ashes, successfully made it all the way down at 67°N ("which is like Iceland on Earth," says one observer).

Over the northern winter of 2008–2009, *Phoenix* found evidence that snow accumulates on the surface and detected what are known as perchlorates (a possible "food" for some microbes) in the soils. It also showed beyond doubt that there is a solid ice layer immediately underneath where *Phoenix* landed. "Nobody really knew how much ice was lurking just below the surface," says Professor Jack Holt, a glaciologist at the University of Arizona. "So that was kind of a surprise to a number of people."

That discovery has, he says, opened the door to discovering that there are much more extensive icy deposits below the surface, which have been remotely detected across whole swaths of the Red Planet. Ice on Mars, the result of changes to the planet's climate over geological time, is no longer confined to the poles. Have these icy deposits always been so prevalent, or are they, as some believe, more of a recent phenomenon?

* * *

Antarctica is as alien as it gets on Earth.

Conditions are scored for extremes. During the summer, the average temperature in coastal regions hovers around freezing point, while it varies between −15°C to −30°C (−5°F to −22°F) inland. In the central plateau, temperatures range from −40°C to −70°C (−40°F to −95°F). The lowest temperature ever recorded on Earth was −89.6°C (−130°F) in the winter of 1983 at the Soviet Vostok research center there. Small wonder it has been referred to as "the Gulag of the South" by those who have willingly stayed at the center in the name of scientific duty.

Though it only covers about 10 percent of the total landmass on Earth, the South Pole contains about 90 percent of the world's supply of ice. The Antarctic continent is shaped like a squat, lopsided letter Q, with the lower squiggle forming the Antarctic Peninsula. It points like a crooked finger toward South America, five hundred miles (some thousand kilometers) distant. Antarctica's ice lies on a foundation of rock, most of which is hidden from view.

Ice flows in strange ways on this southern continent. Around the edge of Antarctica is a ring of mountains through which continental ice is forced to pass. Eventually, it falls into the sea, but first the frozen hulk tends to form ice shelves that are glued to the landmass by the freezing cold. Some ice chunks are as large as small countries. At times these massive sheets break off to form large icebergs; this process has accelerated with recent climate change, which is warming Earth's poles rapidly.

The climate of Antarctica is unique: its air is trapped for most of the year under a giant anticyclone. As a result, winds descend around its outer extremities and flatten as the air flows outward. The winds, immortalized by mariners as the Screaming Sixties, whip up sudden storms and squalls. Thankfully, on the shelf-like coast of the continent around the main ice sheet, a thousand miles from New Zealand, the weather is distinctly better.

It was both the clement conditions and the sheltered inlet of this area that commended it as a stopping-off point for one of the most important voyages of discovery of the nineteenth century. James Clark Ross, a dashing officer in the British Royal Navy, had already discovered the magnetic north pole when he set off to find its southern equivalent in the late 1830s. In the large sailing ships *Erebus* and *Terror*, his expedition ventured farther south than anyone had ever done before.

They happened upon the Antarctic coastline after "a magical journey of towering mountains and shining glaciers," in the memorable phrase of one chronicler of their travels. Ross's own diary records that on January 28, 1841, the sea that now bears his name appeared like a sheet of frozen silver in the uncharacteristically good weather and blue skies. His crew were openmouthed upon finding "a perpendicular cliff of ice between 150 and 200 feet above the level of the sea, perfectly flat and level at the top and without any fissures or promontories on its seaward face."

Because the ice cleared faster here than anywhere else they had happened upon in Antarctica, it became an obvious point of contact for future explorers. Scientists today head for this sheltered sound, which was named after the senior lieutenant of the *Terror*, Archibald McMurdo. Now visitors have the luxury of flying in on modified Hercules transport aircraft on flights from New Zealand. Once they're deposited, the plane often doesn't stick around in case its delicate engine parts freeze over.

Only in the direst of circumstances will the authorities ever attempt a Win-Fly, as the flights in during the dead cold of winter are known. One such situation took place in the austral winter of 2017 to rescue an eighty-seven-year-old man who was experiencing breathing difficulties at the geographical South Pole. Later shown recuperating in a hospital in Christchurch, he smiled for the cameras after his ordeal. That he wore a T-shirt with the phrase GET YOUR ASS TO MARS, made famous in the original *Total Recall* movie, and that he was the second human being to walk on another world had a lot to do with the impression of sangfroid he gave.

But then, Buzz Aldrin has always dreamed of an encore: walking on Mars. "I think we can all say with confidence that we are closer to Mars today than we have ever been," Aldrin had said earlier that same year.

* * *

Antarctica is tough to explore, but it has nothing on Mars.

The Red Planet is roughly 1.5 times farther from the Sun than we are. Mars orbits the Sun at an average distance of 142 million miles (228 million kilometers), compared to 93 million miles (150 million kilometers) for Earth. Its orbit is much more elliptical than Earth's. The Red Planet takes twenty-three of our months—nearly two years—to complete one orbit. It also receives much less heat than Earth. Conditions on Mars would make Vostok station look positively balmy. The average temperature on Mars is about –60°C (–76°F), and though there are places where it can fleetingly hover around the freezing point of water, temperatures can plunge down to –150°C (–240°F) during the polar night.

Atmospheric pressure distinguishes Mars from Antarctica, even though our own southern continent is one of the highest regions on Earth. (The thinner atmosphere there caused the problems with Buzz Aldrin's breathing.) Because of Antarctica's altitude, one Soviet researcher at the same Vostok station where the coldest temperature was measured was astounded to find that potatoes took three hours to cook through. They boiled at 88°C (190°F). On the Red Planet, the average atmospheric pressure is less than a hundredth of that on Earth. There is so little atmosphere on Mars that water molecules would rush out in a mass exodus. If you took a pan of water outside, it would burst outward in a freezing explosion.

Mars has a lower atmospheric pressure because it is roughly half the size of Earth and a tenth of its mass, so its gravitational influence is smaller, roughly 40 percent of ours. Throughout its history, Mars was not able to hold on to its primordial reserves of water, which evaporated or were lost to space. Today, this also means that the Red Planet cannot hold on to as thick an atmosphere as Earth. The average atmospheric pressure on Mars is 6.1 millibars, compared to 1013 millibars at sea level on Earth. The range of pressures on the Red Planet varies, running from nearly 9 millibars at the bottom of the largest basin to 2 millibars at the top of the highest volcanoes.

The atmosphere of Mars is composed almost entirely—95 percent—of carbon dioxide. The gas traps sunlight on the planet's surface, lifting the average temperature there some 5°C (41°F), compared to 35°C (95°F) on Earth. Mars is also almost completely dry. Even more so than Antarctica, it lives up to the nickname of "freezing desert."

* * *

The most Mars-like places in Antarctica are the remarkable Dry Valleys, close by McMurdo Sound. Here the temperature averages –20°C (–4°F) and rarely rises above the freezing point of water. The Dry Valleys receive less annual precipitation than the Gobi Desert. They were discovered by Captain Robert

Falcon Scott on one of his first journeys to the South Pole, in a region known as Victoria Land in honor of the monarch of the time. The Dry Valleys are separated from the remorseless encroachment of Antarctic glaciers by the Transantarctic Mountains.

When Scott and his team happened upon them, they were astounded. "The hillsides were covered with a coarse granitic sand strewn with numerous boulders," he recorded in his diaries. "It was curious to observe that these boulders, from being rounded and sub-angular below, gradually grew to be sharper in outline as they rose in level."

Scott later investigated this area during his more famous, ultimately tragic expedition in 1912. Two of the valleys are named after scientists attached to his expedition, Thomas Griffith Taylor and Charles Wright. The valleys receive at most four inches (ten centimeters) of snow per year, precipitation that is blown away by the harsh winds whistling through the region. They are the coldest and driest places on Earth.

Until the 2000s, scientists had found no trace of life in these harsh valleys. In the early 1970s, when NASA was preparing its first missions to land on Mars, the *Viking* spacecraft, the Dry Valleys were chosen as a test site for some of the life-detecting instruments. If they could find microbes in the Dry Valleys, they would be able to find them on Mars. However, their findings in Antarctica were ambiguous. What resulted was an almighty row between factions within the Viking biology teams, with some claiming that the valleys were entirely sterile and others that they weren't.

Today, cooler heads have prevailed. The original argument was based on biologists' ability to culture any living material from samples of the soil. The greater truth is that nothing could be cultured *from the soils in* the Dry Valleys, hardly surprising given that 90 percent of organisms in any soil cannot be grown in this way. With a sensitive enough probe, though, biologists have subsequently found plenty of evidence for microorganisms throughout the Dry Valleys. Whether that life resulted from material blown in from elsewhere or was indigenous and actually growing there remained a matter of debate until more recent times.

Closer examination reveals thriving microbial ecosystems in the Dry Valleys. Rocks act like little greenhouses and often trap water. Just below the surface of Sun-facing sandstone rocks are layers of lichen and algae that can survive because the dark surface of the rocks is warmed above air temperature. Pores within the rock trap whatever liquid water is available from the occasional snow flurries. The organisms are cocooned from the cold and receive enough sunlight to allow photosynthesis to take place.

At the bottom of the valleys are lakes and ponds, which were also discovered in Scott's time. Some are replete with thick, salty waters that are fed by the annual buildup of snow. Uniquely, they do not drain away. Rather, their liquid content evaporates due to the fearsome winds that constantly blow through the valleys. Around the shorelines of the lakes may be found microbial life in the form of algae, upon which populations of yeast and molds may feed. These microbes support microscopic protozoa, rotifers, and tardigrades, all tiny organisms that congregate at the very base of the food chain.

Whatever their origin, the microorganisms in the Dry Valleys provide clues to ones we may ultimately find on Mars. If we can better understand how such life originally formed here, biologists will be able to get a much better handle on what may have happened on the planet next door. In 2011, when the most powerful camera ever sent to the Red Planet detected what looked like fresh flows of water from orbit, one scientist's comment to the press was especially pertinent: "Mars looks more like the Dry Valleys of Antarctica every day."

* * *

Chris McKay was in graduate school when the Viking missions landed on Mars in 1976. Though they found no evidence for microbial life, he was more intrigued about the absence of organic molecules in the Martian soil. These complex chains of carbon are crucial in the evolution of life as we know it. The singular fact of their complete absence led to an absolute change in scientific opinion about the possibilities for life on Mars. Taken at face value, the lack of organics implied it would be pointless looking for life there. In very simple terms, there was no biochemical backbone on which life could have formed.

"After Viking, there was a general lack of interest in the scientific community," he says. "Viking immediately suggested to many people that there isn't life on Mars, nor could there have ever been. I don't think there was a really objective scientific assessment of the results. The initial disappointment was too much."

Now, in the twenty-first century, the pendulum is swinging back the other way. Over the last eight years, NASA's Curiosity rover has uncovered organics on several occasions on the Martian surface; the latest, in the summer of 2018, were tough organic molecules—"tough" in the sense they had survived for so long—buried in three-billion-year-old rocks that likely had originally accumulated from sediments in a lake.

More recent missions have raised the stakes further, revealing that water once flowed and conditions were probably right for life to have formed in the ancient past. Hematite, a form of iron that is oxidized in the presence of water, has been discovered in various places across the Martian surface. Many rovers, most recently Curiosity, have discerned telltale signs of flooding by dramatic flows of water. From orbit, other spacecraft have observed deeply cut canyons that look like ravines and outflows carved by water. There is also some evidence for an extensive shallow sea in the ancient past that may have covered sizable swaths of the planet's northern hemisphere.

There may even be fresh flows of water on Mars today. Seasonal changes have been observed on the slopes of some Martian craters, although what is causing these dark streaks to appear and disappear is a matter of much debate. Hydrated minerals have been observed in these streaks. Some scientists argue it is because of water from the atmosphere, while others say this might be the result of flowing, briny water. We won't know for sure until some future mission ventures up close to these features, which are called recurring slope lineae.

Scientists are by their very nature conservative, cautious, and slow to adapt. There are times when the shock of the new can cause a radical change in opinion, known as paradigm shifts, but change is rarely sudden. When Alfred Wegener first proposed the notion of plate tectonics on Earth in the 1920s, he was largely ignored. It took nearly four decades for his work to be unequivocally accepted.

Those who study life on Mars cite a similar paradigm shift after the general gloom of Viking's apparent inability to detect life. The shift resulted from a paper that Chris McKay presented with colleagues in 1984 when he was a postgraduate student at the University of Colorado, Boulder. It catalogued the competing theories for the origins of life on Earth and explored whether they would also work on Mars, where many of the same early conditions may have existed. The answer was a resounding yes.

"That's a bit of a weaselly argument," McKay says with a knowing smile today, "because we can't really say what happened on Mars without knowing how life evolved on Earth. But all the theories didn't contradict anything we already know about Mars."

Ever the optimist, McKay says this implicit confirmation increases the likelihood that life could have existed on the Red Planet. Not only do all the theories about the origin of life on Earth apply to Mars, but McKay also points out that life must have evolved pretty quickly. Once life has come about, it is hard to get rid of—no matter what the original conditions were like or how hostile the environment then becomes.

McKay believes that in the ancient past, survival of the fittest would dictate that Martian microbes would have migrated underground. Even if life died pretty quickly after it formed, say within the first billion years of Martian geological history, then its biochemical signature might still be around in the rocks or close below the surface. Certainly, the missions to Mars being launched in 2020 will seek out these markers or "biosignatures" that may still be present. "I think we'll only find organic preserved remains," McKay cautions. "There will be morphological structures, but the organisms themselves will long since have been and gone."

His reasoning may seem obscure, but it comes from the benefit of his own experiences. Chris McKay was involved in an intriguing series of experiments in the permafrost of Siberia, carried out in 1991 by a team of Russian and American astrobiologists. In northeastern Siberia, subzero conditions have persisted for over three million years, a stitch in time compared to Martian geological history. Nevertheless, what they found reveals just how astoundingly hardy life can be in even its lowliest microbial manifestations. Large numbers of bacteria have been effectively freeze-dried. When thawed out, they resume their life functions. The Siberian results show that there are up to 100 million bacteria per gram of frozen soil. Even more remarkably, after being frozen for three million years at a temperature of –10°C to –12°C (10°F to 14°F), they do not seem particularly harmed by the experience.

"They are viable," McKay says. "What keeps them ticking is the natural radioactivity of the rocks, which over ten million years or less is not a problem."

But Mars is a far harsher environment: could microbes have persisted there for such a long period tied up in the permafrost? In principle, yes. Half of the Red Planet has been "dead," geologically speaking, since the intense cratering that immediately followed the birth of the solar system. The southern hemisphere of

Mars bears testimony to ubiquitous bombardment, with surface features that have not been removed by subsequent geological activity. If life ever started, its presence may still be there.

"Those are almost ideal preservation conditions," McKay says. "Dry, frozen, and at low atmospheric pressure. You couldn't ask a curator to do a better job of preserving any samples from that time." He cautions, however, that any microorganisms may have been assaulted by accumulated radiation within the rocks and from space that has bathed the Martian surface, since the atmosphere does a poor job of protection.

Biologists don't have a particularly good understanding of what happens when organisms are frozen for millennia, and the effect of cumulative exposure of microbes to low levels of radiation is largely unknown. The situation is further complicated because the dosage received—on Earth or Mars—would fluctuate over time, making it difficult to extrapolate. Most experiments to date in this field have been with bursts of very high levels of radiation over short periods of time. Today, more precise measurements of radiation are coming from NASA's Curiosity rover in and around Gale Crater. McKay is thus working to better understand how cumulative exposure to natural radioactivity would affect possible Martian microbes.

* * *

If life did begin on Mars three and a half billion years ago, any water would have been locked up as permafrost as its climate subsequently evolved and became cooler. Organic material would have become incorporated into these frozen sediments and remained frozen in place, both physically and chemically.

There is also another factor favoring preservation. Mars does not have plate tectonics, so the upper layers of its surface did not recycle the original material that made up its primordial crust. It has been estimated that erosion and burial rates at the Martian surface are approximately one meter per billion years. Perhaps two-thirds of the Martian surface is older than, and has remained unchanged for, three and a half billion years. This means that even the most ancient of permafrost should be fairly easily accessible to robotic or human explorers. Material a few meters below the surface would also have been protected from dangerous cosmic radiation and ultraviolet rays from the Sun in the epochs since.

Some researchers have even gone so far as to say that underneath the Martian surface, there may well be extant life-forms today. Others are more cautious. Nevertheless, the intriguing possibility remains that there will be compelling evidence for ancient biological activity preserved in the Martian permafrost.

Frozen water is the key. Some Martian craters seem to have formed when they were accompanied by what looks like muddy slurries. Larger impacts were needed to create telltale torrents of mud. If the ground contained significant amounts of water, the craters from the impacts would have been frozen in place thereafter. Closer to the poles, even the smallest of craters seem to be surrounded by telltale signs of ancient muddiness. This suggests that, in these regions at least, the ice is nearer to the surface.

Nearer the equator, however, the relative warmth of the Sun means that the ice migrated farther below the surface. The presence of ground ice becomes apparent from the midlatitudes up toward the poles. "We know there's ice from observations by our radar on spacecraft in orbit," says veteran Mars researcher Richard Zurek of the Jet Propulsion Laboratory (JPL). "There is ice beneath the surface as well as at the polar caps."

Under present conditions, it has been calculated that the surface of Mars is frozen to depths of about a mile (about two kilometers) at the equator and about three miles (six kilometers) at the poles. In 2015, a team led by Ali Bramson, then at the University of Arizona, mapped the location of nearly two hundred craters across an otherwise featureless plain north of the equator, Arcadia Planitia. They were chosen because radar observations from NASA's Mars Reconnaissance Orbiter had measured reflections from under the surface that are characteristic of ice. Many exhibited "terraces" that could only have formed if there was ice under the surface.

To fit the observed "splats" and the radar measurements, the Bramson team estimated that there is a thirteen-story-deep ice sheet buried underneath Arcadia Planitia. A year later another team, led by Cassie Stuurman at the University of Texas at Austin, used a similar radar technique to examine another northern plain called Utopia, where the second *Viking* lander had landed in September 1976. Terrain that has cracked into telltale polygonal shapes and scalloped depressions suggests they were formed by extensive water ice that had sublimated or thawed out. The Stuurman team estimated there were 145,000 square miles (375,000 square kilometers) of ice below the surface.

Taken with the earlier work, these studies show that recent climate change has led to the accumulation of ice at the midlatitudes. Such deep reserves exist because surface ice in the equatorial regions is unstable and will easily sublimate away. At the poles, however, it won't. Higher than 40° latitude in either hemisphere, ground ice will be found closer to the surface. The nearer to the Martian poles you look, the more likely you are to find it. That, in essence, is why the poles could be such a mecca for microbes.

Another tantalizing glimpse of the subsurface water surrounding the south pole of Mars has been inferred from orbit. An unnamed crater in a region called Noachis Terra, in the vicinity of the Martian south pole, reveals the strongest evidence yet for seepage of underground water. Just thirty miles (fifty kilometers) wide, this crater contains dark, tiger-toothed features within its rim that look similar to glacial water seepage seen in Iceland and around Mount St. Helens. This strongly suggests that the water was exposed when the crater was excavated by the impacting body that gouged it out of the surrounding surface.

Not only that, the dark floor of the crater is smooth, which suggests it could well have been covered with a pool of water. The edge of the inside wall of the crater reveals that there are "islands" of material poking up through the floor. The most likely explanation is that water did indeed burst out of the crater rims, and then formed a lake that either evaporated or froze in place.

Certainly, life as we know it depends on water of a certain chemistry—not too salty, for example—so finding life-friendly water could lead to life itself. On Mars, a great deal would have certainly percolated through the soil to form the layer of permafrost or else remained frozen on the surface and thereafter become covered by accumulations of dust.

As such, ice—not least in the form of glaciers—is an important bridge to the past, buttressing theories of what might have happened when the Red Planet was warmer. Could conditions have been clement enough for a large ocean to have covered sizable portions of the planet, as inferred from the geologic evidence? How wet was Mars in its earliest days? Did the planet's ancient climate suddenly change forever, or did it oscillate between warmth and cold?

As ever with the Red Planet, there may well be other factors that are not yet understood. One important insight comes from perhaps the most remarkable discovery ever made about the poles of Mars: the laminated terrain.

* * *

They stretch as far the eye can see in every direction: regular, relentless, and undulating along the perimeters of both polar ice caps on Mars. They look like strange icy layers in a cake created by a cosmic chef with an enhanced sense of aesthetics. Nothing quite like them has been seen elsewhere in the solar system, and that includes the even colder, icier moons of the outer planets.

This is the mysterious layered terrain surrounding the Martian poles that even today defies detailed understanding. The alternating layers of dust and ice are seen at latitudes greater than 80° in both hemispheres and stretch uninterrupted for many hundreds of miles. They appear as fine bands along slopes where the Sun's rays and wind have removed seasonal ice, and they occur in relatively smooth, undulating landforms into which the polar ice has cut steep slopes and scarps. When they were discovered in the 1970s, they caused no end of amazement. Their regularity hints at how the climate has changed in the past. The relative "thickness" of deposited dust and ice has faithfully mimicked climatic conditions throughout Martian history.

Nobody knows how long it would take to deposit an icy layer. Theoretical calculations have shown it could be anywhere from a year to many hundreds of years. Nevertheless, the broader outlines of how the layers formed have come from some remarkable astronomical detective work. By divining the motions of the Red Planet as it has moved through space over recent millennia, climatologists have determined why dust was more likely to be deposited than ice and vice versa at different times in its history.

Planets are not the perfectly defined spheres of childhood drawings, nor are they uniformly dense. Earth, for example, is distinctly pear-shaped. The equator is out of true with the poles by many tens of miles. That irregularity means planets don't spin as perfect spheres would. Mars, too, is oddly shaped, like a slightly squashed egg. It, too, wobbles as it spins, and this has had a pronounced effect on its climate.

When the Martian orbit is at its most elliptical, more sunlight will fall on the hemisphere undergoing summertime, as it is so much closer to the Sun. At present, that is the southern hemisphere. During the southern summer, global dust storms tend to kick up and deposit dust on the northern pole, due to the quirks of atmospheric circulation on today's Red Planet.

This tilt, which is more formally known as obliquity, also has an important effect on climate. As noted earlier, the current Martian tilt (25°) is similar to Earth's (23.5°). Jupiter's immense gravitational influence may have tugged the axis to as high as 46° within the first half billion years of the planet's history. By comparison, Earth's axial tilt has changed only by about one degree over geological time.

Even after the emergence of four enormous volcanoes on the Martian equator, which dampened down the tilt of Mars, the obliquity was reduced to only 35°, still large enough to have had a distinct climactic effect over the succession of seasons. Perhaps today's tilt of 25° is the lowest ever amount by which Mars can roll on its axis. At present, it seems that changes can vary by up to 10° over a period of 100,000 or a million years.

As a result, some have argued that the current deep freeze seen on Mars may just be a temporary phenomenon, so far as geological time is concerned. When the axial tilt is higher, the summer pole would be far hotter. Much less atmospheric carbon dioxide would freeze out on the winter pole in the annual yin and yang of the seasons. More dust would be deposited because there would be less ice around. The opposite would be the case when the planet tilted by much less.

Such changes would obviously contribute to the layering seen in and around the Martian poles. Stronger summer heating at one pole may have released greater amounts of greenhouse gases from the polar caps. Permafrost that is rich in frozen carbon dioxide would have sublimated to allow a temporary greenhouse effect to result. What that also means is that the deep freeze would be temporary in any given region on Mars. The changes to the tilt would cause extensive glaciations, which would then also move down to the midlatitudes or back up to the poles.

If liquid water were available, it could have flowed across the surface before it froze or evaporated. Chemical reactions with the atmosphere would have formed carbonates or salts, reducing the overall atmospheric pressure.

When the planet's tilt was reduced and its orbit became more elliptical, Mars would have been distinctly cooler. At the poles, there would have been a greater chance for snow and glaciers to be more likely. Based on his extensive experience studying glaciers on Earth, Professor Jack Holt, a glaciologist at the University of Arizona, is amazed at the estimated range of ages of the ice which is still present on Mars—anywhere from a few hundred million to several hundred million years. "We don't have anything remotely close to that preserved anywhere on Earth," he says. "On Mars, I think that this ice actually contains a great deal of past climate information that is so far missing from our understanding."

Much more scientific detective work is needed to discern the exact details of the more recent climate history of the Red Planet. While the ice is very old in terms of human experience, it is geologically quite young. Nevertheless, scientists are akin to detectives looking at a crime scene that has long since been altered. They are hampered by not knowing what exactly has taken place. To date, they have inferred details from the equivalent of fleeting snapshots and unreliable witness statements. Worse still, much of the evidence for the evolution of climate on Mars is either circumstantial, contradictory, or plain confusing. One thing most experts agree upon is that the poles of Mars may unlock the mysteries of climate and, hence, the possibility for ancient life. They beckon to us across the interplanetary void, tantalizing and often ambiguous, in the vital information they contain. They will help unravel many of the enduring mysteries about the ancient history of the world next door.

* * *

Since the start of the twenty-first century, ever more detailed suites of instruments have been peering down at the poles of Mars from spacecraft as they orbit overhead. They have been scrutinizing the ice caps in visible light, in the infrared, using laser reflections, and—most revealingly—by radar. Taken altogether, our portrait of the polar ice has been fundamentally redrawn before our very eyes. "We've actually seen exposed cliffs of ice," says Rich Zurek of the Jet Propulsion Laboratory. "And you go, 'Wait a minute, those must be rapidly sublimating into the atmosphere today. So how do you keep this there for any length of time?'" There are many places where there is subsurface ice and a lot of debris surrounding it. "And that insulates the ice and keeps it there," he adds.

One of the persistent riddles is that the northern pole seems warmer than the south yet also seems to have more water ice on display. The explanation is that appearances may be deceptive. The south pole is covered by a frosting of carbon dioxide, which is believed to be quite thin compared to the vast bulk of water ice below.

The Mars Global Surveyor spent most of a decade in orbit around the Red Planet after arriving in September 1997. It accurately scanned the surface below using a laser that fired pulses to build up a detailed map of the terrain. By accurately timing how long the signal took to return, the laser signals revealed the extent of depressions or mountains beneath. If the surface was comparatively lower, the laser pulse took longer to return, and vice versa if it was higher. The Surveyor's orbit was better placed to observe the north pole, and thanks to that, a three-dimensional map of the Martian arctic has been possible.

"Without [the laser instrument], we would have had a lot of trouble with our interpretations because it's our baseline reference for all our current data," admits Jack Holt. His recent work, for example, has used radar that can penetrate below the surface. But making sense of the radar signals requires various corrections that depend on the state of the atmosphere through which they have traveled (such as electrical activity in its outermost layers, which have been found to create aurorae). With a laser instrument, however, what you saw was what you got. Without the laser ranging information, Holt says, "it would have been really impossible to do the level of interpretation that we have."

What laser altimetry has shown, particularly in its greater vertical resolution, is that the north polar cap on Mars is gouged by canyons and troughs that are as deep as 0.6 miles (1 kilometer) beneath the surface. They appear to be carved by the winds and the evaporation of ice.

The statistics stagger when comparing that part of Mars with Greenland on Earth. The shape of the arctic cap on Mars indicates that it is composed primarily of water ice, with a volume of 300,000 cubic miles (1.2 million cubic kilometers) and an average thickness of 0.6 miles (1 kilometer). "Combined together, the Martian polar caps contain about the same amount of water ice as there is in Greenland," notes Dr. Frances Butcher, a planetary glaciologist based at the University of Sheffield in Britain. On Earth, there is much more ice at the poles. Less than 1 percent of Earth's glacial ice is elsewhere; on Mars, 10 percent of its water ice is in the nonpolar regions. What that implies is that ice has played a very important role in the story of Martian climate. To understand that better, researchers need to peel back the outer layers of the ice and peer at what may be lurking below.

* * *

Like Earth, the Martian poles have permanent caps and seasonal ones. The seasonal ice comes from carbon dioxide freezing out onto the winter pole, reducing overall atmospheric pressure in the process (by up to 20 percent as measured on the surface). On the winter pole, the seasonal cap formed by accumulating carbon dioxide is about a meter thick. Depending on how dusty the atmosphere is, the seasonal caps at both poles consist of dirty ice, with dust mixed in. The extent of the mixing depends on how relatively warm or cold it is.

Over the last two decades, new radar instruments have revealed unsuspected details about the structure within the layers. They have also allowed glaciologists to make comprehensive three-dimensional maps of exposed layers of the scarps seen at the poles. In the summer of 2019, researchers in Texas and Arizona presented a new estimate for all the ice currently contained at the north pole of Mars. They examined a deposit of water ice and sand that lies beneath the current permanent cap. These layers formed during many past ice ages on Mars. Hitherto, it had been thought any evidence for this kind of ancient ice would have been lost.

"There is quite a bit more ice [in that basal unit] than people thought," says Professor Jack Holt, who led the study. "We don't have a good age constraint upon it. It probably spans a large range of ages from a billion years up to just five million years."

And it is where some of the missing water might still reside. To their amazement, the researchers estimated there was enough to fill a layer at least five feet (1.5 meters) deep right across the whole planet. Such a large amount of water is not easily explained by current understanding of how climate has evolved on Mars. It could also mean that much of the "unaccounted for" water resides below the surface in a massive frozen aquifer.

In particular, the "shallow radar" aboard the NASA Mars Reconnaissance Orbiter has been able to penetrate down to 1.5 miles (2.5 kilometers) below the surface. The total volume locked up in these buried deposits is about the same as that seen in glaciers and buried ice across the rest of the Red Planet. "There is clearly a process of exchange going back and forth," Holt says. "Or maybe that's just coincidence."

The bottom line shows "there is some process by which significant amounts of ice can get trapped at the pole during these exchanges." It is a one-trick pony:

because the water is frozen, it is not available to flow freely ever again. With each subsequent freezing, less is available for the climate. And that, ultimately, might provide scant comfort in the search for extant life.

"You can have all the right conditions for life," says Stefano Nerozzi, Holt's PhD student who has mapped this ancient ice, "but if most of the water is locked up at the poles, then it becomes difficult to have sufficient amounts of liquid water near to the equator."

* * *

By comparison, the south pole on Mars is very different. One important factor is that it is on average more than three miles (six kilometers) higher than the north pole. That extra height alone might account for the very obvious differences between them. In Jack Holt's estimation, the ice at the south pole appears much older than that at the north. From repeated radar observations, it appears that the wind has had a greater role in shaping the polar caps than had previously been known. Many of the observed patterns of ice and dust at both poles show telltale signs of having been eroded by wind.

But there is a frustrating problem. The highest resolution camera in orbit around Mars today can see down to ten inches (twenty-five centimeters) at the surface (better than Google Earth, in fact). At best, the shallow radar instrument that flies along with it can only see down to thirty-three feet (ten meters). "We can't quite connect them," says Professor Holt. "We haven't been able to positively identify an exposed layer and outcrop that you can trace and connect with [the shallow radar]." Some of these layers can be tracked across the entire polar cap, "which is amazing, and we see all this structure," he says. "The holy grail is still to have a definite correlation between the [changes to the orbit] and the layers."

The way in which the ice has grown and retreated has been modeled at both poles. The ice at the north pole is probably less than five million years old. There has been growth, retreat, and growth. "You see a nice correlation of that in the radar data," Holt says. Some of the layers are cut off mostly around the edges, "where you clearly have had that ice growth and then you have had these periods when you start removing the ice."

In the high-resolution radar observations, "aeolian processes"—as the cumulative effect of wind is known—have played a significant role in shaping the cap. Further work is needed to refine these measurements, with researchers knowing only too well that the finding of finer, thinner bands within the layers may well cause a rethink of how the Martian climate has changed.

* * *

The European Mars Express Orbiter arrived in 2003, three years before NASA's Mars Reconnaissance Orbiter did in 2006. Both have been maneuvered into near-polar orbits around the Red Planet. As they move around Mars, the whole of the planet can be spotted below them (apart from a "collar" of 3° around the geographical poles themselves). Both have radar instruments that have observed distinct differences in the polar regions.

"Our shallow radar works really well for the north pole," says Professor Jack Holt. "It sees lots of internal layering and structure, and you can see all the way down to [the basal unit]."

This instrument has greater resolution at the surface but at the expense of looking far below: "we have shorter wavelengths and higher resolution." The European radar operates at a lower frequency, which means it can penetrate further. "For a reason we don't understand yet," Holt explains, "our signals get scattered by something within the south polar layered deposits. It appears to be a subsurface kind of shallow layer or layers that we just don't understand."

The term they use is "a fog," not in the literal sense but rather a diffuse signal that obscures everything else returned in the radar profile. In other words, there is no useful information that can be interpreted, in much the same way fog obscures a surface seen with the naked eye.

"The European radar does better in the south and not so well in the north," he says. This radar instrument on Mars Express hasn't yet clearly delineated much internal structure within the northern polar layer deposits. It sees the top and the bottom, and, depending on how the information is analyzed, some sort of structures within appear to move up and down.

How the two radars complement each other is interesting, Jack Holt says, although "it's a little frustrating at times." Certainly, that has been the case with one of the more sensational discoveries in recent years: that there is a vast layer of water underneath the south pole of Mars. A team of mostly Italian scientists used the Mars Express radar information to infer that there is a lake-sized reservoir of water with sediment below it. As a likely habitat for biological material, it became front page news in the summer of 2018 when their research was published in the journal *Science*.

"This discovery is changing again the view of the possible presence of liquid water," said one of the team, Elena Pettinelli, a researcher at Roma Tre University in Italy. During dedicated passes between May 2012 and December 2015, the European radar saw "anomalously bright subsurface reflections" beneath the icy Planum Australe region.

Within an area twelve miles (twenty kilometers) wide, something lurks below. The Italian team have interpreted this as a possible sediment-infused body of liquid water buried under the surface, pooling roughly less than a mile (one and a half kilometers) under the frozen ice layers. Other scientists, especially those working with data from the other radar aboard NASA's Mars Reconnaissance Orbiter, remain unconvinced.

Theoretically, the American spacecraft's radar should also be able to see this feature, too. To date, it has not. Everything hinges on the way that the radar "echo" is returned and how it is interpreted. The Mars Express radar instrument can penetrate the top three hundred-plus feet (one hundred meters) below the surface, while the other instrument on the NASA spacecraft sees down to thirty feet (ten meters). There is something strongly attenuating the radar signals, not allowing them to see the bottom half of the ice sheet.

What happens next depends on how these observations are modeled. And, as in so many areas of science, the results depend on just how many assumptions are made at each step along the way. Philosophically, you might say, that depends on just how Earthlike the Martian poles are. "In Antarctica, for example," notes Jack Holt, "when you have a lake, for it to be sustained over any period of time, you have to have melting off the face of the ice sheet."

That will also tend to warp the ice above it, but warping has not been observed in any of the European radar images. If it's an ephemeral feature, it also seems hard to explain. Though salinity (the amount of salt needed to keep the water as liquid) has been inferred, several critics have pointed out that the observations don't seem to match anything known in nature. "There's no amount of salt that can cause an amount of liquid to exist underneath the ice sheet," Holt adds, "even if you made it purely salty."

The simplest way to keep water as liquid is with "a heat anomaly." On Earth, there are such features in the form of volcanoes lying below ice sheets. But on Mars, nobody has a detailed understanding of how much heat is available. Professor Holt points out that if it was some sort of magma chamber that could provide enough warmth, there is no supporting evidence. "The best-case scenario that if there is liquid there, it could be a kind of icy sludge," he says.

The Italian authors, nevertheless, remain confident in their findings as they have taken years to collect the data. As ever with cutting-edge research, it is also a matter of elimination. They looked at other possibilities that might explain the radar reflection before turning to water. They rejected the idea of the layer being due to carbon dioxide or water ice. It is unlikely both could form at the base of the ice cap under those conditions either.

At this stage, it is probably fair to say the jury is still out. A more accurate assessment might come from new information about the interior heat flow within the planet as a whole, which would tell scientists whether pools of liquid water could be possible. For that, measuring the heat flowing just below the surface is crucial. As we will see in the next chapter, attempts to do this are ongoing with the most recent visitor to the surface of Mars, the NASA InSight lander.

Fundamental disagreements are part and parcel of scientific inquiry. Clearly, a deluge of information is helping fine-tune some theories and leading to others being thrown out. One ugly fact, it has often been said, can destroy the beauty of what may appear to be an elegant theory. These are the sorts of riddles which will be addressed by the next missions to Mars. If past experience is anything to go by, the answer will be far stranger than anyone could have ever imagined.

* * *

Liquid water beneath either of the poles has important ramifications for life on Mars. For that reason, the polar ice might provide a new and fertile hunting ground. "The permanent polar caps of Mars are frozen water and would act as a splendid 'cold trap' where organic molecules could condense," wrote one prominent biologist in the early 1980s. "The scavenging oxidizing agents would largely be absent, so that there is an odds-on chance of finding life-forms." Professor Jack

Holt adds, "If there's any life near to the surface of Mars, it's going to be in the icy deposits. But maybe the ice is going to be too young to preserve it."

Certainly, midlatitude glaciers formed where conditions were warmer and may well be easier to reach. And there is another very good reason to examine ice below the surface. Something similar may take place there as has been found on Earth. "Any time we've gone to deep ice on Earth," Jack Holt says, "there are always things living there."

Ultimately, it won't be a single mission that will tell the story of how the polar regions of the Red Planet have changed over time—and whether life has played any part in it. One obvious mission would be to have a small spacecraft land and attempt to take ice cores in the laminated terrain. "With precise dating, you could correlate the precise sequence of how they were lain down," says Professor Holt.

Chris McKay has been working for a number of years on the Icebreaker Life mission concept, which would look for telltale signs of biological activity on the northern plains of Mars. It would be a rerun of the Phoenix mission with a new payload designed to look for any hints of biochemical processes. Though not chosen by NASA in 2015 nor again at the start of 2020, it may eventually be taken up as a project in the years ahead. But there would be little chance for any current technology to actually land farther north on the polar ice caps themselves and survive for any length of time.

The mystery of the Martian poles may not be solved any time soon. According to many researchers, trying to land at either Martian pole is, for now, science fiction. Yet, only a few decades ago, so was establishing a base in the Antarctic or sending a spacecraft to Mars. In the past, the icy poles on Earth were called terra incognita, as they were unknown and unknowable for so long. Thanks to pioneers like Ross, Scott, and, in more recent times, dedicated scientists who live on the ice for months at a time, the poles on Earth are no longer mysterious. The same can be said for Mars.

PLANETARY PROTECTION IN THE NEW SPACE ERA

(2020)

SCIENCE AND GOVERNANCE

Thomas Cheney, Christopher Newman, Karen Olsson-Francis, Scott Steele, Victoria Pearson and Simon Lee

Planetary Protection is the name given to protecting Earth from cross-contamination of biological samples, and vice-versa. In an era of growing Mars exploration, this topic is only becoming more pressing. **Thomas Cheney** is a lecturer in space governance at the Open University. **Christopher Newman** is professor of space law and policy at Northumbria University. **Karen Olsson-Francis** is director of astrobiology at the Open University, **Victoria Pearson** is associate director and **Scott Steele** is a research student. **Simon** Lee is a professor at the Open University Law School.

Committee of Space Research's Planetary Protection Policy is a triumph of technocratic governance in the global sphere. The Policy is produced by a group of scientific experts and subsequently enjoys high regard among the scientific and space community. However, as Committee of Space Research is an independent organization without any legal mandate the Planetary Protection Policy is an example of so-called "soft law" or a non-binding international instrument, in short, no one is under any legal obligation to comply with them. The policy is linked to Article IX of the Outer Space Treaty and its provision calling for the avoidance of "harmful contamination" of the Moon and other celestial bodies. While space activities beyond Earth orbit have been the exclusive preserve of government scientific space agencies this has posed little problem. However as private and "non-science" space activities proliferate and begin to spread their reach beyond Earth orbit, the Planetary Protection Policy is being tested. This paper will examine the challenges of developing and maintaining an effective planetary protection regime in this "New Space" era. This will involve looking at the existing policies, as well as the governance framework they sit within. However, it is also necessary to consider and understand the scientific basis not just for the specifics of the policy itself but the necessity of it. Finally, this paper will consider whether a broader "environmental" framework is needed as space activities diversify in type and location.

Introduction

The Committee of Space Research's (COSPAR) Planetary Protection Policy (PPP)[1] has sought to protect the space environment from "harmful contamination" which would endanger the integrity of the scientific exploration of outer space including the search for life. The PPP predates the Treaty on Principles Governing the Activities of States in the Exploration and Use of Outer Space, Including the Moon and Other Celestial Bodies (Outer Space Treaty/OST).[2] The PPP's non-binding status has allowed a flexible and organic development enabling the Policy to be updated as scientific understanding has developed. Yet, as the geopolitical order shifts, new challenges to the governance of space emerge. This discussion will examine the threat to the consensus that has developed in respect of planetary protection caused by a shift in space exploration. The traditional science-led approach to the exploration of celestial bodies is now being augmented by non-governmental, non-scientific actors looking to pursue commercial activities in outer space. These new actors, some of whom have already expressed skepticism about the current arrangements, will test the way in which planetary protection is enshrined in future missions to other planets.

Summer 2020 has seen the launch of several Mars bound scientific missions, including the first effort from the UAE and China's first independent mission.* Further, the United States has unveiled further details for its Artemis program which intends to return humans to the Moon during the 2020s and involves a commercial lunar payload service in order to open opportunities for private sector activities on the Moon.[3] Elon Musk, Jeff Bezos and others continue to advance their plans for non-governmental activities in outer space. Space governance, and planetary protection, is evolving to deal with these developments, the United States has proposed the Artemis Accords,[4] NASA has updated their planetary protection policy,[5,6] COSPAR also updated the international planetary protection policy in June 2020.[7] Planetary protection is perhaps more important as ever as the number of actors and the diversity of their activities increase. Private and non-governmental space activities present a particular challenge given the status of the COSPAR Planetary Protection Policy in international law and the motivations and intentions of some of these new actors. Non-binding guidelines rely on parties caring about their objectives, however, States have an ability, and a responsibility, to ensure responsible conduct by their nationals and non-governmental entities, in outer space and that includes ensuring adherence to the principles of planetary protection. However, as issues such as space debris demonstrate, it is also necessary to consider whether a broader environmental

* Previous attempt was done in cooperation with Russia.

perspective is needed, whether it is necessary to protect the outer space as an environment, to ensure future use for a range of activities.

This article will start by critiquing the extant planetary protection principles, why they are necessary and their position within international space governance. The discussion will then explore the issues of enforcement in international law, particularly of "non-binding" norms (the so-called 'soft law' provisions), before discussing whether there is a need for a refocusing of international efforts to protect the space environment.

The COSPAR Planetary Protection Policy

The requirement to protect natural celestial bodies has been recognized by both the scientific and legal communities since the early days of space exploration, but this has led to divergence in approach.[8] The concept of planetary protection, while recognized by both communities, is emphasized by the legal community to be concerned with prevention of harmful contamination of celestial bodies. For the scientific community, it refers to the need to ensure that pristine celestial bodies are not contaminated by terrestrial biological (or organic) contamination (forward contamination), specifically in order to avoid compromising the search for extraterrestrial life in the Solar System.[9] It also concerns protection of the Earth's biosphere from the return of (potential) biological extraterrestrial materials (backward contamination). These principles underpin agreed international practices which have evolved as our understanding of the boundaries of life on Earth and elsewhere in the Solar System has progressed[10] and as there has been an increase in commercial activity and exploration, including life detection and sample return missions.

Internationally, technical aspects of planetary protection are developed though deliberations between the space agencies and international scientific organizations, and recommendations are made by COSPAR.[11] The COSPAR Planetary Protection Policy (PPP)[12] defines the specific technical guidelines.[13] The policy is based on two rationales: 1) to ensure that the conduct of scientific investigation of possible extra-terrestrial life forms, precursors, and remnants is not be jeopardized; and 2) the Earth must be protected from the potential hazard posed by extraterrestrial matter carried by a spacecraft returning from an interplanetary mission.[14]

To date, the COSPAR PPP has identified five categories for planetary protection requirements depending on the type of mission, the target body, and the type of scientific investigations involved. For Category I missions, no PP measures are needed, but from II to V, the PP regulations become increasingly stringent depending on the scientific focus of the particular mission on the astrobiological relevance of their individual mission target, as shown in Table 1. These missions include flybys, orbiters and landers to Venus, the Moon, Mars, and the icy moons of the outer Solar System, among other locations as indicated by Table 1. Examples of such missions are the Mars 2020, ExoMars and Mars Sample Return missions. Based on the return of the Apollo samples, samples returned from the moon are not deemed a threat, and as a result the Moon is Category II with unrestricted Earth-return.

Category	Mission type	Target bodies
I	Flyby, orbiter, lander	Undifferentiated, metamorphosed asteroids; lo
II	Flyby, orbiter, lander	Venus; Moon; Comets; Carbonaceous Chondrite Asteroids; Jupiter; Saturn; Uranus; Neptune; Ganymede*; Callisto; Titan*; Triton*; Pluto/Charon*; Ceres; Kuiper-Belt Objects > ½ the size of Pluto*; Kuiper-Belt Objects <½ the size of Pluto; others
III	Flyby, Orbiters	Mars; Europa; Enceladus
IV	Lander Missions	Mars; Europa; Enceladus
V	Any Earth-return mission	—

* *Must be supported by an analysis of the "remote" potential for contamination of the liquid-water environments that may exist beneath their surfaces.*

Table 1. Categories of missions and target as stated in (COSPAR 2020a).

Category IV, however, recognizes that some missions are to bodies where extinct and extant life may exist, e.g., the Martian sub-surface or the sub-surface oceans of icy moons. Indeed, the significance of this is recognized by specific subdivisions of Category IV for Mars (IVa, IVb, and IVc), a prime target for astrobiology investigations. These sub-divisions are dependent on the mission aims and specific target location: Category IVa missions do not have objectives relating to the search for life; Category IVb missions are those investigating the existence of extant life; Category IVc is for missions that target "special regions" (e.g., features such as gullies, subsurface cavities, below 5 m, and Recurring Slope Lineae (RSLs))[15] even if the objective is not related to the search for life. "Martian Special Regions" were created in the 2002 COSPAR PPP for Mars, to recognize regions where terrestrial or native Martian life might flourish. One obvious determinant for these regions is the presence of liquid water. For Category IVc missions, additional restrictions are placed on spacecraft bioburden (terrestrial contamination) and the feasibility of the target site to support the replication of terrestrial microbial life is also taken into consideration. As further information is obtained about proposed targets, the Categories are re-evaluated.

The development of the "Mars Special Regions" highlights the flexibility of the COPSAR PPP and its evolutionary approach, which is essential because we are increasing our understanding of the boundaries within which life can grow on Earth and increasing our awareness of potential habitability elsewhere in the Solar System. Planetary protection is critical for enabling scientists to study the natural environments of celestial bodies without interfering with possible life forms that may have developed there. Most importantly, it also helps to preserve the terrestrial biosphere from possible contamination by extraterrestrial material. The United Nations Committee of the Peaceful Uses of Outer Space (UNCOPUOS), in 2017, noted the long-term standing role of COSPAR in maintaining a Planetary Protection Policy as a reference standard for spacefaring nations and in guiding compliance with Article IX of the Outer Space Treaty.[16]

International Space Law

The space governance regime, of which the COSPAR Planetary Protection Policy is part, is comprised of treaties, such as the foundational Outer Space Treaty, customary international law, and non-binding "soft law" such as UN General Assembly Resolutions and the COSPAR Planetary Protection Policy. As international law the regime directly addresses States. However, private and non-governmental activities are addressed by space law because, while private activities are governed directly by national law these national regimes sit within the framework of international law, specifically the "special regime" (*lex specialis*) of international space law. Space law was first developed through customary international law dating from at least the first instance of state practice, the launch of Sputnik, and strengthened by the Declaration of Legal Principles in 1963[17] but the Outer Space Treaty of 1967 is a milestone as a formally binding or "hard law" instrument which serves as the legal foundation for the space governance regime. The Outer Space Treaty, as its full name implies is a framework treaty laying out key principles. It has been built upon by subsequent treaties focusing on specific aspects, such as elaborating liability and registration provisions. The Agreement Governing the Activities of States on the Moon and Other Celestial Bodies (Moon Agreement)[18] was an attempt to develop a more detailed and specific regime for activities on the Moon and other celestial bodies but despite receiving sufficient ratifications to become an active treaty it has been rejected by the majority of the international community. It is therefore regarded as a "failed" treaty and is of limited relevance. The decades after the Moon Agreement have seen a dearth of "hard law" instruments in space governance. What development there has been has come in the form of customary international law and non-binding instruments or so-called "soft law." The COSPAR Planetary Protection Policy (PPP) is one of the leading examples of "soft law" within space governance.

The key aspects of the Outer Space Treaty[19] for planetary protection are found in Articles I, II, III, VI, VIII, and IX. Article I declares that all States are free to explore and use outer space, including the Moon and other celestial bodies. Such exploration and use should be "carried out for the benefit and in the interest of all countries" and States should "facilitate and encourage international cooperation" in scientific investigation. Article II prohibits the exercise of territorial sovereignty in outer space or on the Moon and other celestial bodies. It also means that the use or occupation of outer space, the Moon and other celestial bodies, gives rise to no sovereign rights over those areas. Article III stipulates that space law is part of international law, and therefore where space law "runs out" international law can be used to "fill the gaps." This is particularly important for enforcement, as it means, among other things, while there is not specific "dispute resolution" mechanism in the Outer Space Treaty states can make use of the International Court of Justice (ICJ) and other such measures. Article VI is vital for modern international space law as it makes States responsible for the activities of their nationals and private corporations in outer space and requires that they "authorize" and "continually supervise" those activities. Similarly, Article VIII establishes that the State on whose registry a space object is carried "retains jurisdiction and control" over the space object and its personnel. Therefore the "state of registry" is able to exercise legal authority over a space object (and its personnel should it have any) despite Article II OST. Article IX is the most directly relevant for this discussion and will therefore be subject to a more detailed examination.

Article IX stipulates that States shall conduct their activities "with due regard to the corresponding interests of all other" parties. They conduct their activities so as to avoid the "harmful contamination" of the Moon and other celestial bodies, as well as changes to the Earth's environment by the introduction of "extraterrestrial matter". Further, states are encouraged to engage in "appropriate international consultations" to avoid "harmful interference" with the activities of other States. As is the case with most of the provisions of the space treaties, "harmful contamination" and "harmful interference" are not defined by the Outer Space Treaty itself. While their ordinary meaning (using the Vienna Convention on the Law of Treaties (VCLT)[20] approach) seems reasonably clear, particularly based on the dictionary definitions,[21] it remains unclear as to what needs to be subject to the harm in question? Environmental law is particularly focused upon harm to "life" whether directly or indirectly. Based on the subsequent practice of the conduct of the various mission to the Moon, and other celestial bodies, the interpretation that can be drawn from State practice is that "harmful contamination" is about biological contamination of celestial bodies particularly where and when there is "potential" for life or related indicators. Therefore, it would be a stretch to interpret "harmful contamination" as being the "contamination" of the celestial body itself *per se*. This also leads to a conclusion that different approaches are in order for different celestial bodies given the different "environmental" conditions on those bodies (the presence of an atmosphere makes Mars different from the Moon for example).[22] However, what standards are necessary to meet the requirements of Article IX are an open question. It seems logical to adopt an "evolutionary approach" to the interpretation of Article IX given the technical nature of the issue. To use, say "planetary protection" standards applied to the Viking landers in 1975 would ignore the developments in both sterilization and the understanding of the outer space environment(s) in the intervening 45 years.

There are planetary protection polices, in addition to the COSPAR PPP, which have been implemented, and can be viewed as "subsequent practice" and therefore help to define the meaning of the terms of the Outer Space Treaty. NASA, the European Space Agency and the Japanese Space Agency (JAXA) all have their own internal policies. These only apply to those agencies and their associated missions, so commercial missions are not necessarily subject to those policies. As mentioned, States are responsible for the activities of their nationals in outer space and required to authorize and supervise those activities. Therefore, non-governmental missions are required to "avoid harmful contamination" as per Article IX OST. The practice of the various space agencies should be viewed as "subsequent practice" and therefore form a baseline of what should be expected of states to require of non-governmental actors in order to secure the necessary authorisations to conduct a space activity. However, unless those States choose to do so there is no obligation to implement COSPAR's PPP. Yet it is important to note the broad acceptance and general adherence to the principles, and indeed, the COSPAR PPP is a respected technically driven international set of guidelines.

As the relevant space agencies operate their own policies and no State has opted to incorporate COSPAR's PPP into their national laws their value as customary international law is limited. However, "soft-law" is a growing part of the space governance infrastructure, and particularly in technical arenas such as "planetary protection" can be a useful approach especially for emerging governance issues. Indeed, Setsuko Aoki argues that soft law, such as the COSPAR PPP, is often best suited for technical guidelines, it reduces the need for compromise, and can be more easily updated than hard law options.[23] COPSAR's PPP is an exemplary example of such technically driven guidelines and should serve as a model for such "soft law" instruments.

New Space and the Problem of Enforcement

Up until the middle years of the 2010s, it was inconceivable that a private company would become seriously engaged in the exploration of celestial bodies other than the Moon within the foreseeable future. The proliferation of private companies seeking to "Mine the Skies"[24] and unlock a prospective trillion-dollar space-resource was greeted with a hopeful but realistic recognition of the difficulties. The size of the investment required coupled with the vast infrastructure needs of space mining meant such ventures would be difficult to bring to fruition, and the troubles experienced by companies like Planetary Resources did little to dampen those doubts.[25]

So, while the exploitation of celestial bodies has been in the zeitgeist of the space community, the aforementioned financial *realpolitik* meant that journeys to celestial bodies remained the province of the scientific and academic actors. The natural corollary of this was that the COSPAR PPP became firmly embedded as the international standard by which contamination of celestial bodies would be avoided. In the United States, for example, no NASA probe will be launched into space without receiving prior approval from the NASA Planetary Protection Officer.[26] Therefore, while COSPAR PPP is entirely voluntary in nature, and it is not enshrined within the OST regime, there has been broad compliance with it by both regulators and mission planners eager not to contaminate a potentially pristine extraterrestrial environment.

This scientific hegemony in respect of planetary protection is not guaranteed to continue in perpetuity, however. The first signs of a serious attempt by the private sector to engage in missions that would require adherence to the principles of planetary protection came at the 2016 International Astronomical Congress, where Elon Musk outlined his plan to colonize Mars using rockets built by his company SpaceX.[27] Unlike some of the other ventures, Musk has the financial resources and SpaceX has evidenced significant technological prowess to make the space community pay serious attention to Musk's proposals. Given that Musk has previously made it clear that he has no great attachment to the planetary protection protocols, believing there is no life on Mars,[28] there a distinct possibility that those looking to explore other celestial bodies for commercial reasons will eschew the restrictive requirements of the COSPAR PPP in favor of a more relaxed approach to planetary protection.

It is not only Musk from the private sector who seeks to challenge the COSPAR regime. In February 2019, the Arch Mission Foundation, a non-profit, independent organization created a "Lunar Library" archive for Israel and secured a launch berth on the Beresheet mission. This was the first attempt at a lunar lander from Israel and was the result of funding from a range of philanthropic organizations and private investment. The mission itself was a failure, with the spacecraft crashing into the lunar surface.[29] However, in August 2019, Nova Spivack, the founder of the Arch Mission Foundation, announced that among the payload (which was, as proposed to the regulator – in this case the UnitedStates' FAA – a DVD-sized archive containing 30 million pages of information and DNA) was a number of tardigrades, the small microscopic lifeforms that are highly resilient to all manner of conditions in space that would normally be considered inimical to most life forms.[30] In a later interview, Spivack further declared that the other launch partners, or the space agencies were unaware of the presence of tardigrades. He claimed that he wanted to take the risk to see how life behaved and would "ask for forgiveness rather than permission".[31]

The lack of disclosure of the existence of the tardigrades casts doubt upon the compliance of the Beresheet mission to the planetary protection guidelines. Missions to the Moon are classed as Category II under the COSPAR PPP. This requires the creation of a planetary protection plan and "a series of reports both pre-launch report, post-launch report, post-encounter report, and an end-of mission report".[32] While the actions of Spivack and the Arch Mission Foundation were clearly aimed at sidestepping the COSPAR guidelines, the reaction of the international community to the seeding of the Moon with tardigrades was a mixture of curiosity and outrage, but little in the way of substantive action.

Recent developments beg the question as to how secure the principles of planetary protection will be if challenged by adventurous entrepreneurs, unfettered by the constraints of the scientific community. The recent NASA Planetary Protection Independent Review Board[33] has led to the announcement of a reframing of the COSPAR guidelines for the Artemis program,[34] but even these revivified proposals may be too cautious for those who are prepared to risk some contamination in exchange for commercial developments. While guidelines provide a suitable mechanism for establishing approved behavior within a specific community of practice, they are ultimately non-binding and not designed to be enforced when actors refuse to adhere to the norms of a group to which they themselves do not belong. In essence, guidelines rely on "carrots" rather than "sticks".[35]

Johnson et al. highlight that although the COSPAR PPP is not binding on any State, that does not mean they cannot be voluntarily adopted by States and then implemented through national legislation or licensing requirements. This is where the legal status of the COSPAR PPP becomes somewhat opaque. Internationally, they do not have the binding force of a treaty, but once a nation decides to incorporate the principles within their licensing regime, it could be unlawful for either an individual or a company to act in a way which is not in accordance with a license (see, for example, United Kingdom Outer Space Act 1986, s3).

Licensing is an inherent requirement of international space law. As stated previously, the five UN space treaties seek to regulate the behavior of States, establishing the obligations and positive rights of States in the peaceful uses and exploration of outer space.[36] Indeed, this is one of the three core principles around which the OST is located; freedom of exploration, the requirement that exploration and use is undertaken for peaceful purposes and that states are responsible for national space activities.[37] It is therefore fundamental to the overarching international regime that State parties to the OST agree to take on responsibility not only for state-sponsored missions, but also non-governmental entities. The provisions of Article VI of the Treaty ensure that the principles of the OST are embedded within all space activity, be it public or private in nature:

> State parties to the Treaty shall bear international responsibility for national space activities in outer space, including the Moon and other celestial bodies, whether such activities are carried out by governmental agencies or by non-governmental entities, and for assuring that national activities are carried out in conformity with the provisions set forth in the present Treaty.

As with much of the OST, this provision represented a compromise between the two dominant powers of the time of drafting: the United States and the Soviet Union (USSR). The USSR at the time believed that only a nation state, "conscious of its intention responsibility should carry on space activities." This is particularly germane when considering notions of planetary exploration as for much of the life of the OST, it was believed that it is only nation states that would either be capable or interested in exploring the Moon or other celestial bodies.

As Johnson et al. point out, the "attribution to the State is direct and automatic. Article VI of the OST makes States internationally responsible for the actions of...private actors such as corporations, non-profit foundations or others not acting in a governmental capacity". It is the second sentence of Article VI that creates the need for a licensing and regulatory framework within the domestic legislation of State Parties to the OST. It states that "The activities of non-governmental entities in outer space, including the Moon and other celestial bodies shall require authorization and continuing supervision by the appropriate State Party to the Treaty". As can be seen, therefore, Article VI creates a positive obligation upon the State to make sure that all non-governmental activities are monitored to ensure compliance with the terms of the OST.

The fact that the principles of planetary protection are not contained within a binding international treaty is, therefore, not fatal to the prospects of ensuring compliance. Indeed, the fluid nature of scientific opinion means that a non-treaty agreement can incorporate changes, such as those proposed by the Stern review,[38] as scientific opinion changes and develops without the need to open a treaty up for renegotiation. As Goh highlights, there is a danger that a poorly constructed and overly restrictive treaty could be prone to premature failure, leaving a significant gap in normative guidance. Non-treaty agreements and guidelines constitute a versatile pre-droit regime that can capture and galvanize developments in the field.

By having States as the guarantors of compliance with the treaty, national regulators have to bear in mind scientific best practice when thinking about the authorization of space missions.[39] The COSPAR PPP is recognized across the world, and – perhaps more importantly – throughout the scientific community as being the best way to capture developments in scientific understanding. This, in turn, will equip national regulators, who have a duty to authorize and oversee all aspects of any mission by their nationals or companies, with the guidance to deal with each mission on a granular level. It is surely logical for nations to incorporate COSPAR guidelines within the extant licensing mechanisms (as the FAA do with their payload reviews and the United Kingdom Space Agency do by virtue of s5 (2) (e) (i) Outer Space Act 1986). Doing this and discharging the supervisory duty under Article VI of the OST, means that approval of missions remains within the competency of States.

What then of the tardigrades and the threat posed to planetary protection by hyper-wealthy entrepreneurs? According to Article VI of the OST, the Beresheet mission was part of a payload launched from the United States (it is important to distinguish the responsibility for the mission from liability under Article VII, however in this case, the FAA in the United States would consider the launch a national space activity and therefore would provide authorization under Article VI as well as being a launching state under Article VII). As the mission involved the activity of an Israeli not-for-profit foundation, both Israel and the United States would be internationally responsible under Article VI of the OST.[40] It is, therefore, for their regulators to decide what action should be taken against the Beresheet mission for the presence of unauthorized tardigrades.

Article VI imposes the positive obligation upon States to license non-governmental actors (of whatever kind) for any area of a space mission. The OST requires consideration to be paid to scientific information, especially regarding the contamination of outer space.[41] That the COSPAR PPP are not binding upon States and merely persuasive is something of a red herring. Any regulator wishing to be seen as responsible ignores the findings of COSPAR at their peril. In the current geopolitical environment, having planetary protection as part of consideration of the granting of a launch license is as close to enforcement as can be realistically hoped for. The burden now passes to individual States to ensure that its nationals and its companies do not engage in exploration that endangers the scientific value of celestial bodies.

Environmental Framework for Outer Space?

In light of the increasing interest in the Moon and other celestial bodies by non-governmental actors, for activities other than scientific investigation, it is worth considering whether there should be a broader "environmental" consideration embodied in the conceptualization of planetary protection, i.e. widen its remit. Increased lunar activities could replicate Earth's "space debris problem" around the Moon (and later Mars); lunar dust, kicked up by activities on the surface could be a significant environmental issue;[42] not to mention the "damage" mining operations will cause to hereto pristine environments. Positioning "planetary protection" as more than just a question of protecting scientific integrity but part of a broader "environmental protection" regime for outer space will strengthen the case for such measures. It will broaden the stakeholders and help guard against an ambivalence toward the objectives of astrobiology.

There has been a clear lack of recognition of the nature of outer space as an environment, doing so provides a framework for understanding and protecting outer space as an arena for human activity. The "Stern Report"[43] draws the boundaries between scientific and commercial space endeavors. There is a recognition of the multi-purpose nature of space activities (commercial, scientific, prestige etc.) but there has not been a sufficiently persuasive argument for the protection of the space environment itself.

While outer space in undoubtedly a unique environment, it is, as discussed, within the scope of international law and therefore under the auspices of the "international community". Central to this international order are the principles set out in the UN Charter of a duty to cooperate, conduct peaceful international relations and a commitment to the rule of law in global governance. The Outer Space Treaty repeatedly reaffirms these principles, particularly the duty to cooperate in the exploration and use of outer space. Article III OST firmly cements space law within the wider general international legal regime with the UN Charter as its foundation. Therefore, it is quite reasonable to draw upon principles in other fields of international law as models for addressing environmental problems in space law.

The United Nations Convention on the Law of the Sea (UNCLOS)[44] is frequently used as an analogue regime for outer space. There are numerous similarities between the law of the sea (particularly as it pertains to the high seas) and the law of outer space. Both the high seas and outer space are "areas beyond territorial jurisdiction" and generally regarded as *res communis* or part of the global commons. A key example of the similarities can be seen between the right of peaceful passage in UNCLOS and Article I of the Outer Space Treaty which provides for a right of free access to all areas of outer space, the Moon and other celestial bodies for all States. A link can even be made to the need to tackle the space debris issue if conceived of as a "hazard to navigation" that potentially impedes the "free access" to outer space.

International environmental law is a useful source for a model or analogue for the development of a more "environmentally" minded space law regime. Similar to outer space, international environmental law has a set of core treaties which have been advanced and developed by non-binding "soft law" instruments, with a limited role for customary international law. While international environmental law is not directly applicable to outer space, it can serve as a model for space law, and the core principles can be transliterated to the space law regime. Further, if these principles of environmental law become further established as customary international law within general international law then they may become applicable to space law.

As it currently stands, with the exception of the "failed" Moon Agreement, Article IX OST is the central pillar of 'space environmental law' and it offers little help. However, it does introduce a requirement of the principle of "appropriate international consultation", and "due regard", but given the paucity of details within Article IX OST there needs to be further elaboration of what is meant by these terms. As discussed above, States can take a proactive role in this, particularly with respect to non-governmental actors such as commercial entities; they can require "environmental impact assessments" or similar measures as part of their national licensing process.

Further, UN guidelines are critical in providing this elaboration and are "a prudent and necessary step toward preserving the outer space environment for future generations".[45] These guidelines should draw on the principles of wider international law, and international environmental law more specifically. Positive action taken at the UN and by the international space community would have the additional advantage of demonstrating intent to "import" or "transliterate" principles that are not necessarily firmly established in customary international law. This would serve the objective of creating an environmental framework to protect space for the future, maintain the viability of scientific endeavors, limit harm to all States and to equally and adequately allow growth of space activity sustainably and economically. This would serve not just to protect the scientific value of outer space as intended by the planetary protection principles espoused by COSPAR and others but would also ensure the sustainability of all outer space activities.

Space is a fragile environment; it lacks many of the "naturally restorative" processes that exist in the terrestrial environment and therefore needs extra care. It is therefore important to place the maintenance and protection of the space environment at the forefront of the space governance regime. Outer space is already an important part of the global economy and will only become more important. Therefore, it is imperative to ensure that the use of space is undertaken on a sustainable footing, to ensure its continued availability. However, the concept of sustainability is a debated one, and there are many differing visions of what "sustainable" means. Inherent to the concept of "sustainable," at least in international environmental law, is that use or exploitation of natural resources can continue however, the needs of future generations and the broader environment must be taken into account.[46] Moreover, sustainability, by definition, includes an ecological, economic, and social component, which seems widely accepted.

As has been explored, even with the provisions of Article IX OST, space law is limited in terms of specific measures to protect the space environment. Space is an environment; a fragile environment in need of greater protection. However, general international law, the Law of the Sea, and international environmental law all provide principles and models for how the space environment can be afforded greater protection. Any international governance measure is only as strong as those that support it, and therefore it is important to build broad support for any measure. That need, combined with a need to allow measures to develop and adapt over time based on new science and technology, lends support for any space environmental instrument to be rooted in "soft law." The COSPAR PPP demonstrate the potential of such measures. States should recall that outer space is a unique and fragile resource. It is imperative that its continued use and development be undertaken in a sustainable manner.

Conclusion

The discovery of extraterrestrial life, in whatever form, will undoubtedly be one of, if not the, most seminal scientific discovery in human history, it is therefore imperative that the integrity of that finding is as unimpeachable as is possible. COSPAR's PPP and allied efforts are central to that. Paradoxically, the inherent responsiveness that is its greatest strength is also potentially its fatal weakness. COSPAR is an expert driven process which generates technical guidelines that are adaptive to circumstances and scientific development and that both makes them an exemplary example of "soft law" and has furnished them with high regard within the scientific community. As non-binding guidelines they are followed because space exploration agencies and mission planners recognize their value as "good practice" for scientists. The Achilles heel in this is that actors who see themselves as outside of this community and who do not care about protecting the scientific integrity of outer space, will not feel obliged to adhere to the guidelines. This situation is not terminal, however, as States have the ability, and the responsibility to ensure respect for the guidelines through Article VI of the OST and the resultant licensing process. Furthermore, as can be seen from the above discussion, States would benefit from promoting the COSPAR PPP as a model. This expert led, adaptive process is a model for other areas of space governance to follow particularly as outer space becoming increasingly commercial and non-governmental. It is also sensible, as this transition occurs, to consider a broader "environmental" framework for outer space to ensure that there is not a repeat of the space debris problem around other celestial bodies, among other issues. The incentive for States to act is clear: there will be significant national prestige for the mission that discovers the existence of extraterrestrial life. The many missions to other worlds have shown that adhering to the COSPAR PPP does not prevent activity but it may be vital to establishing the provenance of any discovery of extraterrestrial life.

COLONIZING MARS

THE MARTIAN CHRONICLES (1946)

OCTOBER 2026: THE MILLION-YEAR PICNIC

Ray Bradbury

Master storyteller **Ray Bradbury** was born in Waukegan, Illinois in 1920. He wrote eleven novels between 1950 and 2006, the first of which was *The Martian Chronicles*. His forte was the short story; he wrote many hundreds of them, most published in anthologies of his work. Although recognized as a novel, *The Martian Chronicles* is itself a collection of short stories that together tell an overarching story. When Bradbury died in 2012, the *New York Times* wrote that he was 'the writer most responsible for bringing modern science fiction into the literary mainstream'.

Somehow the idea was brought up by Mom that perhaps the whole family would enjoy a fishing trip. But they weren't Mom's words; Timothy knew that. They were Dad's words, and Mom used them for him somehow.

Dad shuffled his feet in a clutter of Martian pebbles and agreed. So immediately there was a tumult and a shouting, and very quickly the camp was tucked into capsules and containers, Mom slipped into traveling jumpers and blouse, Dad stuffed his pipe full with trembling hands, his eyes on the Martian sky, and the three boys piled yelling into the motorboat, none of them really keeping an eye on Mom and Dad, except Timothy.

Dad pushed a stud. The water boat sent a humming sound up into the sky. The water shook back and the boat nosed ahead, and the family cried, "Hurrah!"

Timothy sat in the back of the boat with Dad, his small fingers atop Dad's hairy ones, watching the canal twist, leaving the crumbled place behind where they had landed in their small family rocket all the way from Earth. He remembered the night before they left Earth, the hustling and hurrying, the rocket that Dad had found somewhere, somehow, and the talk of a vacation on Mars. A long way to go for a vacation, but Timothy said nothing because of his younger brothers. They came to Mars and now, first thing, or so they said, they were going fishing.

Dad had a funny look in his eyes as the boat went up-canal. A look that Timothy couldn't figure. It was made of strong light and maybe a sort of relief. It made the deep wrinkles laugh instead of worry or cry.

So there went the cooling rocket, around a bend, gone.

"How far are we going?" Robert splashed his hand. It looked like a small crab jumping in the violet water.

Dad exhaled. "A million years."

"Gee," said Robert.

"Look, kids." Mother pointed one soft long arm. "There's a dead city."

They looked with fervent anticipation, and the dead city lay dead for them alone, drowsing in a hot silence of summer made on Mars by a Martian weatherman.

And Dad looked as if he was pleased that it was dead.

It was a futile spread of pink rocks sleeping on a rise of sand, a few tumbled pillars, one lonely shrine, and then the sweep of sand again. Nothing else for miles. A white desert around the canal and a blue desert over it.

Just then a bird flew up. Like a stone thrown across a blue pond, hitting, falling deep, and vanishing.

Dad got a frightened look when he saw it. "I thought it was a rocket."

Timothy looked at the deep ocean sky, trying to see Earth and the war and the ruined cities and the men killing each other since the day he was born. But he saw nothing. The war was as removed and far off as two flies battling to the death in the arch of a great high and silent cathedral. And just as senseless.

William Thomas wiped his forehead and felt the touch of his son's hand on his arm, like a young tarantula, thrilled. He beamed at his son. "How goes it, Timmy?"

"Fine, Dad."

Timothy hadn't quite figured out what was ticking inside the vast adult mechanism beside him. The man with the immense hawk nose, sunburnt, peeling—and the hot blue eyes like agate marbles you play with after school in summer back on Earth, and the long thick columnar legs in the loose riding breeches.

"What are you looking at so hard, Dad?"

"I was looking for Earthian logic, common sense, good government, peace, and responsibility."

"All that up there?"

"No. I didn't find it. It's not there any more. Maybe it'll never be there again. Maybe we fooled ourselves that it was ever there."

"Huh?"

"See the fish," said Dad, pointing.

There rose a soprano clamor from all three boys as they rocked the boat in arching their tender necks to see. They *oohed* and *ahed*. A silver ring fish floated by them, undulating, and closing like an iris, instantly, around food particles, to assimilate them.

Dad looked at it. His voice was deep and quiet.

"Just like war. War swims along, sees food, contracts. A moment later—Earth is gone."

"William," said Mom.

"Sorry," said Dad.

They sat still and felt the canal water rush cool, swift, and glassy. The only sound was the motor hum, the glide of water, the sun expanding the air.

"When do we see the Martians?" cried Michael.

"Quite soon, perhaps," said Father. "Maybe tonight."

"Oh, but the Martians are a dead race now," said Mom.

"No, they're not. I'll show you some Martians, all right," Dad said presently.

Timothy scowled at that but said nothing. Everything was odd now. Vacations and fishing and looks between people.

The other boys were already engaged making shelves of their small hands and peering under them toward the seven-foot stone banks of the canal, watching for Martians.

"What do they look like?" demanded Michael.

"You'll know them when you see them." Dad sort of laughed, and Timothy saw a pulse beating time in his cheek.

Mother was slender and soft, with a woven plait of spun-gold hair over her head in a tiara, and eyes the color of the deep cool canal water where it ran in shadow, almost purple, with flecks of amber caught in it. You could see her thoughts swimming around in her eyes, like fish—some bright, some dark, some fast, quick, some slow and easy, and sometimes, like when she looked up where Earth was, being nothing but color and nothing else. She sat in the boat's prow, one hand resting on the side lip, the other on the lap of her dark blue breeches, and a line of sunburnt soft neck showing where her blouse opened like a white flower.

She kept looking ahead to see what was there, and, not being able to see it clearly enough, she looked backward toward her husband, and through his eyes, reflected then, she saw what was ahead; and since he added part of himself to this reflection, a determined firmness, her face relaxed and she accepted it and she turned back, knowing suddenly what to look for.

Timothy looked too. But all he saw was a straight pencil line of canal going violet through a wide shallow valley penned by low, eroded hills, and on until it fell over the sky's edge. And this canal went on and on, through cities that would have rattled like beetles in a dry skull if you shook them. A hundred or two hundred cities dreaming hot summer-day dreams and cool summer-night dreams ...

They had come millions of miles for this outing—to fish. But there had been a gun on the rocket. This was a vacation. But why all the food, more than enough to last them years and years, left hidden back there near the rocket? Vacation. Just behind the veil of the vacation was not a soft face of laughter, but something hard and bony and perhaps terrifying. Timothy could not lift the veil, and the two other boys were busy being ten and eight years old, respectively.

"No Martians yet. Nuts." Robert put his V-shaped chin on his hands and glared at the canal.

Dad had brought an atomic radio along, strapped to his wrist. It functioned on an old-fashioned principle: you held it against the bones near your ear and it vibrated singing or talking to you. Dad listened to it now. His face looked like one of those fallen Martian cities, caved in, sucked dry, almost dead.

Then he gave it to Mom to listen. Her lips dropped open.

"What——" Timothy started to question, but never finished what he wished to say.

For at that moment there were two titanic, marrow-jolting explosions that grew upon themselves, followed by a half dozen minor concussions.

Jerking his head up, Dad notched the boat speed higher immediately. The boat leaped and jounced and spanked. This shook Robert out of his funk and elicited yelps of frightened but esctatic joy from Michael, who clung to Mom's legs and watched the water pour by his nose in a wet torrent.

Dad swerved the boat, cut speed, and ducked the craft into a little branch canal and under an ancient, crumbling stone wharf that smelled of crab flesh. The boat rammed the wharf hard enough to throw them all forward, but no one was hurt, and Dad was already twisted to see if the ripples on the canal were enough to map their route into hiding. Water lines went across, lapped the stones, and rippled back to meet each other, settling, to be dappled by the sun. It all went away.

Dad listened. So did everybody.

Dad's breathing echoed like fists beating against the cold wet wharf stones. In the shadow, Mom's cat eyes just watched Father for some clue to what next.

Dad relaxed and blew out a breath, laughing at himself.

"The rocket, of course. I'm getting jumpy. The rocket."

Michael said, "What happened, Dad, what happened?"

"Oh, we just blew up our rocket, is all," said Timothy, trying to sound matter-of-fact. "I've heard rockets blown up before. Ours just blew."

"Why did we blow up our rocket?" asked Michael. "Huh, Dad?"

"It's part of the game, silly!" said Timothy.

"A game!" Michael and Robert loved the word.

"Dad fixed it so it would blow up and no one'd know where we landed or went! In case they ever came looking, see?"

"Oh boy, a secret!"

"Scared by my own rocket," admitted Dad to Mom. "I *am* nervous. It's silly to think there'll ever be any more rockets. Except *one*, perhaps, if Edwards and his wife get through with *their* ship."

He put his tiny radio to his ear again. After two mintes he dropped his hand as you would drop a rag.

"It's over at last," he said to Mom. "The radio just went off the atomic beam. Every other world station's gone. They dwindled down to a couple in the last few years. Now the air's completely silent. It'll probably remain silent."

"For how long?" asked Robert.

"Maybe—your great-grandchildren will hear it again," said Dad. He just sat there, and the children were caught in the center of his awe and defeat and resignation and acceptance.

Finally he put the boat out into the canal again, and they continued in the direction in which they had originally started.

It was getting late. Already the sun was down the sky, and a series of dead cities lay ahead of them.

Dad talked very quietly and gently to his sons. Many times in the past he had been brisk, distant, removed from them, but now he patted them on the head with just a word and they felt it.

"Mike, pick a city."

"What, Dad?"

"Pick a city, Son. Any one of these cities we pass."

"All right," said Michael. "How do I pick?"

"Pick the one you like the most. You, too, Robert and Tim. Pick the city you like best."

"I want a city with Martians in it," said Michael.

"You'll have that," said Dad. "I promise." His lips were for the children, but his eyes were for Mom.

They passed six cities in twenty minutes. Dad didn't say anything more about the explosions; he seemed much more interested in having fun with his sons, keeping them happy, than anything else.

Michael liked the first city they passed, but this was vetoed because everyone doubted quick first judgments. The second city nobody liked. It was an Earth Man's settlement, built of wood and already rotting into sawdust. Timothy liked the third city because it was large. The fourth and fifth were too small and the sixth brought acclaim from everyone, including Mother, who joined in the Gees, Goshes, and Look-at-thats!

There were fifty or sixty huge structures still standing, streets were dusty but paved, and you could see one or two old centrifugal fountains still pulsing wetly in the plazas. That was the only life—water leaping in the late sunlight.

"This is the city," said everybody.

Steering the boat to a wharf, Dad jumped out.

"Here we are. This is ours. This is where we live from now on!"

"From now on?" Michael was incredulous. He stood up, looking, and then turned to blink back at where the rocket used to be. "What about the rocket? What about Minnesota?"

"Here," said Dad.

He touched the small radio to Michael's blond head. "Listen."

Michael listened.

"Nothing," he said.

"That's right. Nothing. Nothing at all any more. No more Minneapolis, no more rockets, no more Earth."

Michael considered the lethal revelation and began to sob little dry sobs.

"Wait a moment," said Dad the next instant. "I'm giving you a lot more in exchange, Mike!"

"What?" Michael held off the tears, curious, but quite ready to continue in case Dad's further revelation was as disconcerting as the original.

"I'm giving you this city, Mike. It's yours."

"Mine?"

"For you and Robert and Timothy, all three of you, to own for yourselves."

Timothy bounded from the boat. "Look, guys, all for *us!* All of *that!*" He was playing the game with Dad, playing it large and playing it well. Later, after it was all over and things had settled, he could go off by himself and cry for ten minutes. But now it was still a game, still a family outing, and the other kids must be kept playing.

Mike jumped out with Robert. They helped Mom.

"Be careful of your sister," said Dad, and nobody knew what he meant until later.

They hurried into the great pink-stoned city, whispering among themselves, because dead cities have a way of making you want to whisper, to watch the sun go down.

"In about five days," said Dad quietly, "I'll go back down to where our rocket was and collect the food hidden in the ruins there and bring it here; and I'll hunt for Bert Edwards and his wife and daughters there."

"Daughters?" asked Timothy. "How many?"

"Four."

"I can see that'll cause trouble later." Mom nodded slowly.

"Girls." Michael made a face like an ancient Martian stone image. "Girls."

"Are they coming in a rocket too?"

"Yes. If they make it. Family rockets are made for travel to the Moon, not Mars. We were lucky we got through."

"Where did you get the rocket?" whispered Timothy, for the other boys were running ahead.

"I saved it. I saved it for twenty years, Tim. I had it hidden away, hoping I'd never have to use it. I suppose I should have given it to the government for the war, but I kept thinking about Mars...."

"And a picnic!"

"Right. This is between you and me. When I saw everything was finishing on Earth, after I'd waited until the last moment, I packed us up. Bert Edwards had a ship hidden, too, but we decided it would be safer to take off separately, in case anyone tried to shoot us down."

"Why'd you blow up the rocket, Dad?"

"So we can't go back, ever. And so if any of those evil men ever come to Mars they won't know we're here."

"Is that why you look up all the time?"

"Yes, it's silly. They won't follow us, ever. They haven't anything to follow with. I'm being too careful, is all."

Michael came running back. "Is this really *our* city, Dad?"

"The whole damn planet belongs to us, kids. The whole darn planet."

They stood there, King of the Hill, Top of the Heap, Ruler of All They Surveyed, Unimpeachable Monarchs and Presidents, trying to understand what it meant to own a world and how big a world really was.

Night came quickly in the thin atmosphere, and Dad left them in the square by the pulsing fountain, went down to the boat, and came walking back carrying a stack of paper in his big hands.

He laid the papers in a clutter in an old courtyard and set them afire. To keep warm, they crouched around the blaze and laughed, and Timothy saw the little letters leap like frightened animals when the flames touched and engulfed them. The papers crinkled like an old man's skin, and the cremation surrounded innumerable words:

"GOVERNMENT BONDS; Business Graph, 1999; Religious Prejudice: An Essay; The Science of Logistics; Problems of the Pan-American Unity; Stock Report for July 3, 1998; The War Digest ..."

Dad had insisted on bringing these papers for this purpose. He sat there and fed them into the fire, one by one, with satisfaction, and told his children what it all meant.

"It's time I told you a few things. I don't suppose it was fair, keeping so much from you. I don't know if you'll understand, but I have to talk, even if only part of it gets over to you."

He dropped a leaf in the fire.

"I'm burning a way of life, just like that way of life is being burned clean of Earth right now. Forgive me if I talk like a politician. I am, after all, a former state governor, and I was honest and they hated me for it. Life on Earth never settled down to doing anything very good. Science ran too far ahead of us too quickly, and the people got lost in a mechanical wilderness, like children making over pretty things, gadgets, helicopters, rockets; emphasizing the wrong items, emphasizing machines instead of how to run the machines. Wars got bigger and bigger and finally killed Earth. That's what the silent radio means. That's what we ran away from.

"We were lucky. There aren't any more rockets left. It's time you knew this isn't a fishing trip at all. I put off telling you. Earth is gone. Interplanetary travel won't be back for centuries, maybe never. But that way of life proved itself wrong and strangled itself with its own hands. You're young. I'll tell you this again every day until it sinks in."

He paused to feed more papers to the fire.

"Now we're alone. We and a handful of others who'll land in a few days. Enough to start over. Enough to turn away from all that back on Earth and strike out on a new line——"

The fire leaped up to emphasize his talking. And then all the papers were gone except one. All the laws and beliefs of Earth were burnt into small hot ashes which soon would be carried off in a wind.

Timothy looked at the last thing that Dad tossed in the fire. It was a map of the world, and it wrinkled and distorted itself hotly and went—flimpf—and was gone like a warm, black butterfly. Timothy turned away.

"Now I'm going to show you the Martians," said Dad. "Come on, all of you. Here, Alice." He took her hand.

Michael was crying loudly, and Dad picked him up and carried him, and they walked down through the ruins toward the canal.

The canal. Where tomorrow or the next day their future wives would come up in a boat, small laughing girls now, with their father and mother.

The night came down around them, and there were stars. But Timothy couldn't find Earth. It had already set. That was something to think about.

A night bird called among the ruins as they walked. Dad said, "Your mother and I will try to teach you. Perhaps we'll fail. I hope not. We've had a good lot to see and learn from. We planned this trip years ago, before you were born. Even if there hadn't been a war we would have come to Mars, I think, to live and form our own standard of living. It would have been another century before Mars would have been really poisoned by the Earth civilization. Now, of course—"

They reached the canal. It was long and straight and cool and wet and reflective in the night.

"I've always wanted to see a Martian," said Michael. "Where are they, Dad? You promised."

"There they are," said Dad, and he shifted Michael on his shoulder and pointed straight down.

The Martians were there. Timothy began to shiver.

The Martians were there—in the canal—reflected in the water. Timothy and Michael and Robert and Mom and Dad.

The Martians stared back up at them for a long, long silent time from the rippling water...

THE SANDS OF MARS (1951)

CHAPTER 11

Arthur C. Clarke

Born in 1917, **Arthur C. Clarke** was one of the great authors of science fiction's later golden age. The attention to scientific detail in his story led to *Readers Digest* calling him 'Prophet of the Space Age' in 1969, a few months before the first Moon landing. In 1945, he had famously proposed using geostationary orbit as a location for wireless relay stations, effectively inventing the concept of the modern-day communications satellite. Clarke died in Sri Lanka in 2008.

The amber light was on. Gibson took a last sip of water, cleared his throat gently, and checked that the papers of his script were in the right order. No matter how many times he broadcast, his throat always felt this initial tightness. In the control room, the programme engineer held up her thumb: the amber changed abruptly to red.

'Hello, Earth. This is Martin Gibson speaking to you from Port Lowell, Mars. It's a great day for us here. This morning the new dome was inflated and now the city's increased its size by almost a half. I don't know if I can convey any impression of what a triumph this means, what a feeling of victory it gives to us here in the battle against Mars. But I'll try.

'You all know that it's impossible to breathe the Martian atmosphere – it's far too thin and contains practically no oxygen. Port Lowell, our biggest city, is built under six domes of transparent plastic held up by the pressure of the air inside – air which we can breathe comfortably though it's still much less dense than yours.

'For the last year a seventh dome has been under construction, a dome twice as big as any of the others. I'll describe it as it was yesterday, when I went inside before the inflation started.

'Imagine a great circular space half a kilometre across, surrounded by a thick wall of glass bricks twice as high as a man. Through this wall lead the passages to the other domes, and the exits direct on to the brilliant green Martian landscape all around us. These passages are simply metal tubes with great doors which close automatically if air escapes from any of the domes. On Mars, we don't believe in putting all our eggs in one basket!

'When I entered Dome Seven yesterday, all this great circular space was covered with a thin transparent sheet fastened to the surrounding wall, and lying limp on the ground in huge folds beneath which we had to force our way. If you can imagine being inside a deflated balloon you'll know exactly how I felt. The envelope of the dome is a very strong plastic, almost perfectly transparent and quite flexible – a kind of thick cellophane.

'Of course, I had to wear my breathing mask, for though we were sealed off from the outside there was still practically no air in the dome. It was being pumped in as rapidly as possible, and you could see the great sheets of plastic straining sluggishly as the pressure mounted.

'This went on all through the night. The first thing this morning I went into the dome again, and found that the envelope had now blown itself into a big bubble at the centre, though round the edges it was still lying flat. That huge bubble – it was about a hundred metres across – kept trying to move around like a living creature, and all the time it grew.

'About the middle of the morning it had grown so much that we could see the complete dome taking shape: the envelope had lifted away from the ground everywhere. Pumping was stopped for a while to test for leaks, then resumed again around midday. By now the sun was helping too, warming up the air and making it expand.

'Three hours ago the first stage of the inflation was finished. We took off our masks and let out a great cheer. The air still wasn't really thick enough for comfort, but it was breathable and the engineers could work inside without bothering about masks any more. They'll spend the next few days checking the great envelope for stresses, and looking for leaks. There are bound to be some, of course, but as long as the air loss doesn't exceed a certain value it won't matter.

'So now we feel we've pushed our frontier on Mars back a little farther. Soon the new buildings will be going up under Dome Seven, and we're making plans for a small park and even a lake – the only one on Mars, that will be, for free water can't exist here in the open for any length of time.

'Of course, this is only a beginning, and one day it will seem a very small achievement; but it's a great step forward in our battle – it represents the conquest of another slice of Mars. And it means living space for another thousand people. Are you listening, Earth? Good night.'

The red light faded. For a moment Gibson sat staring at the microphone, musing on the fact that his first words, though travelling at the speed of light, would only now be reaching Earth. Then he gathered up his papers and walked through the padded doors into the control room.

The engineer held up a telephone for him. 'A call's just come through for you, Mr Gibson,' she said. 'Someone's been pretty quick off the mark!'

'They certainly have,' he replied with a grin. 'Hello, Gibson here.'

'This is Hadfield. Congratulations. I've just been listening – it went out over our local station, you know.'

'I'm glad you liked it.'

Hadfield chuckled.

'You've probably guessed that I've read most of your earlier scripts. It's been quite interesting to watch the change of attitude.'

'What change?'

'When you started, we were "they". Now we're "we". Not very well put, perhaps, but I think my point's clear.'

He gave Gibson no time to answer this, but continued without a break.

'I really rang up about this. I've been able to fix your trip to Skia at last. We've got a passenger jet going there on Wednesday, with room for three aboard. Whittaker will give you the details. Goodbye.'

The phone clicked into silence. Very thoughtfully, but not a little pleased, Gibson replaced it on the stand. What the Chief had said was true enough. He had been here for almost a month, and in that time his outlook towards Mars had changed completely. The first schoolboy excitement had lasted no more than a few days: the subsequent disillusionment only a little longer. Now he knew enough to regard the colony with a tempered enthusiasm not wholly based on logic. He was afraid to analyse it, lest it disappear completely. Some part of it, he knew, came from his growing respect for the people around him – his admiration for the keen-eyed competence, the readiness to take well-calculated risks, which had enabled them not merely to survive on this heartbreakingly hostile world, but to lay the foundations of the first extra-terrestrial culture. More than ever before, he felt a longing to identify himself with their work, wherever it might lead.

Meanwhile, his first real chance of seeing Mars on the large scale had arrived. On Wednesday he would be taking off for Port Schiaparelli, the planet's second city, ten thousand kilometres to the east in Trivium Charontis. The trip had been planned a fortnight ago, but every time something had turned up to postpone it. He would have to tell Jimmy and Hilton to get ready – they had been the lucky ones in the draw. Perhaps Jimmy might not be quite so eager to go now as he had been once. No doubt he was now anxiously counting the days left to him on Mars, and would resent anything that took him away from Irene. But if he turned down *this* chance, Gibson would have no sympathy for him at all.

'Neat job, isn't she?' said the pilot proudly. 'There are only six like her on Mars. It's quite a trick designing a jet that can fly in this atmosphere, even with the low gravity to help you.'

Gibson did not know enough about aerodynamics to appreciate the finer points of the aircraft, though he could see that the wing area was abnormally large. The four jet units were neatly buried just outboard of the fuselage, only the slightest of bulges betraying their position. If he had met such a machine on a terrestrial airfield Gibson would not have given it a second thought, though the sturdy tractor undercarriage might have surprised him. This machine was built to fly fast and far – and to land on any surface which was approximately flat.

He climbed in after Jimmy and Hilton and settled himself as comfortably as he could in the rather restricted space. Most of the cabin was taken up by large packing cases securely strapped in position – urgent freight for Skia, he supposed. It hadn't left a great deal of space for the passengers.

The motors accelerated swiftly until their thin whines hovered at the edge of hearing. There was the familiar pause while the pilot checked his instruments and controls: then the jets opened full out and the runway began to slide beneath them. A few seconds later there came the sudden reassuring surge of power as the take-off rockets fired and lifted them effortlessly up into the sky. The aircraft climbed steadily into the south, then swung round to starboard in a great curve that took it over the city. Port Lowell, Gibson thought, had certainly grown since his last view of it from the air. The new dome was still empty, yet already it dominated the city with its promise of more spacious times to come. Near its centre he could glimpse the tiny specks of men and machines at work laying the foundations of the new suburb.

The aircraft levelled out on an easterly course and the great island of Aurorae Sinus sank over the edge of the planet. Apart from a few oases, the open desert now lay ahead for thousands of kilometres.

The pilot switched his controls to automatic and came amidships to talk to his passengers.

'We'll be at Charontis in about four hours,' he said. 'I'm afraid there isn't much to look at on the way, though you'll see some fine colour effects when we go over Euphrates. After that it's more or less uniform desert until we hit the Syrtis Major.'

Gibson did some rapid mental arithmetic.

'Let's see – we're flying east and we started rather late – it'll be dark when we get there.'

'Don't worry about that – we'll pick up the Charontis beacon when we're a couple of hundred kilometres away. Mars is so small that you don't often do a long-distance trip in daylight all the way.'

'How long have you been on Mars?' asked Gibson, who had now ceased taking photos through the observation ports.

'Oh, five years.'

'Flying all the time?'

'Most of it.'

'Wouldn't you prefer being in spaceships?'

'Not likely. No excitement in it – just floating around in nothing for months.' He grinned at Hilton, who smiled amiably but showed no inclination to argue.

'Just what do you mean by "excitement"?' said Gibson anxiously.

'Well, you've got some scenery to look at, you're not away from home for very long, and there's always the chance you may find something new. I've done half a dozen trips over the poles, you know – most of them in summer, but I went across the Mare Boreum last winter. A hundred and fifty degrees below outside! That's the record so far for Mars.'

'I can beat that pretty easily,' said Hilton. 'At night it reaches two hundred below on Titan.' It was the first time Gibson had ever heard him refer to the Saturnian expedition.

'By the way, Fred,' he asked, 'is this rumour true?'

'What rumour?'

'*You* know – that you're going to have another shot at Saturn.'

Hilton shrugged his shoulders.

'It isn't decided – there are a lot of difficulties. But I think it will come off: it would be a pity to miss the chance. You see, if we can leave next year we can go past Jupiter on the way, and have our first really good look at him. Mac's worked out a very interesting orbit for us. We go rather close to Jupiter – right inside *all* the satellites – and let his gravitational field swing us round so that we head out in the right direction for Saturn. It'll need rather accurate navigation to give us just the orbit we want, but it can be done.'

'Then what's holding it up?'

'Money, as usual. The trip will last two and a half years and will cost about fifty million. Mars can't afford it – it would mean doubling the usual deficit! At the moment we're trying to get Earth to foot the bill.'

'It would come to that anyway in the long run,' said Gibson. 'But give me all the facts when we get home and I'll write a blistering *exposé* about cheeseparing terrestrial politicians. You mustn't underestimate the power of the Press.'

The talk then drifted from planet to planet, until Gibson suddenly remembered that he was wasting a magnificent chance of seeing Mars at first hand. Obtaining permission to occupy the pilot's seat – after promising not to touch anything – he went forward and settled himself comfortably behind the controls.

Five kilometres below, the coloured desert was streaking past him to the west. They were flying at what, on Earth, would have been a very low altitude, for the thinness of the Martian air made it essential to keep as near the surface as safety allowed. Gibson had never before received such an impression of sheer speed, for though he had flown in much faster machines on Earth that had always been at heights where the ground was invisible. The nearness of the horizon added to the effect, for an object which appeared over the edge of the planet would be passing beneath a few minutes later.

From time to time the pilot came forward to check the course, though it was a pure formality, as there was nothing he need do until the voyage was nearly over. At mid-point some coffee and light refreshments were produced, and Gibson rejoined his companions in the cabin. Hilton and the pilot were now arguing briskly about Venus – quite a sore point with the Martian colonists, who regarded that peculiar planet as a complete waste of time.

The sun was now very low in the west and even the stunted Martian hills threw long shadows across the desert. Down there the temperature was already below freezing-point, and falling fast. The few hardy plants that had survived in this almost barren waste would have folded their leaves tightly together, conserving warmth and energy against the rigours of the night.

Gibson yawned and stretched himself. The swiftly unfolding landscape had an almost hypnotic effect and it was difficult to keep awake. He decided to catch some sleep in the ninety or so minutes that were left of the voyage.

Some change in the failing light must have woken him. For a moment it was impossible to believe that he was not still dreaming: he could only sit and stare, paralysed with sheer astonishment. No longer was he looking out across a flat, almost featureless landscape meeting the deep blue of the sky at the far horizon. Desert and horizon had both vanished: in their place towered a range of crimson mountains, reaching north and south as far as the eye could follow. The last rays of the setting sun caught their peaks and bequeathed to them its dying glory: already the foothills were lost in the night that was sweeping onwards to the west.

For long seconds the splendour of the scene robbed it of all reality and hence all menace. Then Gibson awoke from his trance, realizing in one dreadful instant that they were flying far too low to clear those Himalayan peaks.

The sense of utter panic lasted only a moment – to be followed at once by a far deeper terror. Gibson had remembered now what the first shock had banished from his mind – the simple fact he should have thought of from the beginning.

There were no mountains on Mars.

Hadfield was dictating an urgent memorandum to the Interplanetary Development Board when the news came through. Port Schiaparelli had waited the regulation fifteen minutes after the aircraft's expected time of arrival, and Port Lowell Control had stood by for another ten before sending out the 'Overdue' signal. One precious aircraft from the tiny Martian fleet was already standing by to search the line of flight as soon as dawn came. The high speed and low altitude essential for flight would make such a search very difficult, but when Phobos rose the telescopes up there could join in with far greater prospects of success.

The news reached Earth an hour later, at a time when there was nothing much else to occupy Press or radio. Gibson would have been well satisfied by the resultant publicity: everywhere people began reading his last articles with a morbid interest. Ruth Goldstein knew nothing about it until an editor she was dealing with arrived waving the evening paper. She immediately sold the second reprint rights of Gibson's latest series for half as much again as her victim had intended to pay, then retired to her private room and wept copiously for a full minute. Both these events would have pleased Gibson enormously.

In a score of newspaper offices, the copy culled from the Morgue began to be set up in type so that no time would be wasted. And in London a publisher who had paid Gibson a rather large advance began to feel very unhappy indeed.

Gibson's shout was still echoing through the cabin when the pilot reached the controls. Then he was flung to the floor as the machine turned over in an almost vertical bank in a desperate attempt to swing round to the north. When Gibson could climb to his feet again, he caught a glimpse of a strangely blurred orange cliff sweeping down upon them from only kilometres away. Even in that moment of panic, he could see that there was something very curious about that swiftly approaching barrier, and suddenly the truth dawned upon him at last. This was no mountain range, but something that might be no less deadly. They were running into a wind-borne wall of sand reaching from the desert almost to the edge of the stratosphere.

The hurricane hit them a second later. Something slapped the machine violently from side to side, and through the insulation of the hull came an angry whistling roar that was the most terrifying sound Gibson had ever heard in his life. Night had come instantly upon them and they were flying helplessly through a howling darkness.

It was all over in five minutes, but it seemed a lifetime. Their sheer speed had saved them, for the ship had cut through the heart of the hurricane like a projectile. There was a sudden burst of deep ruby twilight, the ship ceased to be pounded by a million sledge-hammers, and a ringing silence seemed to fill the little cabin. Through the rear observation port Gibson caught a last glimpse of the storm as it moved westwards, tearing up the desert in its wake.

His legs feeling like jellies, Gibson tottered thankfully into his seat and breathed an enormous sigh of relief. For a moment he wondered if they had been thrown badly off course, then realized that this scarcely mattered considering the navigational aids they carried.

It was only then, when his ears had ceased to be deafened by the storm, that Gibson had his second shock. The motors had stopped.

The little cabin was very tense and still. Then the pilot called out over his shoulder: 'Get your masks on! The hull may crack when we come down.' His fingers feeling very clumsy, Gibson dragged his breathing equipment from under the seat and adjusted it over his head. When he had finished, the ground already seemed very close, though it was hard to judge distances in the failing twilight.

A low hill swept by and was gone into the darkness. The ship banked violently to avoid another, then gave a sudden spasmodic jerk as it touched ground and bounced. A moment later it made contact again and Gibson tensed himself for the inevitable crash.

It was an age before he dared relax, still unable to believe that they were safely down. Then Hilton stretched himself in his seat, removed his mask, and called out to the pilot: 'That was a very nice landing, Skipper. Now how far have we got to walk?'

For a moment there was no reply. Then the pilot called, in a rather strained voice: 'Can anyone light me a cigarette? I've got the twitch.'

'Here you are,' said Hilton, going forward. 'Let's have the cabin lights on now, shall we?'

The warm, comfortable glow did much to raise their spirits by banishing the Martian night, which now lay all around. Everyone began to feel ridiculously cheerful and there was much laughing at quite feeble jokes. The reaction had set in: they were so delighted at still being alive that the thousand kilometres separating them from the nearest base scarcely seemed to matter.

'That was quite a storm,' said Gibson. 'Does this sort of thing happen very often on Mars? And why didn't we get any warning?'

The pilot, now that he had got over his initial shock, was doing some quick thinking, the inevitable court of inquiry obviously looming large in his mind. Even on auto-pilot, he *should* have gone forward more often...

'I've never seen one like it before,' he said, 'though I've done at least fifty trips between Lowell and Skia. The trouble is that we don't know anything about Martian meteorology, even now. And there are only half a dozen met stations on the planet – not enough to give us an accurate picture.'

'What about Phobos? Couldn't they have seen what was happening and warned us?'

The pilot grabbed his almanac and ruffled rapidly through the pages.

'Phobos hasn't risen yet,' he said after a brief calculation. 'I guess the storm blew up suddenly out of Hades – appropriate name, isn't it? – and has probably collapsed again now. I don't suppose it went anywhere near Charontis, so *they* couldn't have warned us either. It was just one of those accidents that's nobody's fault.'

This thought seemed to cheer him considerably, but Gibson found it hard to be so philosophical.

'Meanwhile,' he retorted, 'we're stuck in the middle of nowhere. How long will it take them to find us? Or is there any chance of repairing the ship?'

'Not a hope of that; the jets are ruined. They were made to work on air, not sand, you know!'

'Well, can we radio Skia?'

'Not now we're on the ground. But when Phobos rises in – let's see – an hour's time, we'll be able to call the observatory and they can relay us on. That's the way we've got to do all our long-distance stuff here, you know. The ionosphere's too feeble to bounce signals round the way you do on Earth. Anyway, I'll go and check that the radio is OK.'

He went forward and started tinkering with the ship's transmitter, while Hilton busied himself checking the heaters and cabin air pressure, leaving the two remaining passengers looking at each other a little thoughtfully.

'This is a fine kettle of fish!' exploded Gibson, half in anger and half in amusement. 'I've come safely from Earth to Mars – more than fifty million kilometres – and as soon as I set foot inside a miserable aeroplane *this* is what happens! I'll stick to spaceships in future.'

Jimmy grinned. 'It'll give us something to tell the others when we get back, won't it? Maybe we'll be able to do some real exploring at last.' He peered through the windows, cupping his hands over his eyes to keep out the cabin light. The surrounding landscape was now in complete darkness, apart from the illumination from the ship.

'There seem to be hills all around us: we were lucky to get down in one piece. Good Lord – there's a cliff here on this side – another few metres and we'd have gone smack into it!'

'Any idea where we are?' Gibson called to the pilot.

This tactless remark earned him a very stony stare.

'About 120 east, 20 north. The storm can't have thrown us very far off course.'

'Then we're somewhere in the Aetheria,' said Gibson, bending over the maps. 'Yes – there's a hilly region marked here. Not much information about it.'

'It's the first time anyone's ever landed here – that's why. This part of Mars is almost unexplored: it's been thoroughly mapped from the air, but that's all.'

Gibson was amused to see how Jimmy brightened at this news. There was certainly something exciting about being in a region where no human foot had ever trodden before.

'I hate to cast a gloom over the proceedings,' remarked Hilton, in a tone of voice hinting that this was exactly what he was going to do, 'but I'm not at all sure you'll be able to radio Phobos even when it does rise.'

'What!' yelped the pilot. 'The set's OK – I've just tested it.'

'Yes – but have you noticed where we are? We can't even *see* Phobos. That cliff's due south of us and blocks the view completely. That means that they won't be able to pick up our microwave signals. What's even worse, they won't be able to locate us in their telescopes.'

There was a shocked silence.

'*Now* what do we do?' asked Gibson. He had a horrible vision of a thousand-kilometre trek across the desert to Charontis, but dismissed it from his mind at once. They couldn't possibly carry the oxygen for the trip, still less the food and equipment necessary. And no one could spend the night unprotected on the surface of Mars, even here near the Equator.

'We'll just have to signal in some other way,' said Hilton calmly. 'In the morning we'll climb those hills and have a look round. Meanwhile I suggest we take it easy.' He yawned and stretched himself, filling the cabin from ceiling to floor. 'We've got no immediate worries: there's air for several days, and power in the batteries to keep us warm almost indefinitely. We may get a bit hungry if we're here more than a week, but I don't think that's at all likely to happen.'

By a kind of unspoken mutual consent, Hilton had taken control. Perhaps he was not even consciously aware of the fact, but he was now the leader of the little party. The pilot had delegated his own authority without a second thought.

'Phobos rises in an hour, you said?' asked Hilton.

'Yes.'

'When does it transit? I can never remember what this crazy little moon of yours gets up to.'

'Well, it rises in the west and sets in the east about four hours later.'

'So it'll be due south around midnight?'

'That's right. Oh Lord – that means we won't be able to see it anyway. It'll be eclipsed for at least an hour!'

'*What* a moon!' snorted Gibson. 'When you want it most badly, you can't even see the blasted thing!'

'That doesn't matter,' said Hilton calmly. 'We'll know just where it is, and it won't do any harm to try the radio then. That's all we can do tonight. Has anyone got a pack of cards? No? Then what about entertaining us, Martin, with some of your stories?'

It was a rash remark, and Gibson seized his chance immediately.

'I wouldn't dream of doing that,' he said. '*You're* the one who has the stories to tell.'

Hilton stiffened, and for a moment Gibson wondered if he had offended him. He knew that Hilton seldom talked about the Saturnian expedition, but this was too good an opportunity to miss. The chance would never come again, and, as is true of all great adventures, its telling would do their morale good. Perhaps Hilton realized this too, for presently he relaxed and smiled.

'You've got me nicely cornered, haven't you, Martin? Well, I'll talk – but on one condition.'

'What's that?'

'No direct quotes, please!'

'As if I would!'

'And when you *do* write it up, let me see the manuscript first.'

'Of course.'

This was better than Gibson had dared to hope. He had no immediate intention of writing about Hilton's adventures, but it was nice to know that he could do so if he wished. The possibility that he might never have the chance simply did not cross his mind.

Outside the walls of the ship, the fierce Martian night reigned supreme – a night studded with needle-sharp, unwinking stars. The pale light of Deimos made the surrounding landscape dimly visible, as if lit with a cold phosphorescence. Out of the east Jupiter, the brightest object in the sky, was rising in his glory. But the thoughts of the four men in the crashed aircraft were six hundred million kilometres still farther from the Sun.

It still puzzled many people – the curious fact that man had visited Saturn but not Jupiter, so much closer at hand. But in space-travel, sheer distance is of no importance, and Saturn had been reached because of a single astonishing stroke of luck that still seemed too good to be true. Orbiting Saturn was Titan, the largest satellite in the Solar System – about twice the size of Earth's moon. As far back as 1944 it had been discovered that Titan possessed an atmosphere. It was not an atmosphere one could breathe: it was immensely more valuable than that. For it was an atmosphere of methane, one of the ideal propellants for atomic rockets.

This had given rise to a situation unique in the history of space-flight. For the first time, an expedition could be sent to a strange world with the virtual certainty that refuelling would be possible on arrival.

The *Arcturus* and her crew of six had been launched in space from the orbit of Mars. She had reached the Saturnian system only nine months later, with just enough fuel to land safely on Titan. Then the pumps had been started, and the great tanks replenished from the countless trillions of tons of methane that were there for the taking. Refuelling on Titan whenever necessary, the *Arcturus* had visited every one of Saturn's fifteen known moons, and had even skirted the great ring system itself. In a few months, more was learned about Saturn than in all the previous centuries of telescopic examination.

There had been a price to pay. Two of the crew had died of radiation sickness after emergency repairs to one of the atomic motors. They had been buried on Dione, the fourth moon. And the leader of the expedition, Captain Envers, had been killed by an avalanche of frozen air on Titan: his body had never been found. Hilton had assumed command, and had brought the *Arcturus* safely back to Mars a year later, with only two men to help him.

All these bare facts Gibson knew well enough. He could still remember listening to those radio messages that had come trickling back through space, relayed from world to world. But it was a different thing altogether to hear Hilton telling the story in his quiet, curiously impersonal manner, as if he had been a spectator rather than a participant.

He spoke of Titan and its smaller brethren, the little moons which, circling Saturn, made the planet almost a scale model of the Solar System. He described how at last they had landed on the innermost moon of all, Mimas, only half as far from Saturn as the Moon is from the Earth.

'We came down in a wide valley between a couple of mountains, where we were sure the ground would be pretty solid. We weren't going to make the mistake we did on Rhea! It was a good landing, and we climbed into our suits to go outside. It's funny how impatient you always are to do that, no matter how many times you've set down on a new world.

'Of course, Mimas hasn't much gravity – only a hundredth of Earth's. That was enough to keep us from jumping off into space. I liked it that way: you knew you'd always come down safely again if you waited long enough.

'It was early in the morning when we landed. Mimas has a day a bit shorter than Earth's – it goes round Saturn in twenty-two hours, and as it keeps the same face towards the planet its day and month are the same length – just as they are on the Moon. We'd come down in the northern hemisphere, not far from the equator, and most of Saturn was above the horizon. It looked quite weird – a huge crescent horn sticking up into the sky, like some impossibly bent mountain thousands of miles high.

'Of course you've all seen the films we made – especially the speeded-up colour one showing a complete cycle of Saturn's phases. But I don't think they can give you much idea of what it was like to live with that enormous thing always there in the sky. It was so big, you see, that one couldn't take it in in a single view. If you stood facing it and held your arms wide open, you could just imagine your finger-tips touching the opposite ends of the rings. We couldn't see the rings themselves very well, because they were almost edge-on, but you could always tell they were there by the wide, dusky band of shadow they cast on the planet.

'None of us ever got tired of watching it. It's spinning so fast, you know – the pattern was always changing. The cloud formations, if that's what they were, used to whip round from one side of the disc to the other in a few hours, changing continually as they moved. And there were the most wonderful colours – greens and browns and yellows, chiefly. Now and then there'd be great, slow eruptions, and something as big as Earth would rise up out of the depths and spread itself sluggishly in a huge stain half-way round the planet.

'You could never take your eyes off it for long. Even when it was new and so completely invisible, you could still tell it was there because of the great hole in the stars. And here's a funny thing which I haven't reported because I was never quite sure of it. Once or twice, when we were in the planet's shadow and its disc should have been completely dark, I thought I saw a faint phosphorescent glow coming from the night side. It didn't last long – if it really happened at all. Perhaps it was some kind of chemical reaction going on down there in that spinning cauldron.

'Are you surprised that I want to go to Saturn again? What I'd like to do is to get *really* close this time – and by that I mean within a thousand kilometres. It should be quite safe and wouldn't take much power. All you need do is to go into a parabolic orbit and let yourself fall in like a comet going round the Sun. Of course, you'd only spend a few minutes actually close to Saturn, but you could get a lot of records in that time.

'And I want to land on Mimas again, and see that great shining crescent reaching half-way up the sky. It'll be worth the journey, just to watch Saturn waxing and waning, and to see the storms chasing themselves round his equator.

Yes – it would be worth it, even if *I* didn't get back this time.'

There were no mock heroics in this closing remark. It was merely a simple statement of fact, and Hilton's listeners believed him completely. While the spell lasted, every one of them would be willing to strike the same bargain.

Gibson ended the long silence by going to the cabin window and peering out into the night.

'Can we have the lights off?' he called. Complete darkness fell as the pilot obeyed his request. The others joined him at the window.

'Look,' said Gibson. 'Up there – you can just see it if you crane your neck.'

The cliff against which they were lying was no longer a wall of absolute and unrelieved darkness. On its very topmost peaks a new light was playing, spilling over the broken crags and filtering down into the valley. Phobos had leapt out of the west and was climbing on its meteoric rise towards the south, racing backwards across the sky.

Minute by minute the light grew stronger, and presently the pilot began to send out his signals. He had barely begun when the pale moonlight was snuffed out so suddenly that Gibson gave a cry of astonishment. Phobos had gone hurtling into the shadow of Mars, and though it was still rising it would cease to shine for almost an hour. There was no way of telling whether or not it would peep over the edge of the great cliff and so be in the right position to receive their signals.

They did not give up hope for almost two hours. Suddenly the light reappeared on the peaks, but shining now from the east. Phobos had emerged from its eclipse, and was now dropping down towards the horizon which it would reach in little more than an hour. The pilot switched off his transmitter in disgust.

'It's no good,' he said. 'We'll have to try something else.'

'I know!' Gibson exclaimed excitedly. 'Can't we carry the transmitter up the top of the hill?'

'I'd thought of that, but it would be the devil's own job to get it out without proper tools. The whole thing – aerials and all – is built into the hull.'

'There's nothing more we can do tonight, anyway,' said Hilton. 'I suggest we all get some sleep before dawn. Good night, everybody.'

It was excellent advice, but not easy to follow. Gibson's mind was still racing ahead, making plans for the morrow. Not until Phobos had at last plunged down into the east, and its light had ceased to play mockingly on the cliff above them, did he finally pass into a fitful slumber.

Even then he dreamed that he was trying to fix a belt-drive from the motors to the tractor undercarriage so that they could taxi the last thousand kilometres to Port Schiaparelli...

MARTIAN TIME-SLIP (1964)

CHAPTER 2

Philip K. Dick

Philip K. Dick is perhaps best known as the author of the 1968 novel *Do Androids Dream of Electric Sheep?,* which was adapted into the film *Blade Runner* (1982) by Ridley Scott and starred Harrison Ford. In a career spanning thirty years, Dick wrote forty-four novels and over one hundred short stories. *Martian Time-Slip* was published in 1964, during his most prolific decade; he released twenty-one novels in the 1960s. While it first appears to be a tale of indigenous displacement, it is really a story of alternate perceptions of reality, and the links to mental health. Dick was born in Chicago in 1928 and died in 1982, aged 53.

The ex-plumber, Supreme Goodmember Arnie Kott of the Water Workers' Local, Fourth Planet Branch, rose from his bed at ten in the morning and, as was his custom, strolled directly to the steam bath.

'Hello, Gus.'

'Hi there, Arnie.'

Everybody called him by his first name, and that was good. Arnie Kott nodded to Bill and Eddy and Tom, and they all greeted him. The air, full of steam, condensed around his feet and drained off across the tiles, to be voided. That was a touch which pleased him: the baths had been constructed so as not to preserve the run-off. The water drained out on to the hot sand and disappeared forever. Who else could do that? He thought, Let's see if those rich Jews up in New Israel have a steam bath that wastes water.

Placing himself under a shower, Arnie Kott said to the fellows around him, 'I heard some rumour I want checked on soon as possible. You know that combine from California, those Portugees that originally held title on the FDR Mountain Range, and they tried to extract iron ore there, but it was too low grade, and the cost was way out of line? 1 heard they sold their holdings.'

'Yeah, 1 heard that too.' All the boys nodded. 'I wonder how much they lost. Must have taken a terrible beating.'

Arnie said, 'No, I heard they found a buyer that was willing to put up more than they paid; they made a profit, after all these years. So it paid them to hold out. I wonder who's nuts enough to want that land. I got some mineral rights there, you know. I want you to check into who bought that land and what kind of operation they represent. I want to know what they're doing over there.'

'Good to know those things.' Again everyone nodded, and one man – Fred, it looked like – detached himself from his shower and padded off to dress. 'I'll check into that, Arnie,' Fred said over his shoulder. 'I'll get to it right away.'

Addressing himself to the remaining men, Arnie soaped himself all over and said, 'You know I got to protect my mineral rights; I can't have some smoozer coming in here from Earth and making those mountains into like for instance a national park for picnickers. I tell you what I heard. I know that a bunch of Communist officials from Russia and Hungary, big boys, was over here around a week ago, no doubt looking around. You think because that collective of theirs failed last year they gave up? No. They got the brains of bugs, and like bugs they always come back. Those Reds are aching to establish a successful collective on Mars; it's practically a wet dream of theirs back Home. I wouldn't be surprised if we find out that those Portugees from California sold to Communists, and pretty soon we're seeing the name changed from the FDR Mountains, which is right and proper, to something like the Joe Stalin Mountains.'

The men all laughed appreciatively.

'Now, I got a lot of business ahead of me today to conduct,' Arnie Kott said, washing the soapsuds from him with furious streams of hot water. 'So I can't devote myself to this matter any further; I'm relying on you to dig into it. For example, 1 have been travelling east where we got that melon experiment in progress, and it seems like we're about to be entirely successful in inducing the New England type of melon into growing here in this environment. I know you all have been wondering about that, because everybody likes a good slice of cantaloup in the morning for his breakfast, if it's at all possible.'

'That's true, Arnie,' the boys agreed.

'But,' Arnie said, 'I got more on my mind than melons. We had one of those UN boys visiting us the other day protesting our regulations concerning the niggers. Or maybe I shouldn't say that; maybe I should talk like the UN boys and say "indigenous population remnants", or just Bleekmen. What he had reference to was our licensing the mines owned by our settlement to use Bleekmen at below scale, I mean, below the minimum wage – because even those fairies at the UN don't seriously propose we pay scale to Bleekmen niggers. However, we have this problem that we can't pay any minimum wage to the Bleekmen niggers because their work is so inconsistent that we'd go broke, and we have to use them in mining operations because they're the only ones who can breathe down there, and we can't get oxygen equipment in quantity transported over here at any price less than outrageous. Somebody's making a lot of money back Home on those oxygen tanks and compressors and all that. It's a racket, and we're not going to get gouged, I can tell you.'

Everybody nodded sombrely.

'Now, we can't allow the UN bureaucrats to dictate to us how we'll run our settlement,' Arnie said. 'We set up operations here before the UN was anything here but a flag painted in the sand; we had houses built before they had a pot to piss in anywhere on Mars, including all that disputed area in the south between the US and France.'

'Right, Arnie,' the boys all agreed.

'However,' Arnie said, 'there's the problem that those UN fruits control the waterways, and we got to have water; we need them for conveyance into and out of the settlement and for source of power and to drink and like now, like we're here bathing. I mean, those buggers can cut off our water any time; they've got us by the short hairs.'

He finished his shower and padded across the warm, wet tiles to get a towel from the attendant. Thinking about the UN made his stomach rumble, and his one-time duodenal ulcer began to burn way down in his left side, almost at the groin. Better get some breakfast, he realised.

When he had been dressed by the attendant, in his grey flannel trousers and T-shirt, soft leather boots, and nautical cap, he left the steam bath and crossed the corridor of the Union Hall to his dining-room, where Helio, his Bleekman cook, had his breakfast waiting. Shortly, he sat before a stack of hotcakes and bacon, coffee and a glass of orange juice, and the previous week's *New York Times*, the Sunday edition.

'Good morning, Mr Kott.' In answer to his button-pressing, a secretary from the pool had appeared, a girl he had never seen before. Not too good-looking, he decided after a brief glance; he returned to reading the newspaper. And calling him Mr Kott, too. He sipped his orange juice and read about a ship that had perished in space with all three hundred aboard killed. It was a Japanese merchantman carrying bicycles. That made him laugh. Bicycles in space, and all gone, now; too bad, because on a planet with little mass like Mars, where there was virtually no power source – except the sluggish canal system – and where even kerosene cost a fortune, bicycles were of great economic value. A man could pedal free of cost for hundreds of miles, right over the sand, too. The only people who used kerosene-powered turbine conveyances were vital functionaries, such as the repair and maintenance men, and of course important officials such as himself. There were public transports, of course, such as the tractor-buses which connected one settlement with the next and the outlying residential areas with the world at large ... but they ran irregularly, being dependent on shipments from Earth for their fuel. And personally speaking the buses gave him a case of claustrophobia, they moved so slow.

Reading the *New York Times* made him feel for a little while as if he were back Home again, in South Pasadena; his family had subscribed to the West Coast edition of the *Times*, and as a boy he remembered bringing it in from the mailbox, in from the street lined with apricot trees, the warm, smoggy little street of neat one-storey houses and parked cars and lawns tended from one weekend to the next without fail. It was the lawn, with all its equipment and medicines, that he missed most – the wheelbarrow of fertiliser, the new grass seed, the snippers, the poultry-netting fence in the early spring ... and always the sprinklers at work throughout the long summer, whenever the law allowed. Water shortage there, too. Once his Uncle Paul had been arrested for washing his car on a water-ration day.

Reading further in the paper he came upon an article about a reception at the White House for a Mrs Lizner who, as an official of the Birth Control Agency, had performed eight thousand therapeutic abortions and had thereby set an example for American womanhood. Kind of like a nurse, Arnie Kott decided. Noble occupation for females. He turned the page.

There, in big type, was a quarter-page ad which he himself had helped compose, a glowing come-on to get people to emigrate. Arnie sat back in his chair, folded the paper, felt deep pride as he studied the ad; it looked good, he decided. It would surely attract people, if they had any guts at all and a sincere desire for adventure, as the ad said.

The ad listed all the skills in demand on Mars, and it was a long list, excluding only canary raiser and proctologist, if that. It pointed out how hard it was now for a person with only a master's degree to get a job on Earth, and how on Mars there were good-paying jobs for people with only BAs.

That ought to get them, Arnie thought. He himself had emigrated due to his having only a BA. Every door had been shut to him, and then he had come to Mars as nothing but a union plumber, and within a few short years, look at him. On Earth, a plumber with only a BA would be raking up dead locusts in Africa as part of a US foreign aid work gang. In fact, his brother Phil was doing that right now; he had graduated from the University of California and had never had a chance to practise his profession, that of milk tester. In his class, over a hundred milk testers had been graduated, and for what? There were no opportunities on Earth. You have to come to Mars, Arnie said to himself. We can use you here. Look at the pokey cows on those dairy ranches outside of town. They could use some testing.

But the catch in the ad was simply that, once on Mars, the emigrant was guaranteed nothing, not even the certainty of being able to give up and go home; trips back were much more expensive, due to the inadequate field facilities. Certainly, he was guaranteed nothing in the way of employment. The fault lay with the big powers back Home, China and the US and Russia and West Germany. Instead of properly backing the development of the planets, they had turned their attention to further exploration. Their time and brains and money were all committed to the sidereal projects, such as that frigging flight to Centaurus, which had already wasted billions of dollars and man-hours. Arnie Kott could not see the sidereal projects for beans. Who wanted to take a four-year trip to another solar system which maybe wasn't even there?

And yet at the same time Arnie feared a change in the attitude of the great terrestrial powers. Suppose one morning they woke up and took a new look at the colonies on Mars and Venus? Suppose they eyed the ramshackle developments there and decided something should be done about them? In other words, what became of Arnie Kott when the Great Powers came to their senses? It was a thought to ponder.

However, the Great Powers showed no symptoms of rationality. Their obsessive competitiveness still governed them; right this moment they were locking horns, two light years away, to Arnie's relief.

Reading further in the paper, he came across a brief article having to do with a women's organisation in Berne, Switzerland, which had met to declare once more its anxiety about colonisation.

COLONIAL SAFETY COMMITTEE ALARMED OVER
CONDITIONS OF MARS LANDING FIELDS

The ladies, in a petition presented to the Colonial Department of the UN, had expressed once more their conviction that the fields on Mars at which ships from Earth landed were too remote from habitation and from the water system. Passengers in some cases had been required to trek over a hundred miles of wasteland, and these included women and children and old people. The Colonial Safety Committee wanted the UN to pass a regulation compelling ships to land at fields within twenty-five miles of a major (named) canal.

Do-gooders, Arnie Kott thought as he read the article. Probably not one of them has ever been off Earth; they just know what somebody wrote home in a letter, some aunt retiring to Mars on a pension, living on free UN land and naturally griping. And of course they also depended on their member in residence on Mars, a certain Mrs Anne Esterhazy; she circulated a mimeographed newsletter to other public-spirited ladies throughout the settlements. Arnie received and read her newsletter, *The Auditor Speaks Back*, a title at which he gagged. He gagged, too, at the one- and two-line squibs inserted between longer articles:

> Pray for potable purification! ! Contact
> colony charismatic councillors and witness
> for water filtration we can be proud of!

He could hardly make out the meaning of some of the *Auditor Speaks Back* articles, they were phrased in such special jargon. But evidently the newsletter had attracted an audience of devoted women who grimly took each item to heart and acted out the deeds asked of them. Right now they were undoubtedly complaining, along with the Colonial Safety Committee back on Earth, about the hazardous distances separating most of the landing fields on Mars from water sources and human habitation. They were doing their part in one of the many great fights, and in this particular case, Arnie Kott had managed to gain control of his nausea. For of the twenty or so landing fields on Mars, only one lay within twenty-five miles of a major canal, and that was Samuel Gompers Field, which served his own settlement. If by some chance the pressure of the Colonial Safety Committee was effective, then all incoming passenger ships from Earth would have to land at Arnie Kott's field, with the revenue received going to his settlement.

It was far from accidental that Mrs Esterhazy and her newsletter and organisation on Earth were advocating a cause which would be of economic value to Arnie. Anne Esterhazy was Arnie's ex-wife. They were still good friends, and still owned jointly a number of economic ventures which they had founded or bought into during their marriage. On a number of levels they still worked together, even though on a strictly personal basis they had no common ground whatsoever. He found her aggressive, domineering, overly masculine, a tall and bony female with a long stride, wearing low-heeled shoes and a tweed coat and dark glasses, a huge leather purse slung from a strap over her shoulder ... but she was shrewd and intelligent and a natural executive. As long as he did not have to see her outside of the business context, he could get along with her.

The fact that Anne Esterhazy had once been his wife and that they still had financial ties was not well known. When he wanted to get in touch with her he did not dictate a letter to one of the settlement's stenographers; instead he used a little encoding dictation machine which he kept in his desk, sending the reel of tape over to her by special messenger. The messenger dropped off the tape at an art object shop which Anne owned over in the Israeli settlement, and her answer, if any, was deposited the same way at the office of a cement and gravel works on the Bernard Baruch Canal which belonged to Arnie's brother-in-law, Ed Rockingham, his sister's husband.

A year ago, when Ed Rockingham had built a house for himself and Patricia and their three children, he had acquired the unacquirable: his own canal. He had had it built, in open violation of the law, for his private use, and it drew water from the great common network. Even Arnie had been outraged. But there had been no prosecution, and today the canal, modestly named after Rockingham's eldest child, carried water eighty miles out into the desert, so that Pat Rockingham could live in a lovely spot and have a lawn, a swimming pool, and a fully irrigated flower garden. She grew especially large camellia bushes, which were the only ones that had survived the transplanting to Mars. All during the day, sprinklers revolved and sprayed her bushes, keeping them from drying up and dying.

Twelve huge camellia bushes seemed to Arnie Kott an ostentation. He did not get along very well with his sister or Ed Rockingham. What had they come to Mars for? he asked himself. To live, at incredible expense and effort, as much as possible as they had back Home on Earth. To him it was absurd. Why not remain on Earth? Mars, for Arnie, was a new place, and it meant a new life, lived with a new style. He and the other settlers, both big and small, had made in their time on Mars countless minute adjustments in a process of adaptation through so many stages that they had in fact evolved; they were new creatures, now. Their children born on Mars started out like this, novel and peculiar, in some respects enigmatic to their parents. Two of his own boys – his and Anne's – now lived in a settlement camp at the outskirts of Lewistown. When he visited them he could not make them out; they looked towards him with bleak eyes, as if waiting for him to go away. As near as he could tell, the boys had no sense of humour. And yet they were sensitive; they could talk forever about animals and plants, the landscape itself. Both boys had pets, Martian critters that struck him as horrid: praying mantis types of bugs, as large as donkeys. The damn things were called *boxers*, because they were often seen propped up erect and squaring off at one another in a ritual battle which generally ended up with one killing and eating the other. Bert and Ned had gotten their pet boxers trained to do manual chores of a low calibre, and not to eat each other. And the things were their companions; children on Mars were lonely, partly because there were still so few of them and partly because ... Arnie did not know. The children had a large-eyed, haunted look, as if they were starved for something as yet invisible. They tended to become reclusive, if given half a chance, wandering off to poke about in the wastelands. What they brought back was worthless, to themselves and to the settlements, a few bones or relics of the old nigger civilisation, perhaps. When he flew by 'copter, Arnie always spotted some isolated children, one here and another there, toiling away out in the desert, scratching at the rock and sand as if trying vaguely to pry up the surface of Mars and get underneath...

Unlocking the bottom drawer of his desk, Arnie got out the little battery-powered encoding dictation machine and set it up for use. Into it he said, 'Anne, I'd like to meet with you and talk. That committee has too many women on it, and it's going the wrong way. For example, the last ad in the *Times* worries me because – ' He broke off, for the encoding machine had groaned to a stop. He poked at it, and the reels turned slowly and then once more settled back into silence.

Thought it was fixed, Arnie thought angrily. Can't those jerks fix nothing? Maybe he would have to go to the black market and buy, at an enormous price, another. He winced at the thought.

The not-too-good-looking secretary from the pool, who had been sitting quietly across from him waiting, now responded to his nod. She produced her pencil and pad and began as he dictated.

'Usually,' Arnie Kott said, 'I can understand how hard it is to keep things running, what with no parts hardly, and the way the local weather affects metal and wiring. However, I'm fed up with asking for competent repair service on a vital item like my encoding machine. I just got to have it, that's all. So if you guys can't keep it working, I'm going to disband you and withdraw your franchise to practise the craft of repairing within the settlement, and I'll rely on outside service for our maintenance.' He nodded once more, and the girl ceased writing.

'Shall I take the encoder over to the repair department, Mr Kott?' she asked. 'I'd be happy to, sir.'

'Naw,' Arnie grumbled. 'Just run along.'

As she departed, Arnie once more picked up his *New York Times* and again read. Back home on Earth you could buy a new encoder for almost nothing; in fact, back home you could – hell. Look at the stuff being advertised ... from old Roman coins to fur coats to camping equipment to diamonds to rocket ships to crabgrass poison. Jeez!

However, his immediate problem was how to contact his ex-wife without the use of his encoder. Maybe I can just drop by and see her, Arnie said to himself. Good excuse to get out of the office.

He picked up the telephone and called for a 'copter to be made ready up above him on the roof of the Union Hall, and then he finished off the remains of his breakfast, wiped his mouth hurriedly, and set off for the elevator.

'Hi, Arnie,' the 'copter pilot greeted him, a pleasant-faced young man from the pilot pool.

'Hi, my boy,' Arnie said, as the pilot assisted him into the special leather seat which he had had made at the settlement's fabric and upholstery shop. As the pilot got into the seat ahead of him Arnie leaned back comfortably, crossed his legs, and said, 'Now you just take off and I'll direct you in flight. And take it easy because I'm in no hurry. It looks like a nice day.'

'Real nice day,' the pilot said, as the blades of the 'copter began to rotate. 'Except for that haze over around the FDR Range.'

They had hardly gotten into the air when the 'copter's loudspeaker came on. 'Emergency announcement. There is a small party of Bleekmen out on the open desert at gyrocompass point 4.65003 dying from exposure and lack of water.

Ships north of Lewistown are instructed to direct their flights to that point with all possible speed and give assistance. United Nations law requires all commercial and private ships to respond.' The announcement was repeated in the crisp voice of the UN announcer, speaking from the UN transmitter on the artificial satellite somewhere overhead.

Feeling the 'copter alter its course, Arnie said, 'Aw, come on, my boy.'

'I have to respond, sir,' the pilot said. 'It's the law.'

Chrissake, Arnie thought with disgust. He made a mental note to have the boy sacked or at least suspended as soon as they got back from their trip.

Now they were above the desert, moving at good speed towards the intersect which the UN announcer had given. Bleekmen niggers, Arnie thought. We have to drop everything we're doing to bail them out, the damn fools – can't they trot across their own desert? Haven't they been doing it without our help for five thousand years?

As Jack Bohlen started to lower his Yee Company repairship towards McAuliff's dairy ranch below, he heard the UN announcer come on with the emergency notification, the like of which Bohlen had heard many times before and which never failed to chill him.

'... Party of Bleekmen out on the open desert,' the matter- of-fact voice declared. '... Dying from exposure and lack of water. Ships north of Lewistown –'

I've got it, Jack Bohlen said to himself. He cut his mike on and said, 'Yee Company repairship close by gyrocompass point 4.65003, ready to respond at once. Should reach them in two or three minutes.' He swung his 'copter south, away from McAuliff's ranch, getting a golden-moment sort of satisfaction at the thought of McAuliff's indignation right now as he saw the 'copter swing away and guessed the reason. No one had less use for the Bleekmen than did the big ranchers; the poverty-stricken, nomadic natives were constantly showing up at the ranches for food, water, medical help, and sometimes just a plain old-fashioned handout, and nothing seemed to madden the prosperous dairymen more than to be used by the creatures whose land they had appropriated.

Another 'copter was responding, now. The pilot was saying, 'I am just outside Lewistown at gyrocompass point 4.78995 and will respond as soon as possible. I have rations aboard including fifty gallons of water.' He gave his identification and then rang off.

The dairy ranch with its cows fell away to the north, and Jack Bohlen was gazing intently down at the open desert once more, seeking to catch sight of the party of Bleekmen. Sure enough, there they were. Five of them, in the shade cast by a small hill of stone. They were not moving. Possibly they were already dead. The UN satellite, in its swing across the sky, had discovered them, and yet it could not help them. Their mentors were powerless. And we who can help them – what do we care? Jack thought. The Bleekmen were dying out anyhow, the remnants getting more tattered and despairing every year. They were wards of the UN, protected by them. Some protection, Jack thought.

But what could be done for a waning race? Time had run out for the natives of Mars long before the first Soviet ship had appeared in the sky with its television cameras grinding away, back in the '60s. No human group had conspired to exterminate them; it had not been necessary. And anyhow they had been a vast curiosity, at first. Here was a discovery worth the billions spent in the task of reaching Mars. Here was an extraterrestrial race.

He landed the 'copter on the flat sand close by the party of Bleekmen, switched off the blades, opened the door, and stepped out.

The hot morning sun beat down on him as he walked across the sand towards the unmoving Bleekmen. They were alive; they had their eyes open and were watching him.

'Rains are falling from me on to your valuable persons,' he called to them, the proper Bleekman greeting in the Bleeky dialect.

Close to them now he saw that the party consisted of one wrinkled old couple, a young male and female, no doubt husband and wife, and their infant. A family, obviously, which had set out across the desert alone on foot, probably seeking water or food; perhaps the oasis at which they had been subsisting had dried up. It was typical of the plight of the Bleekmen, this conclusion to their trek. Here they lay, unable to go on any farther; they had withered away to something resembling heaps of dried vegetable matter and they would have died soon had not the UN satellite spotted them.

Rising to his feet slowly, the young Bleekman male genuflected and said in a wavering, frail voice, 'The rains falling from your wonderful presence envigour and restore us, Mister.'

Jack Bohlen tossed his canteen to the young Bleekman, who at once knelt down, unscrewed the cap, and gave it to the supine elderly couple. The old lady seized it and drank from it.

The change in her came at once. She seemed to swell back into life, to change from the muddy grey colour of death before his eyes.

'May we fill our eggshells?' the young Bleekman male asked Jack. Lying upright on the sand were several paka eggs, pale hollow shells which Jack saw were completely empty. The Bleekmen transported water in these shells; their technical ability was so slight that they did not even possess clay pots. And yet, he reflected, their ancestors had constructed the great canal system.

'Sure,' he said. 'There's another ship coming with plenty of water.' He went back to his 'copter and got his lunch pail; returning with it, he handed it to the Bleekman male. 'Food,' he explained. As if they didn't know. Already the elderly couple were on their feet, tottering up with their hands stretched out.

Behind Jack, the roar of a second 'copter grew louder. It was landing, a big two-person 'copter that now coasted up and halted, its blades slowly spinning.

The pilot called down, 'Do you need me? If not, I'll go on.'

'I don't have much water for them,' Jack said.

'OK,' the pilot said, and switched off his blades. He hopped out, lugging a five-gallon can. 'They can have this.'

Together, Jack and the pilot stood watching the Bleekman filling their eggshells from the can of water. Their possessions were not many – a quiver of poisoned arrows, an animal hide for each of them; the two women had their pounding blocks, their sole possessions of value: without the blocks they were not fit women, for on them they prepared either meat or grain, whatever food their hunt might bring. And they had a few cigarettes.

'My passenger,' the young pilot said in a low voice in Jack's ear, 'isn't too keen about the UN being able to compel us to stop like this. But what he doesn't realise is they've got that satellite up there and they can see if you fail to stop. And it's a hell of a big fine.'

Jack turned and looked up into the parked 'copter. He saw seated inside it a heavy-set man with a bald head, a well-fed, self-satisfied-looking man who gazed out sourly, paying no attention to the five Bleekmen.

'You have to comply with the law,' the pilot said in a defensive voice. 'It'd be me who they'd sock with the fine.'

Walking over to the ship, Jack called up to the big bald-headed man seated within, 'Doesn't it make you feel good to know you saved the lives of five people?'

The bald-headed man looked down at him and said, 'Five niggers, you mean. I don't call that saving five people. Do you?'

'Yeah, I do,' Jack said. 'And I intend to continue doing so.'

'Go ahead, call it that,' the bald-headed man said. Flushing, he glanced over at Jack's 'copter, read the markings on it. 'See where it gets you.'

Coming over beside Jack, the young pilot said hurriedly, 'That's Arnie you're talking to. Arnie Kott.' He called up, 'We can leave now, Arnie.' Climbing up, the pilot disappeared inside the 'copter, and once more the blades began to turn.

The 'copter rose into the air, leaving Jack standing alone by the five Bleekmen. They had now finished drinking and were eating from the lunch pail which he had given them. The empty water can lay off to one side. The paka eggshells had been filled and were now stoppered. The Bleekmen did not glance up as the 'copter left. They paid no attention to Jack, either; they murmured among themselves in their dialect.

'What's your destination?' Jack asked them.

The young Bleekman named an oasis very far to the south.

'You think you can make it?' Jack asked. He pointed to the old couple. 'Can they?'

'Yes, Mister,' the young Bleekman answered. 'We can make it now, with the food and water yourself and the other Mister gave us.'

I wonder if they can, Jack said to himself. Naturally they'd say it, even if they knew it wasn't possible. Racial pride, I guess.

'Mister,' the young Bleekman said, 'we have a present for you because you stopped.' He held out something to Jack.

Their possessions were so meagre that he could not believe they had anything to spare. He held his hand out, however, and the young Bleekman put something small and cold into it, a dark, wrinkled, dried bit of substance that looked to Jack like a section of tree root.

'It is a water witch,' the Bleekman said. 'Mister, it will bring you water, the source of life, any time you need.'

'It didn't help you, did it?' Jack said.

With a sly smile the young Bleekman said, 'Mister, it helped; it brought you.'

'What'll you do without it?' Jack asked.

'We have another. Mister, we fashion water witches.' The young Bleekman pointed to the old couple. 'They are authorities.'

More carefully examining the water witch, Jack saw that it had a face and vague limbs. It was mummified, once a living creature of some sort; he made out its drawn-up legs, its ears ... he shivered. The face was oddly human, a wizened, suffering face, as if it had been killed while crying out.

'How does it work?' he asked the young Bleekman.

'Formerly, when one wanted water, one pissed on the water witch, and she came to life. Now we do not do that, Mister; we have learned from you Misters that to piss is wrong. So we spit on her instead, and she hears that, too, almost as well. It wakes her, and she opens her eyes and looks around, and then she opens her mouth and calls the water to her. As she did with you, Mister, and that other Mister, the big one who sat and did not come down, the Mister with no hair on his head.'

'That Mister is a powerful Mister,' Jack said. 'He is monarch of the plumbers' union settlement, and he owns all of Lewistown.'

'That may be,' the young Bleekman said. 'If so, we will not stop at Lewistown, because we could see that the Mister with no hair did not like us. We did not give him a water witch in return for his water, because he did not want to give us water; his heart was not with him in that deed, it came from his hands only.'

Jack said goodbye to the Bleekman and got back into his 'copter. A moment later he was ascending; below him, the Bleekmen waved solemnly.

I'll give the water witch to David, he decided. When I get home at the end of the week. He can piss on it or spit on it, whichever he prefers, to his heart's content.

THE HUMAN EXPLORER

(2004)

WHY SENDING ROBOTS IS NOT ENOUGH

Robert Zubrin

Robert Zubrin is an American aerospace engineer and advocate for the exploration and colonization of Mars. Born in New York City in 1952, he is the creator of the 'Mars Direct' concept, along with Martin Marietta and David Baker. The concept seeks to propose ways to significantly cut the cost of humans exploring Mars. An expanded version of the idea was published by Zubrin in his 1996 book *The Case For Mars*. In 1998, Zubrin and others founded The Mars Society, a non-profit corporation dedicated to the human exploration and settlement of Mars.

President Bush has called for the human exploration of space. His vision changes the orientation of the American manned spaceflight program from one of observing and gathering data on the human experience in space—the medical effects of zero-g and so forth—to a program of going into space to travel across it, to explore worlds.

The president's plan is a step in the right direction, because it gives NASA a much-needed goal. The main reason why NASA's level of achievement in the last three decades has paled in comparison to NASA's level of achievement from 1961 to 1973 is that, in that earlier period, President Kennedy set a clear goal: Reach the moon within a decade. With that specific mission, NASA did Mercury, Gemini, Apollo, and Skylab. It did a host of robotic missions. It developed virtually all the space technologies that we have today, and all the major American space institutions.

But without a goal, NASA's level of achievement has declined, even though NASA's average budget in the 1990s was similar (in inflation-adjusted dollars) to the average NASA budget from 1961 to 1973. The problem has not been a lack of money but an absence of purpose.

So the question arises: What is the right goal for NASA? The right goal is Mars for the following three reasons: Mars is where the science is, Mars is where the challenge is, and Mars is where the future is.

First, the science. A multitude of issues in different scientific disciplines—planetary geology, meteorology, seismology, and other fields—could be answered by putting humans on Mars. But the central questions involve life: Was there, or is there, life on Mars? And if so, what is the nature of that life?

To uncover whether life evolved from chemistry on Mars is the critical experiment for knowing whether the evolution and development of life from chemistry is a general phenomenon in the universe, wherever appropriate physical and chemical conditions exist. Mars appears to have had liquid water for a significant period of time—longer than it took for life to appear in the fossil record on Earth after there was liquid water here. So it's the Rosetta Stone for letting us know the answer to the question of life.

It's also the key for discovering whether all life has to have the same form as life on Earth, since all life on Earth is the same at the biochemical level. It all uses the same RNA and DNA methods of replicating information and so forth—but perhaps it does not have to. Going to Mars can help us find out.

On Earth, fossil-hunting involves hiking long distances through unimproved terrain, and climbing up steep hillsides or cliffs. It involves digging and pickax work, as well as delicate handiwork, like carefully splitting open shales edgewise to reveal the fossils that have been trapped between the pages of rocks pasted together. This is far beyond the ability of robotic rovers like Spirit and Opportunity. If you took one of these robots to a paleontological dig on Earth, the researchers might use it as a platform for putting coffee cups on.

It is true that wearing a spacesuit greatly reduces your situational awareness. Obviously you don't have a sense of smell or a direct sense of touch. But you do retain the ability to pick up samples and manipulate them, and the ability to break rocks open and look inside them. You are able to walk back and forth, looking down at rocks and taking in with your eyes the equivalent of millions of high resolution images.

The human explorer can follow up on all sorts of intuitive clues and observations. Out of the thousands of rocks he has glanced at and the hundreds he has looked at more closely—perhaps he brings ten samples back into the habitat. There he can look at them with a hand-lens; he can thin-section them; he can examine them under a microscope.

Two hundred years after Lewis and Clark, there is not a robot on this planet that you can send to the grocery store to pick up a bag of unbruised apples. If they can't do a trip to the grocery store, how can they explore a planet? How can robots match the intuition, versatility, ingenuity, and common sense of the human explorer?

Now, I'm not putting down robots. It is excellent and important to do robotic missions. But a robot explorer on the surface of a distant planet simply cannot duplicate what a human explorer could do. To find out whether there is life on Mars, we're going to have to set up drilling rigs, drill down into the ground, sample the water, bring it into the lab, and examine the samples under a battery of tests, with a scientist who can react flexibly to the data, in consultation with other scientists back on Earth.

Remember, we've sent robots to Mars to search for life before. In 1976, we sent Viking and did four tests on the Martian soil to determine if there was life. Three suggested there might be life; the other was negative. The meaning of these experiments is still being debated. Viking asked Mars, "Do you have life?" Mars said, "Maybe. Please rephrase the question." If there were humans there, they could immediately have rephrased the question by performing additional experiments. This is the superiority of human exploration.

The second reason to go to Mars is the challenge. It is the chance to do something heroic, to advance humanity on the frontier. A humans-to-Mars program would be a challenge to our entire society. In particular, it would be an inspiration to the next generation: "Learn your science and you can pioneer a new world. Develop your mind and you can be a hero for humanity—doing something that has never been done before, seeing things that no one has seen before, building where no one has built before."

This is the challenge that the youth of my generation got from the Apollo program. As a result of that challenge, I became an engineer. As a result of that challenge, the number of scientists and engineering graduates in this country doubled at every level: high school, college, Ph.D.

And what did those people end up doing? Some went into aerospace, but most of them went off into other scientific ventures: they engaged in medical research, they built Silicon Valley, they created the economic boom of the 1990s. Those 40-year-old techno-nerd billionaires of the 1990s were the 12-year-old boy scientists of the 1960s. It's an investment: If we go to Mars, we'll someday get the payoff that comes from challenging people in a serious way, and by being a society that values great scientific and human achievements.

The final reason to go to Mars is the issue of the future. Imagine you lived 50,000 years ago in Kenya, along with the rest of the human race, and received a proposal from someone who thought humans should colonize Europe or Asia. The skeptics would have said: "Those places are impossible to live in. It's much too cold." If they had robotic probes, they might have sent them to confirm these assumptions with more precision: "Our robotic probes show you could not survive a single winter night in Europe."

But people were able to colonize Europe with the aid of technology: clothing, houses, fire. It is on the basis of our technological ingenuity that humans have left our natural habitat, the Kenyan Rift Valley, and transformed ourselves into a global species with hundreds of nations, languages, and cultural traditions. There has been a vastly richer human experience as a result of the human willingness to leave the known in order to explore and master the unknown. And this is the challenge that Mars holds for us today.

Mars is not just an object of scientific inquiry. It is a world. It is a planet with a surface area equal to all the continents of the Earth put together, with all the resources needed to support not only life but technological civilization, should we choose to exercise our creativity sufficiently to make that possible.

If we do what we can in our day, which is establish that first human foothold on Mars, then five hundred years from now there will be a new branch of human civilization living there. Perhaps many new branches of human civilization will flourish on Mars, with their own cultures, their own languages yet unspoken, their own novel ideas on human social organization, their own traditions of heroic deeds, and their own manifest contributions to technology and invention.

And that is something wonderful. That is something enormously valuable. I wonder if we can even put a price tag on helping to give birth to a new branch of human civilization, one that contributes in unimaginable ways to human progress and the human story.

And not only that, but a branch of civilization whose development shows us that we have the capability to do such things, the capability to engage in yet greater ventures, more daring ventures, further out, toward an unlimited future. And that is the reason why humans should go into space.

WHEN WE DIE ON MARS (2015)

Cassandra Khaw

Cassandra Khaw is a Malaysian author of horror and science fiction, and a scriptwriter for games. Born in 1984, she has had short stories published in *Clarkesworld*, *Fireside Fiction*, *Uncanny* and *Shimmer*. Khaw's novels include *Bearly a Lady*, *A Song for Quiet* and *Nothing But Blackened Teeth*.

"You're all going to die on Mars." This is the first thing he tells us, voice plain, tone sterile. Commander Chien, we eventually learn, is a man not predisposed towards sentimentality.

We stand twelve abreast, six rows deep, bones easy, bodies whetted on a checklist of training regimes. Our answer, military-crisp, converges into a single noise: "Yessir!"

"If at any point before launch, you feel that you cannot commit to this mission: *leave*," Commander Chien stalks our perimeter, gait impossibly supple even with the prosthetic left leg. He bears its presence like a medal, gilled and gleaming with wires, undisguised by fabric. "If at any point you feel like you might jeopardize your comrades: *leave*."

Commander Chien enumerates clauses and conditions without variance in cadence, his face cold and impersonal as the flat of a bayonet. He goes on for minutes, for hours, for seconds, reciting a lexicon of possibilities, an astronautical doomsayer.

At the end of it, there is only silence, viscous, thick as want. No one walks out. We know why we are there, each and every last one of us: to make Mars habitable, hospitable, an asylum for our children so they won't have to die choking on the poison of their inheritance.

Faith, however, is never easy.

It is amoebic, seasonal, vulnerable to circumstance. Faith sways, faith cracks. There are a thousand ways for it to die, to metamorphosize from *yes* to *no, no*, I could *never*.

Gerald and Godfrey go first, both blondes, family men with everything to lose and even more to gain. Gerald leaves after a call with his wife, a poltergeist in the night, clattering with stillborn ambition; Godfrey after witnessing the birth of his daughter third-hand.

We make him name her 'Chance' as a gentle joke, a nod to her significance. Because of her, he'll grow old breathing love instead of red dust. She is his second chance, we laugh, and Godfrey smiles through the salt in his gaze.

"When we die on Mars," I say, as I nestle my hand in the continent of his palm, my heart breaking. "Tell her a fairy tale of our lives. Tell her about how twelve people fought a planet so that billions could live."

His lips twitch. "I will."

He leaves in the morning before any of us wake, his bunk so immaculately made that you would have doubted he was ever there at all.

Five months pass. Ten. Fifteen.

Our lives are ascetic, governed by schedules unerring as the sun's rotation. When we are not honing our trade, we are adopting new ones, exchanging knowledge like cosmic relics under a sky of black metal. Halogen-lit, our existence in the bunker is not unpleasant, only cold, both in fact and in metaphor. Nothing will ever inoculate us against Mars' climate, but we can be taught to endure.

Similarly, chemicals can only do so much to quiet the heart, to beguile it into believing that this is okay, this will be okay. The years on Mars will erode our passion for galaxies, will flense us of wonder, sparing only the longing for affection. When that happens, we must be prepared, must keep strong as loneliness tautens like a noose around the throat.

The understanding of that eventuality weighs hard.

A pair of Thai women, sisters in bearing and intellect if not in blood, depart in the second year. They're followed by an Englishman, rose-cheeked and inexplicably rotund despite fastidious exercise; a willowy boy with deep, memory-bruised eyes; a girl whose real name we never learn, but who sings us to dreaming each night; a mother, a father, a child, a person.

One by one, our group thins, until all that remains is twelve; the last, the best, the most desperate Earth has to give.

"Your turn, Anna. Would you rather give a blowjob to a syphilis-riddled dead billionaire, or eat a kilogram of maggot-infested testicles?"

"Jesus, man!" Hannah, a pretty Latina with double PhDs in astrophysics and aeronautical engineering, shrieks her glee. "What is *wrong* with that head of yours?"

"Nothing!" Randy counters, oil slick smooth. "The medical degree's the problem! Look at enough dead bodies, and everything stops being taboo. I—"

I interrupt, a coy smile slotted in place. "Maggot-infested testicles. Easy."

Both Hannah and Randy guffaw.

"You know syphilis got a cure, right? Why'd you gotta—"

"They're not so bad when you deep-fry them with maple syrup and crushed nuts. Pinch of paprika, dash of star anise. Mmm." It is a fabrication, stitched together from memories of a smoldering New Penang, but I won't tell them that. They deserve this happiness, this harmless grotesquerie, small as it might be.

Hannah jabs a finger in her open mouth, makes a retching noise so absurd that Randy dissolves into laughter. This time, I join in, letting the joy sink down, sink *deep*, catch its teeth on all the hurt snagged between my ribs and drag it all back out. The sound feels good in my lungs, feels *clean*.

A door dilates. Pressurized air hisses out, and Hotaru's silhouette pours in. Of the twelve of us, she's the oldest, a Japanese woman bordering on frail, skin latticed by wrinkles and wartime scars, nose broken so many times that it's just flesh now, shapeless, portentous. When she speaks, everyone listens.

"Everything alright in here?" Her accent rolls, musical and mostly upper-class English save for the way it latches on the 'r's and pulls them stiff.

"Yeah." Randy, long and elegant as his battered old violin, glides out of his seat and stretches. "We're just waiting for Hannah here to check the back-up flight system. Ground control said they found some discrepancies and—"

"You suddenly the medic *and* the engineer, Randy?" Hannah cranes both eyebrows upwards, mouth pinching with mock displeasure. "You want to fly the ship? I'll go sit in the infirmary, if you like. Check out your supply of druuuuuugs."

Randy doesn't quite rise to the bait, only snorts, a grin plucking at the seams of his mouth. He throttles his amusement in an exaggerated cough, and I look away, smiling into the glow of my screen.

Hotaru seems less taken with the exchange, small hands locking behind her back. She waits until we've lapsed into a natural quiet before she speaks again, every word enunciated with a schoolmaster's care.

"If everything is in order, I'll tell Commander Chien that we are prepared to leave." Hotaru's eyes patrol the room, find our gazes one by one. After three years together, it takes no effort at all to read the question buried between each syllable.

"Sounds good," Hannah says, even though the affirmation husks her voice. Her fingers climb to an old-fashioned locket atop her breastbone.

Randy drapes a hand over her shoulder. "Same here."

"Here too," I reply, and try not to linger too long on the ache that tendrils through my chest, a cancer blooming in the dark of artery and tendon. Familial guilt is sometimes heavier than the weight of a rotting world.

Hotaru nods. Like the commander, she will not waste breath on niceties, an efficiency of character I'm learning too well. When your lifespan can be valued in handfuls, every expenditure of time becomes cause for careful evaluation, every act of companionship a hair's width from squander.

"I'll send word then. I imagine we'll have about forty-eight hours to make final preparations," Hotaru pads to the door. She turns at the last instant, skims a look over the precipice of a shoulder and for a moment, I see the woman beneath the skin of legend, stooped from memory and so very tired, a mirror of a mother I'd not seen for decades. "Don't waste them."

"Anna, you awake?"

I yawn into a palm and roll on my side, blink into the phosphor-edged penumbra. "I don't know. Is Malik snoring?"

Hannah whispers a gauzy, sympathetic laugh. She props herself on an elbow, face barely visible, a landscape of thoughtful lines.

"What's up?"

A flash of teeth. She doesn't answer immediately. Instead, she loops a curl about a finger, winds it tight. I wait. There's no rushing Hannah. Under the street-sharpened exterior, she's nervy as an alley cat, quick to flee, to hide behind laughter and slight-of-speech.

"Do you think the radio signal is any good in Mars?"

I shrug. "Not sure if it matters. With the communication delay, we're—"

"—talking about response times of between four to twenty-four minutes. I know, tia. I know," Hannah's voice ebbs. She winds upright, legs crossed, eyes fixed on a place nothing but regret can reach.

An almost-silence; Malik's snoring moving into labored diminuendo.

"Not sure if I ever told ya, but I got a daughter somewhere." Hannah breathes out, every word shrapnel. "Was sixteen when I had her. Way too young. The babydaddy skipped out in the first trimester. He left so fast, you could see dust trails."

A whine of strained laughter, dangerously close to grief, before she hacks it short, swallowing it like a gobbet of bad news.

"My parents wanted me to abort. Said it was for the best. 'Hannah,' they told me. 'This world don't have no God to judge you for choosing reason over guilt.' I refused. I don't even remember why. It's been that long. All I remember was that I wanted to give her a chance out there."

"Did your parents object?" I slink from my bed, cross the ten feet between us to close an arm about her shoulders, press a kiss into the hollow of her cheek. An old sadness reassembles inside me, a thought embedded in biology, not rationality. It's been years since I've spoken to my family. *Isn't it time*, asks a voice that is almost mine, *for you to forgive them*?

Hannah nestles into me and my body bends in reply, curling until we're fitted jigsaw-snug, twins in the womb. "Nah. They weren't that sort. Once I made it clear that it was what I wanted, they went in hundred-and-fifty percent."

I stroke her hair, a storm of dark coils smelling of eucalyptus and mint, a scent that won't keep on Mars.

"They put me into home-schooling, rubbed my feet. Did everything they could to make it easier for me. Nine months later, I had a beautiful little girl. She was perfect, Anna. Ten tiny little toes, cat-gold eyes, hair so soft it was like cotton candy."

"No fingers?"

Hannah pounds knuckles against my sternum. "Very funny."

I trail my fingers over the back of her hand and she lets her fist open, palm warm as we lock grips. "Then what happened?"

"We put her up for adoption."

"And?"

"That was it."

The lie throbs in the air, waiting absolution, release.

"I wish..." Hannah begins, careful, almost too soft to hear, her pulse narrowing. "I wish, sometimes, that I didn't. I mean, kids were never part of my grand plan. But now that we're going? I wonder."

"You could try to call her?"

"How? My parents are dead. I don't know even where to start. It's fine, though," Hannah extracts herself from my arms, pulls her knees close to her chest. There's a new fierceness in her voice, edged both ways, daring me to pry, daring herself to open up. "They told me she went to a good home, a *great* home. That was all I wanted to know then. That's all I need to know now. But."

"Yeah," I don't touch her. Not all places are intended for company. Some agonies you chart alone, walking the length of them until you've domesticated every contour and twinge.

Hannah nods, a jerky little motion, the only one she allows herself. We say nothing, finding instead a noiselessness to share. It is many long minutes before she tips herself backwards and pillows her head on my lap, an arm looping about my hips.

"Stay with me, tia?" Hannah asks and briefly, vividly, I glimpse the sister I'd long excised from daily thought.

"Only if I get a backrub in the morning," I reply, distractedly, drawing circles across her shoulder blades. In my head, a line from a Todd Kern song palpitates on repeat: you can always go home. It could be so easy, so simple. Forgive. Forget.

A tremor undulates through the column of her spine. Laughter or sobs, I can't tell which. "Deal."

"You did what now?" Randy's voice quivers an octave above normalcy, one bad joke away from earnest hysterics.

"I mooned my sister's ex-husband."

"*Why?*"

The shrug in Tuma's rich tenor is almost palpable, like muscles striving under skin. It is also anomalous, out-of-place in a young biologist better remembered for his ponderance than his sense of irreverence. "Why not?"

As expected, Randy cracks up, his laughter melodious, a thing I wish I could scoop into a Petri dish and let grow. I can imagine him in another life, a bluesman with a thimble of whiskey and a room full of worshippers, his eyes alive with their love.

I shake my head, return my attention to the spreadsheets of numbers imprinted in green on my terminal, calculations congregating thick as nebulas. In the corner, a notification pulsates. I ignore it.

"Hi."

We look up as one, fingers retracting from keyboards, faces from screens, to see Stefan's hound-dog frame limned in the doorway, a duffel balanced on one slim shoulder.

"Productive trip?" Tuma asks, swinging around in his chair.

Stefan nods, dislodging his luggage into a pile atop the floor before he drops into an open seat, his face unburdened of ghosts. Not all of them, but enough. "Yeah."

"Your brother finally see the light?" Randy quips, a remark that earns him a fusillade of dirty looks.

"Not exactly. He still thinks we're going against God's will." His eyes shine, illuminated by something sweet. "But he wishes us well. He's happy for me."

"Despite going against God's will?"

Stefan heaves a shrug, mouth curved with secrets, all of them good. "Despite going against God's will."

No one presses for data. Three years teaches you a lot about what a person will allow. From time to time, however, someone makes an excuse to rise, to graze past Stefan and brush fingertips against shoulder or arm, as though contact is enough to transmit a monk's benedictions from brothers to stranger.

On my screen, the icon continues to flash, demanding acknowledgment. Footsteps, like rainfall on metallic tiling. The weight of Randy's arm settles about my shoulders, a barrier against the past.

"You not going to answer that?"

"No." I exhale, hard.

"Why not?"

Because love doesn't grant the right to forgiveness. "Same reason as I said last time."

"You could do like Tuma."

"I'd rather not."

"And why's that?"

"Because screen-capture technology exists," I shoot, hoping that my voice doesn't shake too much, hoping that humor might deflect his curiosity.

And it does. His laugh ricochets through the chamber again, warm, warm, warm. People tilt sly glances over their shoulders. Hannah punches Tuma in the arm, who only chuckles in return, his eyes lidded with delight. When he, with uncharacteristic brazenness, begins expounding on the virtues of his posterior, Randy's laughter becomes epidemic, bouncing from throat to throat. If the sound is a little raw, a little ragged, no one comments. In twelve hours, we give up this planet entirely.

I push from my seat as the sound climbs into a frenzy, and use the diversion to slip out.

In the distance, Hannah's voice, low and thick with aching, echoes, riding that knife-edge between rapture and hurt.

"Henrietta? That's what they're calling you?"

"After my maternal grandmother." A tinny voice, distorted by poor equipment, accent Mid-Western. "Well. You know what I mean."

"Grade school must have been an arena then, chica."

"You have no idea."

I walk into the sleeping hall to see Hannah backlit by a Macbook, its display holding the face of a younger woman, not much older than her teens. Henrietta is paler than her mother, her hair artificially lightened, but she shares the same structural elegance, the same bones.

"I'm really, really glad I got to talk to you," Henrietta declares, after their laughter dims into smiles.

"I'm just happy you don't hate me."

"My biological mother's a literal superhero traveling the universe to save mankind. What's there to hate?" A beat. Henrietta's eyes flick up, over Hannah's shoulder. "Uhm. I think you have company."

The older woman turns slightly, just a glance, before she reverts her attention to the screen. "Yeah. I—"

"It's okay. You can go. I—Galactic penpals?"

"Galactic penpals."

"Sweet." Henrietta quirks her mouth, an expression that has always been indelibly Hannah in my eyes. "And I mean this in the most non-ironic sense of the word ever. I—good luck, mom."

The line cuts and Hannah breathes out, long and slow.

"Is this your fault?" she asks, not turning.

"Mine and Hotaru, really. Hotaru's the one with the necessary clearance—"

"Ass."

"You're welcome."

One hour.

The ship hums like something alive, its vibrations filling our bones, our thoughts. The chatter from mission control is a near-incomprehensible slurry, earmarked by Hotaru's replies, concise and even.

"Final chance for phone calls and other near-instant forms of communication, people!" Hannah roars, flipping switches and levers, a cacophony of motion.

"Everyone I care about in this vessel," remarks Ji-Hyun, stiff, a history of abuse delineated in the margins of her voice.

Everyone I care about in this vessel. The statement tears me open and I breathe the implications deep.

"Anna?" Hannah again.

"I'm going with what Ji-Hyun said. Everyone I care about is already here." And it is not a lie. Not exactly. An almost truth, at worst, that stings to say, but there is no act of healing without hurt.

"Randy." Hotaru's voice cuts through our exchange, before Hannah can press me further.

"Yes, chief?"

"Sing us to Mars, will you?"

The unexpectedness of the request robs Randy of his usual verbosity, but he does not seem to care. Instead, he lifts his gorgeous voice, begins singing a soldier's dirge about going home. Hannah holds my stare for a minute, then lets her expression gentle, looks away. Three years is enough to teach you what people need.

When we die on Mars, it will be a world away from everything we knew, but it won't be alone. We will have each other, and we will have hope.

BEFORE THEY COME (2016)

Sharon Goza

Sharon Goza is a former Integrated Graphics Operations and Analysis Laboratory Manager in the Engineering Directorate, NASA's Johnson Space Centre, Texas. She was involved in developing software simulations and generating animations and 3-D models for many of NASA's robotic systems. Goza now works in the private sector.

I had just installed the last bolt on the habitat connector when an alarm on the downlinked data sounded. Everything in my virtual Mars environment suddenly turned on end, and I found myself looking at the second section of the habitat from my side. Just as I felt my Active Response Gravity Offload and Orientation Unit start to turn me sideways, I felt the jerk of the emergency stop and heard the voice of our CAPCOM, Ann.

"E-stop initiated, R7 Alpha down. Reposition and obtain visuals. R7 Bravo, move to 50 feet from habitat aft. R7 Charlie, move to 50 feet from habitat fore. R7 Delta, move to assess R7 Alpha. I've marked the locations on your maps."

My R7 unit, which was designated Alpha and nicknamed Rama by the team, was locked down. I could either hang horizontally, for possibly hours while the rest of the team got visuals on their targets and the environment was re-scanned, or I could disconnect myself and become a spectator to the unfolding situation. I decided for the later and switched my virtual Mars off, and the quad display showing visuals from each of our R7's on.

"R7 Delta in VR position."
"R7 Bravo in VR position."
"R7 Charlie in VR position."
"Roger, command hold."

Now it was just a matter of waiting until we could assess the situation. Although Earth and Mars were close to being at opposition, we were still 70 plus million miles apart. It took around five minutes for our commands to reach the robots, and five more to receive the visuals and data back. That's why we operated like we did. Months before we started, all of the equipment had arrived on the surface of Mars. The payloads consisted of inflatable habitats, scientific equipment, life support, semi-autonomous rovers, spare parts, and our four R7 humanoid robotic units. First, the rovers were deployed. They were in charge of scanning the landing site. Their data provided us with a three-dimensional model of the entire area, complete with perfectly matched colors and textures. We'd come a long way from early Mars simulations. Now, once you put on your full field helmet and feedback exo-suit, you were there. You could even feel the cracks in the rocks when you touched them. It took a unique operator to be able to detach themselves from the real world and live in a virtual world like it was reality, without getting sick or disorientated. Training for Mars work was even tougher. Advancements in data communication had made it possible to send a huge amount of data back to Earth, but the speed was still dictated by the distance between Earth and Mars. We had to learn not to react. If you received an alarm, as I had, you just stopped. What you were seeing had happened minutes ago, and any reactive movement was not only wasted energy but might send a command that would do more harm to the robot or the environment.

"Mike, from the looks of R7 Alpha's visuals, there may have been a seal failure on the aft endcap. R7 Alpha's trajectory seems to indicate that the force was from the side, rather than the connector," came Ann's voice over the com.

"Understood. Hopefully, he's in one piece and the hab repair will be an easy patch and re-inflate."

"Definitely. We should know more in four."

I thought about putting up a countdown window on my visuals and decided against it. It just made the wait longer. In the meantime, I decided to tap into the public affairs comm line and listen. We may not be the first humans on Mars, but the press considered us the first humanoids on Mars, and they loved a good disaster story. With the networks and the press watching over operations from the catwalk in the Building 9 high-bay at Johnson Space Center, it was hard to appear as if nothing was wrong. We didn't have much to tell them at this point, and the questions I was hearing were getting redundant, so I switched over to engineering.

"Anything you can tell me, Bob?" inquired Ann. By this time, she would have contacted the center director and was probably fishing for something to tell him when he arrived.

"We've gone through the fault tree and known risks. A seal is still the prime suspect, but we didn't expect a seal to break this early in the mission. Thankfully, from the initial scan, the repair container looks perfect, so we don't anticipate any problems getting the required materials to fix it. It'll be quite a bit more difficult for the R7's, though. We're starting up some simulations now so we can try out a few things."

Well, that sounded like fun. I doubted they'd try to mock up anything real and use the ground R6 units to test. It would take too long. So, simulation is all we had. That and us. They would work out the basics of the scenario with ideal conditions and virtual models, but we would be the ones that had to adjust to the real thing.

"R7's in Mars position."

I turned my attention back to my screen and flipped through the visuals one by one. My R7 unit looked fine, except for a detached arm. R7's had plug and play limbs that detached automatically if a force over specs was ever applied to them. It might be inconvenient, but it kept the unit from stripping gears and motors. There weren't any wires either. The sockets had data and power quick connects derived from the old Power Data Grapple Fixtures on the International Space Station. It shouldn't be a problem to put him back together.

The habitat, however, was a lumpy deflated balloon. At 50 feet from the structure, it was hard to see exactly what failed, but it appeared that engineering's suppositions might be correct. Hopefully, none of the pre-packaged equipment within the habitat was harmed, and we could re-inflate it once we repaired the rip.

"R7 Alpha, see if you can stand up, R7 Bravo and Charlie, move in toward the connector and take a closer look. We're deploying the rovers to update the scans as well, but your visuals need to see the details. R7 Delta, go grab a new arm from stores, we'll pick up the other one later."

I re-initialized my virtual Mars based on the latest visual scans, and felt myself re-orient into a horizontal position. I didn't send any commands yet. I wanted to get a feeling of the ground before trying to stand. The R7's control system had automatic correction intelligence, but if I was too far off even that couldn't compensate for the error. I practiced rolling to my belly, pushing up on one arm, and standing until I was confident in my movements, and then connected to the uplink and sent the commands for real. Once again, I disconnected while I waited. I needed to make sure the unit was stable and standing before I sent any other commands.

"R7 Bravo and R7 Charlie, we have a good visual on the endcap. Looks to be a failure near the seam about three inches long. Disconnect and take a break while we complete the new scans. R7 Alpha, you're looking good, data shows a little bit of discrepancy on the locations, but not enough that can't be compensated for. R7 Delta should have your new arm to you in a few minutes. Once you're back in one piece, you two can disconnect and join the rest of the team."

I plopped down in the conference chair and took a drink of my coffee. Our break had turned into a full evening off while the engineering team worked out a solution, and the simulation team mocked it up for us to try. Ann opened the latest charts and began.

"Good news, we've got a plan. Bad news, it's something we've never done before. But, you guys are up for the challenge."

Ann pointed out a slit in the side of the structure.

"We're not exactly sure why the hole went through all the layers of the structure, but we suspect it was almost severed when we set it up, possibly by a part of the descent vehicle. Structures figures the pressure of the inflation finished the job. Interior images don't show a bulge in the area, so we don't think it's something that's still around. However, we've got to repair all the layers if possible, and they're suggesting we sew them together before applying the patches and sealant."

I was skeptical and didn't hesitate to say so. However, Ann ignored me and flipped to the next chart which outlined the procedure.

"First, you'll patch the exterior as you've done many times in the sim. Then comes the hard part. Once the structure is inflated, you'll enter the hab and begin repairing each layer from the inside. Needless to say, the rover doesn't fit in the airlock, so you'll be on your own for environment mapping. It's going to be tedious, and it's going to be slow, but if anyone can do it, you four can. Mike, here are the procedures the engineers came up with, I'd like your team's assessment in an hour. After that, we'll meet with the engineering and simulation teams, tweak the procedures and the sim, and you get the next three hours to practice. We need to seal the hab soon to avoid as much particle contamination as possible."

I sighed, knowing that my day just got a lot longer, but I agreed with Ann. If anyone could do this, we could.

Five hours later we were suited up and back controlling our R7's on Mars. We'd revised the procedures based on our individual skill sets and come up with what we thought was the most feasible. Thankfully our R7's were the closest thing you could get to a human, and all the tools for the soon-to-arrive astronauts would work for us too. The easiest portion was the external patch. The precursor to our glue had been tested early in the Space Shuttle program. Although they scrapped it then, the advanced version worked great. A new application method, as well as advancements in materials, made it simple. Sewing had also been tested in the early 2000's. It was difficult with robotic hands, but not impossible.

"CAPCOM, Patch applied."

"Roger, R7 Delta."

"R7 Bravo, we'll initiate inflate sequence in ten minutes after patch confirmed."

Twenty minutes later, we confirmed we had an inflated habitat. Time for us to enter the habitat, scan the tear from the inside, and get to work. Layer by layer we held, trimmed, folded, sewed, and patched our way to the interior wall. With R7 Bravo, and R7 Charlie holding the seam tight, I was able to use the curved upholstery needle to join the sides of the tear. Once done with the seam, R7 Delta glued a patch over the area. By the time we had worked our way through all 10 layers of various materials, we were pros at the process, and thoroughly exhausted.

#

"Congratulations," said Ann during our briefing the next day. "The pressure is holding and visuals confirm we've got a viable habitat. I've got permission to let you all take the day off if you want. The back-up crew will finish out."

I looked at the rest of my team, and their shaking heads made it unanimous.

"No, thank you. I think we'll finish this out. Suit up, gang, let's finish building the first house on Mars!"

TO SAVE EARTH, GO TO MARS (2016)

Joe Mascaro

Joe Mascaro is an ecologist and science writer based in San Francisco, California. His work has appeared in the *New York Times*, *Los Angeles Times*, *San Francisco Chronicle* and *Space Review*.

Al Gore often ends his presentations on climate change by showing a classic image of the Blue Marble – the full sunlit disk of the Earth, hanging vulnerably against the blackness of space – before adding some cautionary words: 'Don't let anyone tell you we can escape to Mars; we couldn't even evacuate New Orleans. The Earth is the only planet habitable for human beings. We're going to have to make our stand right here.'

This is a rational view of the challenge. We have to deal with famine and disease. An estimated one billion people live on less than $1.90 per day. Climate change is threatening a tenuous global economy. Why would we invest resources on colonising another planet when we need them so urgently on Earth? Such thinking extends up to the highest levels of government. Although it has announced plans to send astronauts to Mars orbit sometime in the 2030s, the Obama administration has with each annual budget request attempted to direct NASA funds away from human spaceflight and planetary science, and toward Earth observation.

But lately I've found myself wondering: what if our societal view of problem-solving is flat-out wrong? More specifically, what if addressing a crisis directly is not always the most effective route to success, and investing resources toward a different, seemingly unrelated challenge could lead to faster, more effective breakthroughs?

It sounds like a wild claim, but this indirect approach, known as lateral innovation, is responsible for many of the most remarkable innovations we encounter in our daily lives, generally with little awareness of where they came from. One of the most dramatic recent examples is the graphics processing unit (GPU). Fast, powerful GPUs were originally developed to meet the demand for increasingly realistic video games, especially ultraviolent first-person shooters. But machine learning, computer vision and neural networks pose very similar computational challenges, and the technology quickly migrated over. Specialised GPUs are now used for diagnostic medicine, facial recognition, self-driving cars and market forecasting. Built on innovations developed for games such as Doom and Grand Theft Auto, machine learning is powering one of the fastest-growing sectors in the world economy.

About a year ago, I joined a company called Planet Labs in San Francisco, and stumbled into a veritable nest of lateral innovation. As an ecologist, I've spent much of my career researching challenges such as deforestation, rising carbon emissions and the loss of biological diversity. In my search for solutions, I've walked the halls of academia and the US government, but at Planet Labs my work has taken me to a place I never expected to be: low Earth orbit. Through the eyes of dozens of microsatellites, each the size of toasters, I watch the world's forests change every day. I collaborate closely with the people who build the satellites, write the control software, and convince skeptical agricultural firms to pay millions for the pixels of data they collect. All of this happens on budgets that are minuscule by NASA standards, using hardware and code that were mostly developed for other purposes.

Watching what my colleagues do, and understanding why they do it, has convinced me that brute force alone will not innovate the technologies that will enable human civilisation to become an effective arbiter of this planet and her resources. The solution requires tapping into the same impractical, impatient, passionate drive that spurred the video-game-fuelled GPU revolution. And although that kind of lateral innovation cannot be instituted forcibly, it can be recognised and fostered.

In short, Gore got it exactly wrong. To save the Earth, we have to go to Mars.

The challenges of surviving on Mars are extreme, but they are hardly alien. Among the most basic steps for colonising Mars are growing food in a cold, arid environment and generating energy without fossil fuels. Today, international development organisations are experimenting with crops that achieve the same basic goal: to produce calories in inhospitable environments. Salt-tolerant potatoes in Pakistan. Drought-tolerant maize in East Africa. Nutrient-rich rice in India. The difference is that the Martian problem requires much more extreme, innovative solutions.

The great agricultural revolution of the previous century was fuelled by lateral innovation. Working in Mexico with the support of the Rockefeller Foundation in the 1940s to 1960s, the biologist Norman Borlaug kicked off a series of world-changing innovations in farming. Among the most important was 'shuttle breeding', a technique Borlaug used to conduct two full growing seasons in the same calendar year. Not everyone was pleased with his unconventional approach. As Mark Stuertz wrote in the *Dallas Observer* in 2012, Borlaug's 'rapid progress drew scowls from fellow wheat breeders and geneticists who maintained wheat strains could only be bred within one specific environment'. Opposition from his boss even led Borlaug to resign from the Rockefeller-backed project.

When a colleague intervened and Borlaug was reinstated, not only did he make good on the plan but, by virtue of the differences in climate and soil at his two test sites in Mexico, Borlaug stumbled into one of his most important innovations. 'As it worked out, in the north, we were planting when the days were getting shorter, at low elevation and high temperature,' he explained. 'Then we'd take the seed from the best plants south and plant it at high elevation, when days were getting longer and there was lots of rain. Soon we had varieties that fit the whole range of conditions. That wasn't supposed to happen by the books.' Problem-solving to breed plants more quickly in Mexico led to better breeds, and to more food. What insights might emerge from engineering potatoes to grow in the cold, arid, dim environment of Mars?

The challenge of engineering Martian crops also attracts a different group of thinkers and occupies a different institutional community than those who traditionally work in agricultural research. Developing salt-tolerant potatoes in Pakistan involves international development programmes, local farmers and land-owners, and private-sector agricultural companies. The laboratory environment used to simulate the surface of Mars is trafficked by astrobiologists and physicists drafting manuscripts for peer review and playing Kerbal Space Program on the side. Different minds see different solutions.

In energy production, too, technology development for a Mars colony could be game-changing for Earth's needs. Solar and nuclear power are the only energy sources that have ever powered craft on Mars: the robotic probes that include the still-active *Opportunity* and *Curiosity* rovers. Mars lacks any fossil-fuel deposits, has no flowing water for hydropower, and its wispy atmosphere cannot push fins around a rotor with much force. It isn't simply that solar and nuclear will be the preferred options for energy on Mars; they are the *only* options. All energy innovations associated with Mars colonisation will need to be long-lasting, and they will be intrinsically carbon-free.

Finally, and perhaps most importantly, living on the Martian surface – and especially returning home from it – will require an ability to manipulate the Martian atmosphere. On Earth, the primary gases in the air we breathe are nitrogen, oxygen, and several trace gases including carbon dioxide. On Mars, the atmosphere is extremely thin, just a little over 1 per cent of the density of Earth's. And although oxygen and nitrogen are essentially absent, carbon dioxide is hyper-abundant, constituting 95 per cent of the Martian atmosphere.

The abundant CO_2 in the Martian atmosphere will be a crucial resource for a self-sustaining Mars colony. By extracting carbon molecules, we can react them with hydrogen to produce methane, a rocket fuel that will send our astronauts home. We can also pull oxygen out of the CO_2 to make water. The basic chemistry involved in these reactions has been known for centuries, but there's been little incentive to innovate the techniques used to manipulate them. Water and oxygen are so bountiful on Earth that we've never had much incentive to develop efficient ways of extracting them from carbon dioxide. But today it is becoming clear that processing CO_2 could be the key to reducing carbon emissions and creating useful materials such as plastics without fossil fuels.

The challenge of colonising Mars shares remarkable DNA with the challenges we face here on Earth. Living on Mars will require mastery of recycling matter and water, producing food from barren and arid soil, generating carbon-free nuclear and solar energy, building advanced batteries and materials, and extracting and storing carbon from atmospheric carbon dioxide – and doing it all at once. The dreamers, thinkers and explorers who decide to go to Mars will, by necessity, fuel unprecedented lateral innovations.

Lateral innovations between Earth and space aren't merely speculation. They are happening right now. In the 1970s, the first *Landsat* mission began returning images of the whole Earth every 16 days. Over the ensuing 45 years, new Landsat sensors have consistently provided that coverage. With each iteration (*Landsat* 8 is flying today), improved detectors have brought more and more of the spectrum into view. The resulting data have transformed scientific awareness of our planet's large-scale ecology and environment. And yet, many fundamental aspects have remained the same over those decades. Landsat captures most of the features of the Earth's surface at 30-metre spatial resolution. At this scale, agricultural fields and large patches of forests are clearly seen, but individual homes and trees are not.

Recently, two private companies have reimagined Earth imaging. Skybox (acquired by Google in 2014) and Planet Labs (which I now call home) have begun to manufacture satellites that are smaller and far less costly than either the Landsat system or most other commercial remote-sensing platforms. Both are developing new approaches and new services – Skybox by producing high-resolution video from space, and Planet Labs by imaging the entire Earth every day.

Lateral innovation is the driving force behind these advances. Pete Klupar, who was director of engineering at NASA's Ames Research Center from 2006 to 2011, has spoken admiringly about the ways the two companies draw on technology from unconventional sources: 'They are using IMUs [inertial measurement units] from video games, radio components from cellphones, processors meant for automobiles and medical devices, reaction wheels meant for dental tools, cameras intended for professional photography and the movies, and open-source software available on the internet.'

The next time you have a cavity drilled, you might want to distract yourself by thinking about orbital mechanics. The drill that your dentist is holding is one of the lateral innovations employed by engineers at Planet Labs. They are relying on a modern version of an old hands-on museum trick in which a docent will seat a guest in a bar stool, place a bike wheel in their hands, and then spin it up. The docent then asks the guest to rotate the bike wheel perpendicularly, and viola! The bar stool starts turning. The effect is caused by conservation of angular momentum. The spinning wheel has a vector of force pushing out of it, and when you try to push that vector it has to have somewhere to go. That is a reaction wheel.

Satellites have for many decades used reaction wheels to point with exquisite accuracy along three axes. The effect works in both zero and Earth-normal gravity, and does not require any propellant. With three reaction wheels (four for redundancy), satellites can stare dead-on target. The planet-hunting Kepler space telescope failed its primary mission after two of its four reaction wheels stopped turning; its reaction wheels, like those on most satellites, were a custom job. For the hundreds of low-cost, tiny satellites to be launched by Planet Labs and a number of other 'small-sat' companies, custom fabrication was not an option. It also was not a necessity, the engineers realised. All they needed was a fast-spinning rotor.

As the medical-devices industry refined components to create a beautiful drill – one that pulls two orders of magnitude faster than a Ferrari – it's safe to say they never imagined the hardware would make its way to space. The decision made by Planet Labs engineers was one of necessity: the only way to build the satellites quickly and cheaply enough was to use off-the-shelf components that could be plugged into a satellite with little modification. Early satellites had bits of measuring tape (the regular hardware-store one with yellow tongues) sticking out to function as radio antennas. Low cost, good performance and high reliability were essential to convince investors that Planet Labs could succeed. Necessity was the mother of lateral innovation.

Since its first demonstration launch in 2013, Planet Labs has successfully deployed 133 satellites. In late 2014, the company lost a flock of 26 satellites when an Antares rocket exploded a few seconds off the launch pad. The Planet Labs team rebounded almost immediately, assembling and shipping out two new satellites in just nine days. We could never have done that without lateral innovation, without pulling drill motors and solid-state drives and CPUs from the shelf.

The lateral flow goes both ways. Just as Planet Labs pulled off-the-shelf innovations into its satellite hardware, innovations developed for spaceflight are benefiting other industries. In 2013, SpaceX hosted 3D metal-printing firms at their warehouse in Hawthorne, California to demonstrate how they 'print' rocket engines for the human-rated version of the Dragon space capsule. Previous human-rated capsules used an escape tower that sits on top of the capsule like a needle. It has to be jettisoned during launch if it isn't used. By contrast, SpaceX built the engines right into the side of the capsule. If the engines aren't used in a rescue operation during launch, then they can be fired later to land the capsule. SpaceX specifically built them to ensure that the Dragon capsule is capable of landing on Mars.

The design of these SuperDraco engines is complex, but not more complex than, say, a device that could suck CO_2 out of the hot gases in a coal-fire smokestack. The technology to capture and contain carbon emissions has eluded sustainability engineers for a quarter century. It's just too costly, according to those who have looked into it. Maybe it's worth taking a stab at innovating it laterally rather than directly. Three-dimensional metal printing of the type used by SpaceX probably provides about the right level of precision. A carbon-capture-and-sequestration instrument is about as complex as a rocket engine. In essence, they do the same thing, just in reverse: a rocket engine supervises a chemical reaction in the additive sense, and a carbon-capture device carries out a chemical subtraction.

The innovations covered here are not spinoffs in the sense that NASA and other government agencies often tout. (Not incidentally, 'Spinoffs' is the name of a report NASA periodically sends to Congress to justify spending on space systems.) We're not talking about commercialising astronaut drinks into Tang.

We're talking about full-fledged, self-sustaining pipelines of innovation that come laterally *to* space hardware (turning dental drills into reaction wheels) or *from* space hardware (3D metal printing, now finding its way into many types of industrial manufacturing). Either way, the innovations were motivated by mission-driven problem-solving. Dental drill reaction wheels exist only because Planet Labs embarked on the enormous task of imaging the entire Earth every day. SpaceX's enhancements in 3D metal printing were overtly driven by the company founder Elon Musk's desire to colonise the Red Planet.

The difference between assuaging skeptical members of Congress with a report on spinoffs and drawing in true innovators can be seen on the factory floor at SpaceX's Hawthorne headquarters. In a report for *60 Minutes* in 2012, the journalist Scott Pelley queried the former NASA astronaut Garrett Reisman about his decision to join SpaceX: 'You have so much background in engineering, such a long and enviable career at NASA. You could have easily gotten a job at Boeing or at Lockheed, but you came here and I wonder why.' Reisman replied: 'If you had a chance to go back in time and work with Howard Hughes when he was creating TWA, if you had a chance to be there at that moment when it was the dawn of a brand new era[...] wouldn't you want to do that? I mean, that's why I'm here.'

Engineers such as Reisman are working hard to take us to Mars. Indeed, SpaceX recently topped a list of tech companies rated based on how much the employees thought their work 'makes the world a better place'. Musk's vision extends even beyond colonising Mars. In the very long run, he envisions 'terraforming' the planet – engineering its climate to be more Earthlike so that it could support life in the open air.

From a conservative, rational perspective, it would be easy to impugn the high-frontier aspects of SpaceX's mission as a waste of talent, time, and technology. Given the urgency of protecting human welfare on the only planet we presently inhabit, it might seem sensible to keep our eyes fixed on the problems immediately ahead of us. But after seeing what lateral innovation can do, I no longer believe that the sensible approach is the correct one. We have to look past the foreground; we have to look beyond our planet. We have to go to Mars.

SURVIVAL (2017)

MARS FICTION AND EXPERIMENTS WITH LIFE ON EARTH

Sabine Höhler

Sabine Höhler is head of the department of philosophy and history at KTH Royal Institute of Technology in Stockholm. Born in 1966 in Germany, she studied undergraduate physics at Karlsruhe University and postgraduate modern history and history of science and technology at Braunschweig University of Technology. In 1999, she completed her doctoral thesis, 'Research and Mythology in Aviation: Scientific Ballooning in Germany, 1880–1910'. In 2011, Höhler became Associate Professor of Science and Technology Studies at KTH, having been a lecturer at Humboldt University of Berlin, International University Bremen (IUB) and Darmstadt University of Technology between 2001 and 2010.

Life on Mars

> The year is 2025 and the research expedition ... has established an outpost on Mars. In the closed atmosphere of their protective ... habitat, the crew utilizes algae to remove carbon dioxide and replace it with oxygen. In the food production module, algae are grown and processed to provide a protein source for the crew. ... This scenario is not science fiction but one of many possible examples of the role of algae in space exploration.[1]

These lines appeared in a 1988 book that envisioned the successful establishment of a "bioregenerative life support system" on Mars.[2] Within just a few decades, so the expectation went, a life support system centered on algal growth for food supply and air revitalization would make human life possible in Mars' extreme environment. The authors of the study *Algae in Space* were biologists and complex systems scientists involved in space research. They took pains to ground their work in scientific fact and to distance themselves from the heated dreams of science fiction. But their ideas of setting up an ecologically balanced system on Mars that maintained life's essential functions were no less visionary.

The US science fiction movie *Red Planet* featured similar ideas of near-future Mars settlement on a planetary scale. In unmanned interplanetary missions humans succeeded in creating an atmosphere on Mars. Water and carbon dioxide were released by melting the Martian polar ice caps. Atmospheric pressure increased and the "greenhouse effect" resulted in a temperature rise. Seeded algae photosynthesized and caused oxygen concentrations to increase and carbon dioxide levels to fall. Like the science book's narrative the movie's story is set in the year 2025, when the first manned research expedition sets out to prepare for a first Mars outpost. The audience learns that humanity is about to abandon an Earth that is thickly polluted and poisoned. With a population of 12 billion people Earth is on the verge of an ecological catastrophe. Like the book the movie relies on the idea of using algae, basic plant life, to produce oxygen and protein, standing in for the basic conditions of human life.

In these two works of science and of fiction David Bowie's 1971 song "Life on Mars?" seems to resound.[3] In the song's refrain "Is there Life on Mars?" Bowie mocks NASA's and other space enthusiasts' ceaseless searches for traces and possibilities of life on the Red Planet. Bowie's song also heavily scorned the twentieth-century US entertainment business as a motor of the capitalist production and consumption cycle. The song and its refrain beautifully capture the essence of exploitative cyclical systems that must tap and feed on new resources continually, be it new markets or new planets. In this view life on Mars became a part and an extension of the rapacious resource economies on Earth. Both the science book and the science fiction film chimed in on this refrain by promoting the creation of living conditions in an extreme environment that was never meant for human life to dwell in—or was it?

Since the early Cold War period, when the hopes of space flight countered the perceptions of industrial pollution, population growth and resource depletion on Earth, ideas of creating a habitable environment on near-Earth planets have gained popularity. Both science and fiction authors were zealously producing and promoting visions of "terraforming," of developing environments with Earth-like qualities in hitherto uninhabitable spaces.[4] It is this relation of Earth and Mars that this paper aims to explore. Mars science and fiction have been understood as an escape route for humans facing the environmental degradation of Earth. But Mars science and fiction can also be read as an eco-technological blueprint for the Earth's environmental restoration. By studying visions of terraforming Mars as practices of shaping and engineering Earth, I aim to show that the Anthropocene, the geological age of humanity, is more than an epoch compromising of the Earth. The Anthropocene has also essentially transformed the understanding of Earthly life to a minimalist principle of survival through infinite metabolic conversions, an understanding that has engendered visions of a future life off Earth. From the 1960s onwards, the notion of immortality based on humankind's survival on another planet conjoined images of recreation and creation, of paradisiacal pasts and eco-technological futures.

In the following I will outline some of the thinking and tinkering with habitable environments on Mars. During the Space Age of the 1960s, experiments with life support technologies literally 'short-circuited' the Earth's life cycles. Humans inserted themselves into the metabolisms of living matter in new ways with the aim of increasing resource efficiency while maintaining self-sufficiency. In basic self-sustaining systems nothing was to be consumed other than what was produced. Which kind of communities emerged when humans recreated nature's metabolic cycles of water, food and energy flows to form symbiotic relationships with living matter?

The first section recalls the aspiration for balance in the political, economic and ecological states of the world in the Cold War period. The search to maintain and recreate equilibrium in materially closed environments went along with new systemic conceptions of life. The second section studies notions of the circulation of matter and biotic mass and the provisioning of the means of exchange, framed by the concept of the biosphere as the Earthly container of processes of life around 1970. The third section introduces closure as the condition and effect of containment: experiments in the 1960s and '70s show that closing the loop was essential for creating viable small-scale systems to circulate human, plant and animal biotic mass. The fourth section explores cultural techniques of counting and converting as crucial for constructing the units and the uniformity needed to upscale the experiment. The fifth section looks at a meso-scale experiment of the 1980s that took the Earth as its object and Mars as its goal. Finally, the sixth section looks at an experiment on the scale of Mars. Thoughts on the recent upswing of Mars exploration will conclude the paper.

1. Equilibrium

Mars is the planet most often referred to when expressing the hope of turning extraterrestrial into terrestrial environments. A rich history of science, technology, and politics of Mars exploration[5] is matched by cultural studies scholarship on science fiction.[6] Best known among the science fiction literature exploring the terraforming of the Red Planet are perhaps Ray Bradbury's *Martian Chronicles*[7] and Kim Stanley Robinson's "Mars Trilogy".[8] Numerous science fiction films addressed Mars colonization projects, among them *Red Planet* (2000), *Total Recall* (1990), *Mission to Mars* (2000) and most recently *The Martian* (2015). State governments, activists and private enterprises campaigned for Mars.[9] Works of science and of fiction share the awareness that reproducing the Earthly environment elsewhere will be no less challenging than sustaining it here. Until the present day the Earth has been held to be unique within the known universe. Earth has developed a nature rich and complex including a climate sufficiently moderate and stable to make human life possible. The problem that the human environment has become increasingly vulnerable to human impact forms the heart of the Anthropocene discourse.[10]

The singularity of the Earth's environment became perhaps most apparent with the exceptional images of Earth taken from space by the Apollo space missions in the 1960s and early 1970s.[11] The Blue Marble's dynamic swirling colors of blue, white, brown and green signified billions of years of converting sunlight into processes of life. The remote view of the Earth as a dynamic and yet well-adjusted planet set a sharp contrast to locally experienced environmental challenges. In public perception space flight both decentered and centered the Earth. Space flight discourse and imagery pushed ambitions of leaving the planet and exploring the possibilities for life elsewhere and also supported the rising environmental movement. Space ecology closely connected approaches of preserving and substituting the fragile Earthly environment.[12] This circumstance has mostly been overlooked in environmental history scholarship. The rise of environmentalism in the latter half of the twentieth century has been studied primarily for its conservationist concerns of safeguarding nature against degradation through industrialization and development.[13] The time period, however, was preoccupied with the stabilization of the Earth's natural systems not only to arrest environmental decline but also to enhance environmental control.[14]

Politically, the Cold-War situation of potential nuclear destruction asked for a new balance of power between the continents. Economically the situation called for balancing wealth between the nations.[15] And ecologically, the threat of environmental degradation appealed for rebalancing the Earth's ecosystems. Balance, political, economic and ecological, became a key concept in international deliberations. Preparing the agenda for the UN Conference on the Human Environment in Stockholm in 1972, Barbara Ward and René Dubos warned that the "two worlds of man—the biosphere of his inheritance, the technosphere of his creation—are out of balance, indeed potentially in deep conflict. And man is in the middle. This is the hinge of history at which we stand".[16] The authors, one a political scientist and economist, the other a biologist and environmentalist, asked for a "coexistence" of nature and technology, of human and non-human life on Earth.[17] In these lines yet another notion of systemic balance becomes visible: the balance of natural creation and technological restitution.

The discourse of balance was not uncontested. Critics of the "myth of ecological equilibrium"[18] reasoned that striving for "ecological balance" led to ill-founded conclusions that concealed existing global imbalances regarding population, pollution and resource consumption.* However, the general equilibrium approach to ecosystems was not based on well-balanced harmony but rested on a systemic concept of stability. Equilibrium encompassed the violent dynamics of a system able to self-regulate its internal environment, be it the Earth's ecosystems or the systems of living organisms. The new framework of ecosystems science and the systemic notion of equilibrium also involved novel scientific definitions of life. Physicist Erwin Schrödinger and physiologist J. B. S. Haldane had radically questioned and refashioned life in the 1940s as a molecular process and an experimental object. Following their lead, life scientists formulated physical and biochemical accounts of life as a homeostatic system of self-identical reproduction.[19] In 1970 the French geneticist Jacques Monod provided a biochemical version of life as an autonomous system constructed not entirely coincidentally, whose actions were directed and nevertheless wholly

* Miguel Ozorio de Almeida, the Brazilian Ambassador to the 1972 Stockholm Conference, argued that on the contrary the problem to be solved was "obtaining the most efficient forms of 'long term ecological imbalance'" (Ozorio de Almeida 1973, 25), applying science and technology to prolong humankind's ability to use natural resources for as long as possible.

contingent in existence, so as to generate an organization capable of self-identical reproduction.[20] The British physicist John Desmond Bernal conceptualized life as an open system of organic reactions favored by catalysts.[21] Bernal based processes of life on thermodynamic principles. The emerging field of biophysics defined life as a highly organized system of material entities slowing down the ever-increasing entropy by means of autoactive and directed organization.[22]

By the 1970s, life had lost all traces of *vis vitalis*, the special properties and powers connected to the living nature that had been prevalent in the late eighteenth and early nineteenth centuries. Twentieth-century ecology couched life instead in the scientific terms and principles of biophysics, biochemistry, and molecular biology. Life's characteristics increasingly concentrated on structural order and functional complexity, on information exchange, metabolic processes, endless reproduction and synthetic reproducibility. This shift of life in the life sciences from an essential natural force to principally synthesizable and reproducible living matter was prerequisite for the visions of exobiologists to emulate eco-technological life forms for extraterrestrial evolution.[23] The new understanding of life was also a precondition for conceptualizing and retaining equilibrium in the Earth's largest ecosystem, the biosphere.

2. Circulation

"Life, the biosphere, creates its own controls and balances".[24] The American ecoscientist and engineer John Allen put forth his understanding of the Earth's biosphere as a system of life's checks and balances in the 1980s. His statement exemplifies the prevalent picture at the time, the picture of the living environment as a cybernetic system. The systems view of the Earth's environment allowed for its perception as a by and large self-regulating environmental system that left plenty of room for improvement through human intervention, regulation and control.

"As man begins to be important in the biosphere, he must consider what system is sufficiently compatible with his inputs and outputs to maintain a viable atmosphere, an adequate water cycle, and other favorable conditions," the renowned American systems ecologist Howard Odum claimed in 1971.[25] Odum's words illustrate the view that the principles of exchange of matter could be scientifically described and operationalized for any materially closed system in order to maintain, optimize, and ultimately also reproduce the system. Odum meticulously computed the power requirements per day to sustain a human being and the input and output levels for "long-range survival," and he concluded: "The design problems are the same as those for the minimal system of man and nature capable of export to the moon for support of human life in space".[26]

Notions of the Earth's biosphere as an ecological "life support system" were directly indebted to the science and technology of spaceflight. And vice versa, the space capsule became a technoscientific model of environmental self-reliance, a blueprint for sufficiency and efficiency solutions to Earth's environmental degradation.[27] Odum's analogy of "total life support," linking the Earthly living space to living in space, could gain such power only in the heyday of the Apollo space missions. But the analogy rested on much older traditions in geology and ecology of thinking about life as confined to certain spaces or habitats. The "biosphere" had first been defined in the nineteenth century as the sphere that encompassed all life on Earth. In 1875 the Austrian geologist Eduard Sueß had reflected on the "zone" on an Earth "formed by spheres" to which organic life was constricted; "on the surface of continents," Sueß asserted, "it is possible to single out a self-contained biosphere".[28] In 1926 the Russian biogeologist Vladimir Vernadsky published his main work *The Biosphere*. His project of a "physics of living matter"[29] added a powerful base to study energy and matter cycles within the environment. Formulating his unified biospheric theory, Vernadsky translated Sueß's "self-contained" biosphere into a "self-maintained" biosphere.[30]

The work of the American ecologist G. Evelyn Hutchinson consolidated the conceptual shift from a phenomenological understanding of the envelope of life to a bio-and geochemical approach to living matter and to the environment as a self-regulating and evolving system sustained by energy, water, oxygen, carbon, nitrogen and minerals. Hutchinson based the "the day-to-day running" of the biosphere on a systems idea of maintaining an "overall reversible cycle".[31] His notion of the biosphere as a circulatory system, whose boundary conditions limited the "amount of life" on Earth, made Hutchinson conclude with respect to its expected life span: "It would seem not unlikely that we are approaching a crisis".[32] In Hutchinson's view, some flaws in the biosphere's composition had allowed for excessive environmental exploitation through humans. Long-term processes of converting organic matter into fossil hydrocarbons had admitted to the extraction of fossil fuels by a pyrotechnical human civilization. The biospheric system had been thrown out of balance and asked for immediate technoscientific intervention.

3. Closure

How should the exchange and circulation of biospheric matter be optimized? Advocates of sufficiency stressed resource limits and the need for stability through careful resource use and complete material recycling. They emphasized the limits of a system that was well capable of continuously reproducing its material form but that was sustained by energy inputs from the sun only. Promoters of efficiency built on scientific progress and proficient technological design as a means of optimizing the Earth's complex life-supporting and regenerating functions. They highlighted the technological possibilities of increasing resource and energy efficiency. Systems dynamics provided the architecture for both perspectives on "the changing metabolic role of man," as Odum phrased the problem.[33] The view of the closed system facilitated the mathematical analysis of global natural processes, or, to quote Odum again: "From an orbiting satellite, the earth's living zone appears to be very simple". The distant view from space facilitated a simplified view of the Earth's biospheric system and the creation of an appropriate experimental setup suited for outer space.

From the early 1960s onwards, scientists on both sides of the Iron Curtain experimented with simple "closed-loop" systems to support astronauts and cosmonauts in space. These experiments went beyond the simple greenhouse spaces in real and imagined spaceships, like the pet plants grown on the Russian Salyut 6 expeditions in the 1970s.[34] The experiments concerned closed self-sufficient exchange systems set up in a scalar analogy to the biospheric processes on Earth. They were designed to be energetically and materially efficient to the highest possible extent. The initial Russian experiments centered on closed cycles of converting carbon dioxide to oxygen and back. Basic structures sustained one human being and a small number of plant organisms on small time scales. The Russian scientist Yevgeny Shepelev self-experimented with Chlorella algae at the Institute of Biomedical Problems in Moscow, managing to spend twenty-four hours in a tightly sealed chamber in which algae regenerated his air and purified his water.

In the 1960s and '70s these algae-based systems were further developed with the BIOS experiments at the Institute of Biophysics in Krasnoyarsk, Siberia. Test facilities Bios-1 and Bios-2 could provide eight square meters of Chlorella per human being. The Bios-3 facility reached the size of a small space station. The Russian scientists attained closure times of up to six months with two- and three-person crews, sustaining a balance of proteins, carbohydrates, vitamins and minerals. Food crops supplied up to 80 percent of the crew's food and provided part of air and water regeneration. Inserting themselves into the metabolic cycles of microalgae and higher plants, the scientists endeavored to establish a holistic system of complete bio-regeneration. These experiments were highly exclusive, carrying a chosen few cosmonauts only. The experiments were also highly selective. They optimized the productivity of food crops and patched the loops for instance by stocking the system with animal proteins.[35]

4. Conversion

Closed ecological [systems] built for space flight literally 'short-circuited' the global biospheric cycles and material flows. The human beings who enclosed themselves in a life support system also repositioned themselves with regard to the imagined and actual place of the human species in the accustomed production and consumption cycles. Leaving their habitual place at the top of the food chain humans inserted themselves at an unfamiliar stage that connected high-modern technology to pre-modern subsistence economy—a combination that has been tried repeatedly in science fiction, for example with large-scale algae farming to offset world population growth.[36] All the while, so the assumption went, humans would remain in control of the cyclical processes. To account for the material exchanges on all scales, ecologists expressed balance in mathematical equations, converting aggregated matter and energy from one form into another. In the small-scale environment of a space capsule it seemed possible to account for the material exchanges by simple experimentation and observation, to testify to the success or the failure of balancing the books.

Applied to the planetary scale the conversion experiments produced bizarre results. They tell a lot about how close science fact is to science fiction with regard to the scope both genres claim they can cover. The American biologist and human ecologist Garrett Hardin for instance calculated that with a 1970 rate of world population growth of 2 percent, within 600 years the world population would amount to 8.27×10^{14} and there would be "standing room only" on all the land areas of the Earth.[37] And he continued: If due to humanity's ceaseless consumptive needs the entire mass of the Earth were finally converted to human flesh, growth would result in a maximum number of 1.33×10^{23} human beings, achieved in only 1,557 years. With his thought experiments Hardin sought to demonstrate the limits of the ecological "carrying capacity" of the Earth or any other intergenerational spaceship, that is, the maximum number of a species that an environment could support indefinitely.[38]

Exactitude is clearly not an attribute to distinguish science from fiction. On the contrary, it is the meticulous precision of their sweeping predictions that make both science fact and science fiction look so fantastic. The example of converting humans and other organic and non-organic matter seamlessly into each other shows how precision and a new organicism went hand in hand in building environmental life support on Earth and in space. Perhaps we must start to think about "biomass" as a new form of biosocial collectivization that around 1970 best matched the ideal of comprehensive regulation but also took up bold ideas of endless possibilities of regeneration, transformation and substitution. Ideas of system maintenance and operability fashioned a prototype of sustainability that was built on utility, functionality and replaceability. As discussed in the following section, the project of Biosphere 2 formed a landmark example of how sufficiency and efficiency regimes worked together to construct a new technological environment, an enclosed living space as modern and as sustainable as a space capsule.

5. Re/Construction

Biosphere 2 was launched in the Arizona desert in the middle of the 1980s as an experiment of unprecedented kind and scale. On a site of three acres, seven interrelated ecosystems simulated planet Earth in miniature. Distinct biomes, from tropical rain forest to ocean (including an artificial coral reef), from savannah and marsh biome to desert, were sealed from the outside world under a huge transparent dome. The dome also housed an "intensive agriculture" biome and a small "city" for eight human beings. Like the Russian Bios experiments, Biosphere 2 was set up as an environmental system materially isolated from the outside world, to be operable at any extreme terrestrial or extraterrestrial site.[39]

The project group of eco-visionaries, among them eco-engineer John Allen, enlisted disciples of architect Buckminster Fuller and ecologist Lynn Margulis, a close collaborator of Gaia originator John Lovelock. Advice was sought from the Russian Bios experimenters and also from NASA representatives who had worked on closed ecological life support. The group endorsed a roughly Darwinian evolutionary thought based on the interplay of environmental need and technoscientific opportunity. The mid-twentieth century had opened a singular window of opportunity in relation to scientific expertise, technological means, energy sources and new materials that would allow the project's architects, for a very short time, to create "biospheric offspring" and release life from its Earth-bound support system of "Biosphere 1".[40] In the prospect of abandoning Planet Earth altogether the architects of Biosphere 2 employed the metaphor of the space capsule: "Why not build a spaceship like the one we've been traveling on?".[41] The ultimate goal was outer space. By miniaturizing Biosphere 1 the experimenters hoped to develop "a prototype for a space colony" and to settle future "Sustainable Communities" on Mars.[42] When closure experiments involving humans began in 1988 the media hailed Biosphere 2 as a "glass spaceship" ready to go.[43]

Biosphere 2 was to evolve life beyond its Earthly origin by harnessing the powers of systems ecology and cybernetics. The biospheric system was designed as "a cybernetic system moderating environmental conditions on our planet for the continuance and spread of life".[44] In short, as Allen and his co-author Mark Nelson saw it, Biosphere 2 was a "Life-Enhancing Feedback System".[45] The newly devised research field of "biospherics"[46] merged holistic and cybernetic approaches to life. Biospherics understood life not as an individual but as an ecological property. "Ecotechnics"[47] aimed at the construction of a "technosphere" that would "stand in" for the geological functions of planet Earth and for its weather and climate systems. An elaborate infrastructure of electrical, mechanical and chemical transmissions formed its fundament. A circuit of pumps and ventilators was responsible for flushing, cleaning, and cooling the air and the water, for moving wind and waves, and for regulating the climate. An intricate network of more than 2,000 sensors, the "nerve system" of the facility, continuously monitored the designated parameters to effect system stability and safety.[48]

Applying the cybernetic ideal of a self-contained and self-sustained system under human auspices, Biosphere 2 also instigated a new ethics for determining what would be useful and what was redundant, what was to be conserved and what was to be discarded in the system. "Sustainability" in the view of the project makers had a purpose and direction: The 4,000 animal and plant species taken on board the "glass ark"[49] were to cooperate to sustain the eight human "biospherians" at the top of the food chain. Biological agents were chosen according to criteria of efficiency. Species could be replaced as long as they would fulfill equivalent biogeochemical functions. "Residents had to earn their keep, performing some useful function in the ecosystem," Allen explained. "Unlike the [biblical] ark, Biosphere 2 welcomed animals to a *web of life*, as participants in the *oikos logos*. From the Greek origins *oikos* (house) and *logos* (governing rules), ecology literally means the 'rules of the house'".[50]

Biosphere 2 was the paradigmatic example of an Anthropocene habitat—an exclusive high-modern living space, clean and sheltered from the inconveniences of nature out there, scientifically organized, technologically monitored and ecologically unsustainable. Neither war nor overpopulation nor pollution nor climate change posed problems within the second biosphere. Instead, an indoor nature "out of control" terminated the project early.[51] Harvest failures and food shortages complicated life under glass. CO_2 concentrations soared and

O_2 concentrations dipped. Biosphere 2 proved functionally unstable owing to its weak biological diversity. Equilibrium, fragile from the onset, could not be retained once violent and disruptive internal dynamics between human, plant, and animal species set in. The rich organic material taken on board metabolized at an unexpectedly high rate and produced an excess of carbon dioxide. A range of pollinating insect species died off, impeding plant fertilization and procreation. Some animal and plant species thrived, however, and fed on the food crops reserved for the human crew.

The American social theorist Jeremy Rifkin[52] critiqued the "engineering approach to the age of biology" which Biosphere 2 pursued, and its aim of "creating a second nature in our image." American political theorist and cultural critic Fredric R. Jameson[53] commented on culture becoming "a veritable 'second nature'" at the moment "the modernization process is complete and nature is gone for good." According to the French sociologist Celine Granjou and the Australian social scientist Jeremy Walker, "nothing exemplifies a cyborg theory of nature better than an infrastructure engineered to put its analytical agenda into effect".[54] Biosphere 2 engineered a veritable second nature by materializing the ecosystems concept itself. Following the American environmental critic Timothy Luke, Biosphere 2 fabricated "an essentially new synthetic ecosystem by mixing and matching various components—soils, plants, and animals—from a wide range of naturally occurring ecosystems to model a simulation, or a copy, of Nature, for which there really is no original".[55] A critique of Biosphere 2 that takes its ecosystems ambitions seriously must note that this system could not handle the contingencies, unforeseen complications and vast expenses of biospheric processes, not even at all costs.*

6. Upscaling

Timothy Luke chose the term "Denature" to describe Biosphere 2's nature that offered space neither for emptiness nor for excess.[56] Biosphere 2 was by no means lifeless. Luke's term of Denature points to the aspirations of the experiment to build an equilibrium that had to exclude all natural traces of chance, disorder, difference, sickness, danger and death—principally, all forms of disturbance. The French philosopher Lucien Sfez[57] looked at Biosphere 2 as a utopia of "perfect health," a nature without decay. Complementing perfect health with a notion of perfect death the French philosopher Jean Baudrillard addressed Biosphere 2 as a "glass coffin".[58] He criticized the project as conserving and musealizing nature during its lifetime.[59] The term "glass coffin" reflects the paradox that the living system of Biosphere 2 aspired to immortality by adhering to a minimalist principle

* In a short note of correspondence in the journal *Nature*, systems ecologist Eugene Odum admitted to the "extremely high cost of providing nature's free life support services" in technologically maintained environments. The monthly electricity bills the crewmembers would have to face if they had to pay for utilities at the U.S. residential rates would be more than $150,000. "At anywhere near this cost, very few of the billions of people on Earth could afford to live in domed cities," Odum concluded. The actual costs of Biosphere 2 are as contended as everything else about the project. Biospherian Jane Poynter speaks $250 million; the *New Scientist* speaks of $150 million.

of "survival" through steady regenerative succession—"just like in paradise".[60] Following Baudrillard the experiment failed because it built on "dissolving, shrinking the metaphor of the living into the metastasis of surviving".[61]

Science fiction offers many clues about how an immortal nature can fail when based on survival as a reduced continuation of life that avoids death. As mentioned previously, the movie *Red Planet* (2000) experiments with the idea of a self-maintained metabolic system on a planetary scale in which implanted photosynthetic microorganisms are to coexist with human organisms on the grounds of calculated gas budgets and energy balances. The aim in this near-future setting is to create not an artificial domed environment but an entire planetary atmosphere. *Red Planet* was praised for its scientific accuracy in studying how terraforming on the planetary scale could get out of hand, spectacularly.

In the movie humans have succeeded in employing algae to create a breathable atmosphere on Mars. Oxygen levels, however, are on the decline and the algae have disappeared. A team of the world's best scientists is sent to find out what went wrong but remains clueless. In a crucial scene the scientist-astronauts discover the reason for the environmental deterioration. Out of the blooming algae an entirely new species has grown that neither the humans nor the algae had reckoned with: A Martian type of cockroaches bred and inserted its mass activity into the closed atmospheric cycle. These cockroaches pose a serious competition to the humans. They feed on the living matter that was meant for the humans to consume (in the movie this flipped human-animal relation goes along with reversing the code of green as the color of life. Rather than being positively attributed, green turns eerie when it becomes associated with living matter seen as pest.) The insects inexplicably produce oxygen from the algae, accounting for the circumstance that the humans can breathe freely. But their parasitical behavior has disrupted the stability of the metabolism. The cockroaches have created an imbalance that causes the fragile atmospheric system on Mars to collapse. Not to mention that they are omnivores, eating plant and human matter alike.

The song "The Tower That Ate People"[62] by Peter Gabriel features on the movie's soundtrack, and it describes the hubris of humans claiming planetary control through their staggering projects. Gabriel sings about a biospheric feeding machine on planetary scale operating in space science and space fiction that eventually turns back on the humans that believed themselves to be in full control.

Do not all closed environments come attached with exclusive "rules of the house" in their endeavor to maintain balanced life support? All of the projects discussed in this paper, whether science fact or science fiction, are fascinating in their own ways, but some turn out to be more abhorrent than attractive, more threatening than protecting.

Life on Mars Revisited

In his Mars song from 1971, David Bowie sang about the flatness and the falseness of turning ordinary dreams into best-selling shows. The spectators take the seat with the clearest view, presented with the silver screen, but then they find they have seen, written and lived the film ten times or more. Throughout the second half of the twentieth and into the twenty-first century this film has been lived again, fueling the production and consumption cycles of living matter on Earth and beyond. This paper has explored some examples of experiments with self-sustaining systems to recreate the Earth's life cycles in outer space. In both small-scale and large-scale experimental settings, the relations of humans to their biosphere were put to the test. The link of biomass, quantifiable amounts of living matter, to biosphere, enclosed controllable living space, has continued to animate thought. In the movie *Red Planet*, two recovered insect specimens act as bearers of the hope of eventually restoring Earth's polluted atmosphere. The mission of terraforming Mars fails but a solution for the environmental problems of the Earth seems at hand. "We triumphed!" exclaims the spaceship commander upon return to Earth. The victory statement points to future amazing projects of mass experimentation that other stories will have to tell.

Most of the Mars stories we are fed are of the unambiguous kind, hailing progress in unquestioned Anthropocene fashion. In 2002 Elon Musk launched his ambitious SpaceX Corporation and since 2013 NASA has been experimenting with a near-future Mars mission on Hawaii.* In 2015 the film *The Martian* celebrated the lone scientist setting up a self-contained and self-sustained atmosphere for plant growth and manure, implanting ideas of the feasibility of creating a livable atmosphere on Mars in large cinema audiences. The Martian survives on minimalist principles. He personifies the Anthropocene as the epoch of humans promising new modes of immortality through infinite metabolic conversion. It is not without irony that in this film as in the other projects described in this paper Earth, the abandoned planet, acts as a standby. Ultimately, Earth provides the living conditions humans believed themselves to have harnessed and recreated. While Earth as a fallback has often been met with disappointment, *The Martian* openly promotes this view, making "Bring him home" its mission. How to value one planet over another? The question of whether "multi-planetary" life will be possible and also desirable might become the most challenging question of the Anthropocene.

* SpaceX, short for Space Exploration Technologies Corporation, was founded in 2002 as a company for space exploration and technology, making a special "Case for Mars" http://www.spacex.com/ (accessed February 28, 2017). HI-SEAS (Hawaii Space Exploration Analog and Simulation) is a project operated by the University of Hawaii, http:// hi-seas.org, and supported by the NASA Human Research Program, https://www.nasa. gov/hrp (accessed February 28, 2017).

THE RETURN OF THE SPACE VISIONARIES (2018)

Rand Simberg

Rand Simberg describes himself as a recovering aerospace engineer with more than forty years of experience in the field. He holds a BSc in applied mathematics and engineering science from the University of Michigan, Ann Arbor and a Masters in Technical Management from West Coast University, Los Angeles. In 2013, he published the book *Safe Is Not an Option: Overcoming the Futile Obsession with 'Getting Everyone Back Alive' That Is Killing Our Expansion into Space*. He is currently Business Development Consultant at SpaceTech Analytics.

In 1969, the year that astronauts first walked on the Moon, Princeton physics professor Gerard K. O'Neill, "almost as a joke," posed a theoretical exercise for his students: Is Planet Earth the best location for a growing techno-industrial civilization? Working through calculations with them, he came to conclude that Earth is indeed not the best location—that other planets, and space itself, would be a better venue for an expanding technological species, offering more energy and raw materials, and risking less pollution of our home planet. "As sometimes happens in the hard sciences," he later explained in an article in *Physics Today*, "what began as a joke had to be taken more seriously when the numbers began to come out right."

Lost on Mars

O'Neill expanded the ideas into the now-classic 1977 book *The High Frontier*. It imagined large spinning habitats built from lunar materials and housing thousands of people. It would be paid for by selling power, using huge arrays, also from lunar materials, to collect sunlight and beam it down to Earth in the form of microwaves. Most industrial activity would be moved off of the home planet, which would become a giant nature park for both inhabitants and tourists visiting from space.

The idea inspired a movement. The L-5 Society—named for one of the stable points equidistant from the Earth and the Moon in the lunar orbital plane, where O'Neill envisioned the habitats might reside—was founded in 1975 to advocate for his vision. Its (now clearly optimistic) slogan was "L5 in '95!" Conferences held at Princeton and a summer study at NASA Ames Research Center, in Mountain View, California, subsequently helped to flesh out concepts on orbital mechanics, how to use the lunar resources recently discovered by Apollo, how to build closed-cycle life-support systems, legal and policy questions, and space-construction techniques.

Against the context of the more measured NASA aspirations we're familiar with—the blink-and-you'll-miss-it moonwalks, astronauts tinkering in low Earth orbit for decades, and far-reaching but uncrewed planetary probes—O'Neill's vision may sound like a pie-in-the-sky aberration. That was indeed how it struck many of his post-Apollo-era contemporaries. Asked about the possibility of federal funding for O'Neill's ideas, Wisconsin Senator William Proxmire famously said, "Not a penny for this nutty fantasy."

Yet O'Neill's vision is strikingly similar to the ones being offered by today's aspiring space tycoons, most notably Elon Musk, founder of Space Exploration Technologies (SpaceX), and Jeff Bezos, founder of Blue Origin. Musk, though focused like a laser on Mars, talks about "Making Humans a Multi-Planetary Species," as the title of a technical article he published last year put it. Bezos, for his part, doesn't confine his ambition even to settling other planets. His stated long-term goal is to get millions of people off of Earth, where they can pursue their own dreams, whatever those may be, whether on other planets or in permanent settlements in space itself. This May, he received the Gerard K. O'Neill Memorial Award for Space Settlement Advocacy from the National Space Society. In an interview given just after accepting the award, he expounded on his O'Neillian vision of a trillion people living in the solar system. In keeping with the theme, he also announced, with some of the show's cast in attendance, that he was saving from cancellation the science-fiction series *The Expanse* by having Amazon Prime take over its production from the SyFy Channel. The series, popular with the science-fiction community, depicts human settlement of the solar system and offers a realistic treatment of how the laws of physics would constrain such an endeavor.

Like O'Neill's movement, democratization is also a key goal of today's space tycoons. Physicist Freeman Dyson, a colleague of O'Neill's at Princeton, argued in 1978 that to avoid being "a luxury that only governments could afford," the cost of space colonization must be lowered to about $40,000 per person, which "would make it comparable to the colonization of America." That figure translates into about $150,000 today. Both the rationale and the dollar figure are strikingly similar to Elon Musk's argument that, in order for mass resettlement to Mars to be viable, the cost per person must become comparable to buying a house—a median of about $200,000 in the United States.

Despite some differences in approach, Musk and Bezos together represent a sharp departure from the conventional approach of America's public space program, which has always been more limited, focused on science and exploration, not human settlement, and has operated on the assumption that only big-government funding and organization could send humans to space. But a look at the history of ideas about space travel, going back much further than O'Neill and Dyson, shows that Musk and Bezos are in fact returning to a longer tradition of dreaming about humanity's future in space—a tradition that, fittingly, is now coming to fruition in America. Musk and Bezos are on the cusp of fulfilling the dreams many others have had of settling space, and of making space travel commercially viable—neither of which Apollo, much less its middling low-Earth-orbit successors, were able to achieve, despite the tragically failed attempt to do so with the space shuttle. Indeed, Bezos and the foreign-born Musk, combining personal dreams with technical prowess and bold entrepreneurship, are much more thoroughly American in their visions than even America's own government-run space program.

With the Trump administration so far appearing more favorable than any previous administration toward the dreams of space entrepreneurs, there are a myriad of signs that the Apollo model—driven as it was by Cold War goals rather than by visions of the human future—will be the exception, not the rule, for how travel to other worlds will be achieved. And our new era in space will look much more like the dreams of early space visionaries, and of more recent supposed fantasists like O'Neill and Dyson, all of whom foresaw a broad, sustained, and prosperous human inhabitance far beyond Earth.

Early Space Visions

While humans have surely been gazing in wonder at the stars since the dawn of our existence, it has been for only a little over a century, after modern telescopes gave us a better sense of what lay in the heavens, that people started to dream of actually leaving the planet to head out toward them, and settle outer space itself.

In 1894, the American inventor and real-estate magnate John Jacob Astor IV (who later died heroically in the sinking of the *Titanic*) wrote the science-fiction novel *Journey in Other Worlds: A Romance of the Future*, depicting life on Saturn and Jupiter in the year 2000. One character in the book enthuses:

> We are all tired of being stuck to this cosmical speck, with its monotonous ocean, leaden sky, and single moon that is useless more than half the time, while its size is so microscopic compared with the universe that we can traverse its great circle in four days. Its possibilities are exhausted; and just as Greece became too small for the civilization of the Greeks, and as reproduction is growth beyond the individual, so it seems to me that the future glory of the human race lies in exploring at least the solar system, without waiting to become shades.

Astor's fictional character probably overstates the number of us who share his viewpoint, but it is one of the earliest known visions of space settlement.

A quarter century later, H. G. Wells saw the conclusion of the First World War as an occasion for optimism. He speculated that humanity might turn away from its recent bloodletting, adopting a global government and using its new technological powers to usher in a new era of history. In *The Outline of History* (1919–1920), Wells prophesied, "Life, for ever dying to be born afresh, for ever young and eager, will presently stand upon earth as upon a footstool, and stretch out its realm amidst the stars."

In 1929, Irish author J. D. Bernal wrote *The World, the Flesh and the Devil*, a work that science-fiction luminary Arthur C. Clarke later called "the most brilliant attempt at scientific prediction ever made." Bernal imagined space settlements on the inside surface of spherical shells, with tens of thousands of people inhabiting each, and speculated—indeed with brilliant prescience—about the use of solar energy, repair bases in space, and multi-stage settlement:

> On earth, even if we should use all the solar energy which we receive, we should still be wasting all but one two-billionths of the energy that the sun gives out. Consequently, when we have learnt to live on this solar energy and also to emancipate ourselves from the earth's surface, the possibilities of the spread of humanity will be multiplied accordingly.

In a subtly warier tone than Wells of a decade before, Bernal foresaw that "at first space navigators, and then scientists whose observations would be best conducted outside the earth, and then finally those who for any reason were dissatisfied with earthly conditions would come to inhabit these bases and found permanent spatial colonies."

Rocket Men

Probably the earliest person to have a vision of space settlement anchored in known physics—and to contribute to its technical realization—was the Russian schoolteacher and engineer Konstantin Tsiolkovsky, born in 1857. He is remembered as one the fathers of rocketry, and for the line "Earth is the cradle of humanity, but one cannot remain in the cradle forever."

Tsiolkovsky envisioned space settlements and, inspired by the radical new steel structure in Paris designed by Gustave Eiffel, space elevators between Earth's equator and a geosynchronous orbit thousands of miles above it. He thought through the issues associated with habitats in a vacuum, and developed the concepts of airlocks and closed-cycle biological environments that could provide food and oxygen for residents of space colonies.

With a lifelong interest in mathematics and physics, he was also one of the first to work out the mathematics of rocketry. In 1903 he derived the exponential rocket equation, since named for him, that dictates how much velocity change one can get for a given amount of propellant, and the associated need for multi-stage rockets.

Tsiolkovsky's work was soon replicated by the American physics professor Robert H. Goddard. After learning physics, Goddard independently derived the rocket equation (Tsiolkovsky's work hadn't yet been disseminated outside of Russia), and published his seminal paper, *A Method of Reaching Extreme Altitudes*, in 1919. As told in Milton Lehman's 1963 biography *This High Man*, Goddard recalled how as a youth he was inspired by a vision on his family's farm in Massachusetts:

> On this day I climbed a tall cherry tree at the back of the barn ... and as I looked toward the fields at the east, I imagined how wonderful it would be to make some device which had even the *possibility* of ascending to Mars, and how it would look on a small scale, if sent up from the meadow at my feet...I was a different boy when I descended the tree from when I ascended. Existence at last seemed very purposive.

Goddard pioneered designs for rockets and actual working models, launching the first liquid-fueled rocket in 1926, a major milestone. Nevertheless, he was ridiculed in 1920 by the *New York Times* editorial board for the foolishness of suggesting rockets could work in space, in ignorance "of the need to have something better than a vacuum against which to react." Long after Goddard's death, three days before the first lunar landing, the astronauts well on their way to the Moon, the *Times* would offer an apology: "It is now definitely established that a rocket can function in a vacuum as well as in an atmosphere. The *Times* regrets the error."

Around the same time as Tsiolkovsky and Goddard were doing their work, similar ideas were being developed by a young physicist named Hermann Oberth, who was born in the Austro-Hungarian Empire but studied for his doctorate in physics in Germany. Inspired by Frenchman Jules Verne's radical new science fiction about voyages to the Moon, Oberth designed his own model rocket in 1909 at the age of fourteen. In the 1920s, he wrote two books exploring his ideas on interplanetary rockets, *The Rocket into Planetary Space* and *Ways to Spaceflight*. The notion had earlier been rejected when he'd submitted it as a thesis proposal for his doctorate.

In Germany, the Verein für Raumschiffahrt—the Society for Space Travel—was founded in 1927, inspired in part by the work of Oberth, who became a member. Another member was the young aristocratic engineering student Wernher von Braun, who had dreams of human voyages to Mars. He and many other amateur rocketeers eventually became swept up in Hitler's program to build the V-2 rockets that bombarded Antwerp and London in 1944.

But von Braun never lost his dream. At the end of the war, he took several of his team and headed west toward Allied lines to escape the advancing Soviets and surrender to the Americans, in hopes that he would have better prospects to achieve his space dreams with them. At first, he was again put to work on ballistic missiles, but with the advent of the U.S. space program in the late 1950s, he was finally able to start applying his engineering and managerial talents, first to satellite launch, then to human spaceflight.

Another German engineer, Krafft Ehricke, was inspired by Fritz Lang's now-classic 1929 film *Woman in the Moon*. Like von Braun, he escaped to the West to avoid capture by the Soviets and to pursue his space dreams, and in America he became part of von Braun's new team, where he pioneered liquid-oxygen and -hydrogen rocket stages, and later ran his own consulting company focused on commercial applications in space. Ehricke developed an idea that he called "The Extraterrestrial Imperative," about the need for humanity to expand into the universe. He characterized life as a "negentropic" process—that is, one that created order, at least temporarily and locally, against a universe tending toward disorder. He thus viewed the expansion of life, first within the solar system and then to other stars and galaxies, as both inevitable and beneficial. (I was one of his students for an extension course he taught on the subject in the early 1980s at California State University, Northridge, toward the end of his life.)

Similar ideas came in the late 1950s from Princeton's Freeman Dyson—before his later work on the economics of space settlements—who wrote "A Space-Traveler's Manifesto." The essay was a paean to space settlement, and to nuclear propulsion, partly to promote his Orion concept of sequentially detonating small nuclear explosives behind a pusher plate to launch a rocket not only into space but across the solar system:

> It is in the long run essential to the growth of any new and high civilization that small groups of men can escape from their neighbors and from their governments, to go and live as they please in the wilderness. A truly isolated, small, and creative society will never again be possible on this planet.

An even more expansive vision came from Dandridge Cole, a physicist and aerospace engineer sadly cut down in the prime of his career by a heart attack. In the early 1960s, he conceived the idea of *Macrolife*, a sort of higher-level life form composed of a self-contained human colony in space, constituting the next step in evolution.

In 1969, it was conventional wisdom to see Neil Armstrong's first step as indeed a giant leap for mankind—not the end of a journey, but just the beginning of a new phase of human history in which the dreams of these space visionaries would at last come to fruition. But it would become clear in short order how wrong this impression was—that these initial dreams were not fulfilled but rather derailed by the Apollo program.

Three Rival Visions of Humanity's Role in Space

Rick Tumlinson, a longtime space activist and co-founder of the Space Frontier Foundation, argues that there are three general categories of space visionaries: the "Saganites," the "von Braunians," and the "O'Neillians"—after Carl, Wernher, and Gerard, respectively.

Saganites view the universe as a precious jewel. How beautiful! "Look at it—but don't touch it!" Tumlinson quips. Space is for scientific inquiry only, and that is best done by investigating it with robots. Later in life Sagan recognized the value of sending humans to other worlds, but as an astrophysicist and planetary scientist, his goals were focused on science, not economic development or settlement.

Since the end of Apollo, this vision has driven what many considered the "Golden Age of Space Exploration," with probes sent to all the known planets in the solar system, even Pluto. The scientific knowledge gained from these inanimate scouts of the space frontier has provided new insight not only into how to further explore the solar system, but also how we might send humans out to settle and develop that frontier.

For example, water has been called the "oil of the solar system," in that it is crucial both for the support of life and as a key constituent for the efficient rocket propellants of oxygen and hydrogen. The relatively recent discovery of significant quantities of water ice on our own moon means that we can make a good start at both "living off the land" and utilizing it as propellant to reduce the cost of venturing beyond the Moon, reducing the amount of payload that must be launched from Earth. In fact, we now know that not just ice but liquid water is much more abundant in the solar system than we previously thought; some of the ocean moons of the gas giants Jupiter and Saturn make the "water planet" Earth a comparative desert. Nevertheless, the Saganite vision is focused on *knowing* about the universe but not *using* it.

What Tumlinson calls, perhaps unfairly, the von Braunian vision is akin to what I have called "Apolloism" ("Getting Over 'Apolloism'," Spring/Summer 2016): The government expends massive resources to send a handful of government employees off to explore another planet. It is what most people continue to consider the normal, perhaps only way to do human spaceflight—though Elon Musk, Jeff Bezos, and others are starting to change that perception.

Many have mistakenly interpreted the Apollo program as a natural derivation from a distinctly American vision for humanity's future in space. But Apollo was motivated largely by the Cold War strategy of beating the Soviets, not by any clear vision of human space exploration or settlement. In truth, Apollo was not a derivation from traditional American values but a deviation: In its rushed, centralized bureaucratic approach, and its hero worship, Apollo was much more a Soviet way of doing things. Its purpose was to win a crucial moral and technological battle in an existential war. It was certainly not a plan for the opening of a new frontier, which would have required a far different, more patient, and more cost-effective approach to getting humans into space and on to other worlds.

The O'Neillian vision is one of massive expansion of humanity into space, and it is much more in line with the visions of all who came before the historical anomaly and disruption of Apollo. But by the accident of history in which we first went into and explored space in the Cold War, the dominant visions have been Apolloistic and Saganite, both government-centric, led by NASA. But now, with reduced launch costs, and the growing interest of billionaires—not only Musk and Bezos but the Russian Yuri Milner, who last year announced his plan to send a privately funded probe to Enceladus, a moon of Saturn that we now know has complex organic molecules under the ice in its oceans—we may be returning to an era of astronomy and space science that is funded privately and philanthropically, as most American observatories were prior to World War II.

The O'Neillian vision could also in theory be driven by government, and one of the many straw-man arguments that opponents of massive space settlement use against it is the assumption that it will be funded by taxpayers. Note again Senator Proxmire's warning, "Not a penny for this nutty fantasy." But it is unrealistic to imagine that there will ever be massive numbers of people going into space at taxpayer expense, at least on any kind of sustained basis in a democracy. It will only happen if they want to go, if they have the financial means to do so, and if, once there, they will be able to continue to pay their way.

In other words, space settlers' activity will of necessity be financially self-sufficient. Despite the criticism of it as outlandishly costly, the radical new approach that O'Neill proposed after Apollo was based on fundamentally economic logic: People living and working in space could provide goods and services that more than compensated for the cost of sending and keeping them there.

The Long Death of Apollo

Because of Apollo's successful proof-of-concept of how to get humans beyond Earth orbit, and the power of its legacy, federal space policy has remained stuck in its mindset, even as the political and budgetary will to back it vanished with Armstrong's first step. The last half-century of NASA plans for voyages to Mars or back to the Moon have thus largely been a series of failed attempts to repeat the 1960s glory days.

In a 1989 speech on the steps of the National Air and Space Museum, President George H. W. Bush laid out a goal of once again sending humans to other bodies in the solar system, most notably Mars. It was called the Space Exploration Initiative (SEI). Its very name carried the seeds of its doom: If the goal is exploration, that can be done much more cost-effectively with robotic space probes.

After the speech, a "90-Day Study" was initiated by NASA. The study's creators solicited input from all of the NASA program administrators on the best way to achieve the goals. Unsurprisingly, the administrators described how each of their programs' work was indispensable, and every existing technology sandbox and project was thrown into the pot. The purpose of the study became not achieving the objective but rather justifying what the agency and its contractors were already doing and, in the face of uncertainty, wanted to continue to do. Cost estimates, as reported by the *New York Times*, put the price tag for going to the Moon and Mars as high as $400 billion. In addition, as Robert Zubrin explained in *The Case for Mars* (1996), NASA leadership itself failed to support the project fully, as it saw the shuttle and space station programs as its only real priorities. And so the initiative was stillborn.

The idea of sending humans to Mars remained off the table through the 1990s. The Clinton administration viewed its space legacy as saving the space station program by a single vote in the House in 1993, in part by deciding to partner with Russia in an attempt to keep Russian engineers out of mischief with countries like Iran and North Korea (which didn't work), and by downscaling the project and changing the name of the space station *Freedom* to the International Space Station (even though it already had Europe, Canada, and Japan as international partners).

U.S. space policy plodded on, visionless and therefore aimless, until February 1, 2003, when the atmosphere sundered the shuttle *Columbia* on entry, scattering the remains of its hull and seven crew over east Texas and Louisiana. It was now clear that the shuttle program had outlived its usefulness, and it was no longer possible to avoid new policy. The following year, President George W. Bush laid one out, called the "Vision for Space Exploration." It was a plan to retire the shuttle in 2010, and to move beyond low Earth orbit for the first time in over three decades, first to the Moon, then on to Mars, with new space transportation systems.

Despite the "e" word in its name, the new plan in fact, for the first time, set policy aims beyond exploration. In a speech in March 2006, the late John Marburger, Bush's science adviser, stated that the vision "subordinates space exploration to the primary goals of scientific, security, and economic interests":

> The Apollo program was what mathematicians call an "existence proof," a demonstration that a problem does have a solution and that efforts to discover its details will not be in vain. Like all firsts, it was unique. No subsequent space endeavor can be quite like it. President Bush's vision also declares the will to lead in space, but it renders the ultimate goal more explicit. And that goal is even grander. The ultimate goal is not to impress others, or merely to explore our planetary system, but to use accessible space for the benefit of humankind. It is a goal that is not confined to a decade or a century. Nor is it confined to a single nearby destination, or to a fleeting dash to plant a flag. The idea is to begin preparing now for a future in which the material trapped in the Sun's vicinity is available for incorporation into our way of life.

This was probably the closest description of the O'Neillian vision in any official statement of U.S. space policy up to that time.

Unfortunately, Michael Griffin, whom Bush had appointed as NASA administrator a year earlier, had different plans. When Griffin rolled out Constellation, his program for implementing the Vision for Space Exploration, he described it as "Apollo on steroids." NASA would build a new giant rocket and capsule to carry out the vision, using legacy hardware from the shuttle and Apollo programs. This returned the Bush vision firmly back to the familiar, with which Griffin probably knew Congress would be more comfortable. But as I wrote in these pages ("A Space Program for the Rest of Us," Summer 2009), the program soon grew out of control, leaving a space-policy mess for the incoming, inexperienced Obama administration to deal with.

Constellation was canceled by the Obama administration in 2010, with a plan to develop commercial capabilities instead. But Congress partially resurrected its aims in the Space Launch System program—a large heavy-lift rocket intended by Congress as a "follow-on launch vehicle to the Space Shuttle"—and in the Orion spacecraft that would take four astronauts beyond low Earth orbit sometime by the early 2020s, according to NASA's current projections. But like Constellation, its replacement remains plagued by budget overruns and schedule slips, and even if it eventually flies, it will do so even more rarely than the space shuttle did, at a cost of billions per flight.

The New Visionaries

The advent of Gerard O'Neill's vision, and the speculation it helped inspire, preceded by just a few years the first flight of the space shuttle in 1981. Many would-be space settlers naively assumed the shuttle would live up to its promise of low-cost routine spaceflight—the original plans called for a shuttle to launch nearly every week. But it was clear by the time *Challenger* was lost in 1986 that it was not going to do so. Fortunately, other developments in the early 1980s would help to sustain the dreams of space settlement, a vision to guide its advocates until they could find the means to realize it.

In 1980 Peter Diamandis—who later went on to found the X Prize Foundation, which offers rewards aimed at spurring major technological breakthroughs—then an M.I.T. undergraduate, co-founded an organization called Students for the Exploration and Development of Space. The following year, he wrote a letter about it to *Omni* magazine, inspiring the founding of chapters at other universities all over the country, including one at O'Neill and Dyson's Princeton. A few years later, a Princeton undergraduate named Jeff Bezos, who was majoring in computer science and electrical engineering and had spoken about space settlements in his high-school valedictorian speech, took courses from O'Neill, and became head of the local chapter.

Meanwhile, in South Africa, an adolescent Elon Musk, fascinated by the planet Mars, was teaching himself to program the new devices called microcomputers. At the age of seventeen, he moved to Canada, and later to the United States, for studies at the University of Pennsylvania, Wharton, and (very briefly) Stanford. In an interview, he once said, "I wasn't born in America—I got here as fast as I could." He became a citizen in 2002.

Bezos was a young child during Apollo; Musk was born as it was coming to an end. Both were part of a generation that felt cheated that they had reached this planet too late to see humans walking on another. But there has been no consensus, at least in government, on the American plan for humanity's future in space. A unified vision has never existed, but it may be starting to coalesce with the ambitions of Bezos and Musk.

Motivated by the vision of thousands or millions of people living and working in space, both men know that this will never happen with expensive government launch systems. Their visions are broadly O'Neillian, even if Musk is obsessed with Mars while Bezos's dreams are destination-agnostic and more expansive.

Establishing human settlements on Mars, and more generally making humanity "a multi-planet species," have always been the stated *raison d'être* of SpaceX. Musk often quips that he wants to die on Mars, "just not on impact." In 2016, at the International Astronautical Congress (IAC) in Guadalajara, Mexico, he laid out in detail for the first time his flight architecture for an extended series of Mars colonization missions—with the first crewed flights beginning as early as 2022. Last year, he published a paper in the journal *New Space* and gave a follow-up presentation at the 2017 IAC in Adelaide, Australia, revising and expanding on the concept. The proposal was for a very large two-stage launcher, using SpaceX's new methane-fueled Raptor engines, dubbed "Big Falcon Rocket" (or BFR—inside the company, they often use a different "F" word). It would dwarf the largest rocket built to date, the Apollo program's Saturn V. He originally called the new overall space-transportation system the Mars Colonial Transporter, but has since renamed it the Interplanetary Transport System to make it seem more widely applicable to the entire solar system.

Bezos started his space company, Blue Origin, in 2000, two years before SpaceX. Until recently, it has been much more secretive about its plans. But while SpaceX has become famously associated with the astonishing feat of repeatedly launching and recovering vertical-takeoff, vertical-landing rockets, since 2015 Blue Origin has done the same with its *New Shepard* vehicle (albeit only to suborbital speeds). Long viewed as the tortoise in the race for the suborbital space market, Blue Origin is probably now firmly in the lead, after the recent bankruptcy of XCOR with its planned Lynx spaceplane, and the continued delays in Virgin Galactic's new SpaceShipTwo vehicle after the catastrophic loss of the first one in 2014, killing one and injuring the other of its test pilots. (Virgin Galactic, however, may be closer to getting back on track after the successful partial-duration supersonic flight test of their new vehicle in May.)

With the repeated successful demonstrations of Blue Origin's rocket over the past couple of years, including a spectacular in-flight abort test—with the crew capsule pushing off from the rocket booster less than a minute after liftoff, which even the company expected would destroy the booster but did not—Blue Origin plans to start flying tourists to space next year. SpaceX, meanwhile, has indefinitely delayed its plan to send two tourists around the Moon, and while Virgin Galactic plans to begin their suborbital space tourism program next year, given the company's spotty track record of meeting previously set deadlines, Blue Origin's plan seems more likely to come to pass.

More importantly, experience gained in the development of the suborbital system is being applied to developing Blue Origin's own orbital launch systems, starting with what they call *New Glenn*, to fly by 2020. They also have plans in the future for a heavy-lift vehicle, which would compete with SpaceX's BFR, that they call *New Armstrong*. (Bezos apparently likes to name his rockets after NASA pioneers, all now dead.)

There has been some contentious but friendly Twitter banter between Musk and Bezos, as the two go head-to-head to help drive down launch prices through continuous improvement and competition. In addition to currently having ample financial resources, both men are technically trained—Musk has a physics degree, Bezos has degrees in electrical engineering and computer science—both have a deep understanding of their space vehicles, both understand the need for a competitive industry with multiple players, and both are driven by their space dreams.

Musk talks about humanity being a multi-planet species, in part as an insurance policy against earthly disasters, whether natural or manmade; he is particularly concerned about the potential danger to humanity posed by artificial intelligence. But he seems to consider two—Earth and Mars—to be a sufficient number for "multi." Musk could in fact be accused of an extension of what Carl Sagan called "planetary chauvinism"—the belief that life can thrive only on planets. But many analysts, I included, don't understand the motivation, once having finally escaped the deep gravity well that has confined us to the planet on which we evolved, to dive down into another, albeit a shallower one.

By contrast, Bezos, as noted, aims for a massive human expansion, and not only to many other planets but to inhabitance of space itself. Wealthier than Musk—on some days he is the world's richest person—Bezos is also more willing to spend his own money. Last November he sold a billion dollars' worth of Amazon stock and claims to intend to do the same every year to provide the rocket company with its annual stipend, and he recently built a large facility at NASA's Kennedy Space Center in Florida to begin the manufacture of his large orbital rockets. Musk, on the other hand, always prefers, if he can find a way, to fund his dreams using OPM—Other People's Money. In the case of Tesla and its subsidiary SolarCity, which sells solar panels for home and commercial use, he's done it with loans from Washington (which he has since paid off), various government subsidies for the production of electric cars, and tax credits offered to people buying electric cars or installing solar panels. In the case of SpaceX, extra funding has come, albeit in this case in exchange for providing a direct service, from NASA and U.S. Air Force contracts.

Over a decade and a half since both men launched their space companies, they have made significant progress in reducing the cost of getting to suborbital and orbital space. If their plans for large reusable launch systems come to fruition in the next few years, with SpaceX's BFR and possibly Blue Origin's *New Armstrong* offering larger payload capacities than NASA's non-reusable Space Launch System, they may well render it obsolete before the full Block 2 version flies. (The planned first flight of the initial Block 1 configuration of SLS has slipped to the end of 2019.) Before its second flight—probably no sooner than a year after its first—it may well be canceled for good, not to be resurrected, perhaps finally putting a stake through the heart of Apolloism.

Change on the Way

Six decades after the formation of NASA, it is finally becoming acceptable to talk about space settlement in polite company. But public policy has yet to catch up with the shift in visions. For instance, the Outer Space Treaty, over half a century old, was written for a different era, when few imagined private activities off the planet. It was modeled on the Antarctic Treaty, for a region whose resources were not to be exploited but only to be studied scientifically—which is perfectly compatible with the Apolloistic and Saganite visions, but not with the O'Neillian one. But change may be on the way.

The Trump administration has reversed course from the Obama plan to skip the Moon, instead refocusing on it. In 2008, President Obama had endorsed "the goal of sending human missions to the Moon by 2020," but revealed a change of mind in a 2010 speech: "I understand that some believe that we should attempt a return to the surface of the Moon first, as previously planned. But I just have to say pretty bluntly here: We've been there before."

By contrast, Jim Bridenstine, the new NASA administrator confirmed by the Senate in April, has experience in the commercial space world and recognizes the value of the Moon and its resources. As a member of Congress, Bridenstine introduced a bill called the American Space Renaissance Act, a grab bag of ideas to make policy more friendly to commercial space activities. In a 2016 speech to the Lunar Exploration Analysis Group, he emphasized the value of lunar water to "service satellites with hydrogen and oxygen ... for a fraction of the cost of launching energy or new satellites from Earth." Although satellites would have to be modified to use this fuel, "it would be a simple economic decision":

> The in-orbit maintenance, servicing, and refueling market, already being planned, could be greatly enhanced by an architecture that includes staging nodes, fuel depots, transit spacecraft and lunar landers. This architecture makes economic sense when considering the cost of building and launching new satellites. And the economics improve when considering the returns from orbital satellite assembly and a new generation of communication satellites with unprecedented bandwidth. To be clear, satellite servicing and assembly requires a lunar program that is permanent to include long term human habitation, machines, rovers, and resource production.

This is a visionary view of the near future that we've heard from no previous NASA administrator.

The space-science community is increasingly recognizing the necessity that their research not just be pure science with potential spin-off technologies on Earth, but have practical applications for the development of the new frontier. At the Lunar Science for Landed Missions Workshop at NASA Ames Research Center in January, the presentations were not just on how to determine the age or origin of lunar samples, but on how to assay the potential for water and other materials that could be used for life support, propellant, construction, and manufacturing. And because the rides for new expeditions may not come from NASA, there were also presentations from a number of private companies planning to offer transportation to the lunar surface, including Blue Origin, with a program they call "Blue Moon."

The Trump administration has also recreated the National Space Council—it had been created during the administration of George H. W. Bush and lasted for only four years. While nominally headed by Vice President Mike Pence (a space-policy tradition going back to Lyndon Johnson's vice presidency under John F. Kennedy), it will actually be run by Scott Pace, an international affairs professor and the director of the Space Policy Institute at George Washington University. Pace is himself no stranger to the concept of space settlements, having long been involved in space activism, including in the L-5 Society and, later, the National

Space Society. He recognizes the need for policy changes in light of changing technology and national goals.

A 'Global Commons'?

In December 2017, at the Galloway Space Law Symposium in Washington, D.C., Scott Pace gave a speech that contains hints of what the future of space policy may hold. In it, he laid out six key policy goals:

- U.S. policy will prioritize the interests of the United States and its friends.
- The United States will be the most friendly jurisdiction for private-sector participation and innovation in space.
- The U.S. government will use legal and diplomatic means to create a stable and peaceful environment for both government and commercial space activities.
- The private sector must have confidence that it will be able to profit from capital investments made in space activities and infrastructure.
- We need to resolve questions and ambiguities in the existing treaty structure about property ownership and rights.
- There is a need to develop international norms through both bottom-up best practices with partners, and top-down, non-binding, confidence-building measures without a need for changing existing treaties and arms agreements.

A key element of the speech was this shot across the bow, implicitly aimed at nations like Russia, with dramatic implications for the future of space development and settlement:

> ... outer space is not a "global commons," not the "common heritage of mankind," not "res communis," nor is it a public good. These concepts are not part of the Outer Space Treaty, and the United States has consistently taken the position that these ideas do not describe the legal status of outer space...[R]eference to these concepts is more distracting than it is helpful. To unlock the promise of space, to expand the economic sphere of human activity beyond the Earth, requires that we not constrain ourselves with legal constructs that do not apply to space.

These are international legal terms of art. The 1967 Outer Space Treaty (OST), to which the United States and all spacefaring nations are signatories, declares space activities "the province of all mankind." That is, it is a region in which all people who wish and can afford to can participate. The 1979 Moon Agreement, to which no spacefaring nation has acceded, uses the phrase "common heritage of mankind," implying that it is a commons that must be shared.* Many people within the space community continue to push behind the scenes for the United States to accede to the Moon Agreement. (It is worth noting that one reason space activists have for years fought U.S. ratification of the United Nations Convention on the Law of the Sea—also known as the Law of the Sea Treaty—part of which would regulate seabed mining, is that the Moon Agreement is modeled on it and, given how little seabed mining has actually occurred as a result, it would be a terrible precedent for the ability to mine space resources.)

Removing ambiguities about property rights is crucial for the dreams of space settlements. There will be no legal issue with private ownership of artificial habitats in space itself; ownership of apartments in them could be dealt with like a condominium on Earth. But if people are to live on planetary bodies, such as the Moon or Mars, it will be difficult for them to do so without clear titled property ownership, including mineral rights, and the ability to transfer ownership, borrow against it, and pass it on to their descendants.

Many space lawyers argue that these rights are either insecure under or outright prohibited by two articles in the Outer Space Treaty. Article II states that "Outer space, including the moon and other celestial bodies, is not subject to national appropriation by claim of sovereignty, by means of use or occupation, or by any other means." In order to enforce a property right, a state must have sovereignty over the land on which the property lies—and this is prohibited by Article II.

Meanwhile, Article VI demands "continuing supervision" by governments of non-governmental activities. Commercial space activities have so far taken place close enough to Earth to allow for a soft interpretation of this clause. But farther from Earth—on the Moon or other bodies—it could well demand a more capacious reading, perhaps requiring a government minder to be physically present with any individual seeking to exploit space resources. This would constitute a practical if not a formal ban on securing private property claims on other planets. (I have previously discussed these issues in greater detail in these pages: "Property Rights in Space," Fall 2012.)

There is also the difficulty of the environmental-protection provisions in Article IX, which requires that states exploring celestial bodies must "avoid their harmful contamination and also adverse changes in the environment of the Earth resulting from the introduction of extraterrestrial matter." Like the phrase "celestial bodies," what counts as "harmful" or "adverse" is not defined, and read even somewhat liberally, these terms would seem prohibitive to any meaningful human settlement of planets or moons, or the use on Earth of resources mined from them.

Scott Pace is well aware of these longstanding issues. His language about becoming friendly to the private sector, and being able to profit from investment, is indicative of a pivot of the new administration from the Apolloistic and

* The Outer Space Treaty is formally known as the Treaty on Principles Governing the Activities of States in the Exploration and Use of Outer Space, Including the Moon and Other Celestial Bodies; the Moon Agreement is formally known as the Agreement Governing the Activities of States on the Moon and Other Celestial Bodies.

Saganite models to a more O'Neillian one that envisions people—Americans and others—living and working in space for their own private purposes.

Pace's call "to expand the economic sphere of human activity beyond the Earth" echoes John Marburger's speech from over a decade earlier. But a president's science adviser has much less clout than the executive director of a National Space Council, particularly one who, like Pace, understands the levers of power in Washington. Reading between the lines of the speech, Pace seems to be saying that, while we believe we can operate within existing treaties, we will not allow them to prevent us from achieving our goals for American defense and private space industry. More recently, during an April Q&A session at the Hudson Institute, he described the OST and other treaties governing U.S. space activities as "broadly permissive" with regard to the stated national objectives. I'd put it a different way: We should interpret the OST, and particularly its "province of all mankind" language, as making the solar system safe for traditional English common law, yet not requiring that it be universally applied.

One possibility Pace may imagine is working with like-minded governments, such as that of Luxembourg, which, with its investments in the companies Deep Space Industries and Planetary Resources, seems determined to become a leader in the use of space resources. Both the United States and Luxembourg have passed national legislation—the U.S. in 2015 with the Commercial Space Launch Competitiveness Act, Luxembourg only last year—allowing commercial use of space resources.

These laws, however, don't guarantee international recognition of space property claims. Many in the international space-law community believe this legislation is incompatible with the OST, and the two nations have been criticized for it. It is particularly important to find international partners to help push hard on this goal because Russia aims to counteract it, curiously continuing to push for an international regime based on the Moon Agreement, despite not having ratified it themselves. As space law expert Frans von der Dunk notes in a recent *Newsweek* article, Russia holds to the "common heritage of mankind" view, in which the harvesting of space resources should be treated as an "international enterprise," with the benefits generally shared among countries.

Securing Property Rights in Space

To work toward an international consensus on the legality of commercial mining in space, I would propose the development of multilateral agreements, both within the Anglosphere and with others of like mind, such as Luxembourg, Japan, the United Arab Emirates, Israel, and any other nation that wanted to participate.

Among other things, this agreement would require that Australia withdraw from the Moon Agreement. Now that it has decided to get more serious about space with the announcement in May that it is creating a space agency, and with Australian mining companies champing at the bit in anticipation of harvesting space resources, it has no good reason to continue to be a party to it, if indeed it ever did.

There is also a strong case to be made that the U.S. State Department ought to officially repudiate the Moon Agreement. The United States, under the Carter administration, was originally a key player in formulating it, and even though it was never ratified by the U.S. or Russia, it continues to carry a great deal of informal weight in the space community. A formal repudiation would finally put a stake through its heart.

Beyond putting the Moon Agreement to rest, a set of multilateral agreements should be based on at least the following principles:

- A clear affirmation that the "province of all mankind" language of the OST is fundamentally incompatible with the "common heritage of all mankind" language of the Moon Agreement. Note that, whereas the "common heritage" in the Moon Agreement is the Moon and celestial bodies themselves, including their natural resources, in the OST the "province of all mankind" is "the exploration and use of outer space." It must be affirmed as logically impossible for states to be parties to both treaties at once, even though many parties to the Moon Agreement are also parties to the OST.

- Formal recognition of the utter impracticality of the view that whoever mines resources in space must "share any benefit with all states," as space lawyer Tosaporn Leepuengtham describes a prevailing interpretation of the "common heritage" principle. Many countries are still pushing for this view, as is clear from the April meeting of the Legal Subcommittee of the U.N. Committee on the Peaceful Uses of Outer Space, where some delegates urged that space mining be "exclusively for the benefit of all countries, regardless of their levels of economic and scientific development," and that discussions begin on "how an international mechanism for the coordination and sharing of space resources could be built." The notion that, say, the sale of liquid oxygen from the Moon to Elon Musk for a trip to Mars should somehow benefit Botswana is absurd. But for imports of space resources to Earth, one way of dealing with the issue could be a tariff that would fund a development bank, from which nations could borrow to fund their own space projects. This would meet the spirit if not the letter of the regime to be established by the failed Moon Agreement.

- A requirement that all parties to the agreements will recognize property claims of individuals from *any* nation, including non-party nations, subject to certain conditions. The U.S. Homestead Act of 1862 could be used as a model, requiring an individual to inhabit a prospective piece of real estate for some designated period of time, and improve it in some sense, in order to gain title. The General Mining Act of 1872 might also be used as a model, regulating mining claims and requiring their purchase for a fee from a governing body, if they are considered to be found on publicly owned land. It is hard to see an argument for such recognition as a "national appropriation," which is what Article II of the OST prohibits.

- A distinction between resources extracted in space for personal use, such as harvesting lunar water for life support; resources extracted in space for space commerce, such as harvesting lunar water to create propellant to sell; and resources brought back to Earth from space and for sale in the terrestrial economy. The latter is the only kind of resource extraction that could justify the sorts of concerns targeted by the Moon Agreement, as it could disrupt commodity markets and disadvantage developing nations.
- A permissive interpretation of Article IX of the OST, which requires avoiding "harmful contamination" of celestial bodies. There is need for a clear interpretation of this clause that would not preclude, say, humans landing on Mars, yet would also ensure the preservation of heritage sites, such as the Apollo landing sites on the Moon or Viking landing sites on Mars.
- A more concerted consideration of establishing civilian national space guards. These agencies would be based on the model of national coast guards, cooperating with each other both for constabulary purposes and to help fulfill the 1968 astronaut rescue agreement, which requires mutual cooperation among nations to aid astronauts in distress. (See James C. Bennett, "Proposing a 'Coast Guard' for Space," Winter 2011.) It may also be desirable for the multilateral agreements to include wording ensuring that the requirement to return astronauts not be used as a means to deny asylum—that is, the space travelers have to want to be returned to their countries of origin.

These proposals, of course, will not be without controversy. There are many, both inside and outside of the United States, who do not in fact share an expansive vision of humanity in space, and some who even find human presence so tainted that they believe it should remain confined to the planet on which it evolved. But if we reject this pessimism in favor of the principles just described, after more than a century of dreams of massive human activity in space, new technologies and wealthy new visionaries may at last enable the most expansive space visions to come to fruition.

WHO OWNS MARS? (2018)

ELON MUSK AND THE GOVERNANCE OF SPACE

Tomás Sidenfaden

Tomás Sidenfaden lives in the San Francisco Bay Area, California. He is the founder and CEO of Nuprizm. He holds a BA in Creative Writing and Music Industry from the University of Southern California. 'Who Owns Mars?' was published on Medium in 2018, where Tomás writes about politics, technology and the future.

On March 11, at the 2018 South by Southwest (SXSW) festival in Austin, Texas, Elon Musk announced his intention to send SpaceX's first "interplanetary ship" to Mars in the first half of 2019.

Musk is known for his aggressive—a kind way to say "wildly optimistic"—timelines, which are themselves a function of his giddiness as an innovator. As a result, observers combine a distrust of the initial timeliness with a full faith in Musk's ability to carry out the project in due course.

Musk has, after all, launched industry-disruptive powerhouses such as PayPal, Tesla, SolarCity, SpaceX, and the Boring Company. Founding any one of these companies would, by itself, represent a historic accomplishment. Place them all within a single Curriculum Vitae and people will start to listen.

Significantly, what Musk has to say isn't limited to such mundane details as quarterly vehicle production forecasts (though he admits they stress him out). Instead, he muses on large topics, such the danger that artificial intelligence poses for humanity, the possibility that we're living in a simulation, and the inevitability of World War III.

Or, like at SXSW, he'll discuss how human beings might get to Mars.

Most of the press surrounding this concept has predictably focused on logistics. Ten years ago we weren't sure we could reliably get humans to the Red Planet. But if Musk is right, we're within just a cosmic blink of an eye from that happening—even if his prediction is slightly off.

But the science of getting there is one thing; what life might look like there—indeed, how society might function there—is another. Musk has discussed both topics.

Life on Mars

Musk envisions a human colony of a million people on Mars, preceded, of course, by an intrepid set of early, capable, and fearless pioneers.

> For the early people that go to Mars, it will be far more dangerous. It kind of reads like Shackleton's ad for Antarctic explorers: Difficult, dangerous, good chance you'll die. Excitement for those who survive.

What's certain is that the first colonizers of the Red Planet aren't likely to raise our collective blood pressure about the intricacies of interplanetary governance. There will be more immediate concerns; considerations of survival, not civilizational durability, will occupy the minds of these early pioneers to Mars.

Even by Elon Musk's standard of cheap ($200K a head), it will still be incredibly expensive to send people there, let alone sustain them once they arrive. And it won't be a tourism trip—those going will have to build all the infrastructure necessary to ongoing survival.

While the first wave of travelers will be focused on securing the most basic conditions for sustenance—air, water, food, shelter—the next wave will utilize expertise and ingenuity to enhance those conditions. These engineers, scientists, doctors, and technologists will attempt to go from making sure Mars is *survivable* to making sure it's *livable*, a subtle though real distinction.

But as the colonial population increases, the colony will presumably start to resemble our planet's major cities. While colonizers won't expect the types of amenities of a New York or a Tokyo, they will begin to expect more services than an Antarctic outpost. Soon, mandatory personnel will give way to—in likely succession—wealthy tourists, entrepreneurs, economic migrants, opportunists, and...criminals.

If that's a particularly bleak perspective on a possible future sequence of events, keep in mind that this assessment assumes the colonization is supported by SpaceX—an American company led by a chief executive who believes personal agency and freedom of action are among civilization's greatest achievements.

Of course, there's no reason to believe Musk's vision is inevitable.

Other Space Racers

Even Musk himself suggests that companies other than SpaceX should be competing to colonize this new planet. But why does Musk envision a corporate competition on Mars? Space flight, until recently, has been the exclusive purview of nation-states—entities large enough to be able to devote the massive resources necessary to equip such missions.

SpaceX may be led by a motivated and capable leader, but it is empowered by dozens of institutions and individual investors that the company has courted over the last 12 years. This type of funding requires dedicated teams to acquire resources, strict timelines for performance, and achievable exit strategies.

But SpaceX and its rivals wouldn't be the only game in town.

For the most part, nation-states are comparatively unencumbered by fund-raising and performance benchmarks. Administrations of democratic societies can hardly expect to devote, say, 25 percent of their tax revenue to space exploration. NASA's entire 2017 budget (Mars exploration aside), for example, was $19.3B, or 0.5 percent of the U.S. federal budget.

China's space program budget is roughly $3B, or less than 0.1 percent of the national budget, according to some observers. Nevertheless, China benefits from a convenient ambiguity between private industry and state-owned enterprises that skews their true investment numbers. In contrast to Western nations, China is comfortable outsourcing innovation to its private sector and rewarding a budding domestic company with the ample power of national tax revenue.

Of course, the U.S. and China aren't the only participants in space. Russia, the E.U., and, most recently, India, are all players on the interplanetary stage.

Russia may struggle to be a long term contender. It's not due to a lack of expertise—they were, after all, the first society to launch us into space. Yet the space program is a government enterprise, and will likely remain so unless and until the system of government changes. Russia's budget is disproportionately dependent on oil and gas. Yet these revenue streams are compromised by the emergence of new oil and gas sources, as well as the global move away from fossil fuels. Thus, Russia's ability to be competitive in space over the long haul depends largely on the ability of its government to diversify its revenue sources.

The European Space Agency (ESA) is, on paper, the most capable entity to compete with NASA and SpaceX. It is a partnership of 22 E.U. member states, responsible for significant contributions to the International Space Station, the Mars Express orbiter, and most recently, Rosetta, the ambitious and partially successful attempt to land a rover on the Comet 67 P/Churyumov-Gerasimenko, over 300 million miles away from Earth. The ESA could achieve remarkable breakthroughs not just scientifically but bureaucratically.

Meanwhile, as the U.S., Russia, China, and the E.U. were all laboring to send payloads into space, India successfully launched a Mars orbiter for just $74 million.

According to Narendra Modi, India's prime minister:

> A one-km auto rickshaw ride in Ahmedabad takes Rs 10 per km and India reached Mars at Rs 7 per km, which is really amazing.

It is also a testament to the capacity of a nation-state to achieve an objective consonant with, and fully reliant upon, all its available resources. India's Mars mission cost less than what it cost Warner Brothers to produce the 2013 space film *Gravity*, which was released in the same year.

Assuming SpaceX, the United States, China, the E.U., Russia, and India are all potential players, there could also be a few other corporations in the mix. SpaceX may be the first pioneer, but other landing outfits will likely follow.

The Biggest Questions Surrounding Interplanetary Colonization

Will they coordinate the logistics of this human colony back on Earth? Will they cluster around the same zones and pool their resources and talents? Will they focus on staking out territory akin to 19th century European colonists? Will they vie for technological, military, and economic dominance? Which model of colonization will each entity promote?

If you ask Musk:

> Most likely the form of government on Mars would be a direct democracy, not representative. So it would be people voting directly on issues. And I think that's probably better, because the potential for corruption is substantially diminished in a direct versus a representative democracy.

But how much thought has he really given this issue of governance in space? He certainly implies he's thought about it a lot. While we terrestrially-minded humans are warring over the existentially meaningless definition of our individual and group identities, or "one sad problem after another," Musk seems preoccupied with what he characterizes as the inspirational goals of greater humanity.

Musk has been vocal about the dangers of artificial intelligence—an amorphous technology the average person knows about mostly through science fiction—though it was just weeks ago that he discovered his companies had profiles on Facebook. Visionaries tend to have interesting blind spots.

Perhaps he really is more concerned with the science of it all. Perhaps he's more preoccupied with missions whose launch dates are nearer than whatever the real launch date for a Mars mission happens to be. Perhaps he is prioritizing what he would *like* to see rather than what is the likeliest governance scenario.

Not to get too bogged down in psychoanalyzing Silicon Valley CEOs, but perhaps all of this stems from an aspirational source: Musk senses the zeitgeist—and the toxicity of our politics—and believes his purpose is uniting us in one of our most fundamental desires: to explore our universe.

Part of Musk's popularity is a result of the care he takes in not being drawn into the politics that now reliably reach into every corner of our lives. But is it overly optimistic to believe we won't export these problems to a new colony millions of miles away?

A good case can be made that the societal framework(s) we set up on Mars will not ultimately be invulnerable to the vicissitudes of Earthly life. Our Martian colonies will likely resemble the tendencies, dispositions, and norms—written and unwritten—found on our own planet. And it will be the events chronologically closest to the Mars mission that will have the greatest impact on the shape of things on the Red Planet.

Let's consider a possible future, a set of developments that could come to pass as Mars is being colonized.

In 2038, the United States' federal budget is almost entirely consumed by three mandatory spending programs: healthcare, retirement benefits, and interest on the debt. Any "discretionary" spending, such as the military, infrastructure, education, science, energy, or space requires issuing new debt. This roils confidence in economies underpinned by dollar-denominated transactions. Investors flee to safer havens, potentially including China's RMB, cryptocurrencies, or commodities. Knock-on effects compound as international markets restructure to account for increased volatility.

In many categories, China exceeds the United States as a global superpower. It has successfully created a middle class for whom basic prosperity is a given and whose focus is naturally oriented toward greater self-determination. Chinese tourists travel the world by the hundreds of millions and are exposed to societies unencumbered by the same censorship and political repression they experience at home. The yawning gap results in domestic political pressure that the Community Party of China works to channel into defensive ends, like nationalism. Domestic anti-establishment efforts, escaping repression at home, manifest common supranational ideologies.

Surveillance is total and virtually unlimited. Privacy is a quaint notion embraced by a small and marginalized group of recluses, modern Luddites, and populations without access to the latest technology. For non-democratic societies, complete transparency is enforced by government decree in the name of security and social cohesion. For democratic societies, business drives surveillance in the interest of actionable market data, while consumers drive it in the name of security. Digital identities are mandatory, required for travel, commerce, and socialization.

Science, medicine, and technology are transcendent, though access to their benefits is unequal. The wealthiest of the world are genetically engineering their children and embedding tech into their persons, while the poorest are still struggling to get access to clean water, roads, and healthcare. Developing economies and remote, undefended territories attract major powers hungry to support increasing energy demands and to ensure the primacy of their own forms of government.

Whole industries, such as freight and transportation, are beginning to be replaced by automation that renders the less technical incapable of reentering the workforce, further exacerbating problems of social cohesion and economic fundamentals.

Conflicts are both national and transnational, continuous, ambiguous and asymmetrical, rarely involving direct military stand-offs, and leave unaligned groups incapable of amassing defensive measures.

Is such a scenario too grim?

Musk has expressed concern about the likelihood of WWIII and the dangers of artificial intelligence. Both of those observations could be considered a distillation of the potential pitfalls mentioned above. Few things are more likely to torpedo humanity's attempt to settle a distant planet like a world war or a hostile takeover by artificial intelligence. The last space race was catalyzed by the race for global military dominance. Will the effort to colonize Mars be a modern replay of that process?

Technology has always been disruptive. But the interconnectedness of modern societies and the proliferation of new tech poses a far greater threat to today's world than the new technologies of the past had on their respective times. The pace of evolution gave former societies dozens, if not hundreds, of years to adapt to the types of change we experience in a matter of years today.

Responsible modern technologists have a duty to consider the impact of their work, whether or not it is ethical, and all the ways in which it may affect our collective future. Mark Zuckerberg's ongoing struggle to manage the power of his product is a perfect example.

When the stakes are large enough, little meaningful engineering occurs without consideration of potential failure modes. Unlike commerce, however, failures of governance are regressive, all-encompassing, and often lethal, setting back progress on all fronts of society by years or decades. In the proper framework, competition breeds excellence—but it can also breed violence and oppression.

To suggest a direct democracy will "most likely" be the system of governance of a future human colony is highly optimistic. What we will need is expert, meticulous planning, not sci-fi musing. Musk is one of the likeliest candidates to propel the human race into Mars, yet this does not mean he is the right person to map out our future there. The organizational and structural aspects of colonizing Mars—as opposed to its scientific and engineering challenges—require serious reflection, not armchair theorizing.

Those Who Ignore The History of Colonization Are Doomed To Repeat It

History is replete with lessons on the pitfalls of colonization. The very word connotes an understanding of the past for which modern civilization has struggled to come to agreement. That's because a defining component of earthly patterns of settlement was the diversity of societies responsible for it: nation-states, races, ethnic groups, migrants, refugees, etc.

All of these were, nevertheless, human. Mars represents an opportunity to grow together as a species. We can arrive, even as separate entities, under a modest framework that accounts for our diverse perspectives and priorities. History informs us of potential alternatives: arms races, open conflict, genocide.

It doesn't have to go that way with interplanetary colonization. But to avoid it, we must encourage—today—meaningful, open, and honest debate about the values and principles that can help humanity survive the heat and pressure necessary to reach the species' true escape velocity.

MARS OR BUST (2018)

HOW SCIENCE FICTION FILMS WILL PROMOTE MARS COLONIZATION REALITY

H Raven Rose

H Raven Rose is a filmmaker, screenwriter and author. She grew up in Georgia's Blue Ridge mountains, and earned a degree in screenwriting from Lesley University, Massachusetts. She recently spent two years living in Wales, attending the Swansea University creative writing research programme as a PhD student. Now back in the United States, she continues her postgraduate studies part time.

Introduction

'Earth is the cradle of the mind, but one cannot live in a cradle forever,' said Russian-Soviet rocket scientist Konstantin E. Tsiolkovsky. Planet Earth is a fragile entity with increasingly reduced resources. Apocalyptic shadows encroach upon humanity. Human extinction or evolution into an unrecognizable form may not be avoidable. Visionaries, Elon Musk, Robert Zubrin, Richard Branson, Jeff Bezos, and others, banking vast intellectual and other resources, powered by passion and dazzling drive, have lucid dreamt a new human destiny. Humanity will colonize Mars. In bygone times, people looked to the glimmering stars and planets of the night sky as a form of wonderlust. As Red Planet colonization becomes possible, wonderlust rapidly becomes wanderlust. Despite the cost, the near-impossibility of the likely dangerous, hard-to-sell, formidable venture, colonizing the glowering red planet, seemingly, is feasible. If this is so, then how can this possible future be made believable, if not inevitable, to all humanity? Global imaginations must be sparked and set on fire through the mesmerizing power of story. Cutting edge Mars colonization mythologies can transform and inspire the collective psyche,[1] and thus take humanity beyond the moon and stars to a new earth on the Red Planet.

Background: Red Planet Siren Call

As ancient human records and narratives reflect, the fiery red ball has long mesmerized Earthlings. 'Mars has become a kind of mythic arena onto which we have projected our Earthly hopes and fears' argued astronomer and Pulitzer Prize-winning author Carl Sagan in his book *Cosmos* (1980). Mars science fiction, whether laughable or idealized, has so far included xenophobic fears of Martian invaders, sole survivor fantasies, or idealistic notions of highly advanced technologically and otherwise enlightened Martians in E.T. utopian paradise. Then there is the grand literary vision of Mars as a human colony. The Red Planet does have all of the resources needed to support human life. Yet, if humanity is to colonize Mars, how will the human race get there from here? 'From a childhood vision to a space-age reality,' said character Tom Chen in the film *The Space Between Us* (2017). Played by Gary Oldman, the character's words are a cinematic summing up of modern Mars mythopoetic power and potential to rewrite reality. Consider the historical evidence that science fiction transmits visions with the power to transform and better human lives and reality.

From Childhood Visions to Space-Age Reality

Imagine a small, freckle-faced red-headed boy born circa 1866. At age ten or eleven, he reads *Twenty Thousand Leagues Under the Sea* by Jules Verne. From then on, the inventive, scrappy little boy, dreamt 'of making voyages under the waters'.[2] His name is Simon Lake. He drew his first plans for a submarine with a diving compartment around age fifteen, then grew up to design, build and successfully test increasingly sophisticated submarines. Prototypes, the *Argonaut Junior* and the *Argonaut*, were seaworthy in 1894 and 1897.[3] The child became the man and fulfilled his boyhood vision of becoming a 'sub-aqueous pioneer' much like the *Twenty Thousand Leagues Under the Sea* story hero who sparked his imagination as a child.[4] The astonishing journeys author—Jules Verne himself—sent a congratulatory cable when *Argonaut* successfully voyaged. In Lake's own words, 'Jules Verne was in a sense the director-general of my life'.[5]

'It's alive! It's alive!' uttered Dr. Frankenstein in the 1931 film[6] adaptation of Mary Shelley's novel *Frankenstein, or the Modern Prometheus* (1818). Those electrifying words led Earl Bakken to his destiny. 'My favorites were those incredible science-fiction films ... in which electricity, usually applied by a "mad" scientist, rendered someone supernaturally strong, invisible, or in some other astonishing way changed. Foremost among those films was *Frankenstein*, the unforgettable story of the learned doctor who, through the "magical" power of electricity, gives life to a collection of inanimate body parts' wrote Bakken in his memoir.[7] Bakken describes seeing the 1931 film *Frankenstein* at the age of eight, 'what intrigued me the most, as I sat through the movie again and again, was not the monster's rampages, but the creative spark of Dr. Frankenstein's electricity. Through the power of his wildly flashing laboratory apparatus, the doctor restored life to the unliving'.[8] Bakken grew up to co-found medical technology company Medtronic

Inc., and himself give life to those who might otherwise be unliving when he created the first wearable, external, battery-powered, transistorized pacemaker in 1957.[9] Medtronic devices have since helped untold numbers of people.

The name Sikorsky is synonymous with helicopters and the technological advances and prowess of the armed forces. Reading science fiction as a ten or eleven-year-old boy stirred Igor Ivanovich Sikorsky, like Simon Lake, to create his remarkable reality. Jules Verne's 1886 novel *Robur The Conqueror*, also known as *The Clipper of the Clouds*, inspired Sikorsky to build a helicopter.[10] In 1923, he founded Sikorsky Aircraft Corporation and built both fixed-wing aircraft and later helicopters. The aeronautic engineer's boyhood flying machine fantasy became a reality years later when he flew the Vought-Sikorsky 300 helicopter in 1939.[11] Igor Sikorsky's son Sergei shared his father's words that the 'helicopter-like vehicle' in *The Clipper of the Clouds* and a Jules Verne quote 'imprinted' on his mind inspired his dream of inventing a helicopter.[12] The quote Sikorsky remembered was 'Anything that one man can imagine, another man can make real.' The word-for-word quote, from Verne's book *The Steam House*, is, 'All that is within the limits of possibility may and shall be accomplished'.[13] Good sci-fi facilitates possibility thinking and directed imagination, or inventiveness. Science fiction is alchemical, it changes minds, and inspires humans to make real whatever they imagine.

Real or Imagined: We Can Remember It for You Wholesale

In *Total Recall*, the 1990 film based on Philip K. Dick's *We Can Remember It for You Wholesale*, Douglas Quaid dreams about the red planet repeatedly then tries to get a Mars virtual vacation memory implant. Quaid increasingly cannot tell the difference between fantasy and reality. Curiously, in reality, humans have a similar inability to distinguish between the imagined and the real. Whether due to the mind-body connection and the function of neural patterning, whether wiring or rewiring, the human brain does not seem to distinguish between fantasy and fact. Whether an experience is virtual or actual, imaginary or real, it impacts the human organism. Pascual-Leone et al. scanned the brains of study participants either playing notes on the piano or imagining doing so and found brain changes of statistical relevance for both groups.[14] A control group, who neither played piano notes nor imagined doing so, had no corresponding brain changes.

The Original Purge: Your Emotions on Story

Well-created content in any genre creates a portal for a narrative consumer, a film or television viewer, story reader or game player, to participate in a transformative rite of passage, enter a liminal space similar to a dream state. Therein, upon identification with characters leading to a satisfying narrative result, the viewer or reader ultimately experiences Aristotle's 'katharsis' or purification.[15]

This common-knowledge vicarious purging of emotions is one reason that many humans love narratives. Ultimately, Aristotle's narrative catharsis transforms the content consumer. Modern stories are meant to entertain and often make people think and feel. In Ed S. Tan's words, 'In general, narration may be seen as the systematic evocation of emotion in an audience, according to a preconceived plan. Narration by means of film is one way of doing this'.[16] As Tan's book title *Emotion and the Structure of Narrative Film: Film as an Emotion* Machine suggests, a film is a feeling-making device. Neuroscience evidence suggests how this may function in humans.

Transformation Through Story

Think of a whole being and whole brain alteration methodology as transformation through story. Narratives can be designed and written to cultivate awareness generally, or promote any other authorial point of view, as well as to specifically prepare humans for spacefaring and the brutal off-world colonization processes which lie ahead. One may devise a story as a compelling rite of passage, or neural-net creating ritual, a virtual somatic journey, with the subliminal potential to inspire and influence society's collective unconsciousness and consciousness.[17] A successful sci-fi or other narrative is thus a vicarious experience which communicates that a fantastical experience is or could be a fact. The liminal in combination with the subliminal becomes supraliminal or conscious and thus real. What does a person experience while consuming such content? Understanding the power of story to drive human reality requires that one consider the human brain and whole being on narrative.

Have Sci-Fi—Will Travel: Fiction as Portal or Gateway

Neural or brain nerve cell wiring and re-wiring is the way that the human brain processes information. The brain creates neural associations or neuronal networks as part of processing and understanding experience, including stories. Good consumer content involves suspension disbelief, both literal and figurative. Compelling stories act as a threshold, or doorway, between the worlds of fantasy and reality. Think of reality as the middle world and a narrative as an actual gateway to the upper or lower worlds of non-ordinary reality. In consuming content, one crosses the threshold, enters and traverses a fictional liminal or Bardo state, undergoes initiation, and thus vicariously experiences an actual mission to or colonization of Mars. Once the film, book, game or other content ends, liminality transcended, the trance ended, the individual awakens into and returns to ordinary reality. In this way, a feature film serves as a neural-net creating somatic ritual with the potential to rewire and transform brains and humanity.

Your Mind on Story: from Fantasy to Reality

Science offers evidence of the movie-mind connection. In the *PsyArt* article 'This is your Brain on Culture: Your Brain on Movies', Dr. Norman Holland summed up the Weizmann Institute findings: 'Specifically viewers' brains behaved alike (high ISC) in the primary visual areas of occipital and temporal cortex, Heschl's gyrus (auditory region), Wernicke's area (language processing), some limbic areas (emotion), the fusiform gyrus (face recognition), and the association cortices that partially integrate primary sensory data'.[18] Holland explains that viewers get 'lost in' movies as passive viewers who surrender or give 'over control' to the film.[19] Holland explains it thus. A movie-goer's brain yields 'control to the movie projector,' shutting off 'reality-testing,' and then they enjoy what he terms the filmic version of Mihaly Csikszentmihalyi's flow state.[20] An interconnection between the psyche and cinema is not surprising. Indeed, in Jungian terms, a human experience, whether dream, fantasy, vision, feature film or content, is an external out-picturing of an internal individual or collective psyche experience. Ira Konigsberg indicates that 'Psychoanalysis and film began at about the same time and the relationship between the two was noticed almost from the start'.[21]

The Good, the Bad, and the Ugly: Cinema as Mind Control!

Uri Hasson and Rafi Malach, with a group of researchers at the Department of Neurobiology, Weizmann Institute of Science, Rehovot, Israel, researched the neuroscience of watching movies. They had five research subjects lie in a fMRI machine—functional magnetic resonance imaging—and watch the opening thirty minutes of the movie *The Good, the Bad, and the Ugly* (1966). What researchers discovered was a high inter-subject correlation (ISC) in study participant brain activity. Some 30% of the subjects' brain areas showed the same activity pattern. Researchers concluded that measuring brain activity during movie watching was useful for 'assessing the cognitive and emotional effectiveness of a movie'.[22] Perhaps the most interesting result found by the researchers, along with the 'high ISC in brain activity,' was 'the same film exerted considerable control over viewers' behavior as measured by tracking their eye movements'.[23] Many subjects, instructed merely to watch the movie, gazed at the same location at the precise cinematic moment. The movie itself directed viewer attention. This result may be perceived as quite Orwellian,[24] as narrative naturally, with or without nefarious intent, is inadvertent mind control.

Neuroscience of Story

It is possible to rewire the brain to rewrite reality. Mirror neurons, brain cells that fire during a human experience or when a person observes another having an experience, inform the transformation through story puzzle.[25] Sympathetic or shared neural activation, when a content consumer identifies with a fictional character and vicariously experiences their story journey, is a function of narrative empathy.[26] Hasson et al. speculate that 'part of the mesmerizing power of movies stems from their ability to take control of viewers' minds, and that viewers often seek and enjoy such control because it allows' deep absorption and mental engagement in a film.[27] Is this taking control of content consumer minds, to facilitate awareness, through motion picture or other content, a form of mind control? Despite valid ethical concerns, good sci-fi stories are undoubtedly useful to rewire the brains of the masses and forward the Mars colonization initiative. Narrative allows the collective consciousness to experience the fantastic as fact. Through topical consumer content, humankind will become comfortable with and committed to the brave new space colonization world which lies before us. We must dream a new reality, one where homo sapiens terraform, and thus green, the planet Mars.

Conclusion: Just the Beginning

Spacefaring and the brutal off-world processes of colonizing and making Mars comfortable will require heroic efforts, much sacrifice and perhaps the loss of many lives. Yet, the valiant human journey and cause will be worth it as, in the words of Joseph Campbell, 'A hero, properly, is one who gives his life to something greater than himself.' Consumer content to promote the mission to Mars enlightenment path has the power and potential to anticipate and influence humanity's future. Topical science fiction summer blockbusters and other popular culture content, including games or television, will play a crucial role in cultivating collective human awareness and dedication to Red Planet colonization. In both fresh, new sci-fi films, games, literature, new media, and other content, and then, in reality, we must go there. To quote Hauser—from the movie *Total Recall*—humankind must '*Get your ass to Mars*'.[28] 'And the rest is, well, it's not history,' as Gary Oldman's character said in *The Space Between Us,* 'it's just the beginning'.[29]

THE MARS DECISION (2019)

HOW TO SHOW THAT AMERICAN DEMOCRACY CAN STILL DO GREAT THINGS

Robert Zubrin

Robert Zubrin is an American aerospace engineer and advocate for the exploration and colonization of Mars. Born in New York City in 1952, he is the creator of the 'Mars Direct' concept, along with Martin Marietta and David Baker. The concept seeks to propose ways to significantly cut the cost of humans exploring Mars. An expanded version of the idea was published by Zubrin in his 1996 book *The Case For Mars*. In 1998, Zubrin and others founded The Mars Society, a non-profit corporation dedicated to the human exploration and settlement of Mars. This is his second contribution to *The Book of Mars*.

There was a bittersweet quality to the recent celebrations of the fiftieth anniversary of the first lunar landing. It was an occasion of justifiable American pride—after all, sending men to the Moon in the late 1960s and early 1970s was not only a feat of human ingenuity and daring but a spectacular national accomplishment, one that, as Jules Verne had sagely predicted a century earlier, only Americans could pull off. But in the half-century since Apollo 11, NASA's human spaceflight program has stagnated. It has had very few memorable successes and certainly performed no comparably glorious feats.

Why not? At the time of the Moon landing, it was generally expected that the United States would quickly go on to Mars. Even several of the Apollo astronauts believed, as they described in their memoirs, that after going to the Moon they might help the United States reach the Red Planet. Apollo 11 astronaut Michael Collins recalled thinking "perhaps I could help them [NASA] plan" a Mars mission. Edgar Mitchell, the sixth man on the Moon, remembered feeling that "it wasn't unreasonable to hope" he'd be assigned to a Mars-bound crew. Gene Cernan, the twelfth and last man on the Moon, recounted with sadness the time that he "finally faced the facts: 'I'm not going to Mars.'"

Many explanations have been offered over the years for why American astronauts have not been sent on to Mars—or anywhere else of note—in the years since Apollo. Three explanations in particular stand out for being widely believed. Each of these explanations is intuitively plausible. Each has a kernel of truth. But these explanations are so incomplete as to be misleading. And taken together, they amount to a profound misunderstanding of how democratic peoples can do great things.

First: popular opinion. It is often argued that American progress in space stalled because of waning public interest. If only (or so the thinking goes) the American people had continued to care about and support NASA as heartily as they had during the 1960s, the space program's record of achievement would have continued unabated.

While this claim does have an obvious basis in fact—of course politicians would be eager to send American astronauts to Mars if voters were clamoring for that, and of course a Mars program would suffer if overwhelmingly opposed by the public—there are two glaring problems with it. It overstates the degree of popular support for the space program in the 1960s; an analysis by historian Roger Launius found that Project Apollo in particular and lunar exploration in general almost never enjoyed majority support in contemporary polls. More importantly, the notion that strong popular support is a prerequisite for an ambitious space program is mistaken in the extreme. Yes, policymakers must take public opinion into account. But public opinion must also be informed and guided by policymakers willing to inspire with their vision, to offer concrete proposals, to make persuasive arguments, and to assemble necessary resources. That is the essence of political leadership in a democratic republic. We did not go to the Moon because of massive, preexisting public support for such an idea, and we haven't failed to reach Mars because of public uninterest.

The second overly simplistic explanation for why the American space program fizzled after Apollo 11 is money: If only (or so the thinking goes) funding for NASA were much higher, we would have been on Mars long ago.

Again, this explanation has a basis in fact. Even before we reached the Moon, NASA's budget was reduced, curtailing plans for Apollo and successor programs. And, as is often pointed out, at the peak of the Apollo program NASA consumed 4 percent of the federal budget, compared to about 0.5 percent today.

But underlying this explanation is the laughably false assumption that more funding results in more accomplishment, that level funding results in comparable accomplishment, and that reduced funding results in less accomplishment. Any student of history could name counterexamples—enterprises, both public and private, in which bigger budgets brought only waste and straitened budgets resulted in urgency and success.

More to the point, blaming low funding for NASA's stagnation is also misleading because, when adjusted for inflation, the agency's funding over the past twenty years has actually been *greater than* its funding was over its first twenty years. Expressed in constant 2018 dollars, NASA's total funding during the period from 1959 (the agency's first full year) through 1978 was $335 billion. The agency's total funding during the period from 1999 through 2018, again expressed in constant 2018 dollars, was $387 billion—an increase of 16 percent.

Now contrast what the agency accomplished during each period. In its first two decades, NASA not only did the Mercury, Gemini, Apollo, Skylab, Ranger,

Mariner, Surveyor, Lunar Orbiter, Viking, Pioneer, and Voyager missions, it developed virtually all the technologies that have enabled space missions then and ever since, including hydrogen/oxygen rocket engines, multi-stage heavy-lift launch vehicles, space life-support systems, spacesuits, lunar rovers, radioisotope generators, space nuclear reactors, deep space navigation and communication technology, space-rendezvous technology, soft landing systems, reentry systems, and most systems that would be used for the space shuttle, and also built the Deep Space Network, the Cape Canaveral launch complex, and most of its centers and testing facilities.

Over the past two decades, however, NASA's accomplishments—with the notable exception of its superb robotic missions of planetary exploration and space astronomy—are not remotely comparable to those of its first two decades. Far from going beyond the Moon, NASA's astronauts have barely flown 0.1 percent of the distance to the Moon. The rate of development of new flight technologies has been near zero. In fact, it has arguably been less than zero in some areas, as exemplified by the failure of NASA's Space Launch System (SLS) program to be able to *redevelop* the J-2 engine that powered the upper stages of the old Saturn V from the Apollo days, leaving SLS, which has about the same takeoff thrust as the Saturn V, with only about half the ability of the Saturn V to throw payloads to the Moon.

The third culprit sometimes blamed for the stalling of the American space program is the fickleness of democratic government. Democracies (or so the thinking goes) are ill suited to the accomplishment of grand undertakings, except in the exigencies of war or war-like circumstances. Absent those conditions, it is held that great things can be accomplished only by private enterprise driven by the promise of profit or by unfree regimes that have a tyrant's constancy of aims.

This explanation, too, has a basis in fact. It is true that the space race against the Soviets—and the larger context of the Cold War—gave a sense of urgency to the Mercury, Gemini, and Apollo projects. It is true that, once the Soviets were beaten and the goal of reaching the Moon was achieved, the purpose of the American space program became less clear. It is also true that presidential administrations have tended to reject their predecessors' plans and timetables for sending humans into space, so that the overall direction of the space program has been erratic and rambling.

The belief that great public works are incompatible with peacetime democracy is also seemingly supported by other evidence from U.S. history outside the space program. There is a reason that the wartime Manhattan Project is cited alongside Apollo as the very model of ambitious publicly funded R&D—as in the refrain "We need a Manhattan Project for X." And some of the biggest American engineering projects were made possible only by flouting democratic norms. "Pure democracy has neither the imagination, nor the energy, nor the disciplined mentality to create major improvements," asserted Raymond Moley in the foreword to a 1970 book by Robert Moses. Moses was the official who, across three decades, built a staggering number of bridges, tunnels, roads, parks, and housing projects in and around New York City. He accomplished all this by amassing quasi-dictatorial powers for himself, running roughshod over his critics, razing whole neighborhoods, and displacing hundreds of thousands of people in the service of his vision of progress. "You have from time to time remarked that I do not have to be elected to office," Moses told one opponent. "Perhaps that is why I am in a position to protect the really long-range public interest."

But this explanation for NASA's stagnation fails like the others. Many great things have been accomplished by democratic means during times of peace in the United States, including massive public works like the Erie Canal, the Hoover Dam, and the Interstate Highway System, and other awesome works that arose from public-private partnership, like the Transcontinental Railroad and much of our modern telecommunications system, including the Internet. Moreover, the undemocratic notion that elected officials cannot protect "the really long-range public interest" perniciously implies that power ought to be handed over to individuals who feel no obligation to answer to the people. And in the case of NASA, blaming the space agency's stagnation on the capriciousness of the American regime is also defeatist, since it implies that unless we give up democratic self-government we must give up all hope of doing great things in space.

Each of these three explanations for why NASA failed to get beyond the Moon—a lack of popular support, a lack of money, and a lack of stable goals in democracies—falls short. But having weighed each of them, we are now prepared to see more clearly the real structural reason for NASA's woes and to figure out, from sound principles of engineering, budgeting, and project management, how to build a human space program that can thrive under American democratic politics.

Purpose-Driven Missions

The main reason for NASA's stagnation—the best explanation for the difference between the rates of accomplishment in the agency's first two decades and the last two decades—is the change in its mode of operation. During the Apollo years, NASA's human spaceflight program was *purpose-driven*. Since then, it has largely been *vendor-driven*. A purpose-driven program spends money to do things. A vendor-driven program does things in order to spend money. In a purpose-driven program, spending is focused and directed toward a well-chosen goal. In a vendor-driven program, spending is unfocused and entropic.

The problem does not affect NASA as a whole. As noted above, the agency's programs for robotic planetary exploration and space astronomy have continued to produce impressive results. This is because they have remained purpose-driven. But without a clear, driving goal, NASA's biggest programs—its human spaceflight effort and associated launch-vehicle development programs—have dispersed hundreds of billions of dollars over the past half-century with very few results worthy of the costs and risks. (One could say virtually no worthwhile results, except for the five space shuttle missions to launch, repair, and upgrade the Hubble Space Telescope—an exception that proves the rule, for in those missions the shuttle was made to serve the purposeful science program.)

Part of the difficulty in moving from a vendor-driven approach to a purpose-driven approach lies outside NASA. The vendor-driven approach is reinforced by certain characteristics of our democratic politics, especially the tendency of members of Congress to prefer and push for projects that bring jobs to their districts. Moving to a purpose-driven approach will sometimes require NASA, with support from the president, to push back against that tendency and all the inefficiency it entails.

To be truly purpose-driven, a space program needs to meet the following conditions:

1. It must have a definite goal.
2. The goal must be assigned a proximate deadline, not a far-off one.
3. Projects must be selected and decisions about technology development must be made with the aim of achieving the goal by the deadline.
4. The projects selected must be pursued as rapidly and efficiently as possible.
5. The goal needs to be rationally chosen to accomplish the most we can.

These five conditions take into account the realities of big engineering projects, including the risks of bloat, sloth, and mission creep. And they take into account the pressures of electoral politics, public opinion, and budgeting. A space program that satisfies all five conditions, a purpose-driven program, is a worthwhile one that has a chance of success. A space program that fails on any of the five conditions is likely to result in waste and stasis and be vulnerable to being killed off by policymakers.

All five of these conditions were fulfilled by the Apollo program. The goal was to reach the Moon by the end of the 1960s. That satisfied conditions 1 and 2. The powerful focus on that goal meant that factions demanding peripheral projects, such as the construction of space stations, were pushed out of the way, thereby satisfying condition 3. Each component of the overall project was pursued expeditiously, in keeping with condition 4. Finally, the goal—by both achieving a major human milestone and winning a geostrategic victory—fully satisfied condition 5.

All this was laid out right from the start. As President Kennedy said in his famous speech before Congress on May 25, 1961, "First, I believe that this nation should commit itself to achieving the goal, before this decade is out, of landing a man on the moon and returning him safely to the Earth. No single space project in this period will be more impressive to mankind, or more important for the long-range exploration of space; and none will be so difficult or expensive to accomplish."

The Graveyard of Purposeless Programs

For most of the period since Apollo, NASA's human spaceflight program has had no goal at all. Some NASA leaders have even explicitly rejected the very idea of having a goal—as when Sean O'Keefe (NASA administrator, 2001–2005) repeatedly declared that the agency should not be "destination-driven."

Even when a nominal goal has been chosen, it has not been treated as authoritative. For example, in a speech on July 20, 1989, the twentieth anniversary of the Apollo 11 Moon landing, President George H.W. Bush announced that the United States would return to the Moon, this time to stay, and then go on to Mars. But that project, dubbed the Space Exploration Initiative, quickly collapsed. NASA at that time was pushing Space Station Freedom as its next major project and the president had unwisely described the space station as a "critical next step in all our space endeavors." So when, three months later, NASA laid out its plans allegedly to accomplish the president's goal, it proposed to send crews to the Moon using massive spaceships assembled on-orbit at a huge space station. The plans were so costly and complex that many veterans of the Apollo program who still filled NASA's ranks at that time could only scratch their heads and wonder, *If we could put a man on the Moon, why can't we put a man on the Moon*? And with the return to the Moon made impossible by the requirement to expand and then use the space station for on-orbit assembly, NASA's even more convoluted Mars mission design was utterly unfeasible. Bush's Space Exploration Initiative died quietly: It had no chance of getting through Congress, since it offered no prospects for attaining meaningful goals within a reasonable schedule and budget.

Sometimes the goal selected for NASA has been made meaningless by not being proximate. This was the fate of the Obama administration's so-called "Journey to Mars" program, which accomplished nothing because it had no specific deadlines to accomplish anything. It thus died without anyone even taking the trouble to kill it, because it had never lived in the first place.

The plan currently on the table, placed there by the Trump administration, calls for returning astronauts to the Moon by 2024. This meets conditions 1 and 2. So far, so good. NASA, however, once again has its heart set on an expensive space station, this time the ironically named "Gateway," to be stationed in lunar orbit. While NASA claims that the Lunar Gateway is needed for missions to the Moon, it is in fact a holdover from an earlier NASA program for capturing asteroids; the plans for the station were laid down by Obama-era NASA administrator Charles Bolden, who openly declared that he had no interest in returning Americans to the Moon.

It gets worse. The problem is not just that NASA is proposing to delay real accomplishment by inserting diversionary programs into the critical path. It is also insisting on undertaking these programs in the slowest and most expensive way possible. NASA administrator Jim Bridenstine has stated that NASA will transport astronauts to the Moon using two technologies it has been developing for many years: the SLS rocket and the Orion crew capsule (along with an actual lunar lander, to be developed at some future date). But the Orion capsule is so heavy that even the long-overdue SLS heavy-lift booster cannot deliver it to low lunar orbit with enough propellant to fly home. So a halfway house is needed not only for the Orion to rendezvous with a lunar lander, but also to refuel—thus the Gateway.

The resulting mission plan requires not only the Gateway, but four launches per mission, involving five different flight elements and six rendezvous operations per mission. This plan is so complex that it would make lunar missions incredibly costly, infrequent, and ineffective, and all but guarantee mission failure.

If we were to build a lunar base on the actual Moon rather than in orbit, refueling could be accomplished using hydrogen/oxygen propellant manufactured from lunar ice. This is the logic behind the Moon Direct plan that I proposed in these pages last year. Not only is that plan much more cost-effective, it also puts our astronauts on the Moon for extended periods, allowing them the time for science and exploration.

In short, the Gateway project is making the technology, rather than the destination, the master—a point hammered home this summer, when, during a White House photo-op for the Moon landing anniversary, President Trump, prompted by Apollo 11 astronaut Mike Collins, asked a pointed question to the NASA administrator: "What about the concept of Mars Direct?" In response, Bridenstine claimed that we cannot go to Mars until we have a lunar orbit space station and a lunar base. As if that were not enough, Bridenstine has also said in an interview that "we're going to need a Gateway-type capability at Mars," implying that he intends to hamper human exploration of the Red Planet with an unnecessary orbiting space station there, too.

The Trump administration's current Moon plans satisfy conditions 1 and 2. But instead of enabling the goal of reaching the Moon by 2024, the Gateway project is disabling it—and likely also harming our chances of getting to Mars anytime soon. NASA is today saying the same thing it said to President George H.W. Bush three decades ago: "You can't do your program until you do my program." Conditions 3 and 4 have gone right out the window.

Wrong Goal, Right Goal

This brings us to condition 5, which holds that it is not enough to have a goal with a proximate deadline, but that the goal must be wisely chosen. Why has the Trump administration picked the goal of sending astronauts back to the Moon by 2024? It is true that by setting a deadline for NASA's human spaceflight program to accomplish its task, the administration has given the agency a very healthy shock, and rousing itself to meet the deadline would certainly restore some of NASA's can-do spirit. But is returning to the Moon the right goal?

In terms of glory and geostrategic influence, the United States is not going to inspire the world with what free people can do by repeating the accomplishments done for the first time a half-century ago by men who are now great-grandfathers. Moreover, while there may be some interesting science and exploration that astronauts could do on the Moon—we could perhaps add to our knowledge about the Moon's origins and about the past of the solar system—there is another obvious destination that holds much greater promise for science: Mars.

Consider the biggest scientific question we could study on Mars: the question of the origins of life. This is one of the greatest mysteries of modern science. We know from fossil evidence dating back at least 3.5 billion years that life appeared on Earth virtually as soon as it could. This suggests that either life evolves quickly and spontaneously when the chemical conditions are suitable, or that life spreads in microbial form across interstellar space and readily takes hold as soon as it finds a habitable environment. If, as many scientists believe, early Mars was warm and wet and thickly enshrouded in carbon dioxide—if, in other words, it was similar to the early Earth—it might very well have hosted life. Methane emissions detected by the Curiosity rover lend support to the suspicion that Mars *still* carries life, protected underground in hydrothermal reservoirs. We need to go to Mars, drill, bring up samples of subsurface water, and see what is there.

And if there is some evidence of present-day or fossilized past life on Mars, the key question becomes *What is its nature?* At the biochemical level, all life on Earth is the same. Whether bacteria, mushrooms, grasshoppers, or people, all terrestrial life uses the same genetic alphabet of DNA and RNA. That is because we all share a common evolutionary ancestor. But what about Martian life? If terrestrial and Martian life both came from a common source, their genetic alphabets will resemble each other, as the English alphabet does that of French. But if each biosphere originated locally, they could be as different as English and Chinese.

The necessary program of drilling, sample-taking, culturing, biochemical analysis, and related observations is far beyond the ability of robotic rovers. Even our three best rovers, Spirit, Opportunity, and Curiosity—marvels of engineering kitted out with a variety of scientific instruments—have been able to discover, in a combined 27 functional years on Mars, just a small fraction of what a single well-equipped scientist could have discovered in a few days. Only human explorers can do the job right. And sending humans to Mars will be worth the cost and risk involved not only because of the potential for answering fundamental questions about the prevalence and diversity of life in the universe but because such a mission would once again astound the world with the daring creative genius of freedom.

In short, unlike NASA's planned lunar venture, which frankly is being done just to have something to do, sending human explorers to Mars would really have a purpose. Therefore, that is what our goal should be.

Mars the Hard Way

Unfortunately, NASA's current thinking about putting astronauts on Mars satisfies none of the five conditions for a purpose-driven program that could succeed. The agency's leaders seem interested in paying only lip service to the goal of humans-to-Mars. They are assigning no deadline for such a mission. And instead of working directly to send people to land on Mars, they are dreaming up a complicated infrastructure whose transparent purpose is to provide a rationale for the Gateway. These are vendor-driven plans.

NASA's putative design for a human mission to Mars would use the Lunar Gateway to support the operations of an interplanetary spaceship called the Deep Space Transport (DST). The DST relies on a slow, unproven electric propulsion

ion drive to travel from the Gateway to Mars and back, with one-way trip times of about 300 days. This contrasts poorly with what chemical rockets can already do, as demonstrated by the robotic missions (Pathfinder, Spirit, Opportunity, and InSight) that have reached Mars in about 200 days after starting from low Earth orbit. If it were starting from low Earth orbit, the DST would take some *600 days* to reach Mars. In other words, the purpose of the Lunar Gateway is to provide a crutch for the feeble DST, allowing a catastrophic choice of propulsion technology to become a merely horrible one.

Furthermore, the xenon intended to be used as a propellant for the DST's ion drive is not available from the Moon, negating all of NASA's claims that the planned outpost on the lunar surface could productively support Mars missions. And using the orbiting Lunar Gateway as a base for the DST will impose massive technical requirements on both systems, since the Gateway will need to include maintenance and propellant-storage facilities, and the DST will need to be made maintainable by astronauts on spacewalks and refuelable on orbit, and all this will need to be backed up by a logistics train transporting propellant and replacement parts from Earth to the Gateway. The requirement for refueling will add to the mission plan numerous critical orbital rendezvous operations that will impose severe timing and coordination constraints, and with them repeated risk of the loss of the mission, vehicle, or crew should any one of them fail to be executed on time.

And all this is needed for what, exactly? The DST does not solve any of the problems that NASA has cited as key obstacles for human Mars exploration, such as cosmic-radiation dosage or health deterioration due to prolonged exposure to zero gravity. On the contrary, it makes these problems significantly worse by greatly increasing the interplanetary transit time over what is otherwise feasible, and requiring a configuration that is inimical to the use of artificial gravity.

But the worst part is that the DST doesn't actually do anything useful. The value of sending human beings to Mars is not in sailing them about in interplanetary space. It is in intensively exploring the surface and searching for evidence of past and present life. The DST does not address those requirements at all. Rather than being derived from a plan to explore Mars, it is a thing in itself, an attempt to realize some science-fiction vision of the interplanetary spaceship.

In a purpose-driven space program, the mission comes first. From the mission comes the plan, from the plan come the vehicle designs, from the vehicle designs come the required technologies. That's how we did Apollo, and how every successful unmanned robotic planetary mission has been done. But the DST effort reverses this logic. NASA wants to employ electric propulsion, so it creates the DST—then insists on imposing the DST on the Mars mission. Instead of the Mars mission being the *reason* for the DST, it must suffer the role of serving as the *rationale* for the DST. The mission is nothing, the vendor contracts are everything.

So: NASA needs the DST to justify the Lunar Gateway, because the Gateway is necessary to prevent the DST from being even worse than it is. It may be nuts, but that's their story and they're sticking to it.

If You Want to Go to Mars, Go to Mars

There is a clear alternative to NASA's series of boondoggles: the Mars Direct plan, which I first proposed in 1990 with my colleague David Baker, and which I have continued to develop and advocate. Under this plan, or others resembling it, necessary payloads are sent on direct trajectories to Mars using the upper stage of a heavy-lift rocket (such as the SLS or the SpaceX Starship system, both now under development, or the currently operational Falcon Heavy). Methane and oxygen propellant can then be produced using Martian water and carbon dioxide even before the crew arrives. For example, in the original Mars Direct plan, an uncrewed Earth Return Vehicle (ERV) is landed on Mars, along with a 100-kilowatt nuclear reactor and a propellant-synthesis unit built into its landing stage. The ERV makes its return propellant by reacting atmospheric carbon dioxide with a small amount of liquid hydrogen it brings with it. So no humans need even launch from the Earth until we know that a vehicle capable of carrying them back to Earth is already awaiting them, fully fueled, on the Martian surface. The crew then launches and is delivered to Mars in a habitation module, which will also serve as their house and laboratory on the Martian surface during their stay of about a year and a half. At the end of their time on Mars, they ride the ERV home, leaving behind their hab module, so that as missions proceed every two years, either a string of small bases or a combined large base is developed.

In the more recent modified Mars Direct plan put forward by SpaceX, a reusable launch vehicle (Starship) lands on Mars and makes its return propellant out of atmospheric carbon dioxide and water ice, so that the same system can serve as both hab module and ERV.

Which approach—some variant of the Mars Direct plan or NASA's Lunar Gateway/DST—is better suited for a Mars program? That depends on the goal of the program. Is the goal to fly around in space, to go further than we have ever gone, perhaps to set a new record for the almanacs? Or is it to bring human explorers to the Red Planet's surface to search for life and develop the technologies required to open the Martian frontier? In a purpose-driven program it clearly must be the latter. The DST concept does nothing toward achieving that objective. Quite the contrary, it inserts the development and support of an entire parallel universe of in-space infrastructure, technologies, and operational capabilities into the Mars mission critical path. As explained above, these new systems do not represent additional Mars mission capabilities. Rather, they are liabilities. Their creation and support imposes additional costs on the Mars program, and if any of them should fail, the Mars mission is off. All this looks very attractive to the vendors, because more complicated components means more contracts, more money, more jobs. But because these components will subtract funds from actual Mars exploration systems and operations, they severely reduce the overall effectiveness of the program.

Put bluntly, the DST is a vehicle for flying around in space. But the purpose of interplanetary travel is not to fly around in space. It is to transit across space to reach, explore, and develop the worlds on the other side of space. This should be done in the simplest way possible. The Mars Direct systems are components of a plan for exploring Mars. Mars Direct would deliver all its payloads to the Martian surface because the surface is where the mission is. For the DST plan any surface activity is at best an afterthought.

It is possible to imagine giving a DST mission some exploration capability by adding to the flight plan the delivery to the Martian surface of a habitation module and descent/ascent vehicle. But in this case the number of mission-critical systems and rendezvous operations would be multiplied, with corresponding added cost and risk. Consider: In the Mars Direct mission, the crew is sent on its way to Mars with a twenty-minute burn of a rocket engine of a type that has been tested and flown hundreds of times before. Once that is done, the crew is on a six-month transit to Mars, and nothing will stop them from reaching their destination unless they choose to abort onto a free-return trajectory that will infallibly take them back to Earth exactly two years after their departure. By contrast, after leaving the Lunar Gateway, the novel DST engines must fire continuously for 300 days for the crew to make it to Mars. If the thrusters, power conditioners, or power system should fail at any point along the way, the crew will be stranded in interplanetary space.

This is not a purpose-driven plan for sending people to Mars. By requiring the development of a wildly complex set of systems, NASA's DST–Gateway planners are cooking up a vendor-driven plan that will accomplish nothing before it is eventually killed off by political realities.

If we want to send human beings to Mars, we don't need a complicated plan involving a lunar space station, a slow-moving spaceship, and many extraneous potential failure points. What we need is a big rocket—a heavy lifter with a payload capacity to the Martian surface of ten tons or more. SpaceX's planned Starship and NASA's planned SLS (with a proper upper stage) could deliver a twenty-ton lander to Mars. (The already-operational Falcon Heavy could send a ten-ton lander.) Even before sending human beings to Mars, we could use such a system to deliver platoons of rovers armed with diverse instruments and tools to reconnoiter regions of interest, demonstrate the systems needed to use Martian resources (including to make propellant), and ultimately prepare a base. Such rover missions would be of scientific value in themselves, and would set the stage, at last, for a human presence on Mars.

Getting Started

At the Oval Office meeting this past summer with the Apollo 11 astronauts, President Trump seemed to sense that he was being given the runaround. But will he take decisive action to address the situation? Will NASA have a purpose-driven plan or a vendor-driven plan? Will we spend money to do great things, or do things in order to spend a great deal of money?

If we allow NASA's human spaceflight program to remain vendor-driven, not only will we not reach Mars by the 2030s, we may not even return to the Moon in any valuable way by then. But if we insist that our space program be purpose-driven, we can reach the Moon by 2024 and Mars before 2030.

This is the choice we face. The stakes are huge: We can prove that democracies in peacetime can still do great things. We can push the United States far to the forefront of technical achievement for a generation or more. We can investigate the very origins of life. We can explore—and even prepare to settle—a new world.

The Moon landing was a grand deed, but a half-century later it is tragic that it remains the peak of our achievement in space. We would be doing far better—and honoring the heroes of Apollo far more appropriately—if this year we were hailing the eighteenth birthday of the first child born on Mars. Let us resolve that by the Moon landing's one hundredth anniversary, Americans will have much more recent epics to acclaim.

EVALUATION OF A HUMAN MISSION TO MARS BY 2033 (2019)

SUMMARY AND CONCLUSIONS

The Institute for Defense Analysis Science and Technology Policy Institute

In August 2017, NASA asked the Institute for Defense Analysis (IDA) Science and Technology Policy Institute (STPI) to conduct an independent assessment of its plans to extend human exploration to the Moon and to Mars. These are the conclusions of that assessment, published in 2019. Although not written for the general public, this report provides a fascinating, and at times sobering, reflection on the commitment it will take to get people to Mars. The IDA is a non-profit corporation that provides objective analysis of national security issues in the United States.

Responding to Section 435 of the NASA Transition Authorization Act of 2017 and NASA's request to evaluate its plans for human spaceflights, in the preceding chapters STPI assessed the human health, technology, schedule, cost, and budget risks associated with a 1,100-day Mars orbital human spaceflight mission to Mars and its precursor missions. This chapter presents our overall findings and conclusions.

The plans presented in this report draw on NASA's *The National Space Exploration Campaign Report*, which establishes a spaceflight program for future human exploration of the Moon and, later, Mars. Additionally, the plans are based on public NASA presentations and documents, internal NASA planning documents, and conversations with NASA personnel. Our overall approach was to assess technology, schedule, and cost risks associated with NASA's current and notional plans to evaluate the feasibility of a Mars human spaceflight mission by 2033. Although the adverse effects associated with each type of risk (e.g., schedule delay, cost overrun) vary by risk category, we generated relative, overall risk rankings (low, medium, or high) based on qualitative assessments of the components of risk: severity of event, likelihood, and consequences. Using this approach, we examine the technology pathway, schedule, and the cost and budget envelope for a 19-year exploration campaign that would require the development and assembly of a cislunar space station in the mid-2020s, five lunar landings in the late 2020s and early 2030s, followed by a crewed mission to Mars orbit later in the 2030s. This chapter summarizes the analyses and conclusions from the preceding six chapters.

A. Assessment
1. Technology Assessment

Accomplishing a human orbital mission to Mars requires the development of four key systems: the SLS and supporting ground systems; Orion; the Gateway; and the DST. These systems are at very different stages of design and technological readiness.

Due to long development programs and ongoing testing, SLS and Orion present low technology risks to a Mars orbital mission in the 2030s. Our assessment identifies a need for 12 SLS launches and 11 Orion launches to support an orbital mission to Mars, including 1 cargo SLS to launch the DST, and 6 SLS and Orion launches to support lunar activities from 2027–2032. SLS Block 2 Advanced Boosters, which would debut on EM-8, present a low-to-medium technology risk, depending on the design that is ultimately selected. Ground system upgrades and refurbishment are nearly complete, and future upgrades do not involve technology innovation, thus ground systems present low technology risk.

Currently, NASA notionally plans to launch crew on the first flight of SLS Block 1B (EM-3) and the schedule presented in this report would launch crew on the debut of SLS Block 2 (EM-8). Although conducting extensive ground testing can reduce the overall risk to astronauts, the notional schedule implicitly accepts a higher level of technology risk than would be the case if new systems were flight tested with uncrewed launches.

The Gateway presents medium technology risk since its current conceptual design largely builds on ISS-heritage technologies. Certain technologies (e.g., xenon refueling and autonomous environmental monitoring) present a medium risk since they have not been previously demonstrated at the scale required by the Gateway. If additional modules are added or mission requirements expand, the Gateway may face higher technology risk. Increasing the complexity and size of the Gateway could require larger power, propulsion, and life support systems, which would add to technology risk. Integrating and testing Gateway elements, whether at its current scale or larger, face medium technology risk.

The DST will require several medium and high-risk technologies. Plans for the DST remain conceptual; many design elements are not yet defined in internal NASA planning documents and NASA currently does not plan to begin design of the system until 2024. In contrast to the Gateway, several new technologies would need to be developed for the DST, especially technologies that need to function without substantial maintenance for the 3 years of the mission. Notably, an ECLSS that meets the performance and reliability requirements of the DST is currently at a low TRL, although NASA plans to test an ECLSS with high oxygen reclamation rates on the ISS starting in 2022. Scaling systems from the Gateway (e.g., 500 kW solar electric propulsion system and reusable in-space engines) presents a medium technology risk. Technologies to transfer cryogenic propellants and prevent unacceptable boil-off losses over long periods of time have not yet been demonstrated at the scales needed for the DST and present a high risk. Integrating each system into a single bus would likely be high risk because a small change to one system could require a series of changes to the rest of the spacecraft due to the scale and interdependent nature of DST systems.

Notional plans for a human orbital mission to Mars rely on several other medium- and high-risk technologies to reach completion. Although spacewalks on the Gateway and DST would be kept to a minimum, deep space spacesuits will be necessary for external repairs. The next generation of spacesuits presents a medium technology risk that must be addressed soon if the spacesuits are to be ready by the launch of the Gateway. Many technologies associated with a Mars surface mission are also high risk, including entry, descent, and landing technologies, a MAV, surface habitation, exploration systems, and in-situ resource utilization.

2. Estimating Schedule

Four complicated elements—SLS, Orion, Gateway, and the DST—would need to be developed and completed to launch a human mission to orbit Mars. These technology developments would be taking place while NASA also designs and launches lunar landers and eventually human astronauts to the Moon's surface. *We find that a Mars 2033 orbital mission cannot be scheduled under NASA's current and notional plans. Our analysis suggests that a Mars orbital mission could not be feasibly carried out until the 2037 orbital window or later.*

Our schedule rests upon two critical assumptions. First, that NASA will choose the architecture of a human Mars orbital mission and begin systems development in 2024 (as indicated in the *Campaign Report*), which drives the development cycle of DST and timeframes for other human spaceflight activities (e.g., lunar landings) in the report. Second, that there is one human mission to the Gateway annually, either to work aboard the Gateway, operate the DST, or to transfer to the lunar surface (as noted by NASA personnel in discussions), which requires NASA to make trade-offs between focusing on lunar landings and preparing the DST for the mission to Mars. Under these assumptions, NASA would begin Phase-A development of the DST in 2024, with final design occurring in 2028 and delivery for integration with SLS in the second half of 2032 for launch and in-space checkout in 2033. In 2034, the DST would depart on a one-year shakedown mission to validate all systems in deep space before returning to the Gateway. Another mission would be launched in the first half of 2036 to refurbish and refuel at the Gateway in preparation for a low-energy transit to Mars in 2037 or later.

For some architectural elements and technologies, current and notional schedules require parallel development efforts. For example, the lunar surface campaign's technology development efforts would have to occur in parallel with development and launch of the Gateway, while the effort to launch astronauts to the lunar surface would occur in parallel with construction of the DST. These parallel development efforts and associated costs and available budgets provide a source of schedule risk.

3. Estimating Cost

Drawing on NASA documents, interviews with NASA, industry studies, and costs of analogous programs, STPI estimates that the total costs for the development and operation of the core architectural elements (SLS, Orion, the Gateway, and the DST) from FY 2019 to expected launch of the Mars orbital mission in FY 2037 is likely to be at least $83 billion in FY 2017 dollars. Including the cost incurred from 2038 to 2040 during the orbital mission, the total estimated cost from FY 2019 onward of a human mission to Mars orbit is $87 billion in FY 2017 dollars. The cost of just the orbital mission to Mars beyond the development of SLS, Orion, and the assembly of the Gateway is $45 billion in FY 2017 dollars.*

Although a human mission to the surface of Mars would not take place until at least 4 years after the first orbital mission to Mars is launched, NASA would need to begin development of the required systems many years in advance. We estimate the development costs for the systems needed for a human landing on the surface of Mars to require an additional $25 billion in FY 2017 dollars. Other human exploration-related costs, including continued support for the ISS, post-ISS flights to another station in LEO, and the costs of the lunar exploration missions would add another $50.9 billion in FY 2017 dollars to the overall cost of the 19-year exploration effort. The aggregate cost of the exploration effort from FY 2019 to FY 2037 is estimated to be $184 billion in FY 2017 dollars.

The elements of a human mission to Mars vary greatly in terms of readiness, greatly affecting the risk of cost overruns in a given program. Orion has already gone through one test launch; the development phases for SLS Block 1 and Orion are almost concluded. Although we believe that there is still risk of additional cost overruns to the SLS and Orion programs in the near term, risk for ongoing cost overruns for these programs over the 19-year period is not high, especially during the period when SLS is being launched on a regular basis.

In contrast to SLS and Orion, NASA has yet to finalize the design of the Gateway. To reduce technology and cost risk, NASA has committed to constructing the Gateway by employing existing technologies. This decision mitigates some cost risk, but risks increase if mission requirements grow for the Gateway, and lead to the addition of more complex systems. Collaborating with international partners also adds to cost risk, as NASA will need to collaborate closely with these partners and integrate their contributions into the overall mission.

* As discussed in Chapter 4, this value includes all R&D and operations costs up to and including the first Mars orbit mission. STPI estimated individual system costs for the Gateway, lunar landing systems, and Mars orbital and surface systems and then added a 30% reserve to account for potential contingencies. We did not include possible cost overruns in this figure.

The DST and the systems needed for a Mars surface mission are only at the conceptual phase, and would employ many new technologies to be used in a new environment. In addition, the energy needed to transport mass from Earth orbit to Mars greatly constrains the mass and, therefore, the design of these elements. Lastly, the cost required to address, mitigate, and retire risks to human health is uncertain and was not estimated in this report, and could be much higher than current NASA budgets for human health research. These uncertainties introduce high risks of cost escalation.

4. Mapping Costs to Budget

NASA was appropriated $20.7 billion for FY 2018 ($20.4 billion in FY 2017 dollars), of which approximately half was devoted to human spaceflight. If NASA continues to receive this amount annually—accounting for inflation—between FY 2019 and FY 2037, it would have a cumulative $192.7 billion in FY 2017 dollars or approximately $10 billion in FY 2017 dollars per year at its disposal for human spaceflight. If NASA were to receive annual budget increases commensurate with annual real GDP growth of an anticipated 1.9 percent, available funds for human spaceflight over the 19-year period could be $233.8 billion in FY 2017 dollars, assuming that the share of the NASA budget devoted to human space exploration remains constant.

Comparing these budgets and our estimates of the costs of systems needed for Mars and lunar exploration with prospective NASA budgets for human space exploration, *we find the aggregate budget under our flat budget scenario ($193 billion FY 2017 dollars) is greater than the aggregate costs for human spaceflight ($184 billion FY 2017 dollars)*. If the human space exploration budget increases at the rate of real GDP growth, giving a cumulative budget of $234 billion FY 2017 dollars, there may be sufficient funding for enhanced exploration (e.g., more Moon landings, funding for the development of other technologies such as nuclear thermal propulsion) or to mitigate the impact of cost overruns. Alternatively, if NASA returns to a more budget-constrained environment such as was envisioned in the 2014 National Academies' Pathways to Exploration report, funding would likely be insufficient to complete all of the exploration activities in our schedule.

While cumulative resources are likely to be sufficient to cover cumulative costs of a human mission to Mars, if the costs of systems are spread over time based on their development cycles, we find that the estimated costs of human space exploration peak at $11.5–$11.8 billion in FY 2017 dollars in FY 2035 and FY 2036, with costs running above $10 billion per year in FY 2017 dollars in FY 2023 and FY 2030–2031 and reaching approximately $10 billion per year in FY 2017 dollars in FY 2021, FY 2022, FY 2025, FY 2026, and FY 2028. *Under flat budgets, a 2037 start date for a mission to Mars is feasible; however, cost peaks will require activities, such as lunar landings, to be rescheduled to ensure that annual appropriations match development costs, which could have implications for the Mars orbital mission launch date. Under a budget that matches real* *growth, budgets always exceed expected costs by $1–6 billion FY 2017 dollars annually, allowing for additional exploration programs, especially after the peak cost years of the DST in the early 2030s.*

We compared our cost estimates associated with a human mission to Mars and the Administration's budget request for FY 2019 to FY 2023. *We find that the current NASA budget sufficiently funds short-term technology development activities for lunar exploration and for the development of the Gateway, though shortfalls may occur late in the budget period as NASA's purchasing power declines over time and Gateway costs approach their peak.*

5. Human Health Risks

Discussions with NASA and a review of internal NASA planning documents and academic literature reveals that the understanding of human health risks associated with an extended human orbital mission to Mars is limited. While no astronauts have died from health-related causes during human spaceflight through 2018, uncertainties remain with respect to threats to human health on an orbital mission to Mars that may not be fully understood given NASA's current and notional plans for ground-based research, ISS missions, Gateway and lunar surface operations, and the DST shakedown cruise. NASA's current Human Research Program Integrated Research Plan to study human health risks associated with long-duration deep space spaceflight lacks sufficient detail in both evidence and strategy to justify the predicted timeline to develop risk mitigation strategies. Further, the document does not present a detailed plan to prioritize NASA's approach to filling in gaps in knowledge, especially on the combined effects of radiation, low-or-micro-gravity, and isolation on astronauts. While clear and detailed in some areas, the current research plan is necessarily vague in others. This vagueness, however, decreases the likelihood that by the 2030s the understanding of human health risks and strategies to mitigate them will be sufficient to meet current standards of risk to astronauts or to ensure crew survival on an extended mission.

Because of the nature of the human health risks and knowledge gaps, developing a clear timeline and specific research plan to adequately address these gaps and create mitigation strategies is difficult and nearly impossible to evaluate at this stage. The primary focus for planned research appears to be on currently known risks as opposed to understanding significant theoretical and unknown risks and holistic human health. Even if NASA could give greater priority to addressing human health risks when planning missions, this focus does not negate the real risks of astronauts dying from health-related factors[†] on an extended orbital mission to Mars, and may require addition risk mitigation efforts prior to the orbital mission.

† As distinct from catastrophic technology failure resulting in death, which has been NASA's primary priority.

B. Feasibility of a Mission to Mars Before 2037

No technology or budget scenario with the current configuration of the Gateway accommodates an orbital mission for Mars departing in 2033. Even under a real growth budget scenario that ends support of the ISS in FY 2025 and forgoes all lunar landings, NASA would face multiple years of budget shortfalls during the concurrent assembly of the Gateway and development of the DST. In order to launch an orbital mission to Mars in 2033, Phase A development for the DST would need to begin at the start of FY 2020 (October 1, 2019), a milestone which would likely be missed. Further, several critical technologies, including thermal management systems, propulsion systems, and ECLSS, are unlikely to be mature by their need-by date for a 2033 mission. Shrinking the DST's formulation and implementation timeline by violating NASA's standard operating procedures for system development would exacerbate budget shortfalls under this scenario throughout the mid-2020s and lead to very high technology, schedule, and cost overrun risk.

A 2035 launch date for the Mars orbit mission would require substantial real increases in NASA's budget, force NASA to reduce the scope of lunar missions, and pose significant technology, schedule, and cost risks. Under flat budgets, NASA would likely need to delay the first lunar landing until after the DST departs in 2035, which would likely delay Mars surface system development and conflict with current and notional plans.

Under either budget scenario, a 2035 launch date would hold high schedule risk due to increased demand for fabrication and testing facilities, and pressure on NASA management during concurrent development and construction of the Gateway and DST. It would also hold high technology risks as some critical technologies for the DST—such as an advanced ECLSS system and propulsion and thermal management technologies—may not be sufficiently mature by the need-by date for DST development. Prolonged delays in any one of many technology development programs would delay the development of the DST, pushing the mission to the next orbital window in 2037.

A 2035 mission would also reduce NASA's ability to mitigate risks to human health, as NASA would be unable to conduct significant human health research in deep space before finalizing the DST's design, and have less time to learn from the operations of the Gateway before starting DST missions. Although budgets may be sufficient under a real growth budget scenario, a 2035 mission to Mars would be high risk from technology and schedule perspectives under NASA's current and notional plans.

C. Implications of Additional Missions for Budget and Schedule

In this section, we consider the budgetary and schedule implications of varying parameters related to the timing of the ISS phaseout and the cadence of Gateway and lunar operations. These variations from our notional schedule represent potential options for NASA to consider.

1. Implications of Changing the Timing of ISS Phaseout

Currently, the Trump Administration plans to end direct support of the ISS by FY 2025. However, NASA personnel have told STPI that NASA requires a space station in LEO through at least 2028 to meet mission needs for extended human health research and technology development that would be unmet by the capabilities offered on the Gateway, and that NASA plans to support a platform in LEO through at least this time. Previous STPI research suggests that an alternative to the ISS will not be available during this time without large subsidies provided by NASA (Crane et al. 2017). As such, large cost savings would likely not manifest if NASA ended support of the ISS in 2025.

STPI's analysis suggests that NASA would have sufficient budget to continue to support the ISS until 2028 while pursuing the lunar and Mars orbit plans outlined in this report. According to our cost and budget analysis, NASA has approximately $1 billion per year in budgetary flexibility during the 2025–2028 period under a real growth budget scenario though not in the flat budget one. Continuing to support the ISS beyond 2028, however, would have substantial budgetary implications. The 2030–2031 period represents the peak of STPI's estimated costs of NASA's human exploration program. Continuing to expend $3 billion FY 2017 dollars per year for ISS support and Commercial Transportation, rather than the assumed $1 billion FY 2017 dollars post-ISS expenditure for astronaut habitation and transportation in LEO, would complicate paying for the final lunar missions and the construction of the DST.

2. Implications of Increasing Gateway Use

The cost of using the Gateway is driven by the costs of the SLS and Orion required to transport a crew, which represent 96 percent of the $1.62 billion FY 2017 dollars additional cost of an additional flight. STPI's budget calculations suggest that there might be sufficient funding for one or at most two additional launches to the Gateway in the 2026–2027 period, were additional launchers and Orion capsules available.

STPI also estimates that the cost of resupplying a crew once transported to the Gateway for a full one-year mission would be less than $500 million FY 2017 dollars for supplies sent on commercial flights. This figure is half of our estimate of $1 billion FY 2017 dollars for astronaut transport costs to LEO that would be required post-ISS. This implies that NASA could potentially consider using the Gateway as a post-ISS outpost where long-term health research could be conducted and training for a future Mars mission might occur without necessarily incurring considerable additional costs—although there may be technological challenges or habitation volume limitations that may restrict continuous operations of the Gateway by a single crew.

3. Implications of the Moon Exploration Missions

Including the cost of development and operation of SLS, Orion, the Gateway, and the lunar landing missions, this report estimates the total projected cost of meeting Administration goals to return to the Moon by the end of the 2020s to be about $62 billion. The base case in this report assumes five landings on the Moon. It is possible, however, that NASA may stay on the Moon beyond the assumptions in the report.

We estimate the cost of an additional lunar exploration mission with a secondary cargo lander and human rover as $3 billion FY 2017 dollars, of which 50 percent is the cost of astronaut transportation using SLS and Orion. The period in which lunar exploration occurs (2028–2032) is the peak-cost period for the construction of the DST. During that time, under a flat budget scenario, there is no budgetary slack for additional lunar missions. However, in 2033 and 2034, funds could be sufficient for one additional lunar exploration mission under the flat budget scenario.

The 2014 National Academies of Sciences' Pathways to Exploration Report assumed a cadence of two lunar missions per year in their campaign rather than the annual missions that STPI's schedule has assumed. Given that in our schedule lunar missions occur in parallel with the peak costs of constructing the DST, increasing the cadence to match the Academies' assumptions would require both transportation and logistics costs to decline substantially. Given that the additional cost of a lunar mission is $2–3 billion FY 2017 dollars depending upon the nature of the mission and the cost of the required supplies and equipment, the costs of crew transportation, supplies, equipment and commercial flights would need to be halved for NASA to be able to support a two-flights-per-year cadence in the flat budget scenario unless construction of the DST were delayed. Additionally, accommodating SLS and Orion production and launch beyond the two per year for lunar missions will require significant infrastructure investment, which has not been estimated.

The schedule presented in this report assumes only five total lunar missions through 2032 and then focuses solely on the missions to Mars orbit and the Martian surface. This assumption, although required to develop and evaluate a possible series of exploration activities leading to a mission to Mars orbit based on NASA's current and notional plans, may be insufficient, especially if future leaders do not want a repeat of the Apollo program, which attempted seven Moon landings before halting human lunar exploration altogether. In the years following 2032, the budget we developed for this report allocated $1 billion FY 2017 dollars a year to support U.S. private (or possibly international) lunar exploration, which could, for example, be used for development purposes or provide a set of lunar landing modules annually. However, this level of investment would likely result in a decrease in the amount of lunar activities for at least the next several years. Creating a sustainable lunar presence for NASA after the missions presented in this report would likely require development of enhanced surface systems, such as permanent lunar habitation systems. Under flat budgets, a trade-off would likely need to be made between lunar surface activities in the late-2030s and Mars surface activities in the early-2040s.

D. Alignment between NASA's Plans for a Human Mission to Mars and its Exploration Strategy Principles

In order to maintain a space exploration campaign over the next two decades through a number of Congresses and Presidential administrations, NASA's plans for human exploration missions will need to be compelling and sustainable. NASA's Human Exploration and Operations Mission Directorate has established eight principles to guide NASA's exploration strategy in such a manner (NASA 2017e). Below we compare the plans for a human mission to Mars with these principles. We find that the mission aligns well with these principles assuming the orbital mission to Mars takes place in 2037 or later, but does not align well with several of them if the mission were to be scheduled earlier.

Principle 1: Fiscal Realism

NASA has set the goal of making a human mission to Mars implementable in the near term under current budgets and in the longer term with budgets commensurate with economic growth. We find that between FY 2019 and FY 2037, the cumulative cost of a human mission to Mars, including a lunar exploration campaign, may fit within cumulative budgeted funds and that, in general, annual budgets meet or exceed annualized costs of human exploration

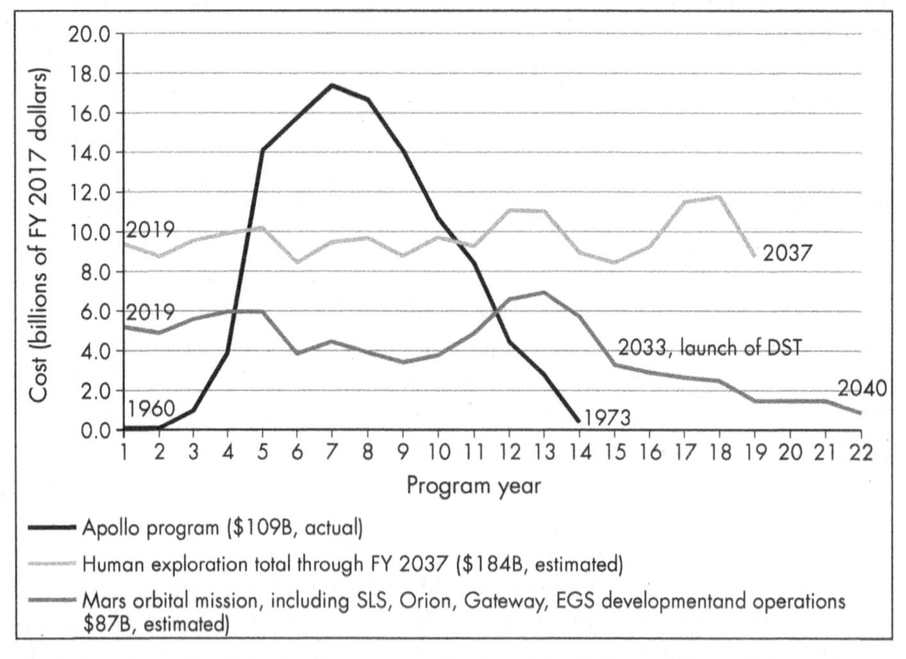

Overlaying the Profile of the Apollo Program with that of the Proposed Mars and Human Spaceflight Activities through 2037.

Source: Ertel and Morse, 1969; Morse and Bays, 1973; Brooks and Ertel, 1973; Ertel and Newkirk, 1978.

missions. The cost of that campaign, however, could not be compressed into a 15-year budget timeframe to FY 2033. Concurrent development of both the Gateway and DST to meet a 2033 mission to Mars could lead to large budget shortfalls throughout the 2020s, even without the development of lunar surface systems. While the cumulative cost of a human mission to Mars could fit within cumulative budgeted funds between FY 2019 and FY 2035, during several years in the late 2020s, projected annual costs would exceed expected budgets, even under a real growth budget scenario.

In comparison to the Apollo program, the series of missions presented in this report represent a more sustainable program from a budget perspective. Figure 24 overlays the total annualized costs of human exploration and the total annualized cost of the orbital mission to Mars and its precursor mission on top of the Apollo program's annualized cost curve. Between 1960 and 1965, NASA's total budget increased 760 percent (calculated from Ertel and Morse, 1969; Morse and Bays, 1973; Brooks and Ertel, 1973; Ertel and Newkirk, 1978). Notably, over the program's 14-year span, the program represented 47 percent of NASA's budget, which is similar to the assumption used in this report that human exploration would compose 50 percent of NASA's budget. However, the Apollo program peaked at 70 percent of NASA's budget in 1967 and had 7 years where it composed more than 50 percent of NASA's budget. This dramatic and rapid budget increase was unsustainable over the long-term. Decisions made during the Nixon Administration changed NASA's funding model to a flatter budget, which has proved sustainable over the following decades. If the United States were to decide that a human mission to Mars orbit before 2037 is a national imperative, NASA's funding model would require a funding profile more akin to that of Apollo. This goal, however, would likely reduce the sustainability of the human spaceflight program in the long run.

Principle 2: Scientific Exploration

NASA states that exploration enables science and science enables exploration. NASA's current plans and programs for robotic missions to the Moon and Mars appear to support exploration well. However, the extent of the benefits to scientific research from NASA's human exploration plans is unclear. The Campaign Report provides limited insight into the scientific findings expected from human lunar and Martian orbital missions. Both the Gateway and DST can play a role as deep space science laboratories. They will have the capability to observe the lunar and Martian surface, respectively, and can also be used to conduct science in deep space. Due to the short time astronauts will stay on the Gateway, however, the duration of human-tended experiments on the platform will be limited. Internal NASA planning documents reviewed by STPI do not adequately justify why many of the scientific activities that may be conducted on the Gateway could not be performed using solely robotic means. The DST does offer some unique possibilities for scientific exploration of Mars, including real-time telerobotic operation of airborne platforms, as well as deep-space human health research. However, the three-year duration of a mission to Mars orbit and inability to resupply during missions may limit the number and type of scientific experiments that could be conducted on the DST during transit or while in orbit.

Principle 3: Technology Pull and Push

Using high TRL technologies for near-term missions, while focusing sustained investments on technologies and capabilities to address challenges of future missions is another NASA goal for human space exploration. Fundamental technologies for the DST—especially those associated with propellant systems and the habitat and its required closed-loop ECLSS technologies—require technologies that are not currently at high TRL. NASA is currently investing in technologies to address DST-related challenges and has plans to advance the technologies through testing on the ISS and Gateway. Scheduling the departure of a mission to Mars for 2033 or 2035 would decrease the alignment with this principle, as the partial overlap between the technology development effort and the decision time frame for the DST design and construction poses technology and schedule risk.

Principle 4: Gradual Buildup of Opportunity

Another NASA goal is near-term mission opportunities with a defined cadence of compelling and integrated human and robotic missions providing for an incremental buildup of capabilities for more complex missions over time. Currently planned, near-term missions have a defined cadence. The Gateway demonstrates deep space operating concepts; lunar landings in the late 2020s and early 2030s demonstrate surface landing and ascent capabilities; and the Mars mission demonstrates the feasibility of humans reaching Mars. The transition from Gateway to DST, however, is considerably more challenging in terms of distance from Earth and mission time. There is a considerable jump from Gateway (~1–2 month sojourns in deep space) to the DST checkout and shakedown missions (6 months and then 1 year in deep space). The leap from the DST checkout mission to the Mars orbital mission is even larger (1 year to 3 years in deep space) with respect to mission time and distance.

Principle 5: Economic Opportunity

NASA is seeking opportunities for U.S. commercial business to further enhance the U.S. space industry through the program for human space exploration. U.S. businesses will gain experience in the development of the Gateway and DST habitats, launch, and in-space propulsion systems. NASA documents, such as the Campaign Report, specifically encourage commercial development in LEO as well as commercial resupply missions to the Gateway and the use of the Gateway as a base for commercial missions to the lunar surface, providing opportunities for the extension of the ISS commercial cargo model to cislunar space.

Principle 6: Architecture Openness and Resilience

Internal planning documents call for a resilient architecture featuring multi-use, evolvable space infrastructure, minimizing unique major developments, with each mission leaving something behind to support subsequent missions. The SLS/Orion/Gateway/DST architectural elements are designed to be configurable for many missions, reusable to the extent feasible, and evolvable as technologies improve.

The current approach to the Gateway, however, does not appear to be optimized to support the Mars mission, although, as explained below, it is aligned to a broader set of space objectives. The mission design for a human mission to Mars uses the Gateway as an in-space dockyard for checkout and refurbishment of the DST. From internal planning documents and conversations with NASA personnel, the justifications for why those operations could not be conducted in high Earth orbit by docking Orion and logistics vehicles directly with the DST are unclear. The trips to the Moon add transportation time and delta-V costs to the operations of the DST. If maintenance is routine and the DST ECLSS can be used to provide life support and living quarters for astronauts during refurbishment then direct docking to the DST could be sufficient. The Gateway, however, provides the possibility of long-term habitation for astronauts in the case of complex repairs that require working with the DST ECLSS off-line for an extended period. In short, the Gateway is used only as an insurance policy in the context of Mars missions.

According to experts we spoke to, the Gateway serves other functions. During our discussions, six other distinct (but not necessarily unique) rationales for the Gateway emerged:‡

- **As a technology testbed for the Mars mission.** The Gateway would test certain technologies needed for the DST, such as Hall thrusters and solar arrays. In addition, the Gateway could test an ECLSS system and radiation hardened electronics in a higher radiation environment than the ISS, which may be necessary to fully understand ECLSS and electronics operations in deep space. Even if the ECLSS system on the Gateway is not as advanced as the one needed for the DST, these data would be useful.
- **As a deep space human habitat laboratory.** The Gateway would host astronauts in deep space for the longest period in spaceflight history until the launch of the DST, providing unique opportunities for research on human health. Although data gathered from 90-day-duration missions may not be fully

‡ As discussed in Subsection 1.B.3, an evaluation of the Gateway was outside the scope of this study.

extensible to a three-year mission to Mars, as discussed in Chapter 6, the environment of the Gateway could allow for novel human health research that cannot be conducted in LEO or on Earth.

- **As a platform to build experience with operations for deep space human missions.** NASA has not operated a human mission in deep space since the Apollo era. The Gateway should help build institutional knowledge for future long-duration deep space missions to the Moon and to Mars.
- **As a deep space science laboratory.** The Gateway could have the capability to observe the lunar surface or deep space from several orbits and conduct science in unique areas of deep space, including telerobotics on the lunar surface.
- **As a transfer station for travel to and from the lunar surface.** Astronauts traveling to the lunar surface via Orion would need to transfer to a lunar descent and ascent vehicle. The Gateway is designed to be used as a transfer port between the two vehicles and a docking station for the descent vehicle between surface missions. The capabilities offered by the Gateway could be extended to private and international lunar surface missions.
- **As an off-ramp for the human exploration program if the journey to Mars is postponed.** The Gateway can be expanded into a more substantial space station with additional modules. If unexpected problems arise in completing a human mission to Mars (e.g., human health concerns for long-duration deep space spaceflight), the Gateway could help sustain the human spaceflight program. Many companies have said they have plans for lunar activities that rely on NASA-sponsored lunar infrastructure. Accordingly, the Gateway may also help spur private cislunar activities.

The design of the Gateway is flexible. This, however, may be a mixed blessing, as space systems without definite mission requirements have historically faced growth in scope, mission, and cost. Internal NASA planning documents contain limited detail with respect to the operations of the Gateway and even fewer details on what it will do between missions related to human missions to the Moon and to Mars when it is to operate autonomously.

Principle 7: Global Collaboration and Leadership

Orion, the Gateway, and potentially the DST provide opportunities for substantial new international and commercial partnerships, leveraging current ISS partnerships and building new cooperative ventures for exploration while maintaining U.S. leadership in the field. International partnerships are envisioned with respect to Orion and the Gateway. International partnerships entail additional costs as well as savings. Commercial partnerships as embodied in the NextSTEP approach are being used to develop the Gateway. Development of the DST is at an early stage, so it is not clear which aspects of the transport or the overall Mars missions would allow for international collaboration.

Principle 8: Continuity of Human Spaceflight

NASA seeks to establish a regular cadence of crewed missions to cislunar space during the lifetime of the ISS for the uninterrupted expansion of human presence into the solar system. Since the ISS's end-of-life may be in 2024 or later, the extent to which this principle is fulfilled remains unclear. It cannot be fulfilled if the ISS is decommissioned in 2024. By the beginning of 2024, at most two short crewed missions (EM-2 and EM-3) will have flown to cislunar space. Our schedule envisions one crewed mission per year from 2025 through 2036 (with the exception of 2035, when the DST is undergoing its one-year shakedown mission), which may comport with "regular cadence." The 2024, 2025, and 2026 flights will be one-month flights associated with the construction of the Gateway, while the 2027 through 2032 missions involve lunar operations. Missions beginning in 2033 are longer than those associated with the initial operations of the Gateway.

E. Observations on the Campaign Report

The *Campaign Report*, used as the basis of this assessment, provides a call to arms for NASA, potential international partners, and the private sector to expand human presence on the Moon and subsequently send humans to Mars. The report also presents a notional description of potential missions and milestones; describes the integration across NASA of the agency's efforts to focus and achieve its objectives; and outlines a framework and set of decisions to meet mission goals in LEO, cislunar space, the lunar surface, and Mars, and evolve its exploration technology capabilities. An important difference from previous NASA strategy documents is the focus on use of open architectures and the development of interface and operational standards, that will enable industry and international partners to contribute to exploration by bringing new capabilities at their own pace and risk.

The *Campaign Report* delineates a host of critical decisions for lunar and Martian exploration technologies that will be made in or around 2024, many of which depend on the timely development of the Gateway. Given the schedule risks outlined in this report, delays in the Gateway could potentially cause disruption to and require redesign of large elements of the future exploration architecture, such as lunar landers, given the dates of decisions for these systems in the *Campaign Report*. Additionally, the 2024 or later dates for several critical decisions, such as the design of the DST and Mars surface systems, will push many of these difficult decisions into another Administration. However, because of the lack of detail and the optimistic assumptions, the report conveys a positive prospect of launching humans to Mars via the proposed architectures. Since the timeline stops at 2024, the question of the report's role in getting NASA to Mars orbit by 2033 cannot be addressed.

The NASA Transition Authorization Act of 2017 specifically calls for NASA to develop a "Human Exploration Roadmap (U.S. Congress 2017)." The *Campaign Report* is not a "roadmap," and commendably does not call itself that. Instead it refers to itself as a "Campaign Report," which is a much more apt title. A roadmap would be expected to include additional details, such as a time-phased sequence of events, decision points, descriptions of all major architectural items, workflows, technology on-ramps and off-ramps, and budgets. Although the *Campaign Report* provides a high-level strategy for the next 7 years of human spaceflight, it is mainly a plan for a plan, and may not ultimately play a substantive role in efforts to place humans in Mars orbit by 2033. Further specificity of NASA's long-term plans in a public document would help Congress and other public policy officials make informed decisions over the coming decades.

F. Overall Assessment and Recommendations

A 2033 launch date for the Mars orbit mission is infeasible given the requirements imposed by NASA's current and notional plans, the National Space Exploration Campaign Report, and the human health, technology, schedule, cost, and budget risks associated with a mission to Mars orbit and its precursor missions. A 2035 launch date for the Mars orbit mission would require substantial real increases in NASA's budget, force NASA to reduce the scope of lunar missions, and pose significant technology, schedule, and cost risks. Our analysis suggests that 2037 is the earliest NASA could feasibly launch the DST to Mars orbit given current and notional plans and budgets *assuming cost peaks in the 2030s can be smoothed*. Delays to the development or testing of the DST or flat budgets with no reduction in lunar missions could push the Mars orbit mission to 2039.

The figure above compares the cost profile of the Mars mission projected in this report to that of Apollo to show the aggressiveness of the effort required. Shifting the DST's construction and launch date to align with a 2037 Mars orbit mission poses some advantages. For example, it allows NASA to learn from the design and operations of the Gateway as part of the design and construction of the DST. Construction of the DST subsequent to the completion of the Gateway also reduces the likelihood of budget shortfalls throughout the 2020s, as SLS and Orion complete development, as the ISS is used for technology testing, and as the Gateway and lunar exploration systems are constructed and operated.

Although limited lunar landings and the 2037 Mars orbital mission are feasible together under notional budgets, limitations imposed by the budget render unlikely a long-term human presence on the Moon in the 2030s concurrent with a 2037 mission to Mars orbit and subsequent Mars landing mission in the early 2040s. The schedule as described in our analysis does not include NASA missions to the Moon after the DST is launched. NASA is constrained both by the limitations of its core architecture—as only one crewed SLS launch is available per year—and by the cost of additional landings relative to the budget available (assuming that the flat budget of approximately $10 billion per year in FY 2017 dollars is available for human space exploration).

The schedule as described in our analysis also does not include R&D towards the development of technologies that could be used for long-term lunar habitation. Instead R&D funds are devoted to the development of technologies intended for long-term stays on Mars, but in the 2040s. Under flat, or even slightly increasing budgets, given the cost of Gateway construction and lunar landing R&D during the 2020s, and DST development and the lunar landings during the early 2030s, cost and budget considerations preclude funding R&D on lunar habitation technologies to allow for long-term lunar habitation or supporting the establishment of commercial lunar activities as part of the 19-year exploration plan we describe. Adding long-term lunar habitation or support of commercial lunar activity to the notional schedule would likely require rising budgets. Alternatively, a second possibility might arise were new launch capabilities to emerge that were less expensive and allowed for a larger number of launches per year than the SLS/Orion combination upon which NASA's current plans—and this notional analysis—are based.

The risks to human health posed by a 1,100-day human orbital mission to Mars are highly uncertain. NASA documents, human research priorities, and the history of human spaceflight suggest that catastrophic technology failure resulting in death is the most important risk to human health, and as such, reducing this risk has been NASA's primary priority. While true, this focus does not negate the real risks of astronauts dying or suffering significant health degradation from spending 3 years in deep space on an extended orbital mission to Mars. The primary focus for planned research appears to be on currently known risks as opposed to understanding significant theoretical and unknown risks and holistic human health. The lack of information on the effects of 3 years in deep space on human health is a significant concern. This information is needed to design the DST to protect human health.

Understanding human variation and its relevance to human health outcomes will require a great deal more data. However, the limited time and resources currently allocated to collecting that data as well as the large knowledge gaps lead to a decreased likelihood of mitigating risks to human health by the time of the first mission to Mars. Although long-duration missions can take place aboard the ISS, limitations to studying synergistic health impacts relevant to deep space travel are inherent to LEO. Future plans for only one- to three-month missions on the Gateway constrain NASA's ability to reduce the knowledge gaps concerning a three-year Mars mission and its impact on human health. The planned one-year shakedown cruise for the DST takes place too late to incorporate changes in design based on information from the shakedown cruise without greatly increasing schedule and cost risk, and even if fully successful, still leaves a two-year gap between those data and the three-year Mars orbital mission.

In light of these findings, we have three high level recommendations. First, given that there is near-certainty that NASA cannot meet the 2033 goal, and 2037 and beyond is a more realistic timeline, NASA has time to consider a mission with value greater than that obtained from just orbiting Mars and returning.

For example, NASA could consider making the first Mars mission a journey to one of the Martian moons, Phobos. This would increase the value of the mission from scientific, exploration, and public interest perspectives without taking on the risk or cost of a surface mission.

Second, regardless of what the specific mission to Mars is, the organizational challenges of managing combined developments and missions to ISS, Gateway, the lunar surface, and Mars are significant and should be addressed. If Congress would like NASA to abide by a specific timeframe in which to reach Mars, a goal not unlike Apollo, there may be value to creating an Associate Administrator position in charge of the Mars missions, with discrete budget authority over the required Mars elements (distinct from Associate Administrator oversight of the Gateway and the ISS).

Lastly, from the point of view of human health risks, given the knowledge gaps, NASA may benefit from developing a unified research plan intended to prioritize its approach to fill in gaps in knowledge, especially on the combined effects of radiation, microgravity, and isolation that may be encountered on a human mission to Mars and precursor missions. We recommend giving human health research a first-order priority in mission planning for the ISS, Gateway and Moon between now and 2037, creating a systematic research plan that addresses synergistic risks of long-duration missions outside of LEO, and emphasizing human research on topics that focus on overall crew health and survival during missions. Given that the uncertainties associated with the human health risks of a Mars orbital mission may not be fully known or mitigated at the time the decision to build a DST is made, NASA should follow the ethical principles laid out in the Academies' report, and discuss publicly and transparently the risk-benefit trade-off associated with NASA's decision to pursue a Mars orbital mission. Clarifying NASA's approach to integrating human health and risk issues into vehicle and mission design may be an important contributor to gaining consensus from stakeholders with respect to the human research approaches NASA is undertaking.

The 20-year plan for human exploration based on NASA's current and notional plans presented in this report represents one of the largest exploration endeavors humanity has ever undertaken. Throughout the 10 Congresses and several Administrations this plan covers, changes are likely to occur to meet the evolving needs of the U.S. Government, U.S. industry, and international partners, requiring NASA's plans to be flexible. During the preparation of this report, we noted the flexibility inherent in NASA's current and notional plans, as NASA has been able to adjust its plans to focus on returning to the Moon to conform to Administration direction. Notably, the designs for the Gateway did not dramatically change whether it was the staging ground for missions to the lunar surface or to Mars. Much of NASA's flexibility comes from the types of missions that compose NASA's longer-term visions for exploration. In particular, current and notional missions for the lunar surface and Mars focus on visiting locations, not staying; although lunar landers and the DST are required to reach a destination, they are not sufficient for establishing long-duration habitation of the Moon or Mars.

Over the coming years, as these initial missions are completed, and plans shift from visiting to staying, organizational inertia, mission creep, and extant infrastructure will reduce NASA's flexibility. For example, by constructing the Gateway, NASA is making a commitment to maintain a presence in cislunar space through at least the lifetime of the Gateway as other NASA systems, such as lunar landers and the DST, will be designed to use a platform like the Gateway. Additionally, the Gateway's role as a dockyard for NASA systems could expand to include private and international lunar surface missions, which may increase system requirements (e.g., adding a propellant depot). Decisions that entrench systems as key components to an overall architecture—such as the ISS—create inertia in the system that may divert plans away from other goals. This is especially true of systems that create locations to stay, as these become permanent destinations—and sometimes the sole destination—because of organizational inertia. However, it is this very organizational inertia that can create sustainable programs that can weather shifting priorities. As future Congresses and Administrations debate the destinations for human spaceflight and the scope of activities to be conducted once there, decision-makers must balance the tension between creating sustainable programs for continuous human habitation in space and the drive to keep setting and meeting explicit horizon goals.

LOST ON MARS (2018)

WHY SPACE COLONIZATION WILL DISAPPOINT YOU

Micah Meadowcroft

Micah Meadowcroft is a writer and editor living in the Washington DC-Baltimore area. He has an MA in Social Science from the University of Chicago and has written for publications such as *New Atlantis*, *Providence*, *Wall Street Journal* and the *Philadelphia Inquirer*. He is currently managing editor of *The American Conservative*.

For the Achaemenid kings of ancient Persia, the world outside their dominion was a desert to turn into a garden. Ahura Mazda, Zoroastrianism's creator and most wise lord, had given them power, and with it the responsibility of regency. In conquest and faithful rule they would undo the drought and disorder made by diabolic Angra Mainyu and bring forth in the dry places fresh springs of water, both verily in walled gardens and metaphorically with truth. What they built with their hands and their laws was to make one paradise.

In Genesis we are told Jehovah planted a garden in the east in Eden, and that there he put the man whom he had formed. He put him in that paradise to dress it and to keep it. For man was made in God's image, after his likeness, male and female. And God said unto them, "Be fruitful, and multiply, and replenish the earth, and subdue it," and gave them dominion over the fish of the sea, and over the fowl of the air, and over the cattle, and over all the earth, and over every creeping thing that creeps upon the earth. But man and woman fell in disobedience. And therefore the Lord God sent man forth from the garden of Eden, to till the ground from whence he was taken.

Later we read that humanity wished to make a name for itself. And accordingly, lest they be scattered abroad upon the face of the whole earth, they set to build a city and a tower, whose top might reach unto heaven. But Jehovah said, "Go to, let us go down, and there confound their language, that they may not understand one another's speech." And did. So men left their tower to ruin, and the name of that place was called Babel, for man's language was confounded, and they were scattered abroad upon the face of the earth.

Today, multiplied and scattered men find themselves seeking new deserts to replenish and subdue. And so they look upward to the stars and planets and dream of reaching unto heaven to make a paradise of its dusts and build cities in its canyons. Whether on the Moon or Mars or Jupiter's Callisto, such new homes would be wastes more bare and wild than any they have known—yet humans yearn for them. But the colonization of space will fail to fulfill our hopes of gardens in that desert, even should we succeed in building towers there or causing plants to grow, unless we first alter ourselves and our purposes.

A New Worldview

Why do we wish to go? It is not just a desire for discovery. That is and still can be accomplished with further and farther voyages of our probes and instruments. We wish to go ourselves, to send humans to space and not merely to bring space to us. There are earthly reasons, which do not explain but justify: We tell ourselves that in exploring other spheres we will learn more of geology and the origins of life, or will derive new techniques and technologies. We say there are minerals to mine, or new means of energy production to harness. Elon Musk, SpaceX's middle-aged boy wonder, gives reason of a slightly higher order when he says Mars must be colonized for the survival of the human race—that it will be a citadel in which the torch of our civilization may be kept alight whatever calamities may come here below, in order to, like Hari Seldon in Asimov's *Foundation*, shorten future dark ages. Musk's rival space-bound billionaire, Amazon's Jeff Bezos, believes that, by moving human industry away from here, the colonization of space will save the Earth from its depredations. Robert Zubrin, president of the Mars Society, has in these pages ("The Human Explorer," Winter 2004) called for manned colonization in grander terms: "It is the chance to do something heroic, to advance humanity on the frontier."

But why are we so eager? Every call for volunteers to be the first to live—and quite probably die—on Mars has been answered with enthusiasm. People are ready to leave brothers and sisters, fathers and mothers, friends and land behind. We want to get out of here, though here will remain a more hospitable place than anything we build in space. There is a subtext of disaster to much of our thought of settling space. Some, like Musk or Christopher Nolan in his 2014 film *Interstellar*, make it explicit: Disaster will strike or has struck and so we must flee. For others, like Zubrin, the beauty of these dreams of space is that they might give us back a sense of hope and destiny, for destiny and hope seem to have been lost. A cadet branch of man on Mars would reassure us "that we have the capability to do such things, the capability to engage in yet greater ventures, more daring ventures, further out, toward an unlimited future." In some psychic sense, then, disaster really has struck. And so we wish to leave.

Hannah Arendt reflected on the significance of humanity's desire for the stars in her 1963 essay "The Conquest of Space and the Stature of Man." Man, she said, has not found a new place for himself after his self-displacement in modernity's rejection of old orders. We stand alienated from ourselves and nature. And as we see deeper into the firmament, and account for more of nature, we only better know the scale of our disorientation, the smallness of our vision. Science cannot bound the cosmos into a comfortable domain, for "this observed universe, the infinitely small no less than the infinitely large, escapes not only the coarseness of human sense perception but even the enormously ingenious instruments that have been built for its refinement." But neither does modern science really wish to find a place for man in all this space, for the questions that would require would produce answers for man that act "as definitions and hence as limitations of his efforts."

This alienation calls for some kind of transcendence—a need to find ourselves and discover where we are. And this is why, Arendt writes, we wish to go not just in spirit and imagination but in body to the stars.

> An actual change of the human world, the conquest of space or whatever we may wish to call it, is achieved only when manned space carriers are shot into the universe, so that man himself can go where up to now only human imagination and its power of abstraction, or human ingenuity and its power of fabrication, could reach.

The erotic drive to get outside, beyond, ourselves requires that we take dramatic action and find a new world. We are looking for home.

In his *Lost in the Cosmos* (1983), Walker Percy said that all this is a sign of our sickness. The optimist may argue that man's search for self-transcendence in such great enterprises is a sign of his freedom, that his stature is unthreatened. But Percy was not an optimist. Man fell in Eden, fell into self-consciousness without self-knowledge, became the being who sees and reads all signs except his self. He can name the other creatures but cannot name himself. Our alienation is complete, our language confounded. As animals, we have an environment, the relevant context of events acting upon us and our reactions to them—the relation of the bell and the food to Pavlov's dogs, or even just that between a body, height, and gravity. But as sign-users—"semiotic" beings, as Percy calls us—we also have a world, a supposed-to-be comprehensive linguistic map of everything we know or could know, an ordered cosmic cartography. In Percy's use, this is the world in your world-view: "All men in all cultures know what is under the earth, what is above the earth, and where the Cosmos came from." You have one whether you like it or not, a symbolically organized world in which everything you can think or experience shows up.

For most human beings who have ever lived, this "world" has involved an understanding of the self in relation to others, to nature, and to a divine transcendence that is beyond the cosmos yet also immanent in it. But, writing of the same sense of alienation and displacement as Arendt, Percy says, "In a post-religious technological society these traditional resources of the self are no longer available, leaving in general only the two options: self conceived as immanent, consumer of the techniques, goods, and services of society; or as transcendent." In the first option, immanent man is mass man, integrated as a part among many into the structures and mechanisms of the social system, accepting his world through passive ignorance of it. In the second option, "the only transcendence open to the self is self-transcendence, that is, the transcending of the world by the self. The available modes of transcendence in such an age are science and art." To broaden the world-view, to actually change his experienced world, seems to require that man secure a position as active observer and interpreter rather than mere participant within a pre-made order. In the outsider role of scientist or artist, man may refuse to be just himself, a consumer conforming to functions made for him by a faceless society. "The pleasure of such transcendence derives not from the recovery of self but from the loss of self. Scientific and artistic transcendence is a partial recovery of Eden, the semiotic Eden, when the self explored the world through signs before falling into self-consciousness." Science's taxonomies and art's sub-creation give scientists and artists a structure or mode in which to explore and position themselves in their world—an occasional detachment from circumstance.

"The environment has gaps," says Percy. "But the world of the sign-user is a totality." What is relevant to us in our circumstances is always partial; we experience little of what is actually there around us, making up our environment. Our world, though, contains even what we have not seen, nor heard, nor touched with our hands, that does not exist apart from our thoughts; it contains ideas and not things only. Space is an intersection of environment and world. For practical purposes—in the environmental realm of causation—the movements of Mars are a blank space; I see or miss its rusty light without effect. Its gravity is to my motions here insignificant. But its *place* in the sky, its presence in thought, is significant. It is near my planet Earth, a new environment relative to my world, and is therefore a part of my world, of my system of signs. I have some idea of it, if only that it is there, another rock in orbit around a great ball of burning gas. But what it is made of is not what it is, and what it means to us has little to do with what it is made of.

Our eagerness to explore is a kind of collapse of world and environment. In discovery of new material conditions—environments—the spiritual might be changed. The colonization of other planets, of *new worlds*, is motivated by hope of a new world-view, a search for a permanent transcendence now only occasionally found. As Charles T. Rubin has written in this journal ("*Thumos* in Space," Fall 2007), "The human explorer manifests his delight, his joy and excitement, at juxtaposing the familiar and the strange; watching, we can, at least in some distant way, feel with him. (Once, merely reading the reports of explorers would have sufficed.)" President Kennedy said in 1962, "We choose to go to the moon in this decade and do the other things, not because they are easy, but because they are hard." But the relationship between expanded environment and world does not by necessity elevate man's dignity and stature, and we may choose to go to Mars not because it is hard—though it is surely the greatest technical challenge we have yet presented ourselves—but because it is easy, easier to be explorer than to find another and better sense of self.

The Self in the New World

And what sense of self has human exploration brought? Before we thought to "explore strange new worlds" with Kirk and Spock and all the rest, early modern man ventured forth in an Age of Discovery. A New World beckoned. And with Europeans' arrival in the Americas came a drastic revision of man's place in the world and thus his relationship to his fellow man, to nature, and to his God. Scholastic figures such as Bartolomé de las Casas and others associated with Francisco de Vitoria's school of Salamanca responded to the discovery and exploration of the Americas with a focus on the alien "Indians"—particularly the obligations Christendom had toward them as fellow creatures of God, bearing rights and deserving justice. But proto-liberals, particularly John Locke, fixated on the vast and seemingly uncultivated expanses of North America, and hastened past cursory acknowledgments of indigenous populations.

In examining the continent from, let's say, an extremely academic remove, Locke found space and scope in which to develop his anthropology of a state of nature, removing the individual from any prior context of relationship, society, or politics. This became new grounds for a fundamental equality of man, in which God played a mostly incidental part: Apart from entrance into voluntary society and communal justice, every individual is judge and executor of the laws of nature, and nature is held in common till some individual mixes his labor with it to create property. And even property creates for him but minimal obligations to his fellow man. In his *Second Treatise on Government*, Locke holds up the bounty of North America as support for his completely individuated pre-political humanity:

> ... for supposing a man, or family, in the state they were at first peopling of the world by the children of *Adam*, or *Noah*; let him plant in some in-land, vacant places of *America*, we shall find that the *possessions* he could make himself, upon the *measures* we have given, would not be very large, nor, even to this day, prejudice the rest of mankind, or give them reason to complain, or think themselves injured by this man's incroachment, though the race of men have now spread themselves to all the corners of the world, and do infinitely exceed the small number was at the beginning.

And elsewhere in his treatise, the father of liberalism connects his world-view even more baldly to his idea of the Western Hemisphere, writing, "Thus in the beginning all the world was *America*."

As English ethicist Oliver O'Donovan reminds us, we can condemn the bizarre ahistoricity of this scheme even simply on the level of ideas, without needing to summon the testimony of pre-Columbian and colonial-era native populations to remind the enlightened doctor they were in fact here first. In *The Desire of the Nations* (1996), O'Donovan writes of "the myth of the social contract" that is at the heart of Locke's political theory—the myth that "society derives from an original free compact of individuals, who have traded in their absolute freedoms for a system of mutual protection and government. So obviously is this myth unhistorical that it is easy to underestimate its hold on the modern mind." No one is ever a free individual among individuals in empty land, without ties to body, family, or culture.

But Locke needed North America to be a vacuous desert for his vision of nature, and consequently his entire project, to hang together. As O'Donovan writes, "Corresponding to the transcendent will is an inert nature, lacking any given order that could make it good prior to the imposition of human purposes upon it." That is, wishing to make an apology for an unbondaged will and construct a new *world*, Locke simply makes America's vast wilderness evidence, regardless of what is actually there. In our exploration, we moderns have muddled environment and world, confusing and distorting in both directions, with worlds projected onto environments, and environments and the people in them altered to fit worlds—think of Manifest Destiny, or the Turner thesis, or so much of the worst of colonialism.

Power Over Nature

Of the modern age, the German Catholic priest-intellectual Romano Guardini said, "What determines its sense of existence is power over nature." Inspirited by Bacon's scientific project—the vexations of nature and discovery of her secrets by the skills of man's *techne*—this world-view seems to both define and demand our efforts toward exploration and self-transcendence. Hannah Arendt assigns it blame for our alienation: "Has not each of the advances of science, since the time of Copernicus, almost automatically resulted in a decrease in his stature?" She laments that the increase in humanity's physical knowledge seems to correlate with further displacement, a loss of cosmic coherence—that, in short, an expanded environment has created an alienated world.

The scientist truly does achieve a kind of self-transcendence. Noting the casualness and speed with which men opened the container that had until then kept the destructive power of nuclear weapons from our childish hands, Arendt observes, "the scientist *qua* scientist does not even care about the survival of the human race on earth or, for that matter, about the survival of the planet itself." The scientific world-view allows the self to transcend the world because it ceases for a moment to consider the questions of interiority. Instead it makes man merely another subject of observation, another material cause and action among many.

There is an irony to this, however, for Arendt writes, "All of this makes it more unlikely every day that man will encounter anything in the world around him that is not man-made and hence is not, in the last analysis, he himself in a different disguise." We are to ourselves the most confusing and inaccessible thing in the cosmos, and so we wish to find ourselves everywhere and in everything even as we seek to escape ourselves. We are our own aliens, whether in space or upon the Earth. The effort at self-transcendence reaches a limit, then, and can free us only so far from our alienation. Even were we to build cities on Europa we would not escape, for

> These new possessions, like all property, would have to be limited, and once the limit is reached and the limitations established, the new world view that may conceivably grow out of it is likely to be once more geocentric and anthropomorphic, although not in the old sense of the earth being the center of the universe and of man being the highest being there is.

That "new" world-view would be just as defined by our desire to know our place as it has been.

While the dominance of a kind of scientific world-view is incontestable, how and why it replaced the religious humanist paradigm of the past remains up for debate. For Arendt, there was an unconscious trade: The growth of the scientific displaced the humane. For Percy, the ascendance of the scientific was incidental: Man has abandoned religion, myth, totems, and all the other old ways, and so now has few options besides science to resort to. Notre Dame's Patrick Deneen sees no accident in this exchange. It is deliberate. Man, by means of science, has in a kind of Faustian bargain replaced religion with the powers of his self. Writes Deneen in these pages ("Nature, Man, and Common Sense," Fall 2007), "The purported aim of lowering human stature is deceptive, inasmuch as its more fundamental motivation lies in displacing the status of the *grantor* of that special status, namely God. By displacing God, humans—increasingly enhanced in power and control by means of science—can occupy the space once occupied by the divine." Man makes himself the subject of experiments that he may no longer be the subject of his God.

Power Over Man

Most of humanity are not scientists. In a scientific and technological age, then, most of humanity will not experience transcendence by science's means. They will not be observer but observed, not experimenter but experiment. Most men are mass man. "If nature is being more and more subjected to the control of man and his works, man himself is also increasingly controlled by those who fit him into 'the system,' even as his work is controlled by the end to which it is directed," Guardini wrote in *Power and Responsibility* (1951). This is the same observation as one made by C. S. Lewis *in The Abolition of Man* (1947), that "Man's power over Nature turns out to be a power exercised by some men over other men," particularly over future generations, as we increasingly turn both our children and the habitat they inherit into engineering projects.

There is obviously something grand about mankind's technological accomplishments, however, and so despite the exclusivity of the scientific caste, its structures become self-justifying, and even invade the language of would-be humanists. Consider this pitch from the Mars Society's Robert Zubrin: "It's an investment: If we go to Mars, we'll someday get the payoff that comes from challenging people in a serious way, and by being a society that values great scientific and human achievements." But the payoff of a system, our system, ordered to power over nature is more power over nature—and hence over other men—for those who wield it, the scientists and technicians. St. Augustine's *libido dominandi* manifests itself in every age.

An antagonist in the first volume of Lewis's sci-fi "Space Trilogy," *Out of the Silent Planet* (1938), justifies shanghaiing the hero Ransom on a journey to Mars by taking modernity's ethic to its dehumanizing conclusion, declaring that small claims must give way to great ones.

> We have learned how to jump off the speck of matter on which our species began; infinity, and therefore perhaps eternity, is being put into the hands of the human race. You cannot be so small-minded as to think that the rights or the life of an individual or of a million individuals are of the slightest importance in comparison with this.

Elon Musk, in a presentation titled "Making Life Interplanetary" on his plans for Mars, said, "Becoming a multi-planet species beats the hell out of being a single-planet species."

We, like Ransom, should respond to such high hopes for a glorious human future with a reminder that we bring human nature with us, even when exploring a new world: "I suppose all that stuff about infinity and eternity means that you think you are justified in doing anything—absolutely anything—here and now, on the off chance that some creatures or other descended from man as we know him may crawl about a few centuries longer in some part of the universe." In this journal's last issue, James Poulos wrote optimistically of the prospect of Mars elevating our eyes and souls ("For the Love of Mars," Spring 2018). To put it in Percy's terms, Poulos hopes man may reform his vision of the world and expand his environment on the sanguine planet simply by lovingly settling it, a united movement of body and imagination. I am not as hopeful.

Again repeating Arendt: "The scientist *qua* scientist does not even care about the survival of the human race on earth or, for that matter, about the survival of the planet itself." In its purity, the quest to vex nature's secrets from her sees human beings as simply something more to deconstruct. Indeed, in *Out of the Silent Planet* and its sequel, *Perelandra*, the danger is never aliens but humans. The promoters of space are not such pure scientists as Arendt's type. But, as lost in the cosmos as the rest of us, they nevertheless seek self-transcendence by means of scientific achievement. We may indeed arrive at a place where the survival of the human race *on Earth* is of less consequence to those possessing power than the furtherance of an abstract humanity that has abandoned much of what made it human.

Are You Happy?

Consider your place in our interplanetary future. Let us speculate. Suppose that SpaceX or some similar enterprise has succeeded, that in the aspirational year 2024 crewed missions settle Mars. Their base is serviced by rockets—Big F***ing Rockets, according to Musk—launched regularly from Earth to Mars. The colonies grow, slowly but seemingly surely. The technical challenges have been immense—but man's resourcefulness greater. The looked-for scientific and technological discoveries materialize with exhilarating speed. Are you happy?

You are Elon Musk. You have achieved your boyhood dream of laying the foundations of towers on other worlds, so that humanity may scatter to the stars lest mankind die upon the Earth. You have made a name for yourself. You have brought forth water in dry places, made gardens where there was only dust and ice. You are dating Grimes. You are very rich. You fight with strangers on Twitter. What is your stature? Where do you fit? Why?

You are a colonist on Mars. You have left behind your country, and your kindred, and your father's house, to go scrape an existence out of the thin infertile regolith of the fourth rock from the Sun. You will live the rest of your life maintaining an assembly-line routine of actions that, along with those of your crewmates, will slowly turn the red around you green. You will die here. Your body will be used for fertilizer. Musk paid off your college debt. You will miss your sister's wedding and your mother's funeral. You will not have children: Cosmic radiation, hardly blocked by the planet's scant atmosphere, causes each rare impregnation to swiftly end in miscarriage, and many crewmembers are now as sterile as the Martian surface. Is it worth it?

You are a scientist on Earth working on the colonization project. You are trying to solve the fertility problem. For now the settlers will simply not be allowed to breed. You may sterilize those still technically fertile until a solution is found. It would be easier. But first there are tumors and leukemias to deal with. We are too fragile, it seems, the god of war too harsh. *Heterocephalus glaber*, the naked mole-rat, would make a far better colonist than the human being. The rodent is cancer- and pain-resistant. It is hive-socialized. So you are breeding mole-rats, tinkering with their DNA using CRISPR-Cas9 genome editing technology to try to find lines of their genetic code that could be useful to you. If man became a little more mole-rat, he might survive and reproduce on other planets. You are addicted to porn. You have not been on a date in months. Does *Heterocephalus glaber* love you?

You are a member of the general public. Humanity has gone to space, is on Mars. You watched all the livestreams. You bought a poster of the first colonists to hang in your second bathroom. It is a print of a painting in the style of Soviet space-race propaganda. You bought your son a tin lunch box shaped like one of Musk's BFRs to take with him to school. You worry your son knows what the F stands for in BFR. You worry he might have ADHD. His teacher, Ms. Perkins, says he is not as well-behaved as his sister was and has suggested you take him to a pediatric psychiatrist. Your insurance will not cover that. Is the achievement of humanity conquering space and colonizing Mars your achievement?

The New World

In Genesis man and woman are called to dominion, to rule the Earth. Guardini writes, "Man's natural God-likeness consists in this capacity for power, in his ability to use it and in his resultant lordship." And elsewhere, "When we examine the motives of human endeavor and the play of forces set in motion by historical decisions, we discover everywhere a basic will at work, the will to dominion." In some sense the pioneer spirit and the desire to colonize space is an expression of that call, something teleological, essential to the human person. But modernity's drive to power over nature is a corruption of what ought to be mankind's dominion over creation.

> Modernity could bask in dreams of yet undiscovered lands, untapped reserves. The concept "colony" was an expression of this. Even the individual peoples and their states embraced, both materially and humanly speaking, unknown, unmeasured possibilities.

Until we have truly taken dominion of the Earth and subdued it as we ought—as lords and stewards placed here by its and our creator—to seek space and the dress and keep of gardens on other worlds is to distract ourselves from our failures. It is the commencement of a new project when we find our first too difficult.

The motives and problems of our desire for the stars that I have sought to explore here are merely those of ourselves. We carry them with us into our post-modern age, and should we venture upward we will carry them into space. The task Guardini set for himself in *Power and Responsibility* is all of ours:

> The core of the new epoch's intellectual task will be to integrate power into life in such a way that man can employ power without forfeiting his humanity. For he will have only two choices: to match the greatness of his power with the strength of his humanity, or to surrender his humanity to power and perish.

To resituate man where he belongs, so that he is no longer lost and alienated, is to situate his power, where it comes from and for whom it has been given. Man can discover his true self in the world when he finds what and how he ought to do, for "the doer is constantly becoming what he does—every doer, from the responsible head of state to office manager or housewife, from scholar to technician, artist to farmer." In doing right—by exercising whatever power he has with responsibility for others—he can embrace his identity as doer. Indeed, he must, for "if the use of power continues to develop as it has, what will happen to those who use it is unimaginable: an ethical dissolution and illness of the soul such as the world has never known."

We continue to fall like the man and woman in the garden because of a failure to see the "fundamental facts of human existence: the essential difference between Creator and created; between Archetype and image; between self-realization through truth and through usurpation; between sovereignty in service and independent sovereignty." Until we regain a cosmic—that is, truly ordered—vision of the world, a chain of being in which to place ourselves and our environment, and until we consider all our placements, even the most mundane, as so situated, we will remain lost.

The new man for the new age Guardini says we require is much like those ancient gardener-kings of Persia, who ruled as regents of Ahura Mazda, under authority. For "this man knows to command as well as how to obey. He respects discipline not as a passive, blind 'being integrated into' a system, but the responsible discipline which stems from his own conscience and personal honor." This humanity needs neither the transcendent system-making of the scientist nor the blind immanence of mass man the consumer. Rather, the future's man takes stock of his dominion, all that his given power gives him responsibility for, and makes a garden of it.

Until we can count on this kind of humanity on Earth, our efforts into space will fail to elevate us. "And, therefore," President Kennedy said, "as we set sail we ask God's blessing on the most hazardous and dangerous and greatest adventure on which man has ever embarked." For whether we stay or go, without it our sickness will only grow worse.

MISSION TO MARS (2019)

THE COMPLETE GUIDE TO GETTING TO THE RED PLANET

Leah Crane

Leah Crane is the physics and space reporter at *New Scientist*. She holds a BA in physics and astronomy from Carleton College. She is based in Chicago.

As rust-coloured dust blows across the empty plains and deep craters of Mars, it just occasionally dances over something made by human hands – perhaps the solar panel of a lander, or the wheel of a rover. The robots we have sent to our neighbouring planet have taught us plenty about it. It hosts the highest mountain in the solar system and probably has underground lakes of liquid water. Long ago it wasn't a freezing desert as it is now, but a warm, wet place. Yet we have never set foot there ourselves.

There's a good reason for that: getting to Mars is hard. Since 1971 there have been 18 attempts to land robots on Mars and 11 of these either crashed, fatally malfunctioned soon after landing or missed the planet altogether. If human lives are at stake, we need better odds of success.

Putting humans on Mars is far from impossible. Doing so is a major goal for NASA, which aims to pull it off in a little over a decade. Elon Musk, the founder of SpaceX, has long said that he wants to build settlements on Mars. China, Russia and India all have their sights set on the planet too.

The most important driver for this Mars rush may be the prestige, but there are good scientific motivations too. While rovers can do marvellous things, they don't have the dexterity, knowledge or intuition that a human would bring to bear on one of the biggest questions our species has asked: are we alone in the universe? Mars is the best place to answer that, says NASA scientist Jennifer Heldmann. "And if you want to really learn about Mars, to answer those fundamental questions, you have to send humans."

To get there we will need to blast off from Earth with more supplies than we have ever put in space before, traverse millions of kilometres of deadly interplanetary nothingness, and land safely at the other end. It is daunting, but it isn't out of the question. Here is our step-by-step guide.

1. Leaving Earth

When Earth and Mars are at their closest, they are about 55 million kilometres apart. That sounds like a lot. But purely in terms of the propulsion systems needed, travelling that distance through space isn't actually too big an ask of our existing rocket technology.

Once you are far enough from Earth, its pull drops considerably and you could cruise to Mars using a reduced thrust. The journey would take about nine months, a little longer than an astronaut's standard six-month stint on the International Space Station (ISS). We don't need to dream up new types of engines or worry about things like solar sails, which accelerate very slowly. All we need is a big rocket pointed in the right direction.

Decades of space exploration have taught us a few things, chief among them being how to build big rockets. There are seven types of rocket in operation that could make it to Mars. The most powerful of these, SpaceX's Falcon Heavy, could shuttle about 18.5 tonnes there. That is more than enough for any lander or rover, but a human mission will be heavier. A crew of six along with food and water to last their journey there and back weighs in at a minimum of 20 tonnes. In 2017, a NASA report estimated that once you factor in scientific equipment and the kit needed to keep explorers alive on the surface – like a power generator and a place to live – a more realistic figure would be about 100 tonnes.

That's not unthinkable. Two rockets that are in development, NASA's Space Launch System (SLS) and SpaceX's Big Falcon Rocket (BFR), are planned to be more powerful than anything that has been launched before. SLS should be able to carry at least 45 tonnes of cargo to Mars, and BFR is expected to haul more than 100 tonnes.

In other words, building bigger, better rockets is something we know how to do. And we could always lighten the load by sending some equipment ahead of the humans. "Everything else is the hard part," says Bruce Jakosky at the University of Colorado, Boulder.

2. In Transit

It might seem as though humans have got to grips with surviving off-planet. After all, the ISS is permanently crewed. But as space exploration goes, visiting the space station is like camping in your back garden. You might feel like you are away from home, but your parents are still bringing you sandwiches. If you are going to Mars, you need to take your own sandwiches.

Except it isn't just food you have got to worry about. If the spacecraft breaks, you must have the spare parts and tools to fix it. If you get sick, you need the right medicine. But packing for every eventuality isn't possible, given that extra weight means more fuel and more expense. What do you do?

Part of the solution will be to take 3D printers that can produce parts on demand. The ISS already has one on board and NASA has been experimenting with it. So a Mars trip could pack a printer and raw material, rather than a bunch of parts that might not be needed.

Stocking the medicine cabinet is more tricky. Our experience on the space station shows germs can thrive in spacecraft. And studies have shown that bacteria growing in simulated microgravity can develop resistance to a broad-spectrum antibiotic, and they retain that resistance for longer than they would on Earth. There are projects in the works to mitigate this, including antibacterial coatings for surfaces that might get dirty, like toilet doors. There is also a suggestion that astronauts could bring along raw pharmaceutical ingredients instead of fully-formulated medications and manufacture their own drugs on demand. A prototype system for automatically synthesising simple medicines has already been tested in space.

Whether or not astronauts get sick, they will definitely feel the physical effects of space travel. Without the pull of gravity to contend with, muscles and bones start to waste away. Studies show that astronauts can lose up to 20 per cent of their muscle mass in under a fortnight, even with daily workouts. The good news is that this may not matter much on Mars because its gravity is so much lower than Earth's – walking on the Red Planet would be far easier. Still, we would want to counteract the effects as much as possible, and astronauts would probably be tasked with hours of daily exercise and special diets. They might also have to wear muscle-compression suits.

As well as missing Earth's gravity, astronauts won't be shielded by its magnetic field, which diverts harmful cosmic radiation. NASA limits radiation exposure for male astronauts to about the equivalent of 286,000 chest X-rays, and around 20 per cent less than this for women, whose bodies may be more susceptible to radiation damage. Astronauts on a Mars mission would hit 60 per cent of that limit on the shortest possible return journey, without taking into account time on the surface. "As it stands right now, every single mission would have to evaluate whether its goals are worth violating the astronaut health standards," says Lucianne Walkowicz at the Adler Planetarium in Chicago. "There's no way to meet the current ones on long-range missions, and there's no way to run the experiment to find out exactly what the risks are without actually doing the mission."

That goes for the mental health risks, too. Being so far from Earth – far enough that home becomes just another point of light in the sky – could be psychologically challenging, says retired NASA astronaut Nicole Stott. "Everything we've done so far, we have had the view of Earth out our window, right there," she says. "It's not just pretty. You have a connection to that place. As long as you can see it present with you, you can maintain that connection." You need a special type of person to cope without it.

3. Who do We Send?

The people we send to Mars will have to meet all the requirements that astronauts do now, including passing strenuous physical and psychological tests. But their skills will have to go beyond that. On the way to Mars, nobody can quit the team and nobody can be added. The handful of people on board will be totally responsible for keeping the mission aloft.

Certain roles like engineers, doctors and scientists will be indispensable. But it won't make sense to look for perfect astronauts, rather the perfect team of astronauts. "You're trying to put together a toolbox, and you wouldn't fill a toolbox with hammers even if they're all the best hammers in the world," says Kim Binsted at the University of Hawaii.

Binsted knows what she's talking about, as chief of the Hawaii Space Exploration Analog and Simulation, in which crews of four to six people live as if they are on Mars. Participants stay for months at a time, donning mock spacesuits when they go outside and enduring a 20-minute communication lag with "Earth".

One thing that consistently causes conflict, says Binsted, is when one or two team members feel different from the others. It could be differences in gender, nationality or even music preference. A crew with three men and one woman, or one person who wants to blast Metallica at all hours, might crumble because the team don't feel like they are all on the same footing. "Given that you're going to have some diversity, you want as much as possible," says Binsted.

Getting a team mission ready will probably involve more intensive group training than astronauts undergo now. The crew will have to learn to deal with each other's personality quirks to defuse even small interpersonal conflicts. "Molehills become mountains in austere environments over time," says NASA psychologist James Picano. "They will have to train as a crew, live together as a crew, simulate those kinds of conditions."

4. Landing and Living on Mars

With nine months of empty space and avoided arguments behind them, the travellers are about to face the most dangerous part of their journey. The trouble with landing on Mars is that its atmosphere is almost non-existent – it is 160 times less dense than Earth's, on average. This means that parachutes don't create enough drag to slow down spacecraft, as they do when landing on Earth. We could use boosters to slow down, like the Apollo astronauts did when they landed on the moon. But because gravity on Mars is stronger than that on the moon, we would need a lot more boosters. This means we will probably need a combination of boosters and something to create drag.

This approach has succeeded for a 1-tonne robot, but it won't be so easy for a heavier craft, which is why researchers are working on finding improved ways to land.

One is NASA's Hypersonic Inflatable Aerodynamic Decelerators, a series of landing devices that use fabric strengthened with Kevlar to form a blow-up structure that is more rigid than a parachute and so creates more drag. The agency has tested small scale models of it on Earth.

Yet the really difficult question isn't how we land, but where. A site near either of the poles would seem the obvious choice because this is where we know there is underground water ice – and possibly an underground lake of liquid water – which would serve as a crucial resource. Humans use a lot of water and it is very heavy, so the amount we could take to Mars would be limited. Plus, many proposed Mars missions involve using water to make rocket fuel to get the explorers home.

The trouble is the pole areas get as cold as −195°C and are prone to storms that make landing even harder. "It's also not a very exciting place. The northern plains of Mars are pretty flat and boring," says Tanya Harrison at Arizona State University. The equatorial region mostly stays above −100°C and can reach 20°C. It also has more sunlight that astronauts could harvest for solar power, rarely gets storms and has all sorts of interesting terrain to explore. But it doesn't seem to have much, if any, accessible water.

It is a tricky problem, but for the first missions, it may be simplest to land somewhere predictable, where rovers have already explored (see "Mars walk"). Once they are down, the explorers will be sticking around for a while. Even if they aren't establishing a permanent settlement, they will have to wait months at a minimum for Earth and Mars to come into alignment again so they can travel home in a matter of months rather than years. There is no visiting Mars without setting up a base.

The base will have to deal with the variety of interesting ways in which Mars can kill you. Apart from the aforementioned gnawing cold, there is the constant risk of being hit by micrometeorites, which often don't burn up in the wispy atmosphere. Then there is the radiation from space, which isn't deflected away because Mars has no planet-wide magnetic field. And with so little atmosphere, the pressure is incredibly low, almost akin to deep space.

The simplest protection from these risks may be the spacecraft that got our explorers here. But the landing craft itself would probably make for cramped quarters. Another option would be to bring their shelter or the materials to build it with them. NASA is running a competition to design 3D-printed habitats, and there have been many entries. A number of them use pieces of the landing craft in their design, but they all also require other building materials, which adds weight to the launch craft. The entries get extra points if they use resources already on the Martian surface, which has inspired plans to make bricks of compressed Martian soil and build igloo-like shelters. NASA has given contracts to several groups studying the best way to make such bricks using precisely engineered replica Mars dust. But even so, building a home on Mars will probably require sending a few packages of building materials on ahead.

It might be possible to rope in the Martian crust itself as a natural radiation shield. One proposal would see humans setting up their habitats in the cylindrical caves created by ancient lava flows. We have seen the entrances to such caves on Mars in satellite images and studied similar structures here. On Earth, these caves are generally about 30 metres wide, but research suggests that on Mars, with its much lower gravity, they could be eight times wider and stretch for miles. One day they could accommodate a whole street of habitats.

The intrepid astronauts will have other pressing needs to think about. Food can be freeze-dried, seeds can be packed, oxygen can be taken in tanks if it can't be scrubbed from the Martian atmosphere once there.

But water is less easy. Even if the astronauts have landed in a location with plenty of it beneath the surface, they will need to have brought heavy mining equipment to reach it. And there's no guarantee that it will be potable once it is out of the ground. "We don't know a lot about the ice that's there," says Harrison. "You don't want to hang your hat on assuming that what's there is all nice, drinkable water because we really don't know." Even if the water is safe to drink, it will be full of fine dust. So the astronauts will have to bring sophisticated filtration systems with them.

That goes for the spacesuits too: they will have to be excellent at keeping dust out, especially as Martian soil may be full of chemicals that can be deadly if inhaled or swallowed. NASA is already working on next-generation spacesuits and special coating materials that would counter the dust problem. "On Mars, there's no margin of error," says Harrison. "Everything has to be working or you die."

5. Home Time

Some people may be hoping that we will settle on Mars permanently in the long term. But all serious Mars mission plans currently involve bringing the explorers back. "It's a very harsh environment and I don't know why we would want to live there," says Jakosky. This means the astronauts need to endure another launch, another nine-month journey, another landing. Luckily, it will be easier the second time. Mars's thin atmosphere and its weaker gravity will mean getting into space won't be as tough. The journey itself will be equally long, but the familiar azure glow of our home world will grow stronger by the day. The landing will be simple, aided by parachutes and Earth's thick atmosphere.

When the explorers peek their heads out of their capsule, they will be splashed by the cool water of our abundant oceans and enveloped in the chatter of other people. They will be home. Back on Mars, the swirling dust will have already covered their footprints. But their habitat will still be standing, ready and waiting for the next visitors.

Mars Walk

There are lots of impressive geological features to explore on the surface of the Red Planet

- The northern hemisphere of Mars is dominated by vast and largely featureless plains such as the Vastitas Borealis. This huge flat area around its north pole is about 4 or 5 kilometres lower than the planet's average elevation.
- Mars's southern hemisphere contains heavily cratered areas like Terra Sirenum. It is a mystery why the northern and southern hemispheres are so starkly different, a characteristic not seen on any other planet.

- Standing almost 13 kilometres high, Elysium Mons is the fourth highest mountain on Mars.
- Olympus Mons is the tallest known mountain in the solar system. Standing nearly 22 kilometres high, it is about two and a half times the height of Everest.
- To the south-east of Olympus Mons are three vast extinct volcanoes, including Pavonis Mons. Hundreds of kilometres wide, the tallest of them peaks at more than 18 kilometres.
- The Valles Marineris is a huge and intricate system of canyons that is more than 4000 kilometres long and up to 7 kilometres deep. Most scientists think this feature is essentially a crack in the planet's crust, which may have formed through plate tectonics.
- Hellas Planitia is an impact basin 3 kilometres deep. It is thought to have formed about 4 billion years ago when a huge asteroid struck Mars.

ONCE UPON A TIME I LIVED ON MARS (2021)

CHAPTER II: ASTRO-GASTRONOMY

Kate Greene

Kate Greene is a New York-based essayist, poet, journalist and former laser physicist whose work has appeared in *Aeon*, *Discover*, *Harvard Review*, *The Economist*, *New Yorker* and *WIRED*, among others. She holds a BSc in chemistry, an MSc in physics and an MFA in poetry, and has taught creative writing at Columbia University, San Francisco State University and the Tennessee Prison for Women. In 2013, she was second-in-command on the first simulated Mars mission for NASA's HI-SEAS project. It is this experience that led to her 2020 book *Once Upon a Time I Lived on Mars*, from which this chapter is extracted.

Let it begin with poutine. Salty, squeaky, messy, and delicious, poutine is a French-Canadian delicacy of french fries topped with cheese curds and brown gravy, currently experiencing international acclaim. Likely invented in a restaurant in Québec in the 1950s, poutine's breakout moment was in 1982 when the *Toronto Star* introduced it by suggesting two types could be found in the eastern part of the country: "regular" and "Italian style," a version made with spaghetti sauce. Shortly thereafter, its popularity exploded. Before 1982, the use of the word "poutine," at least in books digitized by Google, was virtually nonexistent. After 1982, the Google graph of its appearance in text shoots straight into the stratosphere.

There's a decidedly of-the-people feel to poutine. Like a sandwich or a casserole, you can make it your own, dress it up or down, add extravagant sauces and toppings such as ground beef, pickles, kimchi, pork, fennel, curried lentils, a fried egg, pepperoni. In 2001, chef Martin Picard "elevated" poutine by making it with foie gras, now a specialty at his Montreal restaurant, Au Pied de Cochon. And in 2007, in a remote outpost on Canada's Devon Island, 10 degrees north of the Arctic Circle and just west of Greenland across Baffin Bay, poutine went interplanetary.

Devon Island is home to a facility used to conduct analog Mars missions. Called Flashline Mars Arctic Research Station, or FMARS, it mainly consists of a two-story metal can that accommodates crews for long expeditions, allowing them to conduct experiments in a harsh, isolated environment. FMARS comes equipped with ATVs for transportation and shotguns to scare off polar bears, which historically haven't been necessary, but it's good to be prepared.

From May to August 2007, FMARS was home to an international crew of seven people conducting a four-month mission to better understand possible astronaut challenges on Mars.

One of those difficulties turned out to be a homesick Canadian named Simon. To cheer him up, the crew decided to make poutine. While they didn't have french fries or cheese curds, they did have certain dehydrated, shelf-stable, and therefore space-friendly foods such as packets of brown gravy. A start. The crew improvised the rest. Dehydrated scalloped potatoes stood in for fries, and powdered milk hinted at the possibility of cheese curds.

The curds took some time, I learned from Kim Binsted, crewmember on this 2007 FMARS expedition. Like Simon, Binsted was Canadian and therefore familiar with the restorative powers of poutine. The fries also took some finagling. First, crewmembers rehydrated the potatoes, then fried them, and finally baked them for good measure. "And because we were feeling exuberant, we made alternative sauces," Binsted told me. "It took all day, but Simon was extremely happy. We put a Québec flag next to it and took pictures."

The Martian poutine launched dozens of culinary celebrations on Binsted's mission. The crew celebrated birthdays, half birthdays, three-quarter birthdays, any excuse to get creative in the kitchen. All the while, Binsted was reading Ernest Shackleton's diaries from his harrowing 1914 Antarctic journey aboard the *Endurance*. "It was clear that food was important and that [Shackleton's crew] also had special meals, even if it was just seal fat," she said. "I realized food plays an important role in long-duration expeditions and not just in sustaining yourself, but also helping crews bond and reminding people of home."

After the mission, a picture and description of the interplanetary poutine appeared in an academic paper authored by Binsted and others, along with suggestions that meals, in particular celebratory or special meals, might play a crucial role on far-flung space missions:

> The psychosocial preparation and consumption were very clear ... meals eaten en famille provided the social glue that held the crew together. Meal times were an opportunity to discuss the challenges of the day, plan next steps, air complaints, share news, and so on ... Special meals were used to break up the monotony of the long mission, and to mark the passage of time.

Back on Earth, Binsted gave talks about her FMARS experience. In attendance was Jean Hunter. For years, Hunter had been asking questions about space food, including how fermentation might expand the diversity of flavors, textures, and uses of a limited range of space crops, and how omnivorous astronauts might feel, over time, about a diet with vegan menu options. Binsted and Hunter decided to partner on a project—a Mars simulation of their own—to test a problem called menu fatigue, which NASA already knew its astronauts encountered on long-duration missions.

Simply put, crews on the International Space Station tend to eat less over time. And since a healthy diet is crucial to maintaining bone density and overall health in zero gravity, when calories flag, Houston considers it a problem.

I first encountered the concept of menu fatigue, when I myself was fatigued one late February day in 2012, scrolling through Twitter. An NPR headline flashed past: "Why Astronauts Crave Tabasco Sauce."

Why indeed. I clicked. From the article I learned it could be that the lack of gravity shifts fluids in astronauts' bodies. Their sinuses clog; their sense of smell dulls, possibly making food less palatable. And if so, might it be important to reconsider space menus, to make them more appealing?

And then, at the end of the article, the writer added a suggestion from one Kim Binsted that duck fat be included on a Mars mission instead of margarine. It doesn't weigh any more, Binsted noted, it's just as shelf-stable, and it simply tastes so much better.

The article also posted a call for volunteer wannabe astronauts. The piece concluded, "If the idea of pretending you're on Mars for four months is appealing to you, Binsted is still taking applications from people who want to join her simulation."

Well, if you're a certain kind of person, someone who had wanted desperately to go to space camp as a kid but whose parents didn't have the money for it, someone who had geared her whole educational track toward getting the scientific degrees that could qualify her to become an astronaut but who, along the way, had found herself writing about science rather than doing it, which was fine and even at times quite satisfying, but had no plans to get back into science and no big writing assignments or obligations really on the horizon, who was married but childless so the possibility of removing herself from the day-to-day would likely not too drastically upend the lives of others except maybe her wife, her understanding wife, then, especially since being a pretend astronaut matched so closely with her personal hopes and dreams that she had years prior gently stashed on a shelf, you might have been inclined to apply. And so I applied. "Fantasy is hardly an escape from reality," wrote the children's book author Lloyd Alexander. "It's a way of understanding it."

APRIL 7, 2012

Dear HI-SEAS Applicant,

Thank you for your interest in the Hawai'i Space Exploration Analog and Simulation. As you may know, we received almost 700 applications for this mission, for only six crew positions. Because of the huge response, we have had to add one more stage to the process (as originally described in the call for participation). At this point, you are in the "highly qualified candidate" pool of about 150 applicants. However, we will have to narrow that pool down further before moving on to interviews, references and medicals. We expect to be able to notify the 30 semi-finalists by mid-April.

. . .

Thanks again for your application, and for your commitment to human space exploration.

Kim Binsted
HI-SEAS, University of Hawai'i

And so I waited. And as I waited, I dreamt. In one dream, I attended a Mars-simulation tryout where I didn't care much for the other people, my potential crewmates. The guys were competitive and, I suspected, deeply insecure. The women were knowledgeable about all things science and engineering, but were humorless and dull. In this tryout, which spanned days, our every meal was tuna salad sandwiches with a side of tuna salad, a domed, glistening scoop whose pool of watery mayonnaise sogged the bottom of the sandwich. A nightmare, actually.

After a few weeks I still hadn't heard from Binsted. My loud brain told me that it wasn't meant to be. It suggested I read the internet to distract from my disappointment. But my quiet brain said, what if the silence was a mistake? Why not just send an email to see? My quiet brain is often a better friend.

It was a mistake! I should have been contacted with next steps, Binsted told me, and she gave me dates for a Skype interview. I interviewed. Then, a few weeks after that, I was invited to the training workshop with eight others in Ithaca, New York, after which Binsted, Hunter, and the rest of the team would select the crew of six that would participate in the first HI-SEAS mission. I hadn't believed that anything so strange or wonderful, short of actually going to space, could be possible.

The first food in space was dog food. In November of 1957, Laika, the Muscovite street dog, flew on the Soviet satellite Sputnik 2, which had been fitted with a life-support system to prevent carbon-dioxide poisoning, a fan to keep her cool, a bag affixed to her body to collect waste, and powdered meat and bread crumb gelatin to sustain her over the several days she would orbit Earth. She died within hours of launch. Her capsule overheated because part of the rocket failed to separate. This fact wasn't made public until 2002, and for decades the world believed that she died on day six when her oxygen was scheduled to run out, the publicly reported conclusion of the mission.

The second food in space was human food, from a tube, a puree that Yuri Gagarin ate on his historic orbit around the Earth in 1961, the first human space flight. Then along came John Glenn of the NASA Astronaut Corps. In 1962, he circled the Earth three times, sucking down applesauce. Tubes and cubes—food blocks made of meat, vegetables, bread, etc., pressed into bite-size morsels so they wouldn't leave crumbs that could float in an eye or gum up controls or air filters—were standard in the early days of space programs. But engineers were always fiddling with the numbers, trying to find ways to save payload weight and space and compensate for the hassle of eating. In *Packing for Mars*, Mary Roach writes of an unsuccessful proposal in 1964 by a professor at the University of California, Berkeley, to fly a very large astronaut, one with, say, 20 kilograms of fat. By the numbers, the astronaut's adipose would carry 184,000 calories, the researcher claimed, which would supply more than 2,900 calories a day for 90 days. No need to send any food at all.

Today, astronauts who do six-month tours on the International Space Station are not required to subsist off their own bodies and also have more leverage with their meals than astronauts did in the '60s and '70s. There are some two hundred ready-to-eat options. Most ISS foods are sealed in pouches. Many, like the Salisbury steak, require the addition of hot water to rehydrate and heat them. Same goes for the shrimp cocktail and its sauce, a favorite, which features a jolt of horseradish to clear the sinuses. Other foods, like peanut butter and jelly tortilla sandwiches, are ready to go as soon as you smear the condiments. And in fact there is something of a "tortilla culture" on the ISS, which Jean Hunter mentioned early in our HI-SEAS indoctrination, that has allowed some sandwich/wrap/burrito creativity to emerge. For example, a steak and bean burrito isn't technically on the menu, but there's a YouTube video of an astronaut making one. Astronauts are also allowed to send up a personal food treat for their mission, something that reminds them of home. Marshmallow Fluff if they want it.

But what about Mars? What about something longer than six months on the space station? It turns out astronaut food for a Mars mission requires a significant reformulation. Few things in NASA's pantry are designed for the length of such a journey. Nutrients degrade over time, and since the food is prepared well in advance of the mission, it needs to be fortified and palatable for up to seven years.

Binsted and Hunter weren't necessarily interested in developing new, long-lasting foods for Mars, though it is true that bulk ingredients like the ones on HI-SEAS are more shelf-stable and offer the additional benefit of less packaging. Rather, they were more keen on combatting menu fatigue by letting crews be more flexible and creative in their meals. Of course, cooking wouldn't be possible on the actual journey to and from the Red Planet; zero g wouldn't allow for that. Though once on Mars and held to its surface by a gravitational tug about a third as strong as Earth's, a crew with the proper ingredients, utensils, and pots and pans could have any number of gustatory adventures. They could bake and sauté, boil water for pasta, and toss a salad. Soups, latkes, pizza, sushi, beef tagines, apple pie, all made from scratch!

But before project directors, managers, or engineers at NASA would even consider in their mission designs something other than premade pouch food, Binsted and Hunter would need to show the trade-offs between cooking and the resources it eats up, such as water and time, and that less packaging, longer-lasting ingredients, and the creative meals that these ingredients allowed would make a positive difference in a crew's well-being. They'd need to get the data.

On the evening of April 16, 2013, a Tuesday, after a long and winding drive up to an old quarry site on the Hawaiian volcano of Mauna Loa, the six chosen to kick off the HI-SEAS project stepped out of the van. Bags in tow, our boots crunched lava rock like broken plates underfoot to the door of our home for the next four months, a sparsely furnished, newly constructed two-story geodesic dome. The smell of it—off-gassing vinyl from the skin that stretched across the metal frame—was striking and intoxicating, alien and familiar all at once, like driving a new spaceship home, straight off the lot.

Just inside the door was the foyer, which we would use as an "air lock" once we were settled, waiting five minutes before going outside and after coming back in. Straight ahead was a metal shipping container, which housed a workshop with tools and extra supplies. To our right was an archway with white vinyl curtains. Parting them revealed a large common area covered by a thin, blue-gray carpet. Three white rectangular plastic tables, the kind with folding legs, followed the curve of the dome—these would be our workstations. A fourth table was placed near the kitchen, surrounded by black, high-backed rolling chairs, like you might get at Office Depot. This would be our dining room.

In the kitchen was a small fridge, some convection burners, a convection oven, a toaster, a bread maker, a dishwasher, a microwave, standard stuff. Off the kitchen was a door that led to a utility space with a washer and dryer, the control panel for the habitat's electrical system, and another exit. On the other side of the kitchen wall was a small room that acted as a laboratory, and next to it, the ground-floor bathroom. Around the corner from the bathroom were stairs leading to a mezzanine where there was another bathroom and six small rooms where we'd go to sleep each night and wake up every morning.

Our rooms were about the size of small walk-in closets, wedge-shaped like a piece of pie. At crust-edge was a bed. Next to mine was a set of plastic drawers that I used for a nightstand and a small desk and stool. The wall over our sleeping area was curved because of the shape of the structure, making it difficult to sit straight up in bed and adding to a claustrophobic feeling for those prone to it. I'm not. I've always enjoyed small spaces. As a kid, I was always making forts or repurposing large cardboard boxes, rooms within rooms. A little like a hug, maybe. Or just an enclosure that held few surprises and felt completely my own.

When I learned I was selected for the crew, I called my parents. My father, a quiet man, at one point gently said, "I always thought you'd make a good astronaut." At which my mother, a former teacher and very much a talker, reminded me that her high school took part in a national study conducted by Stanford in 1957 to measure the aptitude of American teenagers. While one of her lowest scores was domestic engineering, or housekeeping, one of her highest scores was adaptability to spaceflight, so if such a thing might have a genetic component, she suggested, perhaps I inherited it from her. How the Stanford study's conclusions were drawn, the kind of questions that were asked, what any of it really meant, she didn't really know. But in thinking about my childhood and my relationship to food, to planning and preparing meals, and to cleaning up after, I can absolutely see how it might have been shaped by a mother who was not particularly well suited to the tasks, nor a person who enjoyed them. At an early age, she sold me on the idea of a "meal pill," one of her favorite futuristic concepts. No preparation! No cleanup! Saves time so you can do other things! Our family—my parents, two older brothers, me, and my little sister—ate dinner together most nights without television or other distractions because eating as a family was important, our mother would remind us, though the food that we ate, the meals themselves, were not particularly inspired or inspiring. The canned corn or green beans or salad of iceberg lettuce with ranch alongside spaghetti dressed in RAGÚ, assembled by my mother moving from stovetop to beeping microwave to table, had the mouthfeel of fatigue, of the fact of the need for a dinner for four then five then six people as the family grew. Year after year, every night for decades. Please pass the meal pill.

And now, here I was, dropped inside a project that married my interest and quite possibly genetically bestowed talent for space exploration with my historic ambivalence to food. The six of us wandered around the habitat not knowing what exactly to do because there was so much to do—food inventory, setting up a lab and computers and all other equipment for experiments, figuring out our schedule for meals, chores, exercise, work, free time, correspondences, filling out surveys. The actual science. Eventually, we made our beds, said our good nights, and tried to sleep.

Food is never just food. Consider the Ironman triathlete. She's done the calculations and knows she'll burn 10,000 calories traversing the 140.6 miles over some twelve to fourteen hours, and she needs to eat on schedule so that her body can continue the course she set for it at the beginning of the race or, more accurately, the year before when she signed up for the race, paid the fee, and started in earnest to train. She calls her food "fuel," and it's made of calories and molecules that look like rings and sticks that rearrange themselves in the gut and the bloodstream to spell performance and success.

When I was in my early twenties, I visited a friend who'd moved to Brooklyn right after college. For breakfast one morning, my friend and her roommate, a young woman who worked in the fashion industry, decided we should go to a diner, an authentic, adorable diner nearby that had been around forever. As we looked over our menus, the room-mate asked what kind of cheese we had eaten on our grilled cheese sandwiches as kids. Without hesitation or shame, I said Velveeta processed cheese. It was perhaps one of the most wrong things I could have said in front of my friend who, little did I know, was working to shed indicators that she might have grown up anything less than middle class. My friend instantly covered her mouth in horror and half laughed in disbelief. There was no coming back from my admission. The conversation was over within seconds, yet I'm still struck by how oblivious I was to the true meaning of Velveeta, how quickly I learned my place, and how often I retrieve this memory.

After months ensnared in ice, it finally became clear to Ernest Shackleton, British polar explorer, that his *Endurance* could not be saved. He ordered the crew to abandon ship and set up tent camps on the floes. Unfortunately, they did not have their main food stores, which were trapped below deck and inaccessible due to damage the ship had sustained. Frank Worsley, captain of the *Endurance*, wrote in his diary that despite the move to tents, the crew was in relatively good spirits, though their attitude toward food had significantly changed. "It is scandalous—all we seem to live for and think of now is food. I have never in my life taken half such a keen interest in food as I do now—and we are all alike ... We are ready to eat anything, especially cooked blubber which none of us would tackle before." Eventually, over the course of a week, the crew was able to break through the outside of the ship and retrieve 4.5 tons of food and supplies, an estimated three months of full rations. As time wore on—no one knew how long they'd be adrift on the ice—scarcity conditions revealed fundamental dispositions in the crew. The camp was divided into those who devoured their rations quickly and completely and those who hoarded. The ship's motor expert, Thomas Hans Orde-Lees, had an abject fear of starvation and was the most prominent hoarder, rarely eating an entire ration for any meal. He'd squirrel away a piece of cheese in his clothes, only to rescue it days or even weeks later during leaner times. Food also was used as leverage—to trade out of doing a tedious task like hauling in ice for water or stoking the fire of penguin skins that burned all day. And then there was the sugar pool: each participant gave up one of his three lumps of sugar each day so that on the sixth or seventh day, he might enjoy many returns.

When the celebrated food writer M. F. K. Fisher was asked why she wrote about food rather than about power and security or love, the way other writers do, her short answer was, "like most humans, I am hungry."

My father and I weren't close when I was young. In my early teens, I lamented this fact to my mother, who suggested we make a regular date. Though he had studied Latin and French at Boston College and was trained as a teacher, his job during my formative years was newspaper carrier for *The Kansas City Star*. He owned his own route and bought papers from the *Star*, delivering them with the help of two workers he'd hired to customers in Overland Park, Kansas. Then he'd bill those customers for their subscriptions. It seemed like he was either always billing or sleeping, but our waking schedules overlapped in the early mornings when he came home and before I went to school. We agreed on donuts, then he'd drop me off. I remember sitting on the hard plastic seat at Winchell's Donut House, eating a chocolate-covered long john and drinking milk from the carton, as the strength of the early-morning sun through the windows waned in the first half of the school year and gained in the spring. This is how my father and I began to talk.

M. F. K. Fisher's long response to the question of food writing: "It seems to me that our three basic needs, for food and security and love, are so mixed and mingled and entwined that we cannot straightly think of one without the others. So it happens that when I write of hunger, I am really writing about love and the hunger for it, and warmth and the love of it and the hunger for it ... and then the warmth and richness and fine reality of hunger satisfied ... and it is all one."

In San Francisco, there's a Chinese restaurant called Eric's near the apartment where my wife Jill and I lived for more than a decade. It was our take-out place, it's where we'd bring out-of-town guests, and it's where we'd go when neither of us wanted to cook or the kitchen was still a wreck from previous meals. An order of crab Rangoon, an order of mango chicken, and an order of Eric's spicy eggplant that was not in fact spicy and included shrimp and chicken in a brown sauce, the dish that finally convinced me to appreciate that breed of nightshade. Rice. Fortune cookies. Water and hot tea. Jill and I would talk about our work, what we were reading, what we were thinking about, any of it, all of it. Once, either she or I mentioned our frequency at Eric's to a neighbor who told us about another restaurant a couple blocks over with an identical menu and similar décor. It's called Alice's, she said. Alice and Eric used to be married, the story goes, but they split up. People are usually loyal to one or the other. Jill and I were intrigued. Our next Chinese food outing was to Alice's, about three minutes farther from our apartment and up a slight hill. Walking in, I felt all mixed up. Both restaurants are on a corner, but Alice's faces south and west, while Eric's faces north and east. The space and table layout were nearly identical, but the wall art and the flowers seemed softer somehow. More pink and purple? More irises? Alice's menu also offered our usual dishes (Alice's spicy eggplant, etc.). We tried them. They were similar and good, but ever-so-slightly not the same. We very much wanted to support Alice's, a woman-owned business, and went a few more times, but it was harder than either of us thought to change our habit, to pick up and move to a different place. We ultimately ended up back at Eric's. It's a restaurant special in my spatiotemporal geography, like a you-are-here arrow on a map. It will never not be our place, I think, the way any frequented spot with another person on this planet somehow becomes weightier, holding something that's both of yours forever.

On all of Mars, the place heaviest with memory for me is the dining room table. It's where we ate nearly all of our meals, using nondescript flatware, plain white plates, and bowls on orange plastic place mats, where we drank from our mugs of tea or coffee, where we set up and put away the every-meal line of condiments: mayonnaise, ketchup, soy sauce, mustard, and, absolutely, Tabasco. Each meal included a "hunger and satiety" survey, which we dutifully filled out before and after. To the right of my plate always a notepad and pen.

As a group we had decided before the mission that we'd gather for three daily meals together. As individuals, we were habitual creatures and perhaps somewhat territorial, so we sat in the same seats every day. I think about my seat at the table now, what it meant for me, and what our seats meant for all of us. On the United States' first and only independent space station, Skylab, which orbited the Earth between 1973 and 1979, designers gave the dining table special consideration. They wanted to avoid reinforcing hierarchies. It was the '70s after all, so the table, a small one built for a maximum crew of three, was triangular with equal spacing between seats so no one would automatically, inadvertently or otherwise, preside.

Such was not the case on our mission. At one end of our rectangular table was Angelo Vermeulen, a Belgian artist in his early forties who also held a Ph.D. in biology. Angelo was a Senior TED Fellow recognized for his community art in collaboration with hacker collectives and maker spaces around the world, building sculptures that melded social, biological, and technical elements. He was also our crew commander, leading meetings and acting as the main point of contact between us, the media, and, in some cases, mission support.

To Angelo's right along the long edge of the table, Yajaira Sierra-Sastre, midthirties, chief scientist with a Ph.D. in nanotechnology and material science. She lived in Ithaca, New York, and had grown up in Puerto Rico, where she was becoming something of a celebrity, having done television interviews about her selection as a HI-SEAS crewmember and space education outreach. She managed HI-SEAS's main experiments, which included, among others, testing our noses for NASA's record of how well we could breathe and identify odors over time. She also conducted her own studies of the bacteria in our habitat: bathroom, kitchen, food leftovers, our socks, our sheets, our feet, all of it swabbed and cultured.

Next to Yajaira, Oleg Abramov, a planetary scientist with the United States Geological Survey, in his early thirties, who studied the Late Heavy

Bombardment—the early days of the solar system, when asteroids pummeled the inner rocky planets like the gnarliest game of dodgeball. Oleg, who immigrated to the U.S. from Russia when he was eleven, was the chief geologist and IT specialist who set up and maintained our internet, radios, and computers. Hobbies included backpacking and flying planes, and he was training to be a flight instructor. As such, he was well versed in wilderness first response, which made him a solid choice as our crew's medical officer.

To Oleg's right, opposite Angelo at the other end of the table, was Simon Engler, midthirties, from Calgary, a former combat engineer in the Canadian Army who served in Afghanistan. A driver in a large escort moving VIPs through the Kandahar Province along some of the most dangerous roads in the world, Simon spent his free time while deployed building bomb-sniffing robots and high-altitude balloon drones. On our mission, he was the chief engineer, managing the generator, solar panels, water and waste systems, and 3-D printer, which printed, among other objects, a replacement part for the dishwasher and a hair clip for my bangs. He also conducted a study on robot pets as potential companions for the crew.

To Simon's right was Sian Proctor, a geology professor at South Mountain Community College in Phoenix, Arizona. Sian, in her early forties, was our chief outreach officer, photographer, and producer of a video series during the mission called *Meals for Mars*, in which she prepared recipes submitted by the public for a contest where we, the crew, were the judges. She was also a veteran of the reality survival show called *The Colony* and, in 2009, she made it to the final round of NASA's astronaut selection process, just barely missing the cut but with no hard feelings.

I sat to Sian's right and Angelo's left and with my back to the wall. It was a good spot for me for a couple of reasons. One, I'm rarely comfortable with my back to open spaces, least of which the tall and wide expanse of a geodesic dome. Two, I liked my proximity to Angelo. In addition to running a sleep study and blogging about the mission, I was second-in-command, which was a kind of support role for the commander, though I was still trying to figure out exactly what that meant.

The table was where we ate, and where we had our daily morning meetings, after which Simon would push off dramatically, rolling his chair with Fred Flintstone feet back to his computer. It was where we learned that Sian was highly skilled at making soups or cooking anything, actually, and where Oleg shared his cinematic dreams. Where Angelo, after each meal, happily performed the minor chore of wiping it down with a sponge, no matter who was on dish duty. Where Yajaira's birthday was celebrated on a day when the food experiment allowed for "creative" meals, so we ate a layered Mexican casserole called "enchilasagna," sides of black beans, sweet corn, pita chips, pico de gallo, and a made-from-scratch chocolate cake. It was where, a few days later, we celebrated my birthday on a "non-cooking" day, which meant a just-add-water-and-heat can of macaroni and cheese, leftover rehydrated vegetables, an instant blueberry cheesecake, and Jell-O chocolate mousse decorated with freeze-dried blueberries, cherries, and dehydrated mango, which wasn't terrible, but Yajaira seemed to feel a little bit bad about the contrast.

The table was where, after dinner one night, Oleg asked Sian and me—we were the most familiar with the food inventory—if there was enough Nutella for him to have a spoonful for dessert, but where Angelo didn't hear Oleg's question or our response that there was plenty of Nutella and so loudly objected—Nutella was shared food, after all—just as the glob of hazelnut spread headed toward Oleg's mouth, creating one of the most emotionally charged moments in the mission up to that point. It was where Sian and I recapped Oleg's asking and our response, where Angelo understood his error and apologized, and where Oleg then ate the Nutella, the most bitter Nutella in the solar system, twisted the lid back on the jar, and excused himself, thus ending the conflict that inspired many of us in our mission exit interviews to mention as significant, a conflict that was eventually reported in an article in *The New Yorker* as "the Nutella Incident."

It's where Simon and Oleg played chess; where Sian made a chore chart; where Yajaira and I brainstormed an experiment for mapping the microbiomes of the habitat; where Angelo, Yajaira, and I would linger after meals to talk about our families and childhoods. Where we would sit with a walkie-talkie, and act as Capcom—*roger that; over*—for others when they put on the mock space suits and went outside for rock collecting or photography or mapping caves.

And it's where I discovered one of my favorite Martian rituals: the simple pleasure of the French omelet. Let me tell you about it.

If a creative cooking day fell on a Sunday, our one day off from the regular schedule and the only day when we didn't eat communally until dinner, I would dip into the can of surprisingly high-quality egg powder to make a modest omelet like the kind Julia Child demonstrated on *The French Chef*, season 9, episode 18: thin and quickly cooked, rolled out of the pan onto a plate, seasoned with salt and pepper, some rehydrated shredded cheddar, and parsley flakes. I'd eat it with Finn Crisp crackers spread with reconstituted powdered butter and blackberry jam, sipping Earl Grey from a mug. It was rare for anyone to join me on these mornings, a solitary pleasure.

It's unlikely that there will be French omelets, Julia Child–style, on Mars anytime soon, that is, even assuming astronauts on Mars anytime soon. After the mission, I had a conversation with Grace Douglas, NASA's lead scientist for advanced food study. While the concept of cooking on Mars is a good one, she said, practically speaking, early Mars expeditions will still most likely rely almost entirely on pouch food. Deep-space explorers might also be eating meal-replacement bars—a food item in development by Douglas's team to supply adequate calories and nutrients for at least one meal a day, saving significant payload space and weight. Another type of food system that might ride on the first Mars missions could be a garden. Astronauts on the ISS have successfully grown lettuce, swiss chard, radishes, Chinese cabbage, and peas. And while you can't always count on gardens grown in low gravity, or even on Earth for that matter, Douglas said that astronauts seem to really enjoy eating a fresh, crunchy vegetable every once in a while. It's something NASA takes seriously when considering the psychological effects of food in space.

On our mission, we had sprouts at a few meals. As a general rule, fresh ingredients weren't allowed since they would confound the main food study, but an exception was made for a short period of time. When the sprouts were ready, I did eat one, but just one because they tend to harbor bacteria and I wanted to avoid an upset stomach. Some of the prepackaged soups already disagreed with me, and I didn't want more bad gut. It was nice to smell them, though. A real change of pace. They were sharp and earthy in contrast to the dull and blunted aromas of the rehydrated vegetables we were used to.

I don't recall any particularly severe food cravings during the mission. But I couldn't avoid the effect that the occasional unexpected odor had on me. For one of the nose-related studies, we would sit in front of a computer with a tray of small, covered plastic cups, opaque so you couldn't see what was inside. There were holes in the lids, and when you squeezed the cups, odiferous volatile compounds would puff out. We were to identify the food or nonfood as best we could—it might be soy sauce, eggs, lemon juice, cardboard, or something else—many were often surprisingly hard to name without visual cues—and rate how strong the aroma was and how much we wanted to eat whatever was inside that cup. This was a routine test, conducted every few weeks throughout the mission. In a short time, I thought I knew all the odors the test had to offer. But in early June, only a month and a half in, sitting at the computer in the science lab sniffing cups, I was overcome. Test 9, sample 15: pineapple. Unmistakably fresh pineapple. Something inside me rearranged itself, and a tear slid down my cheek.

In an instant I was back at our pre-mission cookouts in the days before we entered the dome—barbecues with friends, members of the local and scientific community, people who had helped make Mars happen. Grilled pineapple kebabs. Pineapple rings. All the beer, ham sandwiches, burgers, seared ahi tuna, green salads, pies, cakes, cookies. Meals al fresco where I was getting to know everyone and feeling the feelings of anticipation, of jumping into something new and exciting and uncertain. My last phone calls with Jill and my parents before the Mars comms moratorium. I longed for those meals and that time. How was such longing possible? How could food eaten only a month and half prior provoke such nostalgia? What were any of us even doing here? Why Mars? Why weren't we back on Earth with everyone else?

It made me think about how much training Apollo astronauts undertook before going to the moon, so much so it seemed difficult for many of them to fully appreciate the enormity and literal awesomeness of their accomplishment while they were doing it. Landing a capsule, stuffing rocks in a bag, driving a buggy, planting a flag. The simulation was the real thing. These were test pilots. Dealing with emotions like wonder or awe or fear wasn't on the flight plan.

Our crew was in no way trained like real astronauts. Our preparation for the mission was mostly ad hoc and much of what we learned about HI-SEAS and the facilities and scientific studies was on the job. The basics were there, though. For the most part, we were a crew of people who had been selected for our tendency toward analytical thinking and problem solving and could, supposedly, when needed, emotionally detach to complete a task. But then sometimes a different kind of person might slip through the selection process. A person, say, who by all outward appearances is rational and in control and yet, given the right circumstances, is somehow unraveled by the scent of pineapple.

TURNING THE RED PLANET GREEN (2020)

HOW WE'LL GROW CROPS ON MARS

James Romero

James Romero is a London-based science writer interested in planetary geology, astronomy and the space industry. He has contributed to *BBC Science Focus*, *All About Space*, *Physics World* and *Sky and Telescope*, as well as news stories for *space.com*, *astronomy.com*, *New Scientist* and *Sci News*. Outside writing, he organises science festivals and public events that bring together science and the arts.

If humans are to stand a chance of successfully setting up a colony on Mars, we're going to need to figure out a way of producing food on the Red Planet.

In 2016, Wieger Wamelink, a plant ecologist based at Wageningen University, sat down at the New World Hotel in the Netherlands with 50 guests for a one-of-a-kind meal.

Things might have looked ordinary enough from a quick glance at the menu, if maybe a little cheffy – pea puree appetisers to start, followed by potato and nettle soup with rye bread and radish foam, then carrot sorbet to finish.

But the thing that made it such an extraordinary occasion was that all the vegetables used to make the meal had been grown in simulation Martian and lunar soils by Wamelink and his team.

Since then, they have grown an impressive 10 crops, including quinoa, cress, rocket and tomatoes using simulation soils produced using crushed volcanic rocks collected here on Earth. The team produced their simulant soil by grading the particles of rock into different sizes and mixing them in proportions that match rover analyses of the Martian soil.

The soils were initially developed so that rovers and spacesuits could be tested on Earth to see how well they handled the surface materials of Mars and the Moon. Few thought that the soils could ever actually be farmed.

For a start, there were concerns about the texture of the soil, especially after early attempts to farm model lunar soils struggled as a result of tiny, razor-sharp rock fragments that punctured the plants' roots. On Mars, though, the movements of ancient water and ongoing wind erosion have left a far more forgiving surface covering on the planet, and the simulation soils have proved to be successful.

Nutritionally, Wamelink says there's no difference between the 'Martian' crops and those grown in local soils, and when it comes to flavour he was most impressed by the tomatoes' sweetness.

Wamelink and his team are now attempting to improve crop yields by infusing the simulation Mars soil with nitrogen-rich human urine, a resource likely to be readily available on crewed missions to the Red Planet. He also plans to introduce bacteria that will fix more atmospheric nitrogen, and also feed on the toxic perchlorate salts present in Mars soil.

Elsewhere, at Villanova University in Pennsylvania, Prof Ed Guinan and Alicia Eglin are leading the Red Thumbs project, and have had several successes in farming their own Martian simulant.

Initially derived from rocks gathered in the Mojave Desert, the Villanova researchers have augmented their model soil with earthworm farms, due to the animals' ability to release nitrogen from dead organic matter through their burrowing and feeding.

The Red Thumbs project made headlines in 2018 when the international media got excited about the prospect of Martian beer, after Guinan and Eglin's team managed to successfully produce barley and hops.

All the Salad You Want, But No Chips

A couple of years on and Guinan and Eglin have now added tomatoes, garlic, spinach, basil, kale, lettuce, rocket, onion and radishes to their greenhouses. The quality of harvests has varied, but chief among the successes was kale, which actually grew better in the simulant Martian soil than in local soils.

Other crops struggled, such as the much-needed and calorie-dense potatoes. It turns out potatoes prefer more of a loose, uncompacted soil and failed to grow as the simulant soils became heavy and impenetrable when watered, which led to the potatoes being choked out.

Eglin believes that the key to success may be to grow lower yield crops that might enjoy more natural ecosystems than a single-species setup would allow. Even on Earth, agricultural monocultures often suffer over time as nutrients essential for that one plant being grown are progressively depleted and not replaced after each harvest.

To counteract this effect, farmers often introduce secondary species in the same growing area. These wouldn't compete with the main crop, because their root systems are shallower, but they would still offer additional nitrogen fixation to improve soil fertility. Eglin is now planning to test this by growing soybeans, which could prove to be a vital source of protein, and corn alongside pigweed, a leafy vegetable famous for its use in the Caribbean stew callaloo.

But however much success these projects have, we must remember that simulant soils have very real limitations, explains ESA's Christel Paille. She's involved in the Micro-Ecological Life Support System Alternative programme

(MELiSSA), which is exploring a range of technologies for use in long-haul, crewed missions, such as bacterial bioreactors that recycle astronaut waste into air, water and food.

While MELiSSA has provided support to Wamelink, Paille points out that any successes from the model soils must take into account the fact that they're based on limited geographical sampling.

"It's a baseline, but probably not something that we can generalise to any location on the Mars surface. We are always very cautious about a simulant material. It's very difficult in a single simulant to capture all the characteristics [of the Martian surface]," she says.

Perhaps the only way around this is to collect a sample from the surface of Mars and return it to Earth. On 30 July, NASA's Perseverance Rover launched from Cape Canaveral in Florida with its sights set on the ancient river delta deposits in Mars's Jezero Crater. If all goes according to plan, next February the rover will find itself in what's thought to be some of most fertile land on the Red Planet.

Thanks to its plutonium-based power system, the rover should be able to spend up to a decade analysing the surface of Mars. While previous missions have looked for signs of habitable conditions that existed in the past, Perseverance aims to go one step further by searching for signs of past microbial life.

Also, and crucially for those with hopes of growing food on Mars, the rover will collect samples of rocks and soil, and store them in preparation for a potential future robotic mission to return them to Earth for analysis. Until then, the simulation soils are all we have to work with.

There is still much to be learned in the meantime. For instance, rather than commit to individual species, Paille's MELiSSA programme prefers to assess plants within a self-contained, life-supporting ecosystem.

Here, the benefits of edible biomass, oxygen production and even water treatment are balanced against the resources to grow each plant and manage their waste. But predicting crop performance on Mars will require a more fundamental understanding of plant biology.

"It's about going down to the molecular scale," says Paille. "We need to characterise what's happening underground, like in root respiration. How are gases such as oxygen taken up and provided to the root. And how does the carbon dioxide produced actually diffuse out?"

Barriers to Growth

Even if a suitable simulant is developed, there are still other challenges to overcome. Mars is located in an orbit that's around 70 million kilometres further out from the Sun than Earth. As a result, sunlight delivers only 43 per cent as much energy, leaving average temperatures languishing around -60°C. Also because of the planet's tilt and highly elliptical orbit, seasonal variations are extreme.

Another hurdle is the Martian atmosphere, which is much thinner than Earth's and lacking in the nitrogen vital for plant growth. Instead it's dominated by carbon dioxide, which is vital for photosynthesis, but it's at such low concentrations that any plants growing on the surface would struggle to harness enough to spur growth.

The thin atmosphere also exposes the Martian soil to cosmic radiation. This creates a hostile environment for any microorganisms you might introduce to recycle nutrients from dead plant matter.

Also, Jennifer Wadsworth at the UK Centre for Astrobiology has shown that solar radiation can activate chlorine compounds in Martian soil, turning them into toxic perchlorate salts. These are poisonous if eaten and can lead to hypothyroidism, which blocks the release of metabolism-regulating hormones. Poisonous heavy metals such as cadmium, mercury and iron found in the soil also pose their own challenges.

"Everything that's poisonous for people you can think of in terms of heavy metals is in those soils," says Wamelink. "For plants it's not a problem because they'll store it somewhere. But if we eat those plants then it might be [a problem for us]."

Another option may be soil-less techniques already used on Earth. Aeroponics sees plants suspended in the air while their roots are sprayed with a nutrient mist. Alternatively, hydroponics dips the roots into a nutritious liquid.

These approaches can produce larger, faster-growing crops, and have already been used to successfully grow lettuce on the International Space Station (ISS). In fact, the astronauts were so pleased with their harvest, says Wamelink, the scientists back home were disappointed with the amount of lettuce samples returned for analysis after too much was eaten.

Calorie Deficit

Despite the popularity of the ISS lettuce, air or water agriculture alone may not be enough to sustain astronauts on long-haul trips to Mars, thanks again to the problem of growing potatoes.

"It's very difficult to grow potatoes in hydroculture, and just eating lettuce and tomatoes won't be enough because you need calories," says Wamelink. "Potatoes grow much better in soils, where you'll get a lot of harvest per cubic metre and the organic matter that you don't eat can be recycled."

Whether grown in soil, water or misty air, food will likely play much more than a simple nutritional role in any Martian outpost. Sitting down to a proper meal would prove invaluable for the mental health and comfort of any pioneering astronauts living millions of kilometres from home. Who knows, maybe rye bread and radish foam will be on the menu after all.

THE NEXT 500 YEARS: ENGINEERING LIFE TO REACH NEW WORLDS (2021)

CHAPTER 1: THE FIRST GENETIC ASTRONAUTS

Christopher E. Mason

Christopher E. Mason is a geneticist and computational biologist working at Joan & Sanford I. Weill Medical College of Cornell University, New York City. In 2001, he obtained a BSc in genetics and biochemistry from the University of Wisconsin-Madison. A PhD in genetics followed in 2006 from Yale University, Connecticut. Mason was the Principal Investigator for NASA's Twins Study, coordinating ten research teams from across the United States to study the twins Scott and Mark Kelly, and chart the physiological, molecular and cognitive changes that took place when Scott spent a year on the International Space Station while Mark stayed on Earth.

> My skin had not touched anything in 340 days... anything it touched, it felt like it was on fire.
>
> —Astronaut Scott Kelly

Huddled around glowing monitors full of molecular, genetic, and telemetry data, we were united in our bafflement and concern. We simply could not believe our eyes.

"Are these the highest levels ever seen in a human body?" asked Dr. Cem Meydan. "How did he survive?"

It was a crisp December evening in New York City in 2017, at our genetics laboratory at Weill Cornell Medicine. We had just finished the integrated analysis of all the molecular data (DNA, RNA, proteins, small molecules) from Captain Scott Kelly, who had completed the longest-ever NASA mission in space—almost a complete year (340 consecutive days). Kelly's long-duration spaceflight was part of a unique experiment at NASA called the Twins Study, which leveraged identical twin astronauts (Mark and Scott Kelly) to discern what happens to the human body before, during, and after a year in space. The research spanned ten research teams across the United States; our laboratory worked on the genetic, epigenetic, microbial, and gene-expression analyses. We had comprehensive molecular and genetic data from Scott's time in space, which we could compare to Mark's time on Earth. Our job was to (1) assess what happened to Scott during such a long mission, (2) learn about the changes as a guide for Mars missions, and (3) plan for ways to mitigate future risks to other astronauts.

It was clear that his body did not enjoy the return to gravity. Scott himself described the unpleasantness in his book, *Endurance: A Year in Space, a Lifetime of Discovery*. "My ankles swelled up to the size of basketballs," he noted, amazingly with a calm demeanor. "I felt like I needed to go to the emergency room."

Even though he wanted to go to the emergency room, he knew the reason for the body's changes; he had just returned from space! However, this knowledge did not comfort his immune system. He broke out in rashes all over his body, especially where anything touched his skin. His body was even reacting to something as simple as the weight of clothing being pulled down onto his skin by gravity, causing visible irritation. We could see this immune response in the molecular data from his blood work, especially with changes to his proteins and RNA (gene expression). But we all wondered while staring at the monitors ... Was this reaction part of a normal readaptation to gravity? Does this have any impact for the plans to go to Mars?

"These are the highest levels of inflammation markers and cytokine stress I've ever seen," I said. "Let's triple-check the data."

We checked with Dr. Scott Smith at NASA, who leads the biochemistry analysis unit for the twins and other astronauts, and he confirmed that the data were correct. He also noted, "This is the highest we've ever seen, by a long shot." Samples were processed in duplicate, just to be sure, and our measurements and computational analyses matched. While inflammation is a normal part of the body's response to stress, here, Captain Kelly's return to gravity catapulted his inflammation markers to unseen heights (figure 1.1).

Specifically, interleukin receptor antagonist 1 (IL-ra1), which is an important natural anti-inflammatory protein, as well as other cytokines, such as IL-6, IL-10, and C-reactive protein (CRP), were all spiking extremely high upon the return to Earth. CCL-2, which is a cytokine (a type of protein that leaves cells to signal other cells) that recruits immune cells to sites of injury or infection, was also spiking very high.

We quickly searched across the index of all scientific literature and medical journals to see if anyone had ever seen anything close to these levels, especially for IL-ra1 (>10,000 pg/uL). For IL-ra1, the closest we could find was for patients who had just had a myocardial infarction (a kind of heart attack), from a paper in 2004 (by Patti et al.). For IL-10, spikes were found to be associated with patients who had just survived a severe bacterial infection of the blood (called sepsis).

Somehow, even amid this discomfort, when Scott got back to Earth, he jumped right into his swimming pool and went on to live a normal life in the days and years after. However, these markers were not the only thing that dramatically changed. Other changes could be seen across his tissue systems such as his blood and bones,

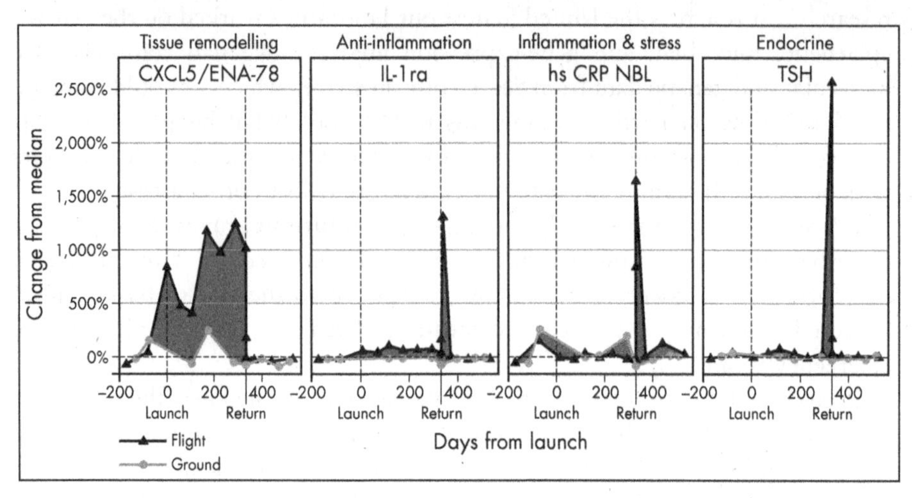

1.1 Many cytokines changed expression during the Twins Study, comparing Scott Kelly's cytokine levels (black) to those of his twin brother, Mark Kelly, who remained on Earth (gray). Dotted lines indicate Scott's launch and return to Earth. Cytokine levels are normalized to their median expression across the analyzed time in both brothers. Some cytokines were elevated throughout the whole mission, such as C-X-C motif chemokine 5 (CXCL5), which plays a role in tissue remodeling. Other molecules primarily spiked upon returning to Earth, such as interleukin-1 receptor antagonist (IL-1 ra) and C-reactive protein (CRP), which deal with inflammation and thyroid-stimulating hormone (TSH).

and we even saw additional molecular changes in his DNA and RNA. We had an unprecedented chance to look at almost everything in the body, from each nucleotide of the genetic code to how cellular responses manifested across Scott's body, resulting in phenotypical changes. Most of these measures were entirely new metrics for any astronaut, including the first complete genetic profiles (genome), as well as other features (figure 1.2) for a spacefaring human. We used all these data to gauge what happened inside the human body during a year in space.

DNA Damage

We first looked at the impact of radiation, which can damage DNA, cells, proteins, and all the regulatory machinery inside cells. Flying at nearly the speed of light are galactic cosmic rays (GCRs), which originate from stars outside our solar system, and solar energetic particles (SEPs), which originate from our sun itself, both sources of radiation that flew through Scott's body. These particles leave a wake of damage like microscopic bullets through the body. GCRs and SEPs are high-energy particles, usually made from protons, helium, and a subset of high-energy ions (HZE ions, which stands for high [H], proton/atom number [Z], and energy [E]). This damage to astronauts was first observed in 1969 and 1970, when Neil Armstrong wore a foil plate around his ankles as he traveled to the moon and back. On this plate, streaks of these HZE particles can be seen displacing the sensor, like marks made by someone drunkenly playing on a high-energy Etch A Sketch or recordings from a nuclear accelerator laboratory after atoms are smashed into each other. Except, in this case, the accelerator is shooting HZE particles, and the laboratory battleground is, unfortunately, the human body.

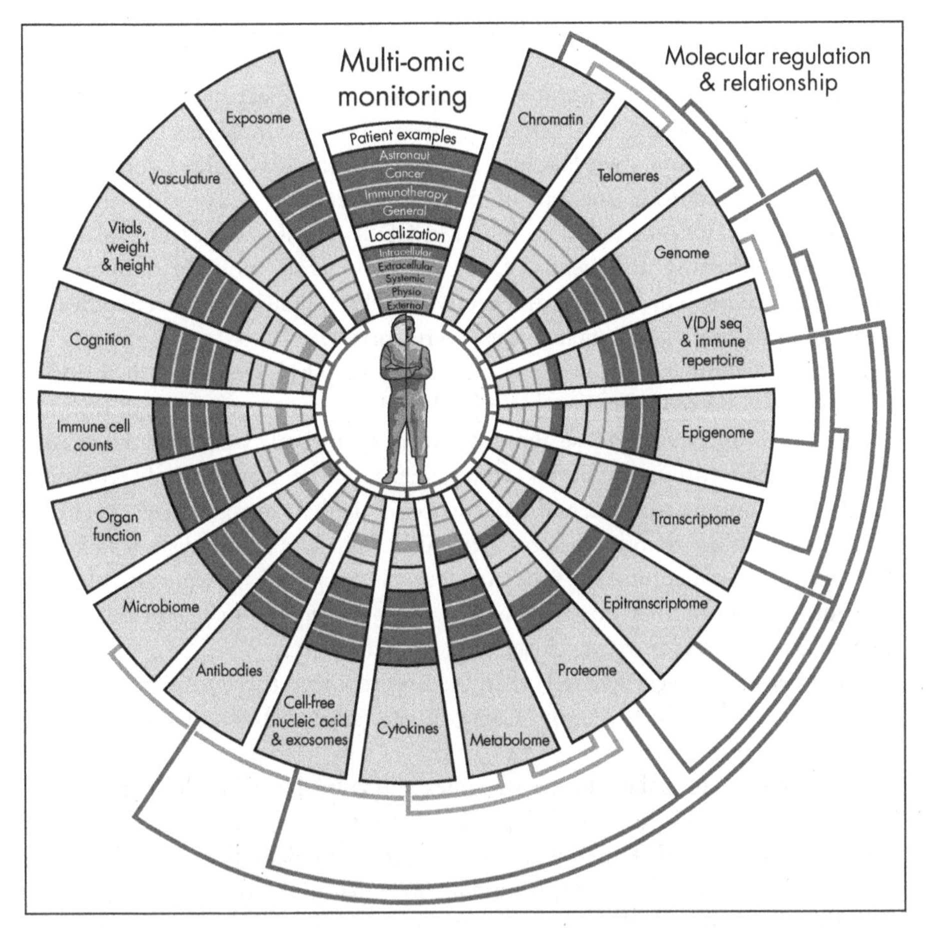

1.2 Multi-omic monitoring platform for astronauts and relation to the clinic: Four monitoring examples are highlighted, including astronauts, cancer patients, immunotherapy patients, and general patients. Each example highlights different-omic data that can be utilized for regular monitoring and follow-up. Molecular interactions between different-omic data demonstrate the need to integrate all these measurements into one platform.

These HZE particles normally go unnoticed during the day, but they can appear in unexpected places. When Scott closed his eyes to go to sleep at night on the International Space Station (ISS), he could see streaks of light, as if there were shooting stars behind his eyelids. These magical displays of light were actually the HZE particles blasting his retinal cells and passing through his eyes, erupting in a lightshow of beautiful, but terrifying, cellular damage as a bedtime story.

Given such reports, we were all worried about what we would find inside Scott after such a long mission. As it turns out, we had several surprises. One of the first things we expected was that his telomeres would probably break down and shrink from radiation and the stresses that accompany spaceflight. Telomeres are the ends of human chromosomes, which normally shrink as you get older, and their lengths are also associated with both diet and stress. As they disappear, the chromosomes become less stable, contributing to the normal molecular process of aging. Dr. Susan Bailey led the research to test this question, and we sent some of our DNA to her lab, and vice versa, to confirm the results.

Unexpected Responses to Spaceflight

Strangely, Scott's telomeres got *longer* when he was in space, which is the opposite of what we expected. We then triple-checked both sample sets of DNA from the Bailey lab and our own lab, and this lengthening was indeed confirmed. It was most pronounced in one type of immune cell called T cells (primarily CD4+ T cells, though evidence was also found in CD8+ T cells), with less evidence of telomere lengthening in B cells (CD19+ cells). Overall, multiple sample replicates, extractions, laboratories, and methods (FISH, PCR, nanopore) confirmed the results, leading us to conclude they were correct.

But then the immediate questions were how and why? We looked at the other data we had collected to make sense of it. Weight loss is associated with telomere maintenance, and Scott did lose about 7 percent of his body weight on the mission because of the rigorous conditions of spaceflight, but he also had daily workouts, nutritionally optimized food, and an absence of alcohol. In some ways, his life in space was healthier than it was on Earth. Also, folic-acid metabolism is linked to telomere maintenance, and the folic-acid levels in Scott's blood were also elevated in flight, adding another possibility. He gained two inches in height during the mission. He also was traveling closer to the speed of light.

Some people got very excited when we first reported these results and asked, "Is space the fountain of youth? Can you get taller and younger if you go to space?" Sort of.

First, we have to isolate all the variables and consider what else happened to him. Scott did travel closer to the speed of light, traveling at an average of 7.68 kilometers per second (km/s), which then enables a calculation using Einstein's relativity and time dilation on a human body. Time dilation occurs when an object moves closer to the speed of light, making time move more slowly for the object in motion relative to the reference frame of other objects. This is dependent on several factors that can be entered into the Einstein/Schwarzschild equation, assuming a few parameters:

(1) A $dr = 0$ (stay at constant radius) and $df = 0$ (same orbital plane);

(2) The ISS orbital speed of 7.68 km/s, with a radius of the ISS at 400 km above the Earth's surface;

(3) The change for Mark Kelly (dt_{MK}) on Earth compared to Scott Kelly (dt_{SK}) on the ISS.

The full equation includes the coordinates of colatitude (theta), the speed of light (c), and the gravitational metric between two spheres (omega), seen here:

$$g = c^2 dr^2 = \left(1 - \frac{r_s}{r}\right) c^2 dt^2 - \left(1 - \frac{r_s}{r}\right)^{-1} dr^2 - r^2 g_{\Omega}$$

Given this equation, Scott became about 0.1 seconds younger than everyone on Earth, including his brother. Since Scott was born 6 minutes after Mark, this made Scott an additional ≈0.1 seconds "younger" than his brother after a year in space. However, even though he is technically younger than what he would have been if he had stayed on Earth, this is not likely a significant factor for his longer telomeres.

We know this because we saw many other modalities of the biology change as well, such as changes in gene expression (off/on or up/down levels of various genes). We all have thousands of genes that change expression every day, so it was not surprising that we could see genes changing when he got to space and when he came back down to Earth. His altered genes' expression included those responsible for DNA repair and cellular respiration. His immune system was also highly activated, including when he received the first-ever flu vaccine in space. Also, we saw evidence of hypercapnia, which is a condition of too much carbon dioxide in the blood and where one can start to feel light-headed and develop a headache; indeed, this irritation was mentioned by Scott in his book. He noted that he got headaches because of the varying carbon dioxide levels, and whenever the CO_2 scrubbers of the space station would break down, he felt as if he had more headaches during these intervals.

We looked at the carbon-dioxide levels on the space station, and though there were some fluctuations, they were not too dramatic and should not have led to physiological changes; we had to look for other causes. As it turns out, breathing in zero gravity is not like breathing on Earth. In particular, every time you breathe out, a small cloud of CO_2 can form in front of your face. This CO_2 minicloud stays by your face, unless you have a fan or move. Thus, some of what we could see in Scott's blood, and likely that of other astronauts, were face-associated, CO_2 miniclouds, more like the atmosphere of Venus than that of Earth.

We also looked at the dynamics in Scott's microbiome, which are the microorganisms (bacteria, viruses, fungi, and other small, nonhuman cells) inside his body. Specifically, we wanted to see what happened to the microbiome during spaceflight. We observed some changes in flight for the ratio of species, specifically for the Firmicutes/Bacteroides (F/B) ratio, using stool data from Drs. Stefan Green, Fred Turek, and Martha Hotz Vitaterna and some of our own data from skin and oral swabs. However, the total diversity was mostly maintained, which is good news. They did eventually return to normal, so there were no big red flags in the microbiome.

But other molecules in Scott's blood did show some unusual features. The mitochondria, which are normally resting inside cells and carrying on cellular respiration to ensure that cells can literally breathe and get energy, were spiking in his blood during the flight—especially when he first got to space. A normal person would have 500 copies of mitochondrial DNA per milliliter (mL) of blood, but Scott showed levels as high as 6,500 copies/mL, based on data from Drs. Kiichi Nakahira and Augustine Choi. We then examined the RNA in the blood, working with Stacy Horner and Nandan Gokhale at Duke University, and there, too, we could see higher levels of mitochondria.

This was an entirely new measure of stress for astronauts, but it has been seen before in other contexts. At Columbia University in New York City, there are laboratories that study extreme variations in mitochondria and even "mitochondrial psychobiology" (in work by Drs. Andrea Baccarelli and Martin Picard), where they have looked in Earth-bound individuals for changes in mtDNA in the blood of people under-going stressful situations. This includes an interesting study of people who gave speeches in a room full of strangers, where the researchers also observed spikes in the blood's mtDNA levels after the talks. Thus, there is ample evidence that mtDNA can appear after general bodily stress, the anxiety of public speaking, or other senses of danger as well.

But—why would human cells start to produce or eject their own means of energy? Here, too, other studies have given clues as to what was happening during a year in space. A 2018 paper (by Irigelsson et al.) showed that white blood cells (lymphocytes) can eject their mtDNA as a way to prime the immune system. These "DNA webs" serve as a warning sign for other immune cells to prepare to fight an infection or defend against a cellular threat, and it seems these webs work in space just as well as they do on Earth. Work from Afshin Beheshti at NASA and our group has now seen the mtDNA stress appear in multiple astronauts, along with other RNA signatures of spaceflight (including small RNAs called miRNAs). All of these surprises, from telomeres, gene expression changes, hypoxic miniclouds, immune stress, mtDNA, and inflammation, happened quickly and seemed to be a rapid, unexpected response to spaceflight, which hopefully would return to normal.

Returning to Earth

Fortunately, almost everything is plastic and malleable about the human body's response to long-term spaceflight. While Scott did gain two inches of height, this gain was just from the lack of compression on his spinal column, and his newfound height disappeared within a few hours of returning to Earth. Also, within forty-eight hours, Scott's telomeres had returned to normal length, and most of his blood and physiological markers were within normal ranges. For his gene-expression dynamics, 91 percent of the changes that occurred while he was in flight returned to normal within six months of returning to Earth.

Thus, most of Scott's spaceflight-induced gene expression returned to normal, but not all. Some genes did carry a "molecular echo" of their time in space, still actively working to continue DNA damage repair and maintain DNA stability. These data also matched what we observed when we examined his chromosomes for other breaks or damage. Even after returning to Earth, Scott showed continual signs of low-level inversions and translocations, which are breaks in the chromosomes, that were continually being healed, replaced with newer cells, and genetically fixed.

Even six months later, some genes were still disrupted in their expression—still adapting—and these are the ones we will cover later in the book, when we discuss the long-term plans for human-genome engineering. The gene expression data showed how the body adapts to space and how, sometimes, it does not completely return to normal. This matches what Scott himself mentioned, that he didn't "feel normal" until seven to eight months after being back on Earth. Also, the work from Dr. Matthias Basner showed that Scott's cognitive speed and accuracy were worse after his return to Earth. In our own work at Cornell with David Lyden, we saw proteins that are normally only in the brain appear in the blood, which matched some of the same genes that created those proteins and indicated a change in the blood-brain barrier. Overall, these molecular changes give us a guide as to which genes may need to be accelerated, decelerated, or otherwise altered to help this response to spaceflight.

Other biological features that could also be tweaked come from clues in the cytokine data, specifically the inflammation markers. Some inflammation markers, like IL-6, went up by thousands of percent on the day he landed, and some even higher two days later. The blood work clearly showed a spike of inflammation cytokines that led to so much pain and is likely why Captain Kelly broke out in rashes. These data were also confirmed with cytokine data from Drs. Tejas Mishra and Michael Snyder from Stanford. When we looked all at the markers together as a pathway, the majority of the functions pointed to muscle regeneration. In short, the pain of using his muscles again was forcing a massive restructuring of the body, with his blood printing the molecular receipt of this expensive physiological purchase. In this amazing event of the human body returning to Earth from space, the blood was screaming out, "Oh crap—gravity! I need to use my muscles again!"

Although landing back on Earth was clearly painful, one good thing about Mars is that it has 38 percent of Earth's gravity. Given that difference, the landing might only constitute 38 percent of an "Oh crap!" moment and 38 percent of a challenge to adapt to the surface when landing on Mars. From these results, it seems that a person could actually survive the trip to Mars, and then likely survive the landing, to begin building a new, rust-hued home.

Future Missions

A large caveat of the Twins Study is that we only had two subjects, derived from a single embryo, with only one in space for a longer duration—so we can only extrapolate these results to others in a limited way. Moreover, spending a year on the ISS is still within Earth's magnetosphere, which extends roughly out to 65,000 km, and still acts as a protective shield from radiation for astronauts. To get a sense of the challenge for a mission to Mars, we can compare other missions to the expected amount of radiation astronauts will incur on the way to the red planet, which is about 300 millisieverts (mSv), as well as a 30-month round-trip mission, which is about 1,000 mSv (figure 1.3). This would be more than six times the amount of radiation Scott saw in his mission. While such radiation is not pleasant, there are ways this can be addressed and protected against, which will be revealed in later chapters.

Indeed, we do not have to accept these radiation risks without defending ourselves against them. Though we do already protect astronauts physically, pharmacologically, and medically, these mitigations need to be improved, and

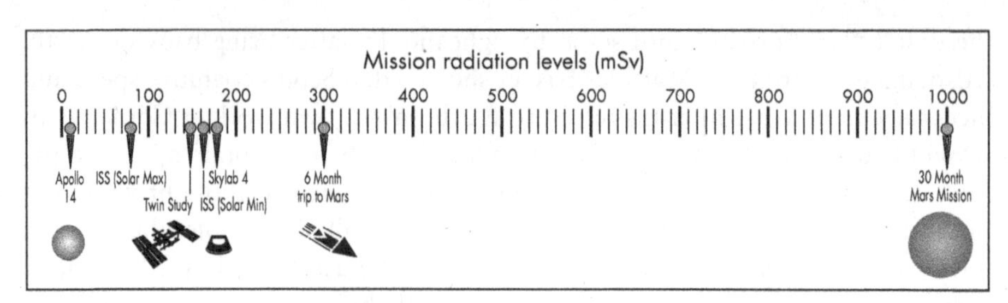

1.3 Radiation metrics for various mission parameters: Estimated and measured radiation metrics for a variety of missions in millisieverts (mSv).

we should further use any other means of protection for them as well. Notably, the one biological defense mechanism that has not yet been implemented for astronauts (though it has been for patients on Earth for a wide range of conditions) is genetic engineering.

Genetic Defenses

Given the clear risks for long-duration missions to other planets (e.g., Mars) and the challenges of later-stage (e.g., interstellar) missions that would put humans in more dangerous environments with more radiation and less ability to create food and maintain proper metabolism, an exploration into our genetic defenses is warranted. In other words, if we can learn the secrets of all other species and craft a series of genetic protections, we would be embarking on not only a needed means of survival, but also a manifestation of our own genetic duty. We do everything we can to keep astronauts safe through engineering their rockets and ships, but could we make some of the protections on the inside, within the astronauts themselves? Should we do such a thing? Is it right to genetically modify astronauts?

Some of these abstract questions became tangible with He Jiankui, who began to genetically modify human embryos using CRISPR (discussed more in later chapters), two of whom were born in 2018. He did all the work in secret and misled the Institutional Review Board (IRB) at his university, kicking off an angry response when he decided to bring gene-edited babies into the world.

Such a process of bringing groundbreaking medical technologies into the world is the absolute worst way to do it—in secret with little oversight—but the idea is no longer hypothetical. The question now is: How do we actually start to regulate genetically engineering embryos or make sure it doesn't go wrong? Numerous examples exist for precision medicine in health and disease, but what is needed to help patients on Earth and future astronauts is more *predictive medicine*. Can a scientist actually engineer something and predict what happens? That is the best test of knowledge.

To this end, the first draft of the 500-year plan was posted on our lab's website in 2011, which included many of the ideas in this book. It was also the first year we submitted the genome and metagenome proposal to NASA, where we had almost none of the information described in this current chapter. Most of the ideas that seemed impossible in 2011 have already become reality, especially the ease with which we can now edit and modify genomes and epigenome (the regulatory landscape of the genome).

But beyond the rapid advancement of science, this plan represents hope and belief in the long-term survival of humans. One of my favorite things about humanity is that we are the only species we know of that can actually create 5-, 500-, or 5,000-year plans, or comprehend any multigenerational plan. Almost all the people who will benefit from such a plan will be born after the death of the plan's creators, yet such plans get made and can serve humanity like an intergenerational Olympic torch, bringing the bright light of past and planned progress to keep hope ignited and eyes looking forward.

The rest of this book will lay out this plan, which addresses the technical, philosophical, and ethical framework for engineering genomes, ecosystems, and planets. While seemingly abstract and almost unbelievable in scope, this large-scale engineering effort is not our first attempt. Mars will, in fact, be the second planet on which we have performed planetary-scale measurements, modeling, and engineering. In 2021, we are doing this planetary-scale engineering on Earth to continue our survival and leave a better planet for the next generations, but, sadly, with scant coordination or planning. We need to do such planetary and biological engineering with far greater precision in the future to fulfill our species' unique role of Shepherds and Guardians. It is no longer a question of "if" we can engineer life—only "how." Engineering life now exists within our generation and will continue to be improved and utilized for generations to come, be it those who exist in 500 years, 5,000 years, or much further into the future.

Engineering is humanity's innate duty, needed to ensure the survival of life.

Notes

Signaling to Mars

1. In his book on the intertwined histories of astronomical observations and literary depictions of Mars, Robert Markley succinctly describes the red planet's orbital opposition, "For a few weeks every twenty-six months, Mars and the Earth are aligned on the same side of the sun in their elliptical orbits. During these periods of opposition, Mars is visible through comparatively small telescopes, and, since the mid-seventeenth century, scientific observations of the planet's surface and atmosphere have clustered during these periods." Robert Markley, *Dying Planet: Mars in Science and the Imagination* (Durham, N.C.: Duke University Press, 2005), 33.
   The occasion for this article, the orbital opposition of 1909, was also the first time Percival Lowell took a successful series of telescopic photographs of Mars in an attempt to prove his theory that a vast infrastructure of canals was built across the Martian surface. For more, see William Graves Hoyt, *Lowell and Mars* (Tucson: University of Arizona Press, 1996); Oliver Morton, *Mapping Mars: Science, Imagination, and the Birth of a World* (New York: Macmillan, 2002); Robert Crossley, *Imagining Mars: A Literary History* (Middletown, Conn.: Wesleyan University Press, 2011).
2. William Henry Pickering was an astronomer with the Harvard Observatory known for discovering Saturn's moon Phoebe, developing new techniques in telescopic photography, and advancing popular knowledge of the surface of Mars. In April 1909, Pickering proposed a plan to communicate with Mars using a massive heliograph. Pickering described the system in a front-page article in the *New York Times*: "My plan of communication would necessitate the use of a series of mirrors so arranged as to present a single reflecting surface toward the planet. Of course one mirror would do as well, but as the area necessary for reflecting the sunlight over 40,000,000 miles would have to be more than a quarter of a mile of glass a single mirror would not be practicable. We would have to use a number of mirrors.
   "These mirrors would all have to be attached to one great axis parallel to the axis of the earth, run by motors, and so timed as to make a complete revolution every twenty-four hours, thus carrying the reflecting surface around with the axis once a day and obviating the necessity of continually readjusting it to allow for the movement of the planets." "Plans Messages to Mars: Prof. Pickering Would Communicate by Series of Mirrors to Cost $10,000,000," *New York Times,* April 19, 1909.
   Signaling to Mars did not mean that Pickering believed there to be life on Mars. A few months later, *Popular Mechanics* magazine noted that Pickering "is also among those who seriously doubt that there are any living beings upon Mars, although he has due respect for the theories of those opposed to him, but he does believe that his scheme of sending messages is the one practical way of finding out, once for all, whether there are such beings, although he admits that if no answering signals were made, it would not disprove the theories that Martians exist." "The Scheme to Signal Mars: Prof. Pickering's Practical Plan," *Popular Mechanics* 12, no. 1 (July 1909): 10.
   A version of Pickering's plan was actually carried out in 1924, when Swiss astronomers "mounted a heliograph in the Alps to flash signals to Mars. The U.S. Navy maintained radio silence for three days to listen to messages from the Martians." Markley, *Dying Planet,* 158.

3. William Preece, engineer-in-chief of the British Post Office and radio experimenter, suggested as early as 1898 that wireless could be used to contact Mars: "If any of the planets be populated with beings like ourselves, then if they could oscillate immense stores of electrical energy to and fro in telegraphic order, it would be possible for us to hold commune by telephone with the people of Mars." William Preece, "Ethereal Telegraphy," *Review of Reviews* 18 (December 1898): 715; quoted in Susan J. Douglas, "Amateur Operators and American Broadcasting: Shaping the Future of Radio," in *Imagining Tomorrow: History, Technology, and the American Future,* ed. Joseph J. Corn (Cambridge, Mass.: MIT Press, 1986). See also Thomas Waller, "Can We Radio a Message to Mars?" *Illustrated World* 33 (April 1920): 242.
4. Gernsback: "Described in the October, 1908, issue of M. E., page 243."
5. Gernsback: "Article in the May, 1908, issue M. E., page 55."
6. Oliver Heaviside proposed the existence of an electrically charged layer in the atmosphere in 1902, but not until 1927 was the existence of the ionosphere confirmed. While Gernsback is correct here that the presence of sunlight affects the transmission of radio waves, it is not the sun's rays themselves but rather their interaction with the ionosphere that causes interference. He could not have known that in 1909, but this was another area where technological achievements outpaced their scientific explanations. Marconi's first successful transatlantic wireless message in 1901, sent *around* the curvature of the earth, was possible because it bounced off the then-undiscovered ionosphere.
7. Gernsback: "'Worlds in the Making.' See Panspermie *[sic]*." This note refers to Svante Arrhenius, *Worlds in the Making: the Evolution of the Universe* (London: Harper & Brothers, 1908). Panspermia is the theory that some form of microscopic life is spread just as evenly throughout the universe as matter itself. Arrhenius, a 1903 Nobel Prize winner in chemistry, writes, "According to this theory life-giving seeds are drifting about in space. They encounter the planets and fill their surfaces with life as soon as the necessary conditions for the existence of organic beings are established" (217). This would imply that "all organic beings in the whole universe should be related to one another and should consist of cells which are built up of carbon, hydrogen, oxygen, and nitrogen" (229).

Enchanting Luna and Militant Mars

1. Letter of 2 January 1953, in C. S. Lewis, *The Collected Letters of C. S. Lewis*, ed. by Walter Hooper, 3 vols. (San Francisco, 2004–7), 3:273.
2. Michael Ward, *Planet Narnia* (New York, 2008), 237.
3. Michael Ward, *The Narnia Code: C. S. Lewis and the Secret of the Seven Heavens* (Wheaton, Illinois, 2010).
4. Robert S. Richardson, "The Day after We Land on Mars," in *The Magazine of Fantasy and Science Fiction* 9, December 1955, 44–52.
5. C. S. Lewis, "The Planets," in *Poems*, ed. by Walter Hooper (New York, 1977), 13–14.
6. C. S. Lewis, "Ministering Angels," in *The Dark Tower and Other Stories*, ed. by Walter Hooper (New York, 1977), 112–23.
7. Lewis, "The Planets," 13–14.
8. C. S. Lewis, *The Discarded Image: An Introduction to Medieval and Renaissance Literature* (Cambridge, 1970), 106.
9. Lewis, *The Discarded Image*, 32.
10. Lewis, *The Discarded Image*, 109.
11. Lewis, "The Planets," 12.
12. Lewis, "The Planets," 12.

13. Letter of 3 September 1927, in Lewis, *The Collected Letters of C. S. Lewis*, 1:730.
14. C. S. Lewis, "Forms of Things Unknown," in *The Dark Tower*, 126.
15. Lewis, "Things Unknown," 127. Pascal's actual line from *Pensées* section III, 206, was *Le silence éternel de ces espaces infinis m'effraie*, "The eternal silence of these infinite spaces frightens me." See Blaise Pascal, *Pascal's Pensées* (New York, 1958), 61.
16. Lewis, "Things Unknown," 132.
17. Ward, *Planet Narnia*, 92.
18. Ward, *Planet Narnia*, 130–1.
19. Ward, *Planet Narnia*, 93.
20. Ward, *Planet Narnia*, 78.
21. Ward, *Planet Narnia*, 139.
22. Ward, *Planet Narnia*, 78.
23. Lewis, "Ministering Angels," 116.
24. Ward, *Planet Narnia*, 92.
25. Lewis, "Ministering Angels," 112.
26. Ward, *Planet Narnia*, 82.
27. From "The Queen of Drum," Canto IV, line 9, in C. S. Lewis, *Narrative Poems*, ed. by Walter Hooper (New York, 1972), 156
28. From "The Queen of Drum" Canto V, line 126, in Lewis, *Narrative Poems*, 170. The full list of lunar deities in the poems of C. S. Lewis is found in Ward, *Planet Narnia*, 126.
29. C. S. Lewis, *Mere Christianity* (San Francisco, 2001), 119. Quoted by Ward, *Planet Narnia*, 93.
30. Ward, *Planet Narnia*, 96.
31. Ward, *Planet Narnia*, 99.
32. Lewis, "Ministering Angels," 123.
33. Ward, *Planet Narnia*, 28.
34. Ursula K. LeGuin, Untitled review, in *New Republic*, 16 April 1977, 29–30
35. Charles A. Brady, "Some Notes on C. S. Lewis's *The Dark Tower and Other Stories*," in *CSL: The Bulletin of the New York C. S. Lewis Society*, 8,11 [95] (1977), 1.
36. Katherine Harper, "C. S. Lewis's Short Fiction and Unpublished Works," in Bruce L. Edwards, ed., *C. S. Lewis: Life, Works, and Legacy*, 4 vols. (Westport, 2007), 2:160.
37. Letter of 20 April 1961, in Lewis, *The Collected Letters of C. S. Lewis*, 3:1258.
38. Anthony Boucher, ed., *The Magazine of Fantasy and Science Fiction* 13, January 1958, 5.
39. Richardson, "The Day after We Land on Mars," 44–52.
40. Boucher, *The Magazine of Fantasy and Science Fiction* 9, December 1955, 44.
41. Boucher, *The Magazine of Fantasy and Science Fiction* 9, December 1955, 52.
42. C. S. Lewis, "The Shoddy Lands," in *The Magazine of Fantasy and Science Fiction* 10, February 1956, 68–74.
43. Richardson, "The Day after We Land on Mars," 47.
44. Richardson, "The Day after We Land on Mars," 47.
45. Richardson, "The Day after We Land on Mars," 48.
46. Richardson, "The Day after We Land on Mars," 49.
47. Richardson, "The Day after We Land on Mars," 49.
48. Richardson, "The Day after We Land on Mars," 49–50.
49. Richardson, "The Day after We Land on Mars," 51.
50. Lewis, "Ministering Angels," 120.

51. C. S. Lewis, "A Reply to Professor Haldane," in *Of Other Worlds: Essays and Stories*, ed. by Walter Hooper (San Diego, 1975), 77.
52. Lewis, "Ministering Angels," 122.
53. Lewis, "A Reply to Professor Haldane," 77.
54. Diana Pavlac Glyer, "'We are *All* Fallen Creatures and *All* Very Hard to Live With': Some thoughts on Lewis and Gender," in *Christian Scholar's Review*, 36, no. 4, Summer 2007 483. For the quotation from Lewis, "We are *all* fallen creature and *all* v. hard to live with," see letter of 8 November 1962. Lewis, *The Collected Letters of C. S. Lewis*, 3:1379.
55. Kathryn Lindskoog, *Sleuthing C. S. Lewis: More Light in the Shadowlands* (Macon, 2001), 104.
56. Lindskoog, *Sleuthing C. S. Lewis*, 106.
57. Lindskoog, *Sleuthing C. S. Lewis*, 105.
58. See for example the lines from Shakespeare's sonnets quoted in C. S. Lewis, *English Literature in the Sixteenth Century, Excluding Drama* (New York, 1954), 489.
59. Brady, "Some Notes on C. S. Lewis's *The Dark Tower and Other Stories*," 1–10.
60. Alastair Fowler, "C. S. Lewis: Supervisor," in Harry Lee Poe and Rebecca Whitten Poe, eds., *C. S. Lewis Remembered* (Grand Rapids, 2006), 105.
61. Letter of 3 February 1953, in Lewis, *The Collected Letters of C. S. Lewis*, 3:288.
62. Fowler, "C. S. Lewis: Supervisor," 105.
63. Walter Hooper, "Preface" in Lewis, *The Dark Tower*, 12.
64. C. S. Lewis, *The Problem with Pain* (New York, 1948), 86.
65. Lewis, "Ministering Angels," 119.
66. Letter of 18 July 1957, in Lewis *The Collected Letters of C. S. Lewis*, 3:872.
67. Lewis, *The Collected Letters of C. S. Lewis*, 2:501.
68. C. S. Lewis, "Hamlet: The Prince or the Poem," in *Selected Literary Essays*, ed. by Walter Hooper (Cambridge, 1969), 99.
69. Lewis, "Things Unknown," 127
70. Letter of 3 February 1954, in Lewis, *The Collected Letters of C. S. Lewis*, 3:423–4.
71. Letter of 9 February 1954, in Lewis, *The Collected Letters of C. S. Lewis*, 3:426.
72. Letter of 20 January 1954, in Lewis, *The Collected Letters of C. S. Lewis*, 3:412.
73. Lewis, "Ministering Angels," 123.

Projecting Landscapes of the Human Mind onto Another World: Changing Faces of an Imaginary Mars

1. This chapter's argument is partly based on Rainer Eisfeld and Wolfgang Jeschke, *Marsfieber*, Munich: Droemer, 2003. Much like Robert Markley's subsequent *Dying Planet. Mars in Science and the Imagination*, Durham: Duke University Press 2005, the book discussed both the imagined Red Planet and the actual Mars progressively unveiled by robotic missions.
2. Wilhelm Beer and Johann Heinrich Maedler, *Beiträge zur physischen Kenntniss der himmlischen Koerper im Sonnensysteme*, Weimar: Bernhard Friedrich Voigt, 1841, VII.
3. Beer and Maedler, Beiträge, ibid.
4. Beer and Maedler, Beiträge, 124, 125.
5. Brian Aldiss and David Wingrove, *Trillion Year Spree. The History of Science Fiction*, London: Paladin, 1988, 603 n. 47; Martin Schwonke, *Vom Staatsroman zur Science Fiction*, Stuttgart: Enke, 1957, 43.
6. Carl Sagan, *Cosmos*, New York/London: Random House, 1980, 106.
7. This brief reference to the British film goes back to *Marsfieber*, 163. *Devil Girl on Mars* was subsequently discussed by Robert Markley, *Dying Planet*, 227–9.

8. Camille Flammarion, *Les Terres du Ciel*, Paris: Marpon & Flammarion, 1884, 208.
9. Camille Flammarion, *Die Mehrheit Bewohnter Welten*, Leipzig: J. J. Weber, 1865, 51-2, 71. Flammarion published the book as a 20 year old.
10. Oliver Morton, *Mapping Mars*, New York: Fourth Estate, 2002, 37-8.
11. Richard A. Proctor, *Other Worlds than Ours*, London: Longmans, Green, 31872 [11870] 85, 109–0.
12. Giovanni Schiaparelli, *Astronomical and Physical Observations of the Axis of Rotation and the Topography of the Planet Mars: First Memoir, 1877–1878*, translated by William Sheehan, MS, Flagstaff: Lowell Observatory Flagstaff (Archives), 1994, 124.
13. Giovanni Schiaparelli, 'Découvertes nouvelles sur la planète Mars', *Révue d'Astronomie populaire* 1.7 (July 1882), 218 ; Camille Flammarion, 'La planète Mars', *Révue d'Astronomie populaire* 1.7 (July 1882), 216.
14. William Sheehan, *The Planet Mars*, Tucson: University of Arizona Press, 1996. 85.
15. Schiaparelli, *Astronomical and Physical Observations*, 123, 124.
16. Percival Lowell, *Mars as the Abode of Life*, New York/London: Macmillan, 1908, 134.
17. Mark R. Hillegas, 'Martians and Mythmakers: 1877–1938', in *Challenges in American Culture*, eds. Ray B. Browne et al., Bowling Green: Bowling Green University Popular Press, 1970, 156.
18. Percival Lowell, *Mars*, Boston/NewYork: Houghton, Mifflin, 1895: 122, 128–9.
19. Lowell, *Mars*, 165, 208–9.
20. Carl Sagan, 'Hypotheses', in: *Mars and the Mind of Man*, New York: Harper & Row, 1973, 13.
21. Percival Lowell, *Mars and its Canals*, New York/London: Macmillan, 1906, 377.
22. Franz Rottensteiner, 'Kurd Lasswitz: A German Pioneer of Science Fiction', in *SF: The Other Side of Realism*, ed. Thomas D. Clareson, Bowling Green: Bowling Green University Popular Press, 1971, 289.
23. Kurd Lasswitz, *Auf zwei Planeten*, Frankfurt: Zweitausendeins, 1979 [11897], 98.
24. Lasswitz, *Planeten*, 875.
25. H. G. Wells, *The War of the Worlds*, New York: Pocket Books, 1953, 2.
26. H. G. Wells, H. G. (1893): 'The Man of the Year Million', in *A Critical Edition of the War of the Worlds*, eds. David Y. Hughes and Harry M. Geduld, Bloomington: Indiana University Press, 1993, Appendix III, 291–2, 293.
27. Frank McConnell, *The Science Fiction of H. G. Wells*, New York/Oxford: Oxford University Press, 1981, 128, 130.
28. McConnell, *Science Fiction*, ibid.
29. McConnell, *Science Fiction*, 132–3.
30. Hadley Cantril, *The Invasion from Mars. A Study in the Psychology of Panic*, Princeton: Princeton University Press, 1940: 47.
31. H. G. Wells, *Star Begotten*, London: Chatto & Windus, 1937: 50–51.
32. Wells, *Star Begotten*, 167–8.
33. Howard E. McCurdy, *Space and the American Imagination*, Washington/London: Smithsonian Institution Press, 1997, 2.
34. McCurdy, *Space*, Washington/London: Smithsonian Institution Press, 1997: 2, 233-4.
35. Benjamin S. Lawson, 'The Time and Place of Edgar Rice Burroughs's Early Martian Trilogy', *Extrapolation* 27.3 (March 1986), 209.
36. Leigh Brackett, *The Secret of Sinharat*, New York: Ace Books, 1964, 8.

37. Richard Slotkin, *Gunfighter Nation. The Myth of the Frontier in Twentieth-Century America*, New York: Atheneum, 1992: 4–5, 6–7, 14, 24.
38. Lawson, 'Time and Place', 213.
39. Edgar Rice Burroughs, *A Princess of Mars*, New York: Random House, 2003 [11912]: XXIII-IV, 14–5.
40. Burroughs, *Princess*, 152.
41. Burroughs, *Princess*, 75.
42. Catherine L. Moore, 'Shambleau', in idem, *Northwest Smith*, New York: Ace Books, 1981 [1933], 2, 3.
43. Leslie A. Fiedler, *The Return of the Vanishing American*, London: Paladin, 1972, 25; Lawson, 'Time and Place', 208.
44. J. Edgar Hoover, *Masters of Deceit. The Story of Communism in America and How to Fight it*, New York: Pocket Books, 1958, VI.
45. Bill Warren, *Keep Watching the Skies. American Science Fiction Movies of the Fifties*, Vol. 1, Jefferson/London: McFarland, IX, 2.
46. Warren, *Keep Watching*, 11.
47. Warren, *Keep Watching*, ibid.
48. Ray Bradbury, quoted in William F. Nolan, 'Bradbury: Prose Poet in the Age of Space', *Magazine of Fantasy & Science Fiction* 24.5 (May 1963), 8.
49. Ray Bradbury, *The Martian Chronicles*, New York: Bantam Books, 1951, 185.
50. Sam Weller, *The Bradbury Chronicles*, New York: William Morrow, 2005, 156, 159.
51. Bradbury, *Martian Chronicles*, 2, 86, 158.
52. Bradbury, *Martian Chronicles*, 172.
53. Ray Bradbury, as quoted in Weller, *Bradbury Chronicles*, 155.
54. Gregory M. Pfitzer, 'The Only Good Alien is a Dead Alien: Science Fiction and the Metaphysics of Indian-Hating on the High Frontier', *Journal of American Culture* 18.1 (Spring 1995), 58.
55. Robert Zubrin, *The Significance of the Martian Frontier*, www.javanet.com/-campr.2/New Mars/Pages/Frontier 1.html, 1998.
56. Carl Sagan, 'Planetary Engineering on Mars', *Icarus* 20 (1973), 513–514.
57. Christopher P. McKay, Owen B. Toon and James F. Kasting, 'Making Mars Habitable", *Nature* 352 (8 August 1991), 489–96; Christopher P. McKay, 'Restoring Mars to Habitable Conditions: Can We? Should We? Will We?', *Journal of the Irish Colleges of Physicians and Surgeons* 22.1 (January 1993), 17–9.
58. Zubrin, *Significance.*
59. Patricia Nelson Limerick, 'Imagined Frontiers: Westward Expansion and the Future of the Space Program', in *Space Policy Alternatives*, ed. Radford Byerly, Jr., Boulder: Westview Press, 1992, 249–262.
60. Paine Commission, http://history.nasa.gov/painerep/parta.html, 1986.
61. Limerick, 'Imagined Frontiers', 253–4, 256–7.

Are the Planets Inhabited?

1. *Radiation in the Solar System: Its Effects on Temperature, and its Pressure on Small Bodies*, by Dr. J. H. Poynting (*Phil. Trans. of the Royal Society*, Vol. 202 A).

The Satellites of Mars: Prediction and Discovery

1. Jonathan Swift, *Gulliver's travels* (London, 1726), part 3, ch. 3.
2. See Marjorie H. Nicolson and Nora M. Mohler, "The scientific background of Swift's Voyage to Laputa" reprinted in Nicolson's *Science and imagination* (Cornell, 1956); Lyle Boyd's "The provenance of Swift" in *The graduate journal,* vii (1965), 235–43, provides the whimsical explanation that Swift was a Martian transported to earth.

3. S. H. Gould, "Gulliver and the moons of Mars", *Journal of the history of ideas,* vi (1945), 91101.
4. Augustus de Morgan, in *A budget of paradoxes* (London, 1872), 133–4, proposes that John Arbuthnot advised Swift on the mathematics.
5. Voltaire, *Micromegas* (1752), ch. 3. Father Louis-Bertrand Castel (1688–1757) was a prolific and determined critic of the new Newtonian science, who looked back to Descartes and Kircher for inspiration (see Donald S. Schier, *Louis Bertrand Castel, Anti-Newtonian scientist* (Cedar Rapids, Iowa, 1941)). A rapid examination of Castel's principal work, *Traité de physique sur la pesanteur universelle des corps* (1724) has failed to locate the argument cited by Voltaire, although I have found a series of curious propositions based on a theory of vortices, such as "109—if the moon is destroyed, it is possible that Mars would take its place; it is possible that Venus and the earth would change places; it is possible that the earth would be overpowered by Venus, or Mars, or Mercury and become its satellites". Perhaps an explicit denial of the existence of Martian satellites is found in one of his 50 articles on the theory of vortices written between 1724 and his death.
6. Bernard le Bovier de Fontenelle, *Entretiens sur la pluralité des mondes* (1866); the quotation here is from p. 211 of an anonymous translation published in London in 1767, but an English translation by Glanvill was already available in 1702.
7. Quoted by Asaph Hall in *Observations and orbits of the satellites of Mars* (Washington, 1878) from David Brewster's "Life of Galileo", pp. 33–4 in *Martyrs of science* (London, 1874). Actually the quotation is from Kepler's *Dissertatio cum nuncio sidereo*—see pp. 14 and 77 of Edward Rosen's *Kepler's Conversation with Galileo's Sidereal messenger* (New York, 1965).
8. J. Kepler, *Narratio de observatis quatuor Jouis satellitibus* (Frankfurt, 1611), p. 3^v; *Johannes Kepler Gesammelte Werke,* iv (Munich, 1941), 319.
9. Edward Stafford Carlos, *The sidereal messenger, and a part of the Preface to Kepler's Dioptrics* (London, 1880), 88.
10. The original letter is unknown except for its publication in Kepler's *Dioptrice* (Augsburg, 1611), 15–16; see F. Hammer's note in *Johannes Kepler Gesammelte Werke,* iv (Munich, 1941), 515.
11. The order of the observations is given explicitly here because of the light it sheds on a statement written by Todd about 30 years later for the *Cosmopolitan magazine* of 11 March 1908, p. 343: "So mine was the first eye that ever saw Phobos recognizing it as a satellite."
12. Asaph Hall, *op. cit.* (n. 7); H. d'Arrest, *Astronomische Nachrichten,* lxiv (1865), 73-4.
13. Simon Newcomb, *The reminiscences of an astronomer* (Boston, 1903), 192.
14. I am indebted to the Hall family for allowing me to quote a portion of this letter; it is currently in the possession of Nancy Hall Denio. Asaph Hall further described his observing method in an unindexed letter in *Monthly notices of the Royal Astronomical Society,* xxxviii (1877), 205-8: "I began to examine the region close to the planet, and within the glare of light that surrounded it. This was done by keeping the planet just outside the field of view, and turning the eye-piece so as to pass completely around the planet."
15. Holden to Rear Admiral John Rogers, Superintendent of the Naval Observatory, Dobbs Ferry, New York, 28 August 1877; the letter is presumably at the U.S. Naval Observatory.
16. Hall to Arthur Searle, Washington, 9 October 1877, in the Harvard University Archives; Hall's greatest potential competition was from the great 48-inch Melbourne reflector, especially since Mars was better placed for southern observers.
17. *Nature,* xvi (1877), 456–7.

18. Simon Newcomb, *op.cit.* (n. 13), 141–2.
19. Hall to Professor Newcomb, South Norfolk, Conn., 23 August 1901, in the Simon Newcomb papers in the Library of Congress.
20. Hall to E. C. Pickering, quoted with permission of the Harvard University Archives.

Mars on Earth: Soil Analogues for Future Mars Missions

1. Lowell, P. 1909. Science 30, 338–340.
2. Klein, H .P. 1979. Rev. Geophys. Space Phys. 17, 1655–1662.
3. Bada, J. L. et al. 2005. A&G 46, 6.26–6.27.
4. Bell, J. et al. 1997. JGR 102, 9109–9124.
5. Parker, D. C. et al. 1999. Icarus 138, 3–19.
6. Bell, J. and Ansty. 2007. Icarus 191, 581–602.
7. Poulet, F. C. et al. 2007. JGR 112, E08S02.
8. Watson, L. et al. 1994. Science 265, 86–90.
9. Grady, M. et al. 1995. Meteorics 30, 511–512.
10. McKay, D. S. et al. 1996. Science 273, 924–930.
11. Bada, J. L. et al. 1998. Science 279, 362–365.
12. Jull, A. et al. 1998. Science 279, 366–369.
13. Becker, L. et al. 1999. EPSL 167, 71–79.
14. Glavin, D. et al. 1999. Proc. Natl Acad. Sci. 96, 8835–8838.
15. Bouvier, A. J. et al. 2005. EPSL 240, 221–233.
16. Jones, K. L. et al. 1979. Science 204, 799–806.
17. Golombek, M. et al. 1999. JGR 104. 8523–8554.
18. Squyres, S. W. et al. 2004. Science 306, 1698–1703.
19. Crisp, J. et al. 2003. JGR 108, ROV 2–1.
20. Cabrol, N. A. et al. 2001. JGR 106, 7785–7806.
21. Sarrazin, P. et al. 2007. 38th LPSC Abstracts, 2147.
22. Gendrin, A. et al. 2005. Science 307, 1587–1591.
23. Morris, R. V. et al. 2006. JGR 111, E02S13.
24. Langevin, Y. et al. 2005. Science 307, 1584–1586.
25. Greeley, R. et al. 1977. JGR 82, 4093–4109.
26. Cabane, M. P. et al. 2004. Advances in Space Research 33, 2240–2245.
27. Mattingly, R. et al. 2004. IEEE Aerospace Conference Proceedings 1, 492.
28. Bada, J. L. et al. 2007. Space Science Reviews 10.1007/s11214-007-9213-3.
29. Farr, T. G. 2004. Planet. and Space Sci. 52, 3–10.
30. Singer, R. B. 1982 JGR 87, 10159–10168.
31. Allen, C. C. et al. 1998a. LPSC XXXI Abstracts, 1287; Allen, C. C. et al. 1998b. LPSC XXIX Abstracts, 1690.
32. Perko, H. et al. 2006. Journal of Aerospace Engineering 19, 169–176.
33. Ming, D. W. et al. 1988. LPSC XIX Abstracts, 780.
34. Ehlmann, B. L. et al. 2007. Seventh International Conference on Mars, 3270.
35. Nornberg, P. et al. 2004. Clay Minerals 39, 85–98.
36. Hansen, A. et al. 2005. Int. J. Astrobiology 4, 135–144.
37. Garry, J. R. et al. 2006. Meteoritics and Planetary Science 41, 391–405.
38. Sutter, B. et al. 2005. LPSCXXXVI Abstracts, 2182.
39. Ewing, S. A. et al. 2006. Geochimica et Cosmochimica. Acta 70, 5293–5322.
40. Navarro-Gonzalez, R. et al. 2003. Science 302, 1018–1021.
41. Quinn, R. et al. 2005. Planet. and Space Sci. 53, 1376–1388.
42. Lester, E. et al. 2007. Soil Biology and Biochemistry 39, 704–708.
43. Buch, A. et al. 2006. Planet. and Space Sci. 54, 1592–1599.
44. Amashukeli, X. et al. 2007. JGR 112, G04S16.
45. Meunier, D. et al. 2007. Advances in Space Research 39, 337–344.

46. Skelley, A. et al. 2007. JGR 112, G04S11.
47. Howard, A., Matsubara, Y. 2007. Eos Trans. AGU 88, P34A-01.
48. Beegle, L. et al. 2007. LPSC XXXVIII Abstracts, 2005.
49. Volpe, R. 1999. International Journal of Robotics Research 18, 669–683.
50. Behar, A. et al. 2004. IEEE Aerospace Conf. 1, 395.
51. Johnson, A. et al. 2005. Proc. 2005 IEEE International Conference on Robotics and Automation, 4463–4469.
52. Beegle, L. et al. 2007. LPSC XXXVIII Abstracts, 2005.
53. Peeters, Z. et al. 2008. In prep.
54. Biemann, K. et al. 1977. JGR 82, 4641–4658.
55. Gonzalez-Toril, E. et al. 2003. Appl. Environ. Microbiol. 69, 4853–4865.
56. Fernandez-Remolar, D. et al. 2004. Planet and Space Sci. 52, 239–248.
57. Klingelhofer, G. et al. 2004. Science 306, 1740–1745.
58. Sarrazin, P. et al. 2007. 38th LPSC Abstracts, 2147.
59. McKay, D. S. et al. 1996. Science 273, 924–930.
60. Anders, E. et al. 1996. Science 274, 2119–2125.
61. Treiman, A. 2001. LPSCXXXII Abstracts, 1304.
62. Golden, D. et al. 2006. LPSC XXXVII Abstracts, 1199.
63. Bada, J. L. et al. 1998. Science 279, 362–365.
64. Sephton, M. et al. 2002. Nat. Prod. Rep. 19, 292–311.
65. Glavin, D. et al. 1999. Proc. Natl Acad. Sci. 96, 8835–8838; Glavin D et al. 2005 LPSC XXXVI Abstracts, 1920.
66. Ehrenfreund, P. et al. 2001. Proc. Natl. Acad. Sci. 98, 2138–2141.
67. Peeters, Z. et al. 2008. In prep.

Physiological and Psychological Aspects of Sending Humans to Mars: Challenges and Recommendations

1. Graham, I. *Space Travel*, 2nd Edition. New York, NY: DK Publishing, 2010.
2. Beatty, J. K., C. C. Peterson, and A. Chaikin. *The New Solar System, 4th Edition.* Cambridge: Sky Publishing, 1999.
3. The National Aeronautics and Space Administration, Apollo Flight Journal, Accessed Feb 19, 2010, http://history.nasa.gov/afj.
4. Angelo, J. A. *The Dictionary of Space Technology*, 3rd Edition, New York, NY: Facts on Files, Inc., 2006.
5. Buckey, J. C. *Space Physiology*. New York: Oxford University Press, 2006.
6. Morphew, M. E. Psychological and Human Factors in Long Duration Spaceflight. NASA Johnson Space Center, 2001.
7. Clement, G. *Fundamentals of Space Medicine*. The Netherlands: Springer, 2005.
8. "Spaceflight Bad for Astronauts' Vision, Study Suggests (2012)." Spaceflight.com, Accessed Dec 6, 2014, http://www.space.com/14876-astronaut-spaceflight-vision-problems.html.
9. Roger D. L., *NASA: A History of the U.S. Civil Space Program*, Malabar, Florida: Krieger Publishing Company, 1994, 55–96.
10. Darling, D. *The Complete Book of Space Flight*. Hoboken, NJ: John Wiley & Sons, 2003.
11. Thirsk, R., A. Kuipers, C. Mukai, and D. Williams. "The Space-flight Environment: The International Space Station and Beyond." *Can. Med. Assoc. J.* 180, (2009): 1216–1220.
12. Stuster, J. *Behavioral Issues Associated with Long Duration Space Expeditions: Review and Analysis of Astronaut Journals*. Santa Barbara, CA: Anacapa Sciences, 2011.

13. Johnson, P. J. "The Roles of NASA, U.S. Astronauts and Their Families in Long-duration Missions." *Acta Astronautica*, 67, no. 5–6 (2010), 561–571.
14. The National Aeronautics and Space Administration, "The International Space Station." Accessed Feb 24, 2010, http://www.nasa.gov.
15. Griffiths, T. *Slicing the Silence: Voyaging to Antarctica*. Sydney, Australia: University of New South Wales Press, 2007.
16. Bishop, S. L. "From Earth Analogs to Space: Getting There From Here." In *Psychology of Space Exploration: Contemporary Research in Historical Perspective*, edited by D. Vakock. NASA History Series, 2011, 47–77.
17. Suedfeld, P. (2010). "Historical space psychology: Early terrestrial explorations as Mars analogues." *Planetary and Space Science*, 58, 639–645.
18. The National Aeronautics and Space Administration, "A Crew Mission to Mars (2012)." http://nssdc.gsfc.nasa.gov/planetary/mars/marsprof.html (accessed Dec 6, 2014).

Water on Mars: A Literature Review

1. Jakosky, B.M.; Haberle, R.M. *The Seasonal Behavior of Water on Mars*; University of Arizona Press: Tocson, AZ, USA, 1992; pp. 969–1016.
2. Martín-Torres, F.J.; Zorzano, M.-P.; Valentín-Serrano, P.; Harri, A.-M.; Genzer, M.; Kemppinen, O.; Rivera-Valentin, E.G.; Jun, I.; Wray, J.J.; Madsen, M.; et al. Transient liquid water and water activity at Gale crater on Mars. *Nat. Geosci.* 2015, *8*, 357–361.
3. Ojha, L.; Wilhelm, M.B.; Murchie, S.L.; McEwen, A.S.; Wray, J.J.; Hanley, J.; Massé, M.; Chojnacki, M.; Chojnacki, M. Spectral evidence for hydrated salts in recurring slope lineae on Mars. *Nat. Geosci.* 2015, *8*, 829–832.
4. Dundas, C.; Byrne, S.; McEwen, A.S.; Mellon, M.T.; Kennedy, M.R.; Daubar, I.J.; Saper, L. HiRISE observations of new impact craters exposing Martian ground ice. *J. Geophys. Res. Planets* 2014, *119*, 109–127.
5. Recurring Martian Streaks: Flowing Sand, Not Water? Available online: http://www.nasa.gov/feature/jpl/recurring-martian-streaks-flowing-sand-not-water (accessed on 27 December 2017).
6. Carr, M.H. *Water on Mars*; Oxford University Press: Oxford, UK; New York, NY, USA, 1996; p. 197.
7. Bibring, J.-P.; Langevin, Y.; Poulet, F.; Gendrin, A.; Gondet, B.; Berthé, M.; Soufflot, A.; Drossart, P.; Combes, M.; Bellucci, G.; et al. Perennial water ice identified in the south polar cap of Mars. *Nature* 2004, *428*, 627–630.
8. Ghosh, J.; Methikkalam, R.R.J.; Bhuin, R.G.; Ragupathy, G.; Choudhary, N.; Kumar, R.; Pradeep, T. Clathrate hydrates in interstellar environment. *Proc. Natl. Acad. Sci. USA* 2019, *116*, 1526–1531.
9. European Space Agency (ESA). Water at Martian South Pole. Available online: http://www.esa.int/Our_Activities/Space_Science/Mars_Express/Water_at_Martian_south_pole (accessed on 17 March 2004).
10. Christensen, P.R. Water at the Poles and in Permafrost Regions of Mars. *Elements* 2006, 2, 151–155.
11. Webster, G.; Brown, D. NASA Mars Spacecraft Reveals a More Dynamic Red Planet. Available online: https://mars.nasa.gov/news/nasa-mars-spacecraft-reveals-a-more-dynamic-red-planet/ (accessed on 10 December 2013).
12. Jones, E.G. Shallow transient liquid water environments on present-day Mars, and their implications for life. *Acta Astronaut.* 2018, *146*, 144–150.
13. Pollack, J.B. Climatic change on the terrestrial planets. *Icarus* 1979, *37*, 479–553.
14. Fairén, A.G. A cold and wet Mars. *Icarus* 2010, *208*, 165–175.

15. Dohm, J.M.; Baker, V.R.; Boynton, W.V.; Fairén, A.G.; Ferris, J.C.; Finch, M.; Furfaro, R.; Hare, T.; Janes, D.M.; Kargel, J.S.; et al. GRS evidence and the possibility of paleooceans on Mars. *Planet. Space Sci.* 2009, *57*, 664–684.
16. Baker, V.R.; Strom, R.G.; Gulick, V.C.; Kargel, J.S.; Komatsu, G.; Kale, V.S. Ancient oceans, ice sheets and the hydrological cycle on Mars. *Nature* 1991, *352*, 589–594.
17. Clifford, S. The Evolution of the Martian Hydrosphere: Implications for the Fate of a Primordial Ocean and the Current State of the Northern Plains. *Icarus* 2001, *154*, 40–79.
18. Ancient Ocean May Have Covered a Third of Mars. Available online: https://www.sciencedaily.com/releases/2010/06/100613181245.htm (accessed on 16 June 2010).
19. Grotzinger, J.P.; Crisp, J.; Vasavada, A.R. Curiosity's Mission of Exploration at Gale Crater, Mars. *Elements* 2015, *11*, 19–26.
20. Choi, C.H. Flashback: Water on Mars Announced 10 Years Ago. Available online: https://www.space.com/8642-flashback-water-mars-announced-10-years.htm (accessed on 16 June 2000).
21. Chang, K. On Mars, an Ancient Lake and Perhaps Life. Available online: https://www.nytimes.com/2013/12/10/science/space/on-mars-an-ancient-lake-and-perhaps-life.html (accessed on 9 December 2013).
22. Moyano-Cambero, C.E.; Trigo-Rodríguez, J.M.; Martín-Torres, F.J. SNC Meteorites: Atmosphere Implantation Ages and the Climatic Evolution of Mars. In *Protection of Materials and Structures from Space Environment*; Springer Science and Business Media LLC: Berlin, Germany, 2013; Volume 35, pp. 165–172.
23. Moyano-Cambero, C.E.; Trigo-Rodríguez, J.M.; Benito, M.I.; Alonso-Azcárate, J.; Lee, M.R.; Mestres, N.; Martínez-Jiménez, M.; Martín-Torres, F.J.; Fraxedas, J. Petrographic and geochemical evidence for multiphase formation of carbonates in the Martian orthopyroxenite Allan Hills 84001. *Meteorit. Planet. Sci.* 2017, *52*, 1030–1047.
24. Scott, E.R.D.; Krot, A.N.; Yamagouchi, A. Carbonates in fractures of Martian meteorite ALH 84001: Petrologic evidence for impact origin. *Meteorit. Planet. Sci.* 1998, *33*, 709–719.
25. Bridges, J.C.; Catling, D.C.; Saxton, J.M.; Swindle, T.; Lyon, I.; Grady, M. Alteration Assemblages in Martian Meteorites: Implications for Near-Surface Processes. *Space Sci. Rev.* 2001, *96*, 365–392.
26. Tomkinson, T.; Lee, M.R.; Mark, D.F.; Smith, C.L. Sequestration of Martian CO2 by mineral carbonation. *Nat. Commun.* 2013, *4*, 2662.
27. Bandfield, J.L.; Glotch, T.D.; Christensen, P.R. Spectroscopic Identification of Carbonate Minerals in the Martian Dust. *Science* 2003, *301*, 1084–1087.
28. Ehlmann, B.L.; Mustard, J.F.; Murchie, S.L.; Poulet, F.; Bishop, J.L.; Brown, A.J.; Calvin, W.M.; Clark, R.N.; Marais, D.J.D.; Milliken, R.E.; et al. Orbital Identification of Carbonate-Bearing Rocks on Mars. *Science* 2008, *322*, 1828–1832.
29. Wray, J.J.; Murchie, S.L.; Ehlmann, B.L.; Milliken, R.E.; Seelos, K.D.; Dobrea, E.Z.N.; Mustard, J.F.; Squyres, S.W. Evidence for regional deeply buried carbonate-bearing rocks on Mars (abstract #2635). In Proceedings of the Lunar and Planetary Science Conference, The Woodlands, TX, USA, 11 March 2011.
30. Glotch, T.D.; Christensen, P.R. Geologic and mineralogic mapping of Aram Chaos: Evidence for a water-rich history. *J. Geophys. Res. Space Phys.* 2005, *110*, 09006.

31. Holt, J.W.; Safaeinili, A.; Plaut, J.J.; Young, D.A.; Head, J.W.; Phillips, R.J.; Campbell, B.A.; Carter, L.M.; Gim, Y.; Seu, R.; et al. Radar Sounding Evidence for Ice within Lobate Debris Aprons near Hellas Basin. In Proceedings of the Mid-Southern Latitudes of Mars. Lunar and Planetary Science Conference, The Woodlands, TX, USA, 10–14 March 2008; Volume 39, p. 2441, Bibcode: 2008LPI. 39.2441H.
32. Amos, J. Old Opportunity Mars Rover Makes Rock Discovery. Available online: https://www.bbc.com/news/science-environment-22832673 (accessed on 10 June 2013).
33. NASA Staff. Mars Rover Opportunity Examines Clay Clues in Rock. Available online: https://www.jpl.nasa.gov/news/news.php?release=2013-167 (accessed on 17 May 2013).
34. Planetary Science Institute. Regional, Not Global, Processes Led to Huge Martian Floods. Available online: http://spaceref.com/mars/regional-not-global-processes-led-to-huge-martian-floods.html (accessed on 12 September 2015).
35. Harrison, K.P.; Grimm, R. Groundwater-controlled valley networks and the decline of surface runoff on early Mars. *J. Geophys. Res. Space Phys.* 2005, *110*.
36. Howard, A.D.; Moore, J.M.; Irwin, R.P. An intense terminal epoch of widespread fluvial activity on early Mars: 1. Valley network incision and associated deposits. *J. Geophys. Res. Space Phys.* 2005, *110*, 12–14.
37. Salese, F.; Di Achille, G.; Neesemann, A.; Ori, G.G.; Hauber, E. Hydrological and sedimentary analyses of well-preserved paleofluvial-paleolacustrine systems at Moa Valles, Mars. *J. Geophys. Res. Planets* 2016, *121*, 194–232.
38. Irwin, R.P.; Howard, A.D.; Craddock, R.A.; Moore, J.M. An intense terminal epoch of widespread fluvial activity on early Mars: 2. Increased runoff and paleolake development. *J. Geophys. Res. Space Phys.* 2005, *110*.
39. Fassett, C.I.; Head, J. Valley network-fed, open-basin lakes on Mars: Distribution and implications for Noachian surface and subsurface hydrology. *Icarus* 2008, *198*, 37–56.
40. Moore, J.M. Hellas as a Possible Site of Ancient Ice-Covered Lakes on Mars. *Icarus* 2001, *154*, 258–276.
41. Weitz, C.; Parker, T. New Evidence That the Valles Marineris Interior Deposits Formed in Standing Bodies of Water. In Proceedings of the Lunar and Planetary Science Conference, Houston, TX, USA, 17 March 2000; p. 1693, Bibcode: 2000LPI. 31.1693W.
42. Thompson, A. New Signs That Ancient Mars Was Wet. Available online: https://www.space.com/6033-signs-ancient-mars-wet.html (accessed on 28 October 2008).
43. Squyres, S.W.; Clifford, S.M.; Kuzmin, R.O.; Zimbelman, J.R.; Costard, F.M. Ice in the Martian Regolith. *Mars* 1992, *1*, 523–554.
44. Head, J.W., III; Marchant, D. Modifications of The Walls of a Noachian Crater in Northern Arabia Terra (24 E, 39 N) During Northern Mid-Latitude Amazonian Glacial Epochs on Mars: Nature and Evolution of Lobate Debris Aprons and Their Relationships to Lineated Valley Fill and Glacial Systems. In Proceedings of the Annual Lunar and Planetary Science Conference, The Woodlands, TX, USA, 13–17 March 2006; p. 1128.
45. Head, J.W.; Nahm, A.L.; Marchant, D.R.; Neukum, G. Modification of the Dichotomy Boundary on Mars by Amazonian Mid-Latitude Regional Glaciation. *Geophys. Res. Lett.* 2006, *33*, 33.

46. Head, J., III; Marchant, D. Evidence for Global-Scale Northern Mid-Latitude Glaciation in The Amazonian Period of Mars: Debris-Covered Glacial and Valley Glacial Deposits in the 30–50 N Latitude Band. In Proceedings of the Annual Lunar and Planetary Science Conference, The Woodlands, TX, USA, 13–17 March 2006; p. 1127.
47. Lewis, R. Glaciers Reveal Martian Climate Has Been Recently Active. Available online: https://www.sciencedaily.com/releases/2008/04/080423131602.htm (accessed on 23 April 2008).
48. Plaut, J.J.; Safaeinili, A.; Holt, J.W.; Phillips, R.J.; Head, J.W.; Seu, R.; Putzig, N.E.; Frigeri, A. Radar evidence for ice in lobate debris aprons in the mid-northern latitudes of Mars. *Geophys. Res. Lett.* 2009, *36*, 36.
49. Wall, M. Q & A with Mars Life-Seeker Chris Carr. Available online: https://www.space.com/11232-mars-life-evolution-carr-interview.html (accessed on 25 March 2011).
50. Dartnell, L.R.; Desorgher, L.; Ward, J.M.; Coates, A.J. Modelling the surface and subsurface Martian radiation environment: Implications for astrobiology. *Geophys. Res. Lett.* 2007, *34*, 02207.
51. NASA Staff. Scalloped Terrain Led to Finding of Buried Ice on Mars. Available online: https://photojournal.jpl.nasa.gov/catalog/PIA21136 (accessed on 23 November 2016).
52. Thomson, L. Lake of Frozen Water the Size of New Mexico Found on Mars – NASA. Available online: https://www.theregister.co.uk/2016/11/22/nasa_finds_ice_under_martian_surface/ (accessed on 23 November 2016).
53. NASA Staff. Mars Ice Deposit Holds as Much Water as Lake Superior. Available online: https://www.jpl.nasa.gov/news/news.php?release=2016-299 (accessed on 22 November 2016).
54. Dartnell, L.R.; Desorgher, L.; Ward, J.M.; Coates, A.J. Martian sub-surface ionising radiation: Biosignatures and geology. *Biogeosciences* 2007, *4*, 545–558.
55. de Morais, A. A Possible Biochemical Model for Mars. In Proceedings of the Lunar and Planetary Science Conference, Rio de Janeiro, Brazil, 19–23 March 2012.
56. Grotzinger, J.P. Habitability, Taphonomy, and the Search for Organic Carbon on Mars. *Science* 2014, *343*, 386–387.
57. Hydrated Minerals—Evidence of Liquid Water on Mars. Available online: https://sci.esa.int/web/mars-express/-/51821-1-hydrated-minerals-ndash-evidence-of-liquid-water-on-mars (accessed on 16 December 2014).
58. Grotzinger, J.P.; Sumner, D.Y.; Kah, L.; Stack, K.; Gupta, S.; Edgar, L.; Rubin, D.M.; Lewis, K.; Schieber, J.; Mangold, N.; et al. A Habitable Fluvio-Lacustrine Environment at Yellowknife Bay, Gale Crater, Mars. *Science* 2013, *343*, 1242777.
59. Rodriguez, J.A.P.; Kargel, J.S.; Baker, V.R.; Gulick, V.C.; Berman, D.C.; Fairén, A.G.; Linares, R.; Zarroca, M.; Yan, J.; Miyamoto, H.; et al. Martian outflow channels: How did their source aquifers form and why did they drain so rapidly? *Sci. Rep.* 2015, *5*, 13404.
60. Space.com Staff. Ancient Mars Water Existed Deep Underground. Available online: https://www.space.com/16335-mars-underground-water-impact-craters.html (accessed on 2 July 2012).
61. Craddock, R.A.; Howard, A.D. The case for rainfall on a warm, wet early Mars. *J. Geophys. Res. Space Phys.* 2002, *107*, 21–31.
62. Head, J.W.; Marchant, D.; Agnew, M.; Fassett, C.I.; Kreslavsky, M. Extensive valley glacier deposits in the northern mid-latitudes of Mars: Evidence for Late Amazonian obliquity-driven climate change. *Earth Planet. Sci. Lett.* 2006, *241*, 663–671.

63. Webster, G.; Martinez, M.; Brown, D. NASA Mars Reconnaissance Orbiter Reveals Details of a Wetter Mars. Available online: http://www.spaceref.com/news/viewpr.html?pid=26817 (accessed on 28 October 2008).
64. Lunine, J.I.; Chambers, J.; Morbidelli, A.; Leshin, L.A. The origin of water on Mars. *Icarus* 2003, *165*, 1–8.
65. Soderblom, L.A.; Bell, J.F. Exploration of the Martian surface: 1992–2007. *Martian Surf.* 2009, 3–19.
66. Ming, D.W.; Morris, R.V.; Clark, R.C. Aqueous Alteration on Mars. In *The Martian Surface: Composition, Mineralogy, and Physical Properties*; Bibcode: 2008mscm.book. B; Cambridge University Press: Cambridge, UK, 2008; pp. 519–540.
67. Lewis, J.S. *Physics and Chemistry of the Solar System*; Academic Press: San Diego, CA, USA, 1997; ISBN 978-0-12-446742-2.
68. Lasue, J.; Mangold, N.; Hauber, E.; Clifford, S.; Feldman, W.; Gasnault, O.; Grima, C.; Maurice, S.; Mousis, O. Quantitative Assessments of the Martian Hydrosphere. *Space Sci. Rev.* 2012, *174*, 155–212.
69. Clark, B.; Morris, R.; McLennan, S.; Gellert, R.; Jolliff, B.; Knoll, A.; Squyres, S.; Lowenstein, T.; Ming, D.; Tosca, N.J.; et al. Chemistry and mineralogy of outcrops at Meridiani Planum. *Earth Planet. Sci. Lett.* 2005, *240*, 73–94.
70. Small, R.J.; Bloom, A.L.; Cliffs, E.; Hempstead, H. Geomorphology: A Systematic Analysis of Late Cenozoic Landforms. *Geogr. J.* 1979, *145*, 485.
71. Boynton, W.V.; Ming, D.W.; Kounaves, S.P.; Young, S.M.M.; Arvidson, R.E.; Hecht, M.H.; Hoffman, J.; Niles, P.B.; Hamara, D.K.; Quinn, R.C.; et al. Evidence for Calcium Carbonate at the Mars Phoenix Landing Site. *Science* 2009, *325*, 61–64.
72. Gooding, J.L.; Arvidson, R.E.; Zolotov, M.Y. Physical and Chemical Weathering. *Mars* 1992, *1*, 626–651.
73. Melosh, H.J. *Planetary Surface Processes by H. Jay Melosh*; Cambridge University Press (CUP): Cambridge, UK, 2011; p. 296.
74. Abramov, O.; Kring, D.A. Impact-induced hydrothermal activity on early Mars. *J. Geophys. Res. Space Phys.* 2005, *110*, E12S09.
75. Schrenk, M.O.; Brazelton, W.J.; Lang, S.Q. Serpentinization, Carbon, and Deep Life. *Rev. Miner. Geochem.* 2013, *75*, 575–606.
76. Hamilton, V.E.; Christensen, P.R. Evidence for extensive, olivine-rich bedrock on Mars. *Geology* 2005, *33*, 433.
77. Ehlmann, B.L.; Mustard, J.; Murchie, S.L. Geologic setting of serpentine deposits on Mars. *Geophys. Res. Lett.* 2010, *37*, 06201.
78. Hartmann, W.K. Martian cratering 8: Isochron refinement and the chronology of Mars. *Icarus* 2005, *174*, 294–320.
79. Ody, A.; Poulet, F.; Bibring, J.-P.; Loizeau, D.; Carter, J.; Gondet, B.; Langevin, Y. Global investigation of olivine on Mars: Insights into crust and mantle compositions. *J. Geophys. Res. Planets* 2013, *118*, 234–262.
80. Swindle, T.; Treiman, A.H.; Lindstrom, D.J.; Burkland, M.K.; Cohen, B.A.; Grier, J.A.; Li, B.; Olson, E.K. Noble gases in iddingsite from the Lafayette meteorite: Evidence for liquid water on Mars in the last few hundred million years. *Meteorit. Planet. Sci.* 2000, *35*, 107–115.
81. Head, J.W.; Kreslavsky, M.A.; Ivanov, M.A.; Hiesinger, H.; Fuller, E.R.; Pratt, S. *Water in Middle Mars History: New Insights from MOLA Data*; P31A–02 INVITED; Bibcode: 2001AGUSM. P31A02H; American Geophysical Union, Spring Meeting Abstracts: Washington, DC, USA, 2001.

82. Head, J.W. *Exploration for Standing Bodies of Water on Mars: When Were They There, Where Did They Go, and What Are the Implications for Astrobiology?* P21C–03, Bibcode: 2001AGUFM.P21C. 03H; AGU Fall Meeting Abstracts: San Francisco, CA, USA, 2001.
83. Meyer, C. The Martian Meteorite Compendium. Available online: http://curator.jsc.nasa.gov/antmet/mmc/ (accessed on 15 March 2017).
84. Shergotty Meteorite—JPL, NASA. Available online: https://www2.jpl.nasa.gov/snc/shergotty.html (accessed on 19 December 2010).
85. Hamilton, V.E.; Christensen, P.R.; McSween, H.Y. Determination of Martian meteorite lithologies and mineralogies using vibrational spectroscopy. *J. Geophys. Res. Space Phys.* 1997, *102*, 25593–25603.
86. Treiman, A.H. The nakhlite meteorites: Augite-rich igneous rocks from Mars. *Geochemistry* 2005, *65*, 203–270.
87. Agee, C.B.; Wilson, N.V.; McCubbin, F.M.; Ziegler, K.; Polyak, V.J.; Sharp, Z.; Asmerom, Y.; Nunn, M.H.; Shaheen, R.; Thiemens, M.; et al. Unique Meteorite from Early Amazonian Mars: Water-Rich Basaltic Breccia Northwest Africa 7034. *Science* 2013, *339*, 780–785.
88. Agree, C. Unique Meteorite from Early Amazonian Mars: Water-Rich Basaltic Breccia Northwest Africa 7034. *Science* 2013, *339*, 780–785.
89. McKay, D.S.; Gibson, E.K.; Thomas-Keprta, K.L.; Vali, H.; Romanek, C.; Clemett, S.J.; Chillier, X.D.F.; Maechling, C.R.; Zare, R.N. Search for Past Life on Mars: Possible Relic Biogenic Activity in Martian Meteorite ALH84001. *Science* 1996, *273*, 924–930.
90. Bada, J.; Glavin, D.P.; McDonald, G.D.; Becker, L. A Search for Endogenous Amino Acids in Martian Meteorite AL84001. *Science* 1998, *279*, 362–365.
91. García-Ruiz, J.M. Morphological behavior of inorganic precipitation systems. *Instrum. Methods Mission. Astrobiol. II* 1999, *3755*, 74–83.
92. Trigo-Rodríguez, J.M.; Moyano-Cambero, C.E.; Benito-Moreno, M.I.; Alonso-Azcárate, J.; Lee, M.R. Ancient Martian Floods in a Plausible Variable Climatic Environment as Revealed from the Sequential Growth of Allan Hills 84001 Carbonate Globules. In Proceedings of the 49th Lunar and Planetary Science Conference 2018 (LPI Contrib. No. 2083), The Woodlands, TX, USA, 19–23 March 2018; University of Glasgow: Scotland, UK, 2018; p. 1448.
93. Schopf, J.W.; Kudryavtsev, A.B.; Czaja, A.D.; Tripathi, A.B. Evidence of Archean life: Stromatolites and microfossils. *Precambrian Res.* 2007, *158*, 141–155.
94. Raeburn, P. *Uncovering the Secrets of the Red Planet: Mars*; National Geographic: Washington, DC, USA, 1998; ISBN13: 9780792273738.
95. Moore, P. *The Atlas of the Solar System*; Chancellor Press: New York, NY, USA, 1990; ISBN13: 9780753700143.
96. Berman, D.C.; Crown, D.A.; Bleamaster, L.F. Degradation of mid-latitude craters on Mars. *Icarus* 2009, *200*, 77–95.
97. Fassett, C.I.; Head, J. The timing of martian valley network activity: Constraints from buffered crater counting. *Icarus* 2008, *195*, 61–89.
98. Malin. An overview of the 1985-2006 Mars Orbiter Camera science investigation. *MARS* 2010, *5*, 1–60.
99. Sinuous Ridges Near Aeolis Mensae. Available online: https://web.archive.org/web/20160305025124/http:/hiroc.lpl.arizona.edu/images/PSP/diafotizo.php?ID=PSP_002279_1735 (accessed on 31 January 2007).
100. Zimbelman, J.; Griffin, L.J. HiRISE images of yardangs and sinuous ridges in the lower member of the Medusae Fossae Formation, Mars. *Icarus* 2010, *205*, 198–210.

101. Newsom, H.; Lanza, N.; Ollila, A.M.; Wiseman, S.M.; Roush, T.L.; Marzo, G.A.; Tornabene, L.L.; Okubo, C.; Osterloo, M.M.; Hamilton, V.E.; et al. Inverted channel deposits on the floor of Miyamoto crater, Mars. *Icarus* 2010, *205*, 64–72.
102. Morgan, A.M.; Howard, A.D.; Hobley, D.E.J.; Moore, J.; Dietrich, W.; Williams, R.; Burr, D.; Grant, J.; Wilson, S.A.; Matsubara, Y. Sedimentology and climatic environment of alluvial fans in the martian Saheki crater and a comparison with terrestrial fans in the Atacama Desert. *Icarus* 2014, *229*, 131–156.
103. Weitz, C.; Milliken, R.; Grant, J.; McEwen, A.; Williams, R.; Bishop, J.L.; Thomson, B. Mars Reconnaissance Orbiter observations of light-toned layered deposits and associated fluvial landforms on the plateaus adjacent to Valles Marineris. *Icarus* 2010, *205*, 73–102.
104. Zendejas, J.; Segura, A.; Raga, A.C. Atmospheric mass loss by stellar wind from planets around main sequence M stars. *Icarus* 2010, *210*, 539–544.
105. Cabrol, N.; Grin, E. *Searching for Lakes on Mars*; Elsevier BV: Amsterdam, The Netherlands, 2010; pp. 1–29.
106. Goldspiel, J.M.; Squyres, S.W. Groundwater Sapping and Valley Formation on Mars. *Icarus* 2000, *148*, 176–192.
107. Conklin, N.; Carr, M.H. The Surface of Mars. *Leon* 1983, *16*, 328.
108. Nedell, S.S.; Squyres, S.W.; Andersen, D.W. Origin and evolution of the layered deposits in the Valles Marineris, Mars. *Icarus* 1987, *70*, 409–441.
109. Matsubara, Y.; Howard, A.D.; Drummond, S.A. Hydrology of early Mars: Lake basins. *J. Geophys. Res. Space Phys.* 2011, *116*, 116.
110. Spectacular Mars Images Reveal Evidence of Ancient Lakes. Available online: https://web.archive.org/web/20160823210537/https:/www.sciencedaily.com/releases/2012/01/100104092452.htm (accessed on 4 January 2010).
111. Warner, N.; Gupta, S.; Kim, J.; Lin, S.-Y.; Muller, J.-P. Hesperian equatorial thermokarst lakes in Ares Vallis as evidence for transient warm conditions on Mars. *Geology* 2010, *38*, 71–74.
112. Palucis, M.C.; Dietrich, W.E.; Hayes, A.; Williams, R.M.; Gupta, S.; Mangold, N.; Newsom, H.; Hardgrove, C.J.; Calef, F.; Sumner, D.Y. The origin and evolution of the Peace Vallis fan system that drains to the Curiosity landing area, Gale Crater, Mars. *J. Geophys. Res. Planets* 2014, *119*, 705–728.
113. Brown, D.; Cole, S.; Webster, G.; Agle, D.C. NASA Rover Finds Old Streambed on Martian Surface. Available online: https://www.nasa.gov/home/hqnews/2012/sep/HQ_12-338_Mars_Water_Stream.html (accessed on 27 September 2012).
114. Lewis, K.W.; Aharonson, O. Stratigraphic analysis of the distributary fan in Eberswalde crater using stereo imagery. *J. Geophys. Res. Space Phys.* 2006, *111*.
115. Chang, A. Mars Rover Curiosity Finds Signs of Ancient Stream. Available online: http://apnews.excite.com/article/20120927/DA1IDOO00.html (accessed on 27 September 2012).
116. Michalski, J.; Dobrea, E.Z.N.; Niles, P.B.; Cuadros, J. Ancient hydrothermal seafloor deposits in Eridania basin on Mars. *Nat. Commun.* 2017, *8*, 15978.
117. Mars Study Yields Clues to Possible Cradle of Life—*Astrobiology Magazine*. Available online: https://www.astrobio.net/also-in-news/mars-study-yields-clues-possible-cradle-life/ (accessed on 8 October 2017).
118. Martin, W.; Baross, J.; Kelley, D.; Russell, M.J. Hydrothermal vents and the origin of life. *Nat. Rev. Genet.* 2008, *6*, 805–814.
119. Irwin, R.P.; Howard, A.D.; Maxwell, T.A. Geomorphology of Ma'adim Vallis, Mars, and associated paleolake basins. *J. Geophys. Res. Space Phys.* 2004, *109*.
120. Hynek, B.M.; Beach, M.; Hoke, M.R.T. Updated global map of Martian valley networks and implications for climate and hydrologic processes. *J. Geophys. Res. Space Phys.* 2010, *115*.

121. Di Achille, G.; Hynek, B.M. Ancient ocean on Mars supported by global distribution of deltas and valleys. *Nat. Geosci.* 2010, *3*, 459–463.
122. Carr, M.H. Formation of Martian flood features by release of water from confined aquifers. *J. Geophys. Res. Space Phys.* 1979, *84*, 2995.
123. Baker, V.R.; Milton, D.J. Erosion by catastrophic floods on Mars and Earth. *Icarus* 1974, *23*, 27–41.
124. Mars Global Surveyor MOC2-862 Release. Available online: http://www.msss.com/mars_images/moc/2004/09/27/ (accessed on 16 January 2012).
125. Irwin, R.P.; Craddock, R.A.; Howard, A.D. Interior channels in Martian valley networks: Discharge and runoff production. *Geology* 2005, *33*, 489.
126. Andrews-Hanna, J.C.; Phillips, R.J.; Zuber, M.T. Meridiani Planum and the global hydrology of Mars. *Nature* 2007, *446*, 163–166.
127. Jakosky, B.M. Water, Climate, and Life. *Science* 1999, *283*, 648–649.
128. Lamb, M.P.; Howard, A.D.; Johnson, J.; Whipple, K.X.; Dietrich, W.E.; Perron, J.T. Can springs cut canyons into rock? *J. Geophys. Res. Space Phys.* 2006, *111*, 111.
129. Grötzinger, J.; Arvidson, R.; Bell, J.; Calvin, W.; Clark, B.; Fike, D.; Golombek, M.; Greeley, R.; Haldemann, A.; Herkenhoff, K.; et al. Stratigraphy and sedimentology of a dry to wet eolian depositional system, Burns formation, Meridiani Planum, Mars. *Earth Planet. Sci. Lett.* 2005, *240*, 11–72.
130. Michalski, J.; Cuadros, J.; Niles, P.B.; Parnell, J.; Rogers, A.D.; Wright, S. Groundwater activity on Mars and implications for a deep biosphere. *Nat. Geosci.* 2013, *6*, 133–138.
131. Zuber, M.T. Planetary science: Mars at the tipping point. *Nature* 2007, *447*, 785–786.
132. Andrews-Hanna, J.C.; Zuber, M.T.; Arvidson, R.E.; Wiseman, S.M. Early Mars hydrology: Meridiani playa deposits and the sedimentary record of Arabia Terra. *J. Geophys. Res. Space Phys.* 2010, *115*.
133. McLennan, S.; Bell, J.; Calvin, W.; Christensen, P.; Clark, B.; De Souza, P.; Farmer, J.; Farrand, W.; Fike, D.; Gellert, R.; et al. Provenance and diagenesis of the evaporite-bearing Burns formation, Meridiani Planum, Mars. *Earth Planet. Sci. Lett.* 2005, *240*, 95–121.
134. Squyres, S.W.; Knoll, A.H. Sedimentary rocks at Meridiani Planum: Origin, diagenesis, and implications for life on Mars. *Earth Planet. Sci. Lett.* 2005, *240*, 1–10.
135. Squyres, S.W.; Knoll, A.H.; Arvidson, R.E.; Clark, B.C.; Grotzinger, J.P.; Jolliff, B.L.; McLennan, S.M.; Tosca, N.J.; Bell, J.F.; Calvin, W.M.; et al. Two Years at Meridiani Planum: Results from the Opportunity Rover. *Science* 2006, *313*, 1403–1407.
136. Wiseman, M.; Andrews-Hanna, J.C.; Arvidson, R.E.; Mustard, J.F.; Zabrusky, K.J. Distribution of Hydrated Sulfates Across Arabia Terra Using CRISM Data: Implications for Martian Hydrology. In Proceedings of the 42nd Lunar and Planetary Science Conference, Woodlands, TX, USA, 7–11 March 2011.
137. Andrews-Hanna, J.C.; Lewis, K.W. Early Mars hydrology: 2. Hydrological evolution in the Noachian and Hesperian epochs. *J. Geophys. Res. Space Phys.* 2011, *116*.
138. Houser, K. First Evidence of "Planet-Wide Groundwater System" on Mars Found. Available online: https://futurism.com/the-byte/mars-groundwater-system-planet-wide (accessed on 28 February 2019).
139. Salese, F.; Pondrelli, M.; Neesemann, A.; Schmidt, G.; Ori, G.G. Geological Evidence of Planet-Wide Groundwater System on Mars. *J. Geophys. Res. Planets* 2019, *124*, 374–395.

140. Mars: Planet-Wide Groundwater System—New Geological Evidence. Available online: https://web.archive.org/web/20120220081803/http:/astrobiology.nasa.gov/articles/mars-ocean-hypothesis-hits-the-shore/ (accessed on 19 February 2019).
141. Brandenburg, J.E. The Paleo-Ocean of Mars. In *MECA Symposium on Mars: Evolution of Its Climate and Atmosphere*; Bibcode: 1987meca.symp. 20B; Lunar and Planetary Institute: Washington, DC, USA, 1987; pp. 20–22.
142. Smith, D.E. The Gravity Field of Mars: Results from Mars Global Surveyor. *Science* 1999, *286*, 94–97.
143. Read, P.L.; Lewis, S.R. *The Martian Climate Revisited: Atmosphere and Environment of a Desert Planet*; Springer: Chichester, UK, 2004; ISBN 978-3-540-40743-0.
144. New Map Bolsters Case for Ancient Ocean on Mars. Available online: https://web.archive.org/web/20120220081803/http:/astrobiology.nasa.gov/articles/mars-ocean-hypothesis-hits-the-shore/ (accessed on 23 November 2009).
145. Carr, M.H.; Head, J. Oceans on Mars: An assessment of the observational evidence and possible fate. *J. Geophys. Res. Space Phys.* 2003, *108*, 5042.
146. Perron, J.T.; Mitrovica, J.X.; Manga, M.; Matsuyama, I.; Richards, M.A. Evidence for an ancient martian ocean in the topography of deformed shorelines. *Nature* 2007, *447*, 840–843.
147. Boynton, W.V.; Feldman, W.C.; Mitrofanov, I.G.; Evans, L.G.; Reedy, R.; Squyres, S.W.; Starr, R.; Trombka, J.; D'Uston, C.; Arnold, J.; et al. The Mars Odyssey Gamma-Ray Spectrometer Instrument Suite. *Space Sci. Rev.* 2004, *110*, 37–83.
148. Ancient Tsunami Evidence on Mars Reveals Life Potential – Astrobiology. Available online: http://astrobiology.com/2016/05/ancient-tsunami-evidence-on-mars-reveals-life-potential.html (accessed on 20 May 2016).
149. Wilson, J.T.; Eke, V.R.; Massey, R.; Elphic, R.C.; Feldman, W.C.; Maurice, S.; Teodoro, L.F. Equatorial locations of water on Mars: Improved resolution maps based on Mars Odyssey Neutron Spectrometer data. *Icarus* 2018, *299*, 148–160.
150. Rodriguez, J.A.P.; Fairén, A.G.; Tanaka, K.L.; Zarroca, M.; Linares, R.; Platz, T.; Komatsu, G.; Miyamoto, H.; Kargel, J.S.; Yan, J.; et al. Tsunami waves extensively resurfaced the shorelines of an early Martian ocean. *Sci. Rep.* 2016, *6*, 25106.
151. Perry, M.R.; Bain, Z.M.; Putzig, N.E.; Morgan, G.A.; Bramson, A.M.; Petersen, E.I.; Smith, I.B. Mars Subsurface Water Ice Mapping (SWIM): The SWIM Equation and Project Infrastructure. In Proceedings of the Lunar and Planetary Science Conference, The Woodlands, TX, USA, 22 March 2019; Volume 50.
152. Evidence of giant tsunami on Mars suggests an early ocean. Available online: https://phys.org/news/2017-03-evidence-giant-tsunami-mars-early.html (accessed on 8 May 2019).
153. Andrews, R.G. When a Mega-Tsunami Drowned Mars, This Spot May Have Been Ground Zero. Available online: https://www.nytimes.com/2019/07/30/science/mars-tsunami-crater.html (accessed on 31 July 2019).
154. Costard, F.; Séjourné, A.; Lagain, A.; Ormö, J.; Rodriguez, J.; Clifford, S.; Bouley, S.; Kelfoun, K.; Lavigne, F. The Lomonosov Crater Impact Event: A Possible Mega-Tsunami Source on Mars. *J. Geophys. Res. Planets* 2019, *124*, 1840–1851.
155. Boynton, W.V.; Taylor, G.J.; Evans, L.G.; Reedy, R.; Starr, R.; Janes, D.M.; Kerry, K.E.; Drake, D.M.; Kim, K.J.; Williams, R.M.S.; et al. Concentration of H, Si, Cl, K, Fe, and Th in the low- and mid-latitude regions of Mars. *J. Geophys. Res. Space Phys.* 2007, *112*.
156. Feldman, W.C.; Prettyman, T.H.; Maurice, S.; Plaut, J.J.; Bish, D.L.; Vaniman, D.T.; Mellon, M.T.; Metzger, A.E.; Squyres, S.W.; Karunatillake, S.; et al. Global distribution of near-surface hydrogen on Mars. *J. Geophys. Res. Space Phys.* 2004, *109*.

157. NASA Staff. Mars' South Pole Ice Deep and Wide. Available online: https://www.nasa.gov/mission_pages/mars/news/mars-20070315.html (accessed on 15 March 2007).
158. Armstrong, J.; Titus, T.N.; Kieffer, H.H. Evidence for subsurface water ice in Korolev crater, Mars. *Icarus* 2005, *174*, 360–372.
159. Plaut, J.J.; Picardi, G.; Safaeinili, A.; Ivanov, A.; Milkovich, S.M.; Cicchetti, A.; Kofman, W.; Mouginot, J.; Farrell, W.M.; Phillips, R.J.; et al. Subsurface Radar Sounding of the South Polar Layered Deposits of Mars. *Science* 2007, *316*, 92–95.
160. Scanlon, K.; Head, J.; Fastook, J.; Wordsworth, R.D. The Dorsa Argentea Formation and the Noachian-Hesperian climate transition. *Icarus* 2018, *299*, 339–363.
161. Head, J.; Pratt, S. Extensive Hesperian-aged south polar ice sheet on Mars: Evidence for massive melting and retreat, and lateral flow and ponding of meltwater. *J. Geophys. Res. Space Phys.* 2001, *106*, 12275–12299.
162. Johnson, J. There's Water on Mars, NASA Confirms. Available online: https://www.latimes.com/archives/la-xpm-2008-aug-01-sci-phoenix1-story.html (accessed on 1 August 2008).
163. On Orbit. Radar Map of Buried Mars Layers Matches Climate Cycles. Available online: https://web.archive.org/web/20101221190147/http:/onorbit.com/node/1524 (accessed on 21 December 2010).
164. Fishbaugh, K.E.; Byrne, S.; Herkenhoff, K.E.; Kirk, R.L.; Fortezzo, C.; Russell, P.S.; McEwen, A. Evaluating the meaning of "layer" in the martian north polar layered deposits and the impact on the climate connection. *Icarus* 2010, *205*, 269–282.
165. German Aerospace Center (DLR). A Winter Wonderland in red and white – Korolev Crater on Mars. Available online: https://www.dlr.de/dlr/en/desktopdefault.aspx/tabid-10081/151_read-31614/#/gallery/33106 (accessed on 20 December 2018).
166. Sample, I. Mars Express Beams Back Images of Ice-Filled Korolev Crater. Available online: https://www.theguardian.com/science/2018/dec/21/mars-express-beams-back-images-of-ice-filled-korolev-crater (accessed on 21 December 2018).
167. Duxbury, N.S.; Zotikov, I.A.; Nealson, K.H.; Romanovsky, V.E.; Carsey, F.D. A numerical model for an alternative origin of Lake Vostok and its exobiological implications for Mars. *J. Geophys. Res. Space Phys.* 2001, *106*, 1453–1462.
168. Orosei, R.; Lauro, S.E.; Pettinelli, E.; Cicchetti, A.; Coradini, M.; Cosciotti, B.; Di Paolo, F.; Flamini, E.; Mattei, E.; Pajola, M.; et al. Radar evidence of subglacial liquid water on Mars. *Science* 2018, *361*, 490–493.
169. Catling, D.C.; Claire, M.W.; Zahnle, K.J.; Quinn, R.C.; Clark, B.C.; Hecht, M.H.; Kounaves, S.P. Atmospheric origins of perchlorate on Mars and in the Atacama. *J. Geophys. Res. Space Phys.* 2010, *115*.
170. Huge Reservoir of Liquid Water Detected Under the Surface of Mars. Available online: https://www.eurekalert.org/pub_releases/2018-07/aaft-hro072318.php (accessed on 25 July 2018).
171. Halton, M. Liquid Water 'Lake' Revealed on Mars. Available online: https://www.bbc.com/news/science-environment-44952710 (accessed on 25 July 2018).
172. Lucchitta, B.K. Mars and Earth: Comparison of cold-climate features. *Icarus* 1981, *45*, 264–303.
173. Grossman, L. Mars (probably) Has A Lake of Liquid Water. Available online: https://www.sciencenews.org/article/mars-may-have-lake-liquid-water-search-life (accessed on 25 July 2018).
174. Giant Liquid Water Lake Found under Martian Ice. Available online: https://www.rte.ie/news/2018/0725/981031-mars-lake/ (accessed on 25 July 2018).

175. Kieffer, H.H.; Jakosky, B.M.; Snyder, C.W.; Matthews, M.S. (Eds.) *Mars*; University of Arizona Press: Tuscon, AZ, USA; London, UK, 2011; p. 1498. ISBN 978-0-8165-1257-7.
176. Howell, E. Water Ice Mystery Found at Martian Equator. Available online: https://www.space.com/38330-water-ice-mystery-at-mars-equator.html (accessed on 2 October 2017).
177. Dundas, C.; Byrne, S.; McEwen, A.S. Modeling the development of martian sublimation thermokarst landforms. *Icarus* 2015, *262*, 154–169.
178. Head, J.W.; Mustard, J.F.; Kreslavsky, M.; Milliken, R.E.; Marchant, D.R. Recent ice ages on Mars. *Nature* 2003, *426*, 797–802.
179. Lefort, A.; Russell, P.; Thomas, N. Scalloped terrains in the Peneus and Amphitrites Paterae region of Mars as observed by HiRISE. *Icarus* 2010, *205*, 259–268.
180. Squyres, S.W. Urey prize lecture: Water on Mars. *Icarus* 1989, *79*, 229–288.
181. Conway, S.J.; Hovius, N.; Barnie, T.; Besserer, J.; Le Mouélic, S.; Orosei, R.; Read, N.A. Climate-driven deposition of water ice and the formation of mounds in craters in Mars' north polar region. *Icarus* 2012, *220*, 174–193.
182. NASA Staff. Steep Slopes on Mars Reveal Structure of Buried Ice. Available online: https://www.jpl.nasa.gov/news/news.php?feature=7038 (accessed on 11 January 2018).
183. Dundas, C.; Bramson, A.M.; Ojha, L.; Wray, J.J.; Mellon, M.T.; Byrne, S.; McEwen, A.S.; Putzig, N.E.; Viola, D.; Sutton, S.S.; et al. Exposed subsurface ice sheets in the Martian mid-latitudes. *Science* 2018, *359*, 199–201.
184. Voosen, P. Ice cliffs spotted on Mars. *Science* 2018.
185. The University of Arizona- HiRISE Dissected Mantled Terrain (PSP_002917_2175). Available online: https://hirise.lpl.arizona.edu/PSP_002917_2175 (accessed on 19 December 2010).
186. Wall, M. Huge Underground Ice Deposit on Mars Is Bigger Than New Mexico. Available online: https://www.space.com/34811-mars-ice-more-water-than-lake-superior.html (accessed on 22 November 2016).
187. Bramson, A.M.; Byrne, S.; Putzig, N.E.; Sutton, S.; Plaut, J.J.; Brothers, T.; Holt, J.W. Widespread excess ice in Arcadia Planitia, Mars. *Geophys. Res. Lett.* 2015, *42*, 6566–6574.
188. Widespread, Thick Water Ice Found in Utopia Planitia, Mars. Available online: https://web.archive.org/web/20161130042608/https:/planetarycassie.com/2016/11/04/widespread-thick-water-ice-found-in-utopia-planitia-mars/ (accessed on 29 November 2016).
189. Stuurman, C.M.; Osinski, G.R.; Holt, J.W.; Levy, J.; Brothers, T.; Kerrigan, M.; Campbell, B.A. SHARAD detection and characterization of subsurface water ice deposits in Utopia Planitia, Mars. *Geophys. Res. Lett.* 2016, *43*, 9484–9491.
190. Byrne, S.; Ingersoll, A.P. A Sublimation Model for the Formation of the Martian Polar Swiss-cheese Features. *Am. Astron. Soc.* 2002, *34*, 837, Bibcode: 2002DPS.34.0301B.
191. Water Ice in Crater at Martian North Pole. Available online: http://www.esa.int/Our_Activities/Space_Science/Mars_Express/Water_ice_in_crater_at_Martian_north_pole (accessed on 27 July 2005).
192. Ice Lake Found on the Red Planet. Available online: http://news.bbc.co.uk/2/hi/science/nature/4727847.stm (accessed on 29 July 2005).
193. Murray, J.B.; The HRSC Co-Investigator Team; Muller, J.-P.; Neukum, G.; Werner, S.C.; Van Gasselt, S.; Hauber, E.; Markiewicz, W.J.; Head, J.W.; Foing, B.H. Evidence from the Mars Express High Resolution Stereo Camera for a frozen sea close to Mars' equator. *Nature* 2005, *434*, 352–356.

194. Orosei, R.; Cartacci, M.; Cicchetti, A.; Noschese, R.; Federico, C.; Frigeri, A.; Flamini, E.; Holt, J.W.; Marinangeli, L.; Pettinelli, E.; et al. Radar subsurface sounding over the putative frozen sea in Cerberus Palus, Mars. In Proceedings of the XIII Internarional Conference on Ground Penetrating Radar, Lecce, Italy, 30 June 2010; Institute of Electrical and Electronics Engineers (IEEE): Piscataway, NJ, USA, 2010.
195. Barlow, N. *Mars: An Introduction to Its Interior, Surface and Atmosphere*; Cambridge University Press (CUP): Cambridge, UK, 2008; ISBN 978-0-521-85226-5.
196. Strom, R.G.; Croft, S.K.; Barlow, N.G. THE MARTIAN IMPACT CRATERING RECORD. *Mars* 2018, *A93-27852 09-91*, 383–423.
197. ESA—Mars Express—Breathtaking Views of Deuteronilus Mensae on Mars. Available online: http://www.esa.int/Our_Activities/Space_Science/Mars_Express/Breathtaking_views_of_Deuteronilus_Mensae_on_Mars (accessed on 14 March 2005).
198. Hauber, E.; The HRSC Co-Investigator Team; Van Gasselt, S.; Ivanov, B.; Werner, S.; Head, J.W.; Neukum, G.; Jaumann, R.; Greeley, R.; Mitchell, K.L. Discovery of a flank caldera and very young glacial activity at Hecates Tholus, Mars. *Nature* 2005, *434*, 356–361.
199. Humayun, M.; Nemchin, A.; Zanda, B.; Hewins, R.H.; Grange, M.; Kennedy, A.; Lorand, J.-P.; Göpel, C.; Fieni, C.; Pont, S.; et al. Origin and age of the earliest Martian crust from meteorite NWA 7533. *Nature* 2013, *503*, 513–516.
200. Shean, D.; Head, J.; Marchant, D.R. Origin and evolution of a cold-based tropical mountain glacier on Mars: The Pavonis Mons fan-shaped deposit. *J. Geophys. Res. Space Phys.* 2005, *110*.
201. Basilevsky, A.T.; Werner, S.C.; Neukum, G.; Head, J.W.; Van Gasselt, S.; Gwinner, K.; Ivanov, B. Geologically recent tectonic, volcanic and fluvial activity on the eastern flank of the Olympus Mons volcano, Mars. *Geophys. Res. Lett.* 2006, *33*, 13201.
202. Milliken, R.E.; Mustard, J.; Goldsby, D.L. Viscous flow features on the surface of Mars: Observations from high-resolution Mars Orbiter Camera (MOC) images. *J. Geophys. Res. Space Phys.* 2003, *108*, 5057.
203. Arfstrom, J.; Hartmann, W.K. Martian flow features, moraine-like ridges, and gullies: Terrestrial analogs and interrelationships. *Icarus* 2005, *174*, 321–335.
204. Head, J.W.; The HRSC Co-Investigator Team; Neukum, G.; Jaumann, R.; Hiesinger, H.; Hauber, E.; Carr, M.; Masson, P.; Foing, B.; Hoffmann, H. Tropical to mid-latitude snow and ice accumulation, flow and glaciation on Mars. *Nature* 2005, *434*, 346–351.
205. Mars' Climate in Flux: Mid-Latitude Glaciers. Available online: http://www.spaceref.com/news/viewpr.html?pid=18050 (accessed on 17 October 2005).
206. Berman, D.C.; Hartmann, W.K.; Crown, D.A.; Baker, V.R. The role of arcuate ridges and gullies in the degradation of craters in the Newton Basin region of Mars. *Icarus* 2005, *178*, 465–486.
207. The University of Arizona. Fretted Terrain Valley Traverse. Available online: https://hirise.lpl.arizona.edu/PSP_009719_2230 (accessed on 16 January 2012).
208. Barnes, J.W.; O'Brien, D.P.; Fortney, J.J.; Hurford, T.A. Superiority of the Lunar and Planetary Laboratory (LPL) over Steward Observatory (SO) at the University of Arizona. *arXiv* 2002, arXiv:astro-ph/0204013.
209. Jakosky, B.; Phillips, R.J. Mars' volatile and climate history. *Nature* 2001, *412*, 237–244.
210. Chaufray, J.; Modolo, R.; Leblanc, F.; Chanteur, G.; Johnson, R.E.; Luhmann, J.G. Mars solar wind interaction: Formation of the Martian corona and atmospheric loss to space. *J. Geophys. Res. Space Phys.* 2007, *112*, 09009.

211. Chevrier, V.; Poulet, F.; Bibring, J.-P. Early geochemical environment of Mars as determined from thermodynamics of phyllosilicates. *Nature* 2007, *448*, 60–63.
212. Morris, R.V.; Golden, D.C.; Ming, D.W.; Shelfer, T.D.; Jørgensen, L.C.; Bell, J.F., III; Graff, T.G.; Mertzman, S.A.; Bishop, J.L.; Ming, D.W.; et al. Mineralogy, composition, and alteration of Mars Pathfinder rocks and soils: Evidence from multispectral, elemental, and magnetic data on terrestrial analogue, SNC meteorite, and Pathfinder samples. *J. Geophys. Res.* 2000, *105*, 1757–1817.
213. Chévrier, V.; Mathé, P.-E.; Rochette, P.; Grauby, O.; Bourrié, G.; Trolard, F. Iron weathering products in a CO2+(H2O or H2O2) atmosphere: Implications for weathering processes on the surface of Mars. *Geochim. et Cosmochim. Acta* 2006, *70*, 4295–4317.
214. Catling, D.C. Mars: Ancient fingerprints in the clay. *Nature* 2007, *448*, 31–32.
215. Bibring, J.-P.; Langevin, Y.; Mustard, J.F.; Poulet, F.; Arvidson, R.; Gendrin, A.; Gondet, B.; Mangold, N.; Pinet, P.; Forget, F.; et al. Global Mineralogical and Aqueous Mars History Derived from OMEGA/Mars Express Data. *Science* 2006, *312*, 400–404.
216. McEwen, A.S.; Hansen, C.J.; Delamere, W.A.; Eliason, E.M.; Herkenhoff, K.E.; Keszthelyi, L.; Gulick, V.C.; Kirk, R.L.; Mellon, M.T.; Grant, J.A.; et al. A Closer Look at Water-Related Geologic Activity on Mars. *Science* 2007, *317*, 1706–1709.
217. Schorghofer, N. Dynamics of ice ages on Mars. *Nature* 2007, *449*, 192–194.
218. Laskar, J.; Levrard, B.; Mustard, J.F. Orbital forcing of the Martian polar layered deposits. *Nature* 2002, *419*, 375–377.
219. Dickson, J.L.; Head, J.W.; Marchant, D.R. Late Amazonian glaciation at the dichotomy boundary on Mars: Evidence for glacial thickness maxima and multiple glacial phases. *Geology* 2008, *36*, 411.
220. Willmes, M.; Reiss, D.; Hiesinger, H.; Zanetti, M. Surface age of the ice–dust mantle deposit in Malea Planum, Mars. *Planet. Space Sci.* 2012, *60*, 199–206.
221. Levrard, B.; Forget, F.; Montmessin, F.; Laskar, J. Recent ice-rich deposits formed at high latitudes on Mars by sublimation of unstable equatorial ice during low obliquity. *Nature* 2004, *431*, 1072–1075.
222. Mars May Be Emerging from An Ice Age. Available online: https://www.sciencedaily.com/releases/2003/12/031218075443.htm (accessed on 18 December 2003).
223. Forget, F.; Haberle, R.M.; Montmessin, F.; Levrard, B.; Head, J.W.; Liou, J.-C.; Johnson, N.L. Formation of Glaciers on Mars by Atmospheric Precipitation at High Obliquity. *Science* 2006, *311*, 368–371.
224. Mustard, J.; Cooper, C.D.; Rifkin, M.K. Evidence for recent climate change on Mars from the identification of youthful near-surface ground ice. *Nature* 2001, *412*, 411–414.
225. Kreslavsky, M.; Head, J.W. Mars: Nature and evolution of young latitude-dependent water-ice-rich mantle. *Geophys. Res. Lett.* 2002, *29*, 14-1–14-4.
226. Beatty, K. Water Ice Found Exposed in Martian Cliffs—Sky & Telescope. Available online: https://www.skyandtelescope.com/astronomy-news/cliffs-reveal-water-ice-on-mars/ (accessed on 23 January 2018).
227. Heldmann, J.L.; Toon, O.B.; Pollard, W.H.; Mellon, M.T.; Pitlick, J.; McKay, C.P.; Andersen, D.T. Formation of Martian gullies by the action of liquid water flowing under current Martian environmental conditions. *J. Geophys. Res. Space Phys.* 2005, *110*.
228. Malin, M.C.; Edgett, K.S.; Posiolova, L.V.; McColley, S.M.; Dobrea, E.Z.N.; Scharlemann, J.P.W.; Laurance, W.F. Present-Day Impact Cratering Rate and Contemporary Gully Activity on Mars. *Science* 2006, *314*, 1573–1577.

229. Head, J.W.; Marchant, D.R.; Kreslavsky, M. Formation of gullies on Mars: Link to recent climate history and insolation microenvironments implicate surface water flow origin. *Proc. Natl. Acad. Sci. USA* 2008, *105*, 13258–13263.
230. Henderson, M. Water Has been Flowing on Mars within the Past Five Years, Nasa says. Available online: https://www.thetimes.co.uk (accessed on 7 December 2006).
231. Malin, M.C.; Edgett, K.S. Evidence for Recent Groundwater Seepage and Surface Runoff on Mars. *Science* 2000, *288*, 2330–2335.
232. Kolb, K.J.; Pelletier, J.D.; McEwen, A.S. Modeling the formation of bright slope deposits associated with gullies in Hale Crater, Mars: Implications for recent liquid water. *Icarus* 2010, *205*, 113–137.
233. Hoffman, N. Active Polar Gullies on Mars and the Role of Carbon Dioxide. *Astrobiology* 2002, *2*, 313–323.
234. Musselwhite, D.S.; Swindle, T.; Lunine, J.I. Liquid CO2 breakout and the formation of recent small gullies on Mars. *Geophys. Res. Lett.* 2001, *28*, 1283–1285.
235. McEwen, A.S.; Ojha, L.; Dundas, C.; Mattson, S.S.; Byrne, S.; Wray, J.J.; Cull, S.C.; Murchie, S.L.; Thomas, N.; Gulick, V.C. Seasonal Flows on Warm Martian Slopes. *Science* 2011, *333*, 740–743.
236. Nepali Scientist Lujendra Ojha Spots Possible Water on Mars. Available online: https://web.archive.org/web/20130604112105/http:/nepaliblogger.com/news/nepali-scientist-lujendra-ojha-spots-possible-water-on-mars/2793/ (accessed on 6 August 2011).
237. NASA Spacecraft Data Suggest Water Flowing on Mars. Available online: https://www.nasa.gov/mission_pages/MRO/news/mro20110804.html (accessed on 4 August 2011).
238. Conrad, P.G.; Archer, D.; Coll, P.; De La Torre, M.; Edgett, K.; Eigenbrode, J.L.; Fisk, M.; Freissenet, C.; Franz, H.; Glavin, D.P.; et al. Habitability Assessment at Gale Crater: Implications from Initial Results. In Proceedings of the Lunar and Planetary Science Conference, The Woodlands, TX, USA, 18–22 March 2013; Bibcode: 2013LPI. 44.2185C. Volume 1719, p. 2185.
239. National Research Council. *Planetary Protection for Mars Missions*; The National Academies Press: Washington, DC, USA, 2007; pp. 95–98. ISBN 978-0-309-10851-5.
240. Daley, J. Mars Surface May Be Too Toxic for Microbial Life—The Combination of UV Radiation and Perchlorates Common on Mars Could be Deadly for Bacteria. Available online: https://www.smithsonianmag.com/smart-news/mars-surface-may-be-toxic-bacteria-180963966/ (accessed on 8 July 2017).
241. Wadsworth, J.; Cockell, C.S. Perchlorates on Mars enhance the bacteriocidal effects of UV light. *Sci. Rep.* 2017, 7, 4662.
242. NASA Staff. NASA Astrobiology Strategy. Available online: https://web.archive.org/web/20161222190306/https:/nai.nasa.gov/media/medialibrary/2015/10/NASA_Astrobiology_Strategy_2015_151008.pdf (accessed on 22 December 2016).
243. Cunningham, G. Mars exploration missions. In Proceedings of the Space Programs and Technologies Conference, Huntsville, AL, USA, 27 September 1990.
244. Viking Orbiter Views of Mars - Channels. Available online: https://history.nasa.gov/SP-441/ch4.htm (accessed on 19 December 2010).
245. Viking Orbiter Views of Mars - Volcanic Features. Available online: https://history.nasa.gov/SP-441/ch5.htm (accessed on 19 December 2010).
246. Viking Orbiter Views of Mars - Craters. Available online: https://history.nasa.gov/SP-441/ch7.htm (accessed on 19 December 2010).

247. Morton, O. *Mapping Mars: Science, Imagination, and the Birth of a World*; Picador: New York, NY, USA, 2002.
248. Arvidson, R.E.; Gooding, J.L.; Moore, H.J. The Martian surface as imaged, sampled, and analyzed by the Viking landers. *Rev. Geophys.* 1989, *27*, 39.
249. Clark, B.C.; Baird, A.K.; Rose, H.J.; Toulmin, P.; Keil, K.; Castro, A.J.; Kelliher, W.C.; Rowe, C.D.; Evans, P.H. Inorganic Analyses of Martian Surface Samples at the Viking Landing Sites. *Science* 1976, *194*, 1283–1288.
250. Keller, J.M.; Boynton, W.V.; Karunatillake, S.; Baker, V.R.; Dohm, J.M.; Evans, L.G.; Finch, M.J.; Hahn, B.C.; Hamara, D.K.; Janes, D.M.; et al. Equatorial and midlatitude distribution of chlorine measured by Mars Odyssey GRS. *J. Geophys. Res. Planets* 2006, *111*.
251. Hoefen, T.M.; Clark, R.N.; Bandfield, J.L.; Smith, M.; Pearl, J.C.; Christensen, P.R. Discovery of Olivine in the Nili Fossae Region of Mars. *Science* 2003, *302*, 627–630.
252. Malin, M.C.; Edgett, K.S. Mars Global Surveyor Mars Orbiter Camera: Interplanetary cruise through primary mission. *J. Geophys. Res. Space Phys.* 2001, *106*, 23429–23570.
253. Golombek, M.P.; Cook, R.A.; Economou, T.; Folkner, W.M.; Haldemann, A.F.C.; Kallemeyn, P.H.; Knudsen, J.M.; Manning, R.M.; Moore, H.J.; Parker, T.J.; et al. Overview of the Mars Pathfinder Mission and Assessment of Landing Site Predictions. *Science* 1997, *278*, 1743–1748.
254. NASA Staff. Mars Odyssey: Newsroom. Available online: https://mars.jpl.nasa.gov/odyssey/newsroom/pressreleases/20020528a.html (accessed on 28 May 2002).
255. Murche, S.; Mustard, J.; Bishop, J.; Head, J.; Pieters, C.; Erard, S. Spatial Variations in the Spectral Properties of Bright Regions on Mars. *Icarus* 1993, *105*, 454–468.
256. Bell, J.F., III. Home Page for Bell (1996) Geochemical Society Paper. Available online: http://marswatch.sese.asu.edu/burns.html (accessed on 19 December 2010).
257. Mitrofanov, I.; Anfimov, D.; Kozyrev, A.S.; Litvak, M.; Sanin, A.; Tret'Yakov, V.; Krylov, A.; Shvetsov, V.; Boynton, W.; Shinohara, C.; et al. Maps of Subsurface Hydrogen from the High Energy Neutron Detector, Mars Odyssey. *Science* 2002, *297*, 78–81.
258. Boynton, W.V.; Feldman, W.C.; Squyres, S.W.; Prettyman, T.H.; Brückner, J.; Evans, L.G.; Reedy, R.; Starr, R.; Arnold, J.R.; Drake, D.M.; et al. Distribution of Hydrogen in the Near Surface of Mars: Evidence for Subsurface Ice Deposits. *Science* 2002, *297*, 81–85.
259. Smith, P.; Tamppari, L.; Arvidson, R.E.; Bass, D.; Blaney, D.; Boynton, W.; Carswell, A.; Catling, D.C.; Clark, B.; Duck, T.; et al. Introduction to special section on the Phoenix Mission: Landing Site Characterization Experiments, Mission Overviews, and Expected Science. *J. Geophys. Res. Space Phys.* 2008, *113*.
260. NASA Staff. NASA Data Shed New Light About Water and Volcanoes on Mars. Available online: https://www.nasa.gov/mission_pages/phoenix/news/phx20100909.html (accessed on 9 September 2010).
261. Mellon, M.T.; Jakosky, B. Geographic variations in the thermal and diffusive stability of ground ice on Mars. *J. Geophys. Res. Space Phys.* 1993, *98*, 3345–3364.
262. NASA Staff. Confirmation of Water on Mars. Available online: https://www.nasa.gov/mission_pages/phoenix/news/phoenix-20080620.html (accessed on 20 June 2008).
263. Martínez, G.M.; Renno, N.O. Water and Brines on Mars: Current Evidence and Implications for MSL. *Space Sci. Rev.* 2013, *175*, 29–51.

264. Renno, N.O.; Bos, B.J.; Catling, D.C.; Clark, B.C.; Drube, L.; Fisher, D.; Goetz, W.; Hviid, S.F.; Keller, H.U.; Kok, J.F.; et al. Possible physical and thermodynamical evidence for liquid water at the Phoenix landing site. *J. Geophys. Res. Space Phys.* 2009, *114*.
265. Chang, K. Blobs in Photos of Mars Lander Stir a Debate: Are They Water? Available online: https://www.nytimes.com/2009/03/17/science/17mars.html?mtrref=undefined&gwh=ED1046D72C9EBD5704CA56E153D00155&gwt=pay&assetType=REGIWALL (accessed on 16 March 2009).
266. Liquid Saltwater Is Likely Present on Mars, New Analysis Shows. Available online: https://www.sciencedaily.com/releases/2009/03/090319232438.htmScienceDaily (accessed on 20 March 2009).
267. Astrobiology Top 10: Too Salty to Freeze. Available online: https://www.astrobio.net/?option=com_retrospection&task=detail&id=3350 (accessed on 19 December 2010).
268. Hecht, M.H.; Kounaves, S.P.; Quinn, R.C.; West, S.J.; Young, S.M.M.; Ming, D.W.; Catling, D.C.; Clark, B.C.; Boynton, W.V.; Hoffman, J.; et al. Detection of Perchlorate and the Soluble Chemistry of Martian Soil at the Phoenix Lander Site. *Science* 2009, *325*, 64–67.
269. Smith, P.; Tamppari, L.K.; Arvidson, R.E.; Bass, D.; Blaney, D.; Boynton, W.V.; Carswell, A.; Catling, D.C.; Clark, B.C.; Duck, T.; et al. H2O at the Phoenix Landing Site. *Science* 2009, *325*, 58–61.
270. Whiteway, J.A.; Komguem, L.; Dickinson, C.; Cook, C.; Illnicki, M.; Seabrook, J.; Popovici, V.; Duck, T.J.; Davy, R.; Taylor, P.A.; et al. Mars Water-Ice Clouds and Precipitation. *Science* 2009, *325*, 68–70.
271. CSA—News Release. Available online: https://web.archive.org/web/20110705011110/http:/www.asc-csa.gc.ca/eng/media/news_releases/2009/0702.asp (accessed on 5 July 2011).
272. NASA Staff. NASA—Mars Rover Spirit Unearths Surprise Evidence of Wetter Past. Available online: https://www.nasa.gov/mission_pages/mer/mer-20070521.html (accessed on 21 May 2007).
273. Bertster, G. *Mars Rover Investigates Signs of Steamy Martian Past*; Press Release. Jet Propulsion Laboratory: Pasadena, CA, USA, 2007.
274. Schroder, C. *"Journal of Geophysical Research" (Abstr.)*. General Assembly: 10254. In Proceedings of the 7th European Geosciences Union, Munich, Germany, 8 April 2005.
275. Morris, R.V.; Klingelhöfer, G.; Schröder, C.; Rodionov, D.S.; Yen, A.; Ming, D.W.; De Souza, P.A.; Fleischer, I.; Wdowiak, T.; Gellert, R.; et al. Mössbauer mineralogy of rock, soil, and dust at Gusev crater, Mars: Spirit's journey through weakly altered olivine basalt on the plains and pervasively altered basalt in the Columbia Hills. *J. Geophys. Res. Space Phys.* 2006, *111*, 111.
276. Ming, D.W.; Mittlefehldt, D.W.; Morris, R.V.; Golden, D.C.; Gellert, R.; Yen, A.; Clark, B.C.; Squyres, S.W.; Farrand, W.H.; Ruff, S.W.; et al. Geochemical and mineralogical indicators for aqueous processes in the Columbia Hills of Gusev crater, Mars. *J. Geophys. Res. Space Phys.* 2006, *111*.
277. Bell, J. (Ed.) *The Martian Surface*; Cambridge University Press: New York, NY, USA, 2008.
278. Morris, R.V.; Ruff, S.W.; Gellert, R.; Ming, D.W.; Arvidson, R.E.; Clark, B.C.; Golden, D.C.; Siebach, K.; Klingelhofer, G.; Schroder, C.; et al. Outcrop of Long-Sought Rare Rock on Mars Found. *Science* 2010, *329*, 421–424.
279. NASA Staff. Opportunity Rover Finds Strong Evidence Meridiani Planum Was Wet. Available online: https://mars.jpl.nasa.gov/mer/newsroom/pressreleases/20040302a.html (accessed on 8 July 2006).

280. Harwood, W. Opportunity Rover Moves Into 10th Year of Mars Operations. Available online: https://spaceflightnow.com/news/n1301/25opportunity/ (accessed on 25 January 2013).
281. Benison, K.C.; LaClair, D.A. Modern and Ancient Extremely Acid Saline Deposits: Terrestrial Analogs for Martian Environments? *Astrobiology* 2003, *3*, 609–618.
282. Benison, K.; Bowen, B. Acid saline lake systems give clues about past environments and the search for life on Mars. *Icarus* 2006, *183*, 225–229.
283. Osterloo, M.M.; Hamilton, V.E.; Bandfield, J.L.; Glotch, T.D.; Baldridge, A.M.; Christensen, P.R.; Tornabene, L.; Anderson, F.S.; Boudreau, B.P.; Arnosti, C.; et al. Chloride-Bearing Materials in the Southern Highlands of Mars. *Science* 2008, *319*, 1651–1654.
284. Grotzinger, J.; Milliken, R. (Eds.) *Sedimentary Geology of Mars*; SEPM (Society for Sedimentary Geology): Tulsa, OK, USA, 2010; pp. 1–48.
285. The University of Arizona. HiRISE—High Resolution Imaging Science Experiment. Available online: https://hirise.lpl.arizona.edu/?PSP_008437_1750 (accessed on 19 December 2010).
286. Target Zone: Nilosyrtis? Available online: http://themis.mars.asu.edu/feature/49 (accessed on 19 December 2010).
287. Mellon, M.T.; Jakosky, B.; Postawko, S.E. The persistence of equatorial ground ice on Mars. *J. Geophys. Res. Space Phys.* 1997, *102*, 19357–19369.
288. Arfstrom, J.D. *A Conceptual Model of Equatorial Ice Sheets on Mars. Comparative Climatology of Terrestrial Planets*; UA Press: Tucson, AZ, USA, 2012; p. 1675.
289. Byrne, S.; Dundas, C.; Kennedy, M.R.; Mellon, M.T.; McEwen, A.S.; Cull, S.C.; Daubar, I.J.; Shean, D.; Seelos, K.; Murchie, S.L.; et al. Distribution of Mid-Latitude Ground Ice on Mars from New Impact Craters. *Science* 2009, *325*, 1674–1676.
290. Thompson, A. Water Ice Exposed in Mars Craters. Available online: https://www.space.com/7333-water-ice-exposed-mars-craters.html (accessed on 19 December 2010).
291. Nerozzi, S.; Holt, J.W. Buried Ice and Sand Caps at the North Pole of Mars: Revealing a Record of Climate Change in the Cavi Unit With SHARAD. *Geophys. Res. Lett.* 2019, *46*, 7278–7286.
292. Ojha, L.; Nerozzi, S.; Lewis, K.W. Compositional Constraints on the North Polar Cap of Mars from Gravity and Topography. *Geophys. Res. Lett.* 2019, *46*, 8671–8679.
293. Brown, D.; Webster, G.; Hoover, R. NASA Rover's First Soil Studies Help Fingerprint Martian Minerals. Available online: https://www.nasa.gov/home/hqnews/2012/oct/HQ_12-383_Curiosity_CheMin.html (accessed on 30 October 2012).
294. Brown, D.; Webster, G.; Neal-Jones, N. NASA Mars Rover Fully Analyzes First Martian Soil Samples. Available online: https://mars.jpl.nasa.gov/msl/news/whatsnew/index.cfm?FuseAction=ShowNews&NewsID=1399 (accessed on 3 December 2012).
295. Chang, K. Mars Rover Discovery Revealed. Available online: https://thelede.blogs.nytimes.com/2012/12/03/mars-rover-discoveryrevealed?mtrref=undefined&gwh=121F367F450884428ECD9DD534D7E7AA&gwt=pay&assetType=REGIWALL (accessed on 3 December 2012).
296. Webster, G.; Brown, D. Curiosity Mars Rover Sees Trend in Water Presence. Available online: https://mars.jpl.nasa.gov/msl/news/whatsnew/index.cfm?FuseAction=ShowNews&NewsID=1446 (accessed on 18 March 2013).
297. Rincon, P. Curiosity Breaks Rock to Reveal Dazzling White Interior. Available online: https://www.bbc.co.uk/news/science-environment-21340279 (accessed on 19 March 2013).

298. NASA Staff. Red Planet Coughs Up A White Rock, and Scientists Freak Out. Available online: https://web.archive.org/web/20130323164757/http:/now.msn.com/white-mars-rock-called-tintina-found-by-curiosity-rover (accessed on 23 March 2013).
299. Lieberman, J. Mars Water Found: Curiosity Rover Uncovers 'Abundant, Easily Accessible' Water in Martian Soil. Available online: http://www.isciencetimes.com/articles/6131/20130926/mars-water-soil-nasa-curiosity-rover-martian.htm (accessed on 26 September 2013).
300. Leshin, L.A.; Mahaffy, P.R.; Webster, C.R.; Cabané, M.; Coll, P.; Conrad, P.G.; Archer, P.D.; Atreya, S.K.; Brunner, A.E.; Buch, A.; et al. Volatile, Isotope, and Organic Analysis of Martian Fines with the Mars Curiosity Rover. *Science* 2013, *341*, 1238937.
301. Grotzinger, J.P. Analysis of Surface Materials by the Curiosity Mars Rover. *Science* 2013, *341*, 1475.
302. Neal-Jones, N.; Zubritsky, E.; Webster, G.; Martialay, M. Curiosity's SAM Instrument Finds Water and More in Surface Sample. Available online: https://www.nasa.gov/content/goddard/curiositys-sam-instrument-finds-water-and-more-in-surface-sample/ (accessed on 26 September 2013).
303. Webster, G.; Brown, D. Science Gains from Diverse Landing Area of Curiosity. Available online: https://www.nasa.gov/mission_pages/msl/news/msl20130926.html (accessed on 26 September 2013).
304. Chang, K. Hitting Pay Dirt on Mars. Available online: https://www.nytimes.com/2013/10/01/science/space/hitting-pay-dirt-on-mars.html (accessed on 1 October 2013).
305. Meslin, P.-Y.; Gasnault, O.; Forni, O.; Schroder, S.; Cousin, A.; Berger, G.; Clegg, S.M.; Lasue, J.; Maurice, S.; Sautter, V.; et al. Soil Diversity and Hydration as Observed by ChemCam at Gale Crater, Mars. *Science* 2013, *341*, 1238670.
306. Stolper, E.M.; Baker, M.B.; Newcombe, M.E.; Schmidt, M.E.; Treiman, A.H.; Cousin, A.; Dyar, M.D.; Fisk, M.R.; Gellert, R.; King, P.L.; et al. The Petrochemistry of Jake_M: A Martian Mugearite. *Science* 2013, *341*, 1239463.
307. Webster, G.; Neal-Jones, N.; Brown, D. NASA Rover Finds Active and Ancient Organic Chemistry on Mars. Available online: https://www.jpl.nasa.gov/news/news.php?release=2014-432 (accessed on 16 December 2014).
308. Chang, K. A Great Moment': Rover Finds Clue That Mars May Harbor Life. Available online: https://www.nytimes.com/2014/12/17/science/a-new-clue-in-the-search-for-life-onmars.html (accessed on 16 December 2014).
309. Mahaffy, P.R.; Webster, C.R.; Stern, J.C.; Brunner, A.E.; Atreya, S.K.; Conrad, P.G.; Domagal-Goldman, S.; Eigenbrode, J.L.; Flesch, G.J.; Christensen, L.E.; et al. The imprint of atmospheric evolution in the D/H of Hesperian clay minerals on Mars. *Science* 2014, *347*, 412–414.
310. Rincon, P. Evidence of Liquid Water Found on Mars. Available online: https://www.bbc.com/news/science-environment-32287609 (accessed on 15 April 2015).

Mars and the Paranormal

1. Strauss, David. *Percival Lowell: The Culture and Science of a Boston Brahmin.* Cambridge: Harvard UP, 2001. 138.
2. Oppenheim, Janet. *The Other World: Spiritualism and Psychical Research in England, 1850–1914.* Cambridge: Cambridge UP, 1985. 137.
3. Strauss, David. *Percival Lowell: The Culture and Science of a Boston Brahmin.* Cambridge: Harvard UP, 2001. 133.

4. Flammarion, Camille. "Spiritualism and Materialism: A Reply to Camille Saint-Saëns." 1900 Rpt. *Haunted Houses*. Trans. E. E. Fournier d'Albe. New York: Appleton, 1924. 13; emphasis in original.
5. Qtd in Haynes, Renée. *The Society for Psychical Research 1882–1982: A History*. London: Macdonald, 1982. 202.
6. Luckhurst, Roger. *The Invention of Telepathy, 1870–1901*. Oxford: Oxford UP, 2002. 10.
7. Viswanathan, Gauri. "The Ordinary Business of Occultism." *Critical Inquiry* 27 (Autumn 2000): 6.
8. Blackwelder, Eliot. "Mars as the Abode of Life." *Our Friends, the Enemy: A Discussion Bearing on Scientific Ethics, with Concrete Illustrations*. New York: American Association for the Advancement of Science, 1910. 3.
9. Viswanathan, Gauri. "The Ordinary Business of Occultism." *Critical Inquiry* 27 (Autumn 2000): 6.
10. Wells, H.G. *A Critical Edition of The War of the Worlds: H.G. Wells's Scientific Romance*. Ed. David Y. Hughes and Harry M. Geduld. Bloomington: Indiana UP, 1993. 152.
11. Burroughs, Edgar Rice. *A Princess of Mars*. 1912. Ed. John Seelye. New York: Penguin, 2007. 12.
12. Gaston, Henry A. *Mars Revealed, or Seven Days in the Spirit World*. San Francisco: A.L. Bancroft, 1880. 28.
13. Ibid., 8.
14. Ibid., 200–201.
15. Ibid., 106.
16. Ibid., 208.
17. Wicks, Mark. *To Mars via the Moon: An Astronomical Story*. Philadelphia: Lippincott, 1911. 204.
18. DuMaurier, George. *The Martian*. 1897. *Novels of George DuMaurier*, London: Pilot Press, 1947. 695.
19. Ibid., 674.
20. Flammarion, Camille. *Uranie*. 1889. Trans. Mary J. Serrano [as *Urania*]. New York: Cassell, 1890. 161.
21. Ibid., 166.
22. Ibid., 175.
23. Ibid., 190.
24. Ibid., 182.
25. Ibid., 185.
26. Ibid., 188–89.
27. Ibid., 185.
28. Ibid., 179.
29. Ibid., 252; emphasis in original.
30. Gratacap, Louis Pope. *The Certainty of a Future Life in Mars*. New York: Brentano, 1903. 61.
31. Ibid., 53.
32. Ibid., 40.
33. Ibid., 158.
34. Ibid., 128.
35. Ibid., 72.
36. Ibid., 79.
37. Ibid., 99.
38. Ibid., 104.
39. Ibid., 183; emphasis in original.

40. Weiss, Sara. *Journeys to the Planet Mars, or Our Mission to Ento.* Rochester: Austin Publishing, 1905. 439.
41. Markley, Robert. *Dying Planet: Mars in Science and the Imagination.* Durham: Duke UP, 2005. 117–18.
42. Weiss, Sara. *Journeys to the Planet Mars, or Our Mission to Ento.* Rochester: Austin Publishing, 1905. 387.
43. Ibid., 475–76.
44. Weiss, Sara. *Decimon Huydas: A Romance of Mars.* Rochester: Austin Publishing, 1906. 154.
45. Kennon, J. L., ed. *The Planet Mars and its Inhabitants: A Psychic Revelation by Iros Urides (A Martian).* Privately printed by Mabel Kean, 1922. n.p.
46. Ibid., n.p.
47. Ibid., 40.
48. Gilbert, J. W., *The Marsian.* New York: Fortuny's, 1940. 12.
49. Ibid., 147.
50. Ibid., 142.
51. Ibid., 108.
52. Stapledon, W. Olaf. *Last and First Men.* London, Meuthen, 1930. 160.
53. Bradbury, Ray. *The Martian Chronicles.* 1950. New York: Bantam, 1979. 21.
54. Shamdasani, Sonu. Introduction. *Des Indes à la Planète Mars: Etude sur un cas de somnambulisme avec glossolalie* by Théodore Flournoy. 1899. Trans. Daniel Vermilye [As *From India to the Planet Mars: A Case of Multiple Personality with Imaginary Languages*]. Princeton: Princeton UP, 1994. xxxi.
55. Flournoy, Théodore. *Des Indes à la Planete Mars: Etude sur un cas de somnambulisme avec glossolalie* by Theodore Flournoy. 1899. Trans. Daniel Vermilye [As *From India to the Planet Mars: A Case of Multiple Personality with Imaginary Languages*]. Ed. Sonu Shamdasani. Princeton: Princeton UP, 1994. 87.
56. Ibid., 120.
57. Ibid., 122.

Planetary Protection in the New Space Era: Science and Governance

1. COSPAR (2020a). COSPAR policy on planetary protection. *Space Res. Today* 208. Available at: https://cosparhq.cnes.fr/assets/uploads/2020/07/PPPolicy-June-2020_Final_Web.pdf (Accessed July 29, 2020).
2. OST (1967). Treaty on principles governing the activities of states in the exploration and use of outer space, including the moon and other celestial bodies. U.N.T.S. 610, at 205.
3. NASA (2020b). The Artemis Accords: principles for a safe, peaceful, and prosperous future. Available at: https://www.nasa.gov/specials/artemis-accords/img/Artemis-Accords_v7_print.pdf (Accessed July 29, 2020).
4. Ibid.
5. NASA (2020c). NASA interim directive: planetary protection categorization for robotic crewed Missions to the Earth's moon. Available at: https://nodis3.gsfc.nasa.gov/OPD_docs/NID_8715_128_.pdf (Accessed July 10, 2020).
6. NASA (2020d). NASA interim directive: biological planetary protection for human missions to Mars. Available at: https://nodis3.gsfc.nasa.gov/OPD_docs/NID_8715_129_.pdf (Accessed July 10, 2020).
7. COSPAR (2020a). COSPAR policy on planetary protection. *Space Res. Today* 208. Available at: https://cosparhq.cnes.fr/assets/uploads/2020/07/PPPolicy June-2020_Final_Web.pdf (Accessed July 29, 2020).
8. Tennen, L. I. (2004). Evolution of the planetary protection policy: conflict of science and jurisprudence? *Adv. Space Res.* 34, 2354–2362. doi:10.1016/j.asr.2004.01.018.

9. Coustenis, A., Kminek, G., and Hedman, N. (2019). The challenge of planetary protection. *Room.* Available at: https://room.eu.com/article/the-challenge-of-planetary-protection.
10. Rummel, J. D., and Billings, L. (2004). Issues in planetary protection: policy, protocol and implementation. *Space Pol.* 20 49–54. doi:10.1016/j.spacepol.2003.11.005.
11. Rettberg, P., Anesio, A. M., Baker, V. R., Baross, J. A., Cady, S. L., Detsis, E., et al. (2016). Planetary protection and Mars special regions: a suggestion for updating the definition. *Astrobiology* 16, 119. doi:10.1089/ast.2016.1472.
12. COSPAR (2020a). COSPAR policy on planetary protection. *Space Res. Today* 208. Available at: https://cosparhq.cnes.fr/assets/uploads/2020/07PPPolicy-June-2020_Final_Web.pdf (Accessed July 29, 2020).
13. COSPAR (2020b). Panel on planetary protection (PPP). Available at: https://cosparhq.cnes.fr/scientific-structure/panels/panel-on-planetary-protection-ppp/ (Accessed July 29, 2020).
14. Coustenis, A., Kminek, G., and Hedman, N. (2019). The challenge of planetary protection. *Room.* Available at: https://room.eu.com/article/the-challenge-of-planetary-protection.
15. Rummel, J. D. (2009). Special regions in Mars exploration: problems and potential. *Acta Astronautica.* 64, 1293–1297. doi:10.1016/j.actaastro.2009.01.006.
16. United Nations (2017). Report of the committee on the peaceful uses of outer space. A/72/20. Available at: https://cms.unov.org/dcpms2/api/finaldocuments?Language=en&Symbol=A/72/20 (Accessed July 28, 2020).
17. United Nations General Assembly. Resolution A/RES/1962 (XVIII) (1962). Declaration of legal principles governing the activities of states in the exploration and use of outer space.
18. Moon Agreement (1979). Agreement governing the activities of states on the moon and other celestial bodies 1979. U.N.T.S. 1363, at 3.
19. OST (1967). Treaty on principles governing the activities of states in the exploration and use of outer space, including the moon and other celestial bodies. U.N.T.S. 610, at 205.
20. VCLT (1969). Vienna convention on the law of treaties 1969. U.N.T.S.1155, at 332.
21. OED (2011). *Concise Oxford English Dictionary.* 12th Edn. Editors A. Stevenson and M. Waite (Oxford, UK: Oxford University Press); 308, 651.
22. Newman, C. J. (2015). Seeking tranquillity: embedding sustainability in lunar exploration policy. *Space Pol.* 33, 29.doi:10.1016/j.spacepol.2015.05.003.
23. Aoki, S. (2012). "The function of 'soft law' in the development of international space law," in: *Soft law in outer space: the function of non-binding norms in international space law.* Editor M. Irmgard (Vienna, Austria: Boehlau Verlag), 57–85.
24. Lewis, J. S. (1997). *Mining the sky: untold riches from the asteroids, comets and planets.* Reading, MA: Helix Books.
25. Foust, J. (2019a). The asteroid mining bubble has burst. *Space Review.* Available at: https://www.thespacereview.com/article/3633/1 (Accessed July 29, 2020).
26. Butler, J. (2006). Unearthly microbes and the laws esigned to resist them. *Georgia Law Rev.* 41, 1355.
27. Newman, C. J. (2015). Seeking tranquillity: embedding sustainability in lunar exploration policy. *Space Pol.* 33, 29. doi:10.1016/j.spacepol.2015.05.003.
28. Berger, E. (2015). When Elon Musk goes to Mars, he won't be overly troubled by planetary protection. *ArsTechnica.* Available at: https://arstechnica.com/science/2015/12/when-elon-musk-goes-to-mars-he-wont-be-troubled-by-planetary-protection/ (Accessed July 29, 2020). doi:10.5771/9783845262635.
29. Foust, J. (2019b). If at first you don't succeed. *Space Review.* Available at: https://www.thespacereview.com/article/3694/1 (Accessed July 29, 2019).

30. Oberhaus, D. (2019). A crashed Israeli lunar lander spilled tardigrades on the moon. *Wired*. Available at: https://www.wired.com/story/a-crashed-israeli-lunar-lander-spilled-tardigrades-on-the-moon/ (Accessed July 29, 2020).
31. Taylor, C. (2019). 'I'm the first space pirate!' how tardigrades were secretly smuggled to the moon. Mashable, United Kingdom. Available at: https://mashable.com/article/smuggled-moon-tardigrade/?europe=true (Accessed July 29, 2020).
32. Johnson, C. D., Porras, D., Christopher, M. H, and O'Sullivan, S. (2019). The curious case of the transgressing tardigrades. *Space Review*. Available at: https://www.thespacereview.com/article/3783/1 (Accessed July 29, 2020). doi:10.1287/5f4e34c9-55a8-4667-b696-4d50c9931b3e.
33. NASA (2019). Report to NASA/SMD: final report. NASA planetary protection independent review board. Available at: https://www.nasa.gov/sites/default/files/atoms/files/planetary_protection_board_report_20191018.pdf (Accessed July 29, 2020).
34. NASA (2020a). NASA "commercial lunar payload services overview". Available at: https://www.nasa.gov/content/commercial-lunar-payload-services-overview (Accessed July 29, 2020).
35. Goh, G. M., and Softly, S. (2008). Catchee monkey: informalism and the quiet development of international space law. *Nebr. Law Rev.* 87, 725.
36. Marboe, I. (2017). "National space law," in *Handbook of Space Law*. Editors F. v. Dunk and F. Tronchetti (Cheltenham, UK: Edward Elgar), 124–204.
37. Masson-Zwann, T. (2017). "Introduction," in *Handbook for New Actors in Space*. Editor C. D. Johnson (Bloomfield, CO: Secure World Foundation), 2–3.
38. NASA (2019). Report to NASA/SMD: final report. NASA planetary protection independent review board. Available at: https://www.nasa.gov/sites/default/files/atoms/files/planetary_protection_board_report_20191018.pdf (Accessed July 29, 2020).
39. Lyall, F., and Larsen, P. (2018). *Space Law*. Abingdon, UK: Routledge.
40. Johnson, C. D., Porras, D., Christopher, M. H, and O'Sullivan, S. (2019). The curious case of the transgressing tardigrades. *Space Review*. Available at: https://www.thespacereview.com/article/3783/1 (Accessed July 29, 2020). doi:10.1287/5f4e34c9-55a8-4667-b696-4d50c9931b3e.
41. Lyall, F., and Larsen, P. (2018). *Space Law*. Abingdon, UK: Routledge.
42. Grush, L. (2019). High-speed lunar dust could cloud the future of human missions to the moon. *Verge*. Available at: https://www.theverge.com/2019/7/17/18663203/apollo-11-anniversary-moon-dust-landing-high-speed (Accessed July 28, 2020).
43. NASA (2019). Report to NASA/SMD: final report. NASA planetary protection independent review board. Available at: https://www.nasa.gov/sites/default/files/atoms/files/planetary_protection_board_report_20191018.pdf (Accessed July 29, 2020).
44. UNCLOS (1982). United nations convention on the law of the sea. U.N.T.S. 1833, at 397.
45. United Nations Committee on the Peaceful Uses of Outer Space (2007). Report of the Committee on the peaceful uses of outer space. A/62/20. Available at: https://www.unoosa.org/pdf/gadocs/A_62_20E.pdf (Accessed July 28, 2020).
46. Dupuy, P.-M., and Vinuales, J. E. (2018). *International Environmental Law*. 2nd Edn. Cambridge, United Kingdom: Cambridge University Press.

SURVIVAL: MARS FICTION AND EXPERIMENTS WITH LIFE ON EARTH

1. Wharton, Robert A., Jr., David T. Smernoff, and Maurice M. Averner. 1988. "Algae in Space." In *Algae and Human Affairs*, edited by Carole A. Lembi and J. Robert Waaland, 485–509. Cambridge: Cambridge University Press.
2. Ibid.
3. Bowie, David. 1971. "Life on Mars?" *Hunky Dory*. New York City, NY: RCA Records. Phonographic Record.
4. Sagan, Carl. 1973. "Planetary Engineering on Mars." *Icarus* 20: 513–514. doi: https://doi.org/10.1016/0019-1035(73)90026-2; Launius, Roger D. 2012. "Venus-Earth-Mars: Comparative Climatology and the Search for Life in the Solar System." *Life* 2: 255–273. doi: https://doi.org/10.3390/life2030255.
5. Conway, Erik M. 2015. *Exploration and Engineering: The Jet Propulsion Laboratory and the Quest for Mars*. Baltimore: Johns Hopkins University Press; Hogan, Thor. 2007. *Mars Wars: The Rise and Fall of the Space Exploration Initiative*. National Aeronautics and Space Administration, Washington, DC: NASA History Division, Office of External Relations; Lambright, W. Henry. 2014. *Why Mars: NASA and the Politics of Space Exploration*. Baltimore, MD: Johns Hopkins University Press.
6. Markley, Robert. 2005. *Dying Planet: Mars in Science and the Imagination*. Durham/London: Duke University Press. doi: https://doi.org/10.1215/9780822387275; McMillan, Gloria, ed. 2013. *Orbiting Ray Bradbury's Mars: Biographical, Anthropological, Literary, Scientific and Other Perspectives*. Jefferson, NC/London: McFarland & Co.
7. Bradbury, Ray. 1950. *The Martian Chronicles*. New York: Doubleday.
8. Robinson, Kim Stanley. 1993. *Red Mars*. New York: Random House; Robinson, Kim Stanley. 1994. *Green Mars*. New York: Random House; Robinson, Kim Stanley. 1996. *Blue Mars*. New York: Random House.
9. Zubrin, Robert. 1996. *The Case for Mars: The Plan to Settle the Red Planet and Why We Must*. New York: The Free Press; Zubrin, Robert. 2003. *Mars on Earth: The Adventures of Space Pioneers in the High Arctic*. New York: Jeremy P. Tarcher/Penguin.
10. Steffen, Will, Paul J. Crutzen, and John R. McNeill. 2007. "The Anthropocene: Are Humans Now Overwhelming the Great Forces of Nature?" *Ambio* 36: 614–621. doi: https://doi.org/10.1579/0044-7447(2007)36[614:TAAHNO]2.0.CO;2
11. Poole, Robert. 2008. *Earthrise: How Man First Saw the Earth*. New Haven/London: Yale University Press.
12. Anker, Peder. 2005a. "The Closed World of Ecological Architecture." *The Journal of Architecture* 10: 527–552. doi: https://doi.org/10.1080/13602360500463230; Anker, Peder. 2005b. "The Ecological Colonization of Space." *Environmental History* 10: 239–268. doi: https://doi.org/10.1093/envhis/10.2.239.
13. Radkau, Joachim. 2014. *The Age of Ecology*. Cambridge: Polity Press; Heise, Ursula K. 2008. *Sense of Place, Sense of Planet: The Environmental Imagination of the Global*. Oxford/New York: Oxford University Press. doi: https://doi.org/10.1093/acprof:oso/9780195335637.001.0001.
14. Hamblin, Jacob Darwin. 2013. *Arming Mother Nature: The Birth of Catastrophic Environmentalism*. Oxford/New York: Oxford University Press; Höhler, Sabine. 2015. *Spaceship Earth in the Environmental Age*, 1960–1990. London: Pickering & Chatto.
15. Ward, Barbara. 1966. *Spaceship Earth*. New York: Columbia University Press.

16. Ward, Barbara, and René Dubos. 1972. *Only One Earth: The Care and Maintenance of a Small Planet.* An Unofficial Report Commissioned by the Secretary-General of the United Nations Conference on the Human Environment, Prepared with the Assistance of a 152-Member Committee of Corresponding Consultants in 58 Countries. New York: Norton.
17. Ibid.
18. Ozorio de Almeida, Miguel A. 1973. "The Myth of Ecological Equilibrium." *The UNESCO Courier* 26: 25–28.
19. Schrödinger, Erwin. 1944. *What is Life? The Physical Aspect of the Living Cell.* Cambridge: Cambridge University Press.
20. Monod, Jacques. 1971. *Chance and Necessity: An Essay on the Natural Philosophy of Modern Biology.* New York: Vintage Books.
21. Bernal, John Desmond. 1967. *The Origin of Life.* London: Weidenfeld & Nicolson.
22. Margulis, Lynn, and Dorian Sagan. 1995. *What is Life?* New York: Simon & Schuster.
23. Lederberg. Joshua. 1966. "Exobiology: Approaches to Life Beyond the Earth." In *Extraterrestrial Life: An Anthology and Bibliography*, compiled by Elie A. Shneour and Eric A. Ottesen, 124–137. Washington, DC: National Academy of Sciences.
24. Allen, John. 1991. *Biosphere 2: The Human Experiment.* New York: Penguin Books.
25. Odum, Howard T. 1971. *Environment, Power, and Society.* New York: Wiley-Interscience.
26. Ibid.
27. Boulding, Kenneth E. 1966. "The Economics of the Coming Spaceship Earth." In *Environmental Quality in a Growing Economy*, Essays from the Sixth RFF Forum on Environmental Quality held in Washington, March 8 and 9, 1966, edited by Henry Jarrett, 3–14. Baltimore: Johns Hopkins Press; Fuller, Richard Buckminster. 1969. *Operating Manual for Spaceship Earth.* Carbondale, IL: Southern Illinois University Press.
28. Sueß, Eduard. 1875. *Die Entstehung der Alpen.* Wien: Braunmüller.
29. Vernadsky, Vladimir Ivanovich. [1926] 1998. *The Biosphere.* New York/Heidelberg: Copernicus/Springer. Revised and annotated English translation. Originally published as *Biosfera* (Leningrad: Nauka, 1926).
30. Ibid.
31. Hutchinson, G. Evelyn. 1970. "The Biosphere." *Scientific American* 223: 44–53. doi: https://doi.org/10.1038/scientificamerican0970-44.
32. Ibid.
33. Odum, Eugene P. 1996. "Cost of Living in Domed Cities." *Nature* 382: 18. doi: https://doi.org/10.1038/382018a0; Hagen, Joel B. 1992. *An Entangled Bank: The Origins of Ecosystem Ecology.* New Brunswick, NJ: Rutgers University Press; Elichirigoity, Fernando. 1999. *Planet Management: Limits to Growth, Computer Simulation, and the Emergence of Global Spaces.* Evanston, IL: Northwestern University Press.
34. Haeuplik-Meusburger, Sandra, Carrie Paterson, Daniel Schubert, and Paul Zabel. 2014. "Greenhouses and their Humanizing Synergies." *Acta Astronautica* 96: 138–150. doi: https://doi.org/10.1016/j.actaastro.2013.11.031
35. Salisbury, F. B. 1994. "Joseph I. Gitelson and the Bios-3 Project." *Life Support and Biosphere Science: International Journal of Earth Space* 1: 69–70.
36. Belasco, Warren. 2006. *Meals to Come: A History of the Future of Food.* Berkeley/Los Angeles: University of California Press.

37. Hardin, Garrett. 1972. *Exploring New Ethics for Survival: The Voyage of the Spaceship Beagle*. New York: The Viking Press.
38. Höhler, Sabine. 2014. "'The Real Problem of a Spaceship Is Its People': Spaceship Earth as Ecological Science Fiction." In *Green Planets: Ecology and Science Fiction*, edited by Gerry Canavan and Kim Stanley Robinson, 99–114. Middletown, CT: Wesleyan University Press.
39. Höhler, Sabine. 2010. "The Environment as a Life Support System: The Case of Biosphere 2." *History and Technology* 26: 39–58. doi: https://doi.org/10.1080/07341510903313048
40. Allen, John, and Mark Nelson. 1989. *Space Biospheres*. Oracle, AZ: Synergetic Press.
41. Allen, John. 1991. *Biosphere 2: The Human Experiment*. New York: Penguin Books.
42. Ibid.
43. Kelly, Kevin. 1990. "Biosphere II: An Autonomous World, Ready to Go." *Whole Earth Review* 67: 2–13; Kelly, Kevin. 1992. "Biosphere 2 at One." *Whole Earth Review* 77: 90–105.
44. Allen, John, and Mark Nelson. 1989. *Space Biospheres*. Oracle, AZ: Synergetic Press.
45. Ibid.
46. Allen, John. 1991. *Biosphere 2: The Human Experiment*. New York: Penguin Books.
47. Ibid.
48. Ibid; Zabel, Bernd, Phil Hawes, Hewitt Stuart, and Bruno D. V. Marino. 1999. "Construction and Engineering of a Created Environment: Overview of the Biosphere 2 Closed System." In *Biosphere 2: Research Past and Present*, edited by B. D. V. Marino and H. T. Odum, 43–63. Amsterdam: Elsevier Science (Ecological Engineering 13, Special Issue "Biosphere 2"). doi: https://doi.org/10.1016/s0925-8574(98)00091-3.
49. Gentry, Linnea, and Karen Liptak. 1991. *The Glass Ark: The Story of Biosphere 2*. New York: Puffin Books/Viking Penguin.
50. Allen, John. 1991. *Biosphere 2: The Human Experiment*. New York: Penguin Books.
51. Kelly, Kevin. 1994. *Out of Control: The New Biology of Machines*, Social Systems and the Economic World. Reading, MA: Addison-Wesley.
52. Rifkin, Jeremy. 1983. *Algeny: A New Word—A New World*. Harmondsworth: Penguin.
53. Jameson, Fredric R. 1995. *Postmodernism, or, the Cultural Logic of Late Capitalism*. Durham, NC: Duke University Press.
54. Granjou, Celine, and Jeremy Walker. 2016. "Promises that Matter: Reconfiguring Ecology in the Ecotrons." *Science and Technology Studies* 29: 49–67.
55. Luke, Timothy W. 1997. "Environmental Emulations: Terraforming Technologies and the Tourist Trade at Biosphere 2." In *Ecocritique: Contesting the Politics of Nature, Economy, and Culture*, 95–114. Minneapolis/London: University of Minnesota Press; Hayles, N. Katherine. 1996. "Simulated Nature and Natural Simulations: Rethinking the Relation Between the Beholder and the World." In *Uncommon Ground: Rethinking the Human Place in Nature*, edited by William Cronon, 409–425. New York/London: W. W. Norton.
56. Ibid.
57. Sfez, Lucien. 1995. *La Santé parfaite: Critique d'une nouvelle utopie*. Paris: Seuil.
58. Baudrillard, Jean. 1994a. "Maleficent Ecology." In *The Illusion of the End*. Translated by Chris Turner, 78–88. Cambridge: Polity Press.

59. Ibid.
60. Baudrillard, Jean. 1994b. "Überleben und Unsterblichkeit." In *Anthropologie nach dem Tode des Menschen: Vervollkommnung und Unverbesserlichkeit*, edited by Dietmar Kamper and Christoph Wulf, 335–354. Frankfurt a. M.: Suhrkamp.
61. Ibid.
62. Gabriel, Peter. 2000. "The Tower That Ate People." *OVO*. Box, Wiltshire: Real World Records. Phonographic Record.

Mars or Bust: How Science Fiction Films will Promote Mars Colonization Reality

1. Jung, 1959. 'The concept of the collective unconscious', *The Collected Works of C.G. Jung*, vol. 9.1. Bollingen Foundation/Pantheon Books.
2. Lake, S., 1938. *Submarine: The Autobiography of Simon Lake*. New York, NY: D. Appleton-Century.
3. Ibid.
4. Ibid.
5. Ibid.
6. Fort, G. & Faragoh, F. E., **1931**. *Frankenstein: Screenplay*. Universal Pictures, p. 32.
7. Bakken, E. E., 1999. *One Man's Full Life. Medtronic, Incorporated.* Retrieved from: http://www.earlbakken.com/content/publications/One.Mans.Full.Life.Book.pdf, p. 28.
8. Ibid., p. 28.
9. Ibid., p. 29.
10. Ryan, B., 1995. 'What Verne Imagined, Sikorsky Made Fly.' *The New York Times*. Retrieved from https://www.nytimes.com/1995/05/07/nyregion/what-verne-imagined-sikorskymade-fly.html
11. Ibid.
12. Ibid.
13. Verne, J., 1881. *The Steam House*. New York, NY: George Munro's Sons, Publishers, p. 6.
14. Pascual-Leone, A., Nguyet, D., Cohen, L. G., Brasil-Neto, J. P., Cammarota, A. and Hallett, M., 1995. "Modulation of muscle responses evoked by transcranial magnetic stimulation during the acquisition of new fine motor skills." *Journal of Neurophysiology*, 74(3), pp 1037–1045.
15. Aristotle, c. 332 B.C.E. *Poetics*.
16. Tan, E. S., 1996. *Emotion and the Structure of Narrative Film: Film as an Emotion Machine*. Routledge, p. 250.
17. Jung, 1959. 'The Concept of the Collective Unconscious', *The Collected Works of C.G. Jung*, vol. 9.1. Bollingen Foundation/Pantheon Books.
18. Holland, N. N., 2012. 'Your Brain on Movies'. *PSYART: A Hyperlink Journal for the Psychological Study of the Arts*. Retrieved from http://psyartjournal.com/article/show/n_holland-your_brain_on_movies
19. Ibid.
20. Csikszentmihalyi, M., 1990. *Flow: The Psychology of Optimal Experience*. New York, NY: Harper & Row.
21. Konigsberg, I., 2007. 'Film Studies and the New Science', *Projections: The Journal for Movies and Mind*, 1(1), 3.
22. Hasson, U., Nir, Y., Levy, I., Fuhrmann, G. and Malach, R., 2004. 'Intersubject synchronization of cortical activity during natural vision'. *Science*, 303(5664), 1634-1640.

23. Ibid.
24. Orwell, G., 1949. *Nineteen Eighty-Four: A Novel.* London, England: Secker & Warburg.
25. Lamm, C. and Majdandži´c, J., 2015. 'The role of shared neural activations, mirror neurons, and morality in empathy–A critical comment'. *Neuroscience Research*, 90, p. 22.
26. Clay, Z. and Lacoboni, M., 2011. 'Mirroring Fictional Others'. *The Aesthetic Mind: Philosophy and Psychology*, 313-331.
27. Hasson, U., Landesman, O., Knappmeyer, B., Vallines, I., Rubin, N. and Heeger, D. J., 2008. 'Neurocinematics: The Neuroscience of Film'. *Projections*, 2(1), 17.
28. *Total Recall*, 1990 [Film]. Directed by Paul Verhoeven. Van Nuys, CA: Carolco Pictures.
29. *The Space Between Us*, 2017 [Film]. Directed by Peter Chelsom. Beijing, China: Huayi Brothers Media Corp.

IMAGE CREDITS

p. 42–3 Figures redrawn by Jeff Edwards
p. 225 Figures redrawn by Jeff Edwards
p. 245 NASA/JPL-Caltech
p. 343 NASA/JPL-Caltech
p. 346 NASA/JPL-Caltech
p. 347 NASA/JPL-Caltech
p. 350 NASA/JPL/University of Arizona and M. T. Lemmon, Texas A&M University
p. 353 NASA/JPL/Univeristy of Arizona, and H. U. Keller and W. Markiewicz, Max Planck Institute for Solar System Research, Katlenburg-Lindau
p. 354 NASA/JPL, Malin Space Science Systems and the University of Arizona
p. 356 NASA/JPL/University of Arizona and M. T. Lemmon, Texas A&M University
p. 393 NASA/JPL-Caltech
p. 394 NASA/JPL-Caltech
p. 396 NASA/JPL/University of Arizona
pp. 404, 406, 409, 410, 417, 421 figures from "Water on Mars: A Literature Review" by M. NAZARI-SHARABIAN, M. AGHABABAEI, M. KARAKOUZIAN & M, KARAMI from Galaxies, Vol. 8, No. 2, 40, 09/05/2020, MPDI. Creative Commons Attribution License (CC BY)
p. 774 Figure redrawn by Jeff Edwards
p. 820 Figure redrawn by Jeff Edwards
p. 821 Figure redrawn by Jeff Edwards
p. 826 Figure redrawn by Jeff Edwards

EXTENDED COPYRIGHT

We are grateful to the following for permission to reproduce copyright material:

MARTIN AMIS: *The Janitor of Mars* from *The New Yorker*, 26/10/1998, copyright © Martin Amis, 1998. Reproduced by permission of The Wylie Agency (UK) Limited;

JEROME BIXBY: "The Holes Around Mars" from *Galaxy Science Fiction*, January 1954. Reproduced by permission of the Estate of the author;

REBECCA BOYLE: "Riding Along With the Mars Rover Drivers" on https://www.popsci.com/technology/article/2012-07/mars-rover-drivers/, 19/07/2012. Reproduced by permission of The YGS Group;

RAY BRADBURY: *The Martian Chronicles*, copyright © Ray Bradbury, 1951. Reproduced by permission of Abner Stein;

WERNHER VON BRAUN: *The Mars Project*, translated by Henry J. White, Apogee Books, 1953, copyright © Board of Trustees of the University of Illinois, 1953, 1962' and Maria von Braun, 1981. Reproduced by permission of the University of Illinois Press;

THOMAS CHENEY, CHRISTOPHER NEWMAN, KAREN OLSSON-FRANCIS, SCOTT STEELE, VICTORIA PEARSON, SIMON LEE: "Planetary Protection in the New Space Era: Science and Governance" from *Frontiers in Astronomy and Space Sciences*, 13/11/2020, copyright © Cheney, Newman, Olsson-Francis, Steele, Pearson and Lee, 2020. This is an open-access article distributed under the terms of the Creative Commons Attribution License (CC BY);

ARTHUR C. CLARKE: *The Sands of Mars*, Chapter 11, Gollancz, copyright © 1951. Reproduced by permission of David Higham Associates;

LEAH CRANE: "Mission to Mars: The complete guide to getting to the Red Planet", *New Scientist*, Issue 3234, 15/06/2019, copyright © New Scientist Ltd, 2019. All rights reserved. Distributed by Tribune Content Agency;

ROBERT CROSSLEY: "Mars and the Paranormal" originally published in *Science Fiction Studies*, Vol. 35, No. 3, November 2008. Reproduced by kind permission of the author;

PHILIP K. DICK: *Martian Time-Slip*, Chapter 2, copyright © Philip K. Dick, 1964. Copyright renewed © 1992, Laura Coelho, Christopher Dick, Isa Hackett, 1992. Reproduced by permission of the Licensor through PLSclear; and The Wylie Agency (UK) Limited;

RAINER EISFELD: "Projecting Landscapes of the Human Mind onto Another World: Changing Faces of an Imaginary Mars" from *Radical Approaches to Political Science: Roads Less Traveled, Verlag Barbara Budrich,* copyright © Verlag Barbara Budrich, 2012. Creative Commons Attribution-ShareAlike 4.0 International License (CC BY-SA 4.0);

JEFF FOUST: "The future of Mars exploration, from sample return to human missions" from *The Space Review*, 07/12/2020. Reproduced by kind permission of the author;

RANDALL GARRETT: "The Man Who Hated Mars" from *Amazing*, September 1956, copyright © Randall Garrett, 1956. Reproduced by permission of JABberwocky Literary Agency, Inc.;

HUGO GERNSBACK: "Münchhausen Departs for the Planet Mars" from *The Perversity of Things,* University of Minnesota Press, copyright © 1909; and "Signaling to Mars" from *The Perversity of Things,* University of Minnesota Press, copyright © 1909. Reproduced by kind permission of the Estate of the author;

OWEN GINGERICH: "The Satellites of Mars: Prediction and Discovery" from *Journal for the History of Astronomy*, SAGE, 01/08/1970, copyright © SAGE Publications, 1970. Permission conveyed through Copyright Clearance Center;

WALTER GOETZ: "Phoenix on Mars: The latest successful landing craft has made new discoveries about water on the red planet", *American Scientist,* Vol. 98, No. 1, Jan-Feb 2010, pp.40-47, Sigma Xi, The Scientific Research Honor Society. Reproduced by permission of the publisher;

SHARON GOZA: "Before they come" on https://roundupreads.jsc.nasa.gov/pages.ashx/525/The, 07/10/2016, copyright © United States Government. Reproduced by permission;

KATE GREENE: *Once Upon A Time I Lived on Mars*, Chapter 2, Icon, 2021, copyright © Kate Greene, 2020. Reproduced by permission of the publisher;

COLIN GREENLAND: *Take Back Plenty*, Orion, 1990, copyright © Colin Greenland, 1990. Reproduced by permission of the Licensor through PLSclear;

NATALIE HAYNES: "Why are we obsessed with Martians?" on https://www.bbc.com/culture/article/20150929-how-were-martians-invented, BBC, 2015, copyright © Natalie Haynes. Reproduced by permission of the author c/o Rogers, Coleridge & White Ltd., 20 Powis Mews, London, W11 1JN; and The BBC;

ANDY HEIL: "The Soviet Mars Shot that almost Everyone Forgets" on https://www.rferl.org/a/soviet-mars-shot-everyone-forgot-space-race/30759023.html, Radio Free Europe/Radio Liberty, 31/07/2020, copyright © RFE/RL, Inc., 2020. Reproduced by permission of kind permission of the author; and Radio Free Europe/Radio Liberty, 1201 Connecticut Ave NW, Ste 400, Washington DC 20036;

SABINE HÖHLER: "Survival: Mars Fiction and Experiments with Life on Earth", from *Environmental Philosophy,* Vol. 14, Issue 1, Spring 2017, pp.83-100, Philosophy Documentation Center, https://doi.org/10.5840/envirophil201731646. Reproduced by permission of the publisher;

ELIZABETH HOWELL, NICOLAS BOOTH: *The Search for Life on Mars*, Arcade Publishing, 2020, pp.1-24. Reproduced by permission of Skyhorse Publishing, Inc.;

DUNCAN HOWITT-MARSHALL: "How the Ancient Greeks Set Us on the Path to Mars" on https://www.greece-is.com/how-the-ancient-greeks-set-us-on-the-path-to-mars/, February 2021. Reproduced with kind permission from the author and greece-is.com;

BRUCE R. JOHNSON: "Enchanting Luna and Militant Mars" from *Sehnsucht: The C.S. Lewis Journal,* Vol. 4, pp.111-124. Wipf and Stock Publishers. Reproduced by permission of the publisher;

PAUL KAROL, DAVID CATLING: "Mars Chronology: Renaissance to the Space Age" on https://www.nasa.gov/audience/forstudents/9-12/features/F_Mars_Chronology.html, NASA, 29/10/2003;

CASSANDRA KHAW: "When we die on Mars", *ClarksWorld Science Fiction and Fantasy Magazine*, Issue 111, December 2015. Reproduced by kind permission of the author;

GEOFFREY A. LANDIS: 'Falling onto Mars' from *Analog Science Fiction and Fact Magazine*, July/August 2002, copyright © Geoffrey A. Landis, 2022. Reproduced by kind permission of the author;

GILBERT V. LEVIN: "I'm Convinced We Found Evidence of Life on Mars in the 1970s" from *Scientific American*, 10/10/2019, copyright © Scientific American, Inc., 2019. Reproduced by permission;

C. S. LEWIS: *Out of the Silent Planet*, chapter 16, copyright © CS Lewis Pte Ltd, 1938. Reproduced by permission;

EVAN LINCK, KEITH W. CRANE, BRIAN L. ZUCKERMAN, BENJAMIN A. CORBIN, ROGER M. MYERS, SHARON R. WILLIAMS, SARA A. CARIOSCIA, RODOLFO GARCIA and BHAVYA LAL: "Evaluation of a Human Mission to Mars by 2033", Institute for Defense Analyses, February 2019, copyright © Institute for Defense Analyses, 2019;

JEFFREY J. MARLOW, ZITA MARTINS, MARK A. SEPHTON: "Mars on Earth: soil analogues for future Mars missions" from *Astronomy & Geophysics*, Vol. 49, Issue 2, April 2008, pp.2.20-2.23, Oxford University Press, https://doi.org/10.1111/j.1468-4004.2008.49220.x. Reproduced by permission of Oxford University Press on behalf of The Royal Astronomical Society;

JOE MASCARO: "To save Earth, go to Mars". Essay originally published by Aeon, May 2016, https://aeon.co/. Reproduced by permission of the publisher;

CHRISTOPHER E. MASON: *The Next 500 Years: Engineering Life to Reach New Worlds*, MIT Press, 2021, pp.1-13, copyright © Christopher E. Mason, 2021. Reproduced by permission of The MIT Press;

CHRISTOPHER P. MCKAY: "Mars: The Case for Life" on http://www.marspapers.org/paper/MAR98003.pdf, March 1998, United States Government, 17 U.S.C. 105. Reproduced by kind permission of the author;

ROBIN MCKEE: "Curiosity rover's descent to Mars - the story so far", *The Observer*, 28/07/2013, copyright © Guardian News & Media Ltd, 2022. Reproduced by permission;

MICAH MEADOWCROFT: "Lost on Mars" from *The New Atlantis*, No. 56, Summer/Fall 2018, pp.69-81. Reproduced by permission of the publisher;

MOHAMMAD NAZARI-SHARABIAN, MOHAMMAD AGHABABAEI, MOSES KARAKOUZIAN, MEHRDAD KARAMI: "Water on Mars-A Literature Review" from *Galaxies*, Vol. 8, No. 2, 40, 09/05/2020, MPDI. Creative Commons Attribution License (CC BY);

ALEXANDER OPARIN: *The Origin of Life*, translated by John Desmond Bernal, Orion, 1967. Originally published as *Proiskhozhdenie zhizni*, Izd. Moskovskii Robochii, 1924;

ANTONIO PARIS: Physiological and Psychological Aspects of Sending Humans to Mars: Challenges and Recommendations" from *Journal of the Washington Academy of Sciences,* Vol. 100, No. 4, Winter 2014, pp.3-20, Washington Academy of Sciences. Reproduced by permission of the publisher;

ROBERT PARK: "The Virtual Astronaut" from *The New Atlantis*, No. 4, Winter 2004, pp.90-93. Reproduced by permission of the publisher;

PAUL PARSONS: "Ingenuity: How the Mars helicopter will fly on another planet", *BBC Science Focus Magazine*, 17/07/2020, copyright © BBC Science Focus Magazine / Ourmedia Ltd. Reproduced by permission;

KARA PLATONI: "The Key to Future Mars Exploration? Precision Landing" from *Air & Space Magazine*, Smithsonian, February 2020, copyright © Smithsonian Institution, 2020.Reprinted by permission of Smithsonian Enterprises. All rights reserved. Reproduction in any medium is strictly prohibited without permission from *Smithsonian* magazine;

JAMES POULOS: "For the Love of Mars" from *The New Atlantis*, No. 55, Spring 2018, pp.43-60. Reproduced by permission of the publisher;

SARA RIGBY: "Mars Sample Return: The mission that will bring home a piece of another planet", *BBC Science Focus Magazine*, 14/10/2021, copyright © BBC Science Focus Magazine / Ourmedia Ltd. Reproduced by permission;

H RAVEN ROSE: "Mars of Bust: How Science Fiction Films Will Promote Mars Colonisation Reality" first published in *Mars Society*, 2018, copyright © H Raven Rose. Reproduced by kind permission of the author;

EDGAR RICE BURROUGHS: *A Princess of Mars*, 1912. Reproduced by kind support of Edgar Rice Burroughs, Inc.;

MARY ROBINETTE KOWAL: "The Lady Astronaut of Mars" on https://www.tor.com/2013/09/11/the-lady-astronaut-of-mars/, copyright © Mary Robinette Kowal. Reproduced by permission of The Gernert Company;

JAMES ROMERRO: "Turning the Red Planet green: How we'll grow crops on Mars", *BBC Science Focus Magazine*, 17/10/2020, copyright © BBC Science Focus Magazine / Ourmedia Ltd. Reproduced by permission;

CHERISE SAYWELL: "Pieces of Mars have fallen to Earth", on https://mslexia.co.uk/magazine/blog/short-story-pieces-of-mars-have-fallen-to-earth, Mslexia, 2015. Reproduced with kind permission of the author; RAND SIMBERG: "The Return of the Space Visionaries" from *The New Atlantis*, No. 56, Summer/Fall 2018, pp.48-68. Reproduced by permission of the publisher;

TOMÁS SIDENFADEN: "Who owns Mars?" on https://medium.com/arc-digital/who-owns-mars-1b03190048fd, 07/05/2018. Reproduced by kind permission of the author;

STEVEN SQUYRES, CLAUDE CANIZARES: "Science Results from the Mars Exploration Rover Mission" in *Bulletin of the American Academy of Arts and Sciences*, Vol. 61, No. 4, Summer, 2008, pp.23-28, American Academy of Arts & Sciences. Reproduced by kind permission of the authors;

SALLY STEPHENS: "The Face on Mars" from *The Universe in the Classroom,* Astronomical Society of the Pacific, No. 25, Fall 1993, copyright © Astronomical Society of the Pacific, 1993. Reproduced courtesy of the Astronomical Society of the Pacific;

ANDY WEIR: *The Martian,* Del Rey, an imprint of Ebury, copyright © Andy Weir, 2014. Reprinted by permission of The Random House Group Limited;

ROGER ZELAZNY: *A Rose for Ecclesiastes*, Rupert Hart-Davis Limited, 1969, copyright © Roger Zelazny. Reproduced by permission of Zeno Agency Ltd;

ROBERT ZUBRIN: "The Human Explorer" from *The New Atlantis*, No. 4, Winter 2004, pp.93-96; and "The Mars Decision" from *The New Atlantis*, No. 60, Fall 2019, pp.46-60. Reproduced by permission of the publisher.

In some instances we have been unable to trace the owners of copyright material, and we would appreciate any information that would enable us to do so.